Tropical Cyclones
Observations and Basic Processes

Developments in Weather and Climate Science

Tropical Cyclones
Observations and Basic Processes

Roger K. Smith
Meteorological Institute
Ludwig-Maximilians University
Munich, Germany

Michael T. Montgomery
Department of Meteorology
Naval Postgraduate School
Monterey, CA, United States

Series Editor
Paul D. Williams

ELSEVIER

Royal Meteorological Society

Elsevier
Radarweg 29, PO Box 211, 1000 AE Amsterdam, Netherlands
The Boulevard, Langford Lane, Kidlington, Oxford OX5 1GB, United Kingdom
50 Hampshire Street, 5th Floor, Cambridge, MA 02139, United States

Notices

Knowledge and best practice in this field are constantly changing. As new research and experience broaden our understanding, changes in research methods, professional practices, or medical treatment may become necessary.

Practitioners and researchers must always rely on their own experience and knowledge in evaluating and using any information, methods, compounds, or experiments described herein. In using such information or methods they should be mindful of their own safety and the safety of others, including parties for whom they have a professional responsibility.

To the fullest extent of the law, neither the Publisher nor the authors, contributors, or editors, assume any liability for any injury and/or damage to persons or property as a matter of products liability, negligence or otherwise, or from any use or operation of any methods, products, instructions, or ideas contained in the material herein.

ISBN: 978-0-443-13449-4

For information on all Elsevier publications
visit our website at https://www.elsevier.com/books-and-journals

Publisher: Candice G. Janco
Acquisitions Editor: Jennette McClain
Editorial Project Manager: Mason Malloy
Production Project Manager: Kumar Anbazhagan
Cover Designer: Victoria Pearson Esser

Typeset by VTeX

I dedicate this book to the memory of my dear daughter, Elena.

Elena Smith (1966–1988)

———

Roger K. Smith

I dedicate this book to my parents, James and June, for their encouragement of lifelong learning and positive thinking and to my wife Leslie, son Michael and daughter Pascale for their love and unwavering support.

———

Michael T. Montgomery

"He thought of how some men feared being out of sight of land in a small boat and knew they were right in the months of sudden bad weather. But now they were in hurricane months and, when there are no hurricanes, the weather of hurricane months is the best of all the year."

Ernest Hemingway, *The Old Man and the Sea*

Contents

10. Tropical cyclone formation and intensification

11. The rotating-convection paradigm

12. Emanuel's intensification theories

13. Emanuel's maximum intensity theory

Companion Website:
https://www.elsevier.com/books-and-journals/book-companion/9780443134494

Preface

It is perhaps ironic that the normally benign tropical atmosphere can produce one of the most powerful and destructive weather systems on earth. These weather systems, known as hurricanes over the Atlantic Ocean, Caribbean Sea, and Eastern Pacific Ocean and typhoons over the Northwestern Pacific Ocean, are born over the warm tropical oceans and have the generic name tropical cyclones, a term that is used for such disturbances in other regions of the tropics. In the Indian Ocean Region they are called simply cyclones. These storms rank with earthquakes as one of the most destructive types of natural disaster. Practically every year, a particular storm makes headlines around the globe for the devastation reeked when it makes landfall. Examples include:

- Severe Tropical Cyclone Tracy (1974), which destroyed the Australian city of Darwin on Christmas Day (Bureau of Meteorology, 1977);
- Hurricane Andrew (1992), which devastated parts of southern Florida (https://en.wikipedia.org/wiki/Hurrikan_Andrew);
- Hurricane Katrina (1995) which caused severe flooding in New Orleans (https://en.wikipedia.org/wiki/Hurrikan_Katrina);
- Cyclone Nargis (2008) which devastated parts of coastal Myanmar (https://en.wikipedia.org/wiki/Cyclone_Nargis);
- Hurricane Sandy (2012), a large hurricane that devastated parts of the East Coast of the United States (https://en.wikipedia.org/wiki/Hurricane_Sandy); and
- Super Typhoon Haiyan (2013), which was one of the strongest tropical cyclones ever recorded (https://en.wikipedia.org/wiki/Typhoon_Haiyan).

The destructive nature of these storms is illustrated by some noteworthy recent storms:

- Hurricane Patricia (2015) was a record setting storm and to our knowledge remains the strongest tropical cyclone on record worldwide in terms of wind speed (245 mph or 394 km/h sustained winds) and the second-most intense on record worldwide in terms of minimum surface pressure (872 hPa). Hurricane Patricia exhibited also a rate of intensification among the most rapid ever observed. "In a 24-hour period, 06:00–06:00 UTC October 22–23, its maximum sustained winds increased from 85 mph (140 km/h) to 205 mph (335 km/h), a record increase of 120 mph (195 km/h)" (https://en.wikipedia.org/wiki/Hurricane_Patricia). These and other record-setting facets of this fascinating storm are presented in Rogers et al. (2017);
- Hurricane Harvey (2017) was a devastating hurricane "that made landfall on Texas and Louisiana in August 2017, causing catastrophic flooding and more than 100 deaths" (https://en.wikipedia.org/wiki/Hurricane_Harvey);
- Hurricane Maria (2017) was a "deadly hurricane that devastated the northeastern Caribbean in September 2017, particularly Dominica, Saint Croix, and Puerto Rico. It is regarded as the worst natural disaster in recorded history to affect those islands". Maria was "the most intense tropical cyclone worldwide" during 2017 and resulted in more than 3000 total deaths (https://en.wikipedia.org/wiki/Hurricane_Maria); and
- Hurricane Ian (2022) "was a large and destructive category four Atlantic hurricane, that was the deadliest hurricane to strike the state of Florida since the 1935 Labor Day hurricane." Ian inflicted "widespread damage across western Cuba and the southeastern United States, especially the states of Florida and South Carolina." At the time of this writing, Ian is known to have "caused at least 150 fatalities" and resulted in "catastrophic damage with losses estimated to be more than $50 billion" (https://en.wikipedia.org/wiki/Hurricane_Ian).

Notwithstanding the obvious practical threat posed by tropical cyclones to populated coastal communities and marine assets, tropical cyclones are fascinating phenomena scientifically, exhibiting a rich interplay between fluid dynamics and thermodynamics and between atmospheric motions on a wide range of scales from deep cumulus clouds to the synoptic-scale disturbances in which they develop. They have a strong coupling also with the ocean. They are a challenge for day-to-day weather forecasting, not only in the tropics, but also in the middle latitudes as a significant fraction of storms move polewards and transition into intense extratropical systems. While track forecasts have seen a dramatic improvement in the last two decades, forecasts of intensity have not. Forecasts of tropical cyclone formation remain a difficult problem also. Gaps in our current understanding of the formation and intensity change processes are rooted in the multi-scale nature of the problem from cloud scale up. Recent work has intimated links between climate change and tropical cyclone behavior (Knutson et al., 2021), though the precise nature of these links remains to be fully articulated.

Despite the importance of tropical cyclones, there have been few textbooks addressing their physics. Apart from various review publications arising from World Meteorological Organization-sponsored workshops, the only books that we are aware of that address the physics of tropical cyclones, at least in English,[1] are those of Anthes (1982) and Emanuel (2005b). The former is now somewhat dated and is limited in the range of physical processes discussed in detail. The latter provides an erudite historical perspective of the influence of hurricanes on art and literature, but the science is limited to an articulation of one particular axially-symmetric paradigm for tropical cyclone behavior by its author.

The aims of this textbook are to present the basic concepts, including the individual physical processes that we believe to be important, and to show how these processes connect. To accomplish these aims we will introduce a number of theoretical models, simple thought experiments and numerical model simulations. Because of the complexity of the overall tropical cyclone problem, we shall rely on numerical model simulations guided strongly by observations. The book is aimed at advanced undergraduate students, graduate students and those in warning services seeking a theoretical basis to underpin their knowledge about tropical cyclone behavior.

The seeds of the textbook were sewn in the summer of 2006 when the two authors rented a holiday cottage in Mittenwald, a small town in the German Alps, with the specific aim of reviewing what was understood and what was not concerning the dynamics of tropical cyclones. After a few days of discussion we had identified large gaps, at least in our understanding, and we set about a comprehensive research program to try to bridge these gaps. A central strategy of this program was to carry out a systematic series of idealized thought experiments, based in part on numerical simulations. The book contains a synthesis of these studies together with a broad survey of relevant basic theory.

An outcome of our research program to date is several review articles (Montgomery and Smith, 2011, 2014, 2017b; Smith and Montgomery, 2015, 2016), that are referred to in the text. However, there appears to be no modern textbook that develops the underlying dynamics and thermodynamics of tropical cyclones from first principles and our book is intended to fill this gap.

A particular aim of the textbook is to provide an understanding of how tropical cyclones form from weak tropical disturbances in a moist, non-saturated environment and how they subsequently intensify, mature, and decay. Like Emanuel's book, we will illustrate many of the processes borrowing from our own research, which has been motivated by a perceived need to build a robust *three-dimensional framework* for interpreting both observations and the results of numerical model simulations of tropical cyclones. Some specific questions addressed are the following:

- Is there a characteristic pathway of formation from weak tropical disturbances?
- Is this formation sequence fundamentally different from the intensification process?
- Why do storms tend to grow in size as they age?
- Are there important differences in tropical cyclone behavior between strictly axisymmetric and three-dimensional representations of the dynamics and thermodynamics?

With computers becoming ever more powerful and numerical forecast models becoming increasingly accurate, one might ask why a theoretical understanding of tropical cyclone behavior is necessary? A similar question was addressed in the context of middle latitude cyclones and anticyclones by Davis and Emanuel (1991). These authors pointed out that

"A proper integration of the equations is not synonymous with a conceptual grasp of the phenomenon being predicted".

In a similar spirit, reflecting on the use of numerical models to generate understanding, James (1994) reminds his readers on page 93 that the whole point of numerical modeling is to distinguish effects that are essential from those that are incidental. Indeed, he goes on to point out that

"comprehensive complexity is no virtue in modeling, but, rather, an admission of failure".

Again, these sentiments could not be more appropriate than when applied to tropical cyclones.

Davis and Emanuel go on to suggest that:

"the emphasis on forecasting may have contributed to an unhealthy separation between observational

---

1. There is another textbook in Russian that presents a state-of-the-art viewpoint from the former Soviet Union in the early 1980s of tropical cyclone theory and modeling including air-sea coupling (Khain and Sutyrin, 1983a,b). The ideas expressed therein are of interest historically, but our limited knowledge of Russian prevents us from providing a deep summary. The topics covered in this two part textbook (index translated into English) appear to be broadly of a similar scope to Anthes (1982), but with additional focus in Part II on the interaction of tropical cyclones with the ocean.

and theoretical work on cyclone dynamics. Observations and theory have yet to be reconciled on some important topics and there has not been enough work to separate the underlying physics of cyclone development from unsystematic details of individual cases. These are necessary if a simple conceptual picture of cyclogenesis is to emerge. A conceptual understanding is not only useful for reconciling theory with observation, but it is valuable also for delineating measurements necessary for accurately integrating forecast models."

While these remarks were aimed at midlatitude cyclones and anticyclones, we believe they are equally pertinent to tropical cyclones. In our experience, it is relatively straightforward in this day and age to download a working numerical model as well as an analysis from some global forecast system to initialize it and then use the model to carry out a case study. It is much harder to interpret the data that the model spits out. Usually the model has all the bells and whistles turned on so that the idea of modeling as a way to separate important processes from those that are incidental is already a challenge. Moreover, in our view, without a simplified conceptual framework at hand, the outcome is unlikely to provide much in the way of understanding.

A book on the theory of tropical cyclones would not be complete without a presentation of observations. We decided that reviewing the observations up front in Chapter 1 would provide motivation for the more theoretical chapters to follow, even though we are aware that a few important concepts in this chapter, such as absolute angular momentum and equivalent potential temperature, will not be addressed until later in the book. We hope that this choice will not be a serious impediment to readers entering the subject for the first time.

To understand the inner workings of tropical cyclones one requires at least an elementary understanding of classical mechanics, fluid mechanics, thermodynamics, and the calculus of multiple variables. We will assume that the reader has some familiarity with fluid dynamics and thermodynamics, although we have provided a brief survey of some of the most important concepts of fluid dynamics and moist thermodynamics in Chapter 2, together with a few key references. Some readers may wish to skip this survey, especially on a first reading. Others may find it useful as reference material.

The existence of large-scale coherent structures in high Reynolds number turbulent flow has been known since approximately 1973. The idea that smaller-scale mixing and chemical reactions may be strongly influenced by these structures has led to new insights with practical implications (e.g., Coles, 1985). Nevertheless, the well-established Kolmogorov model of turbulent energy cascade at smaller scales is not affected at leading order by such flow structures

(Frisch, 1995). On larger planetary scales, however, intense geophysical vortices tend to be quasi-horizontal, long-lived flow structures and occupy a more central role in the system dynamics (e.g., Carnevale et al., 1991).[2] Although fully-formed tropical cyclones are obviously not two-dimensional structures, their longevity and observed resilience to adverse influences naturally invites a quasi two-dimensional analysis of their motion and internal structure evolution. As we shall discover later, however, tropical cyclone vortices are *tightly-coupled* systems in which a wide range of physical processes are involved. In view of the coupling, one must be cautious about cause-and-effect-type arguments. Inspired by the remarks of Ian James cited earlier, our approach will be to try to isolate the more important processes and to understand how they interact. In this approach, the necessary fluid dynamics, moist thermodynamics and boundary layer dynamics need to be considered carefully.

As anticipated in the foregoing discussion, some aspects of tropical cyclone dynamics can be understood to a first approximation in terms of two-dimensional flows. In Chapter 3 we develop an introductory theory for tropical cyclone motion and in Chapter 4 an elementary theory for vortex axisymmetrization, vortex waves and wave-vortex interaction. In these chapters, two dimensional refers to planar horizontal flows, whereas in Chapters 5, 6 and 8, two-dimensional refers to axisymmetric flows in the vertical plane which have a swirling component and transverse component. By their nature, parts of Chapters 3, 4, and 5 call for a familiarity with some advanced mathematical techniques, but we have tried to provide physical interpretations of the results.

In Chapter 5 we develop the axisymmetric theory of vortices in a quiescent environment and in Chapter 6 we examine theories for the frictional boundary layer.

Chapter 7 reviews the determination of various physical parameters required for quantitative theoretical work and in the formulation and implementation of numerical models.

In Chapter 8 we apply some results from Chapters 5 and 6 to explore a prognostic axisymmetric balance theory to understand some fundamental aspects of tropical cyclone evolution and behavior.

The presence of deep convection introduces an intrinsically three-dimensional element that ultimately needs to be confronted. It introduces also a stochastic element that has relevance to the predictability of at least some aspects of storms. To provide a foundation for examining these aspects, Chapter 9 presents a brief review of some of the important aspects of moist convection.

Chapters 10 and 11 focus on flows that are three-dimensional and rely on the ability to solve the governing equations using sophisticated numerical models. In particular, Chapter 10 addresses the genesis problem in a locally favorable environment, while Chapter 11 explores a new

---

2. Frisch (1995) gives a penetrating and comprehensive review of both three-dimensional and two-dimensional turbulent flow problems.

three-dimensional paradigm for understanding tropical cyclone behavior that builds on the classical theory of Ooyama (Ooyama, 1968, 1969, 1982).

Chapters 12 and 13 review and appraise the axisymmetric time dependent and steady-state theories of Emanuel (Emanuel, 1986, 1989, 1995, 1997, 2003, 2012, 2018), respectively, theories that have become firmly implanted in the contemporary literature.

Chapter 14 discusses the budgets of moisture, energy, and angular momentum, the generation of kinetic energy, the thermodynamics of the boundary layer of tropical cyclones, and the constraints required for a steady state vortex, while Chapter 15 presents a novel analysis of an idealized tropical cyclone life-cycle simulation.

Chapter 16 applies the newly-developed rotating-convection paradigm to address a range of basic problems that include tests of the new model using observational data, differences between strictly axisymmetric and three-dimensional dynamics, dependence of vortex evolution on ambient rotation or sea surface temperature, and the dependence of vortex evolution on initial vortex size.

Finally, the epilogue, Chapter 17, highlights the application of some of the ideas developed in the book to examine the formation and evolution of two noteworthy storms during the 2022 Atlantic hurricane season.

We have limited the topics covered in our first book to enable the systematic development of a foundational framework for interpreting the genesis, intensification and decay of an idealized tropical cyclone in a relatively favorable "pouch-like" environment. This material provides a new set of tools to begin a deeper examination of observed tropical cyclone behavior and the convective and air-sea interaction phenomenology that accompanies tropical cyclones in reality. We would like to have covered more advanced topics that encompass the interaction of a tropical cyclone with its atmospheric and oceanic environment, such as an impinging uniform or vertically-sheared flow, the development of a secondary eyewall and eyewall cycle that typically follow the formation of an outer eyewall in major tropical cyclones.

Other topics not addressed include the coupling of the underlying ocean to the evolution of the storm including wind-wave coupling, the interaction with neighboring weather systems including fronts and upper troughs, the extratropical transition when storms move into the middle latitudes, the dynamics of helicity, cloud microphysics, boundary layer rolls and physical processes operative in the surface layer (or emulsion layer) at high wind speeds. Many of these important topics require *inter alia* a more systematic development of the subjects of asymmetric balance dynamics, asymmetric vortex waves, vortex quasi-modes, vortex resilience and wave-vortex interaction. The latter four items are topics that are developed only briefly in primer form using the pedagogical toy model presented in Chapter 4.

When a vortex is subject to strong asymmetric forcing such as an imposed uniform flow, a vertically-sheared flow, or a heterogeneous ocean environment, the approximate Carnot model and related nonlinear boundary layer dynamics developed in Chapters 6 and 13, and the new insight detailed in Chapter 15 involving convective ventilation, need to be revisited and modified suitably, where possible, to consistently account for the ensuing asymmetric wave and convective processes and their coupling to the boundary layer. These topics will be the focus of a planned sequel to this book.

Acknowledgments

My interest in tropical cyclones began while a graduate student at the University of Manchester in the mid-60's when I attended a stimulating course on Geophysical Fluid Dynamics by the late Bruce Morton. Bruce later became my Ph.D. supervisor and a close colleague when I joined him at Monash University in Melbourne, Australia after completing my Ph.D. in Manchester in 1968. A part of my Ph.D. thesis was concerned with the tropical cyclone boundary layer. Bruce was a charismatic mentor and teacher with deep insights in the subject of fluid mechanics. His lectures, while sometimes a little chaotic, were inspirational and stimulating and his inquisitiveness about the subject was infectious. Bruce had a keen interest in all geophysical vortices and his deep insights greatly influenced my early work on tornadoes and dust devils.

My interest in tropical cyclones was rekindled when, in 1986, I was invited by Dr. R. A. (Bob) Abbey to join a United States Office of Naval Research Initiative on Tropical Cyclone Motion. Bob provided generous ONR support to my research group for more than a decade and the work described in Chapter 3 summarizes some of our findings on this topic. In 1988 I moved to a Chair of Meteorology at the Ludwig-Maximilians University in Munich, where much of my research on tropical cyclones was carried out. Contributors to the research that forms the basis for Chapter 3 include Gary Dietachmayer (from Monash University), the late Wolfgang Ulrich, Michael Reeder, Harry Weber, Annetta Kraus, and Arno Glatz.

During the 1990's, the emphasis of the ONR Program shifted from the investigation of tropical cyclone motion to the important problem of storm intensification and the focus of my research group shifted to address this important problem. Sarah Jones and Wolfgang Ulrich played a central role in the group at that time with important contributions also from Dominique Möller, Lloyd Shapiro, Hongyan Zhu, Mai Nguyen, Klaus Dengler, and Reinhardt Hell. Sarah carried out pioneering work on the behavior of vortices in vertical shear and on the extra-tropical transition of tropical cyclones, work that unfortunately we have not been able to cover in this book.

In 2006, I teamed up with my co-author in circumstances described in the Preface and at an early stage we set about trying to consolidate and build on what was understood about tropical cyclone behavior at that time. Important contributions to this endeavor were made by Sang Nguyen, Soeleun Shin, Gerald Thomsen, Steffi Vogl, Ulrike Wissmeier, Christoph Schmidt, Hai Bui, Gerard Kilroy, Nina Črnivec, Shengmin Tang, Sian Steenkamp, Shanghong Wang and Minhee Chang, and their work has had a strong imprint on this book. In particular, Hai Bui, Christoph Schmidt, and Shanghong Wang's research was influential to the work in Chapter 8 and, indeed, in Section 16.1.1. Gerard Kilroy's contributions deserve a special mention as they form the backbone of much of the material presented in Chapters 10, 11, 14, 15 as well as parts of Chapter 16. In summary, I would like to thank all my former students and colleagues for their contributions as well as to Minhee Chang for her perceptive critique of a near final draft of Chapter 9.

Parts of the book were written during regular meetings of the co-authors in a particularly stimulating environment, the Australian Bureau of Meteorology's Regional Forecasting Office in Darwin, Australia. We are extremely grateful to a series of Regional Directors there over the years: Jim Arthur, Geoff Garden, Andrew Tupper, and Todd Smith for encouraging and supporting these visits and to the forecasters "on the bench" for sharing their experience of forecasting weather in the Tropics, especially forecasts of tropical cyclones. It is always humbling as a theoretician to sit with forecasters and to know that, unlike us, they do not enjoy the luxury of having a "too hard basket": they have to make a prediction even when the information available to them is incomplete or even contradictory.

Roger K. Smith
Munich, Germany
October 2022

My interest in tropical cyclones was sparked by two events, both of which occurred while I was a graduate student in the Division of Applied Sciences at Harvard University. The first event was in September 1988 as I watched Hurricane Gilbert, a powerful category-five hurricane, traverse the Caribbean Sea. The second was in the winter of 1990 when I attended an MIT seminar by Frank Marks, Jr., detailing the now-infamous NOAA WP-3D scientific reconnaissance flight through the rapidly intensifying Hurricane Hugo (1989), a storm which ravaged the U.S. Virgin Islands and Puerto Rico and later made a devastating landfall in Charleston, South Carolina. Frank's talk captivated my imagination and made me realize that the Hurricane Problem still had many secrets. (Little did I know then that, twenty years hence, I would be a co-author - with Frank, Peter Black, and Robert Burpee - of a 2010 NOAA-AOML "outstanding research paper" detailing the observations collected during that infamous flight and validating a newly proposed conceptual model of the inner-core evolution of such intense hurricanes!)

At the time, I was studying atmospheric fronts with my Ph.D. supervisor, Brian Farrell. I became intrigued with the problem of the origin of intense tropical cyclones and whether the then–emerging idea of "coherent structures" in fluid dynamics, which I had learned about a few years earlier at a summer school at the National Center for Atmospheric Research, might help advance the state of understanding of tropical cyclone science. I was curious also if the non-geostrophic effects that I was learning about in the dynamics of fronts might play a similar role in the formation of the hurricane eyewall region. In retrospect, our field has learned much since that time about the formation of these dangerous storms, and we have come to better understand the important role of non-gradient effects in these vortices. In addition to Brian, I had the privilege of having several scientific mentors that helped solidify my interest in atmospheric and oceanic science and geophysical vortex dynamics in particular. These include George Carrier, Allan Robinson, Richard Goody, Phillip Marcus, Richard Lindzen, Kerry Emanuel, Joseph Pedlosky, Frank Marks, Jr., Lloyd Shapiro, James McWilliams, Vladimir Vladimirov, Michael McIntyre, Uriel Frisch, James Riley, and James Holton.

Of course, credit must be given to my family of scientific collaborators over the years that helped provide a foundation for this book. Some of the cyclogenesis work summarized in Chapter 1 was carried out while in California, in collaboration with friends Timothy Dunkerton, Chris Davis, Scott Braun, then–student Louis Lussier III and then–postdocs. Zhuo Wang and Blake Rutherford. Some of the work presented in Chapter 4 was developed while in Colorado in collaboration with two of my master's students, Gerald Smith III and Wesley Terwey. Some other work cited in this textbook was carried out over many late-night Skype calls with my always gracious co-author and friend, Roger Smith, and in collaboration with friends James McWilliams, John Persing, Mark Boothe, and Jun Zhang; then–Ph.D. students Gerard Kilroy, Neil Sanger, and Michael Bell; then–master's student Thomas Freismuth; and then–postdoc. Michael Riemer. Other collaborators have been acknowledged above by Roger. Colleagues in California who have provided friendship and collaboration include Tamar and Beny Neta, Pat Harr, and Russ Elsberry; colleagues in Colorado include Wayne Schubert, William Gray, Graeme Stephens, Richard Johnson, William Cotton, Stephen Cox, Thomas Vonder Harr, Dave Randall, Bjorn Stevens, Wen-Chau Lee, Michael Kirby, and George Kiladis; colleagues in Florida include Frank Marks, Jr., Paul Reasor, Xuejin Zhang, S. Gopalakrishnan, Stanley Goldenberg, Howard Friedman, Sim Aberson, and Robert Rogers.

Michael T. Montgomery
Carmel, California
October 2022

The two authors, Roger Smith (left) and Michael Montgomery (right) taken during a joint visit to the UK Meteorological Office in 2016.

Nomenclature

Roman	Meaning	Where first introduced
$3D$	Three dimensional	defined in Section 16.2
a	Mean radius of the Earth, $\approx 6,371$ km	defined in Section 2.10
	Radius of maximum tangential velocity in a Rankine vortex	defined in Sections 4.4.5 and 4.4.8
A	An area within a closed circuit C	defined in Eq. (11.1)
A_n	Projection of area A onto equatorial plane	defined in Section 2.11
AX	Axisymmetric (or axially symmetric)	defined in Section 16.2
b	Buoyancy force per unit mass	defined in Section 2.2
	Specific buoyancy force relative to circular vortex initially in thermal wind balance	defined in Section 5.13.5
B	Baroclinicity of axisymmetric vortex	defined by Eqs. (5.44) and (5.60)
$\mathbf{c}$	Vortex translation velocity	defined in Section 3.2
c_{pd}	Specific heat at constant pressure for dry air (also c_p in Chapters 12 & 13)	defined in Section 2.3.2
c_{vd}	Specific heat at constant volume for dry air	defined in Section 2.3.2
c_{pv}	Specific heat at constant pressure for water vapor	defined in Section 2.5.2
c_w	Specific heat at constant pressure for liquid water	defined in Section 2.5.2
C	The sum of the centrifugal and Coriolis forces per unit mass	defined in Eq. (5.14)
C_D	Surface momentum (drag) coefficient	defined in Section 6.4.2
C_K	Surface enthalpy transfer coefficient	defined in Section 7.5
$\mathbf{c}$	Vortex translation speed	defined in Section 3.3.1
$\mathbf{c}'$	Vortex translation speed relative to the environmental flow	defined in Section 3.3.3
d	Prescribed constant characterizing scale of vorticity dipole	defined in Section 4.1.1
	Small distance to the west of center of circulation	defined in Section 4.3
dA	An area element of a surface	defined in Eq. (11.1)
dh	Moist saturation static enthalpy increment $= c_p dT + L_v dr_v^*$	defined in Eq. (13.12)
$\mathbf{dl}$	Differential vector line segment	defined in Section 2.11
$\mathbf{ds}$	Vector increment along a curve	defined in Eq. (11.1)
D	Discriminant of (dry) Eliassen balance equation	defined by Eq. (5.68)
	Total frictional dissipation, $\overline{[\rho \epsilon_D]}$	defined in Section 14.2.2
$\bar{D}$	Tangential momentum fluxes associated with unresolved horizontal and vertical diffusive processes	defined in Section 14.4
e	Partial pressure of water vapor	defined in Section 2.5
e^*	Saturation vapor pressure	defined in Section 2.5
E	Total moist-saturated energy comprising kinetic, potential and moist saturated enthalpy per unit mass	defined in Eq. (13.25)
E_1	Eddy advection of eddy absolute angular momentum per unit mass	defined in Section 14.4
E_2	Eddy flux divergence of eddy absolute angular momentum per unit mass	defined in Section 14.4
f	The Coriolis parameter $= 2\Omega \sin \phi$	defined in Section 2.10
F_b, F_d	Forcing terms in the Laplacian of p'	defined in Section 9.5
F_r', F_λ', F_z'	Component forcing terms in perturbation momentum equations associated with non-conservative processes, as well as second order mean and eddy advective accelerations	defined in Section 5.13.5
F_r	Radial component of sub grid scale force per unit mass	defined in Section 5.1
F_R	Energy flux at side boundary of cylinder	defined in Eq. (14.25)
F_S	Surface enthalpy flux at base of cylinder	defined in Eq. (14.26)
F_{KEA}	Radial flux of kinetic energy through the side boundary of a cylinder	defined in Section 14.2.3

Roman, continued	Meaning	Where first introduced
F_{KEG}	Radial flux of mechanical energy through the side boundary of a cylinder	defined in Section 14.2.2
F_z	Vertical component of sub grid scale force per unit mass	defined in Section 5.1
F_λ	Azimuthal component of sub grid scale force per unit mass	defined in Section 5.1
$\mathbf{F_E}$	Areally-integrated rate of evaporation from the ocean	defined in Section 14.2.1
$\mathbf{F_{KE}}$	Kinetic energy flux density vector	defined in Section 2.8
$\mathbf{F_{\zeta_a}}$	Horizontal flux of absolute vorticity	defined in Section 11.1
$\mathbf{F_{af}}$	Advective flux of absolute vorticity	defined in Section 11.1
$\mathbf{F_{fri}}$	horizontal force per unit mass due to molecular effects and subgrid-scale eddy momentum fluxes associated with unresolved turbulence	defined in Section 11.1
$\mathbf{F_{naf}}$	Non-advective flux of absolute vorticity	defined in Section 11.1
$\mathbf{F_{tot}}$	Flux of total energy per unit mass	defined in Section 14.2.5
$\mathbf{F_{r_T}}$	Flux of water substance	defined in Section 14.2.1
$\mathbf{F_{rad}}$	Radiative heat flux per unit	defined in Section 14.2.5
g	Magnitude of earth's acceleration due to gravity	defined in Section 2.1
	Also magnitude of effective gravitational acceleration after dropping subscript 'e'	defined in Section 2.1
G	Parameterized deep convective mass flux at middle level in the E89 model	defined in Section 12.1.1
$G(r,p)$	Radial Green function for vorticity source at radius p for azimuthal wavenumber n	defined in Sections 4.4.5 and 4.4.8
$\mathbf{g}$	Acceleration vector due to the earth's gravitational force per unit mass	defined in Section 2.1
$\mathbf{g_e}$	Effective gravitational acceleration vector	defined in Section 2.1
h	The moist static energy $= c_p T + gz + L_v r_v$	defined in Section 2.5.4 and Eq. (14.20)
	The top of vortex boundary layer	defined in Sections 6.5, 12.1.3 and 13.1
h_d	The dry static energy $= c_p T + gz$	defined in Section 2.3.4
H, H_1	Boundary-layer depth parameters	defined in Section 8.1.2
$\mathbf{h}$	The vector $(g, 0, -C)$	defined in Section 16.1.4.1
I^2	Inertial stability	defined in Section 5.7.3
I_g^2	Generalized inertial stability	defined in Sections 5.7.3 and 8.3
$\hat{\mathbf{i}}$	Unit vector in the x-direction	defined in Section 3.3.1
IR	Intensification rate	defined in Section 16.4
$\hat{\mathbf{j}}$	Unit vector in the azimuthal direction	defined in Exercise 16.7
$\hat{\mathbf{k}}$	Unit vector in the z-direction	defined in Section 3.2
k	A local (time-dependent) radial wavenumber	defined in Section 4.4.6
	Azimuthal wavenumber	Fig. 4.21
K_H	Turbulent conductivity of heat	defined in Section 14.2.5
K_M, K_r	Turbulent diffusivities of momentum, and water vapor	defined in Section 14.2.5
K_r, K_z	Radial and vertical turbulent diffusivities of momentum	defined in Section 14.4
$\mathbf{K}$	Curl of specific body force vector associated with sub grid scale motions	defined in Section 2.12
L_v	Latent heat of vaporization	defined in Section 2.5.2
M	Absolute angular momentum per unit mass	defined in Sections 1.2.3 and 5.1
$\bar{M}$	Azimuthally-averaged absolute angular momentum per unit mass	defined in Section 1.2.3
n	Integral azimuthal wavenumber	defined in Chapter 4
N^2	Brunt-Väisälä frequency squared (static stability)	defined in Sections 5.7.3 and 12.2
OW	Okubo-Weiss parameter	defined in Section 1.5.6
p	Pressure	defined in Section 2.1
	Dummy variable of integration	
p_i	Initial pressure of a displaced air parcel	defined in Section 9.2
p_{LCL}	Pressure at the lifting condensation level	defined in Section 9.2
p_{LFC}	Pressure at the level of free convection	defined in Section 9.2
p_{LNB}	Pressure at the level of neutral buoyancy	defined in Section 9.2
p_{ref}	Reference pressure	defined in Sections 9.5 and 14.2.3
p_b', p_d'	Static and dynamic perturbation pressures	defined in Section 9.5
p_h', p_{nh}'	Static and dynamic perturbation pressures	defined in Section 9.11.1
p_{**}	A standard reference pressure	defined in Section 1.2.3
$\hat{p}(r,t)$	Complex Fourier amplitude of perturbation pressure	defined in Section 4.4.8

Roman, continued	Meaning	Where first introduced
P	The Ertel potential vorticity	defined in Sections 2.12 and 8.1.1.3
	Net vertical force per unit mass	defined in Section 14.2.3
P_m^*	Axisymmetric moist saturation potential vorticity	defined in Section 16.1.4.2
P_{rain}	Areally-integrated loss of water by precipitation	defined in Section 14.2.1
q_v	Specific humidity of water vapor (denoted also as q)	defined in Sections 2.5 and 12.2
Q	Normalized heating rate ($= g\dot{\theta}/\theta_0$)	defined in Sections 5.13.3
Q_R	Vertical component of $\mathbf{Q_R}$ at surface	defined in Section 14.2.5
Q_T	Vertical component of $\mathbf{Q_T}$ at surface	defined in Section 14.2.5
$\dot{Q}$	Material heating rate per unit mass ($= DQ/Dt$, where Q is heat input to fluid parcel)	defined in Section 2.3.2
$\mathbf{Q_R}$	Latent heat flux	defined in Section 14.2.5
$\mathbf{Q_T}$	Dry sensible heat flux	defined in Section 14.2.5
r	Radius from center of vortex circulation, O	defined in Sections 3.3.2.1 and 4.3
r'	Radius from center of circulation, O'	defined in Section 4.3
$\mathbf{r}$	Radius vector from center of the Earth	defined in Section 2.1
$\hat{r}$	Unit vector in the radial direction	defined in Section 14.2.5
r_{max}	Radius of maximum asymmetric vorticity	defined in Section 4.1.1
r_o	Radial extent of cyclone at sea level (nominally, radius at which $v_g = 0$)	defined in Section 13.1
r_c	Critical radius	defined in Sections 4.4.4, 4.5.6, and 4.4.8.1
r_b	Radius of M surface at top of subcloud layer	defined in Section 12.1.1
r_{gm}	Radius of maximum gradient wind	defined in Sections 13.1, 13.6.2, and 13.7
r_i	Mixing ratio of ice	defined in Section 2.5
r_L	Mixing ratio of liquid water	defined in Section 2.5
r_m	Radius of maximum tangential wind speed	defined in Sections 3.3.1 and 6.4.5
r_t	Radius of M surface at model tropopause	defined in Section 12.1.1
	Radius where gradient Richardson number first becomes critical	defined in Section 12.3.1
r_v	Mixing ratio of water vapor	defined in Section 2.5
	Local mixing ratio at radius r and boundary layer top	
	$= r_{vs}$ assuming well mixed boundary layer	defined in Section 13.7.3.2
r_v^*	Saturation water vapor mixing ratio	defined in Section 2.5
	Local saturation mixing ratio at sea surface temperature and radius r	defined in Section 13.7.3.2
r_{va}	Mixing ratio of an air parcel in the environment	defined in Section 9.2.4
	Mixing ratio of air parcel in vortex environment and at the boundary layer top	defined in Section 13.7.3.2
	$= r_{vas}$ for a well-mixed boundary layer	
r_{va}^*	Saturation mixing ratio in vortex environment at ambient surface pressure p_{sa} and sea surface temperature T_s	defined in Section 13.7.3.2
r_p	Mixing ratio of a lifted air parcel	defined in Section 9.2.4
r_T	Mixing ratio of total water substance	defined in Section 2.5
r_{vmax}	Radius of maximum azimuthally-averaged tangential wind speed	defined in Section 10.10
R	Potential radius coordinate (defined by $\frac{1}{2}fR^2 = rv + \frac{1}{2}fr^2$)	defined in Section 5.10.4
R_d	Specific gas constant for dry air	defined in Section 2.3.1
Re	Reynolds number	defined in Section 2.6
RH	Relative humidity	defined in Section 2.5
RH_{as}	Relative humidity of ambient air at the top of surface layer	defined in Section 13.1
Ri_c	Critical gradient Richardson number that determines onset of turbulent mixing in the upper-tropospheric outflow	defined in Section 12.3.1
RI	Rapid intensification	defined in Section 10.2
R_v	Specific gas constant for water vapor	defined in Section 2.5
Ro	Rossby number	defined in Section 5.8
R_{umin}	Radius of minimum azimuthally-averaged radial wind speed	defined in Fig. 8.9
R_{vmax}	Radius of maximum azimuthally-averaged tangential wind speed (=RMW)	defined in Section 1.2.3
R_{rwmaxL}	Radius of maximum azimuthally-averaged vertical velocity in the lowest 3 km	defined in Section 15.5.4.1
R_{15}	Average radius of winds equal to 15 m s^{-1}	defined in Section 1.2.6
R_{25}	Average radius of winds equal to 25 m s^{-1}	defined in Section 1.2.6
R_{gales}	Outer radial extent of gale force winds at a height of 2 km or 1 km	defined in Sections 8.5 and 10.10

Roman, continued	Meaning	Where first introduced
R_{galesF}	Outer radial extent of gale force winds at the surface	defined in Section 10.10
R_{nd}	Closest approach of vortex manifold (dividing streamline)	defined in Section 16.11
RAM	Relative Angular Momentum per unit mass about vortex axis ($= rv$)	defined in Section 13.1
s	Laplace transform variable	defined in Section 4.4.4
s_b	Specific moist entropy of boundary layer and anemometer level assuming well-mixed boundary layer	defined in Sections 12.1.1, 12.1.3, 13.1, 13.6.2
s_d	Specific entropy of dry air	defined in Section 2.3.3
s^*	Specific saturation moist entropy	defined in Sections 12.1, 12.8
s_0^*	Specific saturation moist entropy at sea surface temperature and pressure	defined in Section 12.8, 13.1
S	Local radial shear in a circular vortex with basic state angular velocity $\Omega(r)$ ($S = rd\Omega/dr$)	defined in Sections 4.4.5 and 4.5.1
	Spin-up function	defined in Section 14.6
t	Time	defined in Section 2.1
T	Absolute temperature	defined in Section 2.3.1
T	Time in a comoving reference frame	defined in Section 3.2
T_a	Absolute temperature of a fluid parcel's environment	defined in Section 2.2
T_b	Temperature at the top of the inflow layer	defined in Section 12.3.1
T_B	Temperature at top of inflow layer ($= T_b$)	defined in Section 13.1
T_o	Average outflow temperature weighted with the saturation moist entropy of the outflow angular momentum surfaces	defined in Sections 12.3 and 13.1
	$= \overline{T}_{out}$ in Eq. (13.62)	defined in Section 13.7.2
T_d	Dew-point temperature	defined in Section 9.2
T_p	Absolute temperature of a lifted air parcel	defined in Section 9.2.4
T_a	Absolute temperature of the environment of an air parcel	defined in Section 9.2.4
T_{di}	Initial dew-point temperature of a displaced air parcel	defined in Section 9.2
T_{LCL}	Absolute temperature at the lifting condensation level	defined in Section 9.2
T_t	Absolute temperature of outflowing air corresponding to the M-surface passing through the radius of maximum gradient wind, assumed equal to the initial tropopause temperature	
T_v	Virtual temperature	defined in Section 2.5.1
T_{vp}	Virtual temperature of an air parcel	defined in Section 9.2.4
T_{va}	Virtual temperature of an air parcel environment	defined in Section 9.2.4
T_w	Wet-bulb temperature	defined in Section 9.3.4.1
T_ρ	Density temperature	defined in Section 2.5.1
$T_{\rho p}$	Density temperature of a lifted air parcel	defined in Section 9.2.1
$T_{\rho a}$	Density temperature of an air parcel environment	defined in Section 9.2.1
TPW	Total precipitable water per unit horizontal area	defined in Section 14.2.1
TW	Total column water per unit horizontal area	defined in Section 14.2.1
u	zonal component of velocity	defined in Section 2.10
	radial component of velocity	defined in Section 5.1
u_b	Radial component of velocity in boundary layer	defined in Section 13.1
u'	Perturbation radial velocity	defined in Section 4.4.2
u_s'	Perturbation radial velocity associated with smooth (non edge-wave) component	defined in Sections 4.4.8, 4.5
$\mathbf{u}$	Velocity vector	defined in Section 2.1
$\mathbf{u_s}$	Symmetric velocity	defined in Section 3.2
$\mathbf{u_{abs}}$	Absolute velocity vector of a fluid parcel	defined in Section 2.11
$\mathbf{u_h}$	Horizontal velocity vector	defined in Section 2.16
$\hat{u}(r,t)$	Complex Fourier amplitude of perturbation radial velocity	defined in Section 4.4.8
U_a	Zonal velocity component of the vortex environment	defined in Section 3.3.2.1
$U(y)$	Quadratic zonal flow $= U_o + U'y + \frac{1}{2}U''y^2$, where U_o, U', and U'' are constants	defined in Section 3.3.2.1
U	Storm-relative environmental flow	defined in Section 16.11
U', U''	First and second derivatives of zonal flow U	
$\bar{U}$	Zonal flow speed	defined in Section 3.2
U_{max}	Maximum radial velocity	defined in Section 8.1.1.1

Roman, continued	Meaning	Where first introduced
U_{min}	Minimum radial velocity	defined in Section 8.1.1.1
$U_o + iV_o$	Velocity of asymmetric flow across vortex center	defined in Section 4.1.1
U	Velocity vector of the vortex environment	defined in Section 3.2
U$_c$	Flow velocity at the vortex center	defined in Section 3.2
$\bar{\mathbf{U}}$	Zonal flow velocity	defined in Section 3.2
v	Meridional or azimuthal component of velocity	defined in Sections 2.10 and 3.3.1
v_b	Azimuthal (tangential) velocity component in boundary layer	defined in Section 13.1
v_g	Gradient wind - solution to Eq. (4.9) for given pressure & density	defined in Section 4.4.1
v_{gmax}	Maximum gradient wind	defined in Section 13.1
v'	Perturbation tangential (azimuthal) velocity	defined in Section 4.4.2
$V(r)$	Tangential velocity at radius r of an initially circular vortex	defined in Section 4.3
v_m	Maximum tangential wind speed	defined in Section 3.3.1
	Maximum gradient wind	defined in Section 12.3
v_{max}	Maximum gradient wind	defined in Fig. 12.4
v_0	Tangential velocity at top of surface layer (anemometer level, 10 m)	defined in Section 13.1
v$_s$	Surface velocity	defined in Section 8.1.2
$|\vec{\mathbf{V}}|_0$	Horizontal wind speed at anemometer level (10 m)	defined in Section 13.1
V_a	Meridional velocity component of the vortex environment	defined in Section 3.3.2.1
V_{max}	Maximum tangential velocity	defined in Section 8.1.1.1
VT_{max}	Maximum horizontal wind speed	defined in Section 10.3.1
$\hat{v}(r,t)$	Complex Fourier amplitude of perturbation tangential velocity	defined in Section 4.4.8
$\dot{V}$	Azimuthal momentum source per unit mass	defined in Sections 8.1.2 and 16.1
w	Vertical component of velocity	defined in Section 2.1
w_b	Vertical velocity at top of vortex boundary layer	defined in Section 13.2
w_{LFC}	Vertical velocity at the level of free convection	defined in Section 9.2.3
w_{LNB}	Vertical velocity at the level of neutral buoyancy	defined in Section 9.2.3
w_{max}	Maximum local vertical velocity	defined in Section 15.2
W_{max}	Maximum azimuthally-averaged vertical velocity	defined in Section 15.2
x	Zonal component of position	defined in Section 2.10
X	Distance in the zonal direction in a translating coordinate system	defined in Section 3.3.1
X	Vortex track	defined in Section 3.3.2.1
y	Meridional component of position	defined in Section 2.10
Y	Distance in the meridional direction in a translating coordinate system	defined in Section 3.3.1
Y_{smin}	Meridional displacement of the streamfunction center from the vorticity center in a symmetric translating vortex	defined in Section 3.3.1
Y	Minus the non-advective potential vorticity flux	defined in Sections 2.12 and 2.15
z	Height	defined in Sections 2.16, 2.17, and 5.1
z_i	Initial height of a displaced air parcel	defined in Section 9.2
z_{gm}	Height of maximum gradient wind ($=h$)	defined in Section 13.1
Z	Horizontal vorticity flux	defined in Section 2.16
$< ... >$	Three-hour time average and azimuthal average	defined in Section 10.5
$\overline{(...)}$	Azimuthal averaging operator	defined by Eq. (11.7)
$\overline{[...]}$	Averaging operator over a cylinder	defined by Eq. (14.1)

Greek	Meaning	Where first introduced
α	Specific volume of an air parcel ($= 1/\rho$)	defined in Section 2.3.2
	Prescribed constant characterizing orientation of vorticity dipole	defined in Section 4.1.1
	Local concentration/dilution of potential vorticity by the radial vorticity component times radial gradient of diabatic heating	defined in Section 8.1.1.3
	A positive constant characterizing the strength of drag	defined in Section 12.2
	$= R_{nd}/\Gamma U$	defined in Section 16.11
α_d	Specific volume of a dry air parcel	defined in Section 13.1
α_{diss}	Multiplicative factor ($\alpha_{diss} = T_s/T_o$) to include 'dissipative heating' in the extended EPI formulation	defined in Section 13.2
β	Meridional gradient of the Coriolis parameter	defined in Section 2.10
	Local concentration/dilution of potential vorticity by the vertical absolute vorticity times vertical gradient of diabatic heating	defined in Section 8.1.1.3
	Emanuel's 1997 beta parameter for tropical cyclone spin up	defined in Section 12.1.4
	Parameter in E86 formula for square of maximum gradient wind ($= 1 - \epsilon(1 + L_v r_{va}^* RH_{as}/R_d T_s)$)	defined in Section 13.1 & Eq. (13.102)
$\tilde{\beta}$	$= 1 - \epsilon(1 + L_v r_{va}^*/R_d T_s)$	defined by Eq. (13.95)
γ	Complex Fourier amplitude of VR edge-wave	defined in Section 4.4.8
	$= \chi/\rho r$	defined in Sections 5.9 and 16.1.1
γ_0	Initial amplitude of VR edge-wave	defined in Section 4.4.8
γ_r	Real part of $\hat{\gamma}(t)$	defined in Section 4.4.8
γ_i	Imaginary part of $\hat{\gamma}(t)$	defined in Section 4.4.8
Γ	Circulation	defined in Section 2.11
	Circulation of a vortex normalized by 2π	defined in Section 16.11
Γ_a	Absolute circulation	defined in Section 2.11
	Vorticity of the vortex environment	defined in Section 3.2
Γ_1, Γ_2	Contributions to Γ	defined in Eq. (3.11)
δ	boundary layer depth scale	defined in Section 6.4.5
δv	Change in azimuthally-averaged tangential velocity of a vortex	defined in Section 4.5.2
$\delta \mathbf{x}$	Differential line element of fluid parcels	defined in Section 2.9
Δ	Stability discriminant of axisymmetric baroclinic vortex	defined by Eq. (5.48)
	Increment in some quantity	used in Section 14.3.1
	Discriminant of moist-saturated Eliassen balance equation	defined by Eq. (16.11)
ΔM_{flux}	Ventilation diagnostic	defined by Eq. (15.1)
ϵ	The ratio of the specific gas constant for dry air to that for water vapor (R_d/R_v)	defined in Section 2.5
	Small angle between radii r and r' in Fig. 4.7	defined in Section 4.3
	Interface perturbation $= \epsilon(\lambda, t)$ separating rotational and irrotational fluid in a Rankine vortex near $r = a$	defined in Section 4.4.8
	Characteristic amplitude of initial vorticity perturbation in an unbounded Rankine vortex	defined in Section 4.4.8.1
	Thermodynamic efficiency factor ($= (T_B - T_o)/T_B$)	defined in Section 13.1
ϵ_D	Rate of viscous dissipation of kinetic energy	defined in Section 2.8
ζ	Vertical component of relative vertical vorticity	defined in Section 2.10
ζ'	Perturbation vertical vorticity	defined in Section 4.4.2
ζ_a	Vertical component of absolute vertical vorticity	defined in Section 2.9
	Asymmetric vorticity distribution	defined in Section 4.1.1
ζ_b	Vertical component of relative vorticity at top of boundary layer	defined in Eq. (13.38)
ζ_d	Vertical vorticity dipole	defined in Section 4.1.1
ζ_D	Prescribed constant characterizing strength of vortex dipole	defined in Section 4.1.1
ζ_d	Perturbation vorticity of VR edge-wave	defined in Section 4.4.8
	$= \zeta_d'$ in Section 4.5	
$\hat{\zeta}_d(r)$	Complex Fourier vorticity amplitude of VR edge-wave	defined in Section 4.4.8
ζ_{a1}	First order correction to the asymmetric vorticity	defined in Section 3.3.2.3
ζ_s'	Perturbation vorticity associated with smooth (non edge-wave) component	defined in Section 4.5
	$= \zeta_s$ in Section 4.4.8	
$\hat{\zeta}$	Complex Fourier vorticity amplitude of perturbation vorticity	defined in Section 4.5

Greek, continued	Meaning	Where first introduced
$\hat{\zeta}_r$	Real part of $\hat{\zeta}$	defined in Section 4.5
$\hat{\zeta}_i$	Imaginary part of $\hat{\zeta}$	defined in Section 4.5
$\hat{\zeta}_d$	Complex Fourier vorticity amplitude of discrete VR edge wave	defined in Section 4.5
$\hat{\zeta}_{dr}$	Real part of $\hat{\zeta}_d$	defined in Section 4.5
$\hat{\zeta}_{di}$	Imaginary part of $\hat{\zeta}_d$	defined in Section 4.5
$\zeta_{\mathbf{h}}$	Horizontal vorticity component	defined in Section 11.1
ζ_e	Azimuthally-averaged and mean pressure-weighted absolute vorticity gradient of the storm environment	defined in Section 3.4
ζ_s	Symmetric vortex vorticity	defined in Sections 3.3.1 and 4.3
$\hat{\zeta}_0(r, n)$	Radial structure of initial Fourier vorticity amplitude for azimuthal wavenumber n	defined in Section 4.4.5
$\hat{\zeta}_{s0}(r)$	Radial structure of initial smooth Fourier vorticity amplitude	defined in Section 4.4.8
$\hat{\zeta}_s$	Complex Fourier vorticity amplitude of smooth perturbation component	defined in Section 4.5
$\hat{\zeta}_{sr}$	Real part of $\hat{\zeta}_s$	defined in Section 4.5
$\hat{\zeta}_{si}$	Imaginary part of $\hat{\zeta}_s$	defined in Section 4.5
η	Normal displacement of a stretched membrane	defined in Section 2.19
	Azimuthal (or toroidal) component of vorticity	defined in Sections 5.13.1, 13.2 and 13.6
	Absolute vertical vorticity	defined in Section 5.13.3
η_b	Azimuthal component of vorticity at top of vortex boundary layer	defined in Sections 13.2 and 13.6
θ	Potential temperature	defined in Section 2.3.3
	Azimuthal angle from x-axis with origin O in Fig. 4.7	defined in Section 4.3
θ'	Perturbation potential temperature	defined in Section 8.1.1.2
θ_v	Virtual potential temperature	defined in Section 2.5.1
θ_e	Equivalent potential temperature	defined in Eq. (1.2)
	Local equivalent potential temperature at radius r and boundary layer top	defined in Section 13.7.3
θ_{es}	Local equivalent potential temperature at radius r and at the top of surface layer $= \theta_e$ assuming the boundary layer is well mixed between $z = h$ and surface layer	defined in Section 13.7.5
θ_{es}^*	Local saturation equivalent potential temperature at radius r, sea surface temperature T_s and surface pressure p_s	defined in Section 13.7.5
θ_{ea}	Equivalent potential temperature in vortex environment and at boundary layer top (i.e., $r = r_o$ and $z = h$)	defined in Section 13.7.3.2
θ_{esa}	Equivalent potential temperature in vortex environment and at top of surface layer (i.e., $r = r_o$ and $z \approx 10$ m) $= \theta_{ea}$ assuming boundary layer well-mixed between $z = h$ and surface layer	defined in Section 13.7.3.2
θ_e^*	Saturation equivalent potential temperature	defined in Section 9.2.4
θ_{ep}^*	Saturation equivalent potential temperature of a lifted air parcel	defined in Section 9.2.4
θ_{ea}^*	Saturation equivalent potential temperature in environment	defined in Section 9.2.4
	Saturation equivalent potential temperature in vortex environment and boundary layer top (i.e., $r = r_o$ and $z = h$)	defined in Section 13.7.3.2
	$= \theta_{esa}^*$ assuming well-mixed boundary layer	defined in Section 13.7.3.2
θ_w	Wet-bulb potential temperature	defined in Section 9.3.4.1
Θ	Initial potential temperature	defined in Section 10.1
κ	Ratio of dry air gas constant to specific heat at constant pressure (R/c_p)	defined in Section 2.3.2
λ	Azimuthal angle	defined in Section 3.3.2.1
	Azimuthal angle from x-axis with origin O' in Fig. 4.7	defined in Section 4.3
μ	molecular viscosity	defined in Section 2.6
ν	kinematic viscosity (momentum diffusivity)	defined in Section 2.6
ξ	Twice the local absolute rotation rate of a fluid parcel at radius r, $2v/r + f$	defined in Section 5.13.3
π	Ratio of circumference of a circle to its diameter, $\arctan(1) \approx 3.14159265$	
	Exner function, $(p/p_*)^{\kappa}$	defined in Section 2.3.3
π_h	The local value of the Exner function π at radius r and height $z = h$	defined in Section 13.7.3.2
π_{ha}	The environmental value of the Exner function π at $z = h$ and $r = r_o$ (also π_a)	defined in Section 13.7.3.2
π_s	The local value of the Exner function π at radius r and surface $z = 0$	defined in Section 13.7.3.2

Greek, continued	Meaning	Where first introduced
π_{sa}	The environmental value of the Exner function π at $r = r_o$ and $z = 0$	defined in Section 13.7.3.2
ρ	Fluid density	defined in Section 2.1
ρ_d	Density of dry air	defined in Section 13.6
ρ_o	A horizontally-averaged density	defined in Section 2.1
ρ_*	A reference density	defined in Section 2.2
ρ_{ref}	Reference density	defined in Section 9.5
σ	Scaled pressure	defined in Section 2.17
τ	Surface stress	mentioned first in Section 1.4 and defined mathematically in Section 6.4.2
$\tau_{\lambda z}, \tau_{rz}$	Frictional stresses per unit area in the tangential and radial directions	defined in Section 6.4.6
ϕ	Latitude	defined in Section 2.1
	Geopotential in pseudo-height coordinates Z	defined in Section 5.13.3
ϕ_t	Geopotential tendency ($=\partial\phi/\partial t$)	defined in Section 5.13.3
Φ	Effective gravitational potential (or geopotential)	defined in Section 2.1
Φ_v	Dissipation function	defined in Eq. (14.9)
χ	The inverse potential temperature ($1/\theta$)	defined in Section 5.4
χ_1, χ_2	Amplitude functions representing advection of symmetric vorticity by storm-relative asymmetric flow	defined in Section 3.6.3
ψ	Streamfunction for horizontal flow	defined in Section 2.10
	Streamfunction for meridional (secondary) circulation	defined in Section 5.9
ψ'	Perturbation streamfunction	defined in Section 4.4.2
ψ_a	Streamfunction of the vortex environment	defined in Section 3.2
ψ_s	Streamfunction of a symmetric vortex	defined in Section 3.3.1
Ψ_1, Ψ_2	Streamfunction components associated with Γ_1 and Γ_2	defined in Section 3.3.2.1
$\hat{\psi}_n$	Complex Fourier amplitude of perturbation streamfunction for azimuthal wavenumber n	defined in Section 4.4.2
$\hat{\Psi}(r)$	Radial structure of complex perturbation eigen-streamfunction for azimuthal wavenumber n	defined in Section 4.4.3
$\hat{\psi}(r, s)$	Laplace transform of perturbation streamfunction for azimuthal wavenumber n	defined in Section 4.4.4
$\psi_n(r, \lambda, t)$	Complex perturbation streamfunction	defined in Section 4.4.5
$\hat{\psi}$	Complex Fourier amplitude of perturbation streamfunction	defined in Section 4.4.8
$\hat{\psi}_r$	Real part of $\hat{\psi}$	defined in Section 4.5
$\hat{\psi}_i$	Imaginary part of $\hat{\psi}$	defined in Section 4.5
$\hat{\psi}_s$	Complex Fourier streamfunction amplitude of smooth perturbation component	defined in Section 4.4.8
$\hat{\psi}_{sr}$	Real part of $\hat{\psi}_s$	defined in Section 4.5
$\hat{\psi}_{si}$	Imaginary part of $\hat{\psi}_s$	defined in Section 4.5
$\hat{\psi}_d$	Complex Fourier streamfunction amplitude of discrete VR edge-wave component	defined in Section 4.5
$\hat{\psi}_{dr}$	Real part of $\hat{\psi}_d$	defined in Section 4.5
$\hat{\psi}_{di}$	Imaginary part of $\hat{\psi}_d$	defined in Section 4.5
$\hat{\Psi}_1$	Radial structure of Fourier streamfunction amplitude of VR edge-wave component	defined in Section 4.4.8
ω	Eigen-frequency for perturbation flow	defined in Section 4.4.3
	The material derivative of pressure (plays the role of vertical velocity in pressure coordinates)	defined in Section 11.1
ω_r	Real part of eigen-frequency	defined in Section 4.4.3
ω_i	Imaginary part of eigen-frequency	defined in Section 4.4.3
ω_c	Perturbation wave frequency at critical radius r_c: $\omega_c = n\Omega(r_c)$	defined in Sections 4.4.4 and 4.5.6
ω_n	VR edge-wave frequency for integral wavenumber n	defined in Section 4.4.8.1
ω	Three-dimensional vorticity vector	defined in Section 2.9
$\omega_{\mathbf{a}}$	The three-dimensional absolute vorticity vector	defined in Section 2.9
$\mathbf{\Omega}$	The Earth's angular rotation vector or that of a laboratory turntable	defined in Section 2.1
Ω	The magnitude of the angular rotation rate of the Earth or a laboratory turntable	defined in Section 2.1
	Angular velocity of a symmetric vortex about its rotation axis	defined in Section 3.3.1

Acronyms

AMMA	African Monsoon Multidisciplinary Analysis, a NASA field experiment
C130	Lockheed C130 Hercules aircraft
CAPE	Convective available potential energy (defined in Section 9.2.1)
CIN	Convective inhibition (defined in Section 9.2.1)
DC8	McDonnell Douglas DC8 aircraft
DCAPE	Downdraft Convective Available Potential Energy (defined in Section 9.3.4.3)
ECMWF	European Centre for Medium Range Weather Forecasts
ELDORA	Airborne radar mounted on the NRL P3 research aircraft
GATE	Global Atmospheric Research Experiment Atlantic Tropical Experiment
GFS	NOAA Global Forecasting System
GOES	Geostationary Operational Environmental Satellite
GPS	Global Positioning System
GRIP	NASA Genesis and Rapid Intensification experiment
GV	NSF/NCAR Gulfstream-V (G-V) aircraft
HAFS	Hurricane Analysis and Forecast System
HRD	NOAA Hurricane Research Division
IFEX	NOAA Intensity Forecasting Experiment
ITCZ	Inter-Tropical Convergence Zone
LCL	Lifting condensation level (defined in Section 9.1)
LFC	Level of free convection (defined in Section 9.1)
LNB	Level of neutral buoyancy (defined in Section 9.1)
MCS	Mesoscale convective system
minSLP	Minimum sea level pressure
MSLP	Mean sea level pressure
NA	The negative area on an aerological diagram (defined in Section 9.2.1)
NASA	National Aeronautics and Space Administration
NCAR	National Center for Atmospheric Research
NHC	NOAA National Hurricane Center
NOAA	National Oceanographic and Atmospheric Administration
NRL	Naval Research Laboratory
NSF	National Science Foundation
OCS	Outer core strength
OHC	Ocean heat content (defined in Eq. (1.4)
OW	Okubo-Weiss parameter (defined on page 33, footnote 15)
QuikSCAT	Quick Scatterometer
P3	Lockheed WP-3D Orion aircraft
PA	The positive area on an aerological diagram (defined in Section 9.2.1)
PBL	Planetary Boundary Layer
PREDICT	Pre-Depression Investigation of Cloud systems in the Tropics, a NSF field experiment
RAINEX	Hurricane Rainband and Intensity Change Experiment
RMW	Radius of maximum tangential wind
ROCI	Radius of the outermost closed isobar
SAL	Saharan Air Layer
SGS	Sub-grid scale
SST	Sea surface temperature
TCSP	Tropical Cloud Systems and Processes field experiment
TEXMEX	Tropical EXperiment in MEXico
TRMM	Tropical Rainfall Measuring Mission satellite
UK	United Kingdom
U.S.	United States
USAF	United States Air Force
VRW	Vortex Rossby Wave (= VR wave)
WISHE	Wind Induce Surface Heat Exchange (discussed in Sections 12.2 and 16.1.8)
WMO	World Meteorological Organization

Chapter 1

Observations of tropical cyclones

The term *tropical cyclone* refers generically to a class of synoptic-scale, cyclonically-rotating, non-frontal, low-pressure weather systems containing areas of deep cumulus convection that form generally over the warm tropical oceans. Cyclonic means counterclockwise in the Northern Hemisphere and clockwise in the Southern Hemisphere. These weather systems are classified according to their *intensity*, defined as the strength of the maximum sustained near surface wind speed. By convention, *near surface* is taken to be a height of 10 m above mean sea level and *sustained* refers to a 10 min average value, except in the United States and adjoining basins, which adopts a 1 min average. The averaging serves to eliminate temporal turbulent wind gusts that can significantly exceed the average value, especially the 10 min average. The maximum 1 min sustained wind is typically about 14% greater than the maximum 10 min sustained wind.

Tropical cyclones are referred to by different names depending on their geographical location and intensity. When the intensity is below 17 m s^{-1} (60 km h^{-1}, 32 knots), they are called *tropical depressions* or just *tropical lows*. When the intensity lies between 17 m s^{-1} and 32 m s^{-1} the term *tropical storm* is used, except in the Australian region, where they are simply called *tropical cyclones*. The term *severe tropical cyclone* is used in the Australian region, the Southwest Pacific Ocean, west of 160°E, or Southeast Indian Ocean east of 90°E, when the intensity is equal to or exceeds 33 m s^{-1} (120 km h^{-1}, 64 knots). In other regions they are called *hurricanes* over the Atlantic Ocean, the Northeast Pacific Ocean (east of the dateline), and the Caribbean Sea, *typhoons* over the Northwest Pacific Ocean (west of the dateline) and *very severe cyclonic storms* over the North Indian Ocean. An alternative measure of intensity is the minimum sea-level pressure (min-slp), although there is not a unique relation between this and the foregoing wind speed definition for reasons exemplified in Chapter 5, Exercise 5.2.

Forecast offices use also a classification into five categories to indicate the potential wind damage that storms may inflict. The classification scale has some regional variation, the most common being the Saffir-Simpson scale[1]

which is used by the United States National Hurricane Center (NHC). At the lowest end of the scale, a Category 1 Hurricane has maximum sustained winds in the range 33 m s^{-1} to 43 m s^{-1} and is likely to cause roof damage, bring down large tree branches and bring down power lines. At the top end of the scale, a Category 5 Hurricane has maximum sustained winds exceeding 69 m s^{-1} and is expected to produce massive destruction, destroying homes and trees, making an area uninhabitable for weeks or months.

Typically the strongest winds in a tropical cyclone occur in a ring some tens of kilometers from the center and there is a calm region near the center, *the eye*, where winds are light. For moving storms, the wind distribution is distinctly asymmetric with the maximum winds in the forward right quadrant in the Northern Hemisphere and in the forward left quadrant in the Southern Hemisphere. The eye acquires its name because it is often seen in satellite imagery as an approximately circular region free of deep clouds in an otherwise densely overcast region.

The eye is surrounded by a ring of deep convective clouds that slope outwards with height. This ring is called the *eyewall cloud* or simply the *eyewall*. At larger radii from the center, storms usually show spiral bands of deep convective clouds. Fig. 1.1 shows satellite views of the eye and eyewall of a mature typhoon as well as photographs looking out at the eyewall cloud from the eye during aircraft reconnaissance flights into a hurricane and a typhoon.

1.1 Tropical-cyclone tracks

Fig. 1.2 shows the tracks of all tropical cyclones with maximum near-surface winds > 17 m s^{-1} for the period 1979-1988. It shows also the mean direction of all hurricane strength systems during the period indicated for each ocean basin. After forming over the warm tropical oceans, tropical cyclones subsequently move westwards and polewards, although the tracks of individual storms can be quite erratic. To a first approximation tropical cyclones are steered by a mass-weighted average of the broadscale winds through the depth of the troposphere. It is common for storms that reach sufficiently high latitudes to recurve and move eastwards, carried along by the prevailing mid-latitude westerly winds.

Many basic aspects of tropical cyclone motion, including a theory for the poleward and westward drift, can be

---

1. More information about this scale can be found at https://en.wikipedia.org/wiki/Saffir-Simpson_scale.

Tropical Cyclones. https://doi.org/10.1016/B978-0-44-313449-4.00009-6

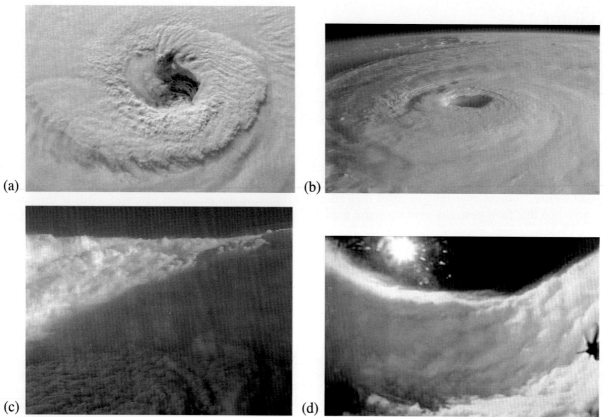

FIGURE 1.1 Eyes of (a) Hurricane Florence (2018) and (b) Hurricane Isabel (2003) from the International Space Station (Courtesy NASA). The lower panels show aerial photographs from reconnaissance aircraft of the eyewall looking out from the eye of a tropical cyclone. (c) Eyewall of Hurricane Georges, 1945 UTC, 19 September 1998 (Photo courtesy of M. Black, NOAA Atlantic Meteorological and Oceanographic Laboratory, Hurricane Research Division (HRD), Miami, Florida, USA), and (d) Typhoon Jangmi (2008) (Photo courtesy N. Sanger).

illustrated in terms of barotropic dynamics in which the vortex structure is assumed to be independent of height. The barotropic dynamics of tropical cyclone motion is the topic of Chapter 3.

Tropical cyclones lose their intensity rapidly when they move over land, but they often continue to produce copious amounts of rain as they move inland. In many cases of landfalling storms, the majority of damage is caused by widespread flooding rather than by strong winds. Near the coast, however, much damage may be caused by high winds and by coastal storm surges.

1.2 Structure

Fig. 1.3 shows a schematic cross section of prominent cloud features in a mature cyclone including the eyewall clouds that surround the largely cloud-free eye at the center of the storm; the spiral bands of deep convective clouds outside the eyewall; and the cirrus cloud canopy in the upper troposphere. The schematic shows also the hub cloud in the center of the eye. The hub cloud is a feature that was first reported by R. Simpson in his historic Typhoon *Marge* flights

in 1952. Also of note was the recognition of the eyewall tilt, which in the early days was not believed to be a real.

To a first approximation, the core of a mature tropical cyclone may be thought of as a horizontal quasi-symmetric circulation on which is superposed a transverse, or overturning "in-up-and-out" circulation in a vertical cross section. These two flow components are often referred to as the *primary circulation* and *secondary circulation*, respectively. When combined, they represent a spiraling motion with inflow in the low to middle troposphere and outflow in the upper troposphere. Air spirals into the storm at low levels and out of the storm in the upper troposphere. The spiraling motions are often evident in cloud patterns seen in satellite imagery and in radar reflectivity displays.

The strongest inflow into the storm is confined to a shallow frictional boundary layer, typically 0.5-1 km deep, but there is generally weak inflow in the lowest few kilometers. As in other parts of the tropics, the air ascends from the boundary layer in the cores of deep convective clouds while in the upper troposphere, ascent occurs also, but at a lesser rate in the more widespread anvils produced by these clouds. The primary circulation is strongest at low levels under the eyewall cloud region and decreases in intensity

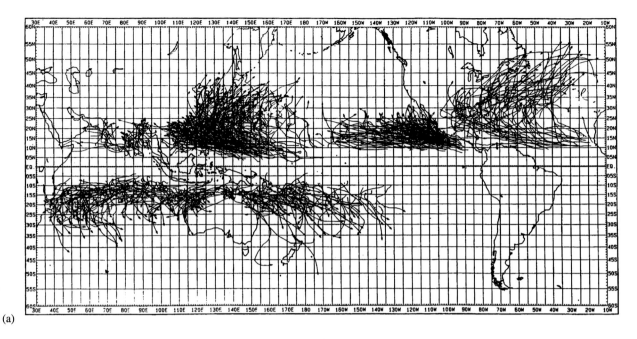

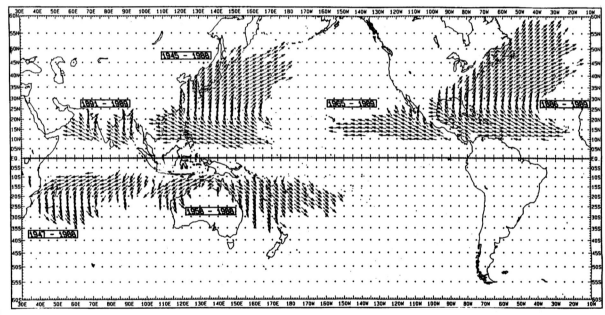

FIGURE 1.2 (a) Tracks of all tropical cyclones (maximum winds > 17 m s^{-1}) for the period 1979-1988. (b) Mean direction of tropical cyclone motion over the periods indicated. From WMO (1993). Republished with permission from WMO.

with both radius and height. In the upper-tropospheric outflow, the primary circulation becomes anticyclonic beyond a radius of a few hundred kilometers.

The secondary circulation is mostly thermally direct, which means that warm air is rising, a process that releases potential energy. However subsidence occurs in the eye, which is the reason why the eye is free of deep clouds. In the eye, the circulation is thermally indirect, a process that requires energy to be supplied. Because of the subsidence, temperatures in the eye are relatively large compared with

those at the same level at larger radii. Outside the eye, most of the temperature excess is found in the upper troposphere.

Tropical cyclones originate and spend much of their lifetime over the tropical oceans where conventional data are sparse. Rarely has it been possible to release radiosondes in the central region of storms and, even then, not in the high wind region of the eyewall. As a result, much of what has been learned about the detailed structure of tropical cyclones has been obtained by instrumented aircraft penetrations into the storms. Such flights began towards the end

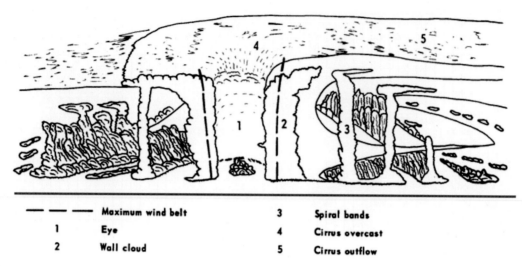

— — — —	Maximum wind belt	**3**	Spiral bands
1	Eye	**4**	Cirrus overcast
2	Wall cloud	**5**	Cirrus outflow

FIGURE 1.3 Schematic cross section of cloud features in a mature tropical cyclone. Vertical scale greatly exaggerated. Source unknown.

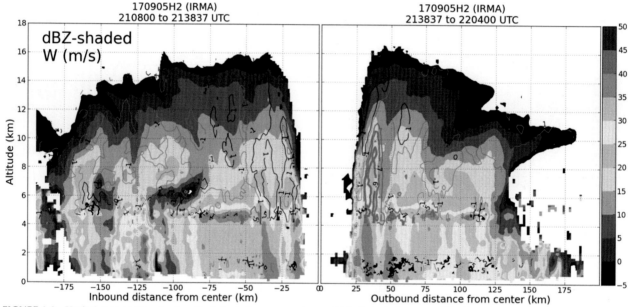

FIGURE 1.4 Vertical cross section of radar reflectivity during a reconnaissance flight through the center of Hurricane Irma (2017). Reflectivity is shown on the color scale in dBZ. The contours show vertical velocity in m s^{-1}. Courtesy Dr. Paul Reasor, NOAA Atlantic Meteorological and Oceanographic Laboratory, Hurricane Research Division, Miami, Florida, USA.

of the Second World War. Until quite recently, the majority of these storm penetrations have been at comparatively low altitudes using turboprop aircraft such as specially instrumented Lockheed WP-3D (P3) Orion aircraft. There have been a few penetrations using higher flying jet aircraft such as the National Aeronautics and Space Administration (NASA) Douglas DC8, but these are unable to overfly the storm completely and only very recently have measurements been made through the full depth of the troposphere and even the lower stratosphere using converted spy planes and pilotless aircraft systems such as the NASA Global Hawk.

Fig. 1.4 shows a vertical cross section of radar reflectivity obtained from the tail radar of a P3 research aircraft dur-

ing its traverse of Atlantic Hurricane Irma (2017). The echo free eye is bounded on each side by a column of strong reflectivity, which marks the eyewall. The edge of the eyewall has a noticeable outward tilt in the upper troposphere. Vertical velocities in the outbound leg are particularly strong, exceeding 9 m s^{-1}. The shallow layer of strong reflectivity at an altitude of about 5 km, often referred to as the "bright band" marks the level at which falling snow and ice crystals melt.

In this chapter we will exemplify many of the salient features of storms using a few particular storms as examples. These include Category 5 Hurricane Patricia (2015) over the Eastern Pacific, which was not only the strongest ever recorded in the Eastern Hemisphere, it was well documented by research aircraft. In particular, dropwindsondes were released from an aircraft flying in the lower stratosphere giving a comparatively high spatial resolution of storm structure below flight level. In addition, Doppler radar and flight level data were obtained by a National Atmospheric and Oceanic Administration (NOAA) P3 aircraft on two days. Other well documented storms that we will show data for are Category 5 Hurricane Isabel (2003) and two weaker storms, Hurricane Earl (2010) and Hurricane Edouard (2014), all from the Atlantic.

1.2.1 Formation and intensification of Hurricane Patricia

Fig. 1.5 shows the track and intensity of Hurricane Patricia and its precursor disturbance taken from the National Hurricane Center best track data, while Fig. 1.6 shows a sequence of visible satellite imagery from 20 Oct. to 23 Oct. during its formation and intensification. As is typical, Hurricane

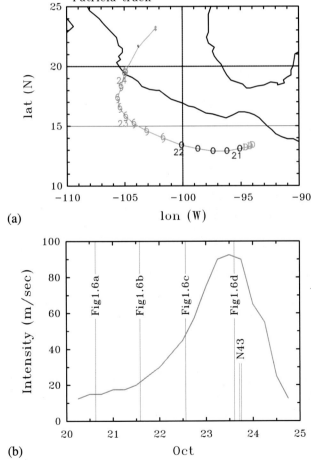

(a)

(b)

FIGURE 1.5 (a) Track, and (b) intensity of Hurricane Patricia in Oct. 2015, based on the best track data from the National Hurricane Center. The single vertical lines indicate the times of the satellite imagery in Fig. 1.6 while the pairs of lines indicate the times of aircraft reconnaissance in the storm.

Patricia developed from a region of active deep cumulus convection, in this case associated with a tropical depression south of the Pacific coast of Mexico (Fig. 1.6a).

At the time of the first image (1500 UTC 20 Oct.), the maximum near surface winds were bordering on tropical storm strength (Fig. 1.5b). Deep convection remained active and on the following day (1400 UTC 21 Oct.) was showing signs of becoming more focused (Fig. 1.6b). Moreover, the near surface winds had strengthened (Fig. 1.5b). In fact, the system was upgraded to tropical storm status by the National Hurricane Center at 0000 UTC on this day.

At about the time of Fig. 1.6b, the system began a period of rapid intensification and less than 12 hours later it was declared a hurricane. Supported by anomalously warm ocean temperatures and a favorable environment with low vertical wind shear, the rapid intensification continued and the hurricane reached Category 5 status with extreme wind speeds of around 100 m s^{-1} at 1200 UTC on 23 Oct. Shortly afterwards, the intensity declined rapidly before the hurricane made landfall.

Some features of the inner core of a tropical cyclone are exemplified by aircraft reconnaissance data obtained during the first NOAA P3 penetration into Hurricane Patricia on 23 Oct. 2015 shown in Fig. 1.7. For comparison, Fig. 1.8 shows similar data from a United States Air Force aircraft reconnaissance flight into Atlantic Hurricane Isabel on 14 Sep. 2003, another well-documented Category 5 storm.

In both cases, the aircraft flew at a pressure altitude of approximately 3 km and therefore, because of the low pressure in the vortex, the aircraft descended as it approached the center and ascended again as it left the center. In the case of Patricia, this descent was slightly more than 1 km (Fig. 1.7a), whereas in Isabel it was only about 500 m (Fig. 1.8b top). The data obtained are summarized below.

1.2.2 Flight level wind structure and temperature structure

The typical wind structure as exemplified by the data for both hurricanes is the progressive increase in wind speed with decreasing radius, followed by a rapid decrease to the center, the location at which the wind speed falls close to zero and where the wind direction changes rapidly. The strongest winds are found beneath the eyewall where vertical velocities are relatively large.

In the case of Patricia, the maximum wind speed, V_{max} is extreme, about 100 m s^{-1}, and it occurs at a radius, R_{Vmax}, of only about 10 km, whereas in Isabel on the day shown, V_{max} is around 70 m s^{-1}, but R_{Vmax} is much larger, about 50 km.

In reality, there can be a significant variation in the size of the inner-core vortex between different storms and, indeed, for the same storm at different stages of its life cycle. This variation is illustrated by the data for Isabel shown in

FIGURE 1.6 Visible imagery from the Geostationary Operational Environmental Satellite (GOES) West satellite showing the formation of Hurricane Patricia in Oct. 2015: (a) The precursor tropical depression at 1500 UTC on 20 Oct.; (b) a well organized convective cluster at 1500 UTC on 21 Oct.; (c) tropical storm Patricia at 1330 UTC on 22 Oct.; (d) Hurricane Patricia at 1430 UTC on 23 Oct. Courtesy Naval Research laboratory (NRL), Monterey, California, USA.

Section 1.2.4. The underlying dynamics controlling inner core size will be examined in Section 10.14.

In Patricia, the temperature rises from approximately 11 °C far from the center to about 30 °C, with much of the rise taking place inside the radius of maximum winds. In Isabel, the temperature rises from approximately 10 °C far from the center to 20 °C. The dew point depression[2] ranges between 5 °C and 10 °C far from the center, falls to zero as the eyewall cloud is penetrated, and increases again inside R_{Vmax}, indicative of subsidence occurring in the eye. Part of the temperature increase in the eye is a result of the aircraft descending to warmer temperatures, but the fall in dew point temperature during descent of the aircraft would have to be an indication of subsidence in the eye as typically the dew-point temperature would have a negative vertical gradient. The dynamics of the eye and the reasons for this subsidence are discussed in Section 5.5.

2. The difference between the temperature and the dew-point temperature.

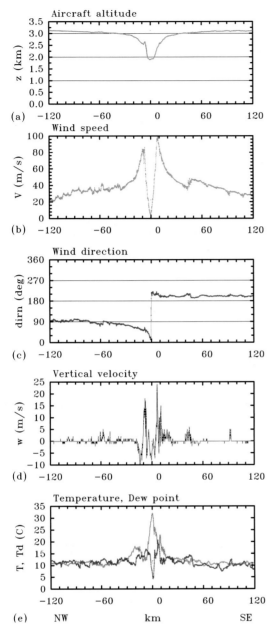

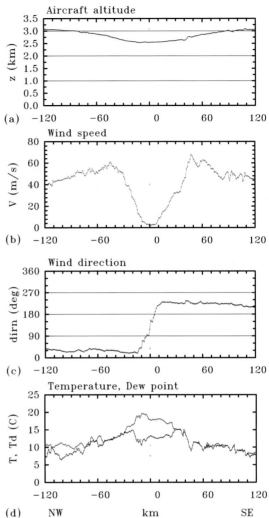

FIGURE 1.8 Flight level data for the USAF C130 penetration of Category 5 Atlantic Hurricane Isabel on 14 Sep. 2003. (a) aircraft altitude; (b) wind speed (*V*), (c) wind direction (*dirn*), (d) temperature (*T*) and dew point temperature (*T_d*). The aircraft track was from the northwest to southeast starting at 1837 UTC and the portion shown took approximately 73 min to complete. The minimum flight level wind speed, the center of the storm is at 0 km radius. Constructed from data made available by NOAA/HRD.

FIGURE 1.7 Flight level data during the first NOAA P3 penetration into for Hurricane Patricia on 23 Oct. 2015: (a) aircraft altitude; (b) wind speed (*V*), (c) wind direction (*dirn*); (d) vertical velocity (*w*); and (e) temperature (*T*) and dew-point temperature (*T_d*). The aircraft track was from the northwest to southeast starting at 1714 UTC and the portion shown took approximately 40 min to complete. The minimum flight level wind speed, the center of the storm, is at 0 km radius. Constructed from data made available by NOAA/HRD.

1.2.3 Vertical cross sections in Hurricane Edouard (2015)

In recent years, it has been possible to obtain vertical cross sections of wind and thermodynamic structure in a few individual storms using the NASA Global Hawk, a high flying unmanned aircraft with the capability to release multiple dropsondes in rapid succession from the lower stratosphere. An example of data so obtained is shown in Fig. 1.9. These data were gathered over a period of 23 hours in Atlantic Hurricane Edouard (2014), shortly after it reached its peak intensity, and are discussed in detail by Smith et al. (2018a). In this case the Global Hawk was flying at an altitude of approximately 18 km.

The storm-relative composite tangential wind component (*v*, Fig. 1.9a) and temperature perturbation (*dT*, Fig. 1.9b) show the classical structure of a warm-cored vortex with the maximum wind in the lower troposphere and the wind decreasing with height, becoming anticyclonic in the upper troposphere. (For the calculation of temper-

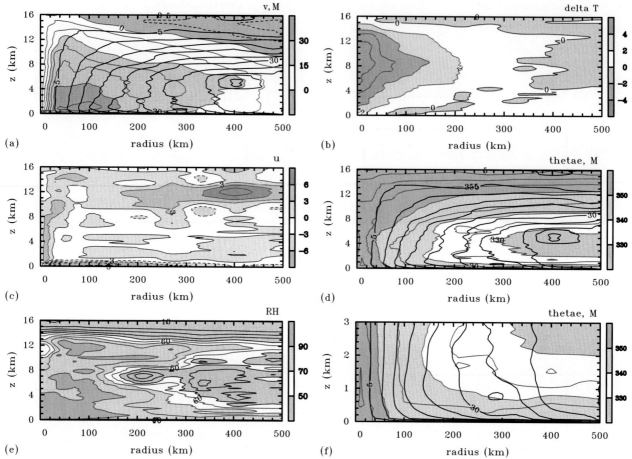

FIGURE 1.9 Radius-height cross sections of selected fields derived from the dropsonde data in Hurricane Edouard (2014): (a) tangential velocity component, contour interval 5 m s^{-1}, shading indicated on the side bar in m s^{-1}, and absolute angular momentum, black lines, contour interval 5 $\times 10^5$ m^2 s^{-1}; (b) temperature perturbation, contour interval 2 K (positive values), 1 K (negative values), shading indicated on the side bar in K; (c) radial velocity component, contour interval 3 m s^{-1}, shading indicated on the side bar in m s^{-1}; (d) pseudo-equivalent potential temperature, contour interval 10 K, shading indicated on the side bar in K, and absolute angular momentum, black lines, contours as in (a); (e) relative humidity, contour interval 10%, shading indicated on the side bar in %; (f) a zoomed in version of panel (d) at heights below 3 km. From Smith et al. (2018a). Republished with permission of Wiley.

ature perturbation, the "environmental temperature" was determined by averaging all 46 dropsondes at radii > 200 km.) The decrease in the tangential velocity component with height corresponds through so-called "thermal-wind balance" with the warm-core structure (see Fig. 1.9b). A theoretical explanation for this structure is given in Chapter 5, specifically in Section 5.2 and in Section 5.3, where the concept of "thermal-wind balance" is explained.

Fig. 1.9a shows also the absolute angular momentum (or M-) surfaces corresponding with the tangential wind component. The quantity M is defined by the equation

$$M = rv + \frac{1}{2}fr^2 \qquad (1.1)$$

where r is the radius, v is the tangential velocity component and f is the Coriolis parameter (a measure of twice the Earth's local rotation rate about a local vertical, Chapter 2).

As shown in Section 5.1, M is theoretically significant because, in an axisymmetric frictionless flow, it is a materially conserved quantity.

The tangential wind structure shows evidence of a weak inner wind maximum near 40 km radius and a height of no more than 1 km, with an outer maximum at a radius of about 100 km at a similar altitude. During the period of observation, the storm was undergoing a so-called eyewall replacement cycle: see Section 1.2.8. The upper-level anticyclone begins at a radius of about 80 km at 16 km altitude and deepens with increasing radius, as does its strength. The maximum anticyclonic flow is found at an altitude between 14 and 15 km at 500 km radius and is clearly increasing beyond this radius.

There is a marked (> 2 °C) positive temperature anomaly inside a radius of about 200 km (Fig. 1.9b). This anomaly has a maximum of nearly 10 °C on the axis of rotation at an altitude of about 8 km. There is a weak cold temperature anomaly at low levels beyond about 60 km radius. The negative temperature anomalies beyond about 400 km radius and those above 13 km may result from the way in which the ambient temperature has been defined and are presumably not significant.

The storm-relative composite radial flow (u, Fig. 1.9c) shows two prominent features of the classical tropical cyclone structure with a layer of strong inflow below about 1 km height extending to large radii as well as a layer of strong outflow in the upper troposphere between about 9 and 14 km, depending on radius. The maximum low-level inflow is about 15 m s^{-1}. The layer of upper tropospheric outflow is a few km deep with a maximum of nearly 12 m s^{-1} at about 12 km altitude and 400 km radius.

In the lower troposphere there are significant regions of outflow above the shallow surface-based inflow layer. This outflow has a local maximum in the inner eyewall (near 20 km radius) and has a layered structure beyond a radius of about 90 km starting near outer eyewall. Calculation of radial flow component from the dropsonde data is somewhat sensitive to the way in which these data are partitioned and some smaller-scale features in Fig. 1.9c are unlikely to be robust (see Smith et al. (2018a) for further discussion). This sensitivity is compounded by an apparent limitation of the assumed steadiness of the storm over the period of data collection discussed above.

While the temperature gives an indication of how warm an air parcel is, another quantity that proves useful for indicating the heat content of a moist air parcel is the pseudomoist equivalent potential temperature, θ_e, which is essentially related to the moist entropy and is defined in Section 2.5.4. In brief, θ_e is determined by the approximate equation

$$\theta_e = T \left(\frac{p_{**}}{p} \right)^\kappa \exp \left(\frac{L_v r_v}{c_{pd} T} \right), \qquad (1.2)$$

where T is the absolute temperature, p is the pressure, p_{**} is the standard pressure, 1000 mb, r_v is the water vapor mixing ratio, L_v is the latent heat of vaporization per unit mass and c_{pd} is the specific heat of dry air at constant pressure per unit mass. The derivation of this equation is discussed in Section 2.5.4 and the significance and use of θ_e are discussed at length in Section 9.2.4 and in subsequent chapters. Suffice it to say here that, like M, θ_e is approximately materially conserved under certain circumstances, a property that makes it useful theoretically.

The cross sections of θ_e in Figs. 1.9d and 1.9f (the latter is a zoomed-in version of the former in the lowest 3 km) shows the structure found in many earlier studies. Principal features are: the mid-tropospheric minimum beyond a radius of about 100 km, increasing in prominence with radius; the tendency for the isopleths of θ_e to become close to vertical in the lower troposphere inside a radius of 100 km;

and the tendency for the isopleths of θ_e to slope outwards and become close to horizontal in the upper tropospheric outflow layer.

With a little imagination, there is an approximate congruence between the θ_e- and M-surfaces in the inner core region and in the upper troposphere, at least out to 250 km radius (the M-surfaces are shown also in Figs. 1.9d and 1.9f). At least in a flow that is quasi-steady, this approximate congruence would be expected if M and θ_e were approximately materially-conserved quantities. The assumed congruence of the M- and θ_e- surfaces is a cornerstone in a theoretical formulation of steady-state and intensifying tropical cyclones reviewed in Chapters 13 and 12, respectively.

Throughout much of the troposphere, θ_e has a negative radial gradient. This is partly a reflection of the radial structure of θ_e in the boundary layer. Below about 600 m, the negative radial gradient of θ_e is apparent only inside a radius of about 100 km and is a result of the presumed increase in surface moisture flux with decreasing radius (Malkus and Riehl, 1960; Ooyama, 1969). This negative radial gradient is the hallmark signature of a mesoscale vortex that is creating its own convectively unstable environment. The relatively high θ_e air in the inner-core boundary layer sustains deep moist convective thermals, the aggregate of which act to draw in air above the frictional boundary layer. Such a localized radial gradient was documented in the classical observational analysis of Hawkins and Imbembo (1976) and has been confirmed by more recent work (Montgomery et al., 2006b; Marks et al., 2008; Bell and Montgomery, 2008; Smith and Montgomery, 2012a).

Maximum values of θ_e here exceed 355 K in the low to mid troposphere near and inside the inner eyewall region. The near surface value is approximately constant at 350 K outside of 100 km radius. The minimum value in the mid to low troposphere falls to values less than 320 K beyond about 300 km radius (the region highlighted in blue (dark gray in print version) in Fig. 1.9d). Further discussion of the dynamical implications of the negative radial gradient of θ_e near and just outside the radius of maximum tangential winds is provided in Sections 12.1.4, 14.3, and 16.1.4.

Values of relative humidity,[3] (RH, panel (d)), exceed 90% inside a radius of 200 km and below a height of about 7 km. At larger radii, values remain relatively high ($> 80\%$) in a shallow near-surface layer, but decrease markedly with height with values of less than 50% through much of the troposphere, especially beyond a radius of about 300 km. These low values are an indication of drying in the subsiding branch of the secondary circulation.

---

3. The relative humidity is calculated relative to water saturation below the freezing level and relative to ice saturation above this level.

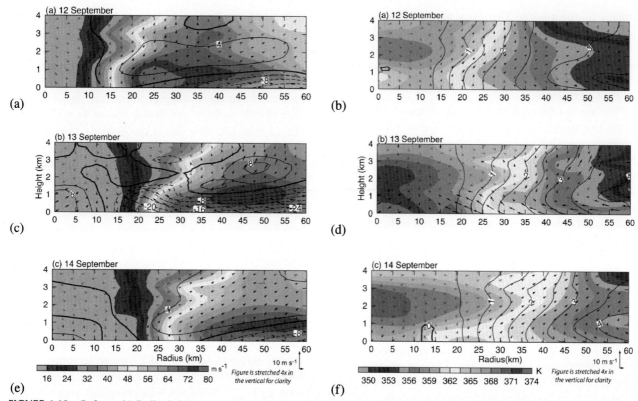

FIGURE 1.10 (Left panels) Radius-height cross sections of azimuthal mean storm-relative tangential wind (color), radial wind (contour), and the secondary circulation (vector) in m s^{-1} derived from GPS dropwindsonde and flight-level data in Hurricane Isabel (2003) on (a) 12, (c) 13, and (e) 14 Sep. 2003. The origin denotes the storm center at the ocean surface. (Right panels) Similar cross sections of θ_e (color; K), absolute angular momentum (contour; m^2 s^{-1} $\times 10^6$), and transverse secondary circulation (vector) on (b) 12, (d) 13, and (f) 14 Sep. 2003. From Bell and Montgomery (2008). Republished with permission of the American Meteorological Society.

1.2.4 Low-level structure of Hurricanes Isabel (2003) and Earl (2010)

Intensive field campaigns using multiple aircraft platforms have enabled researchers to create unprecedented, high-resolution depictions of the low-level, inner-core structure of some notable major hurricanes. The left panels of Fig. 1.10 show radius-height cross sections of the azimuthally-averaged, storm relative, radial u and tangential v wind components in the inner core of Category 5 Hurricane Isabel (2003) on 12, 13 and 14 Sep. 2003 derived from the high-density of dropsonde and flight-level observations using an objective analysis scheme (Bell and Montgomery, 2008). The arrows in these panels depict the transverse secondary circulation.

Inspection of the derived wind structure reveals that the peak mean tangential wind at the radius of maximum tangential wind (RMW) remained very strong on all three days, with the core region of maximum tangential winds decaying from 80 to only 74 m s^{-1}, rising from 500 m to 1 km altitude, and expanding from 25 km to 50 km radius. The low-level radial inflow increased in both depth and intensity from 12 to 13 Sep, but then weakened again on 14 Sep. A persistent region of 5-10 m s^{-1} radial outflow just above the

boundary layer near the RMW is evident on all three days also.

Although Isabel represents only a single case, there are reasons to believe that the inner-core maximum tangential winds would tend to expand radially provided the storm remains in a favorable environment. A physical explanation for the expanding inner-core maximum tangential winds is given in Sections 10.14 and 15.5.2. An explanation for the persistent outflow just above the boundary layer near the maximum winds is presented in Sections 8.1.2 and 15.5.4.

The derived vertical velocity in Fig. 1.10 shows weak vertical motions inside the eye and a maximum updraft nearly collocated with the RMW on each day. Lowest-level (representative of 100 m) radial inflow of 20 m s^{-1}, located at approximately 22-km radius from the center on 13 Sep., suggests significant penetration of air from the eyewall into the eye on this day. This observation of strong inflow inside the eye appears to be robust, but because of limited sampling, it may not be a quantitatively accurate depiction of the axisymmetric-mean inflow at these radii.

The right panels of Fig. 1.10 show the radius-height composite of θ_e, specific absolute angular momentum,[4] M,

---

4. To be discussed in Section 5.6.

and transverse secondary circulation. See caption for details. Some of the most dramatic changes in the inner-core storm structure are illustrated here, with a distinct increase in the low-level θ_e in the eye from 12 to 13 Sep., followed by an increase in θ_e in the outward-sloping eyewall updraft on 14 Sep starting at approximately 2 km altitude. The radial θ_e gradient is generally negative throughout all three days, except for very near the center on the 12 Sep. On the 14 Sep., it appears as if the θ_e has been "mixed out", with relatively low values, typically 365 K, found in the eye, and an increase in the eyewall updraft of up to 5 K. These figures suggest that there were notable changes in the mean θ_e structure over these three days.

Radial profiles of mean potential temperature and water vapor mixing ratio at the lowest composite level are shown in Bell and Montgomery (2008), their Fig. 6. The potential temperature and water vapor mixing ratio are defined in Sections 2.5 and 2.3.3. The radial profiles of these two quantities suggest that the increase in mean θ_e after 12 Sep. was primarily due to increased low-level moisture and occurred despite a rise in central pressure of approximately 10 mb. These data support the hypothesis of persistent latent heat flux from the underlying ocean inside the low-level eye.

The significance of the hypothesized latent heat flux inside the low-level eye and the attendant injection of augmented θ_e from the low-level eye into the eyewall updraft via both mean and eddy advection processes is a topic of scientific interest for three-dimensional hurricane eyewalls (e.g., Zhang et al., 2002; Cram et al., 2007; Persing et al., 2013; Houze, 2014). A notable example is a recent study by Wadler et al. (2021a) who demonstrated (p3517) the quantitative importance of

> *"low-level outflow from inside the eye and eye-eyewall mixing"*

in supporting relatively high θ_e air ascending the eyewall in their Category-5 hurricane simulation experiment. That study supports the hypothesis of a non-negligible influence of eye-eyewall mixing in a three-dimensional, Category-5, hurricane. One of the aims of this book is to develop a more complete understanding of the dynamics and thermodynamics of the inner-core region of an intense storm that may include the injection of augmented θ_e from the low-level eye into the eyewall updraft during parts of its life cycle.

Fig. 1.11 shows composite radius-height cross sections of the azimuthally-averaged radial, u, and tangential, v, wind components, the virtual potential temperature, θ_v, and the pseudo-equivalent potential temperature, θ_e, derived from a large number of dropsonde data collected in Hurricane Earl between 1800 UTC on 29 Aug. and 0600 UTC on 30 Aug 2010. Significantly, the maximum v (marked by a star in panels (a) and (b)) occurs at a low altitude of 0.5 km, within the surface-based layer of inflow, at a radius of

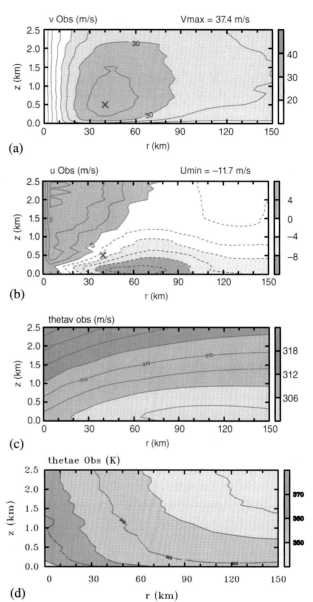

FIGURE 1.11 Radius-height cross sections of (a) azimuthally-averaged tangential wind v; (b) radial wind u; and (c) virtual potential temperature θ_v from dropsonde composites in Hurricane Earl (2010). The data are within the period from 1800 UTC on 29 Aug. to 0600 UTC on 30 Aug. when Earl was undergoing rapid intensification. Contour intervals are 5 m s^{-1} for v, 2 m s^{-1} for u, 2 K for θ_v and 5 K for θ_e. The location of maximum v is shown by a blue (dark gray in print version) cross in (a) and (b). Color shading is as indicated in the side bar. Panels (a) to (c) from Smith et al. (2017), Republished with permission of Wiley. Panel (d) courtesy J. Zhang.

about 40 km. This is not an isolated case; rather the finding that the maximum tangential wind speed occurs in the surface inflow layer is a general one. Since, as shown in Chapter 6, the inflow layer is one in which frictional processes are important, this finding may seem surprising. A theoretical explanation for such behavior is given in Section 6.5.7.

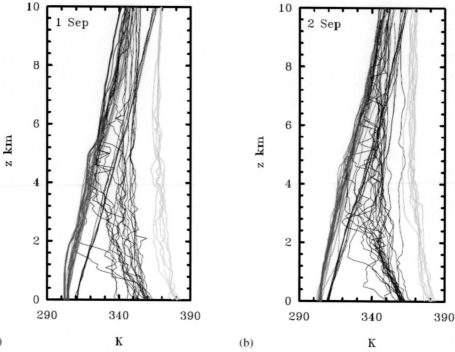

FIGURE 1.12 All soundings of θ_v (red/black curves) and θ_e (blue/green curves) for Hurricane Earl on (a) 1 Sep. and (b) 2 Sep. 2010. The black and green profiles are those for the eye/eyewall region while the red and blue profiles are for soundings made at larger radius. From Smith and Montgomery (2012a). Republished with permission of Wiley.

There is generally inflow outside a radius that slopes outwards with height from about 10 km near the surface and 70 km at 2.5 km. The inflow has a maximum also at a low altitude, within 0.1 km of the surface. This inflow maximum occurs about 20 km beyond the location of the tangential wind maximum. A theoretical explanation for these features is provided in Section 6.5.7 also. The vertical profile of θ_v shows a monotonic increase with height at all radii, which is indicative of a statically-stable boundary layer for vertical displacements of unsaturated air parcels. The structure of θ_e shows a general increase with decreasing radius and a negative vertical gradient, especially beyond a radius of 30 km, indicating that θ_e is not well mixed in the layer of strong inflow. A theoretical explanation for the non well-mixed θ_e in the strong inflow layer is offered in Section 16.7.

1.2.5 Thermodynamic structure of Hurricane Earl's eye and eyewall

Fig. 1.12 shows data from all the dropsonde soundings in Hurricane Earl released at an altitude in the range 10-11 km by the NASA DC8 on 1 and 2 Sep. 2010. There were 25 soundings on 1 Sep. and 29 soundings on 2 Sep. On each day, the soundings fall naturally into two bins: those in the eyewall or eye, and those at larger radii. The former has significantly higher values of θ_e and the soundings in this bin are distinctly warmer than in the latter. The eyewall profiles of θ_e can be distinguished from those in the eye as they are

almost vertical, a feature that is suggestive of moist adiabatic ascent, bearing in mind that the eyewall tends to flare outwards with height. The soundings at larger radii were made within a radius of about 250 km from the storm center. Two representative soundings from the eyewall and eye are shown in Fig. 1.13.

The upper panels of Fig. 1.13 show skew T-log p diagrams of dropsonde observations in the eyewall and in the eye of Hurricane Earl on 1 Sep. 2010, while the lower panels show the corresponding vertical variation of θ_v, θ_e, θ_e^* and water vapor mixing ratio, r_v. (Skew T-log p diagrams are discussed in Section 9.2, the quantity r_v in Section 2.5, the quantity θ_v in Section 2.5.1, and the quantities θ_e and θ_e^* are discussed in Section 2.5.4, and in Section 9.2.4.) As expected, the eyewall sounding (panel (a)) is almost saturated at pressures higher than about 550 mb, the dewpoint depression[5] being close to zero. This structure suggests that the sonde entered the eyewall from above at about this pressure level, the dewpoint depression at larger heights becoming increasingly larger. The eye sounding (panel (b)) is close to saturation at pressures higher than about 790 mb, but is characteristically much drier at lower pressures with a subsidence inversion between about 790 and 750 mb. Similar features of both soundings are evident in a comparison of the corresponding vertical profiles of θ_e and θ_e^*. In both

---

5. The dewpoint depression is the difference between the actual temperature and the dewpoint temperature. The larger its value, the drier the air.

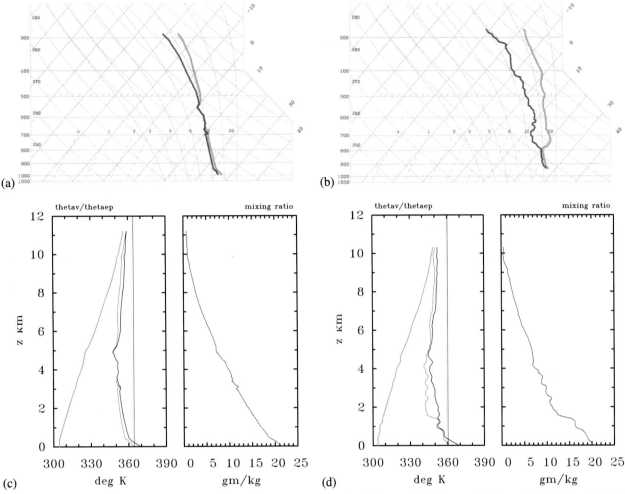

FIGURE 1.13   (Upper panels) Skew T-log p diagrams of dropsonde observations in (a) the eyewall (b) the eye of Hurricane Earl (2010) at 1907 UTC and 2224 UTC, respectively, on 1 Sep. 2010 when Earl was near the end of its period of rapid intensification. In the two top panels, temperature is shown in the red (mid gray in print version) curve and dew point is shown in the blue (dark gray in print version) curve. By definition, the temperature always lies to the right of the dewpoint. Curves that rise nearly vertically are moist adiabats. Horizontal lines are isobars; lines which slope upwards and to the left are dry adiabats; and lines that slope upwards and to the right are isotherms. (Lower panels) The corresponding variation of virtual potential temperature (θ_v, red (mid gray in print version)), pseudo-equivalent potential temperature (θ_e, blue (gray in print version)), saturation pseudo-equivalent potential temperature (θ_e^*, black), and water vapor mixing ratio (r_v) for these soundings. The vertical green (light gray in print version) and black lines to the right of the pseudo-equivalent potential temperature curves depict corresponding conserved values for a moist air parcel lifted from 100 m altitude and the surface, respectively. Constructed from data made available by NOAA/HRD.

soundings, θ_v increases steadily with height while r_v shows a progressive decrease. Both soundings show evidence of convective instability near the surface for air parcels lifted from near the surface. Convective instability is discussed in Chapter 9.

1.2.6   Intensity, strength, and size

Tropical cyclones have a range of sizes and at least two parameters are required to characterize size. One measure of size is the radius of maximum near-surface (10 m altitude) mean wind speed, which provides a scale for the inner core region, i.e., the region of most damaging winds. Another measure is the average radial extent of gale-force winds (17

m s^{-1}), R_{gales}, or the average radius R_{15} of winds equal to 15 m s^{-1}. Either scale would define an outer size for the vortex. Forecasters may be interested also in the mean radius, R_{25} of damaging winds (25 m s^{-1}). An alternative measure of outer size is the radius of the outer closed isobar (ROCI), a metric used by Merrill (1984) and others as it was easier to infer from observations at that time.

Just as the intensity is a measure of the maximum mean near-surface wind speed, one could define a measure for the *strength* of the outer circulation. Some authors define such a measure as a spatially-averaged wind speed over an annulus from 1-2.5°latitude (roughly 110-280 km) from the cyclone center (Weatherford and Gray (1988a,b) refer to this average as the outer-core strength, OCS). The differ-

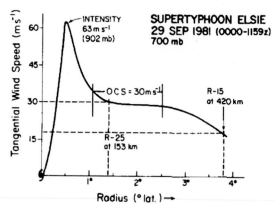

FIGURE 1.14 Illustration of typical cyclone inner-core intensity and outer-core wind strength, plus mean radius of 15 m s^{-1} (R_{15}) and 25 m s^{-1} (R_{25}) winds. From Weatherford and Gray (1988a). Republished with permission of the American Meteorological Society.

ences between intensity and strength are exemplified by the example shown in Fig. 1.14.

A thorough study of reconnaissance aircraft data collected from the western North Pacific by Weatherford and Gray (1988a,b) demonstrated that there is little correlation between intensity and size (e.g., see Fig. 1.15), supporting the earlier findings of Merrill (1984). Extremes of size are exemplified by Cyclone Tracy, which destroyed Darwin, Australia in 1974, and Supertyphoon Tip, which occurred in the western North Pacific in 1979. The peak winds of Tracy were 65 m s^{-1} (Bureau of Meteorology (1977), whereas the radius of gales extended no more than 50 km from the center. On the other hand, Tip had gales extending 1100 km from the center (Dunnavan and Diercks, 1980). This is an extreme difference in size, but even within a single season, tropical cyclones come in all sizes just as they move in a variety of ways and attain a variety of intensities.

Since the early studies of storm size described above, the advent of satellite-borne instruments to measure surface winds over a significant area surrounding storms has made it

possible to construct a more comprehensive climatology of the size and strength of tropical cyclones. One such study is that by Chan and Chan (2012), who used Quick Scatterometer (QuikSCAT) data from a polar-orbiting satellite to study tropical cyclones occurring over the western North Pacific and South China Sea and the North Atlantic, including the Gulf of Mexico and the Caribbean Sea, between 1999 and 2009. Amongst other things, they found that the mean size of typhoons over the Western North Pacific is about 20% larger than for hurricanes over the North Atlantic (240 km compared with 205 km) and their OCS is about 5% (19.6 m s^{-1} compared with 18.6 m s^{-1}).

During the developmental stage, the storm's intensity amplifies to its maximum while the size remains approximately constant. During the mature stage, the storm grows in size, but no longer intensifies. In the decay stage of the storm, the intensity of the inner-core winds decreases while the storm's circulation continues to grow until the deep convective activity collapses and/or the storm transitions to an extra-tropical cyclone.

In Chapters 8, 10, 15 and 16, we examine some pertinent questions related to the foregoing observations including: What dynamical and thermodynamical processes control intensity and size? Why do storms tend to become larger as they mature?

1.2.7 Asymmetries

Normally, only the inner-core region of intense tropical cyclones shows a significant degree of axial symmetry. As shown in Fig. 1.16, the core flow is predominantly axially-symmetric and is generally immersed in a less symmetric exterior flow that blends continuously into the larger-scale synoptic-scale environment. The cyclonic circulation in the lower troposphere may extend 1000 km or more from the center. The flow in the upper tropospheric outflow region is often noticeably asymmetric and, except near the center, the

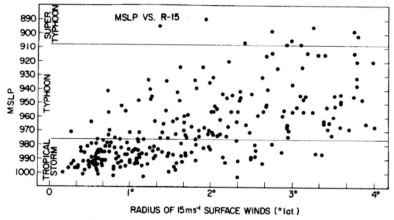

FIGURE 1.15 Scatter plot of minimum sea level pressure MSLP against mean radius of 15 m s^{-1} surface winds. From Weatherford and Gray (1988b). Republished with permission of the American Meteorological Society.

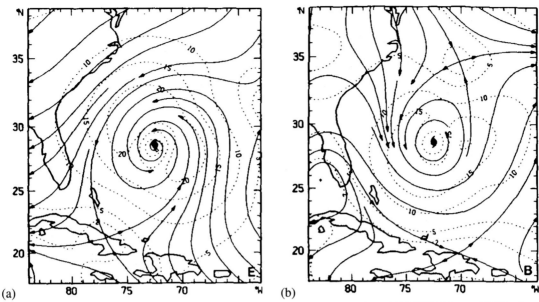

(a)   (b)

FIGURE 1.16 Analyzed wind fields (streamlines/isotachs) for Atlantic Hurricane Josephine (1984) at 0000 UTC 11 Oct. at (a) 900 mb and (b) at 500 mb. The contour interval is 5 m s^{-1}, with an additional contour at 2.5 m s^{-1}. From Franklin (1990). Republished with permission of the American Meteorological Society.

primary circulation there is generally anticyclonic. The transition radius between cyclonic and anticyclonic flow usually slopes downwards with radius, i.e., the upper anticyclone typically deepens with radius. In Chapter 3, the flow asymmetries are shown to have a significant effect on the vortex motion.

Storm asymmetries are usually evident in radar and satellite images of tropical cyclones in the form of spiral-shaped bands of cloud and precipitation. An example is shown in Fig. 1.17, where rainfall patterns obtained from the TRMM[6] microwave imagery are overlaid on the visible imagery. These spiral rainband patterns are typically 5-50 km wide and 100-300 km long. The precipitation-free areas between the bands tend to be somewhat wider than the bands, themselves, presumably, in part, because the velocities of ascent in cloud are typically much larger than the velocities associated with subsidence in the clear air between clouds (cf. Section 9.4). Some cloud bands are composed largely of stratiform clouds with moderate precipitation, while others have localized areas of vigorous deep convection giving rise to heavy precipitation. In tropical cyclones that originate in the monsoon trough, typical over the Western North Pacific and Australian regions, the asymmetric flow may be associated with a band of convection that extends from the cyclone into the trough. An example is shown in Fig. 1.18.

Several types of rainbands have been identified depending on their location relative to the storm center (Willoughby, 1988; Houze, 2010). Fig. 1.19 shows a schematic of these bands, which include the *principal rainband*,

FIGURE 1.17 Severe Tropical Cyclone Monica (2006) reached the highest intensity of any cyclone in the Australian region as it reached the northwest Gulf of Carpentaria, reaching 56 m s^{-1} (110 knots) intensity at 0600 UTC 22 Apr. and then attaining maximum intensity of 69 m s^{-1} (135 knots) 24 hours later about 120 km to the northeast of Nhulunbuy, around the time of this image. The TRMM image shows a pattern of very heavy rain (red) forming an intense symmetric eyewall around a small, complete eye with tightly curved rainbands spiraling into the center - the signature of a mature, intense tropical cyclone. Courtesy Naval Research Laboratory, Monterey, CA, USA. (NRL).

6. The Tropical Rainfall Measuring Mission (TRMM) was a joint project of NASA and Japan's Aerospace Exploration Agency. The mission sought to measure global rainfall using a polar-orbiting, satellite-based rain radar. The satellite was launched in 1997 and ceased operation in 2015.

FIGURE 1.18 Typhoon Winnie 1997. Courtesy NRL, Monterey, CA, USA.

secondary rainbands, and *distant rainbands*. The dashed circle in this figure indicates an approximate boundary between the outer environment of the tropical cyclone and an inner core, which is strongly influenced by the cyclonic vortex circulation. The figure shows also two eyewalls, a feature that is common to mature storms when an older primary eyewall decays and is replaced by a surrounding secondary eyewall. Further details of this process are discussed in Section 1.2.8.

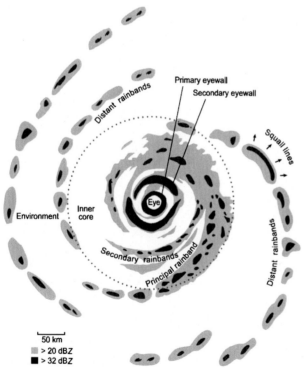

FIGURE 1.19 Schematic illustration of radar reflectivity in a Northern Hemisphere tropical cyclone with a double eyewall. From Houze (2010). Republished with permission of the American Meteorological Society.

The principal rainband is a quasi-stationary feature, presumably associated with flow asymmetries that, in turn, are linked to the vortex motion and possibly the vertical shear of the environment in which the storm is embedded (Willoughby, 1988; Houze, 2010). The vertical structure of convection in the principal band is again strongly influenced by the vortex circulation as discussed below. In some storms, the upwind end of the principal rainband lies outside the inner core region of the storm, in which case the convective elements may be more similar to those in the distant rainbands. These tend to have the character of squall lines, whose vertical structure is at most only weakly affected by the vortex dynamics of the tropical cyclone (Houze, 2010, p327).

Secondary rainbands tend to be smaller and more transient than the principal rainband. These are found inside the inner core region and propagate both radially and azimuthally. The azimuthal component of propagation is typically slower than the local tangential wind component. The behavior of secondary rainbands is consistent with the interpretation that they are vortex Rossby waves. An elementary introduction to the dynamics of these waves is presented in Chapter 4. As shown in Fig. 1.19, secondary rainbands frequently merge with a principal rainband or an eyewall.

Fig. 1.20 shows an idealized plan view of the principal rainband in a mature tropical cyclone with a single eyewall, together with a vertical cross section across the middle portion of the rainband. Much of the band is composed of stratiform precipitation, but cells of deep convection occur on the inner edge of the band (panel (a)). The size of the convective cells in the figure indicates the level of maturity, with the dashed border indicating collapsing cells. An azimuthal jet is located along the axis of the rainband in the lower troposphere, marked by a thick arrow in the figure. The jet is depicted also in the vertical cross section (panel (b)).

Strong reflectivity cores associated with both primary and secondary rainband convection were shown by Barnes et al. (1983) to have a radially outward tilt with height, similar to that of the eyewall. These authors showed also that the flow in a vertical cross section through rainbands has the character of an overturning circulation with radial inflow at heights below 2 km and radial outflow above, the strongest outflow occurring at heights between about 5 and 8 km, somewhat below the outflow associated with the eyewall. Maximum updrafts in rainbands occur in the convective cores (e.g., Barnes et al., 1983; Powell, 1990) and are found typically to be on the order of 4 m s^{-1} (e.g., Black et al., 1996, Fig. 11).

Deep convection in distant rainbands often has more intense updrafts than in rainbands closer to the center and, as noted above, is presumably only weakly affected by the vortical flow and its accompanying overturning circulation.

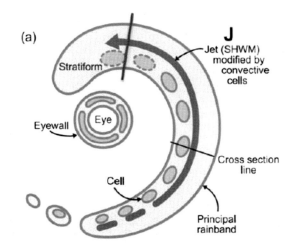

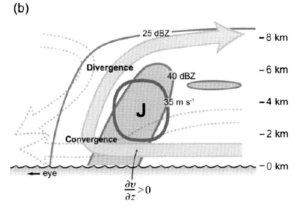

FIGURE 1.20   (a) An idealized view of a principal rainband in a mature tropical cyclone, taken from of Houze (2010). The thick arrow shows the location of the axis of a jet along the rainband in the lower troposphere. The rainband is composed of largely stratified precipitation with embedded convective cells. These cells are located on the radially inward side of the jet. The level of maturity of the cells scales with the size of the cells in the schematic, with the dashed border indicating collapsing cells. Shading indicates precipitation (dBZ; light shading is 25 dBZ and dark shading is 40 dBZ). (b) Cross section across the middle portion of the rainband, emphasizing the updraft structure. The wide arrow indicates the in, up, and out circulation of the convective-scale updraft. Low-level convergence, vertical convergence of vorticity flux, and midlevel divergence regions are indicated. Republished with permission of the American Meteorological Society.

A succinct summary of the current state of knowledge on tropical cyclone rainbands is contained in Moon and Nolan (2015), see their Section 2 and their comprehensive list of references.

1.2.8   Secondary eyewalls

As we have shown above, the eyewall of a mature tropical cyclone is a nearly concentric ring of convection comprising heavy precipitation, the boundary layer roots of which contain the strongest wind speeds in the storm. Beginning with the pioneering study of Willoughby et al. (1982), ob-

servations indicate that once a tropical cyclone has attained a sufficient intensity and the environment remains favorable, a secondary eyewall will form outside the primary one, resulting in one or more concentric rings of convection and a weakening of the primary vortex (Willoughby et al., 1982; Houze et al., 2007). Willoughby and coworkers showed that in intense hurricanes, outer eyewalls contract and act to strangle the original eyewall, halting intensification or causing weakening.

A particularly well observed example of a forming secondary eyewall was documented during the Hurricane Rainband and Intensity Change Experiment (RAINEX) during the mature evolution of Hurricane Rita (2005). The RAINEX flights in Hurricane Rita were the first to extensively observe an eyewall replacement with the high-resolution Electra Doppler radar (ELDORA). Details of the analysis are presented by Bell et al. (2012b). Here we summarize the observed evolution of Rita's eyewall replacement and then highlight some of the pressing questions about the formation process.

Fig. 1.21 shows the airborne radar reflectivity field at 3-km altitude taken by the NRL WP-3D aircraft when a secondary eyewall was starting to form in Hurricane Rita. The left and right panels show the radar reflectivities at the beginning and end of a 23 hour interval starting at 1936 UTC 21 Sep. and ending at 1915 UTC 22 Sep. This time interval spans the interval just before Rita attained peak wind intensity and when the secondary eyewall became organized as a contiguous ring of high reflectivity (exceeding 40 dBZ), indicative of heavy precipitation and relatively strong updrafts.

At around 1900 UTC 21 Sep. the inner eyewall had attained a peak intensity and an outer convective envelope was beginning to emerge in the radial interval 65 to 80 km from the radar reflectivity center. At this time the outer reflectivity field shows an array of disconnected reflectivity features oriented in a downshear configuration with respect to the radial shear of the angular velocity of the swirling flow, $\Omega(r) = V_T(r)/r$, where V_T is the azimuthally averaged tangential velocity of the vortex. Approximately 23 hours later, this outer convective envelope region appears to have contracted inwards approximately 15 km and axisymmetrized into a contiguous and approximately concentric reflectivity structure near 50 km radius.

The bottom panel shows the azimuthally-averaged reflectivity as a function of radius derived independently from the NOAA P3 C-band airborne radar at four consecutive times during this intensive observing period (the first time is at 1943 UTC 21 Sep. and the remaining three times are at 1749, 1913 and 2045 UTC 22 Sep.). Although there are differences between the ELDORA and NOAA radar systems (with differing wavelengths, different beamwidths and different attenuation characteristics), the NOAA radar data broadly confirms the radial regions of the primary and emer-

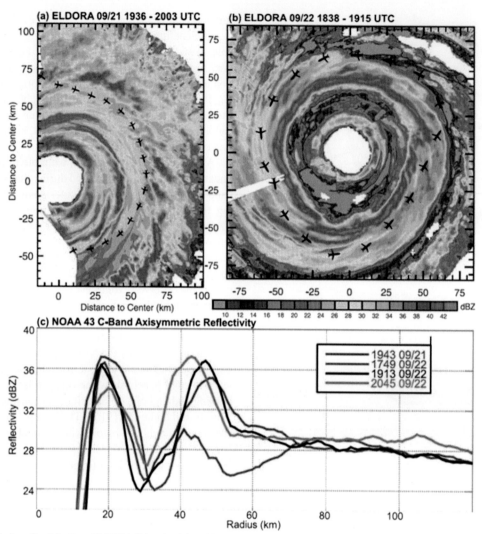

FIGURE 1.21 Radar reflectivity from ELDORA X-band at 3-km altitude during (a) 1936-2003 UTC 21 Sep. and (b) 1838-1915 UTC 22 Sep. 2005, and (c) axisymmetric reflectivity from NOAA C-band at a height of approximately 3 km at four consecutive times shown in inset. From Bell et al. (2012b). Republished with permission of the American Meteorological Society.

gent secondary eyewall depicted by the ELDORA radar during the second time period.

The region of reduced reflectivity between the two eyewalls in Fig. 1.21c is suggestive of a dipole-like pattern of ascent and subsidence. In order to learn more about the evolution of the primary and secondary circulations during this intensive observation period, Fig. 1.22 shows the azimuthally averaged kinematic fields from the ELDORA analysis at the times shown in the first two panels of Fig. 1.21. The top panel (a) shows the radius-height structure of the storm-relative tangential velocity, V_T, relative vertical vorticity, ζ, absolute angular momentum, M, and secondary circulation (vector) - averaged over the interval 1936-2008 UTC on 21 Sep. The middle panel shows the same fields, but averaged over the interval 1838-1914 UTC

on 22 Sep. The third panel shows the difference between these two times, approximately 23 hours apart.

Fig. 1.22a shows a weak overturning circulation associated with the incipient secondary eyewall. The outer overturning circulation subsequently develops into a strong circulation, with peak core updrafts of approximately 6 m s^{-1} at mid-level (Bell et al. (2012b), Figs. 6b and 8). This outer overturning circulation is clearly distinct from that of the primary eyewall. During this time span, the outflow above the boundary layer has completely reversed to inflow below 8 km beyond approximately 50 km radius. The reversal from outflow to inflow in the outer region is associated with an increase in azimuthally-averaged diabatic heating as the outer convective ring becomes more axisymmetric. The formation of an eye-like "moat" is thought to be the result of forced subsidence from both eyewalls.

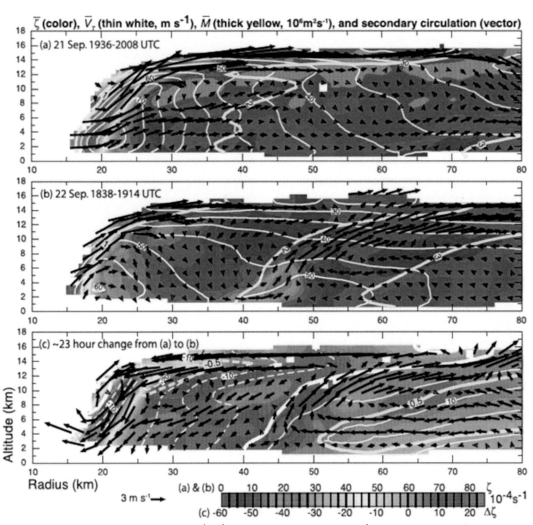

FIGURE 1.22 Axisymmetric relative vorticity (color, 10^{-4} s^{-1}), tangential wind (contour, m s^{-1}), and secondary circulation (vector) derived from ELDORA. (a), (b) Flight patterns and times shown in Fig. 1 of Bell et al. (2012b), and (c) the $\sim$ 23 hour change between the two analysis times. Dashed contours in (c) indicate a negative change of tangential wind and absolute angular momentum, thick contours indicate zero change, and solid contours indicate positive change. Vorticity change in (c) is indicated by the numbers on the bottom of the color bar. From Bell et al. (2012b). Republished with permission of the American Meteorological Society.

The Bell et al. (2012b) study concluded that

"the analyses ... provide new evidence that changes in heating, moist entropy, toroidal and vertical vorticity, and ultimately wind speed are concurrent and highly coupled in a developing secondary eyewall. While the axisymmetric view has been shown to be quite useful for gaining insight into the processes responsible for these changes, the observations suggest also that nonaxisymmetric contributions must be included for a complete understanding of eyewall evolution. Additional observations and further basic research on the topic of secondary eyewall formation and evolution are advocated to improve our understanding and forecasts of the phenomenon and related size expansions of intense tropical cyclones."

Among the open questions on secondary eyewalls, perhaps the most critical is what initiates the formation of the outer eyewall in the first place? In particular, is the conventional (axisymmetric) model of vortex spin up developed later in Chapters 5 and 8 sufficient to understand the initiation process, or do other processes such as nonlinear boundary layer dynamics and/or nonlinear eddy processes intervene in an important way? These questions will be addressed in a planned sequel to this book.

1.3 Surface heat and moisture supply

Tropical cyclones develop over the warm tropical oceans and derive their supply of energy largely in the form of latent heat from the evaporation of sea water. Details of the supply of heat and moisture there as well as the drag exerted by the sea surface on the air above are presented in Chapter 7. As a foretaste and to provide a context for the next section, we consider briefly here the moisture supply.

The evaporation rate of water per unit area of the ocean surface, F_v, is given by the empirical formula

$$F_v = C_E V_{10}(r_v^*(T_s, p_s) - r_{v10}),\qquad(1.3)$$

where V_{10} and r_{v10} are the wind speed and water vapor mixing ratio at a nominal altitude of 10 m, $r_v^*(T_s, p_s)$ is the saturation mixing ratio at the sea surface temperature, T_s, and pressure, p_s, (see Section 2.5), and C_E is a surface exchange coefficient for moisture, normally assumed constant for simplicity with a mean value of about 1.0×10^{-3} (see Section 7.5).

Eq. (1.3) shows that F_v increases with wind speed, but decreases with decreasing moisture disequilibrium, $r_v^*(T_s, p_s) - r_{v10}$, as happens if the near-surface air becomes more humid. Even though at fixed sea surface pressure and temperature, the degree of moisture disequilibrium decreases as the air moistens, $r_v^*(T_s, p_s)$ increases with decreasing pressure so that the disequilibrium may be maintained as the radius from the storm center decreases. Accordingly, as the wind speed increases and pressure decreases, even if the sea temperature remains constant, the moisture disequilibrium can be maintained.

Ooyama (1969) (p15) noted that a sharp decrease of pressure in the inner rain area may help the boundary layer inflow acquire additional moisture and heat shortly before the air ascends in the eyewall updraft.

Assuming that the moisture disequilibrium is approximately maintained, the dependence of F_v on V_{10}, suggests that the strongest water vapor flux from the ocean occurs in the region where the winds are near their peak values. This region is one where breaking waves and whitecaps produce large numbers of sea spray droplets that are carried with the wind. These spray droplets have the same temperature and salinity as the ocean surface and increase the effective surface area of the ocean in contact with the atmosphere. In this way, the spray alters the total sensible and latent heat fluxes in the near-surface layer. The spray drops have an effect also to increase the near-surface drag on the air, while the mass of the spray introduces an additional drag in the vertical momentum equation, tending to stabilize the near-surface layer. Fig. 1.23 shows a photograph of the sea state in wind speeds that are only of marginal hurricane strength.

1.4 Ocean interaction

Tropical cyclones interact with the ocean through the mutual stress exerted at the ocean surface, which at a detailed level will depend on the ocean wave field, and on the surface enthalpy fluxes, i.e., the sum of the sensible and latent heat fluxes. The enthalpy fluxes are influenced also by the

FIGURE 1.23 Ocean surface beneath a strong hurricane. Photograph courtesy of our late friend and NOAA-HRD colleague Michael Black. From Bell and Montgomery (2008). Republished with permission of the American Meteorological Society.

wave field and, in particular, the spray produced by breaking wave crests. The surface stress exerted by the cyclone will generate ocean currents and produce upwelling. If the thermocline is shallow, the upwelling may produce significant cooling of the ocean surface that will affect both the sensible and latent heat fluxes, especially the latter because of the sensitivity of the saturation specific humidity, $q_v^*(T_s, p_s)$, to T_s.

An example of ocean cooling following the passage of Hurricane Harvey (2017) over the Gulf of Mexico in 2017 is shown in Fig. 1.24. Prior to the passage of the hurricane, on 18 Aug., sea surface temperatures (SSTs) along the track were in excess of 29 °C, with some values to the right of the track exceeding 30 °C. Harvey was a relatively slow moving storm and remained in the area where it first made landfall for several days. By 30 Aug., as Harvey had moved inland for the second time and its intensity had declined, SSTs had fallen by up to 3.5 °C in the northern part of the gulf.

The degree of ocean cooling brought about by the passage of a tropical cyclone depends in part on the depth of warm near-surface water, i.e., on the depth of the ocean mixed layer. If the mixed layer is deep, the cooling of the ocean surface brought about by mixing will be minimal, whereas, if the mixed layer is shallow, the cooling may be appreciable. Fig. 1.25 shows vertical profiles of ocean temperature at two locations in the Gulf of Mexico. The profile on the left has a relatively shallow mixed layer, while that on the right has a much deeper one.

The susceptibility of the ocean to cool is measured by the ocean heat content (OHC), defined as the amount of heat through a depth that goes down to the depth of the 26 °C isotherm, $D26$, as indicated by the shaded area in each panel of Fig. 1.25. The formula defining the OHC is given by

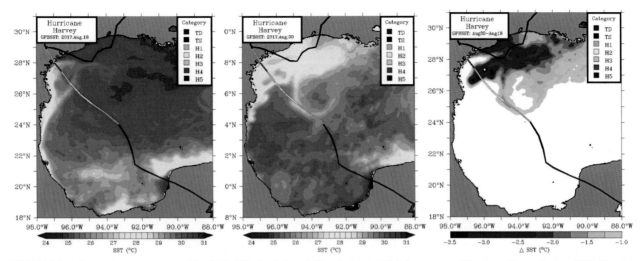

FIGURE 1.24 Sea surface temperature of the Gulf of Mexico before (left) and after (middle) the passage of Hurricane Harvey in Aug. 2017. The right panel shows the cooling that took place over the approximately two-week period between the before and after plots. Courtesy L.K. (Nick) Shay.

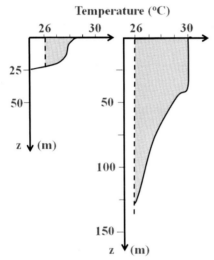

FIGURE 1.25 Sea temperature structure for two stations in the Gulf of Mexico showing examples of high and low OHC. From Leipper and Volgenau (1972). Republished with permission of the American Meteorological Society.

$$OHC = \rho_{sea}c_{sw}\int_0^{D26}(T(z) - 26)dz, \qquad (1.4)$$

where ρ_{sea} is the density of sea water, c_{sw} is the specific heat of sea water and $T(z)$ is the temperature profile above $D26$ in °C. The larger the OHC, the less susceptible will be the ocean to cooling by wind-induced upwelling.

Fig. 1.26 shows the distribution of OHC in the Western Atlantic and Caribbean Sea following the passage of Hurricane Irma (2017) and prior to the passage of Hurricane Maria. Note the low values of OHC brought about by Irma, primarily to the right of Irma's track before the storm reached the northernmost islands of the chain bordering the eastern Caribbean. The more southerly track of Maria en-

abled the storm to avoid this reduction of OHC and Maria intensified rapidly as it reached the high values of OHC near the islands of St. Croix and Puerto Rico.

In the foregoing material, we have shown recent examples of the cold ocean wake trailing moving hurricanes and the cold wake structure of an approximately stationary hurricane. The stationary hurricane resulted ultimately in a markedly reduced sea surface temperature under the high wind region of the storm. Since deep convection relies on warm water to keep the boundary layer θ_e sufficiently high to support its continuation, a reduced sea surface temperature over a region encompassing the eyewall updraft could, depending on the circumstances, limit or altogether suppress further storm development.[7] The moving hurricane cases (Irma and Maria) suggested that most of the cold water was trailing the storm on the right side when viewed in the direction of storm motion.

For the case of a moving storm, an important issue is whether the cold water caused by the stress-induced turbulent upwelling is located underneath the convective eyewall or is displaced well beyond the eyewall. When there is sufficient separation, between the eyewall and the region of cooling, air parcels spiraling inwards can recover from a reduction in θ_e caused by air traversing over a patch of reduced energy content and thermodynamic disequilibrium. Some field experiments have been devoted to addressing this problem. As an example, Fig. 1.27 shows upper ocean

---

7. Based on material to be presented in Chapter 9, if the SST was reduced everywhere by 2°C under the storm, the near-surface θ_e would be reduced by approximately 10 K, assuming a local sea surface pressure of 980 mb. Such a reduction of near-surface θ_e would be enough to offset approximately two-thirds of the enhancement of θ_e that would otherwise occur as air parcels spiral inwards picking up moisture en route before ascending and exiting the boundary layer. In such a scenario, the instability necessary to support the deep convection in the eyewall would disappear and the vortex would spin down rapidly.

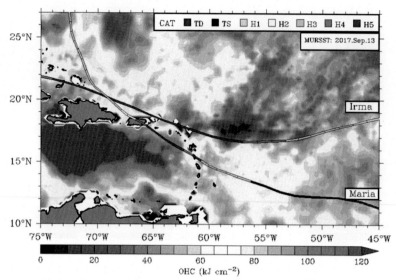

FIGURE 1.26 Ocean heat content (OHC) in a portion of the Western Atlantic and Caribbean Sea on 13 Sep. 2017 following the passage of Hurricane Irma and prior to the passage of Hurricane Maria. The tracks of Irma and Maria are as shown. Courtesy L.K. (Nick) Shay.

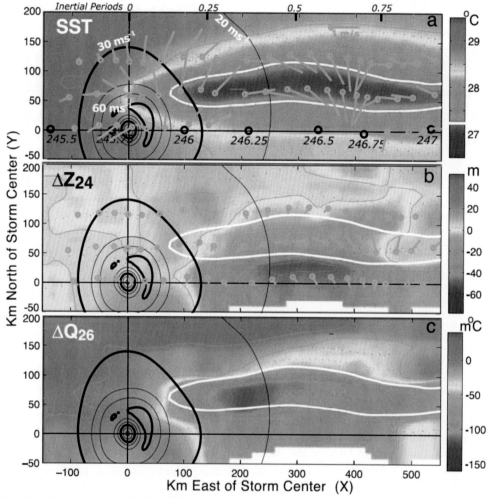

FIGURE 1.27 Plan view of ocean response to Hurricane Frances in a storm-centered coordinate system. The storm moves to the left along the horizontal axis. Wind speed contours (black, white labels) are from the H*WINDS analysis at day 245.75. (a) Sea surface temperature from drifters and floats. Dots are data points; colors and contours show the mapped field. Arrows show mixed layer velocity from the EM-APEX floats. (b) Change in the depth of the 24 °C isotherm from its value before day 245.6 and velocity on this isotherm. Data from floats only. (c) Change in heat warmer than 26 °C, i.e., the Hurricane Heat Content. Data from floats only. From D'Asaro et al. (2007). Republished with permission of American Geophysical Union.

data sampled during the passage of category four Hurricane Frances (2004) over the open ocean. Panel (a) shows sea surface temperature, with dots being data points from drifters and floats. The arrows show mixed layer velocity from the floats. Panel (b) shows the change in the depth of the $24\,^{\circ}C$ isotherm from its value before day 245.6 (the date when the storm passed over the observational array) and the velocity on this isotherm. Panel (c) shows the change in the heat content warmer than $26\,^{\circ}C$. The figure shows clearly a reduction in sea surface temperature by more than $2\,^{\circ}C$, *with the peak reduction located behind and to the right of the highest wind region.*

From their seminal analysis, D'Asaro et al. (2007) found that:

> *"The sea surface temperature (panel a, our insertion) cools behind the storm at all locations. It is largest ($2.2\,^{\circ}C$) in the cold wake, which is clearly offset from both the center of the storm and from the strongest winds. Less cooling occurs on the left-hand side of the storm ($0.8\,^{\circ}C$) and under the central core ($0.4\,^{\circ}C$). The upper layer velocity is also largest in the cold wake, consistent with the cold wake being formed by shear-driven mixing. The upper layer currents and shear will be most efficiently generated on the right-hand side of the storm where the wind stress rotates in a clockwise direction with time … ".*

The $24\,^{\circ}C$ isotherm and change in heat content data shown in panels (b) and (c), respectively, corroborate the cooling of the near surface waters primarily behind and to the right of the storm's path.

The fact that the maximum cooling is azimuthally localized and spatially separated from the strongest wind region suggests that, as intimated above, there is sufficient separation between the coolest water and the eyewall of the storm so that the impact of the shear-induced cooling will be relatively small. In fact, D'Asaro et al. concluded *inter alia* that:

> *"Sea surface temperature under the storm core cools by $0.4\,^{\circ}C$, reducing the air-sea heat flux by about 16%. This negative feedback from ocean mixing to the storm intensity is estimated to decrease the peak winds by about $5\ m\ s^{-1}$ to the observed value of $60\ m\ s^{-1}$."*

That is, the intensity reduction caused by the air-ocean coupling was estimated to be less than 10% of the intensity of the storm.

The foregoing discussion reveals that the air-ocean coupling for a passing hurricane is a complex process involving multiple factors that need to be accounted for in a real-time forecast of an approaching storm. However, for the current textbook, the Frances data suggests that the negative effect of ocean cooling can be safely neglected in a first theoretical treatment of the formation and intensification of a tropical cyclone *provided* there is some modest movement of the developing disturbance or storm over favorable ocean water. We will implicitly assume this to be the case throughout this book and we will pick up the fascinating fluid dynamics and thermodynamics of the air-ocean interaction problem in the planned sequel to this book.

1.5   Tropical cyclone genesis

Over the past decade or so there have been significant strides in improving our understanding of tropical cyclone genesis in the real atmosphere. Before discussing some of these prominent observational advances and supporting theoretical developments, we review first basic knowledge about the formation of tropical cyclones.

Tropical cyclones form in many parts of the world from initial tropical disturbances, often in the form of westward propagating waves in the easterly trade wind flow or from disturbances in the monsoon trough regions. These disturbances have cloud clusters embedded within them, typically comprising multiple mesoscale convective systems, with their accompanying concentrated areas of deep precipitating clouds, widespread stratiform anvils and precipitation regions. The complex, multi-scale nature of the formation process is arguably one of the most intriguing aspects of the problem, and the multi-scale attribute is now becoming more widely recognized. Our observational overview will highlight both classical observations and some new observations adopting the multi-scale perspective.

Unfortunately, a widely accepted definition of tropical cyclogenesis does not exist. However, here is common agreement that the formation of a tropical cyclone, a phenomenon commonly referred to as tropical cyclogenesis, is a process by which some pre-existing, synoptic-scale or mesoscale weather feature in the tropics evolves so as to take on the characteristics of a tropical cyclone as described in the foregoing sections. The physical characteristics adopted widely by operational forecasting centers are as follows: a quasi-circular, closed circulation at or near the surface with the strongest circulating winds within or near the top of the atmospheric boundary layer, the presence of sustained deep moist, precipitating convection near the center of the circulation, and the appearance of outflow in the upper troposphere that turns anticyclonically. These characteristics broadly define a tropical depression, i.e., a nascent hurricane, albeit at a much smaller intensity (sustained winds less than $17\ ms^{-1}$). In this book we align with the tropical cyclone forecasters and define tropical cyclogenesis as the formation of a tropical depression as just described.

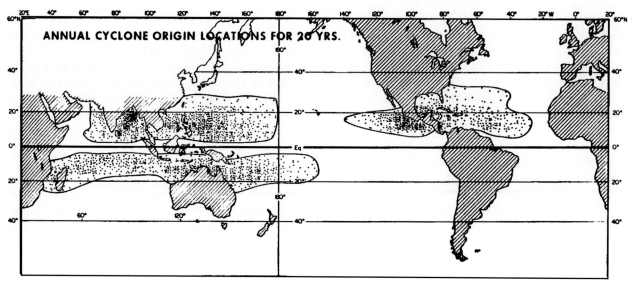

FIGURE 1.28 Locations of tropical cyclone formation over a 20-year period. From Gray (1975). Republished with permission of Colorado State University Libraries.

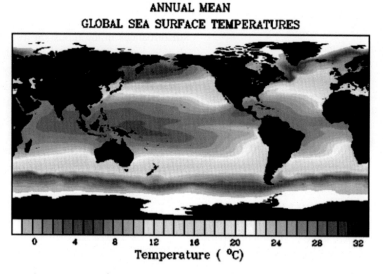

FIGURE 1.29 Annual mean sea-surface temperature (°C). From NOAA, https://www.ospo.noaa.gov/Products/ocean/sst.html.

1.5.1 Formation regions

On average, 86 tropical cyclones form each year throughout the world and about two thirds of these reach the stage of a severe tropical cyclone. The early climatology of initial detection points of each cyclone for a 20-year period by Gray (1975) remains applicable (Fig. 1.28). Significantly, preferred regions of formation are over the tropical oceans that coincide broadly with regions of high SSTs (Fig. 1.29). The warmest waters occur in the Western Pacific, the so-called "warm pool region", while the ocean temperatures in the Southeast Pacific are relatively cold. Indeed, climatological studies by Palmén (1948, 1956) and Gray (1975) have shown that tropical cyclogenesis occurs only in regions where the SST is above 26.5 °C.

There are few formation events within about 2.5°lat. of the Equator, but, contrary to statements in some places, such events do occur (see Section 16.6). Most of the formations (87%) occur between 20°N and 20°S (Fig. 1.30). Another interesting statistic is the frequency of tropical cyclones per 100 years within 140 km of any point as shown in Fig. 1.31.

Of the total number of cyclones that form each year around the globe, the Northern Hemisphere sees nearly two thirds of the total. The Eastern hemisphere has twice as many tropical cyclones as in the Western hemisphere. The disparity is due largely to the absence of tropical cyclones in the South Atlantic and the eastern South Pacific. Since tropical cyclones are generally seasonal weather systems, most basins manifest a maximum frequency of formation

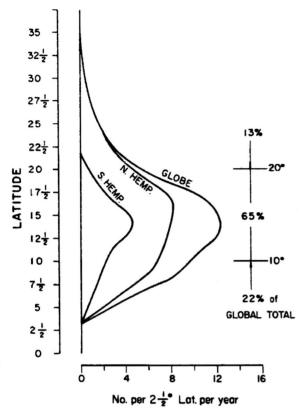

FIGURE 1.30 Latitude of first detection for tropical disturbances that later became tropical cyclones. From Gray (1968). Republished with permission of the Amderican Meteorological Society.

during the late summer to early autumn (fall).[8] The peak in the Southern Hemisphere occurs in January to March. The peak in the Northern Hemisphere occurs from July through September. The most active region is the Northwest Pacific Ocean, where typhoons occur in all seasons.

The seasonal distribution of formation locations is governed by two major factors. One is the association between tropical cyclone formation and SST, with late summer manifesting the highest values of SST. Because of the wind-driven ocean currents of the Kuroshio and Gulf Stream in the North Pacific and North Atlantic basins, respectively, relatively high SST regions extend well northwards of the near equatorial region. Of course, SST is just one factor in the conditions influencing the likelihood of cyclone formation, as illustrated by the dearth of cyclones in the South Atlantic despite similar values of SST during the Southern Hemisphere warm season. The second factor in the seasonal distributions is linked with the seasonal variations in the location of the synoptic-scale flow that supports the favorable conditions and disturbances needed for formation (enumerated specifically below).

8. The prominent exception is the North Indian Ocean, which exhibits a bimodal frequency distribution on account of the Indian monsoon. The first frequency peak typically occurs in May and the second peak occurs in October.

Consider first the case of the monsoon trough areas. As analyzed by Gray (1968), the Inter-Tropical Convergence Zone (ITCZ) extends semi-contiguously around the tropical oceans at low latitudes and is manifest as a convergence line between trade easterlies from the Northern and Southern Hemispheres, or as a convergence zone in westerly monsoon flow. In the latter configuration, the monsoon westerlies usually have trade easterlies on their poleward side. The shear line separating the monsoon westerlies from easterlies has cyclonic shear vorticity and is a climatologically preferred region for tropical cyclone formation. This cyclonic shear region is known as the monsoon trough or monsoon shear line. Characteristic flow patterns for the two ITCZ modes in the upper- and lower-levels are sketched in Fig. 1.32, which is taken from McBride (1995).

Referring to this figure, McBride writes

"The trade convergence line of the ITCZ typically has relatively large vertical wind shear. When monsoon westerlies are present, the low-level monsoon shear line is overlain (in the mean seasonal pattern) by the upper-level subtropical ridge. In the western North Pacific, the ridge above the monsoon trough during the summer is called the subequatorial ridge. This configuration of trade easterlies overlain with westerlies and monsoon westerlies overlain with easterlies gives a (seasonal-mean) vertical wind shear close to zero, with westerly shear on the poleward side and easterly shear on the equatorward side"

The great expanse of the Western Pacific basin allows for a larger variety of synoptic-scale precursors than other basins of the tropical atmosphere. In addition to the monsoon configuration just described, one observes *inter alia* so-called "monsoon gyre" flow regimes that comprise an approximately 2500 km scale cyclonic circulation (or gyre). These gyres occur occasionally during the Boreal summer and last several weeks. The gyres are observed to support the formation of multiple areas of persistent deep convective activity, which sometimes lead to the occurrence of multiple tropical cyclones. In other cases, the monsoon gyre merges with an emergent tropical cyclone, furnishing a giant-sized tropical cyclone (e.g., Supertyphoon Tip as reported in Section 16.5). It turns out that such a regime admits complex flow phenomenology that lies outside the scope of the current textbook. See Lander (1994) and refs. for further details.

Consider next the case of the North Atlantic basin during Northern Hemisphere summer. From right to left, the tropical North Atlantic spans the ocean region starting at the west African coastline and ending in the Caribbean sea, Gulf of Mexico, and Bahamas. Whereas the flow on the southern side of the ITCZ near Africa is often associated with a mon-

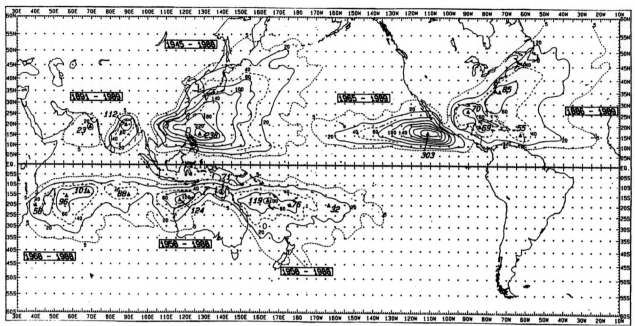

FIGURE 1.31   Frequency of tropical cyclones per 100 years within 140 km of any point. Solid triangles indicate maxima, with values shown. Period of record used is shown in boxes for each basin. From WMO (1993). Republished with permission of the WMO.

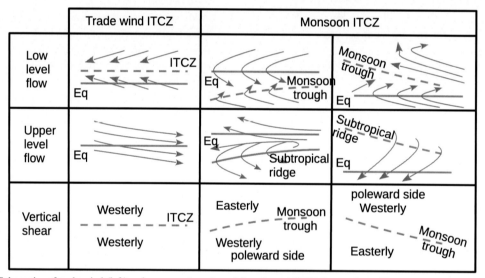

FIGURE 1.32   Schematics of trade-wind (left) and monsoon-type (two right columns) ITCZ flow regimes. The monsoon regimes are subdivided into those typical of the Australian/Southeast Indian Ocean ITCZ during January (middle) and the western North Pacific basin during August (right). Lower panels indicate vertical wind shear between the lower-level and upper-level flow. Adapted from McBride (1995). Republished with permission of the WMO.

soon (westerly) flow, tropical cyclones typically form in the easterly flow north of the ITCZ center line. The juxtaposition of hot and relatively dry air over the Saharan desert with the cooler, moist air associated with the equatorial waters near the African coastline, produces an enhanced trade flow over central Africa known as the African easterly jet.

On the southern flank of the African easterly jet, the localized, westward flowing air mass possesses cyclonic (shear) relative vorticity. This jet is known to be the pro-

genitor of horizontal-shear unstable or nearly neutral, finite amplitude synoptic-scale disturbances, called African easterly waves and illustrated schematically in the forthcoming Fig. 1.35. These waves propagate westwards over the central African continent and out over the tropical North Atlantic ocean. The moderate low pressure and cyclonic relative vorticity regions of these African easterly waves, and related waves excited along the northern side of the ITCZ, are breeding zones for tropical cyclones. In the former case of the easterly jet, the waves have maximum amplitude near, or just below, the jet level of 600 mb. In the latter case associated with the ITCZ, the waves typically have their maximum amplitude near 850 mb, e.g., Wang et al. (2012).

1.5.2 Necessary conditions for formation

As described above, many disturbances travel across the tropical ocean basins during their respective tropical cyclone seasons, and yet relatively few become tropical cyclones. Since the milestone work by Gray (1968), it has become widely accepted by tropical cyclone forecasters and researchers that a number of environmental conditions are necessary for genesis. For a cyclone to form, *at least six preconditions* must be met:

(i) significant values of low-level cyclonic, relative vertical vorticity (similar or larger in value to the local planetary vorticity);

(ii) a location of at least a few degrees poleward of the Equator, giving a relatively significant value of planetary vorticity (but see below);

(iii) weak vertical shear of the horizontal winds (typically less than approximately 12 m s^{-1} spanning the 850 and 200 mb levels);

(iv) sea-surface temperatures (SSTs) exceeding about 26 °C, and a relatively deep thermocline (on the order of 50 m);

(v) An atmosphere which cools fast enough with height (i.e., is "unstable" enough) such that it encourages deep thunderstorm activity.

(vi) large values of relative humidity (say, exceeding roughly 80%) in the low-to-mid troposphere.

Having these conditions met is necessary albeit insufficient since many disturbances that appear to have favorable conditions do not develop. We will discuss below *a new, seventh, necessary condition* that has become established in the tropical North Atlantic and Northeastern Pacific basins. Before doing so, we provide first an elementary interpretation of the foregoing conditions.

The first three factors are functions of the horizontal flow dynamics, while the last three are thermodynamic parameters. The first condition requires an atmospheric region possessing a sufficient cyclonic "spin". Tropical cyclones are not observed to form spontaneously. To develop, they require some persistent convective activity with sufficient cyclonic spin and low-level inflow. The second condition recognizes that some of the Earth's spin is generally needed to provide a background rotation that, in combination with the initial relative cyclonic spin, can be subsequently amplified through the concentration of absolute vertical vorticity (defined below) in a region of horizontally convergent flow (as beautifully illustrated by ice skaters during winter Olympics competitions when adopting a bulk perspective and invoking the conservation of angular momentum).

However, systems can form close to the Equator, or even on the Equator, given a disturbance with sufficient local vertical rotation, but these are relatively rare events (e.g., Kilroy et al., 2020; Steenkamp et al., 2019 and refs.).

The third condition recognizes that large values of vertical wind shear tend to disrupt the persistence and organization of the thunderstorms that are important to the inner part of a developing storm. In addition to the obvious mechanical effect of differential shearing of nascent disturbances (and the convective thermals within), the relative flow between different layers, in combination with dry air in the environment, can produce a source of dry air ("anti-fuel") that will limit or quench the deep convection. The fourth condition reflects the fact that sufficiently warm waters are necessary to support persistent, moist convective instability.

The fifth condition reflects the fact that only a sufficiently rapid decrease of temperature with height (a so-called *conditionally unstable atmosphere*, to be defined later) will permit the heat stored in the ocean waters to be liberated through the troposphere in the form of deep convective clouds. The sixth condition recognizes the fact that dry mid levels are not conducive for the continuing development of widespread thunderstorm activity, due to the *detrimental effects of entrainment of anti-fuel* that dilutes, or altogether vitiates, the buoyancy of moist thermals.

From the ostensibly independent necessary conditions for tropical cyclone formation listed above, Gray defined the dynamic potential for storm development as the product of (i), (ii), and (iii). He similarly defined the product of (iv), (v), and (vi) to be the thermodynamic potential for development. The inferred tropical cyclone formation frequency derived by Gray (1975) using these parameters is consistent with the observed genesis regions in Fig. 1.28.[9] Despite this consistency, some might argue that the above six parameters were adjusted to be in accord with the mean seasonal and geographical distributions of storm formation.

Since the thermodynamic parameters vary slowly in both space and time, Gray (1975) and McBride (1981) argued that the thermodynamic parameters would be expected to remain above suitable threshold values needed for storm genesis throughout the cyclone season. In contrast to the thermodynamic parameters, the dynamic parameters and the corresponding dynamic potential is expected to change more dramatically through synoptic activity in association with easterly waves, monsoon shear lines, monsoon gyres, etc. To address this issue, Gray hypothesized that storms

---

9. The dynamic potential for cyclone development, comprising a multiplication of the foregoing genesis parameters, has been generalized by Emanuel and Nolan (2004) to include non-integer exponents and an energetically-based estimate of a cyclone's potential intensity to be discussed later in Chapter 13. A different approach by Wang and Murakami (2020) employs a logarithmic stepwise regression methodology from 11 dynamic and thermodynamic candidate factors. This method gives an alternative genesis potential index based on only four dynamical factors with no non-integral exponents.

would form only during periods when the dynamic potential attains a magnitude above its regional climatological value during the warm season. This latter notion from Gray will be seen to be complementary with the main tenet of the marsupial model discussed further below.

Although we have so far assumed the six environmental parameters to be independent, Frank (1987) pointed out correctly that some of the parameters are physically linked. For example, regions of high sea-surface temperature in the tropics are usually correlated with conditional instability on account of the weak horizontal temperature gradients in the lower troposphere. Furthermore, high humidity in the middle troposphere tends to prevail in the deep convective regions over warm waters as individual clouds detrain and evaporate. As a result, Frank condensed the list to four parameters by blending (i) and (ii) into a condition on the magnitude of absolute vorticity at low levels and omitting (v).

There have been several observational studies that have proposed parameters pertinent to the potential formation of a tropical cyclone from an individual disturbance. Recent work (described further below) has provided notable advances in this direction by recognizing the essentially Lagrangian nature of the tropical cyclogenesis problem and developing a new overarching theoretical framework that can successfully identify and track potentially favorable pre-depression disturbances (e.g., Montgomery et al., 2012; Tory et al., 2013; Rutherford et al., 2018).

1.5.3 Highlights from field experiments

Serendipitous observations have afforded some unique measurements of a few tropical cyclone genesis events (Ritchie and Holland, 1997; Reasor et al., 2005; Houze et al., 2009). To go beyond this small sample, the past several decades have seen the execution of field experiments aimed in part at further understanding the nature of the genesis process. One prominent study of tropical cyclogenesis in the eastern Pacific basin was that of Bister and Emanuel (1997), which was an outcome of the Tropical EXperiment in MEXico (TEXMEX) that was conducted during the summer of 1991. Aircraft reconnaissance missions were flown out of Acapulco, Mexico, and targeted mesoscale convective systems within easterly disturbances over the nearby Eastern Pacific. This study emphasized the importance of thermodynamical processes within a so-called "mesoscale convective vortex embryo".

Bister and Emanuel proposed that the development of a cool, moist environment resulting from stratiform rain serves as the incubation region for the formation of a low-level, warm-core cyclonic vortex. They argued that sustained precipitation in the stratiform cloud deck together with the evaporation of rain drops below would gradually cool and saturate the layer below cloud base while transporting cyclonic vorticity downwards to the surface. The basic idea is that the processes operating within the vortex embryo will result in an accompanying increase in near surface winds that would increase surface moisture fluxes and lead ultimately to convective destabilization. A subsequent bout of deep convection was hypothesized to induce low-level convergence and vorticity stretching, thereby increasing the low-level tangential winds and "igniting" an amplification process fueled by the increased surface moisture fluxes.

Another prominent field study was conducted during July of 2005, the Tropical Cloud Systems and Processes (TCSP) experiment. Reconnaissance research flights were based out of San Juan, Costa Rica. TCSP was broadly focused on the study of the dynamics and thermodynamics of precipitating cloud systems and tropical cyclones using NASA and NOAA aircraft data and surface remote sensing instrumentation data.

On the question of the origin of tropical cyclones, the genesis of tropical storm Gert that formed in the Gulf of Mexico during the TSCP experiment was studied by several investigators. Of particular scientific interest was the role of deep convection and stratiform convection in the system-scale organization of cyclonic vorticity from the surface to the middle levels. The study by Braun et al. (2010) found *inter alia* that while stratiform precipitation regions may significantly increase the cyclonic circulation at midlevels, convective vortex enhancement at low to midlevels is likely necessary for genesis.

In the late summer of 2008, a field experiment, called Tropical Cyclone Structure (TCS08), was conducted in the Northwestern Pacific basin. The overall objectives were to develop a better understanding of storm-scale processes in the western North Pacific associated with tropical cyclones and to further the understanding of interactions between convective processes and tropical cyclone genesis, structure/evolution, intensity, and predictability (Elsberry and Harr, 2008). In the component of the experiment focused on tropical-cyclone formation, priority was given to developing storms prior to their classification as a tropical depression.

The formation of Typhoon Nuri was arguably the best observed genesis event during the TCS08 experiment. A summary of the large-scale and mesoscale aspects leading to Nuri's genesis was presented in Montgomery et al. (2010a) and Raymond and Carillo (2011). In brief, Montgomery et al. (2010a) used global model analyses, global model forecast data, satellite and dropsonde data to demonstrate that Nuri formed inside the recirculating "cat's eye" region within the critical layer of a westward propagating wave-like disturbance (discussed further below). Raymond and Carillo (2011) presented a complementary study of Nuri's genesis using the airborne ELDORA radar and dropsonde data and implicated the inflow induced by rotating deep convective clouds in spinning up the lower-tropospheric tangential circulation of the system-scale flow.

In the summer of 2010, the Pre-Depression Investigation of Cloud Systems in the Tropics (PREDICT) experiment was carried out in the Atlantic and Caribbean region (Montgomery et al., 2012) to test, inter alia, the marsupial paradigm[10] of tropical cyclogenesis as developed by Dunkerton et al. (2009) and summarized further below. PREDICT was conducted in collaboration with the National Aeronautics and Space Administration's (NASA) Genesis and Rapid Intensification (GRIP) experiment (Braun et al., 2013) and the National Oceanographic and Atmospheric Administration's (NOAA) Intensity Forecasting (IFEX) experiment (Rogers et al., 2006). The scientific scope of the GRIP and IFEX experiments was broader than that of PREDICT, including focus on the intensification process. Some notable scientific findings from GRIP and IFEX related to intensification are touched on in Sections 1.2.4, 1.2.5, 7.1 and discussed in more detail in Section 16.7. With regards the cyclogenesis process, some notable findings from PREDICT stand out.

One finding germane to the thermodynamic discussion above is that the system-average virtual potential temperature for both developing and non-developing disturbances showed a slight warming over the observational period (confirmed by Bell and Montgomery, 2019). The most prominent thermodynamic difference between the non-developing system studied and the two systems that developed was the much larger reduction of θ_e between the surface and a height of 3 km, typically 25 K in the non-developing system, compared with only 17 K in the systems that developed (Smith and Montgomery, 2012b).

Another prominent finding was that the intersection between the parent wave's trough and related critical line/surface acted as a favored region for column moistening and convective aggregation (Wang, 2012). Satellite and related analyses conducted in a co-moving frame of reference by Davis and Ahijevych (2012) and Davis and Ahjevich (2011, personal communication), and related numerical modeling studies (Braun et al., 2010; Montgomery et al., 2010b; Wang et al., 2010) together corroborate the Dunkerton et al. prediction of an underlying focal region for the ensuing deep moist convection and precipitation in favorable environments without excessive vertical wind shear.

The foregoing observational findings cry out for a deeper understanding of the convective-vorticity organization process in tropical waves and related favorable environments. Indeed, a deeper dynamical understanding of these observational findings forms part of the motivation for this book (see e.g., Chapter 10).

1.5.4 The formation of a tropical depression

Having presented the traditional necessary conditions for tropical cyclone formation, together with a summary of some noteworthy field experiments examining this problem, we move now to summarize some modern developments that have been established by recent field campaigns. For simplicity, we will focus here on advances made in the Atlantic and eastern Pacific basins, though many of the results carry over to other basins when tropical waves are the progenitor disturbances (e.g., Montgomery et al., 2010a; Raymond and Carillo, 2011; Montgomery and Smith, 2012; Lussier et al., 2015).

The crux of the genesis problem consists of understanding the processes that result in a tropical depression vortex. It is noteworthy that the PREDICT experiment stands alone over the past three decades in that it was devoted exclusively to examining this transition process (Montgomery et al., 2012).[11]

At the risk of overstating the experience of forecasters, the formation of tropical depressions is tightly linked to synoptic-scale disturbances that come in a variety of forms. In all cases, tropical depressions are not observed to form spontaneously. The most prevalent synoptic-scale disturbances in the Atlantic basin are easterly waves that originate over Africa - the so-called African easterly waves. Typically, they have periods of 3-5 days and wavelengths of 2000-3000 km (e.g., Reed et al., 1977). The parent easterly waves over Africa and the far eastern Atlantic are relatively well studied, as in the landmark GATE[12] campaign in 1974 and more recently in the AMMA[13] campaign in (2006). Sometimes a vigorous, diabatically-activated wave emerging from Africa generates a tropical depression immediately, but in most instances these waves continue their westward course harmlessly over the open ocean, or blend with new waves initiated in the mid-Atlantic ITCZ. In a minority of wave disturbances, the cyclonic vorticity structures they contain become seedlings for the formation of a tropical depression in the central and western Atlantic and farther west.

Fig. 1.33 shows the detection locations of developing and non-developing tropical depressions from 1975 to 2005, based on the work of Bracken and Bosart (2000) and extended to include data from the time frame 1995-2005. It is evident from the figure that there are relatively few Atlantic tropical depressions that fail to become tropical cyclones. It

---

10. In this book we use the term paradigm to describe an overarching theoretical/conceptual framework of ideas.

11. Some of the material presented in the next three subsections is drawn from scientific material written for the PREDICT experiment. The second author's collaborators Tim Dunkerton, Chris Davis and Lance Bosart are acknowledged here for their excellent help in developing that highly successful scientific program.
12. The acronym GATE refers to the WMO Global Atmospheric Research Programme Atlantic Tropical Experiment
13. The acronym AMMA refers to NASA's Saharan Air Layer And The Fate Of African Easterly Waves Field Study of Tropical Cyclogenesis.

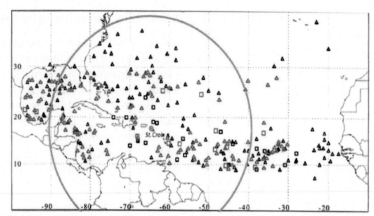

FIGURE 1.33 First-detection locations of developing (triangles) and non-developing (squares) tropical depressions from 1975 to 2005 (1995-2005 in red (mid gray in print version)). The blue (dark gray in print version) circle denotes the approximate PREDICT domain. Courtesy C. Davis.

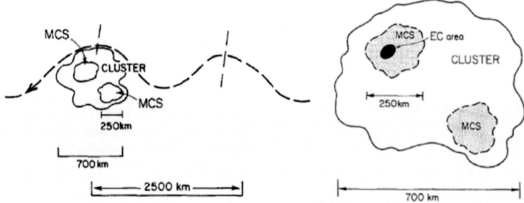

FIGURE 1.34 Schematic (left) of synoptic-scale flow through an easterly waves (dashed) with an embedded cluster of convection in the wave trough. In (right) the cluster is shown to contain mesoscale convective systems (MCSs) and extreme convection (EC, black oval) within one of the MCSs. From Gray (1998). Republished with permission of Meteorology and Atmospheric Physics.

is well known also that most (approximately 80%) tropical waves do not become tropical depressions. This fact is supported by numerous studies (e.g., Frank, 1970; Dunkerton et al., 2009, their footnote 3; Asaadi et al., 2017). Then, the key questions would appear to be the following:

(i) Which tropical waves (or other disturbances) will evolve into a tropical depression?
(ii) What is different about developing waves?
(iii) Can this difference be identified, and on what time scale?
(iv) Why do so few disturbances develop?

1.5.5 The multi-scale nature of genesis in the real world

To answer the foregoing questions, a new approach has been developed in the past decade or so that recognizes the multi-scale nature of tropical cyclogenesis within tropical waves and its accompanying Lagrangian fluid dynamics. Some of the pertinent processes at play are illustrated schematically in Fig. 1.34 (after Gray, 1998). There, two length scales are

illustrated, with a cluster of deep, moist convection confined to the low-pressure trough of the synoptic-scale wave. Within such clusters are individual mesoscale convective systems (MCSs).

Dunkerton et al. (2009) articulated the new approach to the genesis problem, beginning with the realization that (p5588)

> *"the fluid motion is mostly horizontal and quasi-conservative, punctuated by intermittent deep convection, a strongly diabatic and turbulent process."*

The new approach stresses the importance of identifying the intrinsic flow boundaries and advocates analyzing observational data in a manner consistent with the Lagrangian nature of the formation process (p5588):

> *"In order to fully appreciate the transport of vorticity and moist entropy by the flow, their interaction with one another, the impacts of deep convective transport and protection of the proto-vortex from*

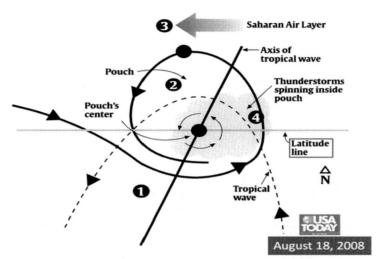

FIGURE 1.35   Idealized schematic of the wave-pouch as presented in USA Today (18 August 2008). Dashed contours represent the easterly wave's streamlines in a "classical" ground-based frame of reference, which is usually open with an inverted-V pattern and trough (1). Solid streamline delineates the wave-pouch as viewed in the "new" frame of reference moving at the same speed with the parent wave. A generally zonal line (gray line) of zero zonal wind speed in this wave-relative framework defines the critical latitude (or "Latitude line" in the sketch). The critical latitude is where the wave's phase velocity matches the local mean flow velocity. See text for detailed description. Graphic with permission from USA TODAY.

hostile influences requires, among other things, an understanding of material surfaces or "Lagrangian boundaries" in the horizontal plane. This viewpoint, although used subconsciously by forecasters, is invisible to researchers working with standard meteorological products in an Eulerian or Earth-relative framework."

The seventh necessary condition, alluded to in Subsection 1.5.2, recognizes the critical importance of a nearly closed circulation region (or *"pouch"*) in the wave-relative frame. For reasons summarized below, a quasi-closed deep pouch spanning the lower troposphere supports the organization of the convection and vorticity around the center of the pouch, the so-called "sweet spot" of the parent wave (defined more precisely below). As shown later in Chapter 10, a further element in the organization process is the subtle but important role of the frictional boundary layer in controlling the location of convective instability.

The scientific basis for the new multi-scale Lagrangian model is detailed in Dunkerton et al. (2009), Montgomery et al. (2010b), Wang (2012) and Asaadi et al. (2016a). Fig. 1.35, published in USA Today, provides a simple conceptual model contrasting the "classical" model (1) with an inverted-V pattern in the low-level earth-relative streamline and pressure tough oriented in a nearly south-north configuration and a "new" model. The new model reveals the advantages of viewing the genesis process in a semi-Lagrangian frame co-moving with a parent wave. As sketched in the figure, a deep pouch (2) can protect the enhanced vortical structures generated by deep convection from the hostile tropical environment. Examples of such an environment are the dry air masses associated with the

Sahara Air Layer (SAL) (3) or environments with strong vertical or horizontal wind shear. Under favorable circumstances, deep convection (4) will tend to be sustained within the pouch. On account of the convectively-generated convergent flow, the wave's pouch may have an opening that permits the influx of vorticity and environmental air on the southwest side of pouch.

The PREDICT experiment was designed principally to test this new model, whose overarching premise was that tropical depression formation is greatly *favored in the critical-layer region* of the synoptic-scale pre-depression wave or subtropical disturbance. The critical layer is a horizontal region of fluid surrounding the arrangement of points where the wave's phase speed coincides with the low-frequency, mean flow. The collection of such points is called the *critical line*. According to this model, the *"sweet spot"*, defined by *the intersection of the wave-trough line and the critical line*, represents an attractor point around which the convection ultimately aggregates and organizes into a tropical depression (e.g., Montgomery et al., 2010b).

Three main hypotheses (H1-H3) were tested in PREDICT. These hypotheses were summarized by Montgomery et al. (2012) in fluid dynamical terms:

"The cat's eye recirculation region within the critical layer of the parent wave is a region of cyclonic rotation and weak straining deformation (H1). This region provides a set of approximately closed material contours. Inside the cat's eye circulation, air tends to be repeatedly moistened by cumulus convection, protected to some degree from lateral intrusion of dry air and deformation by horizontal or vertical shear, and (thanks to its location near the critical level) able

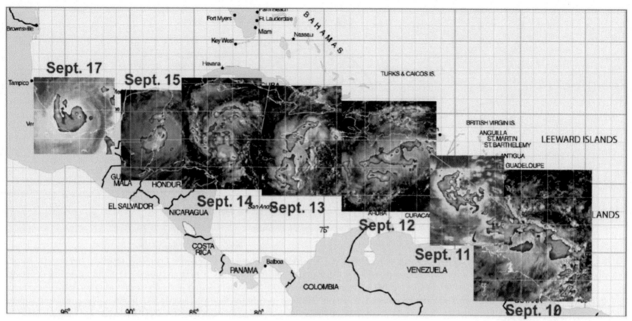

FIGURE 1.36 85-GHz montage (images courtesy of NRL, Monterey) for the active convection periods on each day from 10 to 17 (excluding 16 Sep) during the genesis of Hurricane Karl (PG144). Note the small eye on the Yucatan coast on 15 Sep. From Montgomery et al. (2012). Republished with permission of the American Meteorological Society.

to keep pace with the parent wave until the dominant vortex has strengthened into a self-maintaining entity (H2). During this time the parent wave is maintained and possibly enhanced by diabatically amplified eddies within the wave (proto-vortices on the mesoscale), a process favoured in regions of small intrinsic phase speed (H3)."

The combination of the associated genesis sequence and the overarching framework for describing how such hybrid wave–vortex structures become tropical depressions/storms was likened to the development of a marsupial infant in its mother's pouch, and for this reason was charmingly dubbed the "marsupial paradigm". This model is often colloquially referred to as the "pouch theory".

1.5.6 Practical outcomes

The marsupial model has provided a useful road map for exploration of synoptic-mesoscale connections essential to tropical cyclogenesis. In particular, an important piece of predictive skill derives from the intersection of the wave's trough axis, which is oriented approximately south-north, and the critical line, which is oriented approximately west-east. As noted above, this intersection denotes a unique location along the trough axis around which recirculation occurs and pinpoints the pouch center, or sweet spot as defined above, which has been shown to be the preferred location for the formation of a nonlinear critical layer, or cat's eye circulation, small-scale vorticity aggregation, and ulti-

mately a tropical depression (Dunkerton et al., 2009; Montgomery et al., 2010b; Wang et al., 2010; Montgomery et al., 2012; Davis and Ahijevych, 2012; Asaadi et al., 2016b, 2017). The track of the sweet spot can be predicted accurately several days in advance using the horizontal winds of dynamical models.

As mentioned above, an exciting outcome of the PREDICT observations was the observational confirmation of the theoretical prediction that the sweet spot is the preferred location for cyclogenesis, serving as a "focal point" for an upscale vorticity aggregation process while the proto-vortex and wave pouch move together. An important practical outgrowth of the marsupial model is that the pouch provides a more precise, dynamically relevant target for continuous monitoring that is distinct from the chaotic activity of moist convection scattered throughout the parent wave's trough. Although deep convection exterior to, or along the perimeter of, the pouch can be equally vigorous, it is subject to horizontal strain and/or shear deformation that is inimical to vortical-convective organization.

Tropical cyclone Karl, a well-surveyed formation case during PREDICT, serves as an illustrative example of the foregoing skill using the meteorological pouch products. The GV sampled pre-Karl (called 'PGI44') for five consecutive days from 10 to 14 Sep., 2010.[14] A satellite-derived mosaic of the evolution of the pre-Karl disturbance is given in Fig. 1.36.

---

14. The naming convection for potential disturbances is described in Montgomery et al. (2012).

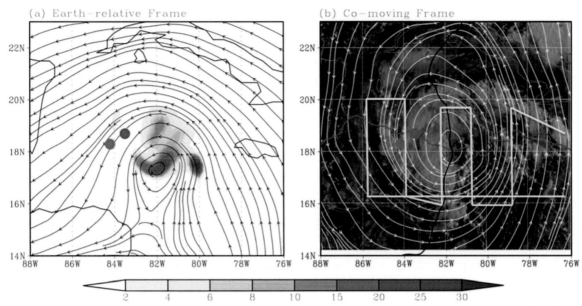

FIGURE 1.37 (Left) ECMWF 36 h forecast of 700-mb Earth-relative streamlines and Okubo-Weiss parameter (shading; units: 10^{-9} s^{-2}) centered on wave pouch PGI44/Al92 (pre-Karl) valid at 1200 UTC 14 Sep. 2010. (Right) Streamlines in the co-moving frame of reference (phase speed of 6.2 m s^{-1} westward), GOES visible imagery at 1225 UTC, and flight pattern of GV aircraft (yellow track). In the right figure, the black curve represents the trough axis and the purple curve the local critical latitude defined by $U = c_x$, where c_x denotes the wave's zonal phase speed. The red dot represents the actual genesis location, and the blue dot is the ECMWF 700-mb predicted sweet spot, defined by the intersection of the trough axis (black curve) and critical latitude (purple curve), at 2100 UTC 14 Sep. 2010. From Montgomery et al. (2012). Republished with permission of the American Meteorological Society.

During the PREDICT midday coordination sessions on 13 Sep., the pouch products were available for planning the next day's flight using the ECMWF forecasts from 0000 UTC 13 Sep. ECMWF 36-hour forecasts depicted a trough located along 82°W at the flight time of 1200 UTC 14 Sep. In an Earth-relative frame, the circulation center as depicted by 700 mb streamlines was at about 17.3°N and located near the southern edge of an extensive region of positive values of the Okubo-Weiss parameter[15] (Fig. 1.37, left). The circulation in the co-moving frame (Fig. 1.37, right) is more clearly defined than in the Earth-relative frame. The 700-mb Earth-relative flow depicts a tropical wave with an inverted-V pattern with only a limited region of weak westerly flow south of the putative center. The circulation center in the co-moving frame of reference is located between two areas of high Okubo-Weiss parameter (Fig. 1.37, left) that appear to be wrapping around the pouch center. From these forecast

products, a large lawnmower flight pattern was constructed that sampled the wave-pouch circulation and a region beyond the central convection.

1.6 Synthesis

In this chapter we have introduced the reader to the basic kinematic and thermodynamic structure of a tropical cyclone vortex during both its intensification, and mature stages. We considered also the conditions necessary to create a tropical cyclone in the real tropical atmosphere and we reviewed new developments in the science of tropical cyclogenesis with new, physically-based, tools to help forecasters on the bench.

The canonical warm-core structure of a mature vortex was revealed by an analysis of airborne dropsondes and flight level data supplied by manned and unmanned reconnaissance aircraft. The rapid intensification of record-setting Hurricane Patricia (2016) was shown to evidence a highly complex and asymmetric distribution of deep cumulus convection focused around a nascent center of circulation. This highly asymmetric configuration of cumulus convection is typical during the spin up stage of a tropical cyclone. As Patricia intensified, the interior region became highly axially symmetric, at least in the core region of the circulation. The progressive transition from a highly asymmetric convective mass to a more circularly symmetric configuration is typical for intensifying systems in favor-

---

15. In simple terms, the Okubo-Weiss (OW) parameter is a frame-independent measure of the tendency for a rotational flow to be shape preserving. Mathematically, assuming a small horizontal divergence in comparison to vertical vorticity in a first approximation, OW is defined here as $OW = \zeta^2 - S_1^2 - S_2^2 = (V_x - U_y)^2 - (U_x - V_y)^2 - (V_x + U_y)^2$ where S_1 and S_2 denote strain deformations, $\zeta = V_x - U_y$ denotes vertical relative vorticity, (U, V) denote zonal and meridional velocity, respectively, and subscripts x and y denote partial differentiation in the zonal or meridional directions (e.g., McWilliams, 1984, p33; Weiss, 1991). In the convention employed here, a negative OW region is unfavorable for rotational organization of deep convection and aggregation of vortical remnants of prior convective activity.

able environments possessing relatively weak vertical wind shear.

Outside the primary eyewall of a mature tropical cyclone, convective rainbands are typically present. These rainbands tend to wrap cyclonically inward towards the center of circulation and consist of a mixture of deep cumulus convection, stratiform convection and wavy azimuthal jets with a corresponding enhanced inflow in the underlying boundary layer. Still to be further understood is the role of asymmetric convection in the intensification process, the role of outer rainbands in the regulation of a storm's peak intensity and formation of an outer secondary eyewall.

The extensive observational data collected in Hurricane Isabel (2003) offers an unprecedented look at the inner-core wind and thermodynamic structure of a Category five storm as the vortex slowly decayed in intensity and expanded in size.

A tropical cyclone acquires its energy primarily through the wind-forced evaporation of water from the underlying ocean. The passage of a tropical cyclone over warm ocean water produces a reduction of the sea surface temperature by wind-forced upwelling of cooler water from below the thermocline in the rear right quadrant of the advancing cyclone in the northern hemisphere.

In the event of a stalled storm, such as occurred in the case of Hurricane Harvey (2017), the upwelling of relatively cool water is no longer confined to the right rear quadrant of the storm. In such an event, the upwelling permeates the high wind region and can lead to a rapid decline in vortex intensity.

The marsupial paradigm offers new understanding into how locally-favorable recirculation ("pouch") regions are generated within synoptic-scale precursor disturbances in the lower tropical troposphere, and has gone a substantial way to providing answers to the itemized questions raised above. On the one hand, these recirculation regions help protect seedling vortices from the detrimental effects of vertical and horizontal shearing deformation and from the lateral entrainment of relatively dry air ("anti-fuel"). On the other hand, these approximate re-circulation regions favor sustained column moistening and low-level cyclonic vorticity enhancement by vortex-tube stretching in association with deep cumulus convection. These ideas have proven useful for bridging the gap between Lagrangian fluid dynamics and the development of pragmatic tools that improve genesis forecasts in the 0-5-day time frame.

To further our understanding of the foregoing observations and related theoretical developments, we believe new researchers to the field would benefit greatly from having a book that develops systematically a basic understanding of the dynamics and thermodynamics of tropical cyclone vortices. The present book is written with this overarching goal in mind. Such knowledge requires a familiarity with basic moist fluid dynamics applied to the tropical atmosphere and

we believe will enable *inter alia* a more in-depth analysis of the theoretical and practical applications of the marsupial paradigm and its operational generalization to cover all basins (e.g., Rutherford et al., 2018, 2015).[16]

---

16. Two additional examples from the 2022 Atlantic Hurricane season are summarized in Chapter 17. These examples illustrate the usefulness of the new tropical cyclogenesis model in combination with the material developed in this textbook.

Chapter 2

Fluid dynamics and moist thermodynamics

In this chapter we review the equations that represent the physical laws governing atmospheric motion. Rigorous derivations of the equations may be found in many text books, such as Gill (1982), Pedlosky (1987), Holton (2004), or McWilliams (2011).

2.1 The equations of motion

The motion of an air parcel is governed by Newton's second law of motion, which states that the acceleration of the parcel equals the force per unit mass that acts upon it. For large-scale geophysical flows on a rotating planet, a rotating coordinate system is used because the fluid as a whole rotates with constant angular velocity. It is the fluid motion relative to the bulk rotation that is most relevant and useful for the prediction of atmospheric flow. In a uniformly rotating frame of reference *pseudo* forces (Coriolis and centrifugal forces) must be added to Newton's equation so as to account for the accelerating (non-inertial) frame of reference. In the rotating reference frame, Newton's second law can be expressed mathematically as follows:

$$\frac{D\mathbf{u}}{Dt} + 2\boldsymbol{\Omega} \wedge \mathbf{u} = -\frac{1}{\rho}\nabla p + \mathbf{g_e} + \mathbf{F}, \qquad (2.1)$$

where $\mathbf{u}$ is the velocity of the air parcel relative to the rotating frame, ρ is its density, p is the pressure, $\boldsymbol{\Omega}$ is the angular rotation vector of the reference frame (e.g., the Earth), $\mathbf{F}$ is a force per unit mass acting on the parcel (e.g., a frictional force), t is the time, and $\mathbf{g_e}$ is the "effective" gravitational acceleration. The "effective" gravitational acceleration is the vector sum of the acceleration due to gravity, $(0, 0, -g)$ and the pseudo force associated with the centripetal acceleration arising from the Earth's rotation

$$\mathbf{g_e} = \mathbf{g} + \Omega^2\mathbf{r_H} = -\nabla\Phi, \qquad (2.2)$$

where $\mathbf{r_H}$ denotes the component of the parcel's position vector $\mathbf{r}$ that is perpendicular to $\boldsymbol{\Omega}$ (see Fig. 2.1) and Φ is the effective gravitational potential, defined as the ordinary gravitational potential minus the gradient of a centrifugal potential, $\Omega^2 r_H^2/2$. The centrifugal action of the rotating body transforms the initially spherical body into an oblate

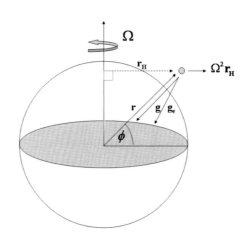

FIGURE 2.1 Effective gravity on the rotating Earth.

spheroid. The surface of the rotating planet is then a surface of constant effective geopotential.

For terrestrial flows, the centrifugal force, $\Omega^2 r_H$, is approximately 0.3% that of true gravity. Hereafter the 'e' subscript will be dropped unless otherwise noted so that the vector $\mathbf{g}$ and scalar g will refer to the effective gravity. It is conventional to take $\Phi = 0$ to correspond with the undisturbed sea surface and to define the local vertical direction to be the unit vector $\hat{\mathbf{k}} = \nabla\Phi/|\nabla\Phi|$. When z denotes the height above the surface in this direction, Φ represents the *potential energy* per unit mass of an air parcel at this level.

The material derivative D/Dt is calculated following the air parcel in the rotating coordinate system and has the form

$$\frac{D\mathbf{u}}{Dt} = \frac{\partial\mathbf{u}}{\partial t} + \mathbf{u}\cdot\nabla\mathbf{u}. \qquad (2.3)$$

In the case of a fluid at rest, $\mathbf{u} = \mathbf{0}$ and the momentum equation reduces to

$$\nabla p = \rho\mathbf{g}. \qquad (2.4)$$

This equation expresses hydrostatic equilibrium and is a form of the *hydrostatic equation*. When expressed in a rectangular or cylindrical coordinate system with z pointing upwards (perpendicular to the local geopotential surface),

Tropical Cyclones. https://doi.org/10.1016/B978-0-44-313449-4.00010-2

$\mathbf{g} = (0, 0, -g)$ and Eq. (2.4) takes the more familiar form

$$\frac{\partial p}{\partial z} = -\rho g. \tag{2.5}$$

Eqs. (2.4) or (2.5) are called the hydrostatic equation because they express a balance between the vertical pressure gradient and the gravitational force, both per unit volume, a state that we call *hydrostatic balance*. The motion of the air described by (2.1) is subject to the constraint that at every point in the flow, mass must be conserved. In general, for a *compressible fluid* such as air, mass conservation is expressed by the equation

$$\frac{D\rho}{Dt} + \rho \nabla \cdot \mathbf{u} = 0. \tag{2.6}$$

We call this the *continuity equation*.

In practice, density fluctuations in the air are rather small and can be neglected unless we want to study sound propagation or buoyancy effects. Usually, shallow layers of air on the order a kilometer deep or less can be considered to be *homogeneous* ($\rho =$ constant) to a first approximation and therefore *incompressible*. In this case the continuity equation reduces to

$$\nabla \cdot \mathbf{u} = 0. \tag{2.7}$$

Formally, this approximation is not accurate for deeper layers since air is compressible under its own weight. For example, the density at a height of 10 km is only about a quarter of that at mean sea level. Even so, for some purposes this *pseudo-incompressible* approximation to the continuity equation often suffices for qualitative insight, even when buoyancy effects are considered. When the density fluctuations are small, their impact in both the continuity and momentum equations can be neglected everywhere, except where they multiply the gravitational acceleration. This is the so-called *Boussinesq approximation*. In situations where further accuracy is desired, one may still neglect the time and horizontal variations of density in the continuity and momentum equations, but retain the vertical variation of the density field. In (2.6) one then sets $\rho = \rho_o(z)$, where $\rho_o(z)$ is the mean density at a particular height z, whereupon (2.6) becomes

$$\nabla \cdot (\rho_o \mathbf{u}) = 0. \tag{2.8}$$

This equation forms part of what is called the *anelastic approximation* in which the density is allowed to vary only with height.

The constraint imposed by (2.7) or (2.8) means that the formulation of fluid flow problems in enclosed regions is intrinsically different to that for problems in rigid body dynamics. In the latter, a typical problem is to specify the forces acting on a body and calculate its resulting acceleration and ultimately its trajectory. *In most fluid flow problems, the force field cannot be imposed arbitrarily: it must produce accelerations that are compatible with ensuring mass continuity at all times. In general, this means that the pressure field cannot be arbitrarily prescribed: it must be determined as part of the solution.* Thus methods for solving fluid flow problems are different also from those required for solving rigid-body problems (see Section 2.7).

For homogeneous fluids, Eqs. (2.1) and (2.7) suffice to determine the flow when suitable initial and boundary conditions are prescribed. However, when density varies with pressure and temperature we need additional equations: a thermodynamic equation and an equation of state (see Section 2.3).

2.2 Buoyancy and perturbation pressure

It is common experience that air that is warmer than its surroundings rises and air that is colder sinks. This tendency may be quantified in terms of the *buoyancy force* per unit mass. The buoyancy force that acts on an air parcel of volume V with density ρ in an environment with density ρ_a is defined as the difference between the weight of air displaced by the parcel, $\rho_a g V$, (the upward thrust according to Archimedes' principle), and the weight of the parcel itself $\rho g V$. This quantity b is normally expressed per unit mass of the air parcel under consideration, i.e.,

$$b = g \frac{(\rho_a - \rho)}{\rho}. \tag{2.9}$$

The calculation of the upward thrust assumes that the pressure within the air parcel is the same as that of its environment *at the same level* and that ρ_a is the density of the environment *at the same height* as the parcel. Using the equation of state (see Section 2.3.1), the buoyancy force may be expressed alternatively in terms of temperature, $b = g(T - T_a)/T_a$, where T is the parcel temperature and T_a is the temperature of its environment. If the air contains moisture, the virtual temperatures need to be used instead, while if the air contains condensate, the density temperature(s) must be used (see Section 2.5.1). It follows that if an air parcel is warmer than its environment, it experiences a positive (i.e. upward) buoyancy force and tends to rise; if it is cooler, the buoyancy force is negative and the parcel tends to sink.

The tendency of warm air to rise is not immediately apparent from the vertical component of forces on the right-hand-side of the momentum equation (2.1). However, we can modify this equation so that a buoyancy term appears explicitly. We simply replace the pressure in this equation by the sum of a reference pressure p_{ref}, a function of height z only, and a perturbation pressure, $p' = p - p_{ref}$. The former is taken to be in hydrostatic balance with a prescribed reference density ρ_{ref}, i.e. it satisfies Eq. (2.4). In idealized flows, the reference density is normally taken as the density profile in the environment, but in real situations, the environmental density is not uniquely defined. In these cases, $\rho_{ref}(z)$ could be taken as the horizontally-averaged density over some large domain surrounding the air parcel. Then Eq. (2.1) becomes

$$\frac{D\mathbf{u}}{Dt} + 2\mathbf{\Omega} \wedge \mathbf{u} = -\frac{1}{\rho}\nabla p' + \mathbf{b} + \mathbf{F}, \qquad (2.10)$$

where $\mathbf{b} = b\hat{\mathbf{k}}$ and b is as defined in (2.9).

A comparison of (2.1) and (2.10) shows that the sum of the vertical pressure gradient force and gravitational force per unit mass acting on an air parcel is equal to the sum of the vertical gradient of perturbation pressure and the buoyancy force, where the latter is calculated from Eq. (2.9) *by comparing densities at constant height*. The scalar part of the expression for $\mathbf{b}$ in (2.10) has the same form as that in (2.9), but with ρ_{ref} in place of ρ_a. However, the derivation circumvents the need to assume that the local (parcel) pressure equals the environmental pressure when calculating b. The pressure deviation is accounted for by the presence of the perturbation pressure gradient force, the first term on the right-hand-side of (2.10).

The foregoing decomposition indicates that, in general, the buoyancy force is not uniquely defined because it depends on the (arbitrary) choice of a reference density. However, the sum of the buoyancy force and the vertical component of perturbation pressure gradient force per unit mass *is* unique. If the motion is hydrostatic, i.e. if the vertical acceleration is small compared with the gravitational acceleration, the perturbation pressure gradient force and the buoyancy force are equal and opposite, but they remain nonunique.

We show in Chapters 5, 9 and 15 that the definition of buoyancy force requires careful consideration and some modification in a rapidly-rotating vortex.

2.3 Thermodynamics

As noted in Section 2.1, when density varies with pressure and temperature we need additional equations including as a minimum a thermodynamic energy equation and an equation of state. These equations are discussed below. When various states of water are present, additional equations governing the rates-of-change of the mixing ratios for water vapor, liquid water, and possibly species of ice as well as conversions between these states are required also. In addition, the thermodynamic equation will need to be modified to account for the latent heat released or consumed during phase changes of water substance. A discussion of moist processes is deferred until Section 2.5.

2.3.1 Equation of state

The *equation of state* is a relationship between the pressure, density, and absolute temperature, T, of an air parcel. For dry air, the ideal gas equation ordinarily suffices to a high level of accuracy for most atmospheric flows

$$p = \rho R_d T, \qquad (2.11)$$

where R_d is the *specific gas constant* for dry air. For moist air, the equation needs to be modified as discussed in Section 2.5.

2.3.2 Thermodynamic energy equation

There are various forms of the *thermodynamic energy equation*, which is an expression of the *first law of thermodynamics*. The law has two parts. The first is that there is a function of state for an air parcel called the internal energy, E. (The internal energy is a macroscopic measure of the average kinetic energy of all the molecules that comprise the fluid parcel.) The second part states that the change in internal energy, dE, is equal to the incremental heat input to the parcel, dQ, plus the incremental work done on the parcel, dW. That is,

$$dE = d^*Q + d^*W. \qquad (2.12)$$

The $*$ notation is a reminder that the heat and work interactions are generally path dependent. That is, these differentials are not exact and the heat and work interactions are generally not state functions. However, for a two parameter system (dry air consisting of a fixed number of molecules is governed by two state variables, such as temperature and pressure), and for a reversible process, d^*W becomes an exact differential, equaling minus the pressure times the change in specific volume, $-pd\alpha$ (positive work is done on system when the volume of the parcel is compressed). The specific volume is the volume per unit mass and is equal to the inverse of the density (i.e., $\alpha = 1/\rho$). Because air behaves approximately as an ideal gas for temperatures well above 0 K, we have $dE = c_{vd}dT$, where c_{vd} is the (constant) specific heat for dry air at constant volume.

The first law then takes the form:

$$c_{vd}dT = d^*Q - pd\alpha. \qquad (2.13)$$

When applied to a moving air parcel undergoing slow time changes in comparison to the molecular time scale governing the relaxation to *local thermodynamic equilibrium*, Eq. (2.13) may be written in the equivalent rate form

$$\frac{DE}{Dt} + p\frac{D\alpha}{Dt} = \frac{DQ}{Dt}, \qquad (2.14)$$

where D/Dt denotes the rate of change with respect to time following a fluid parcel. (This is one form of the thermodynamic energy equation for the parcel.) Hereafter, all differential relations involving Q will be assumed to be exact (dropping star notation). The irreversible (path dependent) part of DQ/Dt may be accounted for by including the viscous dissipation rate, thermal conduction, mixing of constituents and radiative absorption/emission processes.[1] When combined with the un-approximated form of the continuity equation (2.6), this internal energy equation may be written as

$$\rho \frac{DE}{Dt} \equiv \frac{\partial \rho E}{\partial t} + \nabla \cdot (\rho E \mathbf{u}) = -p\rho \frac{D\alpha}{Dt} + \rho \frac{DQ}{Dt}. \quad (2.15)$$

Another commonly used form of the thermodynamic energy equation may be derived as follows. Using the equation of state and the fact that $\alpha = 1/\rho$, Eq. (2.13) may be written in differential form as

$$dQ = c_{pd}dT - \frac{R_d T}{p}dp, \quad (2.16)$$

where $c_{pd} = c_{vd} + R_d$ is the *specific heat at constant pressure* for an ideal gas. Eq. (2.16) may be rewritten more succinctly as

$$d\ln T - \kappa d\ln p = \frac{1}{c_{pd}T}dQ, \quad (2.17)$$

where $\kappa = R_d/c_{pd}$. When applied to a moving air parcel, Eq. (2.17) may be written in the form

$$\frac{D\ln T}{DT} - \kappa \frac{D\ln p}{Dt} = \frac{1}{c_{pd}T}\frac{DQ}{Dt}. \quad (2.18)$$

The total derivative of the heating increment, DQ/Dt, which we abbreviate as $\dot{Q}$, is the *material heating rate per unit mass* with units J kg^{-1} s^{-1}.

For dry air, Eqs. (2.1), (2.6), (2.11), and (2.18) form a complete set to solve for the flow, subject to suitable initial and boundary conditions assuming that $\dot{Q}$ and $\mathbf{F}$ are known.

2.3.3 Potential temperature and specific entropy

Two alternative forms of (2.18) are frequently used. It is common to define the *potential temperature*, θ, as

$$\theta = \frac{T}{\pi}, \quad (2.19)$$

where $\pi(p) = (p_{**}/p)^{\kappa}$ is called the *Exner function* and p_{**} is a constant *reference pressure*, normally taken to be 1000 mb (or 1000 hPa). Then (2.18) becomes

$$\frac{D\theta}{Dt} = \frac{\dot{Q}}{c_{pd}\pi}. \quad (2.20)$$

---

1. See e.g., Gill (1982) for more details. See below for more about viscosity and the dissipation rate.

If there is no heat input into an air parcel during its motion, the right-hand-side of (2.20) is zero, implying that θ remains constant for the parcel in question. A process in which there is no external heat source or sink is called an *adiabatic process*. The potential temperature may be interpreted as the temperature that an air parcel with temperature T at pressure p would have if it were brought adiabatically to a standard reference pressure, p_{**}. *Diabatic processes* are ones in which the material heating rate, $\dot{Q}$, is non zero. The right-hand-side of (2.20) is often referred to as the *diabatic heating rate* and written as $\dot{\theta}$ with units of K s^{-1}. Note that the diabatic heating rate and material heating rate are proportional to one another at constant pressure and we shall often refer to non-zero heating simply as *diabatic heating*.

A second common form of (2.20) is

$$\frac{Ds_d}{Dt} = \frac{\dot{Q}}{T}, \quad (2.21)$$

where $s_d = c_{pd} \ln \theta$ is known as the *specific dry entropy* of the air.

Up to this point, the entropy s_d has been introduced as a variable to alternatively express Eq. (2.20) for the material rate of change of potential temperature. The astute reader will realize that we have formulated the fluid dynamical equations without reference to the second law of thermodynamics. A thermally conducting viscous fluid necessarily includes the operation of irreversible processes associated with heat and momentum diffusion. One consequence of the second law is that irreversible processes increase the entropy of the fluid system, i.e.,

$$\frac{d}{dt}\int \rho s_d dV > 0. \quad (2.22)$$

This inequality implies that the molecular viscosity and molecular thermal conductivity must be positive quantities (Landau and Lifshitz, 1966). The corresponding entropy inventory equation (given by the mass-weighted integral of Eq. (2.21) over the entire fluid) imposes a constraint on the entropy production of a moist precipitating atmosphere with heat sources and sinks, including radiation (Goody, 1995, Section 2.4.1), providing a check on the consistency of numerical simulations of hurricanes or the climate.

At this time, the only study we are aware of that examines the implications of irreversible entropy-production processes on the dynamics of hurricanes is the work by Pauluis and Zhang (2017). Those authors carry out a novel analysis of the impact of some irreversible processes on the maximum intensity of a hurricane vortex as depicted in an idealized, three-dimensional, numerical simulation. Their work finds that unlike the outer rainbands, which are highly inefficient, "the deepest overturning circulation associated with the rising air within the eyewall is an efficient heat engine that produces about 70 % as much kinetic energy as a comparable Carnot cycle."[2]

2.3.4 Static energy

Using the equation of state Eq. (2.11), Eq. (2.16) becomes $dQ = c_{pd}dT - (1/\rho)dp$. Using also the fact that $Dp/Dt = D_h p/Dt + w\partial p/\partial z$ and assuming that the motion is hydrostatic (Eq. (2.5)), this equation can be written for a moving air parcel as

$$\frac{Dh_d}{Dt} - \frac{1}{\rho}\frac{D_h p}{Dt} = \dot{Q}, \qquad (2.23)$$

where $h_d = c_{pd}T + gz = c_{pd}T + \Phi$ is known as the *dry static energy* of the air and D_h/Dt is the material derivative operator following the horizontal wind.

2.4 Prognostic and diagnostic equations

Equations of the type (2.1), (2.6), (2.20), (2.21), and (2.23) that involve time derivatives are called *prediction equations* because they can be integrated in time (perhaps not analytically) to forecast subsequent values of the dependent variable in question. In contrast, equations such as (2.4), (2.5), (2.7), (2.8), and (2.11) are called *diagnostic equations* because they give a relationship between the dependent variables at a particular time, but involve time only implicitly.

2.5 Moist processes

Eqs. (2.18)-(2.23) hold to a good first approximation even when the air contains moisture, in which case a *dry adiabatic process* is one in which the air remains unsaturated. However, tropical cyclones owe their existence to moist processes and the inclusion of these in any theory requires an extension of the foregoing equations to allow for the latent heat that is released when water vapor condenses or liquid water freezes and which is consumed when cloud water evaporates or ice melts. We require also equations characterizing the amount of water substance in the air and its partition between the possible states: water vapor, liquid water, and ice. This section gives only a summary of some of the key thermodynamic results representing moist processes in idealized configurations. More rigorous and complete treatments can be found, e.g., in Emanuel (1994) and Goody (1995).

Water content is frequently expressed by the *mixing ratio*, the mass of water per unit mass of dry air. We denote the mixing ratios of water vapor, liquid water, ice, and total water by r_v, r_L, r_i, and r_T, respectively. Sometimes *specific humidity*, q_v, is used rather than r_v. It is defined as the mass of water vapor per unit mass of moist air and related to r_v by

$q_v = r_v/(1 + r_v)$. Since, in practice, $r_v \ll 1$, the difference between r_v and q_v is not only small, but barely measurable.

An alternative quantity used to characterize moisture in the air is the *partial pressure of water vapor*, e. This quantity is related to r_v by the formula $r_v = \epsilon e/(p - e)$, where p is the air pressure and $\epsilon = R_d/R_v$, i.e., the ratio of R_d to the specific gas constant for water vapor, R_v. Note that $p - e$ in this formula is just the partial pressure of dry air.

When water evaporates from a level water surface into dry air at constant pressure and temperature, the partial pressure e of the air will increase until an equilibrium value e^* is reached. This equilibrium value is an exponentially increasing function of the air temperature only and is called the *saturation vapor pressure*. When $e = e^*$ we say that the air is *saturated*. The corresponding mixing ratio defined by $r_v^* = \epsilon e^*/(p - e^*)$ is called the *saturation mixing ratio*, which can be seen to depend on both the air pressure and temperature. In particular, at a fixed temperature, r_v^* increases with decreasing pressure, a result that has important consequences for tropical cyclones as discussed in Chapters 12 and 13.

When an *unsaturated* air parcel ($e < e^*(T)$) is cooled at constant pressure without water being added or removed, $e^*(T)$ will decrease while e will remain constant and eventually the parcel will become saturated. If the cooling continues, the air parcel would become *supersaturated*, i.e., $e > e^*(T)$. Because of the abundance of tiny impurities in the Earth's atmosphere called aerosols, the degree of supersaturation rarely exceeds more than a fraction of a percent and effectively, the excess water $r_v - r_v^*$ condenses to form tiny cloud drops or *fog*.

A similar situation occurs when an unsaturated air parcel cools when it is lifted or when it rises under its own buoyancy. In this case, its vapor pressure decreases as its pressure decreases, but ignoring any mixing with its environment, its mixing ratio will remain constant. Although the saturation mixing ratio increases with decreasing pressure, the exponential decrease in saturation vapor pressure with decreasing temperature wins and eventually saturation occurs when $r_v = r_v^*$. If the lifting continues, condensation will occur with the formation of cloud (see Chapter 9).

It is sometimes useful to have a measure of the *relative humidity* of the air, i.e., how close the air is to saturation. The relative humidity, RH, is commonly defined as the ratio $e/e^*(T)$ and it is usually expressed as a percentage: $100 \times e/e^*(T)$, but some authors define it as the ratio $r_v/r_v^*(p, T)$. Since p is typically much larger than e, $r_v \approx \epsilon e/p$ so that these definitions are almost equivalent.

2.5.1 Equation of state for moist air

The equation of state for air containing water vapor as well as hydrometeors (liquid water drops, ice crystals) may be written as

---

2. These findings point to additional limitations of the idealized Carnot model discussed in Chapter 13.

$$p = \rho R_d T_\rho. \tag{2.24}$$

where T_ρ is a fictitious temperature, the *density temperature*, defined by

$$T_\rho = T \frac{1 + r_v/\epsilon}{1 + r_T}, \tag{2.25}$$

where, again, $\epsilon = R_d/R_v = 0.622$ is the ratio of the specific gas constant for dry air to that for water vapor, R_v, and r_T is the *mixing ratio of water substance* (i.e. water vapor, liquid water, and ice). The quantity ϵ is equal also to the ratio of the molecular weight of water vapor to the mean molecular weight of dry air.

The use of T_ρ in Eq. (2.24) is an artifice that enables us to retain the gas constant for dry air in the equation. In essence, T_ρ is the temperature of an equivalent dry air parcel with the same density and pressure as the moist air parcel. The alternative would be to use a specific gas constant for the mixture of air and water substance that would depend on the mixing ratios of these quantities (i.e. it would not be a constant!). In the special case when there are no hydrometeors in the air, the density temperature reduces to the *virtual temperature*, T_v, which may be written approximately as $T(1 + 0.61r_v)$.

One can define also the *virtual potential temperature*, $\theta_v = T_v/\pi(p)$, where $\pi(p)$ is the Exner function defined in Section 2.3.3.

2.5.2 Saturation and latent heat release

The introduction of various states of water requires a modification of the equation to account for the latent heat released or consumed during phase changes of water substance. Additional equations governing the rates-of-change of the mixing ratios for water vapor, liquid water, and ice as well as conversions between these states are required also.

As noted above, one measure of the amount of water vapor in the air is the vapor pressure, e. At a given temperature T in the atmosphere, there is an effective limit to the vapor pressure that an air parcel can hold. This value is called the *saturation vapor pressure*, $e^*(T)$ and it decreases rapidly with decreasing temperature. When $e < e^*(T)$, the air parcel is said to be *unsaturated*. Again, the *mixing ratio of water vapor*, r_v is related to e by the formula $r_v = \epsilon e/(p - e)$, where p is the air pressure. The *saturation mixing ratio* r_v^* is therefore a function of both pressure and temperature.

As an initially unsaturated air parcel is cooled isobarically, for example by loss of heat by radiation to space, its vapor pressure remains the same, but the saturation vapor pressure decreases. Eventually, when $e = e^*(T)$ the parcel becomes saturated and if cooling continues, condensation occurs. The conservation of water substance requires that the mixing ratio of the air parcel remains constant as long as the condensed water stays suspended in the air so that the

fractional amount of condensate equals the initial mixing ratio minus the saturation mixing ratio of the cloudy air.

In a similar way, when an air parcel rises adiabatically to lower pressure, its saturation vapor pressure and therefore its saturation mixing ratio decrease, while the mixing ratio r_v remains the same. When the saturation mixing ratio $r_v^*(T, p)$ falls below the initial mixing ratio of the parcel, condensation occurs, the amount of condensate being $dr_v = r_v - r_v^*(T, p)$ per unit mass. The amount of latent heat released is therefore $L_v dr_v$, where $L_v(T)$ is the *latent heat of vaporization*, or *latent heat of condensation*. At $T = 0\,°C$, $L_v = 2.5 \times 10^6$ J kg^{-1} and decreases slowly and approximately linearly with temperature[3] (see e.g., Bohren and Albrecht (1998), p197).

2.5.3 Pseudo-adiabatic ascent

When an ascending air parcel becomes saturated ($r_v = r_v^*(T, p)$), as for example as it rises in a thunderstorm updraft, much of the condensed water falls out of it as rain although a small fraction of condensate remains suspended in the form of minute cloud drops or ice crystals. To a first approximation, we might reasonably assume that all condensate falls out immediately and that the latent heat thereby released goes into raising the temperature of the parcel. We refer to this scenario as *pseudo-adiabatic ascent*. In pseudo-adiabatic ascent, the rate of latent heat release per unit mass of dry air is given by

$$\dot{Q} = -L_v \frac{Dr_v}{Dt}, \tag{2.26}$$

noting that $Dr_v/Dt = 0$ if an air parcel is unsaturated, i.e., unless $r_v = r_v^*(T, p)$.

2.5.4 Equivalent potential temperature, moist entropy, moist static energy

Substitution of (2.26) into (2.20) gives

$$\frac{D}{Dt}\left(\ln\theta + \frac{L_v r_v}{c_{pd} T}\right) = 0, \tag{2.27}$$

with the usual approximation that the material derivative of one over T is small compared with that of r_v, where the latter is bounded above by the saturation value $r_v^*(T, p)$. It follows that the quantity in brackets is materially conserved

---

3. Kirchoff's equation for the rate of change of the latent heat of vaporization with temperature neglecting the pressure dependence is $dL_v/dT = c_{pv} - c_w$ where c_{pv} is the specific heat capacity of water vapor only at constant pressure and c_w is the specific heat capacity of liquid water (Bohren and Albrecht, 1998). Since $c_w > c_{pv}$, the latent heat of condensation will decrease with increasing temperature. More generally, the latent heat of vaporization between vapor and liquid phases must vanish as the temperature approaches the critical point since there is no distinction between liquid and vapor phases there.

in a pseudo-adiabatic process. We define this quantity to be $\ln \theta_e$, whereupon

$$\theta_e = \theta \exp\left(\frac{L_v r_v}{c_{pd} T}\right). \qquad (2.28)$$

The quantity θ_e is called the *pseudo-equivalent potential temperature*, or often just the *equivalent potential temperature*, but the latter may be misleading as there are other definitions of equivalent potential temperature in use. For example, instead of assuming that all the condensate falls out of an ascending air parcel, one might assume that all the condensate is carried with the parcel, an assumption that might be more appropriate for non-precipitating clouds (see Chapter 9). A related quantity to θ_e is the *specific moist entropy*, $s_m = c_{pd} \ln \theta_e$.

Either definition of equivalent potential temperature provides a measure of the moist heat content of the air, even if the air is unsaturated (using r_v instead of r_v^*). Physically, the pseudo-equivalent potential temperature may be thought of as the potential temperature of an air parcel that would be attained when the air parcel is lifted pseudo-adiabatically to a large altitude so that all the liquid water in the parcel is condensed out and the latent heat so released is used to raise the temperature of the parcel (see Chapter 9).

Note that for unsaturated ascent, both θ and r_v are approximately materially conserved until saturation occurs. Thus, for the foregoing physical interpretation to be valid, if the formula (2.28) is used to calculate θ_e at levels where the air parcel is unsaturated, the temperature in the exponential should be strictly the temperature at a level where the parcel first become saturated, the so-called *lifting condensation level* (see Chapter 9). However, for a moist tropical atmosphere, the error in using the absolute temperature is relatively small (on the order of 1 K) because the absolute temperature at which the lifted air parcel becomes saturated is only a few degrees at most below that of the unsaturated air parcel. In particular, the error is less than using a more accurate empirical formula for θ_e such as that developed by Bolton (1980) (e.g., his Eq. (15)).

Substitution of (2.26) into (2.23) gives

$$\frac{D}{Dt}\left(c_{pd}T + \Phi + L_v r_v\right) - \frac{1}{\rho}\frac{D_h p}{Dt} = 0. \qquad (2.29)$$

The quantity $k = c_{pd}T + L_v r_v$ is called the *specific moist enthalpy* and $h = c_{pd}T + \Phi + L_v r_v = k + \Phi$ is known as the *specific moist static energy* of the air. Again, if an air parcel remains unsaturated, r_v is materially conserved, neglecting of course the effects of turbulent mixing. As in the case of θ_e, if the air parcel is unsaturated, the temperature used to evaluate h should be strictly the temperature at a level where the parcel has become saturated.

In situations where $(1/\rho)D_h p/Dt$ can be neglected, h is approximately materially conserved in a pseudo-adiabatic process in which the flow remains in approximate hydrostatic balance.

2.6 Viscosity, diffusion, friction, and turbulence

In general, fluids possess a degree of "stickiness" referred to as viscosity. The viscosity arises from the random Brownian motion of the molecules comprising the fluid. In the presence of macro-scale velocity gradients within some flow, these random motions lead to a transport of momentum down the local velocity gradient which acts over time to smooth out the velocity gradient, a process referred to as momentum diffusion. For smooth, laminar flows of a Newtonian fluid[4] such as air, the effects of viscosity can be represented by taking $\mathbf{F}$ in Eq. (2.1) to be the vector $(1/\rho)\nabla \cdot (\mu \nabla \mathbf{u})$, where μ is the *molecular viscosity*. When variations of μ can be neglected, this expression reduces to $\nu \nabla^2 \mathbf{u}$, where $\nu = \mu/\rho$ is the *kinematic viscosity* of the fluid. Since the effect of viscosity is to diffuse momentum, the coefficient ν is effectively the molecular diffusivity of momentum. For air at room temperature and a pressure of 1000 mb, $\nu = 1.4 \times 10^{-5}$ m^2 s^{-1}. The relative importance of the nonlinear acceleration term to the viscous term is characterized by a non-dimensional *Reynolds number* $Re = UL/\nu$, where U and L are typical velocity and length scales of the flow. Typically, because ν is so small for air, in most situations in the atmosphere, $Re \gg 1$ and the effects of molecular viscosity may be neglected.

Most atmospheric flows are turbulent, especially in regions of large wind shear, or in regions of vigorous buoyant convection. The turbulent "eddies" have a much larger effect on the lateral (i.e. cross-shear) diffusion of momentum than molecular viscosity and it is often necessary to take account of this turbulent diffusion. A crude method for doing this is by likening turbulent diffusion to molecular diffusion, but with a much larger coefficient of viscosity.[5] In this case we take $\mathbf{F}$ in Eq. (2.1) to be the vector $\nabla \cdot (K \nabla \mathbf{u})$, where K is a so-called "eddy diffusivity" or "eddy viscosity" associated with the turbulent fluctuations.[6] If K is taken to be constant, the expression reduces to $K \nabla^2 \mathbf{u}$.

---

4. A Newtonian fluid is one in which the viscous shear stress is linearly related to the local strain. For example, in a unidirectional shear flow $u(z)$, the local shear stress $\tau(z) = \mu du/dz$, where, in this case, du/dz is the rate of strain.

5. The analogy between turbulent eddies and molecular motions in formulations for the turbulent transport of momentum and heat is broadly attributed to L. Prandtl and T. Von Kármán (Schlichting, 1968), though Frisch (1995), Section 9.6, traces the concepts of eddy viscosity and mixing length to nineteenth century researchers.

6. Unlike the formulation for laboratory and engineering flows in which the density is kept explicit in the net force, in large-scale geophysical flows the net force associated with sub-grid-scale (turbulent) motions is usually represented at the outset using momentum diffusivities, wherein explicit density variations are ignored for flow speeds that are much smaller than the speed of sound.

The analogy between eddy viscosity and molecular viscosity is crude because unlike molecular diffusion, where the viscosity is a property of the fluid, K depends on the flow itself and treating it as a constant may be a poor assumption for quantitative calculations. In fact, in some situations, the turbulent diffusion may not even be down the mean velocity gradient, as for example, in the dry convective boundary layer over land. Nevertheless, the constant-K assumption can be useful for elucidating the qualitative effects of turbulent momentum transfer, at least when this transfer is downgradient.

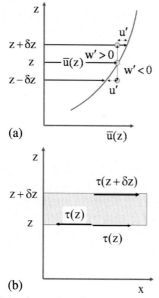

(a)

(b)

FIGURE 2.2 (a) An air parcel moving with speed $\bar{u}(z)$ is displaced to a neighboring level δz, carrying its linear momentum with it. At level $z + \delta z$ it appears as a negative fluctuation, or at $z - \delta z$ it appears as a positive fluctuation. In each case $\rho \overline{u'w'} < 0$. (b) The stress acting on an air layer of thickness δz is $\tau(z + \delta z) - \tau(z)$ per unit area. The stress acting per unit volume is $(\tau(z + \delta z) - \tau(z))/\delta z$.

We illustrate the basic ideas for the case of a mean flow $\bar{u}(z)$ that is one dimensional in the x-direction and constant density (Fig. 2.2a). We assume that the total velocity at any height z is $(\bar{u}(z) + u', v', w')$, where the prime quantities represent turbulent fluctuations that are functions of space and time, but have zero mean in the x-direction. We assume also that $d\bar{u}/dz > 0$. In analogy to molecular diffusion we consider an air parcel at height z moving with the mean flow speed (i.e. $u' = 0$). Suppose that on account of the component of turbulent fluctuation w', it moves to the neighboring level $z + \delta z$, conserving its linear momentum $\rho \bar{u}$. When $w' > 0$, $\delta z > 0$ and the parcel at $z + \delta z$ will have a momentum deficit $\rho u' = \rho(\bar{u}(z) - \bar{u}(z + \delta z))$, whereas, when $w' < 0$, $\delta z < 0$ and it will have a momentum surplus. In both cases the quantity $\rho \overline{u'w'}$ is negative and corresponds to a flux of x-momentum in the negative z-direction, i.e. in the opposite direction to the vertical gradient of $\bar{u}$. Thus, on average in the x-direction, the mean lateral momentum flux,

$\rho \overline{u'w'}$, is negative and hence downgradient, meaning down the gradient of $\bar{u}$. The quantity $\tau = -\rho \overline{u'w'}$ represents a mean turbulent flux of x-momentum in the z-direction and is equivalent to a turbulent shear stress of the fluid above the level z acting on the fluid below the level z. A stress has units of force per unit area.

On the assumption that the mean momentum flux is linearly proportional to the mean shear and that it is downgradient, we define K such that

$$\tau = -\rho \overline{u'w'} = \rho K \frac{d\bar{u}}{dz}. \tag{2.30}$$

Then, as indicated in Fig. 2.2b, the net force per unit area in the x-direction acting on a layer of fluid of thickness δz is $\tau(z + dz) - \tau(z)$. Dividing this net force by δz and letting $\delta z \to 0$ gives the force per unit volume, which is just vertical gradient of the stress, $d\tau/dz$. The net force per unit mass, F_x, is

$$F_x = \frac{1}{\rho}\frac{d\tau}{dz} = \frac{d}{dz}\left(K\frac{d\bar{u}}{dz}\right). \tag{2.31}$$

This expression is a special case of the more general expression in Cartesian coordinates of $\nabla \cdot (K\nabla\mathbf{u})$, or in the case where K is treated as a constant, $K\nabla^2\mathbf{u} = K(\nabla^2 u, \nabla^2 v, \nabla^2 w)$.

The foregoing derivation would apply also to the case of a laminar flow, in which case the velocity fluctuations would be interpreted as those associated with the random motion of the molecules.[7]

Similar expressions to Eq. (2.30) and (2.31) may be obtained for other scalar quantities, c, such as the water vapor mixing ratio. Suppose that $\bar{c}$ has a mean gradient $d\bar{c}/dz$, then turbulent fluctuations give rise to a mean flux of c in the z-direction, $\tau_c = -\rho \overline{c'w'}$. If $d\bar{c}/dz > 0$, then the flux is *normally*[8] in the negative z-direction and we can write

$$\tau_c = -\rho \overline{c'w'} = \rho K_c \frac{d\bar{c}}{dz}, \tag{2.32}$$

where K_c is an eddy diffusivity for the quantity c. The rate of accumulation of $\bar{c}$ in a unit volume, F_c, is simply the divergence of this flux in the volume

$$F_c = \frac{1}{\rho}\frac{\partial \tau_c}{\partial z}. \tag{2.33}$$

Note that waves are associated also with fluctuations about a mean flow and the effects of waves on mean flows

---

7. The analogy between turbulent eddies and molecular motions in the simple "mixing-length" formulation for the turbulent transport of heat (or a scalar) is again broadly attributed to L. Prandtl and T. Von Kármán.

8. Unlike molecular diffusion, it is possible for a turbulent flux to be countergradient, in which case the eddy viscosity would be negative. We will encounter striking examples of this situation in Chapters 4 and 11 when considering the eddy momentum flux characteristics of non-axisymmetric "eddies" during the intensification of a tropical cyclone vortex.

can be analyzed in a similar way to those of turbulent fluctuations. For example, there may be a stress associated with wave motions. Elementary examples of this type will be discussed in Chapter 4 and in Chapter 11 for eddy motions involving both vortex waves and deep moist convection.

2.7 Methods of solution

The momentum equation (2.1), the thermodynamic equation (2.18) or one of its equivalent forms (Eqs. (2.20) or (2.21)), the equation of state (2.11), and the continuity equation (2.6) or one of the approximations to it (Eqs. (2.7) or (2.8)) form a complete set for solving problems of dry air flow. For moist air flows, the equation of state in the form (2.24) must be used and additional equations are necessary to characterize the various species of water substance. The set of equations is referred to as the *primitive equations*.

The primitive equations are rather general and admit a wide range of solutions, including laminar flow, turbulent flow, rotational waves and coherent vortex structures. Because of the nonlinearity of the governing equations, these solution types generally interact with one another, forming a complex tapestry of waves, vortices, and turbulence. If the density is allowed to vary with space and time, the general solutions include motions associated with sound waves (and shock waves) as well as a range of wave types.

The solutions are greatly constrained by the boundaries of the flow domain. For example, at an impervious boundary, there can be no component of flow normal to the boundary.

An *inviscid* fluid is an idealization in which the internal *viscosity* is zero. Such a fluid cannot support shearing stresses and at an impervious boundary the flow adjacent to the boundary must be tangential to it.

If the fluid is *viscous*, there can be no flow at the boundary, but velocity gradients normal to the boundary may be large if the molecular viscosity is accordingly small (more specifically, if the Reynolds number is large compared to unity). In flows that are turbulent, the boundary stress is often related to a mean velocity just above the surface (e.g., at a typical anemometer level of 10 m). An example is discussed in Chapter 6.

At a free surface, such as a water surface, the pressure is often taken to be constant at the surface if surface tension effects can be neglected.

As noted in Section 2.1, the assumption that the fluid is incompressible or anelastic and enclosed by impervious boundaries means that the pressure distribution cannot be imposed arbitrarily, but must be determined as part of the solution. For this reason, arguments for explaining flow behavior on the basis of prescribed forces, as in rigid body dynamics, are usually not possible. However in many flow situations, other methods of solution are possible and some can provide powerful insights into the flow behavior. One such method, used widely in two-dimensional flow problems, is the vorticity-streamfunction approach discussed in Section 2.10.

2.8 Kinetic energy and total energy

In Section 2.1 we derived an expression for the potential energy of an air parcel and in Section 2.3 we derived an equation for the internal energy of an air parcel. An expression for the kinetic energy of an air parcel may be obtained by taking the scalar product of Eq. (2.1) with the velocity vector, $\mathbf{u}$ and using the vector identity $\mathbf{u} \cdot \nabla \mathbf{u} = \nabla(\frac{1}{2}\mathbf{u}^2) + (\nabla \wedge \mathbf{u}) \wedge \mathbf{u}$. Then

$$\rho \frac{D}{Dt}\left(\frac{1}{2}\mathbf{u}^2\right) = -\rho g w - \nabla \cdot (p\mathbf{u}) + p\nabla \cdot \mathbf{u} + \mathbf{u} \cdot \mathbf{F}. \quad (2.34)$$

In the case where the frictional force per unit volume, $\mathbf{F}$, is expressed in terms of molecular viscosity, i.e. $\mathbf{F} = \mu \nabla^2 \mathbf{u}$, where, again, μ is molecular viscosity of the fluid (Gill, 1982, Eq. (4.6.1)) with corresponding momentum diffusivity $\nu = \mu/\rho$,

$$\rho \frac{D}{Dt}\left(\frac{1}{2}\mathbf{u}^2\right) = -\rho g w +$$
$$\nabla \cdot \left[-p\mathbf{u} + \mu \nabla \left(\frac{1}{2}\mathbf{u}^2\right)\right] - \rho \epsilon_D + p\nabla \cdot \mathbf{u}, \quad (2.35)$$

where

$$\epsilon_D = \nu \left[\left(\frac{\partial \mathbf{u}}{\partial x}\right)^2 + \left(\frac{\partial \mathbf{u}}{\partial y}\right)^2 + \left(\frac{\partial \mathbf{u}}{\partial z}\right)^2\right]. \quad (2.36)$$

In this form, the quantity ϵ_D, called the dissipation rate, is always positive and represents the rate of viscous dissipation of kinetic energy, which is manifest as an internal heating of the fluid. Note that, because the Coriolis acceleration is normal to the velocity vector, it does not appear in the kinetic energy equation since it does no work. Following Gill (1982), we note that, using the result of Exercise 2.1, i.e. Eq. (2.71), the local kinetic energy tendency can be written in succinct *flux form*

$$\frac{\partial}{\partial t}\left(\frac{1}{2}\rho\mathbf{u}^2\right) + \nabla \cdot \mathbf{F_{KE}} = -\rho g w - \rho \epsilon_D + p\nabla \cdot \mathbf{u}, \quad (2.37)$$

where

$$\mathbf{F_{KE}} = \left(p + \frac{1}{2}\rho\mathbf{u}^2\right)\mathbf{u} - \mu \nabla\left(\frac{1}{2}\mathbf{u}^2\right). \quad (2.38)$$

The vector field $\mathbf{F_{KE}}$ represents the local rate of energy flow per unit area and is called the *kinetic energy flux density vector*. The definition is not unique because the addition of any non-divergent vector field to it would not alter the validity of Eq. (2.37).

Now combining the kinetic energy equation with the internal energy equation and noting that $\partial \rho \Phi / \partial t + \nabla \cdot (\rho \Phi \mathbf{u}) = \rho w g$ (Exercise 2.2) we obtain the total energy equation

$$\frac{\partial}{\partial t}\left[\rho\left(E + \Phi + \frac{1}{2}\mathbf{u}^2\right)\right] + \nabla \cdot \mathbf{F_{tot}} = \rho \dot{Q}, \qquad (2.39)$$

where

$$\mathbf{F_{tot}} = \rho \mathbf{u}\left(E + \frac{p}{\rho} + \Phi + \frac{1}{2}\mathbf{u}^2\right) - \mu \nabla \left(\frac{1}{2}\mathbf{u}^2\right). \quad (2.40)$$

Note that the kinetic energy consumed by the small-scale turbulence is manifest as thermal energy. Thus, the heating associated with the viscous dissipation ($\rho \epsilon_D$) must be added to the heating rate $\dot{Q}$ in the thermal energy equation and the dissipation term cancels when the kinetic energy and thermal energy equations are combined to give Eq. (2.39). Consequently, the material heating rate, $\dot{Q}$, appearing in Eq. (2.39) represents the heating rate associated with the sum of the divergence of the conductive heat flux, the divergence of the electromagnetic radiation heat flux, the release of latent heat of condensation or fusion or sublimation, the cooling associated with the evaporation of liquid water or ice particles, or the melting of ice, but not that due to viscous dissipation.

The quantity $E + p/\rho$ that occurs in Eq. (2.40) is just the *specific enthalpy of dry air*, $c_{pd}T$ (see Section 2.3.2). The expression $E + p/\rho + \Phi$ is the dry static energy introduced in Section 2.3.4.

Using the result that $\nabla \cdot (p\mathbf{u}) = \rho(D/Dt)(p/\rho) - \partial p/\partial t$ (see Exercise 2.3), Eq. (2.39) may be written in the form

$$\rho \frac{D}{Dt}\left(c_{pd}T + \Phi + \frac{1}{2}\mathbf{u}^2\right)$$
$$- \nabla \cdot \left(\mu \nabla \left(\frac{1}{2}\mathbf{u}^2\right)\right) = \rho \dot{Q} + \frac{\partial p}{\partial t}, \qquad (2.41)$$

In most situations, $\partial p/\partial t$ can be neglected, except when considering acoustic waves. Then, if viscous effects are small and there is no heating, Eq. (2.41) may be written

$$\frac{D}{Dt}\left(c_{pd}T + \Phi + \frac{1}{2}\mathbf{u}^2\right) = 0. \qquad (2.42)$$

This is called *Bernoulli's equation*, which states that for an inviscid, adiabatic flow, the expression in brackets is materially conserved.

For a pseudo-adiabatic, inviscid, process with $\dot{Q}$ given solely by Eq. (2.26), an equation analogous to Eq. (2.42) can be derived which involves the *specific moist static energy*.

$$\frac{D}{Dt}\left(c_{pd}T + \Phi + L_v r_v^* + \frac{1}{2}\mathbf{u}^2\right) = 0, \qquad (2.43)$$

(see Exercise 2.4).

2.9 Vorticity and the vorticity equation

Vorticity is defined as the curl of the velocity vector field and we denote it by $\boldsymbol{\omega}$: i.e.

$$\boldsymbol{\omega} = \nabla \wedge \mathbf{u}. \qquad (2.44)$$

For a volume of fluid rotating as a solid body, the vorticity vector is oriented along the axis of rotation and in the same direction of the rotation vector of the volume; it has a magnitude equal to twice the angular rotation rate of the volume. In general, vorticity is a measure of the local rotation rate of a parcel of fluid. Not surprisingly, vorticity is an important concept for characterizing rapidly rotating flows such as tropical cyclones and other geophysical vortices including tornadoes, waterspouts and dust devils. As will become evident throughout this book, the dynamics of vorticity has far-reaching consequences on the behavior of tropical cyclones.

To help visualize the distribution of the vorticity field it is useful to define a geometric quantity called a *vortex line*. A vortex line is a line (or curve) in the flow that is tangent to the vorticity field at each point along $\boldsymbol{\omega}$. Vortex lines are geometrically analogous to magnetic field lines in physics and, like the magnetic field, are widely thought to obey the "well-known" law that these lines cannot "end" in the fluid and must either close on themselves (like a smoke ring), extend to infinity, or end at a solid or free boundary. However, according to Chorin and Marsden (2000), pages 27-28, the proof of these properties is "hopelessly incomplete" in the fluid dynamics context, due in part to the generally turbulent (and hence chaotic) nature of particle trajectories and the vorticity dynamics in real geophysical flows. Proof of these properties is incomplete, again because of the vague notion surrounding the "ending of vortex lines".

For the most part in this book, we will assume the veracity of these "oft-proved" properties, but remain alert of their potential violation in turbulent flow situations. In distinction with "vortex tubes" (which are comprised of a bundle of vortex lines or filaments - and discussed further below), there is no law saying that the vorticity along a vortex line must remain of constant magnitude. The absence of a constraint on the magnitude of vorticity along a vortex line is analogous to the fact that the magnitude of velocity need not remain constant along a streamline.

We may derive an equation for $\boldsymbol{\omega}$ as follows. First, using again the relationship $\mathbf{u} \cdot \nabla \mathbf{u} = \nabla(\frac{1}{2}\mathbf{u}^2) + \boldsymbol{\omega} \wedge \mathbf{u}$, we write the momentum equation (2.1) as

$$\frac{\partial \mathbf{u}}{\partial t} + (\boldsymbol{\omega} + 2\boldsymbol{\Omega}) \wedge \mathbf{u} + \nabla(\frac{1}{2}\mathbf{u}^2) = -\frac{1}{\rho}\nabla p - \nabla \Phi + \mathbf{F},$$

where $\Phi = g_e z$ is the geopotential. Setting $\boldsymbol{\omega}_a = \boldsymbol{\omega} + 2\boldsymbol{\Omega}$, using the vector formula $\nabla \wedge (\boldsymbol{\omega}_a \wedge \mathbf{u}) = \boldsymbol{\omega}_a(\nabla \cdot \mathbf{u}) - \mathbf{u}(\nabla \cdot \boldsymbol{\omega}_a) + \mathbf{u} \cdot \nabla \boldsymbol{\omega}_a - \boldsymbol{\omega}_a \cdot \nabla \mathbf{u}$ and noting that $\nabla \cdot \boldsymbol{\omega}_a \equiv 0$, the curl of this form of the momentum equation gives

$$\frac{\partial \omega_a}{\partial t} + \mathbf{u} \cdot \nabla \omega_a = (\omega_a \cdot \nabla)\mathbf{u} - \omega_a(\nabla \cdot \mathbf{u})$$
$$+ \frac{1}{\rho^2} \nabla \rho \wedge \nabla p + \nabla \wedge \mathbf{F}. \qquad (2.45)$$

This is called the *Helmholtz vorticity equation*. The quantity $\omega_a = \omega + 2\Omega$ is called the absolute vorticity and is the sum of the relative vorticity, ω, i.e. the curl of the velocity vector relative to a coordinate system rotating with angular velocity Ω, and the *background vorticity*, 2Ω, the vorticity associated with the rotating coordinate system.

The term on the left-hand-side of Eq. (2.45) is simply the material rate-of-change of vorticity, i.e. the rate-of-change following a fluid parcel. The term $(\omega_a \cdot \nabla)\mathbf{u}$ can be interpreted as the rate-of-production of vorticity as a result of the stretching and tilting of absolute vorticity by velocity gradients. The term $\omega_a(\nabla \cdot \mathbf{u})$ represents the rate-of-change of vorticity associated with a material rate-of-change of fluid density and would be zero for a homogeneous fluid. The term $\frac{1}{\rho^2}\nabla\rho \wedge \nabla p$ represents the baroclinic generation of vorticity and the term $\nabla \wedge \mathbf{F}$ is the force generation term.

The continuity equation (2.6) can be used to condense the vorticity equation somewhat as follows

$$\frac{\partial}{\partial t}\left(\frac{\omega_a}{\rho}\right) + \mathbf{u} \cdot \nabla \left(\frac{\omega_a}{\rho}\right) = \left(\frac{\omega_a}{\rho} \cdot \nabla\right)\mathbf{u}$$
$$+ \frac{1}{\rho^3}\nabla\rho \wedge \nabla p + \frac{1}{\rho}\nabla \wedge \mathbf{F}. \qquad (2.46)$$

This is an equation for the vorticity per unit mass ω_a/ρ.

For a homogeneous fluid, or for so-called *barotropic flows* in which the pressure is a function of density alone, the second term on the right-hand-side of (2.46) is identically zero. In such cases, it can be shown, assuming $\mathbf{F} = 0$ also, that specific absolute vorticity lines move with fluid lines, i.e., $\omega_a/\rho \propto \delta \mathbf{x}$, where $\delta \mathbf{x}$ is a differential line element of fluid particles near $\mathbf{x}$ that is initially congruent to the absolute vorticity there. The equation governing the incremental material line of particles $\delta \mathbf{x}$ is given by

$$\frac{\partial \delta \mathbf{x}}{\partial t} + \mathbf{u} \cdot \nabla \delta \mathbf{x} = (\delta \mathbf{x} \cdot \nabla)\mathbf{u}, \qquad (2.47)$$

which is an equation very similar to (2.46), particularly in the event that the solenoidal and diffusion terms vanish.[9]

The significant result that vortex lines move with fluid lines was first proven for a homogeneous inviscid fluid by Helmholtz (1958). In this case, the evolution of vorticity

"is most easily understood through the statement that vortex lines are frozen in the fluid", i.e., vortex lines

are transported with the flow like material curves of fluid particles. ..."

Moffatt (2011). As noted above, this statement holds true for specific absolute vorticity lines in the more general case of inviscid barotropic flow, suggesting again the deep link between material lines and vortex lines.[10]

For the cases of barotropic flow, the vorticity equation does not involve the pressure. This feature has advantages for interpretation because, as noted in Section 2.1, pressure cannot be prescribed arbitrarily, but must be determined as part of the solution. In these cases, provided that a free surface is not involved, such as in theories involving waves at a free surface, interpretations in terms of vorticity circumvent the need to know about the pressure field and are often easier to conceptualize than those based directly on the momentum equation.

An example of the foregoing class of flows is that in which the flow is frictionless and two-dimensional in planes perpendicular to $\hat{\mathbf{k}}$. In such flows, the pertinent vorticity component is the local vertical vorticity and the vorticity equation (2.45) reduces to

$$\frac{D\zeta_a}{Dt} = 0, \qquad (2.48)$$

where the absolute vorticity of air parcels (or in this case air columns), $\zeta_a = \omega_a \cdot \hat{\mathbf{k}}$ is a scalar field and is materially conserved as air parcels move around horizontally. Here, the time derivative D/Dt represents the material derivative operator in the plane perpendicular to $\hat{\mathbf{k}}$ (see Exercise 2.5).

For this class of flows, a knowledge of the vorticity distribution at a particular time enables the flow structure to be determined given suitable boundary conditions on the flow boundaries. This is because the vorticity of the flow is simply the Laplacian of the streamfunction. These results provide a powerful method of predicting the evolution of two-dimensional flows, a method explained in more detail in the next section and exploited in Chapters 3 and 4.

2.10 Vorticity-streamfunction method

The vorticity-streamfunction method is a powerful way of solving two-dimensional flow problems for a homogeneous, incompressible fluid. The method exploits the fact that one can define a streamfunction that satisfies the two-dimensional continuity equation exactly and that, as shown in Section 2.9, for two-dimensional flows there is only one component of absolute vorticity, ζ_a. This component is normal to the plane of flow and may be expressed in terms of a streamfunction. One ends up with just one pair of equations to solve: a prediction equation for the absolute vorticity and a diagnostic equation for the streamfunction, neither of which involves the pressure explicitly.

---

9. The proof that specific absolute vorticity lines move with the flow is essentially as follows. In either the inviscid homogeneous or barotropic case, the solenoidal and diffusion terms in (2.46) vanish. One may then use (2.46) and (2.47) to show that if $\delta \mathbf{x}$ is initially coincident with a portion of ω_a/ρ, it will remain so for all time that a solution exists.

10. See Chorin and Marsden (2000), pages 22-24 for mathematical details.

To illustrate the vorticity-streamfunction method, it is convenient to use a rectangular coordinate system (x, y, z) with the x-axis pointing eastwards, the y-axis pointing northwards and the z-axis pointing in the local vertical direction, $\hat{\mathbf{k}}$, at a particular latitude.

Since tropical cyclones are relatively shallow circulations compared to their horizontal extent and because their horizontal extent is small compared with the radius of the Earth, we can safely approximate the Earth's surface as locally flat. Moreover, it is usually only the local vertical component of the Earth's rotation that is of consequence for the dynamics and it is customary to replace $2\mathbf{\Omega}$ by $\mathbf{f}$, equal to $2|\mathbf{\Omega}| \sin \phi \hat{\mathbf{k}}$, where ϕ is the latitude. The quantity $f = 2|\mathbf{\Omega}| \sin \phi$ is called the *Coriolis parameter*.

When f is treated as a constant, we refer to such a coordinate system as an f-plane. There is a simple and useful generalization of this approximation for flows possessing somewhat larger horizontal length scales in which the variation of f with latitude cannot be neglected, but the explicit curvature of the earth's surface can still be safely ignored. In this circumstance the fluid equations using either Cartesian or cylindrical coordinates may be adopted with the simple modification of retaining a linear dependence of f with latitude. We refer to this coordinate system as a β-plane and write in Cartesian coordinates $f = f_o + \beta y$, where f_o is the value of f at the origin of the coordinates at which $\phi = \phi_0$, and $\beta = df/dy = 2\Omega \cos \phi_0 / a$, where a is the earth's mean radius.

For two-dimensional motion in the (x, y)-plane with velocity components (u, v), the relative vorticity is $\zeta = \partial v / \partial x - \partial u / \partial y$, which satisfies the equation

$$\frac{\partial}{\partial t}(\zeta + f) + u\frac{\partial}{\partial x}(\zeta + f) + v\frac{\partial}{\partial y}(\zeta + f) = 0, \quad (2.49)$$

which is the same as Eq. (2.48), but in component form. For an incompressible fluid, the continuity equation is

$$\frac{\partial u}{\partial x} + \frac{\partial v}{\partial x} = 0, \quad (2.50)$$

and accordingly there exists a streamfunction ψ such that

$$u = -\frac{\partial \psi}{\partial y}, \quad v = \frac{\partial \psi}{\partial x}. \quad (2.51)$$

At any instant of time, the flow is parallel to the lines (or curves) along which ψ is a constant, the so-called streamlines. Substituting the expressions in (2.51) into $\zeta = \partial v / \partial x - \partial u / \partial y$ gives

$$\zeta = \frac{\partial^2 \psi}{\partial x^2} + \frac{\partial^2 \psi}{\partial y^2}. \quad (2.52)$$

Eq. (2.49) is a prediction equation for the *absolute vorticity*, $\zeta + f$, and states that this quantity is conserved fol-lowing columns[11] of air. When ψ is known, u and v can be calculated from the expressions (2.51) and ζ from the formula (2.52). When ζ is known, (2.52) is a two-dimensional, second-order, elliptic partial differential equation (the so-called Poisson equation) for ψ. This can be solved (or "inverted") for ψ subject to suitable boundary conditions on the velocity field on the periphery of the vortex.

A powerful way of understanding what the solutions for ψ are like given the spatial distribution of ζ is to use the *membrane analogy*, which is discussed in Appendix 2.19. We will have occasion to invoke this analogy in several chapters of the book.

In a few simple cases it is possible to obtain an analytical solution of Eqs. (2.49), (2.51), and (2.52), but in general we must resort to numerical methods. The system of equations can be solved numerically using the following steps:

- From a given initial distribution of $\psi(x, y)$ at, say $t = 0$, we can calculate the initial velocity distribution from Eq. (2.51) and the initial vorticity distribution from Eq. (2.52). Alternatively, given the initial vorticity distribution, we can solve Eq. (2.52) for the initial streamfunction distribution ψ and then calculate the initial velocity distribution from Eq. (2.51). From the initial velocity, we can then use Eq. (2.49) to calculate the vorticity tendency $\partial \zeta / \partial t$ at time t = 0.
- We are now in a position to predict the vorticity distribution at a later time, say $t = \Delta t$, using Eq. (2.49), i.e. $\zeta(x, y, \Delta t) \approx \zeta(x, y, 0) + (\partial \zeta / \partial t)_{t=0} \Delta t$.
- Given the vorticity distribution ζ at time Δt we can solve Eq. (2.52) for the streamfunction distribution ψ at time Δt and obtain the new velocity distribution from Eq. (2.51).
- This procedure can be repeated to extend the solution forward to the time $t = 2\Delta t$, and so on.

For flows in which the density fluctuations are important, one needs a prediction equation for the thermodynamic variable as well. Moreover, in this case the vorticity equation involves also the pressure. Nevertheless, methods similar to the vorticity-streamfunction approach can be devised for what are called *balanced flows*. These methods are based on a prediction equation for the potential vorticity, or some asymptotic approximation to it, and diagnostic relationships between streamfunction and/or dynamic pressure (or geopotential) and the potential vorticity as discussed in Section 2.12.

2.11 Circulation

Having summarized the equation governing the evolution of the vorticity vector field, it proves useful to consider also

---

11. As noted earlier, in a two-dimensional flow, there is no dependence of u and v on the z-coordinate and instead of the motion of air parcels, we can think in terms of the motion of thin columns of air in a layer with uniform finite depth.

macroscopic (i.e., integrated) scalar measures of the vorticity field. The tendency of a fluid to circulate around a closed loop C in the fluid is one such measure. The circulation Γ is defined by the line integral of the velocity component tangent to the loop

$$\Gamma = \oint_C \mathbf{u} \cdot \mathbf{dl}, \qquad (2.53)$$

where $\mathbf{u}$ is the earth-relative velocity and $\mathbf{dl}$ is a differential vector line segment in the direction of the loop. This quantity is called the *relative circulation*. Stokes' theorem, relating the line integral of a differentiable vector field round a closed curve to the curl of this vector field over an area enclosed by this loop, then gives

$$\Gamma = \oint_C \mathbf{u} \cdot \mathbf{dl} = \int_A \boldsymbol{\omega} \cdot \hat{\mathbf{n}} dA, \qquad (2.54)$$

where $\hat{\mathbf{n}}$ denotes the unit normal to the surface area element dA.

The relative circulation Γ accordingly equals the so-called *flux of relative vorticity* across the area A. If one defines a vortex tube as the volume of fluid bounded by the vortex lines that thread the fluid and pass through the loop C, then it can be shown that the relative vorticity flux (also called the *strength* of the vortex tube) is constant along the tube. This result hods true at any time. (Exercise 2.10a demonstrates the invariance of vorticity flux along a vortex tube with nonzero strength for a simple (non-turbulent) flow.) How the vorticity flux varies with time requires additional knowledge about the time evolving flow field.

An analogous quantity, the *absolute circulation*, may be defined to account for the motion of the reference frame

$$\Gamma_a = \oint_C \mathbf{u_{abs}} \cdot \mathbf{dl}, \qquad (2.55)$$

where $\mathbf{u_{abs}}$ denotes the absolute velocity of a fluid parcel $(\mathbf{u} + \boldsymbol{\Omega} \wedge \mathbf{r})$ and $\mathbf{r}$ denotes the position vector of the parcel. Stokes' theorem, together with the identity $\nabla \wedge (\boldsymbol{\Omega} \times \mathbf{r}) = 2\boldsymbol{\Omega}$ (Exercise 2.10b) gives

$$\Gamma_a = \int_A \boldsymbol{\omega} \cdot \hat{\mathbf{n}} dA + \int_A 2\boldsymbol{\Omega} \cdot \hat{\mathbf{n}} dA = \Gamma + 2\Omega A_n, \qquad (2.56)$$

where A_n denotes the projection of the area A onto the equatorial plane perpendicular to $\boldsymbol{\Omega}$.

The material rate of change of Γ for a material loop of fluid may be obtained upon materially differentiating the expression for Γ (Eq. (2.53)), and using the vector momentum equation to substitute for $D\mathbf{u}/Dt$. The result (Exercise 2.11) is:

$$\frac{D\Gamma}{Dt} = \oint_C \mathbf{u} \cdot \frac{D\mathbf{dl}}{Dt} + \oint_C (-\frac{1}{\rho}\nabla p - \nabla\Phi - 2\boldsymbol{\Omega} \times \mathbf{u} + \mathbf{F}) \cdot \mathbf{dl}. \qquad (2.57)$$

The first and third integrals on the right hand side of Eq. (2.57) can be shown to vanish (Exercise 2.12). The line integral of the Coriolis force may be expressed as follows

$$-\oint_C 2\boldsymbol{\Omega} \times \mathbf{u} \cdot \mathbf{dl} = -2\Omega\frac{DA_n}{Dt}, \qquad (2.58)$$

where $\Omega = |\boldsymbol{\Omega}|$ (see Exercise 2.13 for details). Stokes' theorem transforms the remaining two line integrals into equivalent area integrals over the area A enclosed by the loop C (Exercise 2.14). Using Eqs. (2.56), (2.57), and (2.58), one may eliminate the material rate of change of A_n in $D\Gamma_a/Dt$ to obtain

$$\frac{D\Gamma_a}{Dt} = \int_A \frac{1}{\rho^2}\nabla\rho \times \nabla p \cdot \hat{\mathbf{n}} dA + \int_A \nabla \times \mathbf{F} \cdot \hat{\mathbf{n}} dA. \qquad (2.59)$$

2.11.1 Kelvin's theorem

If the flow is inviscid and barotropic within A, the two area integrals on the right hand side of (2.59) vanish, implying

$$\frac{D\Gamma_a}{Dt} = 0. \qquad (2.60)$$

That is, if the flow is inviscid and barotropic, then Γ_a is materially conserved following any closed material loop of fluid. This result is known as *Kelvin's circulation theorem* after Lord Kelvin. The theorem represents a generalization of the notion of angular momentum conservation for the small disk of fluid enclosed by the loop C. However, Γ_a is generally not the same as angular momentum and the reader is cautioned to avoid equating these two quantities in three-dimensional vortical flows.[12]

2.11.2 Beyond barotropy

Even if the flow is not barotropic, the first integral on the right hand side of Eq. (2.59) will still vanish for a family of closed material loops if $\nabla\rho \times \nabla p$ is everywhere perpendicular to $\hat{\mathbf{n}}$. Such a situation prevails (i) when the surface $\mathbf{A}$ is chosen to lie within an isentropic surface $\theta = $ const.; (ii) when the flow is adiabatic $(D\theta/Dt = 0)$ (so the material particles comprising the initial surface $\mathbf{A}$ remain on the isentropic surface); and (iii) when the pressure is a function of just density on this family of θ surfaces. These conditions imply the annihilation of the baroclinic generation term on

---

12. A non-circular, disk-like, element will be generally subject to a pressure torque along the periphery of the disk. In such cases, the angular momentum of the disk will not be materially conserved even though Γ_a will be! If, on the other hand, the disk *were to remain circular for all time*, then there would no longer be pressure torques on the periphery of the disk (Why?) and Γ_a would be simply proportional to the component of absolute angular momentum $\mathbf{M}$ that is perpendicular to the material surface $\mathbf{A}$. Since Γ_a is materially conserved, so too will be this angular momentum component. For these reasons vorticity and circulation are sometimes referred to as the spin and spin-momentum variables of fluid dynamics.

the right side of Eq. (2.59) for all time. Therefore, if the flow is inviscid and adiabatic, and pressure is a sole function of *density* on an isentropic surface, then absolute circulation will still be materially conserved for this family of material loops (see Exercise 2.15 for details).

2.12   Potential vorticity

A generalization of the absolute vorticity that proves useful in understanding certain types of three-dimensional flow is the Ertel potential vorticity (PV), P, defined as

$$P = \frac{\boldsymbol{\omega_a} \cdot \nabla\theta}{\rho}. \qquad (2.61)$$

For the general adiabatic and inviscid motion of a rotating stratified fluid, it can be shown that P is materially conserved on fluid parcels, i.e.,

$$\frac{DP}{Dt} = 0. \qquad (2.62)$$

From its definition, we see that P is a projection of the absolute vorticity vector $\boldsymbol{\omega_a}$ onto the gradient of potential temperature $\nabla\theta$, the projection being normalized by the mass density ρ. Thus, at a particular point, P quantifies the mass-normalized spin normal to the isentropic surface, $\theta = $ const.

A heuristic construction of P, and demonstration of its material conservation under the stated conditions, may be achieved by considering the absolute circulation of a small material loop on an isentropic surface and invoking Kelvin's circulation theorem for this loop as the loop is shrunk to an infinitesimal element. Specifically, for adiabatic ($D\theta/Dt = 0$) and inviscid ($\mathbf{F} = 0$) flow, the foregoing Corollary to Kelvin's theorem implies that the absolute circulation Γ_a is conserved for this material loop. Consider now the "pillbox" volume comprising neighboring loops on isentropic surfaces with potential temperatures θ and $\theta + d\theta$ (Fig. 2.3). Since the flow is assumed adiabatic, fluid particles cannot cross the top and bottom surfaces. Furthermore, since the side of the pillbox (∂V) is defined to move with the flow, all surfaces of the pillbox are material surfaces. It follows that fluid cannot enter or exit the pillbox and so the mass of fluid M contained by it must be materially conserved. In the limit as the area and height of the pillbox become infinitesimally small, it can be shown that $\Gamma_a d\theta/M \to P$. Since $DM/Dt = 0$ and $Dd\theta/Dt = 0$, it follows that $DP/Dt = 0$ (see Exercise 2.16 for details). From this simple construction, it is clear that P is essentially the absolute circulation per unit mass per unit Kelvin for a small material loop placed on an isentropic surface.[13]

13. Exercises 2.17 and 2.18 provide alternative, vector calculus, derivations of the equation governing P in the presence of heating/cooling and non-conservative forces. The resulting equation reduces to the material conservation of P in the limit of inviscid and frictionless flow, thus affirming the heuristic construction.

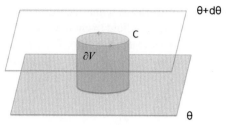

FIGURE 2.3   Lagrangian pillbox volume whose top and bottom are defined by neighboring loops on isentropic surfaces with potential temperatures θ and $\theta + d\theta$. For simplicity, the flow is assumed adiabatic so fluid particles cannot cross the top and bottom surfaces. Since the side of the pillbox (∂V) is defined to move with the flow, all surfaces of the pillbox are material surfaces. As a result, fluid cannot enter or exit the pillbox and thus the mass of fluid M contained within it must be materially conserved.

When diabatic and/or frictional effects are present, P is no longer materially conserved, but is governed by the PV equation

$$\frac{DP}{Dt} = \frac{1}{\rho}\boldsymbol{\omega_a} \cdot \nabla\dot{\theta} + \frac{1}{\rho}\mathbf{K} \cdot \nabla\theta = \frac{1}{\rho}\nabla \cdot \mathbf{Y}, \qquad (2.63)$$

where $\dot{\theta}$ is the diabatic heating rate given by the right-hand-side of (2.20), $\mathbf{K}$ represents the curl of the force $\mathbf{F}$ per unit mass (i.e. $\mathbf{K} = \nabla \wedge \mathbf{F}$) and

$$\mathbf{Y} = \dot{\theta}\boldsymbol{\omega_a} + \nabla\theta \wedge \mathbf{F}. \qquad (2.64)$$

Exercises 2.17 and 2.18 guide the reader in the derivations of Eqs. (2.63) and (2.64). These equations show that diabatic heating in a localized region produces a dipole anomaly of potential vorticity with its axis oriented along the absolute vorticity vector $\boldsymbol{\omega_a}$ and in the opposite direction to it (Fig. 2.4a). Cooling instead of heating reverses the direction of the dipole (Fig. 2.4b).

The action of a localized force $\mathbf{F}$ produces positive relative vorticity to its left and negative relative vorticity to its right (Fig. 2.4c) at a rate $\mathbf{K}$ equal to $\nabla \wedge \mathbf{F}$. Recalling that $P = (1/\rho)(\boldsymbol{\omega_a} \cdot \nabla\theta)$ and remembering that the force generation term in Eq. (2.63) has the form $(1/\rho)\mathbf{K} \cdot \nabla\theta$, we see how the latter contributes to a change in P. Note that like vorticity, a localized force produces a dipole anomaly of potential vorticity. This anomaly is the projection of the dipole anomaly of $\boldsymbol{\omega_a}$, due to $\nabla \wedge \mathbf{F}$, on the potential temperature gradient, $\nabla\theta$.

2.13   Balance dynamics

For situations in which the flow is horizontally dominant ($w \ll |\mathbf{u_h}|$), evolves slowly in time, and remains close to a state of diagnostic force balance (i.e. hydrostatic and gradient wind balance - the latter defined in Chapters 4 and 5), a generalization of the vorticity streamfunction method can be developed using P instead of ζ. Under these circumstances, it is not necessary that P be materially conserved, but the

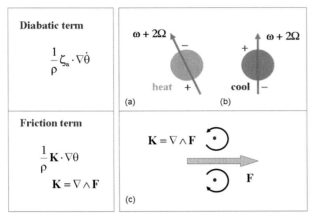

Diabatic term	$\omega + 2\Omega$ $\qquad$ $\omega + 2\Omega$
$\dfrac{1}{\rho}\,\zeta_a \cdot \nabla\dot{\theta}$	heat + $\qquad$ cool $-$
	(a) $\qquad$ (b)
Friction term	$\mathbf{K} = \nabla \wedge \mathbf{F}$
$\dfrac{1}{\rho}\,\mathbf{K} \cdot \nabla\theta$	
$\mathbf{K} = \nabla \wedge \mathbf{F}$	$\mathbf{F}$
	(c)

FIGURE 2.4 PV dipole produced by localized diabatic heating (a), or cooling (b). In (a), the absolute vorticity vector is tilted away from the vertical so as to reflect the orientation of absolute vortex lines, $\boldsymbol{\omega_a}$, in the eyewall of a developing or mature tropical cyclone. In (b), the absolute vortex line is oriented vertically in order to reflect a more typical configuration outside of the primary eyewall region of the vortex where radiative cooling by longwave radiation to space would be prevalent. The plus and minus signs reflect the material tendency of P for the envisaged configuration. Panel (c) shows the force-curl dipole associated with a localized force, $\mathbf{F}$.

forcing from $\dot{\theta}$ and $\mathbf{F}$ must be sufficiently weak to keep the flow evolution suitably "slow" such that the approximation of diagnostic force balance remains well defined.

Like the vorticity-streamfunction method, this generalized method meaningfully allows one to invert the P field for the associated wind and mass fields that are "in balance with" P. Once the wind and mass fields have been deduced from P at a particular time step, given suitable boundary conditions and balance constraints, the P field at the next time step can be obtained by advecting the P field with the winds and modifying P according to Eq. (2.63). This so-called "balance dynamics" approach, in which P and its inverted wind and mass field carries the principal dynamical flow information, retains the geometrical elegance of the vorticity-streamfunction method while including also three-dimensional motions. Illustrative examples of balanced, tropical cyclone evolution will be presented in Chapter 8. These ideas have proven useful in analyses of real-case tropical cyclones (e.g., Möller and Shapiro 2002).[14]

Since moist processes pervade tropical cyclones, an important question arises whether a simple extension of the balance method just described can be developed to include moist variables explicitly. The answer to this question is yes and the solution is to use the density temperature (and corresponding density potential temperature, θ_ρ, which uses T_ρ (Eq. (2.25)) in place of T in the definition of θ) as the appro-

priate scalar function to define P. This choice, along with a suitable balance approximation for rapidly rotating geophysical vortices (e.g., diagnostic force balance as noted above), yields a well-defined and spatially smooth invertibility principle in three dimensions (Schubert et al., 2001 and refs.).[15] Because of the relative smallness of the mixing ratios of water vapor and liquid water, this choice of P is quantitatively very close to the analogous (dry) P based on θ (Eq. (2.61)). The similarity of this moist P to its dry equivalent is strong justification for the common practice of using dry P as a meaningful first approximation for both the axisymmetric and asymmetric balance evolution of tropical cyclone vortices.

The interesting topic of moist PV and balance dynamics *in an axisymmetric context* will be taken up in Chapters 12, 13, and 16.

2.14 PV global constraints

When integrated over a region of space V, Eq. (2.63) gives

$$\frac{D}{Dt}\int_V \rho P\,dV = \int_A (\dot{\theta}\boldsymbol{\omega_a} + \nabla\theta \wedge \mathbf{F})\cdot\hat{\mathbf{n}}\,dS, \qquad (2.65)$$

where A is the surface of V and $\hat{\mathbf{n}}$ is the unit normal to A, outward from V.

An important result that follows from Eq. (2.65) is that purely internal diabatic heating or friction cannot change the mass-weighted potential vorticity distribution within V, i.e., if $\dot{\theta}$ and $\mathbf{K}$ are zero on A, then

$$\frac{D}{Dt}\int_V \rho P\,dV = 0. \qquad (2.66)$$

Thus internal diabatic heating or friction can only redistribute the mass-weighted potential vorticity that is there.

2.15 PV flux form and impermeability theorem

An equivalent form of the PV equation in the presence of heating and momentum forcing follows from Eq. (2.63), in conjunction with the continuity equation, Eq. (2.6):

$$\frac{\partial}{\partial t}(\rho P) + \nabla \cdot (\rho \mathbf{u}P - \mathbf{Y}) = 0. \qquad (2.67)$$

The PV equation is now in the form of an exact conservation relation in the flux sense of Section 2.8, but without an external source or sink term on the right hand side (Exercise 2.20). In this flux form, the first term $\rho \mathbf{u}P$ is referred

---

14. Employing a balance framework, those authors explored factors responsible for the intensification of Hurricane Opal (1995) using three-dimensional PV dynamics, corresponding inversion analyses for both the axisymmetric and asymmetric balance flow components as well as inferred imbalance tendencies.

15. The alternative choice of θ_e as the scalar function is problematic for several reasons in a three-dimensional moist precipitating atmosphere with radiative forcing, including when attempting to "invert" for the balanced flow in tropical cyclone vortices. See Schubert et al. (2001), their Section 5, for details.

to as the *advective flux* because, being proportional to the velocity field, it acts to transport P in the direction of the flow $\mathbf{u}$; the second term $-\mathbf{Y}$ is called the *nonadvective flux* because it is not proportional to the velocity field. The non-advective terms produce dipole-like structures in the ρP tendency in association with local distributions of $\dot{\theta}$, ω_a, and $\mathbf{F}$. Although this form of the PV equation has inspired some to assert an exact analogy with equations for chemical tracers in the atmosphere, P is quite unlike a chemical tracer in the sense that there can be no *net* transport of P across an isentropic surface.

The transport properties of P have been succinctly summarized by McIntyre (2015):

> *"... P can be regarded as the amount per unit mass of a notional chemical substance consisting of charged particles that are permanently trapped on the moving surfaces S."*

Here, S denotes a surface along which θ or a function of θ is constant.

> *"Net charge is conserved: one can have pair production and mutual annihilation, but no net creation or destruction except where a surface S intersects a boundary. In this picture the surfaces S are impermeable to the PV particles even when they are permeable to air undergoing diabatic heating or cooling - a behavior very different from that of a real chemical. The corresponding mathematical statement is sometimes called the impermeability theorem for PV."*

Multiple proofs of the impermeability theorem are presented in Haynes and McIntyre (1987, 1990) and an elegant generalization of the impermeability theorem that includes moist variables is presented in Schubert et al. (2001). (Exercise 2.21 guides the reader through Schubert et al.'s generalization for the simplest case of P based on θ.)

2.16 Vorticity flux equation

It turns out that the tendency equation for vertical vorticity can be put also in a "flux conservation form" similar to the PV equation, Eq. (2.67). In the simple case where θ variations in the horizontal direction are small in comparison to the horizontally-averaged average value (a reasonable first approximation for most tropical flows, at least for tropical storms and weak hurricane/typhoons) the vertical vorticity equation takes the remarkably simple flux conservation form:

$$\frac{\partial \zeta}{\partial t} = -\nabla_h \cdot \mathbf{Z}, \qquad (2.68)$$

where ∇_h is the horizontal gradient operator and

$$\mathbf{Z} = \mathbf{u_h}(\zeta + f) - \boldsymbol{\omega}_h w + \hat{\mathbf{k}} \wedge \mathbf{F}, \qquad (2.69)$$

is the horizontal vorticity flux vector, with $\mathbf{u_h}$ the horizontal velocity vector and $\boldsymbol{\omega}_h$ the horizontal vorticity vector. (Exercise 2.22 considers the general case that is not limited to relatively small horizontal variations of θ. The final result has the same form as Eq. (2.69) albeit with a revised definition of $\mathbf{Z}$.)

The compact form of the vertical vorticity equation (2.68) says that the Eulerian tendency of the vertical component of vorticity equals the *horizontal convergence* of the *vorticity flux* $\mathbf{Z}$. The first term in the flux, $\mathbf{u_h}(\zeta + f)$, is called the "advective vorticity flux", while the sum of the second and third terms, $-\boldsymbol{\omega}_h w + \hat{\mathbf{k}} \wedge \mathbf{F}$, is called the "non-advective vorticity flux". Like the PV equation given above, the first term is called the advective flux because it transports the vertical vorticity by the horizontal flow $\mathbf{u_h}$. Again, the non-advective flux is not proportional to the velocity vector and, in this case, produces dipole-like structures in the vertical vorticity tendency for localized distributions of w, $\boldsymbol{\omega}_h$, or $\mathbf{F}$. Of course, $\mathbf{Z}$ is only unique up to a "do nothing" flux, $\hat{\mathbf{k}} \wedge \nabla\chi$, where χ is an arbitrary scalar field. Nevertheless, Eq. (2.68) makes it immediately clear that there can be no net vertical flux of vertical vorticity across constant height surfaces.

An important deduction about low-level vorticity organization follows immediately from Eq. (2.68): The formation of a mesoscale region of intense cyclonic vorticity in the lower troposphere cannot arise by a downward flux of cyclonic vorticity from above. Rather, the generation of such a vortex can only occur by a sustained horizontal convergence of the flux $\mathbf{Z}$. That is, the formation of a mesoscale region of intense cyclonic vorticity in the lower troposphere can only arise through a horizontal concentration of pre-existing cyclonic vorticity, in combination with a preferential ingestion of cyclonic vorticity (or preferential expulsion of anti-cyclonic vorticity) associated with the horizontal non-advective flux.

In contrast to the material form of the vorticity equation given by Eq. (2.45), which accounts for the rate of change of the vector vorticity following a fluid parcel, the flux form of the vertical vorticity equation proves natural for quantifying changes in the net horizontal circulation within a *fixed*, horizontal, closed loop on account of the advective and non-advective fluxes across the loop.

We will return to the foregoing points later in Chapters 10 and 11 when examining the vorticity dynamics of the formation and intensification of a tropical cyclone in association with deep, cumulonimbus convection over a warm ocean.

2.17 Coordinate systems

In some problems it is convenient to use a vertical coordinate other than height. Common choices are:

- the pressure;
- the logarithm of the pressure;
- the pseudo-height Z, which is a function of pressure that reduces to the actual height in an atmosphere with a uniform potential temperature;
- the potential temperature; or
- σ, which is defined either as the ratio of pressure to its surface value, or the pressure minus some upper-level pressure divided by the surface value minus this upper-level pressure.

Since the potential temperature and dry entropy are uniquely related, coordinates with θ in the vertical are called *isentropic coordinates*. In this case, for example, when the flow is approximately hydrostatic, the potential vorticity takes a particularly simple form (Exercise 2.16):

$$P = -g(\zeta_\theta + f)(\partial\theta/\partial p). \qquad (2.70)$$

Now the scalar vector product no longer appears. Moreover, when expressed in these coordinates, the vertical advection in the derivative D/Dt is proportional to the heating rate (see Eq. (2.20)). For inviscid and adiabatic motion, when $D\theta/Dt = 0$, Exercise 2.16 shows that P is conserved on isentropic surfaces. This turns out to be a very useful result and will be exploited in various places in this book.

2.18 Exercises

Exercise 2.1. Using the full form of the continuity equation (2.6), show that for any scalar field γ,

$$\rho\frac{D\gamma}{DT} = \frac{\partial\rho\gamma}{\partial t} + \nabla\cdot(\rho\gamma\mathbf{u}). \qquad (2.71)$$

Exercise 2.2. Show that

$$\frac{\partial\rho\Phi}{\partial t} + \nabla\cdot(\rho\Phi\mathbf{u}) = \rho g w.$$

Exercise 2.3. Show that

$$\nabla\cdot(p\mathbf{u}) = \rho\frac{D}{Dt}\left(\frac{p}{\rho}\right) - \frac{\partial p}{\partial t}.$$

Exercise 2.4. Show that an equation analogous to Eq. (2.42) for a frictionless pseudo-adiabatic process with $\dot{Q}$ given by Eq. (2.26) is

$$\frac{D}{Dt}\left(E + \frac{p}{\rho} + \Phi + L_v r_v^* + \frac{1}{2}\mathbf{u}^2\right) = 0,$$

assuming that $\partial p/\partial t$ is vanishingly small and can be neglected. What is the physical interpretation of this equation and why is it useful? Discuss.

Exercise 2.5. Consider an inviscid and strictly two-dimensional flow $\mathbf{u} = (u, v, 0)$ in a rectangular β-plane coordinate system (x, y, z) with a latitudinally varying vertical rotation rate $\mathbf{\Omega} = (0, 0, \Omega_0 + \beta y/2)$, where $\Omega_0 = \Omega\sin\phi_0$ is the local vertical rotation rate of the earth at the reference latitude ϕ_0, y is the meridional Cartesian distance along the earth's surface from the reference latitude and z is the local vertical coordinate perpendicular to the modified equipotential surface: $\Phi = \text{const}$.

(a) Show that Eq. (2.45) reduces to the material conservation equations (2.48) (or (2.49)):

$$\frac{D}{Dt}(\zeta + f_0 + \beta y) = 0, \qquad (2.72)$$

where $\zeta = \partial v/\partial x - \partial u/\partial y$ is the relative vertical vorticity, $f_0 = 2\Omega_0$ is the Coriolis frequency at the reference latitude and $\beta = df/dy = 2\Omega\cos\phi_0/a$, where a is the earth's mean radius. Here, D/Dt is the material derivative operator in the two-dimensional β-plane $(x, y, 0)$:

$$\frac{D}{Dt} = \frac{\partial}{\partial t} + u\frac{\partial}{\partial x} + v\frac{\partial}{\partial y}. \qquad (2.73)$$

(b) Next, consider a poleward moving fluid parcel. Show that the relative vertical vorticity must decrease for this parcel and that the local circulation around this parcel must decrease. Offer a physical example of the usefulness of this material conservation equation.

Exercise 2.6. Consider the case of an axisymmetric vortex with tangential velocity distribution $v(r)$, and assume a linear frictional drag force $\mathbf{F} = -\mu v(r)$, with μ (> 0) an effective damping rate associated with the frictional stress at the ground (or sea surface) $z = 0$. Show that the curl of this force per unit mass $\mathbf{K} = \nabla\wedge\mathbf{F}$ equals $\mathbf{K} = -\mu\zeta\mathbf{k}$, where ζ is the vertical component of relative vorticity. Next, show that the Ertel potential vorticity is destroyed locally at the lower boundary at the rate $(1/\rho)(\mathbf{K}\cdot\nabla\theta) = -(1/\rho)\mu\zeta(\partial\theta/\partial z)_{z=0}$. Offer a brief discussion of the potential importance of this frictional force-curl term to the PV dynamics in the vortex boundary layer.

Exercise 2.7. Using the gas law, $p = \rho R_d T_v$, and the usual definition of virtual potential temperature, show that the buoyancy force per unit mass can be written as

$$b = g\left[\frac{(\theta_v - \theta_{vref})}{\theta_{vref}} - (\kappa - 1)\frac{p'}{p_{ref}}\right], \qquad (2.74)$$

where θ_v is the virtual potential temperature of the air parcel in K and θ_{vref} is the corresponding reference value.

Note that the second term on the right-hand-side of Eq. (2.34) is sometimes referred to as the "dynamic buoyancy", but in some sense this is a misnomer since buoyancy depends fundamentally on the density perturbation and this term simply corrects the calculation of the density perturbation based on the virtual potential temperature perturbation.

Exercise 2.8. Show that if the perturbation pressure gradient force term in Eq. (2.28) is written in terms of the Exner function and/or its perturbation, the second term in Eq. (2.34) does not appear in the expression for buoyancy.

Exercise 2.9. The x-momentum equation for the two-dimensional flow of a homogeneous, inviscid fluid in the x-z plane may be written as

$$\frac{\partial u}{\partial t} + u\frac{\partial u}{\partial x} + w\frac{\partial u}{\partial z} = -\frac{1}{\rho}\frac{\partial p}{\partial x}, \qquad (2.75)$$

using standard notation. Assuming that the flow can be decomposed into a zonal (in the x-direction) mean part and a fluctuating part, show that the mean flow satisfies the equation

$$\frac{\partial \bar{u}}{\partial t} = -\frac{\partial}{\partial z}\overline{u'w'}, \qquad (2.76)$$

where an overbar represents a zonal average and a prime represents a deviation (fluctuation) about this average. Physically interpret the resulting mean flow equation.

Exercise 2.10. (a) Show that the relative vorticity flux (or strength) Γ is constant along a vortex tube comprising a bundle of relative vortex lines passing through a material loop C at some time. Conservation of vortex strength along the vortex tube is frequently invoked to show that vortex tubes (and the filaments that comprise them) cannot end in the fluid, but must either close on themselves, extend to infinity, or end at solid or free boundaries. Show this property, assuming a simple configuration for a vortex tube, wherein the vorticity never vanishes along vortex filaments and the vortex tube never splits into two or more sub-tubes.

(b) Show that

$$\oint_C \boldsymbol{\Omega} \wedge \mathbf{r} \cdot \mathbf{dl} = \int_A 2\boldsymbol{\Omega} \cdot \hat{\mathbf{n}}dA = 2\Omega A_n, \qquad (2.77)$$

where A_n denotes the projection of A onto the equatorial plane perpendicular to $\boldsymbol{\Omega}$. That is, $A_n = A\cos\theta$, where θ denotes the angle between the normal vector to the area A and the planet's rotation vector $\boldsymbol{\Omega}$.

Exercise 2.11. Show that

$$\frac{D\Gamma}{Dt} = \oint_C \mathbf{u}\cdot\frac{D\mathbf{dl}}{Dt} + \oint_C (-\frac{1}{\rho}\nabla p - \nabla\Phi - 2\boldsymbol{\Omega}\times\mathbf{u} + \mathbf{F})\cdot\mathbf{dl}, \qquad (2.78)$$

where all variables have their meaning as defined in and around Eq. (2.57) in the main text. Provide a physical interpretation for each of the terms contributing to the tendency of relative circulation.

Exercise 2.12. Show that the first and third integrals on the right hand side of Eq. (2.57) vanish. Provide a supporting physical explanation for why these two integrals vanish.

Exercise 2.13. Show that the line integral of the Coriolis force in Exercise 2.11 may be expressed as the following material derivative of the projected area A_n:

$$-\oint_C 2\boldsymbol{\Omega}\times\mathbf{u}\cdot\mathbf{dl} = -2\Omega\frac{DA_n}{Dt}, \qquad (2.79)$$

where $\Omega = |\boldsymbol{\Omega}|$.

Exercise 2.14. Use Stokes' theorem, together with the foregoing result from Exercise 2.13 between the line-integrated Coriolis force and the material rate of change of A_n, to transform the equation for $D\Gamma/Dt$ into an expression for $D\Gamma_a/Dt$:

$$\frac{D\Gamma_a}{Dt} = \int_A \frac{1}{\rho^2}\nabla\rho\times\nabla p\cdot\hat{\mathbf{n}}dA + \int_A \nabla\times\mathbf{F}\cdot\hat{\mathbf{n}}dA. \quad (2.80)$$

Exercise 2.15. Show that the first integral on the right hand side of Eq. (2.59) will still vanish when the flow is adiabatic ($D\theta/Dt = 0$) and when the surface A is chosen to lie within an isentropic surface $\theta = $ const. In the limit of frictionless ($\mathbf{F} = \mathbf{0}$) flow, show that $D\Gamma_a/Dt = 0$ for these material loops, even though the flow is not generally barotropic. This result establishes the corollary to Kelvin's theorem for an infinite family of material loops that move along an isentropic surface.

Exercise 2.16. Consider a small, closed material loop of fluid particles lying within a particular isentropic surface, defined by $\theta = $ const. Show that the material volume defined by a neighboring surface of potential temperature $\theta + d\theta$ and θ has a constant mass dM for adiabatic flow. Show next, as outlined in the main text, that the absolute circulation of the loop divided by the mass density with respect to θ of the *material* volume is proportional to P. Assuming inviscid and adiabatic flow, show that the mass density, $dM/d\theta$, is materially conserved. Thus, show that when the absolute circulation is normalized by the mass density, the quantity P is materially conserved, i.e., $DP/Dt = 0$.[16] Based on this construction, what, then, is PV and why might PV be useful for understanding vortical flows in stably stratified atmospheres/oceans? Discuss.

Exercise 2.17. Using tensor notation, show that

$$\boldsymbol{\omega}\cdot\nabla(\mathbf{u}\cdot\nabla\chi) = \nabla\chi\cdot(\boldsymbol{\omega}_a\cdot\nabla\mathbf{u}) + \boldsymbol{\omega}_a\cdot(\mathbf{u}\cdot\nabla(\nabla\chi)).$$

Show also that

$$\rho\nabla\chi\cdot\frac{D}{Dt}\left(\frac{\boldsymbol{\omega}_a}{\rho}\right) = \rho\frac{D}{Dt}\left(\frac{\boldsymbol{\omega}_a\cdot\nabla\chi}{\rho}\right) - \boldsymbol{\omega}_a\frac{D}{Dt}(\nabla\chi).$$

---

16. Exercises 2.17 and 2.18 provide alternative, vector calculus, derivations of the equation governing P in the presence of heating/cooling and non-conservative forces. In the limit of inviscid and frictionless flow, the resulting equation reduces to the material conservation of P, consistent with the present construction.

Exercise 2.18. Using the continuity equation for a compressible fluid, $\frac{1}{\rho}\frac{D\rho}{Dt} + \nabla \cdot \mathbf{u} = 0$, and by considering the expression

$$\nabla \chi \cdot \left(\frac{D\boldsymbol{\omega}_a}{Dt} + \boldsymbol{\omega}_a (\nabla \cdot \mathbf{u}) - \boldsymbol{\omega}_a \cdot \nabla \mathbf{u} + \nabla \phi \wedge \frac{1}{\rho}\nabla \rho \right) = 0,$$

show that

$$\rho \frac{D}{Dt}\left(\frac{\boldsymbol{\omega}_a \cdot \nabla \chi}{\rho} \right) + \nabla \chi \cdot \nabla \phi \wedge \frac{1}{\rho}\nabla \rho - \boldsymbol{\omega}_a \cdot \nabla \left(\frac{D\chi}{Dt} \right) = 0,$$

where $\boldsymbol{\omega}_a = \boldsymbol{\omega} + \mathbf{f}$ is the absolute vorticity vector; $\phi = \ln \theta$, θ being the potential temperature; and χ is any scalar function. [Hint: you will find the results of Exercise 2.17 of use in the reduction.]

Show further that the second term in the above equation is zero if χ is a function of state, and that the quantity

$$P = \frac{\boldsymbol{\omega}_a \cdot \nabla \theta}{\rho}$$

is conserved in frictionless motion and when χ is materially conserved, i.e., when $D\chi/Dt = 0$. When χ is the potential temperature, θ, the latter condition is one of adiabatic motion and the quantity P is simply *Ertel's potential vorticity* defined in Section 2.12.

Exercise 2.19. Show that the potential vorticity for the axisymmetric vortex $(0, v(r, z), 0)$ is

$$PV = \frac{1}{\rho}\left(-\frac{\partial v}{\partial z}\frac{\partial \theta}{\partial r} + (\zeta + f)\frac{\partial \theta}{\partial z} \right).$$

Exercise 2.20. Show that in the presence of heating and momentum forcing, the PV equation can be rewritten in exact flux form:

$$\frac{\partial}{\partial t}(\rho P) + \nabla \cdot (\rho \mathbf{u} P - \mathbf{Y}) = 0. \qquad (2.81)$$

Why is this form of the PV equation useful? Discuss.

Exercise 2.21. Show that isentropic surfaces $\theta = \text{const.}$ are impermeable to PV particles even when they are permeable to air undergoing diabatic heating or cooling.

Exercise 2.22. (a) Show that when moisture effects are neglected in the horizontal momentum equations and when potential temperature deviations are small in comparison to a horizontally averaged potential temperature, the horizontal component of the vector momentum equation (2.1) may be put in equivalent rotational form

$$\frac{\partial \mathbf{u_h}}{\partial t} + \hat{\mathbf{k}} \wedge \mathbf{Z} + \nabla_h \left(\frac{1}{2}\mathbf{u}^2 + c_p \pi \bar{\theta} \right) = 0, \qquad (2.82)$$

where $\mathbf{u_h}$ is the horizontal velocity vector, $\bar{\theta}$ is the horizontal average of dry potential temperature θ, $\Pi = c_p(p/p_{**})^\kappa$ is the Exner function, and $\mathbf{Z}$ is the horizontal vector field

$$\mathbf{Z} = \mathbf{u_h}(\zeta + f) - \boldsymbol{\omega}_h w + \hat{\mathbf{k}} \wedge \mathbf{F}, \qquad (2.83)$$

where ζ is the relative vertical vorticity and w is the vertical velocity component. Show next that by taking the vertical component of the curl of (2.82), the vertical vorticity equation takes the exact flux-divergence form given by:

$$\frac{\partial \zeta}{\partial t} = -\nabla_h \cdot \mathbf{Z}. \qquad (2.84)$$

This equation demonstrates that the Eulerian tendency of the vertical component of absolute vorticity is given by the *horizontal convergence* of the *flux* $\mathbf{Z}$. The astute reader may ask why might this form of the vorticity equation be useful in understanding the formation of an intense vortex such as a hurricane, tornado, water spout or dust devil? Discuss.

(b) Show next that the foregoing result may be generalized to include horizontal variations in θ by adding $-\hat{\mathbf{k}} \wedge c_{pd}\theta'\nabla_h \Pi$ to the vector $\mathbf{Z}$, viz.,

$$\mathbf{Z} = \mathbf{u_h}(\zeta + f) - \boldsymbol{\omega}_h w + \mathbf{k} \wedge \mathbf{F} - \mathbf{k} \wedge c_{pd}\theta'\nabla_h \Pi, \quad (2.85)$$

where $\theta' = \theta - \bar{\theta}$ is the deviation of the dry potential temperature from its horizontal average. Show that the vertical component of the curl of the updated (2.82) results again in (2.84) using the updated flux given by (2.85).

(c) Finally, sketch a generalization of the foregoing vorticity equation to include moisture variables.

2.19 Appendix: the membrane analogy

In Section 2.10, we stated that the vorticity-streamfunction method is a powerful way of solving for the evolution of two-dimensional flows of a homogeneous, incompressible fluid. As noted there, one step in this method requires the solution of a Poisson type of partial differential equation for the streamfunction ψ given the spatial distribution of vorticity, ζ at a given time, i.e.,

$$\frac{\partial^2 \psi}{\partial x^2} + \frac{\partial^2 \psi}{\partial y^2} = \zeta(x, y). \qquad (2.86)$$

Except for a few simple distributions of ζ (see e.g., Chapter 3), it is necessary to resort to numerical methods to solve this equation and there are several such methods available. However, a powerful way of seeing what the qualitative structure of the solutions must look like is to invoke the so-called *membrane analogy*. One of the several prototype problems in classical physics where the Poisson equation arises is that for the equilibrium displacement $\eta(x, y)$ of a stretched membrane, or drum skin, that is subject to a prescribed force distribution, $F(x, y)$, where x and y are rectangular Cartesian coordinates. If the displacement is sufficiently small, the equation for the displacement is simply

$$\frac{\partial^2 \eta}{\partial x^2} + \frac{\partial^2 \eta}{\partial y^2} = -F. \tag{2.87}$$

Intuition about how such a membrane would deform for simple spatial distributions of F can be used to infer what the solution for $\eta(x, y)$ should look like given appropriate conditions for η along the boundary of some domain. For example, we would expect the maximum displacement of the membrane to be where the applied force is a maximum and the sign of this displacement to have the same sign as the force.

In the case of Eq. (2.86), the vorticity would be analogous to minus the force acting on the membrane and the streamfunction would be analogous to the membrane displacement. Other applications of the membrane analogy will be discussed in Chapters 3-5 and 8-9.

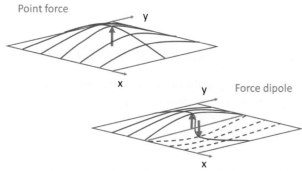

FIGURE 2.5 Response of a membrane to an upward point force and force dipole. Negative membrane displacements indicated by dashed lines.

A particular example would be the displacement induced by a point force directed upwards when the displacement is constrained to be zero along the boundary (Fig. 2.5, top left). Another example would be the displacement brought about by a point dipole of force, i.e. two point forces located infinitesimally close together, but acting in opposite directions (Fig. 2.5, bottom right). In the case of the point force, the displacement in the interior of the domain is everywhere upward and is a maximum at the location of forcing. In the case of a dipole, the displacement is positive on the side of the dipole where the forcing is positive and negative on the side where the forcing is negative, but it is zero at the dipole itself.

Chapter 3

Tropical cyclone motion

3.1 The observations to be explained

In Section 1.1, we showed that, while at low latitudes, tropical cyclones tend to move westwards and polewards on average, except over the southwest Pacific Ocean, where the average motion is eastwards and polewards. To a reasonable first approximation, the motion can be understood in terms a so-called "steering concept" in which a tropical cyclone is simply carried along by some mass-weighted vertical mean of its larger-scale environmental flow. An appropriate mean might extend in depth from, say 850 mb to 200 mb, avoiding the strong inflow layer near the surface and the outflow layer at the top of the cyclone. While the steering idea accounts for perhaps 80% of the motion, the deviations of the motion vector from that of the so-called *mean steering current* are important in a forecasting context over several days.

Perhaps, because the flow above some shallow surface friction layer and below the outflow layer is for the most part quasi-horizontal (except, of course, in the eyewall region), many important dynamical aspects of motion may be understood in terms of a simple barotropic theory in which the flow is assumed to be independent of height. In this chapter we review the barotropic dynamics of vortex motion as it relates to the steering concept, showing how deviations from the mean steering current arise. In a proposed companion volume we will discuss the additional effects of baroclinicity including the effects of vertical wind shear on tropical cyclone motion.

3.2 The partitioning problem

An important issue that arises in the development of a theory for tropical cyclone motion in a particular environment is the so-called partitioning problem. To be able to articulate how a tropical cyclone interacts with its environment, we need to define first what we mean by "the tropical cyclone" and what we mean by its "environment". Since Nature makes no distinction between the cyclone and its environment, any partitioning we choose to analyze this interaction is necessarily artificial and non-unique. Moreover, interpretations of the motion will depend to some degree on the chosen partitioning scheme. An early and insightful paper exploring the issues of partitioning is that of Kasahara and Platzmann (1963).

Various schemes have been proposed to partition the cyclone and its environment. One obvious possibility is to define the cyclone as the azimuthally-averaged flow about the vortex center and the residual flow (i.e., the asymmetric component) as "the environment". Then the question arises: which location should be chosen as "the cyclone center"? Possibilities might include the location of the minimum surface pressure, the location of the maximum vorticity and the center of the vortex circulation (i.e., the minimum streamfunction) at some level, or the minimum wind speed at that level. We show in Section 3.3.1 that, in general, these locations are not coincident.

Many idealized theoretical studies consider the motion of an initial, analytically-prescribed, axisymmetric vortex in an analytically-prescribed environmental flow or in a quiescent environment. Some early papers on this topic are those of Chan and Williams (1987), Fiorino and Elsberry (1987), Shapiro and Ooyama (1990), Smith et al. (1990), and Smith and Ulrich (1990). If the flow is barotropic, there is no mechanism to change the absolute vorticity of a moving air column. Since the relative vorticity of the vortex itself tends to greatly exceed the local planetary vorticity or that of its environment, it would seem reasonable to assume that, to a first approximation, the vortex vorticity will be largely carried along with (i.e., advected by) the environmental flow. Therefore, a reasonable definition of "the vortex" at a later time would be the relative vorticity distribution of the *initial* vortex, appropriately relocated. Then, all the flow changes accompanying the vortex motion reside in the residual flow, which serves to define the new "vortex environment". With these definitions, the location of the relative vorticity maximum would be a suitable choice for the vortex center location, at least for a monopole vortex,[1] notwithstanding the fact that it is strictly absolute vorticity that is conserved and not relative vorticity (see Section 3.3.2 for more on this point).

---

1. A monopole vortex has maximum vorticity at its center. In circumstances when the maximum vorticity occurs at a finite radius from the circulation center, as for example in a vortex ring located on the inside of the hurricane eyewall, one can use an integrated measure of the vorticity distribution, such as the vorticity centroid. In such vortices, barotropic shear instabilities and trochoidal motions associated with the ring vortex configuration act on shorter (local advective) time scales (Nolan and Montgomery, 2000; Nolan et al., 2001). These motions are approximately independent of the synoptic-scale steering flow.

Tropical Cyclones. https://doi.org/10.1016/B978-0-44-313449-4.00011-4
Copyright © 2023 Elsevier Inc. All rights reserved.

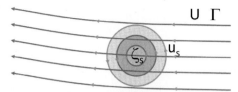

FIGURE 3.1 Schematic illustrating a partitioning between a vortex and its environment. The vortex is defined by an axisymmetric velocity distribution $\mathbf{u_s}$ with corresponding relative vorticity ζ_s. The vortex environment is characterized by the velocity distribution $\mathbf{U} = \mathbf{u} - \mathbf{u_s}$ where $\mathbf{u}$ is the total velocity vector. The relative vorticity of the environment is Γ. See text for further discussion.

An advantage of the foregoing partitioning scheme is that all the subsequent flow changes are contained in one component of the partition, the "environment" while "the vortex" remains unchanged at its new location. Moreover, starting with a vortex that is "well-behaved" in the sense of having zero circulation at large radial distances and hence finite integrated kinetic energy, the vortex remains well behaved. Further, one does not have to be concerned with vorticity transfer between the symmetric vortex and the environment as this transfer is zero, by definition. The partitioning scheme has advantages also for understanding the motion of initially asymmetric vortices discussed in Chapter 4.

The partitioning scheme is best illustrated mathematically in a barotropic context, based on the vorticity-streamfunction ideas outlined in Section 2.10. Let the total wind be expressed as $\mathbf{u} = \mathbf{u_s} + \mathbf{U}$, where $\mathbf{u_s}$ denotes the axisymmetric velocity field and $\mathbf{U}$ is the velocity field of the vortex environment, and define $\zeta_s = \hat{\mathbf{k}} \cdot \nabla \wedge \mathbf{u_s}$ and $\Gamma = \hat{\mathbf{k}} \cdot \nabla \wedge \mathbf{U}$, where $\hat{\mathbf{k}}$ is the unit vector in the vertical (see Fig. 3.1). Then the barotropic vorticity equation (Eq. (2.49)) in vector form can be partitioned into two equations:

$$\frac{\partial \zeta_s}{\partial t} + \mathbf{c} \cdot \nabla \zeta_s = 0, \qquad (3.1)$$

and

$$\frac{\partial \Gamma}{\partial t} = -\mathbf{u_s} \cdot \nabla(\Gamma + f) - (\mathbf{U} - \mathbf{c}) \cdot \nabla \zeta_s - \mathbf{U} \cdot \nabla(\Gamma + f), \quad (3.2)$$

where $\mathbf{c}(t)$ is the vortex translation speed and t is the time. Note that $\mathbf{u_s} \cdot \nabla \zeta_s = 0$, because for a symmetric vortex $\mathbf{u_s}$ is normal to $\nabla \zeta_s$.

Eq. (3.1) states simply that the symmetric vortex translates with velocity $\mathbf{c}$ and Eq. (3.2), which is obtained by subtracting Eq. (3.1) from the full vorticity equation (2.49), is an equation for the evolution of the vorticity of the vortex environment.

Having solved Eq. (3.2) for $\Gamma(x, y, t)$, we can obtain the corresponding environmental streamfunction, ψ_a, by solving Eq. (2.52) in the form $\nabla^2 \psi_a = \Gamma$. Assuming that "the vortex" is carried along by the flow across its center, the

vortex translation velocity $\mathbf{c}$ may be obtained by calculating the flow velocity $\mathbf{U_c} = \mathbf{k} \wedge \nabla \psi_a$ at the vortex center and the vortex track can be obtained by integrating the flow at the vortex center with time. The degree of approximation involved with this assumption is discussed further in Section 3.3, where a range of examples is presented.

3.3 Prototype problems

3.3.1 Symmetric vortex in a uniform flow

Consider a barotropic vortex with an axisymmetric relative vorticity distribution, ζ_s, embedded in a uniform zonal air stream $U\hat{\mathbf{i}}$ on an f-plane. Here $\hat{\mathbf{i}}$ is a unit vector in the x-direction. It is easy to verify that the total streamfunction for this flow has the form:

$$\psi(x, y, t) = -Uy + \psi_s(x, y, t), \qquad (3.3)$$

where $\psi_s(x, y, t)$ is a function only of radius r from the vortex center given by $r^2 = (x - Ut)^2 + y^2$, and the initial location of the vortex is at the origin, $(0, 0)$. The corresponding total velocity field is

$$\mathbf{u} = (U, 0) + \left(-\frac{\partial \psi_s}{\partial y}, \frac{\partial \psi_s}{\partial x} \right), \qquad (3.4)$$

and the relative vorticity distribution, $\zeta_s = \nabla^2 \psi_s$. From the partitioning perspective in Section 3.2, all the relative vorticity is ascribed to "the vortex" and "the environment", characterized by the flow vector $U\hat{\mathbf{i}}$, has zero vorticity. Since the flow at the center of an axisymmetric vortex in isolation is zero, it follows from Eq. (3.4) that the flow at the vortex center, $\mathbf{U_c} = U\hat{\mathbf{i}}$. Thus, $\mathbf{c} = U\hat{\mathbf{i}}$ and the vortex translates with speed U in the x-direction.

The foregoing result is consistent with the mathematics of the previous section. First note that in Eq. (3.2), $\mathbf{c} = \mathbf{U}$ so that this equation is trivially satisfied and, with $\mathbf{c} = U\hat{\mathbf{i}}$, Eq. (3.1) becomes simply

$$\frac{\partial \zeta_s}{\partial t} + U \frac{\partial \zeta_s}{\partial x} = 0. \qquad (3.5)$$

This is a linear, first-order, partial differential equation with the characteristics given by $dx/dt = U$. Thus Eq. (3.5) is a mathematical statement of the vorticity field being simply translated by the uniform flow.

Even though the relative vorticity in this uniform flow remains symmetric about the moving vortex center, neither the streamfunction distribution $\psi(x, y, t)$, nor the pressure distribution $p(x, y, t)$, are symmetric. Moreover, in general, the locations of the minimum central pressure, maximum relative vorticity, and minimum streamfunction (where $\mathbf{u} = \mathbf{0}$) do not coincide.

Specifically, there are three important deductions from Eq. (3.4):

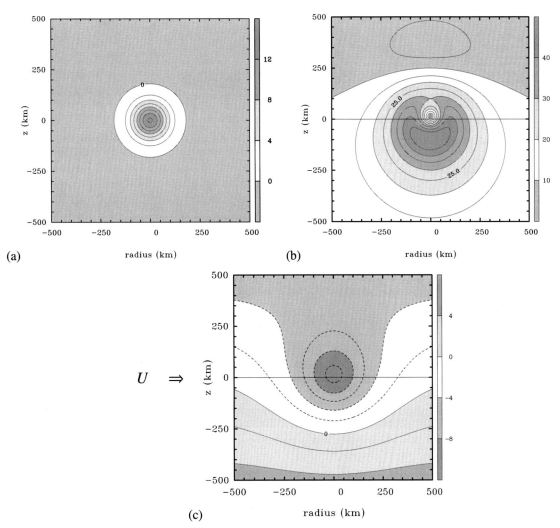

FIGURE 3.2 Contour plots of (a) relative vorticity, (b) total wind speed, and (c) streamlines, for a vortex with a symmetric relative vorticity distribution and maximum tangential wind speed of 40 m s^{-1} in a uniform westerly flow, U, with speed 10 m s^{-1} on an f-plane. The maximum tangential wind speed occurs at a radius of 100 km. Contour intervals: 4×10^{-4} s^{-1} for relative vorticity; 5 m s^{-1} for wind speeds; 4×10^{6} m^2 s^{-1} for the streamfunction. From Callaghan and Smith (1998). Republished with permission of the Australian Bureau of Meteorology.

- In Earth-relative coordinates, the total velocity field of the translating vortex is not symmetric about the axis of circulation of the vortex.
- The maximum wind speed is simply the arithmetic sum of the large scale steering flow, U, and the maximum tangential wind speed of the symmetric vortex, $V_m = (\partial \psi_s / \partial r)_{max}$.
- The maximum wind speed occurs on the right-hand-side of the vortex in the direction of motion in the northern hemisphere and on the left-hand-side in the southern hemisphere.

Fig. 3.2 shows an example of the vorticity, streamfunction and total wind speed distribution for the cyclonic, tropical-cyclone-scale vortex (maximum tangential velocity

40 m s^{-1} at a radius[2] of 100 km) translating in a uniform westerly current of 10 m s^{-1}. This tangential wind profile is shown in Fig. 3.3.

Because the vorticity field is Galilean invariant while the pressure field and streamfunction fields are not, it is advantageous to define the vortex center as the location of maximum relative vorticity and to transform the equations of motion to a coordinate system $(X, Y) = (x - Ut, y)$, whose origin is at this center (see Section 3.6.1).

In this frame of reference, the streamfunction center, i.e., the location of the minimum streamfunction) is at the point $(0, Y_{smin})$, where (see Exercise 3.1)

$$U - \Omega(Y_{smin})Y_{smin} = 0. \qquad (3.6)$$

2. While a little too large for a tropical cyclone, this radius is used in this chapter so that numerical calculations in support of the analytic theories do not require excessive spatial resolution to represent the vortex.

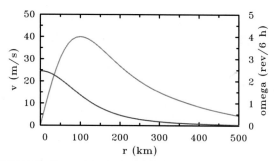

FIGURE 3.3 Tangential velocity profile $V(r)$ and angular velocity profile $\Omega(r) = V(r)/r$ for the symmetric vortex.

This point is to the left of the vorticity center in the direction of motion in the Northern Hemisphere. In the moving coordinate system, the momentum equations may be written in the form

$$\nabla p = -\rho f[\Omega(R)(-Y, X) + (U, 0)]. \qquad (3.7)$$

(see Exercise 3.1). The minimum surface pressure occurs where $|\nabla p| = 0$, which from Eq. (3.7) is at the point $(0, Y_{pmin})$ where

$$Y_{pmin}\Omega(Y_{pmin}) = U. \qquad (3.8)$$

Comparing Eqs. (3.6) and (3.8), it follows that Y_{pmin} and Y_{smin} are equal, but not zero. However, in practice they are relatively small compared with the radius of maximum tangential wind speed.

Consider, for example, the case where the inner core is in solid body rotation out to the radius r_m, of maximum tangential wind speed v_m. In this case, $\Omega = v_m/r_m$ so that $Y_{smin} = (U/v_m)r_m$. Taking typical values: $U = 5$ m s^{-1}, $v_m = 50$ m s^{-1}, $r_m = 30$ km yields $Y_{smin} = 3$ km, the latter being much smaller than r_m.

Clearly, for weak vortices (small v_m) and/or strong background flows (large U), the values of Y_{smin}/r_m and Y_{pmin}/r_m are comparatively large and the difference between the various centers may be more significant quantitatively.

3.3.2 Vortex motion on a β-plane

Another prototype problem for tropical-cyclone motion considers the evolution of an initially-symmetric barotropic vortex on a Northern Hemisphere β-plane[3] in a quiescent environment. In this problem, the initial absolute vorticity distribution, $\zeta + f$, is not symmetric about the vortex center: an air parcel at a distance y_o polewards of the vortex center will have a larger absolute vorticity than one at the same distance equatorwards of the center. Moreover, Eq. (2.49) tells us that $\zeta + f$ is conserved following air parcels.

3. The idea of a β-plane is discussed in Section 2.10.

At early times, at least, air parcels will move in circular trajectories about the vortex center. Those parcels lying initially to the west of the center will move equatorwards while those on the eastward side will move polewards. Since the planetary vorticity decreases for parcels moving equatorwards, their relative vorticity must increase as their planetary vorticity decreases. Conversely, the relative vorticity of air parcels moving polewards decreases. Thus we expect to find a growing cyclonic vorticity anomaly to the west of the vortex and an anticyclonic anomaly to the east. In the framework of Section 3.2, we regard this as part of the environment (i.e., Γ) and note that it has an associated streamfunction asymmetry with a nonzero flow across the vortex center. Thus the vortex center will move approximately with the asymmetric flow across it, approximately because as noted earlier, we are taking the vortex center to be characterized by the maximum relative vorticity and not the maximum absolute vorticity.

If the asymmetric streamfunction is approximately uniform across the vortex, the whole vortex will translate with the velocity of the flow across the vortex core and, to a first approximation, we can determine the evolution of the vorticity asymmetry by assuming that the flow remains circular relative to the moving vortex center. This assumption is strictly justified only for radii at which the tangential velocity component greatly exceeds the speed of vortex motion. However, since typical motion speeds of tropical cyclones are often not more than several m s^{-1} and typical radii of gales (wind speed = 17 m s^{-1}) may be 200 km or more (see, e.g., Section 1.2.6), the assumption should provide a useful starting point in the development of a tractable analytic theory for tropical cyclone motion. We show in the next subsections that this is, indeed, the case.

In this and subsequent sections, it is advantageous to transform the equations of motion into a frame of reference (X, Y, T) moving with the vortex center. The details of this transformation are provided in Section 3.6.1 (see especially Eq. (3.32)). Then Eq. (3.1) becomes $\partial \zeta_s/\partial T \equiv 0$ and the environmental vorticity equation (3.2) becomes

$$\frac{\partial \Gamma}{\partial T} = -\mathbf{u_s} \cdot \nabla(\Gamma + f) - \mathbf{U} \cdot \nabla \zeta_s$$
$$-\mathbf{U} \cdot \nabla(\Gamma + f) - \mathbf{c} \cdot \nabla f, \qquad (3.9)$$

where, in particular, $\mathbf{U}$ is now the flow velocity relative to the translating frame. The last term $-\mathbf{c} \cdot \nabla f$ characterizes the rate-of-change of planetary vorticity due to the meridional motion of the reference frame. In the present section, the environmental flow is assumed to be quiescent so that $\mathbf{U}$ represents only the perturbation flow associated with vorticity asymmetries induced by the moving vortex, itself.

To begin with, we adopt a heuristic approach to develop a zero-order theory and a first-order correction to it that highlights the basic physics. Later, in Section 3.3.3.4, we

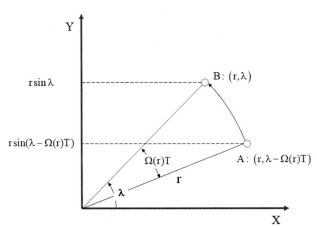

FIGURE 3.4 An air parcel moving in a circular orbit of radius r with angular velocity $\Omega(r)$ is located at the point B with polar coordinates (r, λ) at time T. At time $T = 0$ the parcel was located at point A with coordinates $(r, \lambda - \Omega(r)T)$. During this time it undergoes a meridional displacement $r[\sin\lambda - \sin(\lambda - \Omega(r)T)]$. Adapted from Smith (1991). Republished with permission of Wiley.

point out that the theory can be formalized mathematically as an expansion in a small parameter.

3.3.2.1 Zero-order theory

Consider a column of air that at time T is at the point with polar coordinate (r, λ) *located at the (moving) vortex center* (Fig. 3.4). This column would have been at the position $(r, \lambda - \Omega(r)T)$ at the initial instant, where $\Omega(r) = V(r)/r$ is the angular velocity at radius r and $V(r)$ is the tangential wind speed at that radius. The initial absolute vorticity of the parcel is $\zeta_s(r) + f_0 + \beta r \sin(\lambda - \Omega(r)T)$, while the absolute vorticity of the column at its current location is $\zeta_s(r) + \Gamma(r, \lambda, T) + f_0 + \beta r \sin\lambda$, where $\Gamma(r, \lambda, T)$ is the relative vorticity perturbation at the point (r, λ) at time T. Since the absolute vorticity of the air column is conserved, the vorticity perturbation is obtained by equating these expressions to give

$$\Gamma(r, \lambda, T) = -\beta r[\sin\lambda - \sin(\lambda - \Omega(r)T)]$$

or

$$\Gamma(r, \lambda) = \Gamma_1(r, T)\cos\lambda + \Gamma_2(r, T)\sin\lambda, \qquad (3.10)$$

where

$$\Gamma_1(r, T) = -\beta r \sin(\Omega(r)T),$$
$$\Gamma_2(r, T) = -\beta r[1 - \cos(\Omega(r)T)]. \qquad (3.11)$$

The solution for Γ in Eq. (3.10) is equivalent to that obtained by solving Eq. (3.9) with only the first term retained on the right-hand-side. In cylindrical coordinates centered at the vortex center, the approximated equation (3.9) may be written as

$$\frac{\partial\Gamma}{\partial T} + \Omega(r)\frac{\partial}{\partial\lambda}(\Gamma + f) = 0. \qquad (3.12)$$

In essence, this is a form of Eq. (3.9) when only the first term on the right-hand-side is retained (see Exercise 3.2). At this stage, **U** and **c** have not been calculated yet.

From the solution for Γ in Eq. (3.10), we can calculate the asymmetric streamfunction $\psi_a(r, \lambda, T)$ corresponding to this asymmetry using Eq. (2.52). It turns out to be easier at first to calculate the instantaneous streamfunction, ψ_a, at any time T, in non-translating Cartesian coordinates at the vortex center. In this non-translating frame, the streamfunction should satisfy the boundary condition that ψ_a is at least bounded as $r \to \infty$ so that the asymmetric perturbation velocity tends to zero at large radii. From the form of Γ, it is reasonable to expect that ψ_a will have the form:

$$\psi_a(r, \lambda) = \Psi_1(r, T)\cos\lambda + \Psi_2(r, T)\sin\lambda, \qquad (3.13)$$

whereupon, as shown in Section 3.6.2,

$$\Psi_n(r, T) = -\frac{r}{2}\int_r^\infty \Gamma_n(p, T)\,dp$$
$$-\frac{1}{2r}\int_0^r p^2\Gamma_n(p, T)\,dp \quad (n = 1, 2). \qquad (3.14)$$

The Cartesian velocity components *in the non-translating frame*, $(U_a, V_a) = (-\partial\psi_a/\partial y, \partial\psi_a/\partial x)$ are given in (r, λ) coordinates by

$$U_a = \cos\lambda\sin\lambda\left[\frac{\Psi_1}{r} - \frac{\partial\Psi_1}{\partial r}\right] - \sin^2\lambda\frac{\partial\Psi_2}{\partial r} - \cos^2\lambda\frac{\Psi_2}{r}, \qquad (3.15)$$

$$V_a = \cos^2\lambda\frac{\partial\Psi_1}{\partial r} + \sin^2\lambda\frac{\Psi_1}{r} - \cos\lambda\sin\lambda\left[\frac{\Psi_2}{r} - \frac{\partial\Psi_2}{\partial r}\right]. \qquad (3.16)$$

In order that these expressions give a unique velocity at the instantaneous origin, they must be independent of λ as $r \to 0$, in which case

$$\left.\frac{\partial\Psi_n}{\partial r}\right|_{r=0} = \lim_{r\to 0}\frac{\Psi_n}{r}, \quad (n = 1, 2).$$

Then

$$(U_a, V_a)_{r=0} = \left(-\left.\frac{\partial\Psi_2}{\partial r}\right|_{r=0}, \left.\frac{\partial\Psi_1}{\partial r}\right|_{r=0}\right), \qquad (3.17)$$

and using (3.14) it follows that

$$\left.\frac{\partial\Psi_n}{\partial r}\right|_{r=0} = -\frac{1}{2}\int_0^\infty \Gamma_n(p, T)\,dp. \qquad (3.18)$$

Assuming that the symmetric vortex moves with the velocity of the asymmetric flow across its center as discussed in Section 3.2, the vortex speed is simply

$$\mathbf{c}(t) = \left[-\left. \frac{\partial \Psi_2}{\partial r} \right|_{r=0}, \left. \frac{\partial \Psi_1}{\partial r} \right|_{r=0} \right], \qquad (3.19)$$

which can be evaluated using Eqs. (3.18) and (3.11).

The vortex track in a reference frame fixed in space, $\mathbf{x}(t) = [x(t), y(t)]$ may be obtained by integrating the equation $d\mathbf{x}/dt = \mathbf{c}(t)$. Using Eqs. (3.18) and (3.19) gives

$$\begin{bmatrix} x(t) \\ y(t) \end{bmatrix} = \begin{bmatrix} \frac{1}{2} \int_0^\infty \left\{ \int_0^t \Gamma_2(p, t') dt' \right\} dp \\ -\frac{1}{2} \int_0^\infty \left\{ \int_0^t \Gamma_1(p, t') dt' \right\} dp \end{bmatrix}. \qquad (3.20)$$

Finally, with the expressions for Γ_n from Eq. (3.11), Eq. (3.20) reduces to

$$\begin{bmatrix} x(t) \\ y(t) \end{bmatrix} = \begin{bmatrix} -\frac{1}{2}\beta \int_0^\infty r \left[t - \frac{\sin(\Omega(r)t)}{\Omega(r)} \right] dr \\ \frac{1}{2}\beta \int_0^\infty r \left[\frac{1 - \cos(\Omega(r)t)}{\Omega(r)} \right] dr \end{bmatrix}. \qquad (3.21)$$

This formula determines the vortex track in terms of the initial angular velocity profile of the vortex, $\Omega(r)$. For simplicity, we consider only cases where $\Omega(r)$ is positive at all radii to avoid potential singularities in integrals such as Eq. (3.21).

To illustrate the solutions, we use the tangential velocity profile, $V(r)$ and the corresponding profile of $\Omega(r)$ shown in Fig. 3.3. The maximum wind speed of 40 m s^{-1} occurs at a radius[4] of 100 km and the region of approximate gale force winds (> 17 m s^{-1}) extends to about 280 km. The angular velocity has a maximum at the vortex center and decreases monotonically with radius. Finally, the vortex has finite kinetic energy (see Exercise 3.3).

Fig. 3.5 shows the asymmetric vorticity field calculated from Eqs. (3.10) and (3.11) and the corresponding streamfunction field calculated from Eq. (3.13) at selected times. The value for β is taken to be 2.23×10^{-11} m^{-1} s^{-1} corresponding with a latitude of 12.5°N. The integrals involved are calculated using simple quadrature.

After five minutes the asymmetric vorticity and streamfunction fields show an east-west oriented dipole pattern. The vorticity maxima and minima occur at the radius of maximum tangential wind and there is a southerly component of the asymmetric flow across the vortex center (Fig. 3.5a). As time proceeds, the vortex asymmetry is rotated by the symmetric vortex circulation and its strength and scale increase. The rotation is characterized by the first

4. As noted earlier, while this radius is somewhat large for an intense tropical cyclone, the choice enables one to carry out a numerical solution of the problem in a relatively large domain with adequate resolution for comparison with the analytical theory.

term on the right-hand-side of Eq. (3.9). The reasons for the increase in strength and scale are discussed below. In the inner core of the vortex (typically $r < 200$ km), the asymmetry is rapidly sheared by the relatively large radial gradient of Ω (Fig. 3.3). In response to these changes in asymmetric vorticity, the streamfunction dipole strengthens and rotates also, whereupon the asymmetric flow across the vortex center increases in strength and rotates northwestwards.

The structure of the streamfunction fields in Fig. 3.5 may be qualitatively understood in terms of the membrane analogy by referring to the corresponding vorticity fields, recalling that minus the vorticity is analogous to a force acting normal to the membrane and the streamfunction is analogous to the membrane displacement. Note that, in general, areas of positive vorticity correspond with negative values of streamfunction and conversely. Note that the streamfunction fields are generally smoother than the vorticity fields, consistent with the idea that the inverse of the Laplacian operator ∇^2 acts like a smoothing operator.

3.3.2.2 Comparison with numerical solutions

The first four panels of Fig. 3.6 compare the zero-order analytical solutions with numerical solutions at 24 h. Even at this time, the asymmetric vorticity and streamfunction patterns show remarkable similarity to those diagnosed from the complete numerical solution of Eq. (2.49), which can be regarded as the control calculation. The numerical calculation was performed on a 2000 km × 2000 km domain with a 20 km grid size and for the same value of β. Further details of this calculation may be found in Smith and Ulrich (1990).

Despite the similarities between the analytically and numerically calculated vorticity patterns in Fig. 3.6, the noticeable difference in their orientation is manifest in a more westerly oriented stream flow across the vortex center in the analytical solution and this is reflected in differences in the vortex tracks shown in Fig. 3.7. It follows that the analytical solution gives a track that is too far westward, but the average speed of motion is comparable with, although a fraction smaller than in the control case for this entire period. Even so, it is apparent that even this zero-order analytical solution captures much of the dynamics of the full numerical solution.

3.3.2.3 A first-order correction

The analytical theory can be considerably improved by taking account of the contribution to the vorticity asymmetry, Γ_a, by the relative advection of symmetric vortex vorticity, ζ_s. This contribution is represented by the term $-\mathbf{U} \cdot \nabla \zeta_s$ in Eq. (3.9) (the second term on the right-hand-side) where $\mathbf{U}$ is now just $\mathbf{U_a} - \mathbf{c}$. Here, the components of $\mathbf{U_a}$, U_a and V_a, are estimated using the zero-order solution obtained in

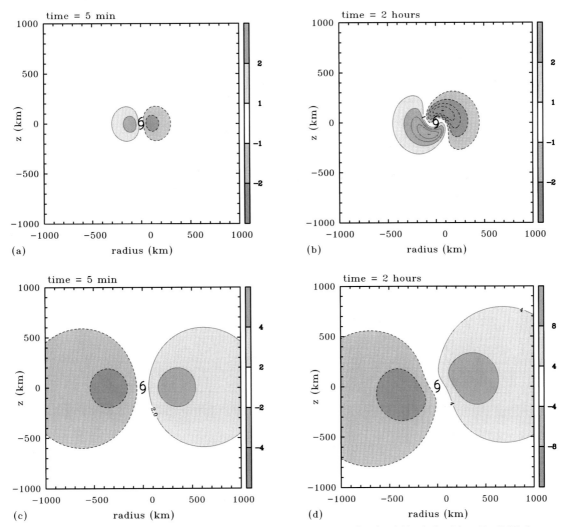

FIGURE 3.5 Asymmetric vorticity field calculated from Eq. (3.10) (top panels) and streamfunction field calculated from Eq. (3.13) (bottom panels) at selected times: (left panels) 5 min, (right panels) 2 h. The corresponding plots for 24 h are shown in Fig. 3.6. Contour intervals for ζ_a are: 1×10^{-7} s^{-1} in (a), and 1×10^{-6} s^{-1} in (b). Contour intervals for ψ_a are: 2×10^3 m^2 s^{-1} in (c), and 4×10^4 m^2 s^{-1} in (d). Adapted from Smith and Ulrich (1990). Republished with permission of the American Meteorological Society.

Section 3.3.2.1 and **c** is the vortex translation velocity given by Eq. (3.19).

Again, with the assumption that air parcels move in circular orbits about the vortex center while conserving their absolute vorticity, Γ_{ac} satisfies the equation:

$$\frac{\partial \Gamma_{ac}}{\partial T} + \Omega(r)\frac{\partial \Gamma_{ac}}{\partial \lambda} = -(\mathbf{U_a} - \mathbf{c}) \cdot \nabla \zeta_s, \qquad (3.22)$$

where the components of $\mathbf{U_a}$ are given by Eqs. (3.15) and (3.16). Further details of the solution for Γ_{ac} are given in Section 3.6.3.

The vorticity correction Γ_{ac} is exemplified by the structure at 24 h shown in Fig. 3.8. This structure consists of a pair of azimuthal wavenumber-one gyres, in which the maximum amplitude of the asymmetry is located about 600 km from the vortex center. At this radius, the tangential wind

speed of the vortex is barely a tenth of its maximum value (Fig. 3.3).

The sum of the zero-order and first-order vorticity asymmetries at 24 h is shown in Fig. 3.6e and the corresponding streamfunction asymmetry is shown in Fig. 3.6f. Comparing these panels of Fig. 3.6 with the corresponding panels (c) and (d) shows a much improved agreement between the analytical solution and the numerical solution. Now, the corrected analytically-calculated track is much closer to that calculated numerically as shown in Fig. 3.7 (compare the tracks A1 and N).

3.3.2.4 Vortex-relative flow

Fig. 3.9 shows the flow relative to the moving vortex at 24 h, based on the improved analytic theory discussed above. The relative flow at large radial distances is 2.3 m s^{-1} from

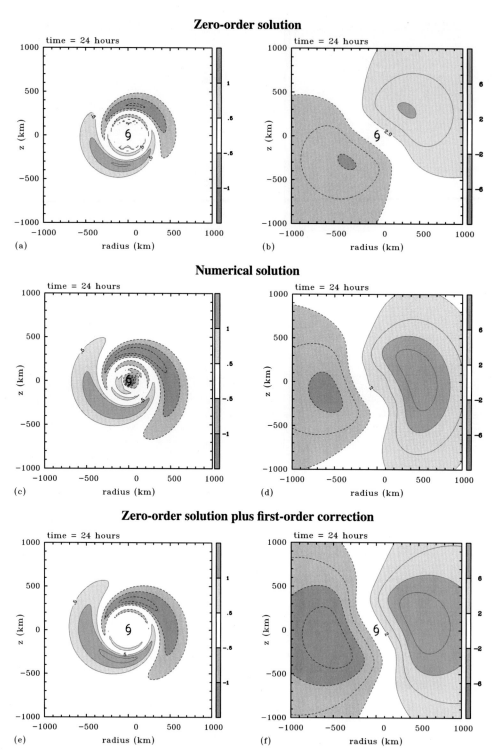

FIGURE 3.6 Comparison of the analytically-computed asymmetric vorticity (left panels) and streamfunction fields (right panels) with those for the corresponding numerical solutions at 24 h (middle row). Only the inner part of the numerical domain, centered on the vortex center, is shown (the calculations were carried out on a 4000 km × 4000 km domain with impervious boundary conditions at the north and south boundaries and periodic boundary conditions at zonal boundaries). Panel (a) and (b) show the zero-order vorticity and streamfunction fields, while panels (e) and (f) show the corresponding fields with next-order correction added as discussed in the text. Contour intervals for vorticity 5×10^{-6} s^{-1} and for streamfunction 2×10^{5} m^{2} s^{-1}. The tropical cyclone symbol represents the vortex center. Adapted from Smith and Ulrich (1990). Republished with permission of the American Meteorological Society.

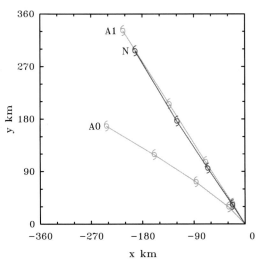

FIGURE 3.7 Comparison of the analytically calculated vortex track from the zero-order theory (denoted by A0) compared with that for the corresponding numerical solution (denoted by N). The track by A1 is the analytically corrected track referred to in the text. From Smith and Ulrich (1990). Republished with permission of the American Meteorological Society.

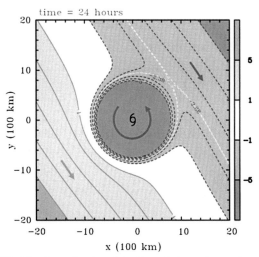

FIGURE 3.9 Streamfunction of the flow relative to the moving vortex based on the zero- and first-order analytical calculation. Contour interval 1×10^6 m^2 s^{-1}. The white dashed contours have the values -2.38×10^6 m^2 s^{-1} and -2.36×10^6 m^2 s^{-1}. These show approximately the location of the dividing streamline that bounds the recirculating flow on the northeastern side of the vortex. Adapted from Smith and Ulrich (1990). Republished with permission of the American Meteorological Society.

north-northeast. Because this flow is relatively weak compared with the maximum wind speed of the vortex (40 m s^{-1}), there is a broad region of closed relative circulation extending out to a radius on the order of 700 km. In real storms, such regions are important when they extend over an appreciable depth as they provide protection from the intrusion of dry air from the storm environment, allowing for deep convection to progressively moisten the region (e.g., Figs. 1.16 and 1.35; see also Chapter 8 and Section 16.11).

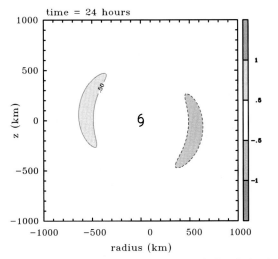

FIGURE 3.8 The first-order correction to the analytically-calculated vorticity field, Γ_{ac}, at 24 h. Contour intervals 5×10^{-6} s^{-1}. The tropical cyclone symbol represents the vortex center. From Smith and Ulrich (1990). Republished with permission of the American Meteorological Society.

3.3.2.5 Evolution of the β-gyres

As time proceeds, the strength of the vorticity and streamfunction asymmetries and the radii at which the maxima occur continue to increase until about 60 h when the radius of the maximum stabilizes. This increase in the strength and scale of the gyres in the model is easy to understand if we assume that air parcels move in circular orbits around the moving vortex center (Fig. 3.10). As shown in Subsection 3.3.2.1, the change in relative vorticity of an air parcel circulating around the vortex is equal to its displacement in the direction of the absolute vorticity gradient times the magnitude of that gradient. For an air parcel at radius r the maximum possible meridional displacement is $2r$, which limits the size of the maximum asymmetry at this radius. The time for this displacement to be achieved is $\pi/\Omega(r)$, where $\Omega(r)$ is the angular velocity of a fluid parcel at radius r. Since Ω is largest at small radii, air parcels there attain their maximum displacement relatively quickly, while air parcels at larger radii require more time. At a given time, the maximum displacement of any air parcel is proportional to $r\Omega(r)$, i.e., $V(r)$, so that at early times, this maximum displacement occurs at the radius r_{max} of the maximum tangential wind (Fig. 3.10a). Here, early times refer to times less than $\pi/\Omega(r_{max})$.

Although air parcels at larger radii are rotating more slowly, given sufficient time they have the ability to achieve much larger meridional displacements than those at small radii and this is exactly what happens (Fig. 3.10b). Ultimately, of course, if $\Omega(r)$ decreases monotonically to zero, there is a finite radius beyond which the tangential wind speed is less than the translation speed of the vortex. As

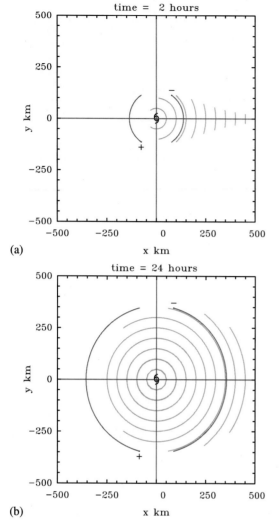

(a)

(b)

FIGURE 3.10 Approximate trajectories of air parcels which, for a given radius, give the maximum meridional displacement and therefore the maximum asymmetric vorticity contribution at that radius. The figures refer to the case of motion of an initially-symmetric vortex on a β-plane with zero basic flow at (a) 2 hours, (b) 24 hours. The particles are assumed to follow circular paths about the vortex center with angular velocity $\Omega(r)$, where Ω decreases monotonically with radius r. Solid lines denote trajectories at 50 km radial intervals. Curved arcs marked '+' and '-' represent the trajectories giving the overall largest meridional displacements and therefore the overall axisymmetric vorticity maxima and minima, respectively. These maxima (+) and minima (-) occur at the positive and negative ends of the relevant arcs. From Smith and Ulrich (1993). Republished with permission of Wiley.

the maximum in the asymmetry approaches this radius, the effect of the vortex motion on the asymmetry can no longer be ignored. This feature is illustrated in Fig. 3.11, which shows the trajectories of selected particles relative to the vortex center calculated from the numerical solution described earlier. As the coherent part of the wavenumber-one asymmetries reaches the radius where the tangential wind speed is not appreciably greater than the vortex translation speed, the analytical theory must become progressively in-

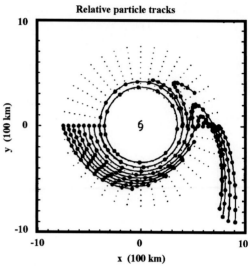

FIGURE 3.11 Selected particle trajectories relative to the moving vortex in the numerical calculation described in the text. The trajectories start along the x-axis at the initial time. Thick dots indicate 6 hourly positions. Thin dots mark circular paths at the initial radii of particles. From Smith and Ulrich (1990). Republished with permission of the American Meteorological Society.

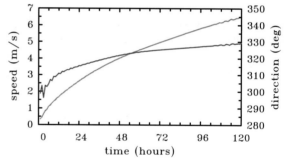

FIGURE 3.12 Vortex translation speed (red (mid gray in print version) curve) and direction of motion (blue (dark gray in print version) curve) as a function of time calculated from the zero-order theory encapsulated by Eqs. (3.19) and (3.21).

accurate, but the comparison of the analytically-calculated tracks with those from a numerical solution in Fig. 3.7 indicates that the analytical theory remains reasonably accurate for many days, at least for the outer size of vortex used in the calculation. Based on the foregoing ideas, one would expect the agreement to deteriorate more rapidly in time as the outer size of the vortex decreases, a result that was verified by Smith and Ulrich (1990).

3.3.2.6 Vortex translation velocity

The growth in size of the β-gyres and the increase in their strength leads to a progressive increase in the speed of vortex translation, but the rate of increase steadily decreases (Fig. 3.12) as the tangential component of flow at the radius of maximum vorticity asymmetry becomes comparable to the poleward component of vortex motion. The rotation

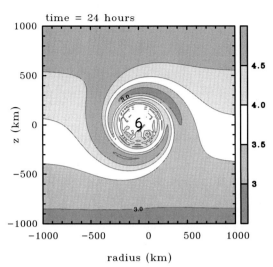

time = 24 hours

FIGURE 3.13 Analytically calculated absolute vorticity distribution at 24 h corresponding with the vorticity asymmetry in Fig. 3.7a.

of the vorticity asymmetry is accompanied by a change in the flow orientation between the streamfunction gyres and hence in the flow across the vortex center. As shown also in Fig. 3.12, these changes lead to a change in the direction of vortex movement from towards about 300 deg. at the initial time to about 330 deg. during the first five days of integration.

3.3.2.7 Vorticity filamentation

Since the absolute vorticity is the conserved quantity in the barotropic flow problem, it is instructive to examine the evolution of the isolines of this quantity as the flow evolves. At the initial time, the contours are very close to circular near the vortex center and are oriented zonally far from the center. The pattern after 24 h, shown in Fig. 3.13, illustrates how contours are progressively wound around the vortex with those nearest the center drawn out into long filaments. This filamentation process is associated with the strong differential rotation associated with the rapidly decreasing angular velocity with radius (see Fig. 3.3).

In reality, the strong gradients associated with the small-scale filamentary vorticity structures would be removed by diffusive processes. The filamentation is comparatively slow at larger radial distances so that coherent vorticity asymmetries occur outside the rapidly-rotating and strongly-sheared core. One consequence of these processes is that the larger-scale asymmetries have the main effect on the vortex motion. On account of the filamentation process, there is a natural tendency for vortices to axisymmetrize disturbances in their cores.

The effects of filamentation can be understood also in terms of the membrane analogy. Intuition would suggest that the largest displacements of a membrane would occur in regions where the normal force acting on the membrane

is of a single sign. In contrast, in regions where the force oscillates in sign over short distances, the membrane displacement would be comparatively smaller. Using this analogy explains why the largest contributions to the streamfunction come from coherent regions of like signed vorticity. This explains why it is the larger-scale asymmetries that have the main effect on the vortex motion.

The axisymmetrization process in rapidly-rotating vortices is analyzed in more detail in Section 4.1.1.

3.3.3 The effects of an environmental flow

The analytical theory can be extended also to vortex motion in an environmental flow. Using the partitioning scheme of Section 3.2, we write the horizontal velocity vector $\mathbf{u} = \mathbf{u_s} + \mathbf{U}$, where $\mathbf{u_s}$ denotes the velocity vector of an axisymmetric vortex that translates unchanged in structure with vector velocity $\mathbf{c}$ and $\mathbf{U}$ is the velocity field of the vortex environment, which includes any flow asymmetries generated by the vortex.

Let $\zeta_s = \hat{\mathbf{k}} \cdot \nabla \wedge \mathbf{u_s}$ and $\Gamma = \hat{\mathbf{k}} \cdot \nabla \wedge \mathbf{U}$, where $\hat{\mathbf{k}}$ is the unit vector in the vertical. Then, as in Section 3.2, $\mathbf{u_s} \cdot \nabla \zeta_s = 0$ and in a reference frame moving with the vortex, $\partial \zeta_s / \partial T = 0$ and Γ satisfies the equation

$$\frac{\partial \Gamma}{\partial T} = -\mathbf{u_s} \cdot \nabla (\Gamma + f) - \mathbf{U} \cdot \nabla \zeta_s$$
$$- \mathbf{U} \cdot \nabla (\Gamma + f) - \mathbf{c} \cdot \nabla f. \quad (3.23)$$

This is just a repeat of Eq. (3.9). Now, suppose that, in the absence of the vortex, the environment would remain steady with velocity field $\mathbf{U_i}$ and relative vorticity field Γ_i. Examples of such flows are presented below. Then we may write $\mathbf{U} = \mathbf{U_i} + \mathbf{U_{as}}$, where $\mathbf{U_{as}}$ is the velocity field induced by vortex. We refer to $\mathbf{U_i}$ below as the basic flow and $\mathbf{U_{as}}$ as the perturbation flow. Then, following the same approximate procedure as in Section 3.3.2.3, Eq. (3.23) may be written as one for the local tendency of asymmetric vorticity Γ_{as}:

$$\frac{\partial \Gamma_{as}}{\partial T} + \mathbf{u_s} \cdot \nabla \Gamma_{as} = \underbrace{-\mathbf{u_s} \cdot \nabla (\Gamma_i + f)}_{\partial \Gamma_1 / \partial T} \underbrace{-\mathbf{U_{as}} \cdot \nabla \zeta_s}_{\partial \Gamma_2 / \partial T}$$
$$\underbrace{-\mathbf{U_i} \cdot \nabla \zeta_s}_{\partial \Gamma_3 / \partial T} \underbrace{-(\mathbf{U_i} + \mathbf{U_{as}}) \cdot \nabla (\Gamma_i + \Gamma_{as} + f)}_{\partial \Gamma_4 / \partial T} \underbrace{-\mathbf{c} \cdot \nabla f}_{\partial \Gamma_5 / \partial T}.$$
$$(3.24)$$

Here, the operator on the left-hand side may be written as $\partial \Gamma_{as} / \partial T + \Omega(r) \partial \Gamma_{as} / \partial \lambda$ as in Eqs. (3.12) and (3.22), and $\Gamma_n (n = 1 - 5)$ represents the contribution to Γ arising from $\partial \Gamma_n / \partial T$. Moreover, as in Section 3.3.2.3, $\mathbf{U_{as}}$ and Γ_{as} on the right-hand side are estimated from the zero-order theory in which these terms are set to zero. The five terms on the right-hand-side of Eq. (3.24) may be interpreted as follows:

- $\partial\Gamma_1/\partial T$ is the asymmetric vorticity tendency associated with the advection of the initial absolute vorticity of the basic flow by the symmetric vortex circulation;
- $\partial\Gamma_2/\partial T$ is the asymmetric vorticity tendency associated with the perturbation flow acting on the symmetric vortex;
- $\partial\Gamma_3/\partial T$ is the asymmetric vorticity tendency associated with the relative initial flow acting on the symmetric vortex;
- $\partial\Gamma_4/\partial T$ is the asymmetric vorticity tendency associated with the advection of the absolute vorticity of the basic flow plus perturbation flow by the total environmental flow, the relative initial flow plus the perturbation flow.
- $\partial\Gamma_5/\partial T$ is the asymmetric vorticity tendency associated with the poleward component of the vortex motion.

Eq. (3.24) may be solved in the same way as Eq. (3.22) when $\mathbf{u_s}$ and $\mathbf{U_i}$ are specified. At each time, the vortex translation speed may be obtained in the same way as in Section 3.3.2.1.

3.3.3.1 Case I: linear zonal wind profile

For a zonal flow with uniform shear, $\mathbf{U_i} = U'y\hat{\mathbf{i}}$, where U' is a constant and $\Gamma_i = -U'$, a constant, where $\hat{\mathbf{i}}$ is a unit vector in the zonal direction. Then Eq. (3.24) becomes

$$\frac{\partial\Gamma_{as}}{\partial T} + \Omega(r)\frac{\partial\Gamma_{as}}{\partial\lambda} = -\mathbf{u_s}\cdot\nabla f - \mathbf{U_{as}}\cdot\nabla\zeta_s$$
$$- \mathbf{U_{as}}\cdot\nabla(\Gamma_{as}+f)\underbrace{-U'Y\hat{\mathbf{i}}\cdot\nabla\zeta_s}_{\partial\Gamma_3/\partial T} - U'Y\hat{\mathbf{i}}\cdot\nabla\Gamma_{as} - \mathbf{c}\cdot\nabla f,$$

$$(3.25)$$

noting that $\hat{\mathbf{i}}\cdot\nabla f = 0$. The effects of uniform shear are contained in the last two terms of this equation, the dominant term, at least at early times being the first of these, which represents the vorticity tendency associated with the shearing of the vortex by the environmental flow. Thus, the main difference compared to the calculation in Section 3.3.2 is the emergence of an azimuthal wavenumber-2 vorticity asymmetry from this term.

This result is easy to understand by reference to Fig. 3.15. The vorticity gradient of the symmetric vortex is negative inside a radius of 255 km (say r_o) and positive outside this radius (Fig. 3.14). Therefore, $\partial\zeta_s/\partial X$ is positive for $X > 0$ and $r > r_o$ and negative for $X < 0$ and $r < r_o$. If $U = U'Y$, $-U'Y\hat{\mathbf{i}}\cdot\nabla\zeta_s = -U'Y\partial\zeta_s/\partial X$, which is negative in the first and third quadrants for $r > r_o$ and positive in the second and fourth quadrants (Fig. 3.15). For $r < r_o$, the signs are reversed.

Fig. 3.16 shows the calculation of Γ_3 obtained by solving Eq. (3.25) at 24 h when $U' = 5$ m s^{-1} per 1000 km with just the term $\partial\Gamma/\partial T$ on the right hand side. This vorticity asymmetry has extrema at two radii, one at a small radius ($< r_o$) and a one at a larger radius ($> r_o$), where the magnitude of the asymmetry is largest (1.1×10^{-5} s^{-1} at 24

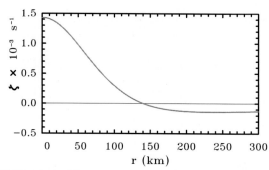

FIGURE 3.14 Radial profile of vortex vorticity, $\zeta(r)$, corresponding with the tangential wind profile in Fig. 3.3. From Smith (1991). Republished with permission of Wiley.

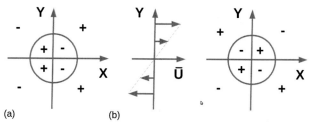

FIGURE 3.15 Schematic depiction of the azimuthal wavenumber-2 vorticity tendency arising from the term $-\mathbf{U}\cdot\nabla\zeta_s = -U'Y\partial\zeta_s/\partial X$ in Eq. (3.25) in the case of a uniform zonal shear $U = U'Y$. (a) shows the sign of the vorticity gradient $\partial\zeta_s/\partial X$ in each quadrant for $0 < r < r_o$ and $r_o < r$ where r_o is the radius at which the vorticity gradient $d\zeta_s/dr$ changes sign (see Fig. 3.15). (b) shows the vorticity tendency $-U\partial\zeta_s/\partial X$ in the eight regions. From Smith (1991). Republished with permission of Wiley.

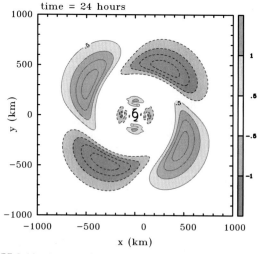

FIGURE 3.16 Asymmetric vorticity contribution for the case of a uniform zonal shear with $U' = 5$ m s^{-1} per 1000 km calculated as discussed in the text. Contour interval is 5×10^{-6} s^{-1}. Dashed lines indicate negative values. The vortex center is marked by a cyclone symbol. From Smith (1991). Republished with permission of Wiley.

h). Note that all azimuthal wavenumber asymmetries other than wavenumber-1 have zero flow at the vortex center and therefore have no effect on the vortex motion.

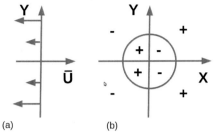

(a)   (b)

FIGURE 3.17 Schematic depiction of the wavenumber-1 vorticity tendency arising from the term $-\mathbf{U} \cdot \nabla \zeta_s = -U \partial \zeta_s / \partial X$ in the case of a parabolic basic shear flow $U = \frac{1}{2} U'' Y^2$, with $U'' < 0$. (a) shows the profile $U(Y)$ and (b) shows the vorticity tendency $-U \partial \zeta_s / \partial X$, in the eight regions defined in Fig. 3.15. The sign of $\partial \zeta_s / \partial X$ in these regions is shown in Fig. 3.15a. From Smith (1991). Republished with permission of Wiley.

In the case of uniform shear, there is a small wavenumber-1 contribution to the flow asymmetry resulting from the penultimate term in Eq. (3.25) and this term influences the vortex track (see Subsection 3.3.3.3).

3.3.3.2 Case II: parabolic zonal wind profile

We consider now the case of a parabolic velocity profile (i.e., linear shear) in which U' is taken to be zero, but U'' and therefore $\partial \Gamma / \partial Y = -U''$ are nonzero. Linear shear has two particularly important effects that lead to a wavenumber-1 asymmetry, thereby affecting the vortex track at zero order. The first is characterized by the contribution to the absolute-vorticity gradient of the basic flow (the first term on the right-hand-side of Eq. (3.24), which directly affects the zero-order vorticity asymmetry, Γ_1. The second is associated with the distortion of the vortex vorticity as depicted in Fig. 3.17 and represented mathematically by Γ_3, which originates from the third term on the right-hand-side of Eq. (3.24). For further details of the asymmetries, the reader is referred to the study by Smith (1991).

3.3.3.3 Vortex tracks

Fig. 3.18 shows the vortex tracks calculated from the analytical theory with corresponding numerical calculations. The tracks for three calculations with uniform shear are shown in panel (a): one in which the shear is positive with $U' = 5$ m s^{-1} per 1000 km (designated LP1); one where it is negative, but with the same magnitude (designated LN); and one where it is positive, but with double the shear in LP1 (designated LP2). These are compared with the tracks for the case with zero basic flow (designated ZBF).

There is broad agreement between the analytical and numerical calculations in all cases, indicating that the analytical theory captures the essence of the dynamics involving shear, even though the analytically-calculated motion is a little too fast. The eastward or westward displacement in the cases with zonal shear are in accordance with expectations that the vortex is advected by the basic flow and the different meridional displacements are attributed to the wavenumber-

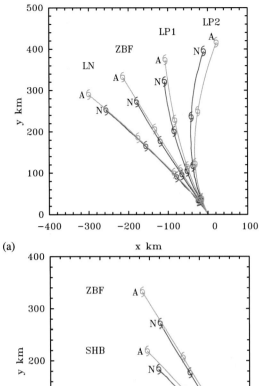

(a)

(b)

FIGURE 3.18 Analytically calculated vortex track (denoted by A) compared with the corresponding numerical solution (denoted by N): (a) uniform shear flow cases and (b) linear shear flow cases. Each panel includes the analytically and numerically calculated track for the case of zero basic; flow (denoted ZBF). Cyclone symbols mark the vortex position at 12-h intervals. (See text for explanation of other letters.) From Smith (1991). Republished with permission of Wiley.

1 asymmetry arising from the tendency term $-U'Y\hat{\mathbf{i}} \cdot \nabla \Gamma_{as}$ in Eq. (3.25) (see Smith, 1991).

Panel (b) of Fig. 3.18 shows a similar comparison for two cases of a parabolic shear flow: SNB with $U'' = -\beta_o$ and $\beta = 0$ and SHB $U'' = -\frac{1}{2}\beta_o$ and $\beta = \frac{1}{2}\beta_o$. Again, the tracks are compared with the case of zero basic flow (ZBF) with $\beta = \beta_o$. Here β_o is the standard value of β used earlier. These three calculations have the same absolute vorticity gradient, β_o, but the relative contribution to it from U'' and β is different. Note that the poleward displacement is reduced as U'' increases in magnitude. This effect can be attributed to the wavenumber-one asymmetry arising from the tendency term $\partial \Gamma_4 / \partial T$ in Eq. (3.24).

In the calculations just described, U'' is negative and the relative environmental flow is everywhere easterly

(Fig. 3.17). A case in which U'' is everywhere positive with easterly flow in the south and westerly flow to the north is examined by Ulrich and Smith (1991), albeit on the basis of a numerical simulation and its analysis (see their Figs. 5-7 and accompanying discussion).

3.3.3.4 Higher-order theory and its limitations

The analytical theory described above can be extended to account for higher-order corrections to the vorticity asymmetry. Smith and Weber (1993) showed that the theory can be derived from a formal power-series expansion of variables representing the vortex asymmetry in terms of a single nondimensional parameter $\epsilon = BL^2/U$, where B is the meridional gradient of basic state absolute vorticity and L and U are length and velocity scales, respectively, for the outer part of the vortex. The derivation requires *inter alia* that $\epsilon \ll 1$ and that the magnitude of the basic flow scales with ϵ. The corrections involve higher-order azimuthal wavenumber asymmetries. Mathematically an azimuthal wavenumber-n vorticity asymmetry has the form

$$\Gamma(r, \lambda, n) = \Gamma_1(r, T)\cos(n\lambda) + \Gamma_2(r, T)\sin(n\lambda)$$
$$(n = 1, 2, \ldots),$$

which may be written

$$\Gamma(r, \lambda, n) = \Gamma_n(r, T)\cos(n\lambda + \alpha), \qquad (3.26)$$

with α being a phase shift function. The associated streamfunction asymmetry has a similar form:

$$\psi_a(r, \lambda, n) = \psi_n(r, T)\cos(n\lambda + \alpha),$$

where (see Section 3.6.2)

$$\psi_n = -\frac{1}{2n}\left[r^n \int_r^\infty p^{1-n}\Gamma_n(p, T)dp \right.$$
$$\left. + r^{-n}\int_0^r p^{1+n}\Gamma_n(p, T)dp \right], \quad (n \neq 0). \quad (3.27)$$

For the vortex shown in Fig. 3.4, the tracks obtained from the higher-order analytical theory show improved agreement with those obtained from a numerical solution of the problem to at least 72 h, indicating that the theory captures the essential features of the dynamics of vortex motion on the β-plane. The reader is referred to Smith and Weber for additional details. The higher-order theory is still limited by the issues discussed in Subsection 3.3.2.5 and can be expected to break down sooner for vortices in which the vortex velocity profile decays rapidly with radius.

3.3.4 More general environmental flows

Insightful thought experiments can be devised to provide examples of vortex motion in non-zonal environmental flows, possibly with zonal as well as meridional gradients of environmental absolute vorticity. Two such examples are discussed below.

3.3.4.1 Stationary Rossby wave

An example of a steady environmental flow that is not purely zonal and has a variable absolute vorticity gradient across it is a stationary finite amplitude Rossby wave. Vortex motion in such flows was examined by Ulrich and Smith (1991), Smith and Ulrich (1993), and Kraus et al. (1995). As an example, Fig. 3.19 shows the track of a symmetric vortex within a Rossby wave with meridional wavenumber one. The vortex is located initially at the point $(-500 \text{ km}, -1000 \text{ km})$, where the environmental flow has an easterly component. As in the simpler flow configurations discussed earlier in this chapter, the motion is governed by Eq. (3.24) and is due *inter alia* to advection by the background flow and to a drift in the direction and to the left of the local absolute vorticity gradient.

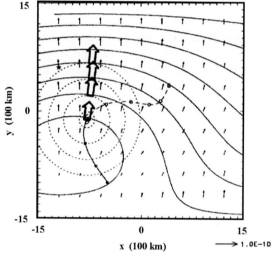

FIGURE 3.19 Absolute vorticity gradient (arrows) and streamlines for the calculation in Subsection 3.3.4.1: (a) imposed environment and vortex track. The vortex position at 72 hours is marked by a cyclone symbol and at 24-hour intervals thereto by solid dots. The subsequent track is shown by the dashed line and open circles. The thick arrows show the relative magnitude and direction of the imposed environment averaged over annular regions 1-3, 3-5, and 5-7 degrees latitude compared with the analytic value at the vortex center. The three annular regions are denoted by dashed circles. For clarity, the thick arrows are scaled differently from the smaller ones, those at the vortex center denoting the same vector. From Smith and Ulrich (1993). Republished with permission of Wiley.

Detailed analyses of the vortex motion in such more general environmental flows can be used to investigate possible relationships between the motion, the flow in which it is embedded and the local absolute vorticity gradient. They can be used also to investigate strategies for estimating the environmental flow from observations, by, for example, averaging the flow surrounding a vortex over annular regions centered on the vortex. Such annuli are highlighted by dashed circles in Fig. 3.19. Averaging strategies are discussed in more depth in Section 3.4.

3.3.4.2 Flow with strong horizontal deformation

Another idealized flow configuration that is pertinent to fully understanding tropical cyclone motion in a general environmental flow is that provided by a horizontal deformation flow field (Kraus et al., 1995). Such a flow field is exemplified by a stationary Rossby wave with meridional wavenumber two (Fig. 3.20). In this case, the x-axis to the left of the two regions of closed streamlines on opposite sides of the x-axis and centered at $x = 1000$ km[5] is an axis of contraction. Clearly, if the vortex motion were governed solely by advection, one could imagine that this axis would be a sensitive location for determining the subsequent vortex track. For a vortex located initially just above this axis, the motion would be polewards around the local high, whereas for one located just below this axis, the motion would be equatorwards around the region of closed streamfunction.

In reality, a small error in locating the vortex center in such a situation could lead subsequently to a large track error. However, in the Rossby wave there is an additional effect to consider because the non-zero environmental vorticity gradient would lead to a drift of the vortex relative to the deformation flow.

Kraus et al. (1995) applied the analytic theory of vortex motion sketched out earlier to study vortex motion in this Rossby wave. The study was complemented by corresponding numerical solutions of the problem.

Fig. 3.20 shows the vortex tracks calculated numerically and analytically for nine initial vortex positions. In broad terms, the agreement is good, evidence that the theory captures the essential processes involved in the motion. For the initial vortex position (500 km, 500 km), the difference between the two tracks is as small as 25 km after 48 hours. A similar result is obtained in the case (0 km, −500 km) (41 km), whereas for the position (−500 km, 500 km), the worst case, the difference between the vortex positions is 75 km after 24 hours and 296 km after 48 hours. The comparison suggests that the analytical theory provides an accurate representation of the vortex motion for at least 24 hours and can be used therefore to interpret the numerical calculations.

As in the simpler flows considered herein, the vortex track is determined primarily by two effects, both of which are stronger west of the center of deformation than in the east. Firstly, the symmetric vortex is simply advected by the initially imposed large-scale flow. This advection depends strongly on the position of the vortex relative to the axis of contraction of the deformation field. For example,

5. This streamfunction dipole has an anticyclonic gyre to the north of the x-axis and a cyclonic gyre to the south of this axis.

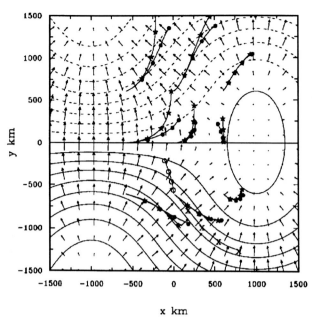

FIGURE 3.20 Vortex tracks calculated analytically (symbols ★) and numerically (symbols ●), where the symbols denote the vortex center position every 12 hours. Solid (positive) and dashed lines (negative) are streamfunctions of the planetary-wave field; the contour interval is 1×10^6 m² s⁻¹. Small arrows represent the absolute vorticity gradient $\nabla(\Gamma + f)$, where the length of the vector is proportional to $|\nabla(\Gamma + f)|$. Additional tracks are shown starting at the initial vortex position (0 km, −500 km). These are calculated from the components of the vortex motion which are associated with the developing vorticity asymmetries only (symbol O) and assuming advection by the flow associated with the deformation field only (symbol X). Note that the x-axis lies north of and not at the Equator. From Kraus et al. (1995). Republished with permission of the Deutsche Meteorologische Gesellschaft.

the eastward component of the large-scale flow reaches a strength of 8.33 m s⁻¹ 1600 km west of the deformation field center, but is only 0.24 m s⁻¹ 600 km east of that point. Secondly, the initially symmetric vortex develops an asymmetry in the presence, *inter alia*, of a basic-state absolute vorticity gradient. The flow across the vortex center associated with the azimuthal wavenumber-one component of this asymmetry gives an additional contribution to the motion. The analytical theory shows that the main contribution to this component is proportional to $\nabla(\Gamma + f)$, whereupon the asymmetry is largest in the region where $|\nabla(\Gamma + f)|$ is largest, i.e., west of the center of deformation. For all initial vortex positions shown in Fig. 3.20, these two effects partly cancel each other, so that the speed $|\mathbf{c}|$ of the vortex center relative to the stationary Rossby wave is smaller than the advection speed at the vortex center. Fig. 3.20 shows the two effects separately for the initial vortex position (0 km, −500 km).

As noted above, if the vortex motion were determined by advection only, the line of contraction of the deformation field could be expected to be a line along which the subsequent vortex motion has a certain sensitivity. Vor-

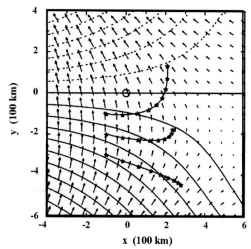

FIGURE 3.21 Vortex tracks calculated numerically with the initial vortex positions 100 km west and 100 km, 200 km, and 300 km south of the axis of contraction of the deformation field. The star symbols (★) denote the vortex center position every 12 hours. Solid (positive) and dashed lines (negative) are streamfunctions of the planetary-wave field; the contour interval is 1×10^6 m^2 s^{-1}. Small arrows represent the absolute vorticity gradient $\nabla(\Gamma + f)$, where the length of the vector is proportional to $|\nabla(\Gamma + f)|$. From Kraus et al. (1995). Republished with permission of the Deutsche Meteorologische Gesellschaft.

tices initialized a fraction to the north of this line would subsequently track polewards, while those initially a fraction to the south would subsequently track equatorwards. It is evident that the existence of such lines (or regions) of sensitivity vis-á-vis the subsequent vortex motion have important implications for forecasting tropical cyclone motion because the initial positions are always inaccurate to some degree. In the present model calculations, the line of sensitivity is shifted equatorwards on account of the contribution to the vortex motion from the vortex asymmetry. This feature is illustrated in Fig. 3.21, which shows the tracks of three vortices initialized 100 km, 200 km, and 300 km to the south of the axis of contraction of the deformation field. Note that the first two eventually track polewards, while only the third continues to track equatorwards.

3.4   Observations of the β-gyres

There have been numerous attempts to detect the β-gyres or their effects in observational data (e.g., Carr and Elsberry, 1990; Franklin, 1990; Franklin et al., 1996; Glatz and Smith, 1996; Kim et al., 2009), but there are numerous difficulties in doing this. For one thing, the real atmosphere is not barotropic, the amplitude of the gyres is relatively weak, the environmental flow is not uniform and the data are mostly too sparse (Reeder et al., 1991). In the 1980's, the Hurricane Research Division of the NOAA/Atlantic Oceanographical and Meteorological Laboratory conducted a series of "synoptic flow" experiments designed to collect wind and thermodynamic data in the environment of

tropical cyclones in the Atlantic basin to assess their impact on track forecasts. These data provided an opportunity to assess the barotropic theory of tropical cyclone motion.

One particularly well documented storm was Hurricane Josephine (1984), the data for which were subsequently analyzed by Franklin (1990) and Glatz and Smith (1996). The data coverage obtained during one day of observations for this storm is shown in Fig. 3.22. One issue that arises immediately in trying to appraise the theory is the partitioning problem and, in particular, how to extract the "vortex environment" from the data. Simply removing an azimuthally-averaged circulation about the storm center to define "the vortex" and treating "the environment" as the residual would still leave any vortex-induced asymmetries such as the β-gyres as part of this environment. This decomposition is different from that used to develop the theory in Section 3.3.3 as there, "the environment" and the vortex-induced asymmetries are clearly separated, the environment being prescribed at the initial time. Other difficulties in isolating the β-gyres from observations are that, in reality, the horizontal flow structure varies with height and the vorticity and flow structures associated with the gyres are expected to be relatively weak.

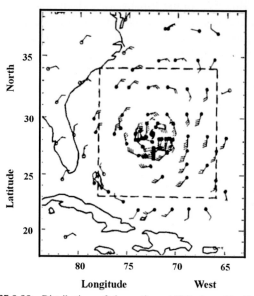

FIGURE 3.22 Distributions of observations at 700 mb used by Franklin (1990) for the objective analyses of Hurricane Josephine on 11 October 1984. The boundaries of an inner nested domain are indicated by the dashed lines; the horizontal filtering parameter was set to 6° latitude outside this area, and 4° inside. Josephine's location is indicated by the tropical storm /hurricane symbol. Wind barbs follow the normal convention with a full barb equivalent to 5 m s^{-1}. Observation types are identified as follows: ODW (0), rawinsonde (0), USAF aircraft (W), WP-3D aircraft (C1). From Franklin (1990). Republished with permission of the American Meteorological Society.

Early attempts to define the environmental flow of tropical cyclones were based on an azimuthal average of objectively analyzed flow fields over some distant annulus surrounding a storm (e.g. an annulus from 5°-7° latitude) and as a basis for comparing observations with barotropic vortex theory, a deep layer pressure-weighted mean was constructed, typically from 800 mb to 300 mb. Because of data scarcity, prior to 1990, flow fields were based on composite data sets constructed from many storms (e.g. Carr and Elsberry, 1990).

For a range of storm composites stratified according to the direction of storm motion, the latitude, the intensity and the speed, Carr and Elsberry (1990) showed that the storm motion vector tended to have a westward and poleward component relative to the environmental flow as defined above and offered this finding as support for the theory of "β-drift".

Franklin (1990) carried out a similar analysis, but for a single storm, Hurricane Josephine (1984). The environment of this storm was relatively well sampled on three consecutive days. He used a slightly different deep layer mean from 850 mb to 300 mb and considered the total wind field to be the sum of two components, an axisymmetric vortex and a residual flow, which he referred to as the vortex environment. The structure of the environmental flow on the three days on which the storm was sampled (his Fig. 8) did not show evidence of the β-gyres found in theoretical models such as those shown in Figs. 3.5 and 3.6.

Franklin tried computing the azimuthally-averaged and mean pressure-weighted absolute vorticity gradient of the storm environment (his $\nabla(\zeta_e + f)$) for annular bands from 3°-5° latitude and 5°-7° latitude as well as the azimuthally-averaged environmental winds in each of these bands (his V_p). He found that

"On each day, the 5°-7° mean vorticity gradient is oriented to the north, while the so-called propagation vector, V_p, is oriented toward the west-northwest or northwest. Furthermore, as the magnitude of $\nabla(\zeta_e + f)$ increases over the three days, the magnitude of V_p, increases proportionally".

However, he found that the same relationship does not hold for the 3°-5° band. Moreover he noted that it is not obvious why,

"even if the basic current could be estimated by the 5°-7° band average flow, this same radial band would be the appropriate one for averaging the environmental vorticity gradient".

Franklin noted a strong variability of Josephine's environmental flow (averaged over some annular region) with distance from the storm center and with pressure, noting that Josephine moved at nearly right angles to the 5°-7° band azimuthally averaged wind at 500 mb, whereas the vortex motion was more consistent with the flow near 700 mb.

He concluded that, in general, single-level mid-tropospheric data may not be representative of a hurricane's net environmental flow.

Glatz and Smith (1996) analyzed the same data for Hurricane Josephine, but assumed a three-way partitioning of the deep-layer mean flow, consistent with the partitioning in Section 3.3.3. They assumed the "vortex environment" to have a linear variation in both the zonal and meridional direction. This structure was obtained by fitting a linear function to the horizontal wind components over the largest circular domain centered on the storm that could be constructed from the data. The "vortex" was obtained as an azimuthal average of the residual flow after this "environmental flow" had been removed and the wavenumber-one asymmetry of the residual flow was computed also. These wavenumber-one asymmetries are shown in Fig. 3.23, both for a mean pressure-weighted flow between 850 mb and 500 mb and one between the surface and 100 mb.

On 10 and 12 October, the wavenumber one asymmetries in the 850-500 mb layer mean are similar in scale structure and magnitude to that arising in the prototype problem shown in Fig. 3.6. However, there are major changes in the orientation of the gyres over the three days of observation. For example, the axis joining the maxima and minima of the gyres (the outer gyres on 11 October) moved by more than 90° in a clockwise direction between 10 October and 11 October, only to move a similar angle counterclockwise by 12 October. Bearing in mind the mechanism for the establishment of the gyres portrayed in Fig. 3.10, it is difficult to explain such rapid changes in orientation of the gyres in terms of barotropic theory. Notwithstanding possible limitations of the data, it is possible that baroclinic effects may have been responsible for such changes, such as changes in vertical shear. In support of this interpretation, the changes in orientation in the deep-layer-mean wavenumber-one asymmetry shown also in Fig. 3.23 were less than in the lower tropospheric mean, although the data above 400 mb used to construct the deep layer mean were based on the United States National Meteorological Centre's assimilated analysis: there were no additional data from aircraft-released dropsondes at these levels.

While suggestive, neither the Franklin study or that by Glatz and Smith provided conclusive evidence for the existence of β-gyres in Nature, but they did outline a range of difficulties that would be faced in providing such evidence. In a further study, Franklin et al. (1996) presented an analysis of 16 datasets from synoptic flow experiments collected in 10 Atlantic basin hurricanes and tropical storms, but the results fell short of providing clear evidence for the β-gyre circulations, apparently "due to the presence of other synoptic features in the vortex environment". However, Franklin et al. do say that "while it is impossible to conclusively interpret any particular vorticity anomaly as a β-gyre, the tendencies over the sample strongly suggest that the β ef-

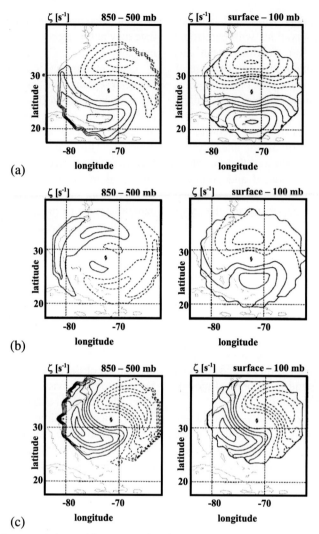

FIGURE 3.23 Azimuthal wavenumber one relative vorticity asymmetries of the 850-500 mb layer mean flow (left panels) and surface to 100 mb deep-layer-mean flow (right panels) in Hurricane Josephine on (a) 10 October, (b) 11 October, and (c) 12 October 1984. Contour interval 2.5×10^{-6} s^{-1}. Dashed lines indicate negative values. From Glatz and Smith (1996). Republished with permission of Wiley.

fect was involved in the formation and evolution of many of them".

Kim et al. (2009) claim to have positively identified β-gyres in global analyses of some typhoons. Like Glatz and Smith (1996), they use a three component partitioning of the flow, but the large-scale "environment" is obtained by applying a low-pass filter to the data. Again, the "vortex" is defined as an azimuthal average of the residual flow with the environment removed. The residual flow is then examined for β-gyre-like structures.

An example of the analyzed vortex and the residual flow with the vortex removed is shown in Fig. 3.24. While the asymmetry shown has the key features of a β-gyre with two counter-rotating circulations straddling the vortex center and a nearly-uniform flow between these circulations that is towards the northwest, it is not clear whether a gyre with this orientation should be expected as the authors do

not compute the environmental vorticity gradient and, in particular, its orientation. Moreover, they do not demonstrate that these gyres have temporal persistence. In fact, they do note that using any particular global model,

"the beta gyre may be well defined at some analysis times but not at other times",

suggesting that, in general, the gyres do not have the slowly-varying temporal continuity that is seen in barotropic models.

It may be pertinent to reiterate a conclusion of Ulrich and Smith (1991) that:

"In all the basic flows considered, the environmental flow across the vortex centre provided an accurate steering current for the vortex. Furthermore, the vor-

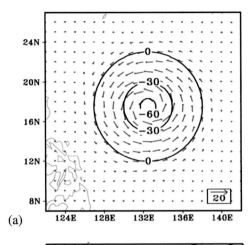

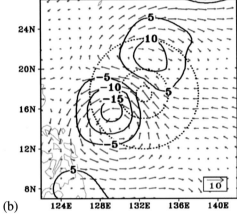

FIGURE 3.24 Analyzed axisymmetric (top panel) and residual (bottom panel) 850 mb wind and perturbation geopotential height fields for Typhoon Sonca (2005). Contour interval for geopotential is 30 m. The dotted circles in the bottom panel are the symmetric height fields in the top panel. From Kim et al. (2009). Republished with permission of Springer.

tex asymmetries were concealed by the presence of the basic flow and they became apparent only when the latter was removed. It follows that the so-called beta gyres may be exceedingly difficult to observe in nature."

Indeed, in reality, any partitioning of the basic flow, the vortex and the vorticity asymmetries from the total flow is necessarily nonunique.

3.5 Exercises

Exercise 3.1. Referring to the analysis of an axisymmetric vortex in a uniform flow in Section 3.3.1, show that the location of the minimum streamfunction) is displaced a meridional distance Y_{smin} from the vorticity center, where $U - \Omega(Y_{smin})Y_{smin} = 0$ (see Eq. (3.6)).

Exercise 3.2. Starting from Eq. (2.50) and the assumptions that air parcels move in circular orbits about the vortex cen-

ter while conserving their absolute vorticity and that the relative advection of vortex vorticity is small, show that the asymmetric vorticity approximately satisfies the equation:

$$\frac{\partial \Gamma}{\partial t} + \Omega(r)\frac{\partial \Gamma}{\partial \lambda} = -\beta r \Omega(r) \sin \lambda. \qquad (3.28)$$

Exercise 3.3. Show that the equation

$$\frac{\partial X}{\partial t} + \Omega(r, t)\frac{\partial X}{\partial \lambda} = -\beta r \Omega(r, t) \cos \lambda,$$

has the solution

$$X = -\beta r(\sin \lambda - \sin(\lambda - \omega)),$$

where

$$\omega = \int_0^t \Omega(r, t')dt'.$$

3.6 Appendices

3.6.1 Appendix 1: transformation of the momentum equation to an accelerating frame of reference

The vector momentum equation may be written

$$\frac{\partial \mathbf{u}}{\partial t} + \mathbf{u} \cdot \nabla \mathbf{u} + \mathbf{f} \wedge \mathbf{u} = -\frac{1}{\rho}\nabla p.$$

Now

$$\mathbf{u} \cdot \nabla \mathbf{u} = \boldsymbol{\omega} \wedge \mathbf{u} + \nabla(\frac{1}{2}\mathbf{u}^2).$$

Therefore

$$\frac{\partial \mathbf{u}}{\partial t} + (\boldsymbol{\omega} + \mathbf{f}) \wedge \mathbf{u} = -\frac{1}{\rho}\nabla\left(p + \frac{1}{2}\mathbf{u}^2\right).$$

Now transform the stationary coordinate system $(\mathbf{x}, t)$ to the moving coordinate system $(\mathbf{X}, T)$ (see Fig. 3.25), where

$$\mathbf{x} = \mathbf{X} + \mathbf{x_c(t)}, \quad t = T,$$

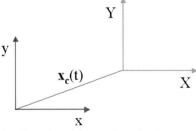

FIGURE 3.25 Illustration of the transformation from a stationary coordinate system (x, y, t) to a moving coordinate system (X, Y, T).

and

$$\frac{d\mathbf{x_c}}{dt} = \mathbf{c}(t) = (c_1(t), c_2(t)).$$

Then

$$\frac{\partial}{\partial x} = \frac{\partial}{\partial X}\underbrace{\frac{\partial X}{\partial x}}_{=1} + \frac{\partial}{\partial Y}\underbrace{\frac{\partial Y}{\partial x}}_{=0} + \frac{\partial}{\partial T}\underbrace{\frac{\partial T}{\partial x}}_{=0},$$

$$\frac{\partial}{\partial y} = \frac{\partial}{\partial X}\underbrace{\frac{\partial X}{\partial y}}_{=0} + \frac{\partial}{\partial Y}\underbrace{\frac{\partial Y}{\partial y}}_{=1} + \frac{\partial}{\partial T}\underbrace{\frac{\partial T}{\partial y}}_{=0},$$

$$\frac{\partial}{\partial t} = \frac{\partial}{\partial X}\underbrace{\frac{\partial X}{\partial t}}_{=-c_1} + \frac{\partial}{\partial Y}\underbrace{\frac{\partial Y}{\partial t}}_{=-c_2} + \frac{\partial}{\partial T}\underbrace{\frac{\partial T}{\partial t}}_{=1},$$

or

$$\begin{pmatrix} \frac{\partial}{\partial x} \\ \frac{\partial}{\partial y} \\ \frac{\partial}{\partial t} \end{pmatrix} = \begin{pmatrix} 1 & 0 & 0 \\ 0 & 1 & 0 \\ -c_1 & -c_2 & 1 \end{pmatrix} \begin{pmatrix} \frac{\partial}{\partial X} \\ \frac{\partial}{\partial Y} \\ \frac{\partial}{\partial T} \end{pmatrix},$$

or, in vector notation,

$$\frac{\partial}{\partial t} = \frac{\partial}{\partial T} - \mathbf{c} \cdot \nabla_X, \quad \nabla_\mathbf{x} = \nabla_\mathbf{X}.$$

Note that although $T = t$, $\partial/\partial T \neq \partial/\partial t$ because the former holds $\mathbf{x}$ constant while the latter holds $\mathbf{X}$ constant.

Let $\mathbf{U}$ be the velocity in the moving frame, i.e., $\mathbf{U} = \mathbf{u} - \mathbf{c}$. Then the momentum equation transforms to

$$\left(\frac{\partial}{\partial T} - \mathbf{c} \cdot \nabla\right)(\mathbf{U} + \mathbf{c}) + (\boldsymbol{\omega} + \mathbf{f}) \wedge (\mathbf{U} + \mathbf{c}) =$$
$$-\frac{1}{\rho}\nabla p - \nabla\left(\frac{1}{2}(\mathbf{U} + \mathbf{c})^2\right), \quad (3.29)$$

where

$$\boldsymbol{\omega} = \nabla_\mathbf{x} \wedge \mathbf{u} = \nabla_\mathbf{X} \wedge \mathbf{U}.$$

Further reduction gives

$$\frac{\partial \mathbf{U}}{\partial T} + (\boldsymbol{\omega} + \mathbf{f}) \wedge \mathbf{U} = -\frac{1}{\rho}\nabla p - \nabla(\frac{1}{2}\mathbf{U}^2)$$
$$-\frac{\partial \mathbf{c}}{\partial T} + \mathbf{c} \cdot \nabla \mathbf{U} - (\boldsymbol{\omega} + \mathbf{f}) \wedge \mathbf{c} - \nabla(\mathbf{U} \cdot \mathbf{c}), \quad (3.30)$$

using the fact that $\nabla \mathbf{c} = \mathbf{0}$ because $\mathbf{c} = \mathbf{c}(t)$. Now

$$\nabla(\mathbf{U} \cdot \mathbf{c}) = \mathbf{U} \cdot \underbrace{\nabla \mathbf{c}}_{=0} + \mathbf{c} \cdot \nabla \mathbf{U} + \mathbf{U} \wedge \underbrace{(\nabla \wedge \mathbf{c})}_{=0} + \mathbf{c} \wedge \underbrace{(\nabla \wedge \mathbf{U})}_{=\omega}$$

Therefore

$$\boxed{\frac{\partial \mathbf{U}}{\partial T} + (\boldsymbol{\omega} + f) \wedge \mathbf{U} = -\frac{1}{\rho}\nabla(p + \frac{1}{2}\rho\mathbf{U}^2) - \mathbf{f} \wedge \mathbf{c} - \frac{d\mathbf{c}}{dt}.}$$
$$(3.31)$$

Using

$$\nabla \wedge (\mathbf{f} \wedge \mathbf{c}) = \mathbf{f}\underbrace{(\nabla \cdot \mathbf{c})}_{=0} - \underbrace{(\nabla \cdot \mathbf{f})}_{=0}\mathbf{c} + \mathbf{c} \cdot \nabla\mathbf{f} - \underbrace{\mathbf{f} \cdot \nabla \mathbf{c}}_{=0},$$

the corresponding vorticity equation takes the form

$$\boxed{\frac{\partial \boldsymbol{\omega}}{\partial T} + \mathbf{U} \cdot \nabla(\boldsymbol{\omega} + \mathbf{f}) = (\boldsymbol{\omega} + \mathbf{f}) \cdot \nabla \mathbf{U} - \mathbf{c} \cdot \nabla f.} \quad (3.32)$$

3.6.2 Appendix 2: derivation of Eq. (3.14)

In general, we require the solution of $\nabla^2 \psi = \zeta$, when

$$\zeta(r, \lambda, T) = \Gamma_1(r, T)\cos n\lambda + \Gamma_2(r, T)\sin n\lambda. \quad (3.33)$$

Then, in polar coordinates,

$$\nabla^2\psi = \frac{\partial^2\psi}{\partial r^2} + \frac{1}{r}\frac{\partial\psi}{\partial r} + \frac{1}{r^2}\frac{\partial^2\psi}{\partial\lambda^2} =$$
$$\Gamma_1(r, T)\cos n\lambda + \Gamma_2(r, T)\sin n\lambda. \quad (3.34)$$

Assuming a solution of the form $\psi = \hat{\psi}_1(r, T)\cos n\lambda + \hat{\psi}_2(r, T)\sin n\lambda$, this equation becomes

$$\frac{d^2\hat{\psi}_k}{dr^2} + \frac{1}{r}\frac{d\hat{\psi}_k}{dr} - \frac{n^2}{r^2}\hat{\psi}_k = \Gamma_k(r, T), (k = 1, 2). \quad (3.35)$$

A solution of this equation may be obtained by the method of variation of parameters (e.g. Boyce and DiPrima, 1986). As T can be treated as a constant in this solution, we omit writing the T dependence in the derivation that follows. It is easy to check that the homogeneous form of Eq. (3.35) has two solutions $\psi_{h1} = r^n$ and $\psi_{h2} = r^{-n}$. A particular solution of the nonhomogeneous equation is sought for in the form $\psi_k = u_1(r)\psi_{h1} + u_2(r)\psi_{h2}$. Then

$$\frac{d\psi_k}{dr} = \psi_{h1}\frac{du_1}{dr} + \psi_{h2}\frac{du_2}{dr} + u_1\frac{d\psi_{h1}}{dr} + u_2\frac{d\psi_{h2}}{dr}. \quad (3.36)$$

The functions $u_1(r)$ and $u_2(r)$ are as yet unknown. We choose one relation between these to satisfy

$$\psi_{h1}\frac{du_1}{dr} + \psi_{h2}\frac{du_2}{dr} = 0. \quad (3.37)$$

Then

$$\frac{d^2\psi_k}{dr^2} = \frac{d\psi_{h1}}{dr}\frac{du_1}{dr} + \frac{d\psi_{h2}}{dr}\frac{du_2}{dr} + u_1\frac{d^2\psi_{h1}}{dr^2} + u_2\frac{d^2\psi_{h2}}{dr^2}.$$
$$(3.38)$$

Substitution of the derivatives (3.36) and (3.38) into Eq. (3.35) then gives with a little algebra, using the fact that ψ_{h1} and ψ_{h2} satisfy the homogeneous form of Eq. (3.35),

$$\frac{d\psi_{h1}}{dr}\frac{du_1}{dr} + \frac{d\psi_{h2}}{dr}\frac{du_2}{dr} = \Gamma_k(r), \quad (k = 1, 2). \quad (3.39)$$

Solving Eqs. (3.37) and (3.39) for the derivatives of $u_1(r)$ and $u_2(r)$ gives

$$\frac{du_1}{dr} = -\frac{\psi_{h2}}{W(\psi_{h1}, \psi_{h2})}\Gamma_k(r),$$

$$\frac{du_2}{dr} = \frac{\psi_{h1}}{W(\psi_{h1}, \psi_{h2})}\Gamma_k(r) \quad (3.40)$$

where $W(\psi_{h1}, \psi_{h2}) = \psi_{h1}d\psi_{h2}/dr - \psi_{h2}d\psi_{h1}/dr$ is called the Wronskian of the independent solutions ψ_{h1} and ψ_{h2}. In the present case, $W(r) = -2n/r$ and

$$\frac{du_1}{dr} = \frac{r^{1-n}}{2n}\Gamma_k(r) \quad \text{and} \quad \frac{du_2}{dr} = -\frac{r^{n+1}}{2n}\Gamma_k(r). \quad (3.41)$$

Integration with respect to r gives

$$u_1 = \frac{1}{2n}\int_a^r p^{1-n}\Gamma_k(p)dp$$

and

$$u_2 = -\frac{1}{2n}\int_a^r p^{n+1}\Gamma_k(p)dp \quad (3.42)$$

and the general solution of Eq. (3.35) is

$$\psi_k = \frac{r^n}{2n}\int_a^r p^{1-n}\Gamma_k(p)dp$$
$$- \frac{r^{-n}}{2n}\int_a^r p^{n+1}\Gamma_k(p)dp + Ar^n + Br^{-n}, \quad (3.43)$$

where A and B are constants. These constants are determined by suitable boundary conditions.

In anticipation of an application of the theory in Chapter 4, we assume for the present that the domain of integration is finite from a to b and that $\psi_k(a) = 0$ and $\psi_k(b) = 0$. The first of these conditions gives $B = -Aa^{2n}$ and the second gives

$$A(b^{2n} - a^{2n}) = -\frac{b^{2n}}{2n}\int_a^b p^{1-n}\Gamma_k(p)dp$$
$$+ \frac{1}{2n}\int_a^b p^{n+1}\Gamma_k(p)dp. \quad (3.44)$$

Finally, after a little algebra,

$$\psi_k = \frac{r^n}{2n} \times$$
$$\left[\int_a^r p^{1-n}\Gamma_k(p)dp - \frac{1}{r^{2n}}\int_a^r p^{n+1}\Gamma_k(p)dp\right.$$
$$+ \frac{1 - \frac{a^{2n}}{r^{2n}}}{1 - \frac{a^{2n}}{b^{2n}}} \times$$
$$\left.\left(-\int_a^b p^{1-n}\Gamma_k(p)dp + \frac{1}{b^{2n}}\int_a^b p^{n+1}\Gamma_k(p)dp\right)\right]$$
$$(k = 1, 2), \quad (3.45)$$

In the special case in this chapter, $a = 0$ and $b \to \infty$, whereupon

$$\psi_k = -\frac{1}{2nr^n}\int_0^r p^{n+1}\Gamma_k(p)dp - \frac{r^n}{2n}\int_r^\infty p^{1-n}\Gamma_k(p)dp$$
$$(k = 1, 2), \quad (3.46)$$

When $n = 1$, this formula is identical to Eq. (3.15). In the general case it is equal to Eq. (3.27).

3.6.3 Appendix 3: solution of Eq. (3.22)

The asymmetric flow $\mathbf{U_a}$ is obtained from Eqs. (3.15) and (3.16) and $\mathbf{c}$ is obtained from Eq. (3.19). We can calculate the streamfunction ψ_a' of the *vortex-relative* flow $\mathbf{U_a} - \mathbf{c}$, from $\psi_a' = \psi_a - \psi_c$, where

$$\psi_c = r\left(V_a\cos\lambda - U_a\sin\lambda\right)|_{r=0}$$
$$= r\left(\left.\frac{\partial\Psi_1}{\partial r}\right|_{r=0}\cos\lambda + \left.\frac{\partial\Psi_2}{\partial r}\right|_{r=0}\sin\lambda\right). \quad (3.47)$$

Then using Eqs. (3.13), (3.14), (3.17), and (3.47) we obtain

$$\psi_a' = \Psi_1'(r, T)\cos\lambda + \Psi_2'(r, T)\sin\lambda, \quad (3.48)$$

where

$$\Psi_n'(r, t) = \Psi_n - r\left[\frac{\partial\Psi_n}{\partial r}\right]_{r=0}, \quad (n = 1, 2)$$
$$= \frac{1}{2}r\int_0^r\left(1 - \frac{p^2}{r^2}\right)\Gamma_n(p, T)dp. \quad (3.49)$$

After a little more algebra it follows using Eqs. (3.15), (3.16), (3.18), and (3.49) that

$$-(\mathbf{U_a} - \mathbf{c})\cdot\nabla\zeta_s = \chi_1(r, T)\cos\lambda + \chi_2(r, T)\sin\lambda, \quad (3.50)$$

where

$$\begin{bmatrix}\chi_1(r, T)\\\chi_2(r, T)\end{bmatrix} = \frac{1}{r}\frac{d\zeta_s}{dr} \times \begin{bmatrix}\Psi_2'(r, T)\\-\Psi_1'(r, T)\end{bmatrix}. \quad (3.51)$$

Now using Eqs. (3.50) and (3.51), Eq. (3.49) can be written as

$$\frac{d\Gamma_{ac}}{dT} = \frac{1}{r}\frac{d\zeta_s}{dr}(\Psi_2'(r,T)\cos\lambda - \Psi_1'(r,T)\sin\lambda),$$

where d/dT denotes differentiation following a fluid parcel moving in a circular path of radius r about the vortex center with angular velocity $\Omega(r)$. It follows that

$$\Gamma_{ac} = \frac{1}{r}\frac{d\zeta_s}{dr}\int_0^T (\Psi_2'(r,t')\cos\lambda(t') -$$
$$\Psi_1'(r,t')\sin\lambda(t'))dt', \qquad (3.52)$$

where $\lambda(t') = \lambda - \Omega(r)(T - t')$. Using Eq. (3.49), this expression becomes

$$\Gamma_{ac} = \frac{1}{2}\frac{d\zeta_s}{dr}\int_0^T\int_0^r \left(1 - \frac{p^2}{r^2}\right) \times$$
$$(\Gamma_2(p,t')\cos\lambda(t') - \Gamma_1(p,t')\sin\lambda(t'))dp\,dt',$$

and it reduces further on substitution for Γ_n from Eq. (3.11) and the above expression for $\lambda(t')$ giving

$$\Gamma_{ac} = \frac{1}{2}\beta\frac{d\zeta_s}{dr}\int_0^r p\left(1 - \frac{p^2}{r^2}\right)\int_0^T (...)dt'\,dp,$$

where $(...) = (\cos\{\lambda - \Omega(r)(T - t')\} - \cos\{\lambda - \Omega(r)(T - t') - \Omega(p)t'\})$.

On integration with respect to t' we obtain

$$\Gamma_{ac}(r,\theta,T) = \Gamma_{11}(r,T)\cos\lambda + \Gamma_{12}(r,T)\sin\lambda, \quad (3.53)$$

where

$$\Gamma_{1n}(r,T) = -\frac{1}{2}\beta\frac{d\zeta_s}{dr}\int_0^r p\left(1 - \frac{p^2}{r^2}\right)\eta_n(r,p,T)dp,$$
$$(3.54)$$

and

$$\eta_1(r,p,T) = \frac{\sin(\Omega(r)T)}{\Omega(r)} - \frac{\sin(\Omega(r)T) - \sin(\Omega(p)T)}{\Omega(r) - \Omega(p)},$$
$$(3.55)$$

$$\eta_2(r,p,T) = \frac{1 - \cos(\Omega(r)T)}{\Omega(r)} + \frac{\cos(\Omega(r)T) - \cos(\Omega(p)T)}{\Omega(r) - \Omega(p)}.$$
$$(3.56)$$

The integrals in (3.54) can be readily evaluated using quadrature. Recall from Section 3.3.2.1 that we are considering only situations where $\Omega(r)$ does not vanish to avoid possible singularities in the foregoing integrals.

Chapter 4

Vortex axisymmetrization, waves and wave-vortex interaction

Observations of tropical cyclones indicate that storms are generally asymmetric and only the inner-core region of the more intense storms can be reasonably approximated as axisymmetric (e.g., Fig. 4.1). The outer region of storms is invariably asymmetric and weaker storms undergoing intensification are usually highly asymmetric (e.g., Figs. 1.6a,b and 1.16). As a quantitative illustration of this geometric property in the case of a mature hurricane, Fig. 4.2 shows the relative magnitude of the wind asymmetries in comparison to the azimuthally averaged tangential velocity for the particularly well-observed Atlantic Hurricane Gloria (1985).[1] The Gloria winds are Fourier decomposed into azimuthal wavenumber components (n), with the symmetric component ($n = 0$) and the asymmetric components ($n = 1$ to 4), in a coordinate system translating with the hurricane. The figure shows the radial variation of the 700 hPa (aircraft reconnaissance level) wind-amplitude components in the hurricane. These radial profiles are constructed from analyses using Doppler-radar derived winds within 75 km of the storm center, and Omega dropwindsondes in the storm's environment (detailed in Franklin et al., 1993). The symmetric radial wind is relatively small in comparison at this level and is not shown.

As is evident from Fig. 4.2, near the storm center, the asymmetries are much weaker than the symmetric tangential wind. Moreover, the wavenumber 2, 3, and 4 asymmetries are generally much weaker than the wavenumber one asymmetry.[2] Thus, as a first step towards understanding the dynamics of asymmetric motions in a realistic tropical cyclone vortex above the boundary layer, it is defensible to make a small-amplitude (i.e., linear) approximation in the

FIGURE 4.1 A view of Hurricane Gloria (1985) from space on September 25, 1985. At this time, the storm was near its maximum intensity of 145 knots and the eye is located northeast of the Bahamas. The space image features a small eye and large convective bands. Source: https://en.wikipedia.org/wiki/Hurricane_Gloria.

rapidly rotating core region of a mature hurricane within a few hundred kilometers of the storm center.[3]

As shown in Chapter 3, the mere presence of the β effect would contribute to an asymmetry in the flow even if the environment were quiescent, which, in reality it is not. Another source of flow asymmetry arises simply as an imprint of the storm environment, which in nature is never symmetric. For example, there is generally variability in the broadscale trade-wind environment in the lower troposphere and frequently variability in the upper-tropospheric flow associ-

---

1. This type of Fourier analysis has been affirmed by others, e.g., Reasor et al. (2000) (their Fig. 1). The Hurricane Gloria analysis remains one of the most comprehensive data sets for a developed storm that covers both the synoptic-scale environment and the inner-core region of a mature hurricane.

2. As noted by Shapiro and Montgomery (1993): "There is a region near 40 km where all asymmetric wind components, including $n = 1$, are very small. The asymmetric wind near that radius is dominated by a uniform, wavenumber one, flow relative to the ground that approximates the translation of the hurricane. The relative wind, in the translating system, is thus very small. The 5 m s^{-1} maximum in the wavenumber one component at $r = 0$ is likely due to local, convectively-induced asymmetries (Franklin et al., 1993; Shapiro and Ooyama, 1990)."

3. The near-axisymmetric property of a well-developed tropical cyclone in its rapidly rotating core region can be anticipated also invoking the Okubo-Weiss (OW) parameter summarized in Chapter 1. In this case the strong azimuthal shearing process outside the rapidly spinning core dominates the vorticity there. When the shearing dominates the vorticity, asymmetric vorticity perturbations tend to be shredded to oblivion (e.g. Rozoff et al., 2006). Astrophysicists use the term "spaghettification" to describe an analogous process that operates near the vicinity of the event horizon of a rotating black hole. Coherent vortices are likened to the "black holes of turbulence" (Haller and Beron-Vera, 2013).

Tropical Cyclones. https://doi.org/10.1016/B978-0-44-313449-4.00012-6

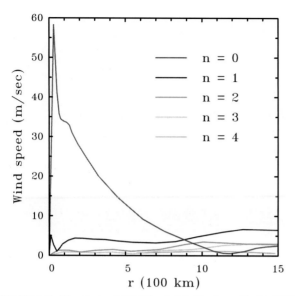

FIGURE 4.2 Magnitude of 700-hPa wind in Hurricane Gloria (1985). Azimuthal average of tangential wind speed ($n = 0$) as well as magnitude of azimuthal wavenumber, $n = 1, 2, 3$, and 4 asymmetric wind components are shown. The winds are computed in a coordinate system translating with the hurricane. Adapted from Shapiro and Montgomery (1993) (their Fig. 1). Republished with permission of the American Meteorological Society.

ated for example by migratory upper-level troughs or seasonal ridges. On the cyclone scale, near-surface frictional effects in a moving vortex and environmental vertical wind shear are a major source of flow asymmetries. On an even smaller scale, an important source of flow asymmetry is associated with deep cumulus convection, which as shown later, has a stochastic element. Internal shear instabilities of the swirling flow are an additional source of flow asymmetries.

It will be shown later in this chapter that high azimuthal wavenumbers, such as those associated with deep convection, have a tendency to be preferentially damped compared to low azimuthal wavenumbers in the core of intense vortices. This tendency is a consequence of a process called "vortex axisymmetrization". It turns out that axisymmetrization is essential to understanding the observed coherence and persistence of large-scale vortices in the atmosphere and oceans (McWilliams, 2011 and refs.). Therefore, some extended discussion of this topic here will prove useful in subsequent chapters and also in the interpretation of some observed features noted previously in the Introduction.

4.1 Illustration of flow asymmetries

In the previous chapter we saw how vortex flow asymmetries, whether they are considered a part of the vortex or a part of its environment, can influence the vortex motion. In the next section we study first the motion of initially

asymmetric vortices on an f-plane. For simplicity, as motivated by the foregoing observations, we focus primarily on low azimuthal wavenumber perturbations and consider moderate- or small-amplitude wind (and vorticity) perturbations to the main vortex. The issues to be addressed here are relevant to the problem of initializing tropical cyclone forecast models as well as to an understanding of possible track changes as cyclones develop new flow asymmetries or as existing flow asymmetries evolve in time.

Asymmetries have implications not only for tropical cyclone motion, but also for intensification and/or vortex structure change. The processes involved can be quite complex and are often intimately tied up with vorticity-wave motions, within or on the periphery of the vortex. Therefore, in later sections we examine the dynamics of vortex waves using an insightful toy model of a strong localized vortex immersed in a weak environment of vorticity perturbations. Again, in accord with the foregoing observations, for the most part we limit the mathematical analysis to that of small-amplitude perturbations on an intense circular vortex, together with their estimated change to the mean vortex via the temporal integration of suitable eddy covariance terms. Some aspects of the nonlinear dynamics of eddy-vorticity perturbations on such a strong vortex are touched on briefly. Such advanced topics must be deferred until later, where one must contend also with the presence of deep moist convection and the frictional boundary layer in a developing or mature cyclone.

4.1.1 Examples of vortex axisymmetrization

We construct first an asymmetric vortex by adding a dipole of vorticity to the initial vorticity distribution shown in Fig. 3.11. The vorticity dipole has the form

$$\Gamma_d(r, \lambda) = \zeta_D (r/d)^2 \exp(-r^2/d^2) \cos(\lambda - \alpha), \quad (4.1)$$

where ζ_D, d, and α are prescribed constants characterizing the dipole strength, scale, and orientation. Thus the vorticity maximum and minimum of the dipole occur at (d, α) and $(d, \pi + \alpha)$, respectively.

We consider four calculations with $\alpha = 0$ so that the dipole is oriented west-east. In the first calculation, S1, $d = \sqrt{2}$ and $\zeta_D = 0.2\zeta_o$, where ζ_o is the maximum value of ζ in the symmetric vortex. In the second calculation, S2, $d = 2\sqrt{2}$ and $\zeta_D = 0.1\zeta_o$ so that the velocity at the origin associated with the dipole is the same as in S1. These two calculations are carried out on an f-plane. The third and fourth calculations, S3 and S4, are the same as S1 and S2, but are for a β-plane. The calculations are carried out numerically by a direct integration of Eq. (2.49) with the initial vorticity distribution (symmetric vortex plus dipole) described above. The Kasahara-Platzman partitioning scheme is used to analyze the subsequent vortex evolution. In this scheme, the

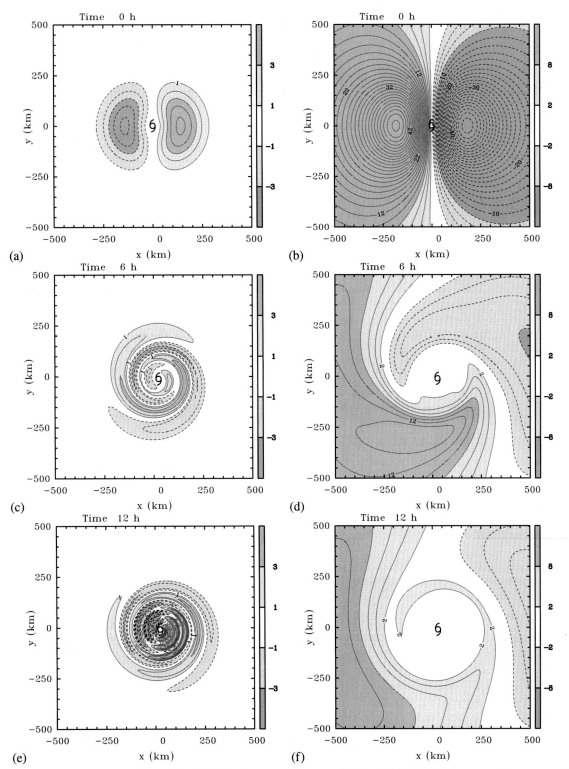

FIGURE 4.3 Evolution of the asymmetric vorticity field (ζ_a) and corresponding streamfunction field for the initially asymmetric vortex on an f-plane in the case of small-scale asymmetry (calculation S1). Shown are (a) the initial fields, and the fields at (b) 6 h and (c) 12 h. Note that only one quarter of the total flow domain is shown. Contour intervals are 2×10^{-5} s^{-1} for ζ_a and 5×10^4 m^2 s^{-1} for ψ_a. Positive values solid in red (mid gray in print version), negative values dashed in blue (dark gray in print version). Zero contours have been excluded. Adapted from Smith et al. (1990). Republished with permission of Wiley.

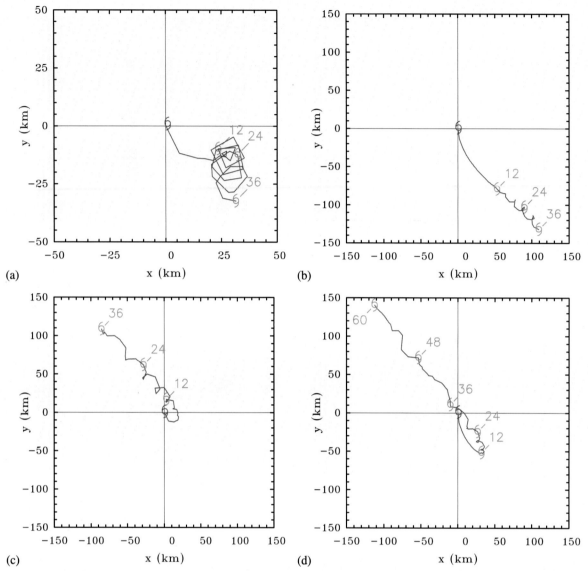

FIGURE 4.4 Tracks of initially asymmetric vortices in the calculations S1 to S4 defined in the text. (a) Small asymmetry, f-plane; (b) large asymmetry, f-plane; (c) small asymmetry, β-plane; and (d) large asymmetry, β-plane. Adapted from Smith et al. (1990). Republished with permission of Wiley.

asymmetric component of the vortex is regarded as a part of the environment, even at the initial instant.

Fig. 4.3 shows the evolution of the asymmetric vorticity component and associated streamfunction at selected times for the calculation S1. It can be seen that within a circle of radius about $2r_{max}$ centered on the vortex, the asymmetric vorticity field undergoes rapid distortion due to the relatively large (radial) shear of the tangential wind field in this region. For example, for the flow parameters chosen, the angular velocity of the symmetric vortex decreases monotonically with radius (see Fig. 3.3) so that in 6 h an air parcel of 20 km radius completes approximately 2.5 revolutions compared with 1.4 revolutions at 100 km (i.e., r_{max}) and 0.5 revolutions at 200 km. Outside this circle, the distortion of the asymmetry proceeds more slowly. Initially, the asym-

metric flow across the vortex is towards the south Fig. 4.3a, but its direction rotates counterclockwise with the gyres of the asymmetric streamfunction as the vorticity asymmetry is rotated. Therefore the vortex track forms a counterclockwise arc as shown in Fig. 4.4a.

As the asymmetric vorticity distribution is wound around the vortex by the angular shear of the tangential wind, the associated flow is reduced in strength and after about 12 h, the vortex essentially stalls. The reduction in strength of the asymmetric flow as the asymmetric vorticity field suffers angular shear can be understood in terms of an approximate analytical solution for the problem in a frame of reference moving with the basic vortex. Then, in the same spirit as the calculation in Subsection 3.3.2.1, we can show that the asymmetric vorticity distribution at time T is given by

$$\Gamma_d(r, \lambda, T) = \zeta_D(r/d)^2 \exp(-r^2/d^2) \cos(\lambda - \Omega(r)T)$$
$$= \Gamma_1(r, T) \cos \lambda + \Gamma_2(r, T) \sin \lambda, \qquad (4.2)$$

where now

$$\begin{pmatrix} \Gamma_1 \\ \Gamma_2 \end{pmatrix} = \zeta_D(r/d)^2 \exp(-r^2/d^2) \times \begin{pmatrix} \cos(\Omega(r)T) \\ \sin(\Omega(r)T) \end{pmatrix}.$$
$$(4.3)$$

Note that Eq. (4.2) has the same form as Eq. (3.10).

For an unbounded domain, we can solve the Poisson equation for the associated streamfunction, $\psi(r, \lambda, T)$ using the method outlined in Subsection 3.3.2.1. Writing $\psi(r, \lambda, T) = \Psi_1(r, T) \cos \lambda + \Psi_2(r, T) \sin \lambda$, the functions $\Psi_1(r, T)$ and $\Psi_2(r, T)$ are given in terms of $\Gamma_1(r, T)$ and $\Gamma_2(r, T)$. Moreover, the Cartesian velocity of the asymmetric flow across the vortex center, (U_a, V_a) is given by Eq. (3.17), using Eq. (3.18).

The solution for (U_a, V_a) may be obtained succinctly by writing this vector in complex notation, $U_a + iV_a$, where $i^2 = -1$, and by considering the complex function

$$\Gamma_1(r, T) + i\Gamma_2(r, T) = \zeta_D \frac{r^2}{d^2} \exp\left(-\frac{r^2}{d^2} + i\Omega(r)T\right).$$
$$(4.4)$$

Then, using Eq. (3.17) and Eq. (3.18), one obtains

$$U_a + iV_a = -\frac{i}{2}\zeta_D \int_0^\infty \frac{r^2}{d^2} \exp\left(-\frac{r^2}{d^2} + i\Omega(r)T\right) dr.$$
$$(4.5)$$

As T increases, the integrand in Eq. (4.5) oscillates more and more rapidly. Then, as a result of cancellation, the integral itself decreases monotonically in value. This explains why the vortex motion eventually stalls.

Fig. 4.5 shows the evolution of the asymmetric vorticity field for calculation S2 and Fig. 4.4b shows the vortex track in this simulation. As expected, since the maximum asymmetry is at a larger radius than S1, it is less rapidly wound up by the radial shear of the basic vortex. Accordingly, the asymmetric component of flow across the streamfunction center rotates less rapidly and decays less rapidly in strength with time. As a result, the vortex moves farther from its initial position than in S1 and its track rotates only slowly towards the east after the first three hours. As might be anticipated from the results of Section 3.3.2, the effect of a nonzero β would be to induce an east-west vorticity tendency in addition to the existing asymmetry. This is confirmed by the calculations S3 and S4, the vortex tracks for which are shown in panels (c) and (d) of Fig. 4.4.

In S3 the vortex no longer stalls after 12 h, but recurves to move along a north-westwards track as the beta-induced asymmetries begin to dominate. In S4, the beta effect becomes important also, but not so rapidly, and again the track turns north-westwards as it does so. These calculations show that the importance of vortex asymmetry on the track depends strongly on the scale of the asymmetry. The larger this scale, the less rapidly is the asymmetry wound up by the vortex circulation and the more persistent is the effect of the asymmetry. It is clear that initial asymmetries concentrated outside the radius of maximum tangential wind can have a significant effect on subsequent vortex positions and would need to be represented in tropical-cyclone forecast models.

4.2 Vortex merger and separation, Fujiwhara effect

So far we have considered the effects of a vortex asymmetry on the motion of a single vortex. But suppose there is more than one vortex in a particular flow field. A case of this type was considered in Chapter 3, where we examined the motion of a relatively-small scale vortex in a planetary-scale vortex that was part of a stationary Rossby wave. In that case, the vorticity distribution of the smaller-scale vortex is distorted by the flow of the planetary wave leading, in part, to an azimuthal wavenumber-one asymmetry flow component about the vortex that contributes to the drift of the vortex relative to the flow of the planetary wave, itself, at the vortex location.

Another situation of some interest is when the total flow arises from a single pair of vortices on an f-plane. We consider the case here where the two vortices are both cyclonic and have equal strength. From what we have learned in Chapter 3, we can imagine that each vortex will become deformed by the flow field of each other and that they will be, in part, advected by the other vortex. If the vortices are both narrow and if they are sufficiently far apart so that the angular shear experienced by the remote vortex is minimal, the effect on the motion by deformation in the flow field of the remote vortex is weak compared with the advection by that flow field, the vortices will move cyclonically in an approximately circular orbit about each other. Sometimes, two tropical cyclones form close enough together that this tendency to rotate about each other becomes an important component of their tracks. This effect is referred to as the *Fujiwhara effect*.

In the more general case, when the distortion of each vortex by the flow field of the other becomes appreciable, or when the two vortices have different strengths, other interesting effects occur. When the vortices have a similar strength and size, and if they are close enough together, they merge and axisymmetrize to form a single vortex. If they are not sufficiently close, they separate, possibly with considerable distortion, depending on their separation, while rotating cyclonically about each other.

An example of the foregoing behavior is shown in Fig. 4.6 for the case of initially identical, axisymmetric, barotropic vortices separated by a certain distance

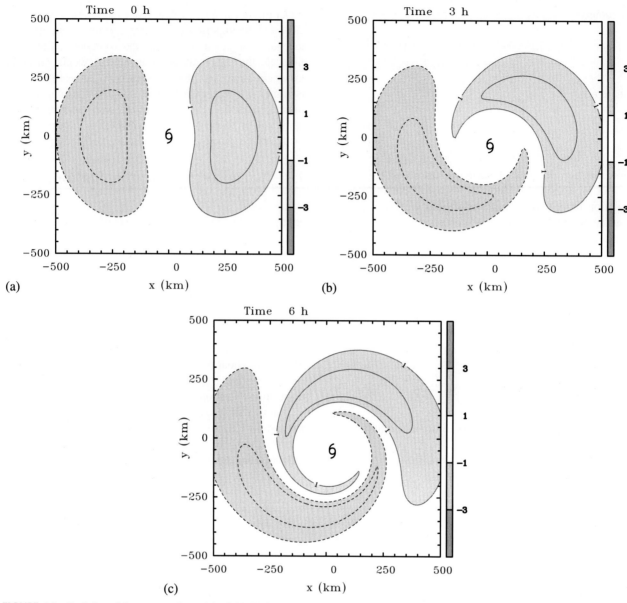

FIGURE 4.5 Evolution of the asymmetric vorticity field (Γ) for the initially asymmetric vortex on an f-plane in the case of large-scale asymmetry (simulation S2). Shown are (a) the initial field, and the fields at (b) 6 h and (c) 12 h. Contour interval is 1×10^{-5} s^{-1}. Positive values solid in red (mid gray in print version), negative values dashed in blue (dark gray in print version). Zero contours have been excluded. Adapted from Smith et al. (1990). Reproduced with permission of Wiley.

d. Each vortex has the tangential wind profile $v(r) = \omega_o r \exp\left(-\frac{1}{2}r^2/r_m^2\right)$, where r is the radius from its center, r_m is the radius of maximum tangential wind, v_m and $\omega_o = e^{\frac{1}{2}} v_m/r_m$ is a constant. Each vortex has a concentrated core of positive relative vorticity, which is surrounded by a weak field of negative vorticity.[4] The vortices are situated initially at points $(-\frac{1}{2}d, 0)$ and $(\frac{1}{2}d, 0)$ in the (x, y)

plane. In the calculation shown, $v_m = 20$ m s^{-1} and $r_m = 75$ km. The subsequent flow is solved numerically using the methodology described in Section 2.10 on a domain 2000 km $\times$ 2000 km for three different values of d: 180 km, 200 km, and 220 km. The model grid spacing is 5 km. The numerical method, itself, is the same as that used by Smith et al. (1990). In particular, it applies a weak biharmonic filter on the vorticity field at each time step to control aliasing, but no explicit diffusion.

---

4. This structure is a necessary consequence of the requirement that the vortices have a finite kinetic energy with zero circulation at infinite radius.

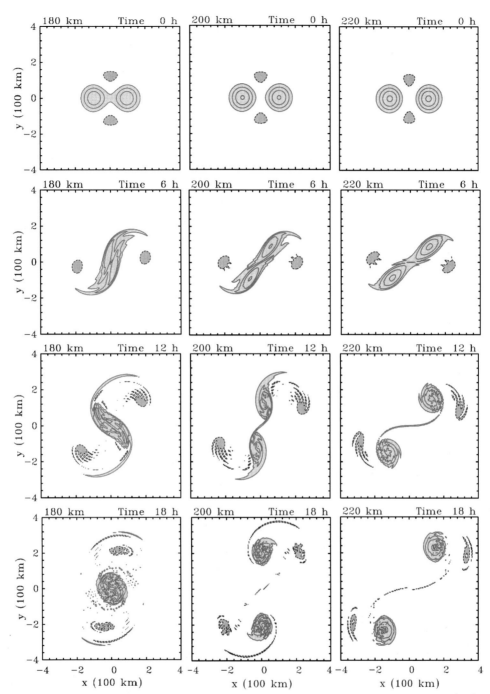

FIGURE 4.6 Evolution of relative vorticity in the two-vortex calculations described in Section 4.2. Contour interval: 2×10^{-4} s^{-1} Positive values solid in red (mid gray in print version), negative values dashed in blue (dark gray in print version).

The left panels of Fig. 4.6 show the evolution of the vorticity field[5] from the initial time at 6 hour intervals to 18 hours when $d = 180$ km. For the chosen contour intervals

for positive and negative vorticity, there is a slight overlap of the two vortex cores at the initial time as well as two small regions to the north and south where the negative vorticity of the two vortices reinforce each other enough to feature in the plots. These two vortices are close enough together so that, after 6 hours, they have virtually amalgamated as a single elongated patch of vorticity with an axis running approximately from the south-southwest to the north-northeast. As

5. We have not sought to remove the initial vorticity, suitably relocated, from these figures as in the partitioning scheme of Section 3.2 as we restrict attention here to a qualitative discussion of the total flow behavior without a detailed analysis of the vortex asymmetries and their influence on the motion of the individual vortices.

time proceeds, this patch of vorticity rotates cyclonically, becoming progressively more circular in shape as the patch axisymmetrizes, a feature exemplified by the vorticity at 18 hours. The noisiness of the fields at later times is, in part, because of the filamentation processes taking place and, in part, because the vorticity structures are becoming comparable in scale with the resolution (an example of the filamentation process is shown later in the upper panels of Fig. 4.17).

The middle column of Fig. 4.6 shows the vortex evolution for $d = 200$ km. Now the two vortex cores do not overlap for the chosen contour intervals for positive vorticity at the initial time, but the two small regions to the north and south where the regions of negative vorticity of the two vortices reinforce remain virtually unchanged. The pattern of vorticity at 6 hours is only slightly different than for the case with $d = 180$ km, but the elongated patch of positive vorticity has not rotated cyclonically so far, having now a roughly southwest to northeast orientation. In addition, the two centers of locally maximum vorticity in this pattern are further apart.

The subsequent evolution for the case with $d = 200$ km is now quite different. By 12 hours, the elongated vorticity patch of cyclonic vorticity has begun to split into two, one moving north-northeast and the other moving south-southwest while continuing to rotate slowly cyclonically about each other. This rotation diminishes as the two cyclonic patches separate and by 18 hours, the axis joining them is still not quite oriented south to north.

As the separation of the initial vortices increases further (right column of Fig. 4.6), the trends seen between the left and middle panels at a given time continue. At 6 hours, the elongated patch of vorticity has not rotated cyclonically so far, and the two centers of locally maximum vorticity in this pattern are further apart than for smaller d. By 12 hours, the two vortices have essentially separated and this separation continues to 18 hours. Of course, as the vortices move apart, the mutual cyclonic rotation slows markedly.

So far we have summarized the evolution of the positive vorticity regions, but it is of interest to examine briefly the evolution of negative regions. The negative vorticity regions, highlighted with the chosen contour interval, are seen to rotate cyclonically as well and to become partially filamented in all three experiments. For $d = 180$ km, the two surviving negative vorticity patches rotate cyclonically about the amalgamated cyclonic vortex. For $d = 200$ km and $d = 220$ km, the two surviving negative vorticity patches each rotate about the un-merged vortices as satellite features. However, it should be remembered that these negative vorticity patches are part of a sea of weak negative vorticity. Animations of the vorticity fields for a longer time indicate that the flow induced by this field of negative vorticity has an important effect on the motion of the un-merged vortices.

A case in which one of the vortices is much stronger than the other is examined by Smith et al. (1990): see their Fig. 14 and accompanying discussion. In this case, the small and weaker vortex becomes rapidly drawn out into a vortex filament by the angular shear of the strong one, becoming progressively axisymmetrized. As this happens, the effect of the smaller vortex on the cyclonically-spiraling track of the larger one rapidly diminishes. The related case of a strong, but small-scale, cyclonic vortex that is immersed in the cyclonic vorticity field of the primary vortex is summarized in Section 4.7.2, where it is shown that vortex waves play an important role in the merger process.

The processes described in this section are relevant to understanding aspects of tropical cyclogenesis and the early stages of vortex intensification described in Chapters 10, 11, 15 and 16.

4.3 The pseudo-mode

The pseudo-mode is a particular solution for an asymmetric flow structure that occupies a relatively unique position in the theory of vortex motion. For this reason, it merits some discussion before proceeding to an analysis of vortex waves. There are two alternative, but equivalent interpretations of the pseudo mode. The first is that, as shown in Section 4.1.1, if an azimuthal wavenumber-one disturbance is added to the initial axisymmetric circulation, the interaction between the disturbance and vortex will generally lead to a displacement of the initial vortex. Alternatively, the pseudo-mode is an apparent asymmetry of an axisymmetric vortex in a coordinate system that is displaced relative to the vortex center. This alternative interpretation suggests a simple method to determine the mathematical form of the vortex displacement.

Imagine an axisymmetric vortex with tangential velocity distribution $V(r')$ where r' is the radius from the center of circulation, O'. Suppose we analyze its structure in a polar coordinate system (r, θ) with its origin, O, situated a small distance d to the west of the center of circulation (see Fig. 4.7). Define angles θ, λ and ϵ as shown in the figure, noting from the geometry of the figure that $\lambda = \theta + \epsilon$. In the (r, θ) coordinate system, the tangential wind $\mathbf{V}(\mathbf{r}')$ has components $(-V(r') \sin \epsilon, V(r') \cos \epsilon)$.

It is clear from the figure that $r' \sin \epsilon = d \sin \theta$ and $r = d \cos \theta + r' \cos \epsilon$. Assuming that $d \ll r'$, $\epsilon \ll 1$, whereupon, to first order, $\cos \epsilon \approx 1$ so that $r' = r - d \cos \theta$. Then, using a Taylor series expansion, it follows that, for sufficiently small d, $V(r') = V(r - d \cos \theta) \approx V(r) - d \cos \theta (dV(r)/dr)$. Using these relationships, it is straightforward to show that

$$\mathbf{V} = \left(-V(r) \frac{d}{r} \sin \theta, \ V(r) - d \cos \theta \frac{dV}{dr} \right) + O(d^2).$$

$$(4.6)$$

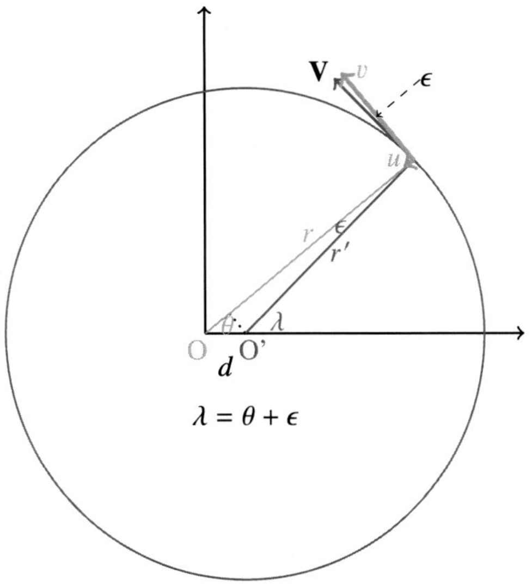

FIGURE 4.7 Geometrical configuration for calculating the structure of the pseudo-mode.

This formula is an expression of the tangential wind vector about O' in the (r, θ) coordinate system. The details are left to Exercise 4.1. It is straightforward to show also that the corresponding streamfunction of the vortex, $\psi(r, \theta)$, is

$$\psi = \int_0^r V(s)ds - d\cos\theta \, V(r) \qquad (4.7)$$

and the relative vorticity of the vortex is

$$\zeta(r, \theta) = \frac{1}{r}\frac{d}{dr}(rV) - \frac{d\cos\theta}{r}\left[\frac{d}{dr}\left(r\frac{dV}{dr}\right) - \frac{V}{r}\right]$$
$$= \zeta_s(r) - d\cos\theta \frac{d\zeta_s}{dr}, \qquad (4.8)$$

where $\zeta_s(r)$ is the relative vorticity of the symmetric vortex about its true center. It is seen from the last two equations that both ψ and ζ have an axisymmetric component as well as an azimuthal wavenumber-one component. *The azimuthal wavenumber-one components in Eqs. (4.7) and (4.8) define the pseudo mode.* In the case of the streamfunction, the pseudo-mode has amplitude $d \times V(r)$ and the maximum occurs at the radius of maximum tangential wind speed. In the case of the vorticity, the pseudo-mode has amplitude $|d\zeta_s/dr|$ and the maximum occurs where the radial vorticity gradient is largest.

The pseudo-mode is important because it is almost always a feature in analyses of data from numerical model simulations or observational data as a result of errors in determining the true vortex center. In fact, its existence can be exploited to determine the vortex center, the idea being to vary the location of an approximate center in such a way as to minimize the amplitude of the pseudo mode (e.g., Weber and Smith, 1995, p638, Section (b)). However, the

pseudo-mode has a uniquely different interpretation from other types of wavenumber-one asymmetries.

In Chapter 3 and earlier in the present chapter, we interpreted the streamflow across the vortex center as the primary cause of vortex motion. Clearly, this interpretation does not apply to the pseudo-mode as the direction of the inferred streamflow is at right-angles to the direction of vortex motion (or displacement). This feature is highlighted in Fig. 4.8a, which shows the streamfunction structure of the pseudo-mode in the horizontal $x - y$ plane for the tangential velocity profile in Fig. 3.3 when the origin of coordinates is displaced a distance $d = 10$ km to the west of the actual vortex center. Since the westward displacement of the origin of coordinates puts this origin in a northerly flow with strength $-V(d)$, the streamfunction of the pseudo-mode must have a positive gradient to the west across the origin, which explains the sign of the streamfunction dipole seen in Fig. 4.8a.

It is seen that the vorticity asymmetry in Fig. 4.8b is concentrated inside the radius of maximum tangential wind of the true vortex, while, consistent with the ideas in Section 2.19, the streamfunction asymmetry extends well beyond this radius. In a similar way, since the relative vorticity of the vortex has a maximum at the true center of vortex circulation, the vorticity of the pseudo-mode, which is correcting for the misplaced vortex center, must have a positive anomaly to the east of the displaced center with a maximum at the radius of maximum $|d\zeta/dr|$. Because of the mathematical approximations made in the foregoing derivation of the pseudo-mode, the analysis of the pseudo-mode in vorticity is formally inaccurate in the vicinity of the vortex center if, as in the present case, the radial gradient of vorticity there is zero. Consideration of the lateral displacement of a symmetric bell-shaped vorticity profile to the west indicates that the center of the pseudo-mode dipole is located a distance $\frac{1}{2}d$ to the west of the true location of maximum vorticity. The details are left to the reader as an exercise.

4.4 Vortex shear waves and vortex Rossby waves

The asymmetries we investigated analytically in Chapter 3 and in Section 4.1 were associated wholly with advective processes, since we made the assumption that to a first approximation, the vorticity perturbation is advected by the tangential velocity of the initial axisymmetric vortex. With this assumption, waves that, in general, are able to propagate on the vorticity gradient of the basic vortex are excluded from the analysis. However, in some situations, wave motions are important to the vortex dynamics also and we review now some elementary aspects of vortex waves on an initially circular vortex.

For reasons given at the beginning of this chapter, we consider now the linear theory of two-dimensional waves on

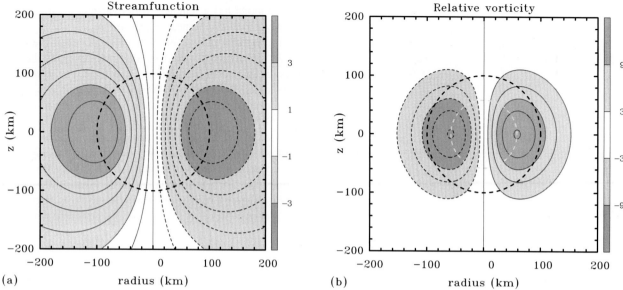

FIGURE 4.8 Structure of the pseudo-mode in the horizontal $x - y$ plane for the tangential velocity profile in Fig. 3.3 and for a westward displacement of the vortex center with magnitude $d = 10$ km: (a) streamfunction, ψ, (b) relative vorticity, ζ. The dashed black circle shows the radius of maximum tangential wind and the dashed yellow (light gray in print version) circle in (b) shows the radius of maximum $|d\zeta_s/dr|$, where r is the radius. Contour intervals: for $\zeta = 3 \times 10^{-5}$ s^{-1}; for $\psi = 5 \times 10^4$ m^2 s^{-1}. Shading as indicated on the sidebar for streamfunction is ± 1, $\pm 3 \times 10^5$ m^2 s^{-1}. Shading as indicated on the sidebar for vorticity is ± 3, $\pm 9 \times 10^{-5}$ s^{-1}. Positive values solid in red (mid gray in print version), negative values dashed in blue (dark gray in print version).

an initially circular vortex in gradient wind balance. However, before discussing the waves, it is first necessary to define the basic state about which the waves oscillate.

4.4.1 Force balances in a circular vortex

For a purely axially-symmetric vortex flow relative to a stationary circulation center, gradient wind balance[6] consists of a balance in the radial direction between the Coriolis, centrifugal and radial pressure gradient forces per unit mass, respectively:

$$\frac{v^2}{r} + fv - \frac{1}{\rho}\frac{\partial p}{\partial r} = 0, \qquad (4.9)$$

where r is radius in a cylindrical-polar coordinate system, ρ is the constant density in the incompressible approximation, v is the tangential velocity, and other notation is as defined in prior sections. (Exercise 4.2 derives this force balance for steady, axisymmetric, inviscid swirling flow and then obtains the solution and limiting cases discussed below.)

When the centrifugal force dominates the Coriolis force, i.e., $v^2/r \gg fv$, the limiting relation

$$\frac{v^2}{r} - \frac{1}{\rho}\frac{\partial p}{\partial r} = 0, \qquad (4.10)$$

---

6. Referred to also as "cyclogeostrophic balance".

is called *cyclostrophic balance*. In this case, the dominant radial force balance is between the centrifugal and radial pressure gradient forces. Conversely, when the centrifugal force is small compared to the Coriolis force, the limiting relation is called *geostrophic balance*

$$fv - \frac{1}{\rho}\frac{\partial p}{\partial r} = 0, \qquad (4.11)$$

a balance between the radial pressure gradient force and the Coriolis force acting on the tangential flow. In a typical mature hurricane (e.g., Fig. 2 of Shapiro and Montgomery, 1993), the axisymmetric tangential flow is close to cyclostrophic balance in the high wind zone of the vortex ($r < 200$ km), and close to geostrophic balance in the far outer region of the vortex ($r > 700$ km). At intermediate radii ($200 < r < 600$ km), all terms in Eq. (4.9) are of similar order of magnitude.

In summary, gradient wind balance links the mass (pressure) and tangential wind fields in a circular vortex and serves as a valid, zero-order, horizontal diagnostic force balance at all radii in a hurricane. The uniform validity of gradient wind balance as a first approximation for the swirling circulation in a hurricane excludes the region within the frictional boundary layer and in the upper tropospheric outflow layer of the vortex where gradient wind balance is generally not a valid first approximation (see Chapters 6 and 11 for further discussion).

4.4.2 Vortex waves and instabilities

In a stationary cylindrical coordinate-system, the linearized vorticity-equation for incompressible, non-divergent motion is

$$\left(\frac{\partial}{\partial t} + \frac{v}{r}\frac{\partial}{\partial \lambda}\right)\zeta' - \frac{1}{r}\frac{\partial \psi'}{\partial \lambda}\frac{d\zeta_a}{dr} = 0, \qquad (4.12)$$

where ψ' denotes the small-amplitude, perturbation streamfunction, $\zeta' = \nabla^2 \psi'$ is the perturbation vertical vorticity, $v(r)$ the basic-state tangential velocity at radius r, and $\zeta_a(r) = f + \zeta = f + (1/r)d(rv)/dr$ the basic-state absolute vorticity at radius r. On an f-plane, f does not appear explicitly in the problem for the perturbation streamfunction. When Eq. (4.12) has been solved for ψ', the perturbation radial and azimuthal winds are obtained in the usual way (see Section 2.10) from

$$u' = -\frac{1}{r}\frac{\partial \psi'}{\partial \lambda}, \quad v' = \frac{\partial \psi'}{\partial r}. \qquad (4.13)$$

Again, the streamfunction representation ensures that the continuity equation is exactly satisfied for incompressible, non-divergent flow.

The solution to Eq. (4.12) may be obtained by an azimuthal Fourier analysis. Specifically, let $\psi' = \hat{\psi}_n(r,t)e^{in\lambda}$, where $\hat{\psi}_n(r,t)$ denotes the Fourier amplitude of the azimuthal wave-number n and i is the imaginary number ($i^2 = -1$). Also, let $\Omega(r) = v(r)/r$ be the local angular rotation rate of the basic-state vortex. The linearized vorticity equation in Fourier space then becomes (Exercise 4.3):

$$\left(\frac{\partial}{\partial t} + in\Omega\right)\left[\frac{1}{r}\frac{\partial}{\partial r}\left(r\frac{\partial \hat{\psi}_n}{\partial r}\right) - \frac{n^2}{r^2}\hat{\psi}_n\right]$$
$$-\frac{in}{r}\frac{d\zeta}{dr}\hat{\psi}_n = 0. \qquad (4.14)$$

Under certain circumstances, the streamfunction amplitude $\hat{\psi}_n(r,t)$ can grow exponentially in time in the linearized formulation. In such cases, there exist exponentially unstable solutions.

On an f-plane, a necessary condition for the existence of exponentially unstable solutions is that the radial gradient of the basic state relative vorticity, $d\zeta/dr$, changes sign at least once within the flow. This instability theorem is a curvilinear generalization of Lord Rayleigh's 1880 "inflection point" theorem for straight-line, basic state, shear flows. A second necessary condition for exponential instability, together with a supporting physical interpretation for both necessary conditions will be sketched below for completeness. Detailed consideration of barotropically (and/or baroclinically) unstable vortices and the nonlinear life cycle that accompanies these instabilities in the tropical cyclone context is presented in Schubert et al. (1999), Nolan and Montgomery (2000), Nolan et al. (2001), Nolan and Montgomery

(2002), Schecter and Montgomery (2006), Hendricks et al. (2009), and Hendricks and Schubert (2010).

4.4.3 Generalized Rayleigh and Fjortoft instability theorems

Here we sketch the derivation of the two necessary conditions for the existence of exponentially unstable solutions to Eq. (4.14) and develop also a physical interpretation of these two necessary conditions using the anticipated properties of localized vortex Rossby (VR) waves in a variable vorticity gradient mean vortex flow as discussed in Section 4.4.6.

Since the angular velocity and relative vertical vorticity of the basic state vortex are generally functions of radius, but not of azimuth or time, we may assume separable exponential solutions to Eq. (4.14) in azimuth and time. However, because of the radial dependence of the basic state flow, the radial dependence of the solution may not be assumed to have an exponential form. Accordingly, solutions to the eigen-streamfunction must be sought in the form:

$$\psi'(r,\lambda,t) = \hat{\Psi}(r)\exp i(n\lambda - \omega t), \qquad (4.15)$$

where $\hat{\Psi}(r)$ is the radial structure function for the Fourier amplitude of the eigen-streamfunction, λ is the azimuthal angle ($0 \leq \lambda \leq 2\pi$), i is the imaginary number and ω is the eigen-frequency. It is understood also that the real part of the eigen-solution (4.15) and related derivatives must be taken when analyzing the wave's eddy angular momentum flux or eddy energy flux properties. The eigen-frequency may be purely real or have both a real and an imaginary part. In the latter case, the solution may be shown to consist of a complex conjugate pair of an exponentially growing and an exponentially decaying eigenmode in time.

We assume next the existence of a non-trivial unstable eigenmode and write $\omega = \omega_r + i\omega_i$, where the r-subscript denotes the real part and the i-subscript denotes the imaginary part of the eigen-frequency. It follows that exponentially growing modes have $\omega_i > 0$ (and conjugate decaying modes have $\omega_i < 0$). Using the eigenmode structure given by Eq. (4.15), the following radial structure equation for $\hat{\Psi}(r)$ may be derived (after canceling common factors):

$$(\omega - n\,\Omega(r))\left[\frac{1}{r}\frac{\partial}{\partial r}\left(r\frac{\partial}{\partial r}\right) - \frac{n^2}{r^2}\right]\hat{\Psi}(r)$$
$$+\frac{n}{r}\frac{d\zeta}{dr}\hat{\Psi}(r) = 0. \qquad (4.16)$$

Because $\omega_i > 0$, it is evident that $\omega - n\Omega(r)$ never vanishes in the interval $0 \leq r < \infty$. In this case, one may multiply the radial structure equation (4.16) by $r\hat{\Psi}^*(r)/(\omega - n\Omega(r))$, where $\hat{\Psi}^*(r)$ denotes the complex conjugate of the eigen-streamfunction amplitude at radius r. Now, integrate the resulting equation over the flow domain, integrate by parts where appropriate and apply the boundary conditions $\hat{\Psi} \rightarrow$

0 as $r \to 0$ and $r \to \infty$ to eliminate boundary terms. When performing the latter steps, the following identity is used:

$$\frac{1}{(\omega - n\Omega(r))} = \frac{(\omega_r - n\Omega(r)) - i\omega_i}{|\omega - n\Omega(r)|^2}. \quad (4.17)$$

The resulting integral equation is:

$$-\int_0^\infty \left(\left| \frac{d\hat{\Psi}}{dr} \right|^2 + \frac{n^2}{r^2} |\hat{\Psi}|^2 \right) r\, dr$$

$$+ \int_0^\infty \frac{n(\omega_r - n\Omega)|\hat{\Psi}|^2}{|\omega - n\Omega|^2} \frac{d\zeta}{dr} dr$$

$$- i\omega_i \int_0^\infty \frac{n|\hat{\Psi}|^2}{|\omega - n\Omega|^2} \frac{d\zeta}{dr} dr = 0. \quad (4.18)$$

We separate now the integral equation (4.18) into real and imaginary parts and deduce two (necessary) conditions that must be satisfied for the existence of an exponentially growing eigenmode. Since there is only one imaginary term in the integral equation, the integral multiplying the unstable frequency ω_i (> 0) must vanish, i.e.,

$$\int_0^\infty \frac{n|\hat{\Psi}|^2}{|\omega - n\Omega|^2} \frac{d\zeta}{dr} dr = 0. \quad (4.19)$$

After using Eq. (4.19), the remaining integral equation (4.18) becomes:

$$\int_0^\infty \left(\left| \frac{d\hat{\Psi}}{dr} \right|^2 + \frac{n^2}{r^2} |\hat{\Psi}|^2 \right) r\, dr = -\int_0^\infty \frac{n^2 \Omega |\hat{\Psi}|^2}{|\omega - n\Omega|^2} \frac{d\zeta}{dr} dr. \quad (4.20)$$

The first condition (4.19) leads immediately to a necessary condition for shear instability. Apart from the mean radial vorticity gradient $d\zeta/dr$, the other terms comprising the integrand in Eq. (4.19) are non-negative. *A vanishing integral necessarily implies that the mean vorticity gradient change sign at least once in the flow. This is the generalized Rayleigh theorem for vortex shear instability.*[7]

The second condition (4.20) leads to a second necessary condition for shear instability. *Eq. (4.20) implies that $\Omega(r)$*

and $d\zeta/dr$ must be negatively correlated on average, with a weighting function equal to $n^2|\hat{\Psi}|^2/|\omega - n\Omega|^2$. This theorem, referred to here as a generalized Fjørtoft theorem, gives an additional constraint on the existence of asymmetric shear instabilities in a rapidly rotating, incompressible, non-divergent circular vortex. For although a given vortex may satisfy Rayleigh's necessary condition for shear instability, if $\Omega \times d\zeta/dr$ is everywhere positive, the vortex will be exponentially stable.[8]

4.4.3.1   Physical interpretation

In light of now accepted interpretations of shear instability as the phase locking of counter-propagating Rossby-like waves (e.g., Schubert et al., 1999, their Section 2), it is insightful to examine the physical content of the above necessary conditions. Given that sheared disturbances in a variable vorticity environment behave as VR waves as discussed in Section 4.4.6, we may interpret the *generalized Rayleigh theorem as a necessary condition for the existence of counter-propagating VR waves relative to the local mean swirling flow.* Similarly, *the generalized Fjørtoft theorems represent necessary conditions for these counter-propagating VR waves to phase lock and grow in strength.*[9]

4.4.4   Solution to initial value problems

A formal solution to the general initial value problem for Eq. (4.14) may be obtained using Laplace transform techniques. The Laplace transform of an arbitrary function $\chi(r, t)$ is defined by

$$\hat{\chi}(r, s) = \int_0^\infty e^{-st} \chi(r, t) dt, \quad (4.21)$$

and if $\hat{\chi}(r, s)$ is known, the inverse transform is obtained as a contour integral in the complex s-plane:

$$\chi(r, t) = \frac{1}{2\pi i} \int_{c-i\infty}^{c+i\infty} e^{st} \hat{\chi}(r, s) ds, \quad (4.22)$$

---

7. In the analogous problem of a basic state zonal shear flow, $U(y)$, the mean vorticity gradient is equal to minus the second (meridional) derivative of the zonal flow, $d\zeta/dy = -U''(y)$, where $'$ denotes a meridional derivative. In this case, the requisite sign change of the mean vorticity gradient is equivalent to a sign change in the concavity of the mean flow. For a smooth zonal flow, a change of concavity can only occur if there is (at least) one inflection point in the velocity profile. For this reason the theorem is sometimes referred to as "Rayleigh's inflection point theorem" (Rayleigh, 1880). However, the curvilinear case is somewhat more general since the vanishing of $d\zeta/dr$ is *not* equivalent to the vanishing of U''. For this reason we regard the theorem as a generalization of Rayleigh's theorem. Note that this theorem is distinct from Rayleigh's 1916 theorem on the axisymmetric centrifugal instability of curvilinear shear flow (Rayleigh, 1916, see Section 5.7 for details).

8. In this normal-mode context, the Fjørtoft theorem can be relaxed further. Let r_s denote a radius at which $d\zeta/dr = 0$. Then (4.19), together with (4.20), implies that $(\Omega - \Omega(r_s))d\zeta/dr < 0$ somewhere in $0 < r < \infty$ is necessary also for instability. If, on the other hand, $(\Omega - \Omega(r_s))d\zeta/dr > 0$ everywhere in the flow, then the vortex is exponentially stable. Finally, if $\Omega(r)$ is monotonic and $d\zeta/dr$ changes sign only once, a tighter necessary condition is that $(\Omega - \Omega(r_s))d\zeta/dr \leq 0$, for all radii, with equality occurring only at r_s. See Gent and McWilliams (1986) for details.

9. These necessary conditions, and supporting physical interpretation in terms of counter-propagating, phase-locked VR waves, have been generalized to three-dimensional, quasi-balanced motions in a stably stratified tropical cyclone vortex by Montgomery and Shapiro (1995). These instability theorems have been extended also to generalized stability theorems involving both regular normal-mode (discrete) and singular normal-mode (sheared) VR waves by Ren (1999). In a planned sequel to this book, we will return to investigate VR waves, VR instabilities, and their nonlinear life cycles for some canonical tropical cyclone configurations.

where c is a constant chosen so that the contour of integration in the complex s-plane lies to the right of all singularities of $\hat{\chi}(r, s)$. It is easy to show that the Laplace transform of $\partial\chi/\partial t$ is $s\hat{\chi}(r, s) - \chi(r, 0)$ and it follows that the Laplace transform of Eq. (4.14) satisfies the ordinary differential equation

$$(s + in\Omega(r))\left[\frac{1}{r}\frac{\partial}{\partial r}\left(r\frac{\partial}{\partial r}\right) - \frac{n^2}{r^2}\right]\hat{\psi}(r, s) - \frac{in}{r}\frac{d\zeta}{dr}\hat{\psi}(r, s) = \left[\frac{1}{r}\frac{\partial}{\partial r}\left(r\frac{\partial}{\partial r}\right) - \frac{n^2}{r^2}\right]\hat{\psi}_n(r, t = 0).$$

(4.23)

Dividing by $s + in\Omega(r)$ and noting that the right-hand-side of Eq. (4.23) is the initial vorticity $\hat{\zeta}_o$ of the n-th Fourier component, we obtain

$$\left[\frac{1}{r}\frac{\partial}{\partial r}\left(r\frac{\partial}{\partial r}\right) - \frac{n^2}{r^2}\right]\hat{\psi}(r, s)$$
$$- \frac{in}{r}\frac{d\zeta}{dr}\frac{\hat{\psi}(r, s)}{(s + in\Omega(r))} = \frac{\hat{\zeta}_0(r, n)}{(s + in\Omega(r))}.$$

(4.24)

In principle, when this equation has been solved for $\hat{\psi}(r, s)$, the inverse transform can be obtained for $\psi(r, t)$.

From the calculus of residues, we know that the general solution of Eq. (4.24) consists of a sum of discrete exponential solutions associated with the zeros of the Wronskian,[10] together with an integral along branch cuts associated with each zero of $(s + in\Omega(r))$ (purely imaginary s) that characterizes the continuous spectrum. The set of discrete exponential solutions correspond with "regular normal modes" (including the exponentially unstable solutions summarized in the prior section) while the set of integrals along branch cuts correspond with "singular normal modes" (see e.g. Section 4.4.5) since the associated elemental solutions to the vortex wave equation possess a singular structure in radius. This second set of solutions comprise a continuum of *critical radii*. The solutions associated with each *critical radius* rotate azimuthally with a frequency ω_c equal to n times the corresponding angular velocity of the mean vortex at the critical radius, $\Omega(r_c)$, i.e., $\omega_c = n\Omega(r_c)$.

Whereas the continuum solutions, when considered in isolation (via Eq. (4.15)), possess singular behavior in the vicinity of each critical radius, superposition of neighboring continuum solutions often permits an escape from the singular structure constrained by their elemental form (e.g., see Section 4.4.6 and refs.). Although explicit analytical solutions to Eqs. (4.14) and (4.24) for arbitrary initial conditions have been obtained only in a handful of cases (Carr and Williams, 1989; Sutyrin, 1989; Smith and Rosenbluth,

1990; Montgomery and Kallenbach, 1997; Nolan and Montgomery, 2000; Brunet and Montgomery, 2002), a brief examination of a subset of these cases proves instructive as a foundation for a more complete understanding of the evolution of small-amplitude vortex asymmetries in tropical cyclones, and the axisymmetrization process in particular.

In the next section we consider arguably the simplest case of a Rankine wind profile outside the radius of maximum tangential wind. Within the realm of barotropic, linearized, non-divergent dynamics, this specific vortex possesses only a single-signed vorticity gradient and, according to the generalized Rayleigh theorem discussed above, is exponentially stable, supporting only sheared waves that are ultimately axisymmetrized by the differential rotation of the mean vortex.

4.4.5 Case I: bounded Rankine vortex: $V = \Gamma/r$, $\Omega = \Gamma/r^2$, $\Gamma = $ constant, $a \leq r \leq b$

In this case $d\zeta/dr = 0$ and Eq. (4.14) becomes

$$\left(\frac{\partial}{\partial t} + in\Omega\right)\left[\frac{1}{r}\frac{\partial}{\partial r}\left(r\frac{\partial\hat{\psi}_n}{\partial r}\right) - \frac{n^2}{r^2}\hat{\psi}_n\right] = 0. \quad (4.25)$$

The inverse Laplace transform of Eq. (4.25) is obtained readily in this case owing to simple poles in the complex s-plane associated with the vanishing of $s + in\Omega(r)$ at critical radii (without branch cuts),

$$\left[\frac{\partial}{\partial r}\left(r\frac{\partial}{\partial r}\right) - \frac{n^2}{r}\right]\hat{\psi}_n(r, t) = r\hat{\zeta}_0(r, n)e^{-in\Omega(r)t}. \quad (4.26)$$

The solution of this equation in the domain $a \leq r \leq b$ is

$$\hat{\psi}_n(r, t) = \int_a^b G(r, p)\hat{\zeta}_0(p, n)e^{-in\Omega(p)t}pdp, \quad (4.27)$$

where the Green function is given by

$$G(r, p) = \frac{1}{2nr^n(a^{2n} - b^{2n})} \times$$
$$\begin{cases} (p^n - b^{2n}p^{-n})(a^{2n} - r^{2n}), & a \leq r \leq p \\ (p^n - a^{2n}p^{-n})(b^{2n} - r^{2n}), & p \leq r \leq b \end{cases} \quad (4.28)$$

Finally, the Fourier inversion for azimuthal wavenumber n is

$$\psi_n(r, \lambda, t) = e^{in\lambda}\int_a^b G(r, p)\hat{\zeta}_0(p, n)e^{-in\Omega(p)t}pdp. \quad (4.29)$$

Some solutions for this case are shown in Figs. 4.9 and 4.10 for various initial distributions of $\hat{\zeta}_0(r, n)$ and n. Fig. 4.9 shows the evolution of the vorticity field and

---

10. See Appendix 3.6.2.

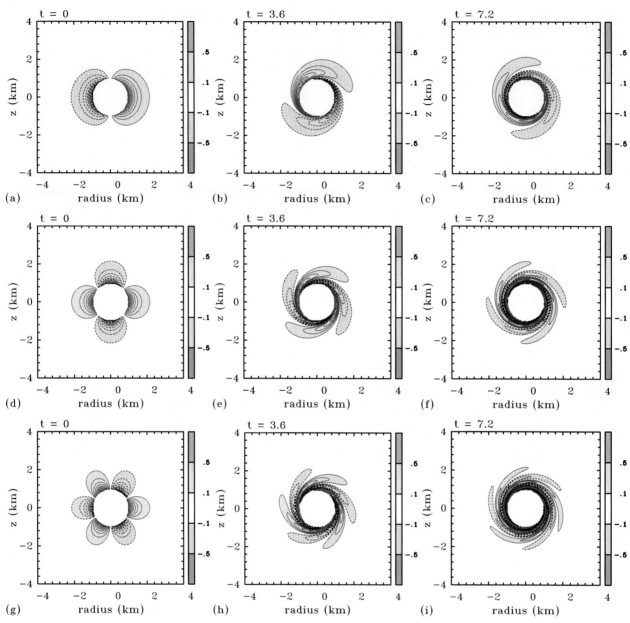

FIGURE 4.9 Nondimensional perturbation vorticity field for azimuthal wavenumbers $n = 1, 2$, and 3 (by row) in Case I for the initial vorticity $\hat{\zeta}_0(r) = 1/r^3$ and at nondimensional times $t = 0$, $t = 3.6$, and $t = 7.2$ (by column as indicated). Contour interval is 1×10^{-1} units. Shading as indicated on the sidebar in the same units. Positive values solid in red (mid gray in print version), negative values dashed in blue (dark gray in print version). Adapted from Smith and Montgomery (1995). Republished with permission of Wiley.

Fig. 4.10 shows the corresponding perturbation streamfunction fields for a distribution of initial vorticity $\hat{\zeta}_0(r, n) = 1/r^3$ that is non-tilted with respect to the azimuthal shear. In the figures, the rows represent azimuthal wavenumbers $n = 1, 2$, and 3 respectively, while the columns designate nondimensional times $t = 0$, $t = 3.6$, and $t = 7.2$, respectively. The figures show how the initial disturbances are sheared out preferentially in the inner region of the vortex by the large angular shear of the vortex.

This behavior is reminiscent of that shown earlier in Chapter 3 and in Section 4.1.1, wherein the perturbation vorticity is differentially advected around the vortex center, just like that of a passive tracer. Indeed, the method of solution developed in Chapter 3 can be applied to solve Case I (note that Eq. (4.26) is equivalent to Eq. (3.35)), but it cannot be generalized to cases in which the vortex has a radial gradient of vorticity. Note also that, as shown in Section 4.1.1, the case of azimuthal wavenumber one is not quite consistent with the assumption of a stationary vortex

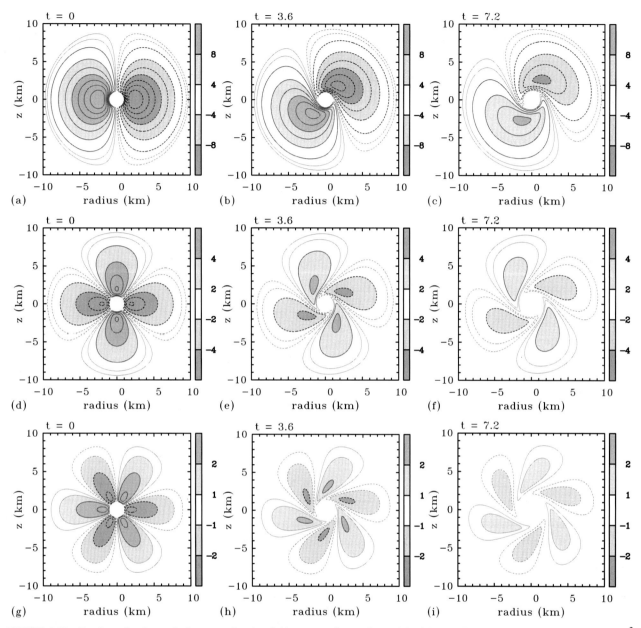

FIGURE 4.10 Nondimensional perturbation streamfunction field corresponding to the vorticity fields in Fig. 4.9. The contour interval is 2×10^{-2} Shading as on the side bar. Positive values solid in red (mid gray in print version), negative values dashed in blue (dark gray in print version). Adapted from Smith and Montgomery (1995). Republished with permission of Wiley.

because, then there is a flow across the vortex center that would cause the vortex to translate. However, the results of Chapter 3 suggest that the stationary vortex assumption may be justified for small perturbations to asymmetry where the flow across the vortex center is weak, or if it is assumed that the analysis applies to a vortex that is translating with an approximately uniform velocity.

Again as in Section 4.1.1, the radial scale of the perturbation vorticity decreases with time and consequently the local radial wavenumber of the perturbation increases linearly with time. Because the local radial wavenumber can

be shown to generally increase linearly with time (see below for details), the conservation of perturbation vorticity following the mean angular velocity implies that the amplitude of the perturbation streamfunction (illustrated in Fig. 4.10) must ultimately decay quadratically toward zero with time. The algebraic growth of perturbation streamfunction for an initially upshear-tilted perturbation and the algebraic decay of perturbation streamfunction for a downshear-tilted perturbation is a simple illustration of the growth and decay phases, respectively, of the "Orr mechanism" for simple sheared-vorticity disturbances (e.g., Lindzen, 1990).

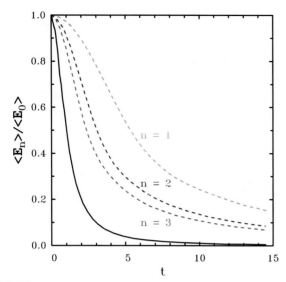

FIGURE 4.11 Normalized perturbation kinetic energy for the bounded Rankine vortex (Case I) with $\hat{\zeta}_0(r) = r^{-3}$; a = 1, b = 10. The dashed curves show the energy decay for: n = 1 (green (gray in print version)), n = 2 (blue (dark gray in print version)), and n = 3 (red (mid gray in print version)). The solid curve shows the limiting energy decay given by $1/[1 + (S_{eff}t)^2]$, where $S_{eff} = -1.01$. Adapted from Smith and Montgomery (1995). Republished with permission of Wiley.

Physically pertinent properties of these illustrated solutions, such as the dependence of integrated kinetic energy on azimuthal wavenumber, and the preferential damping of high azimuthal wavenumber disturbances over low azimuthal wavenumber disturbances, even in the absence of viscous damping, are examined and explained in Smith and Montgomery (1995), their Sections 2c, 3d, and 3e.

Fig. 4.11 shows an example of the dependence of the integrated perturbation kinetic energy on time and azimuthal wavenumber for the three perturbation vorticity fields displayed in Fig. 4.9. Here the kinetic energy of the disturbance is integrated over the fluid annulus and is normalized by its initial value. The lowest curve in the figure depicts a limiting energy decay given by $1/[1 + (S_{eff}t)^2]$ where S_{eff} is an effective radial shear:

$$S_{eff} = \frac{\int_a^b r\frac{d\Omega}{dr}|\hat{\zeta}_0(r)|dr}{\int_a^b |\hat{\zeta}_0(r,n)(r)|dr}, \qquad (4.30)$$

the integrals being over the initial vorticity disturbance. The fact that high azimuthal wavenumber perturbations tend to be preferentially damped over low wavenumber perturbations provides a plausible explanation of the observation of nearly axisymmetric flow in the near-core region of intense tropical cyclones and, in particular, the predominance of very low wavenumber wind asymmetries, primarily n = 0 and n = 1, as illustrated in Fig. 4.2.

4.4.6 More on vortex waves

Unlike the foregoing calculations, and those presented in Chapter 3, the vortex wave dynamics become strongly dispersive when the term $ud\zeta/dr$ in Eq. (4.12) is comparable with the other terms. In the absence of discrete exponentially-unstable modes, this term can temporarily forestall the algebraic decay of the perturbation streamfunction noted above. With the $ud\zeta/dr$ term present, perturbation vorticity is no longer conserved following the mean angular velocity Ω. For the case of a negative mean radial vorticity gradient, i.e. $d\zeta/dr < 0$, the vorticity anomalies tend to propagate azimuthally more slowly than the local mean angular velocity. In this case, the vorticity waves are retrograde relative to the local mean angular velocity.

The aforementioned properties may be inferred from the local dispersion relation for tightly-spiraled wave packets whose radial scale is small compared to the characteristic radial scale of the mean vortex as defined by v. For basic state circular vortices possessing a smooth-in-radius vorticity distribution, sheared wave packets are governed by the local dispersion relation (Montgomery and Kallenbach, 1997; McWilliams et al., 2003):

$$\omega = n\Omega + \frac{\frac{n}{r}\frac{d\zeta}{dr}}{k^2 + \frac{n^2}{r^2}}, \qquad (4.31)$$

where ω is the wave packet's local frequency, $\Omega = v(r)/r$ is the local mean angular velocity as defined above, ζ is the local mean relative vertical vorticity of the basic state vortex, k is the local (time-dependent) radial wavenumber whose time derivative obeys the relation $dk/dt = -nd\Omega/dr$, and n is the (constant) azimuthal wavenumber.

To gain perspective on the foregoing dispersion relation, it is helpful to recall the dispersion relation for planetary Rossby waves on a mid-latitude beta plane in a spatially uniform zonal mean flow U:

$$\omega = kU - \frac{k\beta}{k^2 + l^2}, \qquad (4.32)$$

where β is the meridional gradient of planetary vorticity (df/dy), k is here the zonal wavenumber, l is the meridional wavenumber and ω represents the planetary Rossby wave frequency (e.g., Holton, 2004, Section 7.7).

A comparison between Eqs. (4.31) and (4.32) indicates that n/r is the cylindrical analog of the zonal wavenumber k and the mean angular velocity Ω is the cylindrical analog of the zonal mean flow U. It follows that minus the radial gradient of the mean relative vorticity, $-d\zeta/dr$, is analogous to the meridional gradient of planetary vorticity, $\beta = df/dy$. For these reasons the ensuing vorticity waves with non-negligible $ud\zeta/dr$ are accordingly called "vortex Rossby waves" (referred to above and hereafter as

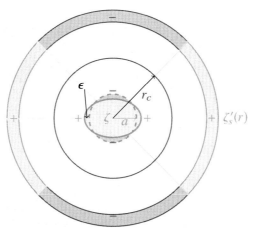

FIGURE 4.12 Toy wave-vortex model developed in Sections 4.4.8 and 4.5. An initially circular basic state vortex with core vorticity ζ is perturbed by a smooth-in-radius vorticity perturbation, $\hat{\zeta}_s(r)\exp(i[n\lambda - \omega_n t]) + $ c.c., where n is the azimuthal wavenumber and c.c. denotes complex conjugate. Here the basic state is a Rankine vortex, which consists of constant relative vorticity inside $r = a$, and zero relative vorticity outside $r = a$. Since the basic Rankine flow is irrotational beyond the edge of the core region ($r > a$), the vorticity perturbation outside the core evolves as a sheared disturbance with no radial propagation, just like that of a passive tracer in the azimuthal swirling flow $\Omega(r)$. Because of the discontinuity in the basic state vorticity at $r = a$, the perturbation vorticity of the vortex edge-wave, $\hat{\zeta}_d$, will have a singular (Dirac delta function) structure at $r = a$. The perturbed interface, $\epsilon(\lambda, t)$, depicted as a wavenumber-2 disturbance in the figure, marks the boundary between the perturbed rotational 'core' fluid ($0 < r < a + \epsilon$) and the perturbed irrotational 'environment' fluid ($a + \epsilon < r < \infty$). The ensuing vorticity response to the initial perturbation is the sum of a sheared vorticity disturbance plus a discrete, vortex Rossby (VR) edge wave, $\hat{\zeta}_d\exp(i[n\lambda - \omega_n t]) + $ c.c, whose amplitude is proportional to the product of the smooth initial vorticity amplitude and the Dirac delta function, $\delta(r - a)$. The azimuthal propagation of the discrete VR edge wave is retrograde relative to the tangential velocity of the mean vortex at $r = a$. The critical radius r_c denotes the radius where the angular frequency of the discrete VR edge wave equals n times the mean angular velocity of the vortex: $\omega_n = n\Omega(r_c)$, where the VR edge wave frequency is given by Eq. (4.55). When the smooth vorticity perturbation is initialized close to r_c, the sheared vorticity disturbance rotates azimuthally at a similar rotation rate to the discrete VR edge wave and a type of "resonant response" occurs in both the perturbation solution and the wave-vortex interaction. The ensuing nonlinear dynamics involves the stirring of perturbation vorticity around $r = r_c$ in a so-called "critical layer" region. See text for further details.

VR waves).[11] In general, the dynamics of VR waves on smooth vorticity basic states are more complex than the simple shear waves examined so far using the Rankine vortex model (e.g., Montgomery and Kallenbach, 1997; Nolan and Montgomery, 2000; Brunet and Montgomery, 2002; Montgomery and Brunet, 2002; McWilliams et al., 2003; Nikitina and Campbell, 2015a,b).

4.4.7   Relevance to tropical cyclones

At this point it seems to be worth reviewing the possible applications of this family of wave-vortex models and the types of questions one might wish to answer in the context of tropical cyclones. One question that arises from observa-

tions is how does the inner-core vortex respond to intermittent episodes of deep convection within or outside the vortex core? Another question that arises is how do wind asymmetries generated through the interaction of the vortex with its local environment impact the mean vortex as it moves through this environment? While it is beyond the scope of this book to give a complete answer to these questions, the discussion that follows provides a tractable analytic framework for obtaining such answers, at least for sufficiently weak perturbations. In this context, it is intended to provide a stepping stone to enable the reader to further explore the extensive literature on the topic of vortex waves and their impact on storm structure and behavior.

As a prelude to a more complete theoretical treatment involving both sheared and discrete VR waves, we limit our focus here to the dynamics of waves in the Rankine vortex model. In this limiting model, the VR waves are mathematically *distinct* from the sheared vorticity disturbances (see Fig. 4.12). In the remaining sections we use this limiting Rankine vortex model to highlight some interesting properties of the sheared vorticity waves and their implied wave-vortex interaction at second order in wave amplitude.

---

11. In the fluid dynamics literature, when using the Rankine vortex as the basic state, the asymmetric vorticity waves in non-divergent flow are called Kelvin waves after Lord Kelvin who first discovered them (Kelvin, 1880). However, in geophysical fluid dynamics the term "Kelvin waves" is reserved for either a class of eastward propagating waves in the equatorial atmosphere (or ocean) or with waves propagating parallel to a topographic boundary, such as a north-south oriented coastline (e.g., Gill, 1982). In the general case of a basic state vortex with a radially smooth vorticity distribution, the local dispersion relation is indeed mathematically isomorphic to the corresponding dispersion relation for planetary Rossby waves in zonal shear flow. Therefore, we adhere to the convention of calling these waves VR waves.

4.4.8 Case II: unbounded Rankine vortex

In an unbounded domain, the Rankine vortex is defined by the tangential velocity profile

$$v(r) = \begin{cases} r/a, & r \leq a, \\ a/r, & a \leq r, \end{cases} \qquad (4.33)$$

where a is the radius of maximum winds. This profile is shown in Fig. 4.13. The corresponding profile of basic-state vorticity is

$$\zeta(r) = \begin{cases} 2/a, & r < a, \\ 0, & a < r, \end{cases} \qquad (4.34)$$

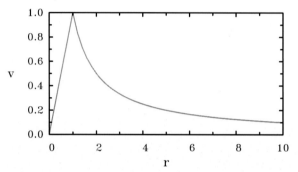

FIGURE 4.13 Radial tangential wind profile, $v(r)$, for the full Rankine vortex.

Here, as in Case I, the vorticity has been scaled by V_m/R_m and the tangential velocity has been scaled by V_m, where V_m denotes the maximum tangential velocity of the basic state vortex and R_m denotes the radius of this maximum.

The discontinuity in the mean-state vorticity at $r = a$ effectively introduces another boundary to the system. Since this boundary lies in the interior of the fluid, kinematic and dynamic boundary conditions must be satisfied at the disturbed interface defined by $r = a + \epsilon$, where $\epsilon(\lambda, t)$ is the interfacial displacement field in the neighborhood of $r = a$. The kinematic boundary condition requires that the normal velocity be continuous at $r = a + \epsilon$, while the dynamic boundary condition requires that the perturbation pressure be continuous at this radius.[12] Consistent with the linearization of the horizontal momentum equations about the basic state vortex, these two matching conditions may be evaluated at $r = a$. Once u is determined, the evolution of the

---

12. Strictly speaking, the dynamic boundary condition requires that the total pressure be continuous at $r = a + \epsilon(\lambda, t)$. At the level of linear theory, this condition is effected by employing a Taylor series expansion for the total pressure in powers of ϵ and neglecting second order terms in perturbation quantities. Since the basic state tangential velocity in the Rankine vortex is continuous at $r = a$, the gradient wind balance equation (4.9) implies that the basic state pressure and its radial gradient are continuous at $r = a$. As a result, the basic state pressure and its radial gradient drop out of the linearized dynamic boundary condition for the Rankine vortex.

disturbed interface, at leading order in wave amplitude, may be found by integrating the linearized interfacial equation

$$\left(\frac{\partial}{\partial t} + \Omega(r)\frac{\partial}{\partial \lambda} \right)\epsilon = u \qquad (4.35)$$

for ϵ at $r = a$.

For the full Rankine vorticity profile (4.34), the linearized vorticity equation (4.12) is modified to

$$\left(\frac{\partial}{\partial t} + \Omega(r)\frac{\partial}{\partial \lambda} \right)\zeta' = 0 \quad r \neq a. \qquad (4.36)$$

To solve Eq. (4.36), the discontinuity in the basic-state vorticity at $r = a$ must be accounted for. Since the wave problem is linear, the superposition principle may be used to separate the solution into two parts by letting $\zeta' = \zeta_s + \zeta_d$, where ζ_s is defined to be smooth for all r and ζ_d accounts for the discontinuity in the basic-state vorticity at $r = a$. The vorticity equation (4.36) is then split into two parts

$$\left(\frac{\partial}{\partial t} + \Omega(r)\frac{\partial}{\partial \lambda} \right)\zeta_s = 0, \quad \forall r \qquad (4.37)$$

$$\left(\frac{\partial}{\partial t} + \Omega(r)\frac{\partial}{\partial \lambda} \right)\zeta_d = 0 \quad r \neq a. \qquad (4.38)$$

Eq. (4.37) is formally identical to the system solved in Case I, but with the modified boundary conditions given above. The corresponding solution in Fourier space (see Exercise 4.4) is then

$$\hat{\psi}_s(r, t) = \int_0^\infty G(r, p)\hat{\zeta}_{s0}(p)e^{-int\Omega(p)}p\,dp, \qquad (4.39)$$

where the appropriate Green function is

$$G(r, t) = -\frac{1}{2n} \begin{cases} p^{-n}r^n, & 0 \leq r \leq p \\ p^n r^{-n}, & p \leq r \leq \infty \end{cases} \qquad (4.40)$$

and $\hat{\zeta}_{s0}(p)$ is the smooth component of $\hat{\zeta}$ for azimuthal wavenumber n at time $t = 0$. (Here, to avoid clutter, the dependence of $\hat{\zeta}_{s0}$ on azimuthal wavenumber is notationally suppressed.)

The Fourier-space equivalent to Eq. (4.38) is

$$\left(\frac{\partial}{\partial t} + in\Omega(r) \right)\hat{\zeta}_d = 0, \quad r \neq a. \qquad (4.41)$$

Anticipating that the solution to Eq. (4.41) will yield the discrete wave modes, which are irrotational on both sides of the mean-state vorticity discontinuity, $\hat{\zeta}_d$ is assumed to be separable and of the form

$$\hat{\zeta}_d = \gamma(t)\delta(r - a). \qquad (4.42)$$

Here, γ is an undetermined temporal multiplier for $\hat{\zeta}_d$, and $\delta(r - a)$ is the one-dimensional Dirac delta function. As is usual in this type of formulation, expressions involving the delta function should be interpreted as a generalized function. That is, the delta function is a limiting process of a narrow function around $r = a$ whose amplitude grows commensurately with the spatial shrinking of the nonzero values of the function, but whose integral containing $r = a$ remains equal to unity.

The delta function arises naturally from the implied radial derivative of the Heaviside step function of the basic state vorticity; *i.e.*, from the radial derivative of the discontinuous basic state vorticity at $r = a$. This delta function term is associated with the radial advection of the basic Rankine vorticity by the perturbation radial velocity in Eq. (4.14). On kinematic grounds, the radial velocity must be finite and continuous at all radii. In terms of the streamfunction for this edge-wave component, $\hat{\psi}_d$, Eq. (4.42) becomes

$$\nabla^2 \hat{\psi}_d = \gamma(t)\delta(r - a). \qquad (4.43)$$

This perturbation streamfunction is assumed to be separable also and to be of the form $\hat{\psi}_1 = \gamma(t)\hat{\Psi}_d(r)$. Thus, Eq. (4.43) becomes

$$\left[\frac{1}{r}\frac{d}{dr}\left(r\frac{d}{dr}\right) - \frac{n^2}{r^2}\right]\hat{\Psi}_d = \delta(r - a). \qquad (4.44)$$

For $r \neq a$, Eq. (4.44) is Euler's equation. Two conditions are needed to match the solutions in each region across $r = a$. The first is the kinematic boundary condition requiring that the radial velocity u be continuous at $r = a$. Consequently, the Fourier streamfunction-amplitude must be continuous across $r = a$. The second condition on the discrete mode streamfunction results from integrating Eq. (4.44) over a small radial interval that includes $r = a$. This yields a jump condition for $d\hat{\psi}_d/dr$:

$$\frac{d}{dr}\hat{\Psi}_d(a^+) - \frac{d}{dr}\hat{\Psi}_d(a^-) = 1. \qquad (4.45)$$

Applying the boundary conditions, and the continuity and jump conditions at $r = a$, yields

$$\hat{\Psi}_d = -\frac{a}{2n}\begin{cases} a^{-n}r^n, 0 \leq r \leq a \\ a^n r^{-n}, a \leq r \leq \infty. \end{cases} \qquad (4.46)$$

To complete the solution, $\gamma(t)$ must be determined. The remaining constraint is the dynamic boundary condition at $r = a + \epsilon$, which for the Rankine vortex requires that the perturbation pressure be continuous there. In Fourier space, the azimuthal-momentum equation is given by

$$\frac{\partial \hat{v}}{\partial t} + in\Omega(r)\hat{v} + \zeta_a \hat{u} = -\frac{in}{r}\hat{p}, \qquad (4.47)$$

where $\zeta_a = f + \zeta$ is the absolute vorticity of the mean vortex.

Consistent with the accuracy of linearized dynamics, we may evaluate Eq. (4.47) on each side of $r = a$. Subtracting these two evaluations from one another and imposing continuity of perturbation pressure at $r = a$ then gives

$$\frac{\partial}{\partial t}\left(\hat{v}(a^+) - \hat{v}(a^-)\right) + \frac{in}{a}\left(\hat{v}(a^+) - \hat{v}(a^-)\right)$$
$$-\frac{2}{a}\hat{u}(a) = 0, \qquad (4.48)$$

where "+" denotes evaluation just outside $r = a$ and "-" denotes evaluation just inside $r = a$.

In terms of the *total* perturbation streamfunction amplitude, Eq. (4.48) becomes

$$\frac{\partial}{\partial t}\left(\frac{\partial \hat{\psi}}{\partial r}(a^+) - \frac{\partial \hat{\psi}}{\partial r}(a^-)\right) +$$
$$\frac{in}{a}\left(\frac{\partial \hat{\psi}}{\partial r}(a^+) - \frac{\partial \hat{\psi}}{\partial r}(a^-)\right) + \frac{2in}{a^2}\hat{\psi}(a) = 0. \qquad (4.49)$$

Now, from the superposition principle, $\hat{\psi} = \hat{\psi}_s + \gamma\hat{\Psi}_d$, where ψ_s and its derivatives are everywhere smooth by construction. Since Ψ_d is continuous, but has a unit jump in its derivative across $r = a$, Eq. (4.49) simplifies (Exercise 4.5) to

$$\frac{d\gamma}{dt} + \frac{i}{a}(n-1)\gamma = -\frac{2in}{a^2}\hat{\psi}_s(a, t), \qquad (4.50)$$

a first-order linear differential equation for $\gamma(t)$. Multiplying this equation by the integrating factor $\exp\{i(n-1)t/a\}$ and substituting for $\hat{\psi}_s(a, t)$ using Eq. (4.39), Eq. (4.50) becomes

$$\frac{d}{dt}(\gamma e^{i(n-1)t/a}) =$$
$$-\frac{2in}{a^2}\int_0^\infty G(a, p)\hat{\zeta}_{s0}e^{\{i(n-1)/a - in\Omega(p)\}t}p\,dp. \qquad (4.51)$$

Integrating Eq. (4.51) in time and then multiplying through by $\exp\{-i(n-1)t/a\}$ gives

$$\gamma(t) = -\frac{2n}{a}\int_0^\infty \frac{G(a, p)\hat{\zeta}_{s0}}{(n-1-an\Omega(p))}e^{-int\Omega(p)}p\,dp$$
$$+c_1 e^{-i(n-1)t/a}, \qquad (4.52)$$

where c_1 is the constant of integration at $t = 0$. Eq. (4.52) then yields

$$\gamma(t) = -\frac{2n}{a}\int_0^\infty \frac{G(a, p)\hat{\zeta}_{s0}}{(n-1-an\Omega(p))}e^{-in\Omega(p)t}p\,dp$$
$$+\left(\gamma_0 + \frac{2n}{a}\int_0^\infty \frac{G(a, p)\hat{\zeta}_{s0}}{(n-1-an\Omega(p))}p\,dp\right)e^{-i(n-1)t/a}, \qquad (4.53)$$

where γ_0 is the initial amplitude of the VR edge-wave associated with the radial discontinuity in the vorticity of the mean Rankine vortex. The total perturbation streamfunction-amplitude, $\hat{\psi}$, is thus (Exercise 4.6):

$$
\begin{aligned}
\hat{\psi} = &\int_0^\infty G(r,p)\hat{\zeta}_{s0}e^{-in\Omega(p)t}\,p\,dp \\
&- \frac{2n}{a}\hat{\Psi}_d \int_0^\infty \frac{G(a,p)\hat{\zeta}_{s0}}{(n-1-an\Omega(p))}e^{-in\Omega(p)t}\,p\,dp \\
&+ \hat{\Psi}_d\left(\gamma_0 + \frac{2n}{a}\int_0^\infty \frac{G(a,p)\hat{\zeta}_{s0}}{(n-1-an\Omega(p))}p\,dp\right) \\
&\times e^{-i(n-1)t/a},
\end{aligned}
\tag{4.54}
$$

where G is given by Eq. (4.40) and $\hat{\Psi}_d$ is given by Eq. (4.46). To obtain the physical-space streamfunction, the inverse Fourier transform must be applied to Eq. (4.54).

4.4.8.1 Interpretations

Eq. (4.54) can be given a relatively simple interpretation. Setting aside for the moment the possibility of an apparent singularity for $n > 1$ in the second and third integrals associated with a vanishing denominator and nonzero perturbation vorticity there, it is clear that the first term in (4.54) may be identified solely with the continuous-spectrum solution and represents the unbounded analog of the solution presented in Case I. The second and third terms together represent a conversion term that transfers a portion of the kinetic energy from the continuous-spectrum solution into the discrete (VR wave) mode: the second term represents a transition between the continuous spectrum solution and the discrete mode; the third term furnishes the projection of the initial condition onto the discrete VR wave. Note that even with no initial VR edge wave mode, i.e. $\gamma_0 = 0$, the continuous-spectrum solution always projects onto the VR edge wave mode at later times provided the corresponding radial integral does not vanish.

For $n > 1$, the discrete Rossby modes rotate more slowly than the vortex, and, as alluded to above, represent retrograde VR edge-waves at $r = a + \epsilon(\lambda, t)$, where as defined above ϵ is the amplitude of the perturbed interface between the rotational core and irrotational environment of the vortex.

In this mathematical idealization, which assumes a singular basic state vorticity gradient at $r = a$, the VR wave dispersion relation differs from the continuous relation given by Eq. (4.31) in that the wave frequency is proportional to the jump in basic-state vorticity between the core and environment ($\Delta = \zeta + f - f = \zeta$) rather than the radial derivative of the basic-state vorticity ($d\zeta/dr$). The dimensional dispersion relation for the singular case (without sheared vorticity disturbances) can be obtained from the general equation (4.50) on setting the sheared solution

component to zero (i.e., setting $\hat{\psi}_s = 0$), solving for γ and re-dimensionalizing the wave's frequency. The dimensional frequency ω_n for the VR edge wave is thus given by[13]

$$
\omega_n = \frac{1}{2}\zeta(n-1),
\tag{4.55}
$$

where n is the integral azimuthal wavenumber and ζ is the constant vorticity of the vortex core (for more on the geometry of these VR edge waves - see Fig. 4.12). Apart from the non-dimensionalization introduced earlier, the exponent in the term given by $\exp[-i(n-1)t/a]$ in Eq. (4.53) for the discrete mode is just the VR edge wave frequency given by Eq. (4.55).

For $n = 1$, it may be shown that, in the absence of perturbation vorticity inside the radius of maximum winds, as $t \to \infty$ only the non-rotating edge-wave component remains in the perturbation field. That is, as $t \to \infty$, the first and second terms in Eq. (4.54) tend to zero as t^{-2}, leaving just the third term:

$$
\hat{\psi} \to \hat{\Psi}_d\left(\gamma_0 - \frac{2}{a^2}\int_0^\infty \frac{G(a,p)\hat{\zeta}_{s0}}{\Omega(p)}p\,dp\right).
\tag{4.56}
$$

Recall from Eq. (4.46) that for $n = 1$, $\hat{\Psi}_d(r)$ is proportional to $v(r)$ as given by Eq. (4.33). Thus the remaining term (4.56) is proportional to the non-rotating normal-mode associated with a translation of the basic-state vortex (i.e., the pseudo-mode discussed in Section 4.3). This result is as expected based on the azimuthal wavenumber one component of Eq. (4.7). The constant of proportionality is derived from the limiting Green function integral in Eq. (4.56) weighted by the structure of the initial condition $\hat{\zeta}_{s0}$ and represents an angular rotation of the pseudo-mode. (At the time shown in Fig. 4.14b, the second term in Eq. (4.54) has not yet become small in comparison to the corresponding third term.)

For non-zero perturbation vorticity inside the radius of maximum winds, the asymptotic solution for $n = 1$ has a non-vanishing oscillatory component that rotates around the center of the vortex with the maximum angular velocity of the basic state vortex. In a mature tropical cyclone, where the basic-state angular velocity has a small, but generally non-zero radial gradient inside the radius of maximum winds, as $t \to \infty$ the latter effect would probably be of little significance since these perturbations would ultimately axisymmetrize also. At early times, however, the transient wavenumber-one component could produce a cycloidal track in the fully non-linear formulation - as demonstrated near the beginning of this chapter (Fig. 4.4a,b).

For $n > 1$ and when the smooth perturbation vorticity $\hat{\zeta}_{s0}$ is nonzero at radii immediately surrounding p_c, where the denominator in the integrals associated with the function $n - 1 - an\Omega(p)$ vanishes, the solution for $(\gamma(t), \hat{\psi}(r,t))$

---

13. Kelvin (1880).

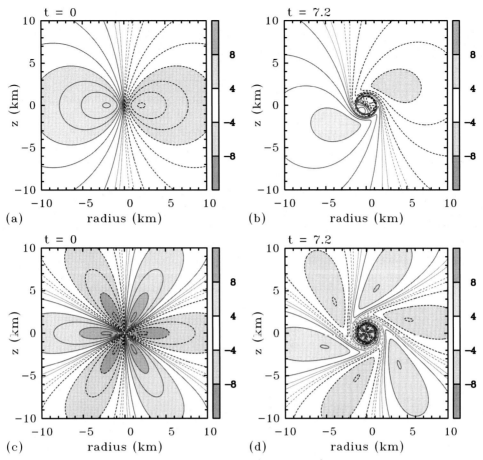

FIGURE 4.14 Perturbation streamfunction forced by initial vorticity-profile $\hat{\zeta}_0 = 1/(r+1)^3$. Panels (a) and (b) are for wave number $n = 1$ and the contour interval is 2×10^{-2}. Panels (c) and (d) are for $n = 3$ and the contour interval is 2×10^{-3} The columns are for times $t = 0.0$ and $t = 7.2$, respectively. Solid lines denote $\psi \geq 0$; dashed lines denote $\psi < 0$. Adapted from Smith and Montgomery (1995). Republished with permission of Wiley.

given by Eqs. (4.53) and (4.54), respectively, is more readily assessed by combining the two corresponding integrals. For long times, it suffices to consider the expression (4.53) for $\gamma(t)$ governing the time dependence of the discrete VR wave:

$$
\gamma(t) = -\frac{2n}{a} \int_0^\infty G(a, p) \, \hat{\zeta}_{s0} \times
$$
$$
\frac{[e^{-in\Omega(p)t} - e^{-i(n-1)t/a}]}{n - 1 - an\Omega(p)} p \, dp
$$
$$
+ \gamma_0 \, e^{-i(n-1)t/a}. \quad (4.57)
$$

In this form it is easy to see that the singularity associated with the vanishing of $n - 1 - an\Omega(p)$ in the denominator of the integrand in (4.57) is a removable singularity since the corresponding numerator vanishes there also and L'Hospital's rule gives a finite limit as $p \to p_c$. However, the time-asymptotic solution must generally account for the possible "resonant growth" associated with the frequency matching of the discrete VR wave with the perturbation vorticity ζ_s at the critical radii $p = p_c$.

For discrete radii where $n - 1 - an\Omega(p)$ vanishes (see Section 4.5.6), the integral (4.57) can be shown to grow algebraically no faster than $\sim t$. Algebraic growth of the discrete VR wave in these cases can be anticipated physically because the forcing on the right hand side of Eq. (4.50) includes a component with the same (non-dimensional) frequency as that of the discrete VR wave (i.e., $\omega_n = (n - 1)/a$).[14]

Eventually for $t \sim O(\epsilon^{-1})$, with ϵ here characterizing the amplitude of the initial disturbance, the discrete VR wave grows progressively to a sufficient amplitude that the previously neglected nonlinear terms become comparable to the linear terms and must be retained in the vicinity of $r = a$ and $r = r_c$ to render the solution physically consistent at leading order. Section 4.5.6 provides further details.

---

14. The simplest demonstration is to consider the limiting initial condition $\hat{\zeta}_{s0} \sim \delta(r - r_c)$, where, again, δ denotes Dirac's delta function. The radial integral in (4.57) then collapses at $r = r_c$ and L'Hospital's rule furnishes the limiting value of the integral, which is finite, but whose magnitude grows with time as t.

4.4.8.2 Summary

Explicit monochromatic wave solutions have been constructed for an unbounded Rankine-vortex, wherein ζ experiences a finite jump at the radius of maximum tangential winds, but is otherwise uniform inside and outside this radius. In this model, the perturbation solution is a superposition of a shear-wave (continuous spectrum) component and a VR edge-wave (discrete spectrum) component for $n > 1$, the latter of which propagates azimuthally, at a slower angular speed than the mean vortex at $r = a$. However, because these edge waves do not propagate radially, they are unable to transport energy out of the vortex core.

In cases where the initial perturbation vorticity is non-zero surrounding the critical radii associated with the VR edge-waves, the long-time behavior of the VR edge-wave contains generally an algebraically growing component associated with the "resonant frequency matching" between the initial perturbation vorticity and the VR edge-wave at its critical radii.

4.4.9 Case III: unbounded Rankine-like vortex with multiple discontinuities in ζ

An extension of the unbounded-vortex model of the prior section allows multiple discontinuities in ζ. With more than one discontinuity, wave interference effects can arise. As an example, consider the simplest case of a three-region model in which the innermost vorticity ζ_1, is greater than the intermediate vorticity ζ_2, which is greater than the outermost vorticity ζ_3 (Fig. 4.15). Such a distribution can be regarded physically as a three-region approximation of a vortex monopole (or tropical cyclone) possessing a finite transition region between its rapidly rotating core and its slowly rotating environment.

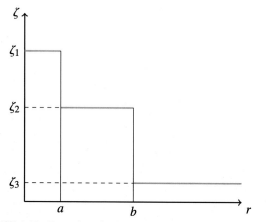

FIGURE 4.15 Extension of the unbounded-vortex model to allow for multiple discontinuities in ζ as illustrated by the red (mid gray in print version) lines with downward jumps at radii a and b.

Because the generalized radial gradient of vorticity is everywhere non-positive, the vortex is exponentially sta-

ble.[15] If the initial condition consists of no exterior edge-wave, but rather a smooth disturbance-vorticity in $a \leq r \leq b$, some portion of the energy of the initial disturbance will be permanently transferred to the exterior edge-wave. Sheared disturbances play an essential role in this transfer process, for without them the disturbance energy, as $t \to \infty$, resides solely in the interior edge-wave.

Bearing in mind the limitations of the three-region model, the foregoing summary of VR waves on smooth basic state vortices suggests that one may expect that the continuous model possesses an analogous mechanism for transferring energy outwards. This expectation was demonstrated by Montgomery and Kallenbach (1997).

4.5 Wave-vortex interaction

In order for the excited waves to cause a permanent change to the mean vortex, the waves must irreversibly alter the distribution of vorticity in the fluid. In two-dimensional incompressible fluid dynamics, the change in the mean vortex is governed by the mean tangential velocity tendency equation:

$$\frac{\partial}{\partial t} \langle v \rangle = -\langle u' \zeta' \rangle, \qquad (4.58)$$

where u' denotes the perturbation radial velocity, ζ' denotes the perturbation relative vorticity and the bracket symbol $\langle ... \rangle$ denotes an azimuthal average at radius r around the vortex center. This mean equation is obtained by azimuthally averaging the tangential momentum equation and taking into account the mean and perturbation contributions of the flow. The mean equation shows that *the mean tangential velocity will increase in time at radius r if there is a net inward flux of eddy cyclonic vorticity across this radius*, i.e.,

$$\langle u' \zeta' \rangle < 0. \qquad (4.59)$$

Conversely, *the mean tangential velocity will decrease at r if there is a net outward flux of eddy cyclonic vorticity there.*

The foregoing results are simply a rephrasing of Stokes' theorem when applied to a circular loop of radius r centered on the vortex. In fluid dynamics, Stokes' theorem says that the line integral of tangential velocity round the loop equals the area integral of vorticity contained within the loop. If cyclonic vorticity is advected inside the loop by irreversible wave action, the area integral of vorticity will increase and the tangential velocity at radius r will increase in time accordingly.

The following identity between the azimuthally-averaged eddy vorticity flux and the azimuthally-averaged eddy momentum flux divergence proves useful:

---

15. See prior summary of the generalized Rayleigh's theorem for exponential stability of circular shear flow.

$$\frac{1}{r^2}\frac{\partial}{\partial r}\langle r^2 u'v'\rangle = \langle u'\zeta'\rangle, \qquad (4.60)$$

where v' is the perturbation tangential velocity. This (Taylor) identity (Exercise 4.7) allows the mean vortex equation to be written in an equivalent eddy, angular momentum flux form:

$$\frac{\partial}{\partial t}\langle v\rangle = -\frac{1}{r^2}\frac{\partial}{\partial r}\langle r^2 u'v'\rangle. \qquad (4.61)$$

This form of the mean vortex equation shows that a convergence of eddy angular momentum flux is required to strengthen the mean tangential flow. After integrating Eq. (4.61) over the entire flow domain A, it follows that the integrated change in specific angular momentum of the vortex, $\delta M_v = \int r\delta v dA$, must vanish. Here, δM_v denotes the change in specific angular momentum of the mean vortex and δv denotes the local change in mean tangential velocity obtained by time-integrating the mean tangential acceleration in Eq. (4.61). For if there are no external torques acting on the fluid, the Rossby/shear waves redistribute angular momentum locally, but cannot change the angular momentum of the vortex.

Having constructed the analytical solution for the perturbation streamfunction and vorticity as a function of time in the unbounded Rankine vortex of Case II, this solution may be used as a first approximation to evaluate the eddy vorticity (and momentum) fluxes. With the fluxes in hand, one may then estimate the change in the mean vortex tangential velocity, $\delta v(r)$, by time integrating the wave-induced force per unit mass. The methodology of using the first-order linear solution to estimate the second-order-in-amplitude wave fluxes is called "the quasi-linear approximation".

An explicit expression for the eddy vorticity flux may be obtained as follows. Recall that the perturbation streamfunction and vorticity amplitudes for azimuthal wavenumber n may be represented as follows:

$$\psi' = Re\left[\hat{\psi}(r,t)e^{in\lambda}\right], \qquad (4.62)$$

$$\zeta' = Re\left[\hat{\zeta}(r,t)e^{in\lambda}\right], \qquad (4.63)$$

where 'Re' denotes the real part of the expression. The complex Fourier amplitudes $\hat{\psi}$ and $\hat{\zeta}$ can then be split into their component real and imaginary parts

$$\hat{\psi} = \hat{\psi}_r + i\hat{\psi}_i, \qquad (4.64)$$

$$\hat{\zeta} = \hat{\zeta}_r + i\hat{\zeta}_i. \qquad (4.65)$$

Hence,

$$u' = -\frac{1}{r}\frac{\partial}{\partial\lambda}\left[\text{Real}\left\{\hat{\psi}\,\text{exp}^{in\lambda}\right\}\right]$$

$$= -\frac{1}{r}\frac{\partial}{\partial\lambda}\left[\hat{\psi}_r\cos(n\lambda) - \hat{\psi}_i\sin(n\lambda)\right]$$

$$= \frac{n}{r}\left[\hat{\psi}_r\sin(n\lambda) + \hat{\psi}_i\cos(n\lambda)\right], \qquad (4.66)$$

and similarly

$$\zeta' = \hat{\zeta}_r\cos(n\lambda) - \hat{\zeta}_i\sin(n\lambda). \qquad (4.67)$$

Thus

$$u'\zeta' = \frac{n}{r}\left[\hat{\psi}_r\hat{\zeta}_r\cos(n\lambda)\sin(n\lambda) - \hat{\psi}_r\hat{\zeta}_i\sin^2(n\lambda)\right.$$
$$\left. +\hat{\psi}_i\hat{\zeta}_r\cos^2(n\lambda) - \hat{\psi}_i\hat{\zeta}_i\cos(n\lambda)\sin(n\lambda)\right], \qquad (4.68)$$

which, after azimuthally averaging and recalling the orthogonality of the sine and cosine functions, gives

$$\langle u'\zeta'\rangle = \frac{n}{2r}\left[\hat{\psi}_i\hat{\zeta}_r - \hat{\psi}_r\hat{\zeta}_i\right]. \qquad (4.69)$$

Recalling the explicit wave solution for the unbounded Rankine vortex model (Case II), we can express the perturbation streamfunction and vorticity amplitude functions as follows:

$$\hat{\psi}(r,t) = \hat{\psi}_d(r,t) + \hat{\psi}_s(r,t) = \gamma(t)\hat{\Psi}_d(r) + \hat{\psi}_s(r,t), \qquad (4.70)$$

$$\hat{\zeta}(r,t) = \hat{\zeta}_d(r,t) + \hat{\zeta}_s(r,t) = \gamma(t)\delta(r-a) + \hat{\zeta}_s, \quad (4.71)$$

where $\hat{\psi}_d$ and $\hat{\zeta}_d$ are the discrete VR wave perturbation streamfunction and vorticity components, respectively, and $\hat{\psi}_s$ and $\hat{\zeta}_s$ are the smooth streamfunction and vorticity components associated with the sheared disturbances, respectively. As noted under Case II, the radial structure of the discrete streamfunction is denoted by $\hat{\Psi}_d(r)$ and the time-dependence of this mode is represented by $\gamma(t)$. As in the foregoing section, $\delta(r-a)$ denotes the one-dimensional Dirac delta function centered at $r=a$.

Making use of the real and imaginary parts of these Fourier components, and substituting the complete solution for the waves into the azimuthally-averaged vorticity flux equation (4.58), an explicit expression for the wave-induced acceleration (Exercise 4.8) may be obtained that comprises four distinct eddy vorticity flux components:

$$\frac{\partial}{\partial t}\langle v\rangle = -\langle u'\zeta'\rangle$$
$$= -\langle u_d'\zeta_d'\rangle - \langle u_d'\zeta_s'\rangle - \langle u_s'\zeta_d'\rangle - \langle u_s'\zeta_s'\rangle$$
$$= -\frac{n}{2r}\left[\left(\hat{\psi}_{di}\hat{\zeta}_{dr} - \hat{\psi}_{dr}\hat{\zeta}_{di}\right) + \left(\hat{\psi}_{di}\hat{\zeta}_{sr} - \hat{\psi}_{dr}\hat{\zeta}_{si}\right)\right.$$
$$\left. + \left(\hat{\psi}_{si}\hat{\zeta}_{dr} - \hat{\psi}_{sr}\hat{\zeta}_{di}\right) + \left(\hat{\psi}_{si}\hat{\zeta}_{sr} - \hat{\psi}_{sr}\hat{\zeta}_{si}\right)\right].$$
$$(4.72)$$

4.5.1 Effect of discrete VR wave only

Let us focus first on the $\langle -u_d'\zeta_d'\rangle$ term, i.e., the eddy vorticity flux due entirely to the discrete neutral VR wave component. Inserting the solution for this neutral wave into the eddy vorticity flux (Exercise 4.9) gives

$$\langle u_d'\zeta_d'\rangle = -\frac{n}{2r}\left(\hat{\psi}_{di}\hat{\zeta}_{dr} - \hat{\psi}_{dr}\hat{\zeta}_{di}\right)$$
$$= -\frac{n}{2r}\left[\gamma_i\hat{\Psi}_d\gamma_r - \gamma_r\hat{\Psi}_d\gamma_i\right]\delta(r-a) = 0. \quad (4.73)$$

Eq. (4.73) is a reminder that a neutral VR wave, by itself, will not alter the mean vortex.[16]

The non-acceleration by steady VR waves is an example of a *non-acceleration theorem*. The non-acceleration result can be understood also by recalling Stokes' theorem discussed above. *If the discrete VR wave reversibly undulates the vorticity contour comprising the edge of the Rankine vortex, there will be no net transfer of the wave's perturbation vorticity across any radius and hence the tangential velocity of the mean vortex will not change.*

4.5.2 Effect of exterior disturbance on outer flow

We consider next the potential wave, vortex interaction associated with an initially smooth vorticity disturbance, as represented by the term $\langle -u_s'\zeta_s'\rangle$. Here the initial disturbance is assumed to be located sufficiently far from the core of the vortex, i.e., $r \gg a$, so the two interaction terms $\langle u_s'\zeta_d'\rangle$ and $\langle u_d'\zeta_s'\rangle$ are negligible.

The ensuing wave, mean-flow interaction for the smooth perturbation can be understood qualitatively as follows. The differential rotation of the mean angular velocity causes the vorticity perturbation to become progressively sheared into a trailing spiral pattern around the vortex center. For a cyclonic vortex, the sheared pattern spirals cyclonically inwards. The sheared disturbance generates an eddy angular momentum flux that is convergent on the inward side of the shear wave packet, and divergent on the outward side. The radial divergence of this averaged eddy momentum flux implies an increase in v on the inward side of the wave packet and a decrease on the outward side. The net result, after integrating over the wave's lifecycle, is a force dipole, as illustrated schematically in Fig. 4.16.

For an initially smooth vorticity perturbation centered at a large radius, the positive inner peak of $\delta v(r)$ will be slightly larger than the magnitude of the negative outer peak. That this must be so follows from the result noted previously that the change in angular momentum density, $r\delta v(r)$, must have a zero integral after integrating over the entire fluid. Thus, while the sheared waves locally redistribute angular momentum of the vortex, the net angular momentum is left unchanged. This and other properties can be inferred from the foregoing analytical solution.

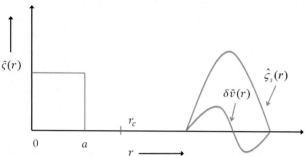

FIGURE 4.16 Schematic of wave-induced vortex interaction near the initial smooth vorticity perturbation $\hat{\zeta}_{s0}(r)$. Other solution components in Eq. (4.72) are omitted for clarity. Shown are the mean Rankine vortex, the initially upright smooth vorticity perturbation $\hat{\zeta}_{s0}(r)$, and the change in the mean tangential velocity of the vortex, $\delta v(r)$, induced by the differential shearing of the smooth vorticity perturbation after a suitably long time interval $t \gtrsim S^{-1}$, where $S = rd\Omega/dr$ is the local (radial) shear of the mean vortex at radius r. The amplitude of the initial vorticity perturbation and wave-induced, mean-flow change, the latter being proportional to the square of the perturbation amplitude, are exaggerated for clarity. Other notation is as defined in Fig. 4.12.

The change in the mean tangential velocity $\delta v(r)$ is furnished by substituting the smooth solution component into the expression for the wave-induced force $\langle -u_s'\zeta_s'\rangle$ and integrating this expression in time from $t = 0$ to $t = \infty$. The result is (Exercise 4.10):

$$\delta v(r) = -\frac{n}{2r}\hat{\zeta}_{s0}(r)\times$$
$$\left[\int_0^\infty \sin(n\Omega(r)t)\int_0^\infty G(r,p)\hat{\zeta}_{s0}(p)\cos(n\Omega(p)t)\,pdpdt\right.$$
$$\left. - \int_0^\infty \cos(n\Omega(r)t)\int_0^\infty G(r,p)\hat{\zeta}_{s0}(p)\sin(n\Omega(p)t)\,pdpdt\right],$$
$$(4.74)$$

where, again, the Green function $G(r,p)$ is given by Eq. (4.40) and $\hat{\zeta}_{s0}$ is the smooth, monochromatic, initial vorticity perturbation. For simplicity, $\hat{\zeta}_{s0}$ is assumed to be real-valued, corresponding to a perturbation with no initial tilt with respect to the angular shear.

As time increases from zero to suitably large values,[17] the expression for δv predicts an asymmetric "reverse S" pattern, with a comparatively large positive δv interior to the smooth perturbation vorticity maximum in relation to the negative δv outside of the perturbation maximum (see Fig. 4.16 and Fig. 4.18c below from Nolan and Farrell, 1999). This shear-induced, force dipole, is opposite to what one would obtain if the axisymmetrizing waves acted diffusively (i.e., downgradient).

For a given cyclonic vortex monopole, $\Omega(r)$ is a maximum at the vortex center and the corresponding radial gradient is directed inwards, i.e., towards the vortex center. A

---

16. This result holds true for initially circular vortex basic states with any number of constant vorticity regions and their corresponding discrete neutral VR waves. This non-acceleration theorem assumes that there are no exponential instabilities and no sheared vorticity components.

---

17. By suitably large values we mean $t \gtrsim S^{-1}$, where $S = rd\Omega/dr$ is the local (radial) shear of the mean vortex at radius r.

diffusive closure for the eddy momentum fluxes would be downgradient, i.e., outwards. However, the strengthening of the mean vortex, interior to the perturbation vorticity maximum ($\delta v > 0$), is the result of the inward-directed, eddy momentum flux and its associated radial convergence as the perturbation vorticity is ultimately sheared into a trailing spiral pattern.

This analytical model succinctly illustrates the anti-diffusive nature of sheared vorticity perturbations evolving like the Orr mechanism at long times. The *peripheral spin up of the outer circulation* predicted by the quasi-linear approximation is consistent qualitatively with the findings of Sutyrin and Radko (2019) and Sutyrin (2019) regarding the peripheral intensification of vortices when subject to small-amplitude random vorticity perturbations outside the vortex core region.

4.5.3 Effect of exterior disturbances on v_{max}

Since one of the aims of this book is to understand how vortex asymmetries can influence the maximum tangential velocity of the mean vortex, v_{max}, it is pertinent to determine whether the smooth vorticity perturbation exterior to $r = a$ will strengthen or weaken the maximum tangential wind at $r = a$? We can answer this question in the context of our toy model by examining the third term in the mean vortex equation, i.e., the term comprising the mean eddy flux of the discrete VR wave vorticity by the sheared radial wind component, $\langle -u_s' \zeta_d' \rangle$. Because, by construction, we have assumed that the smooth vorticity perturbation is nonzero only outside $r = a$, the other interaction term in the mean vortex equation, $\langle -u_d' \zeta_s' \rangle$, vanishes at $r = a$.

From the mean vortex equation (4.72), the change in the mean tangential wind at $r = a$ by the exterior vorticity perturbation may be estimated by substituting the discrete VR wave solution, $\gamma(t)\delta(r - a)$, into the mean vortex equation. The result is:

$$
\begin{aligned}
\frac{\partial}{\partial t} \langle v \rangle \bigg|_{r=a} &= -\langle u_s'(a,t)\zeta_d'(a,t) \rangle \\
&= -\frac{n}{2a}\left(\hat{\psi}_{si}(a,t)\hat{\zeta}_{dr}(a,t) - \hat{\psi}_{sr}(a,t)\hat{\zeta}_{di}(a,t) \right) \\
&= -\frac{n}{2a}\left(\gamma_r(t)\hat{\psi}_{si}(a,t) - \gamma_i(t)\hat{\psi}_{sr}(a,t) \right)\delta(r - a).
\end{aligned}
$$

$$(4.75)$$

As before, the subscripts 'r' and 'i' denote the real and imaginary parts, respectively. An explicit expression for the smooth streamfunction perturbation at $r = a$ is required in order to time integrate this equation. Recalling the solution from case II, an explicit expression for $\hat{\psi}_s(a,t)$ was obtained in terms of $\gamma(t)$ from the dynamic boundary condition at $r = a$. On substituting the inferred expression for $\hat{\psi}_s(a,t)$ from Eq. (4.50) into Eq. (4.75), grouping real and

imaginary terms, and simplifying yields a compact equation for the tendency of mean tangential velocity at $r = a$ in terms of the amplitude of the discrete VR wave:

$$
\frac{\partial}{\partial t} \langle v \rangle \bigg|_{r=a} = -\frac{a}{8} \frac{\partial |\gamma|^2}{\partial t} \delta(r - a), \qquad (4.76)
$$

where $|\gamma|^2$ denotes the square of the absolute value of γ.

Integrating Eq. (4.76) in time, from $t = 0$ to $t = \infty$, gives (Exercise 4.11) the net change in the mean tangential velocity at $r = a$ forced by the interaction between the sheared vorticity solution outside the core and the discrete VR wave:

$$
\delta v|_{r=a} = -\frac{a}{8} |\gamma|^2 (t \to \infty)\delta(r - a). \qquad (4.77)
$$

Here, as in prior subsections, a is the radius of maximum tangential wind of the basic state vortex. For simplicity, we have assumed that $\gamma(t = 0) = 0$, i.e., there is no discrete VR wave at time $t = 0$.[18]

The equation for δv shows that the net effect of the outer-core vorticity perturbation is to cause a weakening of v_{max}. As is evident from Eq. (4.77), the weakening of the tangential wind maximum is of second order in the amplitude of the emergent VR wave. The weakening may be anticipated physically by considering the angular momentum constraint as discussed above. Equivalently, from an energetics perspective, as the sheared vorticity wave disturbance in the outer region excites the inner-core discrete VR wave, the inner-core VR wave acts like an energy trap for the mean vortex. In this sense, the discrete VR wave acts as a "negative energy wave", weakening the core vortex.

4.5.4 Effect of near-core disturbances on v_{max}

The prior subsection demonstrated that outer-core vorticity perturbations ultimately cause a weakening of v_{max}. We ask now whether the outcome is the same for near-core vorticity perturbations? To address this question, we consider the mean flow interaction resulting from vorticity perturbations excited initially near $r = a$. The answer to this question is not as simple as it might first appear. To the extent that a weakening effect of the discrete VR wave as found above persists, the weakening effect may be countered by the spin up effect caused by the irreversible, inward, cyclonic eddy vorticity flux due to the sheared vorticity component as illustrated above.

---

18. As previously noted, the delta function $\delta(r - a)$ represents a generalized function defined by a limiting process of a narrow function around $r = a$ whose amplitude grows commensurately with the spatial shrinking of the nonzero values of the function, but whose integrated value around $r = a$ remains constant. In the case of Eq. (4.77), a radial integration of the expression for δv across a small interval containing $r = a$ implies a small, but nonzero, weakening of v at $r = a$.

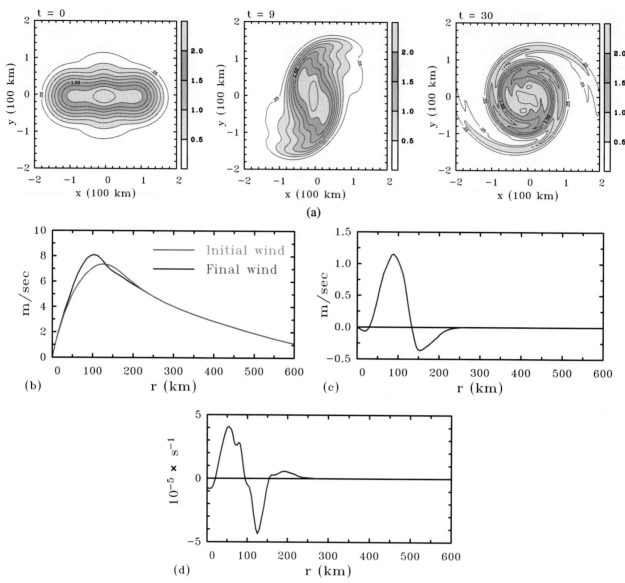

FIGURE 4.17 Typical vortex axisymmetrization for inner-core vorticity perturbation on a cyclonic monopole vortex. (a) Time evolution of the absolute vorticity field spanning 30 h for a basic state plus nominal double-cluster cyclonic vorticity anomaly. Only the innermost 400 km × 400 km of the model domain is shown. Contour interval 0.25×10^{-4} s^{-1} and units on color bar ×10^{-4} s^{-1}. (b) Azimuthal mean tangential wind v at $t = 0$ and $t = 49$ h (1.6 τ_{eddy}); (c) change in azimuthal mean tangential wind δv over the 49-h period, and (d) change in azimuthal mean vorticity $\delta\zeta$ over 49 h for the double-cluster convective experiment. Only the innermost 600 km of the model domain is shown. Adapted from Enagonio and Montgomery (2001). Republished with permission of the American Meteorological Society.

To fix ideas, imagine that convective activity has generated an azimuthal wavenumber two vorticity perturbation near $r = a$. An example of the wave and mean vortex evolution for such a vorticity perturbation is illustrated in Fig. 4.17. In this case, the ensuing sheared VR waves are found to increase and sharpen the mean tangential velocity near where the waves are excited. This example, and other calculations like this, demonstrate that the *strengthening/sharpening of the mean vortex by the sheared VR waves* is distinct from what was found above for exterior perturbations in regards to their effect on v_{max}.

Since momentum diffusion would tend to reduce both the maximum velocity and its gradient, this example again illustrates the anti-diffusive property of sheared VR waves. That is, sheared VR waves excited near the radius of maximum tangential wind on an inviscid barotropic vortex ultimately behave anti-diffusively and will contribute to intensifying the vortex.

4.5.5 Model limitations applied to smooth vortices: quasi-modes

In the foregoing example, the initial mean vortex had a smooth vorticity distribution. However, the toy model derived above does not strictly apply in this case. Given a mean vorticity distribution that is smooth and monotonic decreasing with radius, a discrete VR wave eigenmode is not part of the analytical solution. From the viewpoint of the toy model, the discrete VR wave eigenmode disappears into the continuous spectrum of sheared waves, formally leaving just one term on the right hand side of the mean vortex equation (4.72). That is, the mean vortex equation simplifies to:

$$\frac{\partial}{\partial t}\langle v \rangle = -\langle u_s' \zeta_s' \rangle. \tag{4.78}$$

Under these circumstances, the generalized waves become sheared VR waves, possessing both irreversible shearing and Rossby wave dispersion effects. Although the mathematical solution no longer explicitly contains a discrete VR wave eigenmode, the solution for a smooth vortex nonetheless retains some discrete VR wave characteristics.

A quasi-mode is not a true eigenmode of the linearized vortex Rossby wave equation. It can be shown to be a superposition of sheared vorticity waves with similar angular phase speeds (e.g., Reasor and Montgomery, 2001 and refs.). In this respect, the quasi-mode resembles a discrete VR wave mode for a finite time until the differential rotation of the mean vortex ultimately shears the initial vorticity perturbation into fine-scale vorticity spirals around the vortex, like that of a passive scalar (the long time limit of the Orr mechanism).

The dynamics of a quasi-mode is generally more complex than the analytical solution considered so far. An analysis of the quasi-mode and its contribution to vortex axisymmetrization, vortex alignment, and vortex resilience more generally, is beyond the scope of the present chapter.

4.5.6 Resonant wave, vortex interaction

To conclude this primer on wave, vortex interaction theory, we return briefly to the toy Rankine model and its implied wave-vortex interaction in the quasi-linear approximation.

As already intimated in Section 4.4.8.1, an interesting phenomenon occurs in this model when the outer-core vorticity perturbation approaches the radius where the discrete VR wave's frequency ω_n equals the angular velocity of the mean vortex. In this circumstance, the sheared disturbance (ζ_s') rotates at the same rotation rate as the discrete VR wave (ζ_d') and a type of "resonant response" occurs in the streamfunction solution (4.54) and the mean vortex interaction (4.58). The specific radius where this resonance occurs is the so-called "critical radius", r_c, defined dimensionally by $\omega_n = n\Omega(r_c)$, where ω_n is the temporal frequency of the

discrete VR wave (given by Eq. (4.55), see also below) and, again, $\Omega = v(r)/r$ is the angular velocity of the basic state vortex.[19]

As summarized in Section 4.4.8.1, this resonance phenomenon is associated with the *apparent* singularity in the second and third integrals on the right hand side of (4.54) where the denominator vanishes (the non-dimensional critical radius) and occurs when the perturbation vorticity does not vanish there. As intimated there, for times $t \sim O(\epsilon^{-1})$, the nonlinear terms are expected to become comparable to the linear terms. Such mathematical details are beyond the scope of this chapter. Here we sketch briefly the qualitative nature of the nonlinear solution whose quantitative characteristics have been explored elsewhere by Terwey and Montgomery (2003) and in a companion formulation of a related problem as summarized in Section 4.7.3.

For the toy Rankine model, the dimensional dispersion relation for the discrete VR eigenmodes was given previously

$$\omega_n = \frac{1}{2}\zeta(n-1). \tag{4.79}$$

This relation shows that azimuthally propagating, discrete, VR waves in this model exist only for $n \geq 2$. For such modes, the non-dimensional critical radius (Exercise 4.12) is given by

$$r_c = a/\sqrt{(1 - 1/n)}. \tag{4.80}$$

The outermost critical radius accordingly occurs for $n = 2$ wherein $r_c = \sqrt{2}a$. For higher azimuthal wavenumbers, r_c monotonically decreases with wavenumber and asymptotically approaches the radius of maximum wind $r = a$ as $n \to \infty$.

As the outer initial vorticity perturbation is placed progressively inward towards the critical radius, δv is found to be significantly amplified relative to its counterpart response for an exterior perturbation sufficiently distant from r_c. In particular, the amplification of δv near r_c forces a commensurately magnified weakening of the mean vortex in the vicinity of $r = a$ (e.g., Figs. 5.6 and 5.7 of Terwey and Montgomery, 2003). The weakening of v_{max} can be understood qualitatively as a consequence of the conservation of net angular momentum.

As intimated above, the quasi-linear theory is expected to progressively break down in the neighborhood of r_c and nonlinear effects, that have so far been neglected, must be retained to furnish a physically consistent solution (Section 5.5 of Terwey and Montgomery, 2003). The fully nonlinear solution entails the stirring of the perturbation vorticity in a horizontal layer encompassing the critical radius.

---

19. This resonance phenomenon is similar in some respects to the dynamics of the quasi-mode summarized above. We plan to explore these topics further in a sequel to this textbook after a basic conceptual model for tropical cyclone evolution is developed.

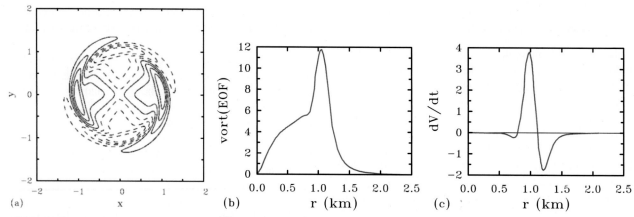

FIGURE 4.18 Ensemble mean response of the $r^{-1/2}$ vortex under "convective" stochastic forcing for $k = 2$, where k is the azimuthal wavenumber: (a) Primary Empirical Orthogonal Function (EOF) perturbation vorticity response, (b) complex magnitude of the primary EOF shown in (a), and (c) mean eddy flux divergence (and implied change in azimuthal mean tangential velocity). All quantities plotted are non-dimensional. Adapted from Nolan and Farrell, 1999. Republished with permission of the American Meteorological Society.

The stirring region is called a nonlinear critical layer.[20] Although the magnitude of the wave-induced, vortex interaction in the nonlinear solution near both r_c and $r = a$ is reduced relative to its quasi-linear prediction, a resonant-type response persists nonetheless.

4.6 Synthesis

We have considered the problem of inviscid vortex axisymmetrization as a foundation for understanding the observed near-axisymmetry of mature tropical cyclones as discussed in Chapter 1 and in the beginning of this chapter. Various examples using both nonlinear and linear non-divergent vorticity models illustrate the axisymmetrization process when applied to a tropical cyclone-like vortex.

A toy model using a Rankine basic state vortex was developed next to understand some hitherto unappreciated properties of the "surf zone" of sheared vorticity disturbances outside an intense vortex core as well as their associated wave, vortex interaction in the presence of discrete VR waves on the vortex's edge. The toy model offers a useful framework for understanding the anti-diffusive behavior of vortex shear waves and their interaction with discrete VR wave modes, though intrinsic limitations of the model as applied to more realistic smooth vortices need to be kept in mind.

By design, the results of this chapter have been obtained by neglecting diabatic processes and boundary layer processes. To proceed further, suitable theoretical tools need to be developed that confront these important elements be-

fore re-integrating axisymmetrization, VR wave, and wave, vortex interaction processes into an overarching conceptual model of tropical cyclone evolution.

4.7 Enrichment topics

4.7.1 Vortex intensification by stochastic forcing with secondary circulation

Up to this point we have omitted the secondary circulation, which is superimposed on the primary circulation. Above the frictional boundary layer, the inflow is relatively weak and so neglecting the secondary circulation seems an apt approximation for developing a basic understanding of vortex axisymmetrization applied to tropical cyclones. However, even in the limit of weak secondary flow, concerns could be raised about the barotropic nondivergent formulation for its neglect of vortex tube stretching and vortex tube compression effects in localized regions of convergence and divergence, respectively.

The dynamics of vortex intensification by small-amplitude, stochastic, vorticity perturbations in the presence of primary and secondary circulation was explored in a pioneering study by Nolan and Farrell (1999), e.g., Fig. 4.18. The model developed by them appears useful when aiming to understand the evolution of vorticity asymmetries and their interaction with the primary and secondary circulations in realistic tropical cyclone prediction models provided suitable modification is made to account for possible "resonant wave, vortex interaction" effects as summarized above (and in Section 4.7.3) and related influences of a superposed secondary circulation (Caillol, 2017; Caillol, 2019).

---

20. The zone where the irreversible stirring of vorticity is confined is called a "Kelvin cat's eye", named after Lord Kelvin who first considered this type of recirculating flow in the vicinity of a critical level in shear flows perturbed by neutral (non-growing) wave modes. This critical layer is not unlike the critical layer occurring within tropical easterly waves discussed in Chapter 1.

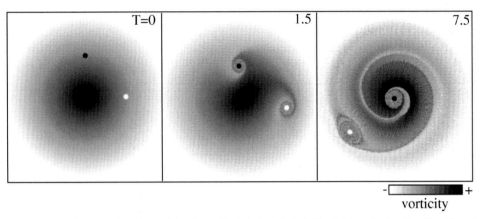

FIGURE 4.19 Gradient-driven radial separation of a vorticity clump (black dot) and hole (white dot) in a circular shear flow. From Schecter and Dubin (1999). Republished with permission of the American Physical Society.

4.7.2 Point vortex analog of wave, vortex model

In Sections 4.4 and 4.5 we considered the problem of vortex axisymmetrization and wave, vortex interaction primarily in the limit where the vorticity perturbations to the master vortex are of sufficiently small amplitude that nonlinear advective terms can be neglected in a first approximation. As demonstrated at the beginning of this chapter, this is a defensible first approximation for a mature tropical cyclone/typhoon vortex. However, when the convectively-induced vorticity becomes comparable to that of the background vortex, the linear approximation no longer applies as a first approximation and an alternative description becomes necessary. The need for an alternative approach becomes especially acute during tropical cyclogenesis and the early stages of vortex intensification when the winds of the background vortex are not dominant and the convectively-induced vorticity anomalies possess large amplitudes in comparison to that of the azimuthally-averaged flow (e.g., Kilroy et al., 2017c). Related examples illustrating the interaction between two separate tropical storm strength vortices were presented in Section 4.2.

If the vorticity perturbations are strong compared to that of the background flow, the perturbation vorticity can be considered as dominant locally (i.e., with an associated positive OW parameter amidst a surrounding sea of negative OW). That is, the wave-like description of the vorticity evolution developed in this chapter should be replaced by a particle-like description. In such a localized limit, the small amplitude approximation is replaced by an approximation in which the net local circulation of the vorticity anomaly is small in comparison to that of the background vortex.

The limiting case of locally infinite vorticity, which is confined to a vanishingly small area, is the so-called "point-vortex" limit. The fluid dynamics of point vortices has a rich and interesting scientific history (Aref, 1983). When considering the interaction of one (or more) point vortices that occurs within a background vortex which, itself, supports

VR waves and possesses radial shear, special consideration needs to be given to the interaction of the point vortices with the background vortex's critical radii (introduced in Section 4.5.6).

The specific formulation of a point vortex immersed in the background vorticity gradient of a parent vortex (Fig. 4.19) furnishes a particularly elegant model of the *asymmetric vortex merger problem* (Schecter and Dubin, 1999). For the case of a locally intense "vorticity clump" possessing the same parity as the background vortex, the asymptotic model correctly predicts the inward movement of the "clump" and its eventual arrest near the center of the background vortex (merger). Equally interesting is that locally intense "vorticity holes" with opposite parity as the background vortex are correctly predicted to be expelled from the background vortex. It would seem an extension of this model to the highly-asymmetric genesis stage within a meso-α scale pouch (e.g., Chapter 1) might prove useful theoretically, especially if the effects of a weak secondary circulation (such as Nolan and Farrell (1999) and Caillol (2019)) driven by the latent heat release of the aggregate clouds could be included also.

4.7.3 VR wave pathway to secondary eyewall formation?

A novel model has been developed by Nikitina and Campbell (2015a,b) to explore a hypothesized VR wave pathway to the formation of a secondary eyewall in a tropical cyclone. The model shares some similarity with the toy wave, vortex model developed in this chapter, but no longer assumes a singular vorticity waveguide. Following Brunet and Montgomery (2002), the model assumes a smooth basic state vorticity distribution that approximates a Gaussian distribution in the vortex core region.

The model is formulated in a barotropic nondivergent context and poses the secondary eyewall formation problem as the long-time response to a forced VR wave that propagates along the waveguide of the vortex's primary eye-wall. The specific forcing consists of an imposed vorticity wave of fixed amplitude at a radius ($r = r_1$) corresponding to the primary eyewall of the tropical cyclone (Fig. 4.20). The study examines both the steady-state and transient linear wave response, as well as the quasi-linear, wave-vortex interaction induced by the imposed forcing. The study determines the response in the region exterior to the forced VR wave, which includes the critical radius ($r = r_c$) of the forced wave, where the wave-vortex interaction is expected to be greatest (Fig. 4.21), as foreshadowed in Sections 4.4.8.1 and 4.5.6.

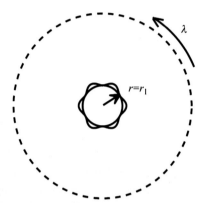

FIGURE 4.20 The flow domain in the forced linear vortex wave model of Nikitina and Campbell (2015a). The domain is given in terms of polar coordinates r and λ by $r_1 \leq r < \infty$, $0 \leq \lambda < 2\pi$ with $r = 0$ corresponding to the vortex center. The waves are generated at the primary eyewall $r = r_1$. From Nikitina and Campbell (2015a). Reproduced from Studies in Applied Mathematics with permission of Wiley publisher.

Although this model is fairly advanced (but certainly accessible to advanced undergraduate or graduate students in physics or applied mathematics), the work is well deserving of thorough study for its mathematical rigor and potential generalization to more realistic formulations that include plausible representations of surface friction and cumulus convection.

4.8 Exercises

Exercise 4.1. Verify Eqs. (4.7) and (4.8). Recall that in polar coordinates (r, λ), the streamfunction, ψ, is related to the radial and tangential velocity components, u and v, by the formulae

$$ru = -\frac{\partial \psi}{\partial \lambda} \quad \text{and} \quad v = \frac{\partial \psi}{\partial r},$$

in order to satisfy the continuity equation $\partial(ru)/\partial r + \partial v/\partial \lambda = 0$.

Exercise 4.2. Assume a steady, inviscid and axisymmetric swirling flow in cylindrical polar coordinates ($\mathbf{U} = (0, V(r), 0)$) about a circulation center at the origin. From Newton's second law, derive the horizontal force balance

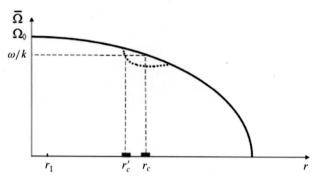

FIGURE 4.21 Time evolution of mean angular velocity near the critical radius in the weakly nonlinear asymptotic model of Nikitina and Campbell (2015b). The initial critical radius $r = r_c$ is given by $\Omega(r_c) = \omega/k$, where $\Omega(r)$ is the mean angular velocity of the initial vortex. If the angular velocity decreases in the critical layer near $r = r_c$, as shown by the dashed line, the critical radius moves toward the vortex center, from location $r = r_c$ to $r = r_c'$ ($< r_c$). From Nikitina and Campbell (2015b). Reproduced from Studies in Applied Mathematics with permission of Wiley publisher.

known as gradient wind balance given by Eq. (4.10). Assuming the radial pressure field $p(r)$ is known from observations, solve this equation for $V(r)$. Next, develop a general solution classification scheme that includes the "normal" and so-called "anomalous" solutions. Physically, what is different about the anomalous solutions? Explain why the anomalous solutions generally do not exist for a localized vortex whose tangential velocity vanishes as $r \to \infty$.

Exercise 4.3. Verify the derivation of the linearized vorticity equation given by Eq. (4.14), which governs small-amplitude vortex shear and Rossby waves.

Exercise 4.4. After taking a Fourier transform in azimuth, use the method of Green functions in one dimension to obtain the solution for the smooth streamfunction perturbation given by Eq. (4.39) for the vortex wave model in an unbounded domain. Verify explicitly that Eq. (4.39) solves the differential equation (4.37) for the smooth perturbation component. (The solution given by Eq. (4.39) can be obtained also using the methodology presented in Section 3.6.2, see Eq. (3.46).)

Exercise 4.5. Follow the steps outlined in the main text to obtain the differential equation for $\gamma(t)$ given by Eq. (4.50). Explain how this differential equation emerges from the dynamic boundary condition at the edge of the vortex core $r = a$ when applied to the total perturbation streamfunction for the unbounded domain.

Exercise 4.6. Follow the steps outlined in the main text to obtain the general Fourier mode solution for the total perturbation streamfunction amplitude given by Eq. (4.54). Physically interpret the three terms comprising the derived solution.

Exercise 4.7. Verify the so-called Taylor identity given by Eq. (4.60). Why is this relationship useful physically? Discuss.

Exercise 4.8. Follow the steps outlined in the main text to obtain Eq. (4.72), the wave-induced acceleration comprising four distinct eddy vorticity flux components. Physically interpret each contribution to the acceleration of the mean vortex.

Exercise 4.9. Obtain Eq. (4.73). Interpret the result using the idea of "non-acceleration" by the steady, discrete VR edge wave.

Exercise 4.10. Follow the steps outlined in the main text to obtain Eqs. (4.74). Explain physically how this solution is anti-diffusive at long times.

Exercise 4.11. Follow the steps outlined in the main text to obtain Eq. (4.77). Interpret the result physically. What does this result imply about the sign of the corresponding eddy angular momentum flux near the radius of maximum tangential wind? In particular, discuss how one might misidentify this result as "downgradient diffusion". Finally, speculate how Eq. (4.77) might be used to help interpret the evolution of a realistic (observed or simulated) tropical cyclone.

Exercise 4.12. Obtain Eq. (4.80) and interpret the result physically.

Chapter 5

Axisymmetric vortex theory fundamentals

As explained in Chapters 1 and 4, to a reasonable first approximation, a mature tropical cyclone consists of a horizontal, quasi-axisymmetric circulation (the primary circulation) on which is superposed a transverse (overturning) circulation (the secondary circulation). The primary circulation refers to the tangential or swirling flow rotating about the central axis, and the secondary circulation to the "in-up-and-out circulation", which has low and middle level inflow and upper-level outflow. In this chapter we examine some basic aspects of the axisymmetric dynamics of these circulation components of tropical cyclones on the basis of the physical laws governing fluid motion and the thermodynamic processes that occur. For simplicity, we examine here the dynamics of a stationary (non-moving) axisymmetric tropical-cyclone-like vortex. First we introduce the governing equations without making the assumption of axial symmetry. Then we go on to develop an approximate theory for the primary circulation, followed by one for the secondary circulation.

5.1 Equations of motion in rotating cylindrical polar coordinates

As already discussed in Chapter 2, some basic properties of vortical flows of variable density that are relevant to tropical cyclones may be elucidated by considering the full equations of motion in a rotating coordinate system in cylindrical polar coordinates, (r, λ, z) on an f-plane.[1] The gravitational force per unit mass $\mathbf{g} = (0, 0, -g)$ is taken to point in the direction opposite to z, where g is the effective gravita-

tional acceleration[2] due to gravity. The basic configuration is sketched in Fig. 5.1.

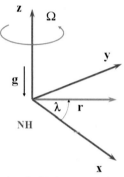

FIGURE 5.1 Schematic cylindrical polar coordinate system in which the $x - y$- or $r - \lambda$-plane is normal to the Earth's effective gravity at the surface at latitude ϕ. This coordinate system rotates about the z-axis with angular rotation rate $\Omega = \frac{1}{2} f$.

The traditional primitive equations of motion comprising the momentum equations, the continuity equation, the thermodynamic equation, and the equation of state for motion on the f-plane may be expressed as follows:

$$\frac{\partial u}{\partial t} + u \frac{\partial u}{\partial r} + \frac{v}{r} \frac{\partial u}{\partial \lambda} + w \frac{\partial u}{\partial z} - \frac{v^2}{r} - fv = -\frac{1}{\rho} \frac{\partial p}{\partial r} + F_r, \tag{5.1}$$

$$\frac{\partial v}{\partial t} + u \frac{\partial v}{\partial r} + \frac{v}{r} \frac{\partial v}{\partial \lambda} + w \frac{\partial v}{\partial z} + \frac{uv}{r} + fu = -\frac{1}{\rho r} \frac{\partial p}{\partial \lambda} + F_\lambda, \tag{5.2}$$

$$\frac{\partial w}{\partial t} + u \frac{\partial w}{\partial r} + \frac{v}{r} \frac{\partial w}{\partial \lambda} + w \frac{\partial w}{\partial z} = -\frac{1}{\rho} \frac{\partial p}{\partial z} - g + F_z, \tag{5.3}$$

$$\frac{\partial \rho}{\partial t} + \frac{1}{r} \frac{\partial \rho r u}{\partial r} + \frac{1}{r} \frac{\partial \rho v}{\partial \lambda} + \frac{\partial \rho w}{\partial z} = 0, \tag{5.4}$$

$$\frac{\partial \theta}{\partial t} + u \frac{\partial \theta}{\partial r} + \frac{v}{r} \frac{\partial \theta}{\partial \lambda} + w \frac{\partial \theta}{\partial z} = \dot{\theta}, \tag{5.5}$$

$$\rho = p_* \pi^{\frac{1}{\kappa}-1} / (R_d \theta), \tag{5.6}$$

where u, v, w are the radial, azimuthal and vertical velocity components in the three orthogonal coordinate directions, (F_r, F_λ, F_z) is the divergence of the sub-grid-scale turbulent momentum fluxes and/or the frictional force per unit mass

---

1. On an f-plane, the coordinate system is assumed to rotate with a uniform angular velocity $\Omega = \frac{1}{2} f$ equal to the *local vertical* rotation rate of the earth at latitude ϕ, where $f = 2\Omega \sin \phi$ and ϕ the constant reference latitude. In this formulation, we follow standard practice and neglect the influence of the Coriolis force associated with the *local horizontal* rotation rate of the earth, i.e., the Coriolis terms proportional to the cosine of latitude. The validity of this approximation is generally justified for large-scale tropical motions in a thin fluid envelope, but the vigor of convective updraughts during the spin up and mature phases of a cyclone's lifecycle raises some questions in principle about the potential system-scale impact of these non-traditional Coriolis terms. At the present time, we have no reason to expect these non-traditional terms will lead to any significant effect. In any event, we leave this matter for others and proceed with a traditional formulation of the moist atmosphere, retaining only Coriolis terms proportional to the sine of latitude.

2. Effective gravity is explained in Chapter 2.

Tropical Cyclones. https://doi.org/10.1016/B978-0-44-313449-4.00013-8

and $\dot{\theta}$ is the diabatic heating rate. Temperature in K is given by $T = \pi\theta$, where θ is the potential temperature and π is the *Exner function* (see Section 2.3.3).

Multiplication of Eq. (5.2) by r and recalling that $u = Dr/Dt$ leads to the equation

$$\frac{\partial M}{\partial t} + u\frac{\partial M}{\partial r} + \frac{v}{r}\frac{\partial M}{\partial \lambda} + w\frac{\partial M}{\partial z} = -\frac{1}{\rho}\frac{\partial p}{\partial \lambda} + rF_\lambda, \quad (5.7)$$

where

$$M = rv + \frac{1}{2}fr^2, \quad (5.8)$$

is the *absolute angular momentum* per unit mass of an air parcel about the rotation axis. In this equation, rv is the *relative angular momentum* in the rotating coordinate system and $\frac{1}{2}fr^2$ is the angular momentum of an air parcel associated with the spinning Earth at a radius r about the local vertical. The latter term is often referred to as the *planetary angular momentum*.

If the flow is axisymmetric ($\partial p/\partial\lambda = 0$) and frictionless ($F_\lambda = 0$), the right-hand-side of Eq. (5.7) is zero and the absolute angular momentum is materially conserved as rings of air move radially and/or vertically, i.e.,

$$\frac{DM}{Dt} = 0 \quad (5.9)$$

where

$$\frac{D}{Dt} = \frac{\partial}{\partial t} + u\frac{\partial}{\partial r} + w\frac{\partial}{\partial z} \quad (5.10)$$

is the material derivative operator for axisymmetric flow. Apart from a factor of 2π, M is equivalent to Kelvin's circulation, Γ_a, for a horizontally oriented circle C of radius r enclosing the center of circulation, i.e., $2\pi M = \Gamma_a = \int_C \mathbf{u_{abs}} \cdot \mathbf{dl}$, where $\mathbf{u_{abs}}$ is the absolute velocity and $\mathbf{dl}$ is a differential horizontal line segment along the circle. For an axisymmetric and inviscid flow, the material conservation of M is equivalent to Kelvin's circulation theorem (Section 2.11.1).

5.2 The primary circulation

In the absence of diabatic heating ($\dot{\theta} = 0$) and friction, ($F_r, F_\lambda, F_z) = (0, 0, 0)$, Eqs. (5.1) and (5.3) have an exact solution for a steady ($\partial/\partial t \equiv 0$), axisymmetric ($\partial/\partial\lambda \equiv 0$), freely spinning vortex $v(r, 0, z)$ in which the secondary circulation is absent ($u = 0$, $w = 0$). As shown in Section 5.8, this solution provides a simple theory for the primary circulation of a tropical cyclone, even when this circulation evolves slowly with time. With the foregoing assumptions, Eqs. (5.1) and (5.3) give,

$$\frac{\partial p}{\partial r} = \rho\left(\frac{v^2}{r} + fv\right), \quad (5.11)$$

and

$$\frac{\partial p}{\partial z} = -\rho g, \quad (5.12)$$

while Eqs. (5.2), (5.4) and (5.5) are identically satisfied. Eq. (5.11) expresses a balance between the radial pressure gradient force and the centrifugal and Coriolis forces, each per unit volume, a state that we refer to hereafter as *gradient wind balance*. We call this equation the *gradient wind equation*. As discussed in Section 4.4.1, if the background rotation is weak, i.e., $f \ll v/r$, the second term on the right-hand-side of Eq. (5.11) may be neglected and we speak then of *cyclostrophic balance*. Eq. (5.12) is just the *hydrostatic equation*, Eq. (2.5). The radial forces in gradient wind balance are shown schematically in Fig. 5.2.

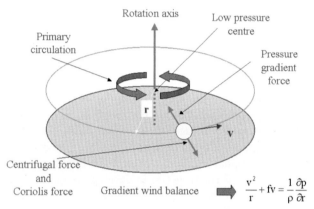

FIGURE 5.2 Schematic diagram illustrating the gradient wind force balance per unit mass in the primary circulation of a tropical cyclone.

A scale analysis carried out later (see Section 5.8) shows that the gradient wind equation and hydrostatic equation provide a good first approximation for representing the primary circulation of tropical cyclones at levels where frictional effects can be neglected, even in the presence of modest diabatic heating and a nonzero, but relatively weak secondary circulation. Support for approximate gradient wind balance throughout much of the free troposphere is found in analyses of aircraft observations in mature hurricanes (Willoughby, 1990).

Taking ($\partial/\partial z$) [Eq. (5.11)] and ($\partial/\partial r$) [Eq. (5.12)] to eliminate the pressure we obtain

the *thermal wind equation*

$$g\frac{\partial \ln\rho}{\partial r} + C\frac{\partial \ln\rho}{\partial z} = -\frac{\partial C}{\partial z}, \quad (5.13)$$

where

$$C = \frac{v^2}{r} + fv \quad (5.14)$$

denotes the sum of the centrifugal and Coriolis forces per unit mass. Eq. (5.13) is a linear first-order partial differential equation for $\ln\rho$.

The characteristics of the equation satisfy

$$\frac{dz}{dr} = \frac{C}{g}. \qquad (5.15)$$

The characteristics coincide with the isobaric surfaces because a small displacement (dr, dz) along an isobaric surface satisfies $(\partial p/\partial r)dr + (\partial p/\partial z)dz = 0$. Then, using the equations for hydrostatic balance $(\partial p/\partial z = -\rho g)$ and gradient wind balance $(\partial p/\partial r = \rho C)$ gives the equation for the characteristics (Eq. (5.15)). The density variation along a characteristic is governed by the equation

$$\frac{d}{dr} \ln \rho = -\frac{1}{g}\frac{\partial C}{\partial z}, \qquad (5.16)$$

which is another form of Eq. (5.13).

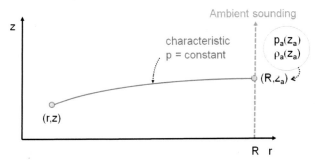

FIGURE 5.3 Characteristics of Eq. (5.13) as given by Eq. (5.15). See text for discussion of solution method given an ambient sounding for pressure and density $(p_a(z), \rho_a(z))$ as a function of height at radius R.

Given the vertical density profile, $\rho_a(z)$, and a prescribed tangential wind distribution, $v(r, z)$, Eqs. (5.15) and (5.16) can be integrated along the isobars to obtain the balanced axisymmetric density and pressure distributions for the wind distribution. The method is indicated in Fig. 5.3. Starting at a chosen point (r, z), Eq. (5.15) is integrated outwards with respect to r to determine the characteristic (i.e., isobar) through (r, z). Assuming that an ambient sounding of pressure $p_a(z)$ or density $\rho_a(z)$ is known[3] at radius R, the pressure at (r, z) is simply $p_a(z_R)$, where z_R is the height of the characteristic at $r = R$. Knowing the density $\rho_a(z_R)$ at height (z_R), Eq. (5.16) may be integrated backwards along the characteristic to determine the density at (r, z).

Since $\partial C/\partial z = (2v/r + f)(\partial v/\partial z)$, it follows from Eq. (5.16) that, for a barotropic vortex $(\partial v/\partial z = 0)$, ρ is constant along an isobaric surface, i.e. $\rho = \rho(p)$, whereupon, from the equation of state (2.11), T is a constant also. It follows that, at constant height, all the quantities p, ρ, and T, will decrease with decreasing radius because these quantities decrease with increasing height and the characteristics slope downwards with decreasing radius. Thus a barotropic vortex is cold-cored.

3. Actually, if only one of $p_a(z)$ and $\rho_a(z)$ are known, the other may be determined using the hydrostatic equation, i.e., $\rho_a(z) = -(1/g)dp_a/dz$ or $p_a(z) = \int_z^\infty g\rho_a dz$.

5.3 Interpretation of the thermal wind equation

The thermal wind equation is a simplified form of the toroidal vorticity equation (the equation for the tangential component of vorticity: see Section 5.13.1; Eq. 5.79) and can be interpreted as a balance between the tendency of three effects (Fig. 5.4).

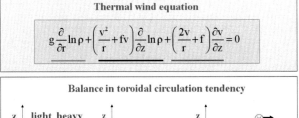

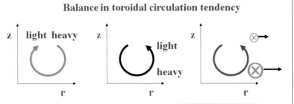

FIGURE 5.4 Schematic diagram illustrating the balance of three tendencies in the toroidal vorticity equation expressed by the thermal wind equation for a warm-cored vortex.

In terms of the buoyancy, Coriolis and centrifugal force fields, the first term in Eq. (5.13) is a positive azimuthal vorticity tendency associated with the propensity of warm (light) air in the core of the vortex to rise and cool (heavy) air on the periphery to sink in the presence of a gravitational force field acting downwards. The second term is a negative vorticity tendency associated with the propensity of light air to be centrifuged radially inwards and heavy air to be centrifuged radially outwards in the presence of the centrifugal force field acting outwards. The third term is a negative vorticity tendency associated with the decrease of the outward force C with height associated with the negative vertical shear of the tangential wind component.

The thermal wind equation (5.13), or equivalently Eq. (5.16), shows that in a cyclonic vortex in the Northern Hemisphere $(v > 0)$, in which the tangential wind speed decreases with height $(\partial v/\partial z < 0)$, $\ln \rho$ and hence ρ decreases with decreasing radius along the isobaric surface. Thus the temperature and potential temperature increase as r decreases and the vortex is warm cored (i.e. $\partial T/\partial r < 0$). This prediction of the thermal wind equation *is consistent with* the observation that tropical cyclones are warm-cored systems and that the tangential wind speed decreases with altitude above the frictional boundary layer (see e.g., Section 1.2.3). If the tangential wind speed were to increase with height $(\partial v/\partial z > 0)$, the vortex would be cold cored. Note that the characteristics, or isobars, dip down as the axis is approached on account of Eq. (5.15).

Fig. 5.5 shows an example of a calculation using the foregoing theory. The tangential wind field in Fig. 5.5a is

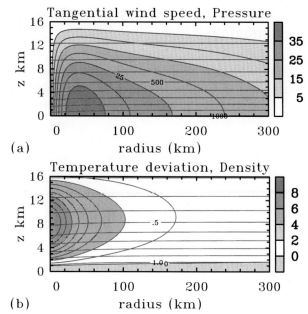

(a)

(b)

FIGURE 5.5 The primary circulation of a tropical cyclone. (a) isotachs of tangential wind speed V (contour interval 5 m s^{-1}) in red (mid gray in print version), shaded and isobaric surfaces in magenta (dark gray in print version) (contour interval 100 mb). (b) temperature deviation δT (contour interval 1 °C) in red (mid gray in print version), shaded (positive deviations) and light blue (dark gray in print version) (negative deviations), and density surfaces, ρ, in magenta (contour interval 0.1 kg m^{-3}).

broadly typical of that in a tropical cyclone. It is used to calculate the pressure field, shown also in this figure, as well as the density and temperature fields, assuming for illustrative purposes that $f = 5 \times 10^{-5}$ s^{-1} corresponding with a latitude of about 20°N. The density field and the difference in temperature from that at large radii are shown in Fig. 5.5b. Note that the isobars as well as the density contours dip down near the vortex axis. Temperatures in the vortex core region are mostly warmer than in the environment at the same height, except at low levels where they are cooler as the vortex in Fig. 5.5 is approximately barotropic near the surface.

Fig. 5.6 shows the contours of absolute angular momentum, M, corresponding to the tangential wind field in Fig. 5.5 and isentropes, θ, in balance with this velocity distribution. Typically, the M-surfaces slope outwards with height and values of M generally increase with increasing radius (see Section 5.7). The isentropes are approximately horizontal at larger radii, but dip downwards in the central region of the vortex. The values of θ increase with height, characteristic of a stably-stratified vortex (see Subsection 5.7.3).

The observational evidence that the primary circulation of a hurricane is approximately in gradient wind balance and hydrostatic balance makes the foregoing analysis a good start in understanding the structure of this circulation. However, the solution of the thermal wind equation neglects

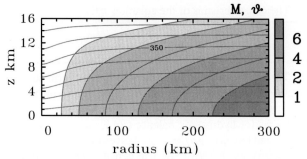

FIGURE 5.6 Isopleths of absolute angular momentum, M, corresponding to the tangential velocity in Fig. 5.5 and isentropes, θ = constant, in balance with this velocity distribution. Contour interval for M: 1×10^6 m^2 s^{-1}, shading levels indicated on the side bar; for θ 10 K.

the secondary circulation associated with nonzero u and w and it neglects the effects of condensational heating, radiative cooling and of friction near the sea surface. These are topics of subsequent subsections.

5.4 Generalized buoyancy

For an axisymmetric vortex in thermal wind balance, the local pressure gradient force per unit mass is $(1/\rho)(\partial p/\partial r, 0, \partial p/\partial z)$ when expressed in cylindrical coordinates (r, λ, z). Using Eqs. (5.11) and (5.12), this pressure gradient force may be written as $(C, 0, -g)$, which defines a *generalized gravitational vector*, $\mathbf{g}_*$. Thus, the isobars in this vortex are locally normal to the vector $\mathbf{g}_*$: see Fig. 5.7.

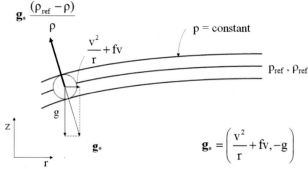

FIGURE 5.7 Schematic radial-height cross section of isobaric surfaces in a cyclonically-rotating vortex showing the forces on an air parcel including the gravitational force g, per unit mass, and the sum of the centrifugal and Coriolis forces $C = v^2/r + fv$, per unit mass. Note that the isobaric surfaces are normal to the local generalized gravitational force $\mathbf{g}_* = (C, 0, -g)$. For an air parcel of density ρ, the Archimedes force acting on it (the weight of displaced fluid per unit volume) is $-\mathbf{g}_* \rho_{ref}$, where ρ_{ref} is the local density of the vortex, while the weight of the air parcel is $\mathbf{g}_* \rho$. Therefore, the net buoyancy force, $\mathbf{b}$, acting on the parcel is $\mathbf{g}_*(\rho - \rho_{ref})/\rho$ per unit mass. For a cyclonically-rotating vortex and an air parcel with $\rho < \rho_{ref}$, $\mathbf{b}$ points upwards and inwards. Adapted from Smith et al. (2005). Republished with permission of Elsevier.

The expression for buoyancy (2.9) is valid also in such a vortex, but then there exists a radial component of buoyancy as well. The Archimedes force acting on the parcel is $-\rho_{ref}\mathbf{g}_*$ and the effective weight of the parcel is $\rho\mathbf{g}_*$, where ρ_{ref} is now the local reference density of the air parcel's environment at radius r. Accordingly, we may define a *generalized buoyancy force per unit mass* as the difference between these forces per unit mass:

$$\mathbf{b} = \mathbf{g}_* \frac{\rho - \rho_{ref}}{\rho}, \qquad (5.17)$$

analogous to the derivation of Eq. (2.9).

In a cyclonically-rotating vortex, air parcels that are lighter than their environment have an inward-directed component of generalized buoyancy force as well as an upward component, while heavier parcels have an outward component as well as a downward component. This result provides the theoretical basis for a *centrifuge*. A centrifuge is a device used to separate lighter from heavier fluids by rotating them rapidly in some container.

5.4.1 Exercises

Exercise 5.1. Assuming the most general form of the mass conservation equation:

$$\frac{\partial \rho}{\partial t} + \frac{1}{r}\frac{\partial(\rho r u)}{\partial r} + \frac{1}{r}\frac{\partial(\rho v)}{\partial \lambda} + \frac{\partial(\rho w)}{\partial z} = 0,$$

show that for inviscid flow, the absolute angular momentum per unit volume,

$$M_v = \rho\left(rv + \frac{1}{2}fr^2\right),$$

satisfies the equation:

$$\frac{\partial M_v}{\partial t} + \frac{1}{r}\frac{\partial(ruM_v)}{\partial r} + \frac{1}{r}\frac{\partial(vM_v)}{\partial \lambda} + \frac{\partial(wM_v)}{\partial z} = -\frac{\partial p}{\partial \lambda}.$$

Exercise 5.2. (a) Show that if the tangential circulation of a tropical cyclone is in gradient wind balance, there is strictly not a unique relationship between central pressure relative to the far environment and the maximum tangential wind speed.[4]

(b) Using a modified Rankine vortex as an example, show that for a given maximum tangential wind speed, the central pressure decreases as the tangential wind profile becomes broader. [Hint, the *modified Rankine vortex* is a circular vortex whose tangential wind distribution has a solid body "core" and a power-law decay outside the core. The

wind field is defined by the formula

$$v(r) = v_m \times \begin{cases} \dfrac{r}{r_m} & \text{for} \quad r \le r_m, \\ \left(\dfrac{r_m}{r}\right)^\alpha & \text{for} \quad r \ge r_m, \end{cases} \qquad (5.18)$$

where v_m is the maximum wind speed, r is the radius and α is a positive decay exponent.]

(c) Show that, for a finite pressure difference, α must be larger than unity for this profile.

Exercise 5.3. Show that at a point (r_{vmax}, z_{vmax}) where the tangential wind speed is a maximum, v_{max}, the radial gradient of M is $v_{max} + fr_{vmax}$ and the vertical gradient of M is zero. Show that for sufficiently strong vortices, $v_{max} \approx \frac{\partial M}{\partial r}|_{(r_{vmax}, z_{vmax})}$.

Exercise 5.4. Show that ρ/χ is a function of p only. Hence show that the thermal wind equation (5.13) may be reformulated as

$$g\frac{\partial(\ln\chi)}{\partial r} + C\frac{\partial(\ln\chi)}{\partial z} = -\frac{\partial C}{\partial z},$$

or alternatively as

$$g\frac{\partial\chi}{\partial r} = -\frac{\partial}{\partial z}(\chi C), \qquad (5.19)$$

where $\chi = 1/\theta$.

Exercise 5.5. Show that in terms of the Exner function, Eqs. (5.11) and (5.12) may be written as

$$\chi C = c_p\frac{\partial\pi}{\partial r} \quad \text{and} \quad -\chi g = c_p\frac{\partial\pi}{\partial z}, \qquad (5.20)$$

respectively, where $\chi = 1/\theta$. Use these relationships to verify Eq. (5.19), which is the thermal wind equation for the vortex $v(r,z)$ using the variables χ and C.

Exercise 5.6. Show that for a steady, axisymmetric, inviscid flow, the velocity vector (u, w) in the meridional plane is tangential to the M-surfaces, i.e., the streamlines of the overturning circulation are congruent to the M-surfaces.

Exercise 5.7. The M-surfaces in a steady axisymmetric vortex are inclined, sloping upwards and outwards. Show that the tangential wind of an air parcel descending along one of these surfaces is increasing although the tangential wind at a fixed height and radius is not. Show next that if the M-surfaces are moving, $\partial v/\partial t = -(1/r)\mathbf{u_s}\cdot\nabla M$, where $\mathbf{u_s}$ is the transverse velocity component in a vertical (r, z)-plane.

5.5 The tropical cyclone eye

Observations show that the eye is a cloud free region surrounding the storm axis where the air temperatures are

---

4. Note that, despite this lack of uniqueness, tropical cyclone forecasters find it useful to have an approximate empirical relationship between the central pressure and maximum tangential wind speed (see e.g., Callaghan and Smith, 1998; Kossin, 2015 and refs.).

warmest. Therefore, it would be reasonable to surmise that the air within it has undergone descent during the formative stages of the cyclone, and that possibly it continues to descend. The question then is: why doesn't the inflowing air spiral in as far as the axis of rotation. We address this question in Chapter 6, but note here that eye formation is consistent with other observed features of the tropical cyclone circulation. Assuming that the primary circulation is in gradient wind balance, we may integrate Eq. (5.11) with radius to obtain a relationship between the pressure perturbation at a given height z on the axis and the tangential wind field distribution, i.e.,

$$p(0, z) = p_a(z) - \int_0^\infty \rho \left(\frac{v^2}{r} + fv \right) dr, \qquad (5.21)$$

where $p_a(z) = p(\infty, z)$ is the environmental pressure at the same height. Differentiating Eq. (5.21) with respect to height and dividing by the density gives the perturbation pressure gradient force per unit mass along the vortex axis:

$$-\frac{1}{\rho} \frac{\partial(p - p_a)}{\partial z} = \frac{1}{\rho} \frac{\partial}{\partial z} \int_0^\infty \rho \left(\frac{v^2}{r} + fv \right) dr. \qquad (5.22)$$

Observations in tropical cyclones show that the tangential wind speed generally decreases with height above the boundary layer (e.g., Fig. 1.9a, Reasor et al., 2000, Fig. 4.) and that the vortex broadens with height in the sense that the radius of the maximum tangential wind speed increases with altitude (see Fig. 1.9a). This behavior, which is consistent with outward-sloping absolute angular momentum surfaces as discussed above, implies that the integral on the right-hand-side of Eq. (5.22) decreases with height. Then, Eq. (5.22) shows that there must be a downward-directed perturbation pressure gradient force per unit mass along the vortex axis. If unopposed, this perturbation pressure gradient force would drive subsidence along and near to the axis to form the eye. However, as air subsides, it is compressed and warms relative to air at the same level outside the eye and thereby becomes locally buoyant (i.e., relative to the air outside the eye). In reality, this upward buoyancy approximately balances the downward directed (perturbation) pressure gradient force so that the actual subsidence results from a small residual force. In essence the subsiding flow remains close to hydrostatic balance.

As the vortex intensifies, the downward pressure gradient force given by Eq. (5.22) must increase and the residual force must be downwards to drive further subsidence. On the other hand, if the vortex weakens, the residual force must be upwards, allowing the air to re-ascend. In the steady state, the residual force must be zero and there is no longer a need for up- or down motion in the eye, although, in reality there may be motion in the eye associated with turbulent mixing across the eyewall or with asymmetric instabilities within the eye.

It is not possible to directly measure the vertical velocity that occurs in the eye, but one can make certain inferences about the origin of air parcels in the eye from their thermodynamic characteristics, which can be measured.

The above arguments are incomplete as they are based on the assumption of exact balance and cannot account for the residual force. The theory based on the Eliassen equation in Section 5.9 provides a more complete way of understanding the formation of the eye (see Subsection 5.10.5).

5.6 Spin up of the primary circulation

The key element of vortex spin up in an axisymmetric setting can be illustrated from the equation for absolute angular momentum per unit mass, Eq. (5.7). If the frictional torque per unit mass, rF_λ, acting on a fluid parcel is zero, then M is materially conserved (see Eq. (5.9)). Since M is related to the tangential velocity by the formula (5.8):

$$v = \frac{M}{r} - \frac{1}{2}fr, \qquad (5.23)$$

and when M is materially-conserved, both terms in this expression lead to an increase in v as r decreases and to a decrease in v when r increases. Thus *a prerequisite for spin up in an inviscid axisymmetric flow is the existence of radial inflow.*

The above result underpins *the classical spin up mechanism for tropical cyclones* (see Section 8.2). This mechanism calls for convergence of air parcels above the frictional boundary layer to produce spin up. The processes capable of producing such inflow in a tropical cyclone are discussed later.

The converse result explains why the upper-tropospheric outflow of a tropical cyclone becomes anticyclonic beyond a certain radius, since if air parcels move outwards, they spin more slowly and if the outward displacement is large enough, the direction of spin will change.

An alternative, but equivalent interpretation for the material rate-of-change of the mean tangential wind in an axisymmetric inviscid flow follows directly from Newton's second law of motion (see Eq. (5.2)) in which the sole force is the generalized Coriolis force, $-u(v/r + f)$. Either viewpoint shows that spin up or spin down is related to the radial motion of air parcels and therefore depends on the secondary circulation, which will be examined in Section 5.9.

5.7 Stability

Having shown in Section 5.2 that any steady vortical flow with velocity field $\mathbf{u} = (0, v(r, z), 0)$ is an exact solution of the inviscid equations without secondary circulation or heating, it is appropriate to ask under what circumstances such solutions are stable. We consider here the problem of local axisymmetric stability based on simple parcel arguments.

5.7.1 Barotropic vortices

For simplicity, we begin by considering a barotropic vortex rotating with tangential velocity $v(r)$ in an incompressible fluid. Solid body rotation is the special case $v(r) = \Omega r$, where Ω is the constant angular velocity. The approach is to investigate the forces acting on a fluid parcel that is displaced radially outwards from A at radius r_1 to B at radius r_2 in Fig. 5.8. The fluid parcel at A has the tangential velocity v_1. Assuming frictional torques can be neglected, the parcel conserves its absolute angular momentum M so that its velocity v' at B is given by

$$r_2 v' + \frac{1}{2} f r_2^2 = r_1 v_1 + \frac{1}{2} f r_1^2,$$

or

$$v' = \frac{r_1}{r_2} v_1 + \frac{1}{2} \frac{f}{r_2} (r_1^2 - r_2^2) \qquad (5.24)$$

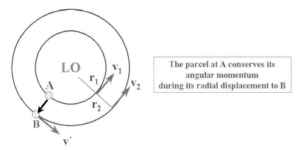

FIGURE 5.8 Radial displacement of an air parcel in a vortex on an f-plane.

Other parcels at the same radius as B have a velocity v_2 that is different, in general, from v'. In equilibrium, before the displacement, these parcels are in a balanced state in which the radially-inward pressure gradient force they experience is exactly balanced by the outward force C; i.e.,

$$\frac{1}{\rho} \frac{dp}{dr} \bigg]_{r=r_2} = \frac{v_2^2}{r_2} + f v_2. \qquad (5.25)$$

It is assumed that the disturbance in pressure brought about by the displaced parcel at radius r_2 can be neglected. Consequently, the displaced parcel will experience the same radial pressure gradient force as other parcels at radius r_2. Now, since the displaced parcel rotates with velocity v', the sum of the centrifugal and Coriolis forces acting on it is $v'^2/r_2 + f v'$. Therefore, the displaced parcel experiences a net *outward* force per unit mass, $F = $ centrifugal $+$ Coriolis force $-$ radial pressure gradient force

$$= \frac{v'^2}{r_2} + f v' - \frac{1}{\rho} \frac{\partial p}{\partial r} \bigg]_{r=r_2}.$$

Using Eqs. (5.24) and (5.25), this expression can be written after a little algebra (see Exercise 5.8)

$$F = \frac{1}{r_2^3} \left[\left(r_1 v_1 + \frac{1}{2} r_1^2 f \right)^2 - \left(r_2 v_2 + \frac{1}{2} r_2^2 f \right)^2 \right]. \qquad (5.26)$$

Recognizing the terms inside the inner parentheses as the absolute angular momentum, we may rewrite Eq. (5.26) as follows

$$F = -\frac{1}{r_2^3} \left[M^2(r_2) - M^2(r_1) \right]. \qquad (5.27)$$

Therefore, an air parcel displaced outwards experiences an inward force (i.e., a restoring force per unit mass) that is proportional to the difference in M^2 between the two radii. We see immediately that if the square of M increases with radius, the restoring force will be inward. In this case, the vortex is said to be *centrifugally stable*.

For small displacements from the initial equilibrium position one may expand the expression for the net outward force as a Taylor series, retaining only the first order term in the displacement $r' = r_2 - r_1$. The net outward force per unit mass acting on the displaced parcel is then approximated by

$$F = -\frac{1}{r^3} \frac{dM^2}{dr} r', \qquad (5.28)$$

where it is understood that, to first order in r', the equilibrium quantities above are evaluated at the initial position. This expression enables us to establish a local criterion for the stability of a general rotating flow $v(r)$, analogous to the criterion in terms of $\text{sgn}(N^2)$ for the vertical stability of a density stratified fluid at rest (see, e.g., Subsection 5.7.3 and Eq. (5.42)). It follows from Eq. (5.28) that *a general swirling flow $v(r)$ is axisymmetrically stable, neutral or unstable if the radial gradient of the square of the absolute angular momentum is positive, zero, or negative.*[5] The height-radius distribution of M^2 for the *baroclinic* vortex in Fig. 5.5 is shown in Fig. 5.9.

For the special case of solid body rotation, $v = \Omega r$, and for a small displacement from radius $r_1 = r$ to $r_2 = r + r'$, Eq. (5.28) gives

$$F = -4 \left(\Omega + \frac{1}{2} f \right)^2 r'. \qquad (5.29)$$

---

5. This is the second Rayleigh theorem on vortex stability as mentioned in Section 4.4.3 (Rayleigh, 1916). In the fluid dynamics literature, the quantity $(1/r^3) dM^2/dr$ is called Rayleigh's discriminant. To keep with meteorological custom, this quantity will be referred to as the centrifugal (or inertial) stability parameter. As a point of historical interest, the 1916 derivation employed an energetics-based argument using the device of interchanging two concentric rings of fluid and computing the change in kinetic energy of the two rings after interchange under the constraint of the material conservation of angular momentum. The derivation employed here (and in the forthcoming application to a baroclinic vortex) is force-based, in the spirit of Kármán (1956).

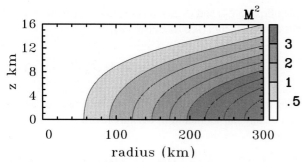

FIGURE 5.9 Height-radius distribution of M^2 for the vortex in Fig. 5.5. (contour interval: 1×10^{13} m⁴s⁻²). Shading levels indicated on the side bar.

Thus, an air parcel displaced outwards experiences an inward restoring force, proportional to the displacement and to the square of *the absolute angular frequency* $\Omega + \frac{1}{2}f$. Note that if the rotation rate $\Omega \gg \frac{1}{2}f$, such as in a rotating tank in the laboratory, the rotation of the coordinate system can be neglected and $F = -4\Omega^2 r'$.

The presence of a restoring force when an air parcel is displaced radially in a rapidly-rotating fluid leads to the concept of *centrifugal* or *inertial stiffness*. Broadly speaking, the larger the rotation rate, the harder it is to displace an air parcel radially. A notable exception to this "rule of thumb" occurs for tangential wind profiles exhibiting a $1/r$ decay law with radius (see Exercise 5.10).

5.7.2 Exercises

Exercise 5.8. Verify that

$$\left.\frac{v'^2}{r_2} + fv' - \frac{1}{\rho}\frac{\partial p}{\partial r}\right]_{r=r_2} = \frac{1}{r_2^3}\left[(r_1v_1 + \frac{1}{2}r_1^2 f)^2 - (r_2v_2 + \frac{1}{2}r_2^2 f)^2\right],$$

and show that in the case of solid body rotation, $v = \Omega r$, the expression for F in Eq. (5.26) reduces to that on the right-hand-side of Eq. (5.29).

Exercise 5.9. Show that, in an axisymmetric flow, the absolute vorticity, $\zeta + f$, and absolute angular momentum, M, are related by

$$\zeta + f = \frac{1}{r}\frac{dM}{dr}. \tag{5.30}$$

Exercise 5.10. Consider the centrifugal (inertial) stability of the modified Rankine vortex defined in Exercise 5.2 (see Eq. (5.18)). For several values of α between zero and unity and for α greater than unity, sketch the *r*elative angular momentum distribution. When is the vortex centrifugally stable or unstable? Consider next the corresponding absolute angular momentum distribution for cases in which the Earth's

rotation is significant and the Coriolis parameter has the same sign as the core rotation rate ("cyclonic case") and the opposite sign as the core rotation ("anticyclonic case"). When is the vortex centrifugally stable? Does the Earth's rotation always stabilize the vortex? Explain your answer physically. Now, consider the limiting case of $\alpha = 1$. Why is $\alpha = 1$ a limiting case? Explain. Is the 'rule of thumb' given in the text generally correct? Discuss.

5.7.3 Baroclinic vortices

The foregoing axisymmetric stability theory can be readily extended to the case of a balanced baroclinic vortex in a stably-stratified atmosphere. In this case we consider the displacement of an air parcel from a point $\mathbf{r} = (r, z)$ to a neighboring point with coordinates $\mathbf{r} + \mathbf{r}' = (r + r', z + z')$ (see Fig. 5.10a). Again we assume that M is materially conserved during the displacement, but now, in general, the density of the parcel is allowed to change also. Nevertheless, in the situation where the displacement of the parcel is dry adiabatic, the potential temperature, θ, is materially conserved. As in the previous example, we examine the net force on the parcel at the point $\mathbf{r} + \mathbf{r}'$, assuming that the parcel has the same pressure as that of its surroundings at that point. Let the volume of the air parcel at this point be V_p.

The net vertical force per unit mass acting on the parcel at $\mathbf{r} + \mathbf{r}'$ is the weight of the air displaced (the Archimedes force), $g\rho V_p$, minus the weight of the parcel itself, $g\rho_p V_p$, where ρ_p is the density of the parcel at $\mathbf{r} + \mathbf{r}'$ (Fig. 5.10a). The vertical force per unit mass is obtained by dividing this difference by the parcel mass $\rho_p V_p$, i.e.,

$$F_{vertical} = g\frac{\rho - \rho_p}{\rho_p}. \tag{5.31}$$

Since $\rho = p/R\pi\theta$ and the parcel materially conserves its potential temperature during its displacement ($\theta = \theta_1$), Eq. (5.31) may be written

$$F_{vertical} = g\frac{\theta_1 - \theta_2}{\theta_2}, \tag{5.32}$$

where the subscripts 1 and 2 refer to equilibrium values at the points $\mathbf{r}$ and $\mathbf{r} + \mathbf{r}'$, respectively. For a small vector displacement $\mathbf{r}'$ (i.e., $|\mathbf{r}'| \ll |\mathbf{r_i}|$), the difference $\theta_2 - \theta_1 = \theta(r_1 + r', z_1 + z') - \theta(r_1, z_1)$ can be approximated in terms of partial derivatives, whereupon

$$F_{vertical} = -\frac{g}{\theta}\frac{\partial\theta}{\partial r}r' - \frac{g}{\theta}\frac{\partial\theta}{\partial z}z', \tag{5.33}$$

where the partial derivatives are evaluated at the point $\mathbf{r}$.

The net radial force acting on the air parcel may be obtained in a similar way, but with the sum of the centrifugal and Coriolis forces, C, playing the role analogous to the gravitational force. Here, unlike the case of the gravitational

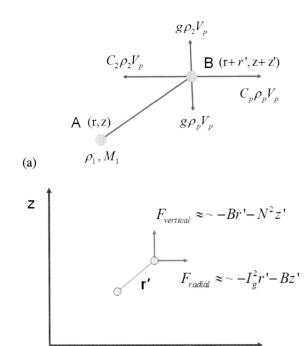

$$F_{radial} = \left[\left(\frac{M_1^2}{r_2^3} - \frac{r_2 f^2}{4} \right) - \left(\frac{M_2^2}{r_2^3} - \frac{r_2 f^2}{4} \right) \frac{\theta_1}{\theta_2} \right]. \tag{5.35}$$

This equation may be rewritten as

$$F_{radial} = \left[\frac{1}{r_2^3} \left(M_1^2 - M_2^2 \frac{\theta_1}{\theta_2} \right) + \frac{r_2 f^2}{4} \left(\frac{\theta_1}{\theta_2} - 1 \right) \right]. \tag{5.36}$$

Then, expressed in terms of differentials at the point (r, z) as before, the expression becomes

$$F_{radial} = -\theta \left[\frac{1}{r^3} \frac{\partial}{\partial r} \left(\frac{M^2}{\theta} \right) - \frac{r f^2}{4} \frac{\partial}{\partial r} \left(\frac{1}{\theta} \right) \right] r'$$
$$-\theta \left[\frac{1}{r^3} \frac{\partial}{\partial z} \left(\frac{M^2}{\theta} \right) - \frac{r f^2}{4} \frac{\partial}{\partial z} \left(\frac{1}{\theta} \right) \right] z'. \tag{5.37}$$

The radial and vertical components of Newton's second law of motion for the air parcel then take the following vector form

$$\left(\frac{\partial^2 r'}{\partial t^2}, \frac{\partial^2 z'}{\partial t^2} \right) = (-R_1 r' - R_2 z', -Z_1 r' - Z_2 z'), \tag{5.38}$$

where, in terms of χ, it can be shown that

$$R_1 = \xi(\zeta + f) + \frac{C}{\chi} \frac{\partial \chi}{\partial r}, \qquad Z_1 = -\frac{g}{\chi} \frac{\partial \chi}{\partial r}, \tag{5.39}$$
$$R_2 = \frac{1}{\chi} \frac{\partial}{\partial z}(C\chi), \qquad Z_2 = -\frac{g}{\chi} \frac{\partial \chi}{\partial z}. \tag{5.40}$$

Note that $Z_2 = N^2$, the Brunt-Väisälä frequency squared, and on account of the thermal wind equation (5.19) in the form $g \partial \chi / \partial r = -\partial(C\chi)/\partial z$ (cf. Eq. (5.19)), $Z_1 = R_2$, which henceforth we denote by B. The quantity R_1 is the generalized inertial (centrifugal) stability[6] (cf. Eq. (5.28) and we will write it as I_g^2. Note that I_g^2 reduces to

> the *inertial stability*
> $$I^2 = \xi(\zeta + f), \tag{5.41}$$

if χ is independent of r. The force perturbations represented by Eq. (5.38) are shown in Fig. 5.10b. In the analysis below, it will be assumed that the vortex is both statically and centrifugally stable, i.e., $N^2 > 0$ and $I_g^2 > 0$.

In summary, the local stability of a baroclinic vortex depends on the three spatially varying parameters characterizing:

FIGURE 5.10 (a) Forces acting on an air parcel displaced radially and vertically from a point with coordinates $\mathbf{r} = (r, z)$ to a neighboring point with coordinates $\mathbf{r} + \mathbf{r}' = (r + r', z + z')$ in an axisymmetric stably-stratified vortex. (b) The net forces expressed in terms of the displacement $\mathbf{r}' = (r', z')$. Here C is the sum of the centrifugal and Coriolis forces, M is the absolute angular momentum and ρ is the density. Subscripts '1' and '2' refer to conditions at points $\mathbf{r}$ and $\mathbf{r} + \mathbf{r}'$ and subscript 'p' refers to the displaced air parcel at $\mathbf{r} + \mathbf{r}'$. For a vortex in which the tangential wind speed decreases with height, the quantity B defined by Eq. (5.44) is typically negative (see e.g., Fig. 5.11c). In this case, the forces involving B are in the directions of the component displacements. Under certain circumstances, this behavior can lead to instability, even when the familiar static stability (on r-surfaces) and centrifugal stability (on z-surfaces) are both positive so that the vortex might naively be expected to be stable. See text and exercises for further discussion.

force field, the radial force field is not spatially uniform within the vortex. The net radial force on the air parcel per unit mass is then

$$F_{radial} = \frac{C_p \rho_p - C_2 \rho_2}{\rho_p}, \tag{5.34}$$

where C_p, the value of C for the air parcel (or ring of air) at (r, z), is determined by the assumption that M is conserved during the parcel displacement, as in the barotropic case (Fig. 5.8a). Now, however, the parcel density changes as well, but again θ is materially conserved. Using the relationship between C and M, i.e. $C = M^2/r^3 - \frac{1}{4}rf^2$, replacing ρ by the corresponding value of $p/R\pi\theta$, and assuming that the parcel pressure at B is the same as the equilibrium pressure, Eq. (5.34) may be written

6. The adjective "inertial" owes its origin to the rotational analogue of Newton's first law of motion in which a rotating body will preserve its angular momentum unless acted upon by a torque in the direction of rotation.

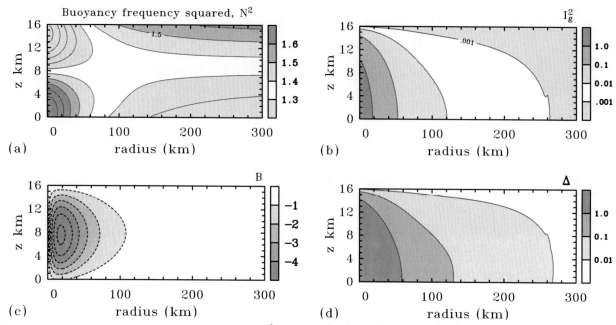

FIGURE 5.11 Height-radius distributions of the quantities (a) N^2 (Unit $\times 10^{-4}$ s^{-2}); (b) I_g^2 (Unit $\times 10^{-5}$ s^{-2}; shown are contours 1, 0.1, 0.01, 0.001 Units); (c) B (Unit $\times 10^{-6}$ s^{-2}); (d) the discriminant, Δ, for the vortex in Fig. 5.5 (Unit $\times 10^{-10}$ s^{-2}; shown are contours 1, 0.1, 0.01 Units). In each panel, shading values on the side bar.

the *static stability*

$$N^2 = -\frac{g}{\chi}\frac{\partial \chi}{\partial z}; \qquad (5.42)$$

the *generalized inertial stability*

$$I_g^2 = I^2 + \frac{C}{\chi}\frac{\partial \chi}{\partial r}; \qquad (5.43)$$

the *baroclinicity*

$$B = \frac{1}{\chi}\frac{\partial}{\partial z}(C\chi) = \xi\frac{\partial v}{\partial z} + \frac{C}{\chi}\frac{\partial \chi}{\partial z}. \qquad (5.44)$$

The height-radius distributions of these parameters for the vortex of Fig. 5.5 are shown in the first three panels of Fig. 5.11. Note that for a vortex in which the tangential wind speed decreases with height, B is typically negative and contributes to force perturbations in the direction of the displacement as seen in Fig. 5.10b.

If the quantities N^2, I_g^2, and B are appreciably constant on the scale of a parcel displacement, Eq. (5.38) has solutions of the form $(r', z') = (r_0', z_0')\exp(i\omega t)$, where r_0', z_0', and the frequency ω must satisfy the pair of eigenvector equations:

$$(I_g^2 - \omega^2)r_0' + Bz_0' = 0,$$
$$Br_0' + (N^2 - \omega^2)z_0' = 0. \qquad (5.45)$$

These equations have a nontrivial solution for (r_0', z_0') only if

$$(N^2 - \omega^2)(I_g^2 - \omega^2) - B^2 = 0,$$

i.e., if

$$\omega^4 - \omega^2(N^2 + I_g^2) + N^2 I_g^2 - B^2 = 0, \qquad (5.46)$$

then

$$\omega^2 = \frac{1}{2}(N^2 + I_g^2) \pm \frac{1}{2}\sqrt{\{(N^2 + I_g^2)^2 - 4(N^2 I_g^2 - B^2)\}}. \qquad (5.47)$$

Note that the expression under the square-root sign can be rewritten as $(N^2 - I_g^2)^2 + 4B^2$ and it is always positive. Then the solutions for ω^2 are both positive provided that the magnitude of $(N^2 + I_g^2)$ is larger than the square root term, i.e., if the discriminant Δ defined by

$$\Delta = \frac{1}{4}[(N^2 + I_g^2)^2 - (N^2 - I_g^2)^2 - 4B^2]$$
$$= N^2 I_g^2 - B^2 > 0 \qquad (5.48)$$

In this case the parcel oscillates along the displacement eigenvector (r_0', z_0') with frequency ω given by the square root of Eq. (5.47) and the flow is symmetrically *stable*. The stable solutions may be easily interpreted in the special case of small baroclinicity, that is, $B^2 \ll (N^2 - I_g^2)^2$. Taylor expanding the frequency equation (5.47) in the limit of small B^2 gives $\omega^2 \approx N^2 + B^2/(N^2 - I_g^2)$ for the case of the positive root and $\omega^2 \approx I_g^2 - B^2/(N^2 - I_g^2)$ for the case of the negative root. The former solution corresponds to shear-modified buoyancy oscillations that materially conserve potential temperature; the latter solution corresponds to shear-modified centrifugal oscillations that materially conserve absolute angular momentum per unit mass. Providing $N^2 > I_g^2$, which is typically the case for a tropical cyclone, the solution given by Eq. (5.47) implies that as B^2 increases from zero the shear-modified buoyancy frequency will increase, while the shear-modified centrifugal frequency will decrease relative to the barotropic counterpart $B^2 = 0$.

If, on the other hand, $\Delta < 0$, the negative root solution for ω^2 is negative. In this case, two solutions for ω are imaginary and one of these gives an exponentially growing solution for the parcel displacement. In the growing solution case, a displaced parcel continues to move away from its initial position and the displacement is therefore unstable. We refer to this situation as *symmetric instability*. Based on these results, it follows that a statically stable ($N^2 > 0$) and centrifugally stable ($I_g^2 > 0$) vortex will be symmetrically unstable when $\Delta < 0$ (Exercise 5.15). It is left for the reader to show that Δ is proportional to the *potential vorticity*, P, multiplied by the *absolute angular rotation rate* $\frac{1}{2}\xi = V/r + \frac{1}{2}f$ (see Exercise 5.14). The height-radius distributions of the discriminant Δ for the vortex in Fig. 5.5 is shown in Fig. 5.11d.

The trajectory of a particle can be obtained as follows. Suppose that the flow is symmetrically stable and that the initial parcel displacement is to the point (r_0', z_0'). Let ω_1 and ω_2 be the square of the eigenfrequencies given by Eq. (5.47). From (5.45), the corresponding eigenvectors are: $(1, (\omega_1^2 - I_g^2)/B)$ and $(1, (\omega_2^2 - I_g^2)/B)$. Then the parcel displacement at time t may be written as

$$(r'(t), z'(t)) = \left(1, \frac{\omega_1^2 - I_g^2}{B}\right)(c_{1+}e^{i\omega_1 t} + c_{1-}e^{-i\omega_1 t})$$
$$+ \left(1, \frac{\omega_2^2 - I_g^2}{B}\right)(c_{2+}e^{i\omega_2 t} + c_{2-}e^{-i\omega_2 t}),$$
$$(5.49)$$

where $c_{1\pm}$ and $c_{2\pm}$ are four constants. At $t = 0$, $(r'(t), z'(t)) = (r_0', z_0')$ and we assume that the particle has no initial motion: $(dr'/dt, dz'/dt) = (0, 0)$. These conditions determine $c_{1\pm}$ and $c_{2\pm}$. The second condition gives us $c_{1-} = c_{1+} = \frac{1}{2}c_1$ and $c_{2-} = c_{2+} = \frac{1}{2}c_2$ so that if both values of ω^2 are positive, Eq. (5.49) becomes

$$(r'(t), z'(t)) = c_1 \left(1, \frac{\omega_1^2 - I_g^2}{B}\right) \cos(\omega_1 t)$$
$$+ c_2 \left(1, \frac{\omega_2^2 - I_g^2}{B}\right) \cos(\omega_2 t). \quad (5.50)$$

Evaluating this equation at $t = 0$ gives $c_1 + c_2 = r_0'$ and $c_1(\omega_1^2 - I_g^2) + c_2(\omega_2^2 - I_g^2) = Bz_0'$, from which,

$$c_1 = [(\omega_2^2 - I_g^2)r_0' - Bz_0']/(\omega_2^2 - \omega_1^2) \quad (5.51)$$

and

$$c_2 = [Bz_0' - (\omega_1^2 - I_g^2)r_0']/(\omega_2^2 - \omega_1^2). \quad (5.52)$$

Finally

$$r'(t) = \frac{(\omega_2^2 - I_g^2)r_0' - Bz_0'}{\omega_2^2 - \omega_1^2}\cos(\omega_1 t) +$$
$$\frac{Bz_0' - (\omega_1^2 - I_g^2)r_0'}{\omega_2^2 - \omega_1^2}\cos(\omega_2 t) \quad (5.53)$$

and

$$z'(t) = \frac{(\omega_2^2 - I_g^2)}{B}\frac{(\omega_1^2 - I_g^2)r_0' - Bz_0'}{(\omega_2^2 - \omega_1^2)}\cos(\omega_1 t) +$$
$$\frac{(\omega_2^2 - I_g^2)}{B}\frac{Bz_0' - (\omega_1^2 - I_g^2)r_0'}{(\omega_2^2 - \omega_1^2)}\cos(\omega_2 t). \quad (5.54)$$

An example of an air-parcel trajectory in a symmetrically-stable rotating shear flow is shown in Fig. 5.12a and one that is in a symmetrically-unstable flow is shown in Fig. 5.12b.

5.7.4 Exercises

Exercise 5.11. Verify that

$$M^2 = r^3 C + \frac{1}{2}f^2 r^4$$

Exercise 5.12. Verify Eq. (5.38) and the functional forms for R_1, R_2, Z_1, and Z_2 in Eq. (5.39) and Eq. (5.40).

Exercise 5.13. Show that I_g^2 defined in Eq. (5.43) can be written as

$$I_g^2 = I^2 + \frac{C}{\chi}\frac{\partial\chi}{\partial r}, \quad (5.55)$$

where $I^2 = \xi(\zeta + f)$. Show also that

$$B = \xi\frac{\partial v}{\partial z} + \frac{C}{\chi}\frac{\partial\chi}{\partial z} = g\frac{\partial}{\partial r}\ln\theta, \quad (5.56)$$

where $\xi = 2v/r + f$ denotes twice the absolute rotation rate at radius r as defined above.

The second term on the right-hand side of Eq. (5.55) characterizes the effects of a radially-varying density on the inertial stability (recall that ρ is proportional to χ at constant pressure). For a warm-cored vortex with ρ increasing with radius, this term increases the inertial stability. Similarly, the second term in the middle expression of Eq. (5.56)

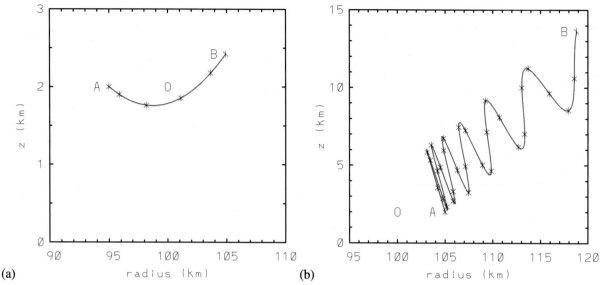

FIGURE 5.12 Displacement of an air parcel in a rotating shear flow that is (a) symmetrically-stable, and (b) symmetrically-unstable. In each case the parcel is initially displaced from O to A. The subsequent track is the blue (dark gray in print version) line from A to B. The '*' marks show locations every two minutes. For these calculations we take N^2, I^2, and B to be constants. In each case $N = 1.0 \times 10^{-2}$ s^{-1} and $I_g = 5.0 \times 10^{-3}$ s^{-1}. In (a), $B = 3.27 \times 10^{-6}$ s^{-2}, and in (b) $B = -1.01 \times NI$.

characterizes the effects of a vertically-varying density on the inertial stability. For a stably-stratified vortex with ρ decreasing with height, this term is negative and for a warm-cored vortex with v decreasing with height, it increases the magnitude of the baroclinicity term, B.

An important deduction from Eq. (5.56) is that a warm-cored vortex with $\partial\theta/\partial r < 0$, (i.e., $\partial\chi/\partial r > 0$) implies that $B < 0$.

Exercise 5.14. Recalling the result of Exercise 2.19, that the Ertel potential vorticity (Eq. (2.23)) for the vortex $v(r, z)$ is

$$P = \frac{1}{\rho}\left(-\frac{\partial v}{\partial z}\frac{\partial\theta}{\partial r} + (\zeta + f)\frac{\partial\theta}{\partial z}\right), \quad (5.57)$$

deduce that

$$P = \frac{1}{\rho g \xi \chi}(I_g^2 N^2 - B^2). \quad (5.58)$$

Exercise 5.15. Show that $\nabla\theta \wedge \nabla M$ is proportional to the potential vorticity, i.e., to $I_g^2 N^2 - B^2$. Show that this quantity is negative when the θ-surfaces are more steeply inclined to the horizontal than the M-surfaces. Note that this configuration of M- and θ-surfaces is therefore a condition for symmetric instability. Show next that an alignment of the M- and θ-surfaces implies zero potential vorticity.

Exercise 5.16. Show that in the symmetrically-unstable case, the terms involving $\cos(\omega_2 t)$ in Eqs. (5.53) and (5.54) must be replaced with $\cosh(\omega_2^* t)$, where $\omega_2^{*2} = -\omega_2^2$.

5.8 Scale analysis

Since the theory for the steady primary circulation in Section 5.2 was derived on the assumption that there was no secondary circulation ($u \equiv 0$, $w \equiv 0$), it is necessary to examine the applicability of this theory when a secondary circulation exists. To this end we perform a scale analysis of the dynamical equations (5.1)-(5.3) and continuity equation (5.4). For the present, it suffices to assume that the air is homogeneous and frictionless and that the motion is axisymmetric. In the next chapter we repeat this analysis for flows in which friction is important. We define velocity scales (U, V, W) for (u, v, w), length scales (R, Z) for (r, z), a time scale T for t, and scales Δp_r and Δp_z for changes in pressure, p, in the radial and vertical directions, respectively. We define two nondimensional parameters: a *swirl ratio* $S_r = V/U$ and a *Rossby number* $Ro = V/fR$. We assume also an *advective time scale* for the secondary circulation $T = R/U$ (for axisymmetric motions the azimuthal advective time-scale V/R is not relevant). Then the terms in Eqs. (5.1) to (5.4) have scales as shown in Table 5.1.

5.8.1 Continuity equation

From line (1) in Table 5.1 we see that the continuity equation (5.4) implies that $W/Z \sim U/R$. This useful result can be employed to simplify the scalings.

TABLE 5.1 Scaling of the terms in Eqs. (5.1) to (5.4), beginning with the last of these. Here $S_r = V/U$ is the swirl ratio, $A = Z/R$ is the aspect ratio of the flow, and $Fr = U/\sqrt{gZ}$ is the Froude number.

Continuity

$\dfrac{1}{r}\dfrac{\partial \rho u}{\partial r}$	$+\dfrac{\partial \rho w}{\partial z}$	$=0$	(5.4)
$\rho\dfrac{U}{R}$	$\rho\dfrac{W}{Z}$		(5.4)a

Radial momentum

$\dfrac{\partial u}{\partial t}$	$+u\dfrac{\partial u}{\partial r}$	$+w\dfrac{\partial u}{\partial z}$	$-\dfrac{v^2}{r}$	$-fv$	$= -\dfrac{1}{\rho}\dfrac{\partial p}{\partial r}$	(5.1)
$\dfrac{U}{T}$	$\dfrac{U^2}{R}$	$W\dfrac{U}{Z}$	$\dfrac{V^2}{R}$	fV	$\dfrac{\Delta p_r}{\rho R}$	(5.1)a
S_r^{-2}	S_r^{-2}	S_r^{-2}	1	$\dfrac{1}{Ro}$	$\dfrac{\Delta p_r}{\rho V^2}$	(5.1)b

Tangential momentum

$\dfrac{\partial v}{\partial t}$	$+u\dfrac{\partial v}{\partial r}$	$+w\dfrac{\partial v}{\partial z}$	$+\dfrac{uv}{r}$	$+fu$	$=0$	(5.2)
$\dfrac{V}{T}$	$U\dfrac{V}{R}$	$\dfrac{WV}{Z}$	$\dfrac{UV}{R}$	fU		(5.2)a
S_r^{-2}	S_r^{-2}	S_r^{-2}	S_r^{-2}	$S_r^{-1}Ro^{-1}$		(5.2)b

Vertical momentum

$\dfrac{\partial w}{\partial t}$	$+u\dfrac{\partial w}{\partial r}$	$+w\dfrac{\partial w}{\partial z}$	$= -\dfrac{1}{\rho}\dfrac{\partial p}{\partial z}$	$-g$	(5.3)
$\dfrac{W}{T}$	$U\dfrac{W}{R}$	$W\dfrac{W}{Z}$	$\dfrac{\Delta p_z}{\rho Z}$	g	(5.3)a
$A^2 Fr^2$	$A^2 Fr^2$	$A^2 Fr^2$	$\dfrac{\Delta p_z}{\rho g Z}$	1	(5.3)b

5.8.2 Momentum equations

Using the fact that $W \sim UZ/R$, we divide terms in lines (5.1)a and (5.2)a by V^2/R to obtain those in lines (5.1)b and (5.2)b, respectively. The terms in line (5.3)a are divided by g to obtain those in line (5.3)b. The terms in lines (b) of each equation are then nondimensional.

5.8.2.1 Gradient wind balance

It follows from line (5.1)b in Table 5.1 that, for high swirl ratio squared ($S_r^2 \gg 1$) and $Ro \sim 1$, $\Delta p_r \sim \rho V^2$ and that the radial momentum equation is closely approximated by the gradient wind equation. (Note that if the pressure term had a larger order of magnitude than the sum of the centrifugal and Coriolis terms, it would drive a radial acceleration so that the radial velocity scale would eventually violate the assumption that $S_r \gg 1$).

Line (5.2)b in the table shows that all the nonlinear terms in the azimuthal momentum equation are of the same order of magnitude. These are dominant if $Ro \gg 1$, but the Coriolis term is of equal importance[7] if $Ro \ll 1$.

5.8.2.2 Hydrostatic balance

Line (5.3)b in Table 5.1 shows that the ratio of the acceleration terms in the vertical momentum equation to the gravitational acceleration is $WU/Rg = A^2 Fr^2$, where $A = Z/R$ is the *aspect ratio* for the flow and $Fr = U/\sqrt{gZ}$ has the form of a *Froude number*.

In the free troposphere, typical scales for a tropical cyclone vortex are $U \sim 5$ m s^{-1}, $W \sim 0.1$ m s^{-1}, $R \sim 200$ km, $Z \sim 10$ km, and $g \sim 10$ m s^{-2}. Then $A \sim 0.05$, $Fr \sim 1.6 \times 10^{-2}$ and $WU/Rg \sim 2.5 \times 10^{-7}$. We conclude that $\Delta p_z \sim \rho g Z$ and that the motion on these scales is very close to hydrostatic balance with the vertical pressure gradient force term in the momentum equation balancing the gravitational acceleration, i.e., Eq. (5.3) is closely approximated by Eq. (5.12).

---

7. In this axisymmetric configuration, there is no azimuthal pressure gradient force, by definition. Thus, we cannot argue that the Coriolis term is dominant, since then the other terms would be negligible, but then we would conclude that $fu = 0$!

TABLE 5.2 Scaling of the terms in perturbation form of the hydrostatic equation. Here $A = Z/R$ is the aspect ratio and $Fr' = U/\sqrt{g_r Z}$ is the modified Froude number, based on the reduced gravity, g_r.

Vertical momentum: perturbation form					
$\dfrac{\partial w}{\partial t}$	$+u\dfrac{\partial w}{\partial r}$	$+w\dfrac{\partial w}{\partial z}$	$= \quad -\dfrac{1}{\rho}\dfrac{\partial p'}{\partial z}$	$+b$	(5.3′)
$\dfrac{W}{T}$	$U\dfrac{W}{R}$	$W\dfrac{W}{Z}$	$\dfrac{\Delta p'_z}{\rho Z^*}$	$\dfrac{g\Delta T}{T^*}$	(5.3′)a
$A^2 Fr'^2$	$A^2 Fr'^2$	$A^2 Fr'^2$	$\dfrac{\Delta p'_z}{\rho g_r Z^*}$	1	(5.3′)b

Even though the motions on the foregoing scales are hydrostatic, the question remains as to whether disturbances to the large scales are themselves hydrostatic. In other words, when we subtract the reference pressure p_{ref} from p, is it still legitimate to neglect Dw/Dt? To answer this question we must carry out a scale analysis of the vertical component of Eq. (5.3) with the reference pressure removed. In essence, this is just the vertical component of Eq. (2.10) with $\mathbf{F} = 0$. The scaling is as indicated in Table 5.2. In this table, $Fr' = U/\sqrt{g_r Z^*}$ is a *modified Froude number*, based on the reduced gravity, $g_r = g\Delta T/T_*$, T is the temperature, T_* is a reference temperature and Z^* is a height scale for the perturbation pressure difference in the vertical direction, $\Delta p'_z$. A typical scale for $\Delta p'_z$ might be 10 mb for a weak tropical low and perhaps 50 mb or more for an intense tropical cyclone. For a disturbance confined to the troposphere, it is reasonable to assume that $Z^* \leq Z$.

Typical temperature differences in a tropical cyclone are about 10 K whereas T_* is typically 300 K, giving typical values of about 0.3 m s^{-2} for g_r and 9×10^{-2} for Fr'. Again, it is clear that the terms on the right-hand side of Eq. (5.3) must balance and hence, in tropical cyclones, the perturbations, themselves, are in close hydrostatic balance on the system scale. We deduce that to a very good approximation,

$$0 = -\frac{1}{\rho}\frac{\partial p'}{\partial z} + b. \qquad (5.59)$$

The validity of the gradient wind balance and hydrostatic balance means that the theory for the primary circulation worked out in Section 5.2 should be a good approximation to reality, even in the presence of a nonzero, but relatively weak secondary circulation.

5.8.2.3 *Internal consistency of balance arguments*

Invoking the approximations of gradient wind balance and hydrostatic balance permits enormous simplifications in dynamical studies of large-scale vortices. However, it is as well to remember that, from Newton's second law, it is small

TABLE 5.3 Scaling of the terms in the thermodynamic equation. Here, Θ is a scale for θ, $\Delta\Theta_1$ and $\Delta\Theta_2$ are typical variations of Θ over the radial scale R and vertical scale Z, and $\dot{\Theta}^*$ is a scale for the diabatic heating rate. The quantities B_* and N_*^2 are typical scales for the baroclinicity, B, a measure of the radial potential temperature gradient (see Exercise 5.17), and the static stability, N^2, respectively.

Thermodynamic equation				
$\dfrac{\partial\theta}{\partial t}$	$+u\dfrac{\partial\theta}{\partial r}$	$+w\dfrac{\partial\theta}{\partial z}$	$= \quad \dot{\theta}$	(5.5)
$\dfrac{\Delta\Theta_1}{T}$	$U\dfrac{\Delta\Theta_1}{R}$	$W\dfrac{\Delta\Theta_2}{Z}$	$\dot{\Theta}^*$	(5.5)a
$\dfrac{\Delta\Theta_1}{T}$	$U\dfrac{B_*}{g}\Theta$	$W\dfrac{N_*^2}{g}\Theta$	$\dot{\Theta}^*$	(5.5)b

departures from balance that actually drive the weak radial and vertical accelerations in systems of this scale. The subtlety is that, despite this fact, if one constructs a balance theory based on the foregoing scaling approximations as in Section 5.9 below, one can no longer argue that *in this theory the flow is being driven by the force imbalances*. See Section 5.9 for more on this point.

5.8.3 Thermodynamic equation

The scaling of the thermodynamic equation (5.5) is shown in Table 5.3. Again, using the scaling $U/R = W/Z$ from the continuity equation, the ratio of the radial advection term to the vertical advection term is seen to be on order of $\Delta\Theta_1/\Delta\Theta_2$, where the scales $\Delta\Theta_1$ and $\Delta\Theta_2$ are defined in the table caption.[8] This ratio is typically of order unity (Exercise 5.17), so that, in general, no approximation can be made to the thermodynamic equation. Nevertheless, it is instructive to note some limiting cases:

---

8. Note that the scales R and Z are taken to be appropriate to variations in kinematic quantities. In general, they need not be appropriate also as scales for variations in potential temperature.

(1) For a steady vertical plume in a barotropic flow ($\partial\theta/\partial r = 0$), the thermodynamic equation may be written as

$$0 = \dot{\theta} - w\frac{\partial\theta}{\partial z}.$$

In this situation the rate of diabatic heating of an ascending air parcel is exactly balanced by the rate of adiabatic cooling as the parcel expands. The potential temperature of the parcel does not change.

(2) For an unsteady vertical plume in a weakly baroclinic flow ($B_* \ll AN_*^2$), the thermodynamic equation may be written approximately as

$$\frac{\partial\theta}{\partial t} = \dot{\theta} - w\frac{\partial\theta}{\partial z},$$

whereupon the local rate of increase in θ is the difference between the rate of diabatic heating and the rate of adiabatic cooling. In atmospheric situations, for example in a cumulonimbus updraft, this difference is typically small so that relatively large heating rates lead only to small local increases in temperature. In the tropical cyclone eyewall, temperature perturbations are only a degree or two in excess of surrounding values.

(3) When the baroclinicity is important, i.e., $B_* \approx N_*^2 Z/R$, the rate of heating or cooling by radial advection makes an important contribution to the local rate of change of θ.

5.8.4 Exercise

Exercise 5.17. Using Eq. (5.19), show that the baroclinicity B, defined by Eq. (5.44), can be written as

$$B = g\frac{\partial}{\partial r}\ln\theta. \tag{5.60}$$

Hence show that with the scales defined in this section ($\Delta\Theta_1 = 10$ K, $R = 200$ km, $Z = 10$ km) and assuming typical tropospheric scale of 10^{-4} s^{-2} for N_*^2, the ratio of the radial and vertical advection terms in the thermodynamic equation (5.5) is proportional to $\Delta\Theta_1/\Delta\Theta_2$ and show that this ratio is ~ 1.

5.9 The secondary circulation

If the vortex is axisymmetric and in approximate gradient wind and hydrostatic balance, we can derive an equation for the streamfunction, ψ, of the secondary circulation, i.e., the circulation in a vertical plane. This streamfunction is such that

$$u = -\frac{1}{r\rho}\frac{\partial\psi}{\partial z}, \qquad w = \frac{1}{r\rho}\frac{\partial\psi}{\partial r}, \tag{5.61}$$

which ensures that the quasi-steady form of the continuity equation (5.4), is satisfied. The equation for ψ follows

by differentiating the thermal wind equation in the form of Eq. (5.19) with respect to time t and using the azimuthal momentum and thermodynamic equations to eliminate the local time derivatives. It is convenient to write the last two equations in the following form

$$\frac{\partial v}{\partial t} + u(\zeta + f) + wS = -\dot{V} \tag{5.62}$$

and

$$\frac{\partial\chi}{\partial t} + u\frac{\partial\chi}{\partial r} + w\frac{\partial\chi}{\partial z} = -\chi^2\dot{\theta} \tag{5.63}$$

where $\zeta = (1/r)\partial(rv)/\partial r$ is the relative vertical vorticity and $S = \partial v/\partial z$ is the vertical shear of the tangential wind. Note that we have added a momentum sink term $F_\lambda = -\dot{V}$ in the former equation to represent the effect of surface friction on the tangential wind component of the vortex in the balance formulation. Here the term $\dot{V}$ represents a distributed body force and is confined to a thin layer adjacent to the lower boundary. Further consideration of this thin friction layer will be developed in Chapter 6 and its consequences on the slow evolution of the vortex will be examined in Chapter 8.

In the absence of a secondary circulation, for general forcing terms $\dot{V}$ and $\dot{\theta}$, v and χ given by Eqs. (5.62) and (5.63) will change at rates that will destroy thermal wind balance. To ensure that the vortex remains in both gradient wind and hydrostatic balance as time proceeds, the time derivative of the thermal wind equation, Eq. (5.19), must be satisfied at all times, i.e., $\partial v/\partial t$ and $\partial\chi/\partial t$ must satisfy the equation:

$$g\frac{\partial}{\partial r}\left(\frac{\partial\chi}{\partial t}\right) + \frac{\partial}{\partial z}\left(C\frac{\partial\chi}{\partial t} + \chi\frac{\partial C}{\partial t}\right) = 0. \tag{5.64}$$

Substituting the time derivatives from Eqs. (5.62) and (5.63) in Eq. (5.64) gives

$$-g\frac{\partial}{\partial r}\left(u\frac{\partial\chi}{\partial r} + w\frac{\partial\chi}{\partial z} + \dot{\Theta}\right)$$
$$-\frac{\partial}{\partial z}\left[C\left(u\frac{\partial\chi}{\partial r} + w\frac{\partial\chi}{\partial z} + \dot{\Theta}\right)\right.$$
$$\left. + \chi\xi\left(u(\zeta + f) + wS + \dot{V}\right)\right] = 0$$

where $\chi = 1/\theta$ and $\dot{\Theta} = \chi^2\dot{\theta}$. We will refer to $\dot{\Theta}$ as the heating function. Then

$$\frac{\partial}{\partial r}\left[-g\frac{\partial\chi}{\partial z}w - g\frac{\partial\chi}{\partial r}u\right] -$$
$$\frac{\partial}{\partial z}\left[\left(\chi\xi(\zeta + f) + C\frac{\partial\chi}{\partial r}\right)u + \frac{\partial}{\partial z}(\chi C)w\right] =$$
$$g\frac{\partial\dot{\Theta}}{\partial r} + \frac{\partial}{\partial z}(C\dot{\Theta}) + \frac{\partial}{\partial z}(\chi\xi\dot{V})$$

or, using Eq. (5.19),

$$\frac{\partial}{\partial r}\left[-g\frac{\partial \chi}{\partial z}w + \frac{\partial}{\partial z}(\chi C)u\right] -$$
$$\frac{\partial}{\partial z}\left[(\chi\xi(\zeta + f) + C\frac{\partial \chi}{\partial r})u + \frac{\partial}{\partial z}(\chi C)w\right]$$
$$= g\frac{\partial \dot{\Theta}}{\partial r} + \frac{\partial}{\partial z}(C\dot{\Theta}) + \frac{\partial}{\partial z}(\chi\xi\dot{V}).$$

(5.65)

Then substitution for u and w from Eqs. (5.61) into Eq. (5.65) gives

$$\frac{\partial}{\partial r}\left[-g\frac{\partial \chi}{\partial z}\frac{1}{\rho r}\frac{\partial \psi}{\partial r} - \frac{\partial}{\partial z}(\chi C)\frac{1}{\rho r}\frac{\partial \psi}{\partial z}\right] +$$
$$\frac{\partial}{\partial z}\left[\left(\chi\xi(\zeta + f) + C\frac{\partial \chi}{\partial r}\right)\frac{1}{\rho r}\frac{\partial \psi}{\partial z} - \frac{\partial}{\partial z}(\chi C)\frac{1}{\rho r}\frac{\partial \psi}{\partial r}\right]$$
$$= g\frac{\partial \dot{\Theta}}{\partial r} + \frac{\partial}{\partial z}(C\dot{\Theta}) + \frac{\partial}{\partial z}(\chi\xi\dot{V}).$$

(5.66)

This linear second-order partial differential equation is called the *Eliassen equation*[9]. In essence, *the equation determines the streamfunction of the secondary circulation that is required to maintain the vortex in balance in the presence of processes that would otherwise force it out of balance*. As foreshadowed in Subsection 5.8.2.3, in a strict balance theory, this interpretation replaces that in which the balanced secondary circulation is attributed to unbalanced forces.

In terms of the parameters introduced in Section 5.7.3, the equation may be written finally as

the Eliassen equation

$$\frac{\partial}{\partial r}\left[\gamma N^2 \frac{\partial \psi}{\partial r} - \gamma B\frac{\partial \psi}{\partial z}\right] +$$
$$\frac{\partial}{\partial z}\left[\gamma I_g^2 \frac{\partial \psi}{\partial z} - \gamma B\frac{\partial \psi}{\partial r}\right] =$$
$$g\frac{\partial \dot{\Theta}}{\partial r} + \frac{\partial}{\partial z}(C\dot{\Theta}) + \frac{\partial}{\partial z}(\chi\xi\dot{V})$$

(5.67)

where $\gamma = \chi/(\rho r)$. This equation contains the same three spatially-varying parameters N^2, I_g^2, and B that arise in the linear stability analysis of a baroclinic vortex (Section 5.7.3).

The discriminant of the Eliassen equation is

$$D = 4\gamma^2(I_g^2 N^2 - B^2),$$

(5.68)

which is proportional to Δ, where Δ is defined by Eq. (5.48) (Exercise 5.18). Recall that a baroclinic vortex is symmetrically stable if $\Delta > 0$. It can be shown that Eq. (5.67) is

elliptic if $D > 0$ and *hyperbolic* if $D < 0$ (see e.g., Exercise 5.18). Thus the symmetric stability of the vortex globally is a requirement that the Eliassen equation is elliptic globally, which, in turn, is a requirement that the equation may be solved as a diagnostic equation for the streamfunction ψ.

5.9.1 Exercises

Exercise 5.18. Show that the Eliassen equation can be written in the form

$$\bar{A}\frac{\partial^2 \psi}{\partial r^2} + 2\bar{B}\frac{\partial^2 \psi}{\partial r \partial z} + \bar{C}\frac{\partial^2 \psi}{\partial z^2} + \bar{D}\frac{\partial \psi}{\partial r} + \bar{E}\frac{\partial \psi}{\partial z}$$
$$= g\frac{\partial \dot{\Theta}}{\partial r} + \frac{\partial}{\partial z}(C\dot{\Theta}) + \frac{\partial}{\partial z}(\chi\xi\dot{V}),$$

(5.69)

where

$$\bar{A} = -g\frac{\partial \chi}{\partial z}\frac{1}{\rho r} = \gamma N^2,$$

$$\bar{B} = -\frac{\partial}{\partial z}(\chi C)\frac{1}{\rho r} = -\gamma B,$$

$$\bar{C} = \left(\chi\xi(\zeta + f) + C\frac{\partial \chi}{\partial r}\right)\frac{1}{\rho r} = \gamma I_g^2,$$

$$\bar{D} = \frac{\partial \bar{A}}{\partial r} + \frac{\partial \bar{B}}{\partial z},$$

and

$$\bar{E} = \frac{\partial \bar{B}}{\partial r} + \frac{\partial \bar{C}}{\partial z}.$$

Given that the discriminant of this equation is $4(\bar{A}\bar{C} - \bar{B}^2)$ (see e.g., Sneddon, 1957, p108[10]), verify Eq. (5.68).

Exercise 5.19. Show now that the quantity D in Eq. (5.68) equals $\rho g\xi\chi\gamma^2 P$, where P is the Ertel potential vorticity, P, in the form (5.57) given above. Hint: Show first that for a symmetric vortex with tangential wind speed distribution $v(r, z)$, $\boldsymbol{\omega} + \mathbf{f} = -(\partial v/\partial z)\hat{\mathbf{r}} + (\zeta + f)\hat{\mathbf{z}}$ and $\nabla\theta = -(1/\chi^2)\nabla\chi = -(1/\chi^2)[(\partial \chi/\partial r)\hat{\mathbf{r}} + (\partial \chi/\partial z)\hat{\mathbf{z}}]$ so that

$$P = \frac{1}{\rho\chi^2}\left[\frac{\partial v}{\partial z}\frac{\partial \chi}{\partial r} - (\zeta + f)\frac{\partial \chi}{\partial z}\right].$$

(5.70)

Exercise 5.20. Show that the forcing term for ψ involving diabatic heating in Eq. (5.67) can be expressed in terms of the material generation rate of generalized buoyancy, where the generalized buoyancy force **b** per unit mass is defined by Eq. (5.17). Hint: Show first that the diabatic heating rate

---

9. Or Sawyer-Eliassen equation following the pioneering work of these two authors in the theory of fronts and large-scale circulations.

10. Sneddon uses $\bar{B}$ as the coefficient of the mixed derivative term instead of $2\bar{B}$ used here and defines the discriminant as $\bar{B}^2 - 4\bar{A}\bar{C}$. We have chosen here to reverse the sign so that positive P corresponds with positive discriminant.

term in Eq. (5.67), comprising the first two terms on the right-hand side, can be written:

$$F = \frac{\partial}{\partial r}\left(\frac{g}{\theta^2}\frac{D\theta}{Dt}\right) + \frac{\partial}{\partial z}\left(\frac{C}{\theta^2}\frac{D\theta}{Dt}\right)$$

and deduce that this term is approximately equal to

$$F \approx \frac{1}{\theta_*}\hat{\mathbf{j}}\cdot\nabla\wedge\frac{D\mathbf{b}}{Dt}, \qquad (5.71)$$

where θ_* is an average potential temperature over the vortex and $\hat{\mathbf{j}}$ is a unit vector in the tangential direction.

Exercise 5.21. Starting from the Boussinesq system of equations, show that the Eliassen equation takes the simplified form

$$\frac{\partial}{\partial r}\left[\left(N^2 + \frac{\partial b}{\partial z}\right)\frac{1}{r}\frac{\partial\psi}{\partial r} - \frac{S\xi}{r}\frac{\partial\psi}{\partial z}\right] +$$
$$\frac{\partial}{\partial r}\left[\frac{I^2}{r}\frac{\partial\psi}{\partial z} - \frac{S\xi}{r}\frac{\partial\psi}{\partial r}\right] = +\frac{\partial\dot{b}}{\partial r} + \frac{\partial}{\partial z}(\xi\dot{V}),$$

$$(5.72)$$

where N^2 is here the ambient static stability, $b = -g\rho'/\rho_0 \approx +g\theta'/\theta_0$ is the buoyancy force per unit mass in the Boussinesq approximation, ρ' is the density anomaly, ρ_0 is the average density over the domain, θ' is the potential temperature anomaly, θ_0 is the ambient surface potential temperature, $\dot{b}$ is the buoyancy generation rate in the Boussinesq form of the thermodynamic equation ($= g\dot{\theta}/\theta_0$), $S = \partial v/\partial z$ is the vertical shear of the gradient wind, I^2 is defined in Eq. (5.41) and the remaining notation is as defined in the text.

Exercise 5.22. Show that for an axisymmetric flow, the agradient wind $v' = v - v_g$, where v is the tangential wind and v_g is the gradient wind, is given by the quadratic equation

$$v'^2 + r\xi_g v' - r\left(\frac{Du}{Dt} - F_r\right) = 0.$$

Hence show that

$$v' = -\frac{1}{2}r\xi_g + \sqrt{\frac{1}{4}r^2\xi_g^2 + r\left(\frac{Du}{Dt} - F_r\right)}. \qquad (5.73)$$

Explain why the positive sign needs to be chosen before the square root sign. Derive an approximate expression for v' when $Du/Dt - F_r$ is sufficiently small and derive a criterion for "sufficiently small".

Note the implications of this exercise, which shows that the agradient tangential wind can be estimated from a full prognostic balance solution (see e.g., Chapter 8) including the secondary circulation. The prognostic solution would be required to estimate Du/Dt.

5.10 Solutions of the Eliassen equation

The Eliassen equation is a seemingly complex partial differential equation and it is helpful to develop an intuitive way of understanding what the solutions look like. The simplest elliptic partial differential equation similar in structure to the Eliassen equation is Poisson's equation, which, as we saw in Appendix 2.19, arises in the equilibrium displacement of a stretched membrane subject to a normal force distribution acting on it.

Two particular examples shown in Fig. 2.5 are the membrane displacement induced by a point force directed upwards when the displacement is constrained to be zero along the boundary and the displacement brought about by a point dipole of force, i.e., two point forces located infinitesimally close together, but acting in opposite directions. In the case of the point force, the displacement in the interior of the domain is everywhere upward and is a maximum at the location of forcing. In the case of a dipole, the displacement is positive on the side of the dipole where the forcing is positive and negative on the side where the forcing is negative, but it is zero at the dipole itself.

5.10.1 Boundary effects in the membrane analogy

Let us compare two specific examples of a membrane, the first in which a point force, $\delta(x - \frac{1}{2})\delta(y - \frac{1}{2})$ is applied at the center $(\frac{1}{2}, \frac{1}{2})$ of a square membrane $0 \leq x \leq 1, 0 \leq y \leq 1$, and the second in which the point force is at the center of a rectangular region $0 \leq x \leq 3, 0 \leq y \leq 1$. Here the function $\delta(x)$ denotes the one-dimensional Dirac delta function as defined in Section 4.4.8. The isopleths of membrane displacement in these cases, scaled so that the maximum membrane displacement is unity in each case, are illustrated in panels (a) and (b) of Fig. 5.13, respectively. In the first case, the displacement isopleths are symmetrical, being close to circular near the center of the domain and becoming more square-like as the membrane is progressively influenced by the zero displacement condition at the domain boundaries.

In the case of the rectangular domain, the scale of the membrane displacement is set by the nearest boundaries, i.e., those at $y = 0$ and $y = 1$ and some of the symmetry is lost, the displacement being somewhat elongated in the x-direction compared with the y-direction. This elongation is an indication that the membrane is feeling the lateral boundaries more weakly than those at $y = 0$ and $y = 1$. As the rectangular domain becomes further stretched in the x-direction, one can guess that the boundaries in this direction will have progressively less effect on the pattern of displacement.

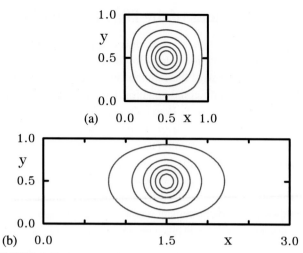

FIGURE 5.13   Isopleths of membrane displacement subject to a point force (approximated as a sharp Gaussian profile) at the center point of (a) a square ($\frac{1}{2}, \frac{1}{2}$), and (b) a rectangle ($\frac{3}{2}, \frac{3}{2}$). In each case, the displacement isopleths are scaled to give unit displacement at that point.

5.10.2   Scale effects in the membrane analogy

In the derivation of the Poisson equation (2.87) that relates to the simple membrane, it is assumed that the membrane properties are isotropic (i.e., that the local response to forcing is independent of direction), but one could envisage a situation where, for example, the local response is larger in one direction than another. In this case the coefficients of the second derivatives in the membrane equation would be different constants. An example of relevance in the meteorological context is the equation

$$N^2 \frac{\partial^2 \psi}{\partial x^2} + f^2 \frac{\partial^2 \psi}{\partial z^2} = -F(x, z), \qquad (5.74)$$

where x and z are coordinates in the horizontal and vertical, respectively, f and N are constants and the function $F(x, z)$ is prescribed. The unknown ψ might refer to a streamfunction of some overturning circulation in a vertical plane. Suppose the equation applies to the rectangular domain $0 \leq x \leq L$ and $0 \leq z \leq H$ and for simplicity that $F(x, z)$ represents a point forcing proportional to $\delta(x - \frac{1}{2}L)\delta(z - \frac{1}{2}H)$. Again, $\delta(x)$ is the one-dimensional Dirac delta function.

Eq. (5.74) may be transformed to one with unit coefficients by defining new independent coordinates $X = x/N$ and $Z = z/f$, whereupon, using the property that $\delta(ax) = (1/a)\delta(x)$,

$$\frac{\partial^2 \psi}{\partial X^2} + \frac{\partial^2 \psi}{\partial Z^2} = -\frac{1}{fN} \delta\left(X - \frac{L}{2N}\right) \delta\left(Z - \frac{H}{2f}\right). \quad (5.75)$$

In this form, in the absence of boundaries, the membrane response, in this case ψ, would be isotropic. Invoking the results of Section 5.10.1, in the transformed rectangular domain ($0 \leq L/N, 0 \leq H/f$), the pattern of membrane displacement will be influenced by the minimum distance,

$\min(L/N, H/f)$. These two distances are the same when $L = L_R$, where $L_R = NH/f$, a quantity called the *Rossby length*.

In some meteorological applications, H may be taken as the depth of the troposphere, N is typically the Brunt-Väisälä frequency on the order of 0.01 s^{-1} and f is typically the Coriolis parameter, typically 10^{-4} s^{-1} at middle latitudes and even smaller at tropical latitudes. At middle latitudes, L_R is on the order 10^3 km and it is even larger in the Tropics. Typically, in either case, $H \ll L$.

The point of making the transformation of vertical coordinate is that *the response to the forcing in* Eq. (5.75) *is isotropic* as in Eq. (2.87) with the length scale L_R in both coordinate directions. Thus, *in physical coordinates* (x, z), a response with vertical scale H will have a horizontal scale L_R. Then the question is, how does the response vary with the ratio of L to L_R in typical geophysical flows?

The situation where $H/f \leq L/N$, i.e., $L_R \leq L$, which is usually the case for large-scale geophysical flows, is illustrated in Fig. 5.14 by numerical solutions of Eq. (5.74) in the region where $L = 2000$ km and $H = 10$ km for four different values of L_R. Again, the same sharply-pointed Gaussian forcing function F is used to approximate the Delta function numerically at the point ($\frac{1}{2}L, \frac{1}{2}H$) and the isopleths of "membrane displacement" are normalized so that the maximum "displacement", here ψ, is unity. Note how the horizontal scale of response decreases as L_R decreases, i.e., when the rotational stiffness, characterized here by the Coriolis frequency, f increases relative to the buoyancy frequency, characterized by N.

In less common cases where $H/f \geq L/N$, the horizontal scale of response will be set by L and the vertical scale of response is then Lf/N, a quantity called *the Rossby depth*.

If other boundary conditions are imposed along all or part of the domain boundary, the foregoing ideas may have to be modified.

5.10.3   Other anisotropic effects in the membrane analogy

One could envisage a situation where the local response to a point force depends on the spatial location. In that case, the coefficients of the derivatives would vary spatially and a cross-derivative term $\partial^2 \eta / \partial x \partial y$ might appear also. The Eliassen equation is an example of such an equation. Despite the additional complexity, the membrane analogy is still useful in understanding the broad structure of solutions in this more general case.

5.10.4   Exercise

Exercise 5.23.   The quantity R defined by $\frac{1}{2} f R^2 = rv + \frac{1}{2} f r^2$ is called the *potential radius*. It is the radius at which the tangential wind of an air parcel that is displaced radially

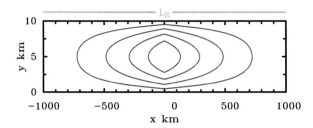

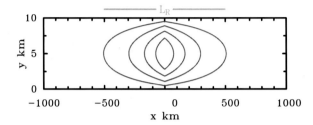

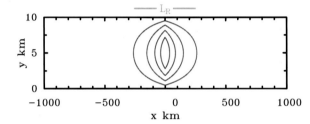

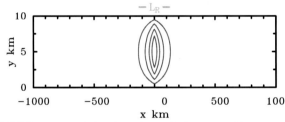

FIGURE 5.14 Isopleths of membrane displacement subject to a point force (approximated as a sharp Gaussian profile at the center point $(\frac{1}{2}L, \frac{1}{2}H)$ of a rectangular domain (L, H) with aspect ratio $H \ll L$ for various Rossby lengths $L_R \leq L$ as indicated. In each case, the displacement isopleths are scaled to give unit displacement at that point.

while conserving its absolute angular momentum is zero. Show that

$$R^2 = 2M/f. \qquad (5.76)$$

Show next that in a transformation from (r, z) coordinates to (R, χ) coordinates,

$$\frac{\partial}{\partial r} = \frac{\partial R}{\partial r}\frac{\partial}{\partial R} + \frac{\partial \chi}{\partial r}\frac{\partial}{\partial \chi}, \quad \frac{\partial}{\partial z} = \frac{\partial R}{\partial z}\frac{\partial}{\partial R} + \frac{\partial \chi}{\partial z}\frac{\partial}{\partial \chi},$$

where

$$\frac{\partial R}{\partial r} = \frac{r\zeta_a}{Rf}, \quad \frac{\partial R}{\partial z} = \frac{r}{fR\xi}\left(B - \frac{C}{\chi}\frac{\partial \chi}{\partial z}\right),$$

$$\frac{\partial \chi}{\partial r} = -\frac{\chi B}{g}, \quad \frac{\partial \chi}{\partial z} = -\frac{\chi N^2}{g}.$$

Here, $\zeta_a = \zeta + f$, is the absolute vorticity and $\xi = 2v/r + f$ is twice the absolute angular velocity. Hint: the expression for $\partial R/\partial z$ requires use of the thermal wind equation.

5.10.5 Point source solutions in an unbounded domain

Before presenting solutions of the Eliassen equation in a bounded vortex, it is instructive to examine solutions for point (ring) sources of heat and azimuthal momentum in an unbounded domain. Because the forcing term on the right-hand-side of the Eliassen equation representing a heat source appears as radial and vertical derivatives of the diabatic heating rate $\dot{\theta}$, the relevant membrane analogy is the situation involving a force dipole. In the case of angular momentum forcing, the forcing involves a vertical derivative of $\dot{V}$.

Solutions corresponding to point sources of heat and azimuthal momentum are shown schematically in Fig. 5.15. The vortex axis lies to the left of the figure. For reasons discussed below, the flow through the heat source is upwards approximately along a constant absolute angular momentum (M- or R-) surface,[11] while the flow through an absolute angular momentum source is outwards approximately along an isentropic (θ- or χ-) surface. For heat and angular momentum sinks, the flow direction is reversed. In barotropic vortices (panels (a), (b), (d), and (e)), the M-surfaces are nearly vertical and the θ-surfaces are nearly horizontal. In warm-core vortices (panels (c) and (f)), the M- and θ-surfaces tilt outwards with height as shown in Fig. 5.6.

As shown in Sections 5.7.1 and 5.7.3, the generalized inertial stability I_g^2 acts to resists radial motion in the same way that the static stability N^2 acts to resist vertical motion. As can be inferred from the discussion in Section 5.10.2, the ratio of vertical scale to radial scale of the point-source-induced gyres in Fig. 5.15 is proportional to the ratio I_g/N. If $I_g \ll N$, the radial scale will be large compared with the vertical scale as indicated in panels (a) and (d) and the opposite is true when $N \ll I_g$ as indicated in panels (b) and (e). Vertical gradients of M associated with the vertical shear of the primary circulation imply a slope of the M surface and therefore a slope of the updraft through the point sources of heat (panel (c)). Likewise, horizontal gradients of θ associated with the vertical shear imply a slope of the θ surfaces and thereby a tilt of the flow through an angular momentum source (panel (f)).

The reason why the flow through a heat source is approximately along an M-surface and that through an angular momentum source is approximately along a θ-surface is as follows. The Eliassen equation determines the secondary circulation consistent with the tangential momentum equation (5.62) and potential temperature equation (equivalent to

---

11. Here R refers to the potential radius defined in Exercise 5.23.

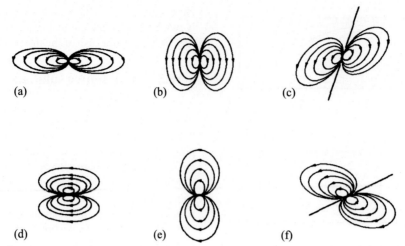

FIGURE 5.15 Streamfunction responses to point sources of: (a) heat in a barotropic vortex with weak inertial stability, (b) heat in a barotropic vortex with strong inertial stability, (c) heat in a warm-core baroclinic vortex, (d) momentum in a barotropic vortex with weak inertial stability, (e) momentum in a barotropic vortex with strong inertial stability, and (f) momentum in a warm-core baroclinic vortex. (Based on Figs. 8, 9, 11, and 12 of Eliassen, 1951.) From Shapiro and Willoughby (1982). Republished with permission of the American Meteorological Society.

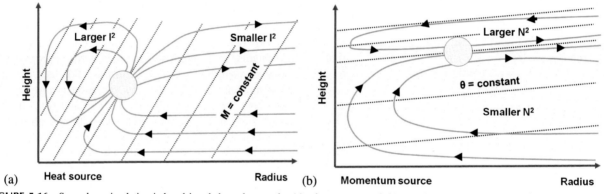

FIGURE 5.16 Secondary circulation induced in a balanced vortex by (a) a heat source and (b) a cyclonic angular momentum source showing the distortion induced by variation in inertial stability, I_g^2, static stability, N^2, and baroclinicity B^2. The strong flow through the heat source is approximately along constant M surfaces and that through the angular momentum source is approximately along constant θ surfaces. (Adapted from Willoughby, 1995.) Republished with permission of WMO.

Eq. (5.63)). If there is only a heat source, i.e., $\dot{V} = 0$, then Eq. (5.62) implies that M (or equivalently R) is materially conserved. Likewise, if there is only a momentum source, i.e., $\dot{\theta} = 0$, then Eq. (5.62) implies that θ (or equivalently χ) is materially conserved. Although the flow associated with a heat source lies generally along the M-surface and that associated with a momentum source lies generally along a θ-surface, in each case the flow does have a small component across the respective surface. *It is the advection by this flow component that causes evolution of the primary circulation* (see Exercise 5.6).

5.10.6 Point source solutions in a partially bounded domain

Fig. 5.16 shows the secondary circulation induced by point sources of heat and absolute angular momentum in a balanced, symmetrically stable, tropical-cyclone-like vortex in a partially bounded domain. These solutions can be obtained, for example, using the so-called method of images. Again, the streamline patterns show two counter-rotating cells of circulation (or gyres) that extend beyond the source. There is a relatively strong flow between these gyres and a weaker return flow outside them. The secondary circulation locally through a heat source is approximately along an M-surface, and that through an angular momentum source is approximately along a θ surface. The flow emerging from the heat source spreads outwards above the source and subsequently subsides, flowing back towards the source from below. In essence, a heat-induced and typically outward-sloping updraft is surrounded by compensating subsidence. Similarly, compensating inflow is found above and below momentum-induced outflow. Recalling that the radial scale of the gyres is given by the local Rossby length NH/I_g, the local ratio of radial to vertical scale is then N/I_g. As shown in Fig. 5.11b, there is typically a large radial variation of I_g^2 across the vortex, with the largest values in the inner core region.

It is found that the effect of the induced secondary circulation in a balanced flow tends to cancel the direct effect of forcing. For example, the work done by expansion in the updraft induced by a positive heat source nearly balances the actual heating so that the increase in temperature is comparatively small (cf. Subsection 5.8.3, item (2)). Similarly, a positive momentum source produces outflow that advects compensating low values of absolute angular momentum from the central region of the vortex.

5.11   Representation of the diabatic heating rate, $\dot{\theta}$

In the Eliassen balance framework, the effects of inner-core deep convection in forcing the secondary circulation are represented by spatial gradients of the diabatic heating rate $\dot{\theta}$, which is proportional to the material heating rate $\dot{Q}$ (see Section 2.5.3). In a simple explicit formulation of diabatic heating in which condensation occurs when the saturation mixing ratio of water vapor, r_v^*, decreases as an air parcel expands and cools, the diabatic heating rate is just the latent heat of condensation, L_v, times the material rate of change of condensation, $-Dr_v^*/Dt$, i.e., $\dot{Q} = -L_v Dr_v^*/Dt$ (see Section 2.5.3).

When air parcels are ascending more or less vertically and steadily in a convective cloud, the material heating rate may be approximated by the vertical advection so that, to a first approximation,

$$\dot{Q} \approx -L_v w \frac{\partial r_v^*}{\partial z} \qquad (5.77)$$

where, as usual, w is the vertical velocity and z is the height. The quantity r_v^* is a monotonically increasing function of saturation pseudo-equivalent potential temperature, θ_e^* (see Exercise 9.4). In other words, to a first approximation, $\dot{Q}$ depends on the vertical velocity of the ascending air and on its thermodynamic properties embodied in r_v^*. However, in the Eliassen balance framework, w is determined by knowing $\dot{Q}$. Therefore, if the diabatic heating rate were to be related to the explicit condensation using Eq. (5.77), the determination of both w and $\dot{Q}$ would have to be part of the solution. In particular, $\dot{Q}$ could not be prescribed independently. While it proves instructive to examine solutions of the Eliassen equation for prescribed heating distributions, the limitations of doing this, as for example in Chapter 8, need to be borne in mind. We return to examine the use of Eq. (5.77) further in Section 16.1.4.

5.12   Buoyancy relative to a balanced vortex

Tropical cyclones are rapidly-rotating warm-cored vortices and the warm core is therefore positively buoyant relative to the vertical temperature distribution in the environment. However, we have seen that, on the cyclone scale, hydrostatic and gradient wind balances exist to a good approximation and the radial density (or buoyancy) gradient is related by the thermal wind equation to the decline in the mean tangential velocity and density with height. If there were strict balance, the generalized buoyancy force relative to an ambient sounding would be exactly balanced by the perturbation pressure gradient force.

In general, hydrostatic balance is a poor approximation in individual convective clouds, in which the updraft is attributed to a positive buoyancy force within it (Chapter 9). A pertinent question is then, how does one characterize the local buoyancy of clouds when they are embedded in a warm-cored vortex such as a tropical cyclone? For example, the warmest temperatures in a tropical cyclone are found in the eye, in which case the eyewall clouds would have negative buoyancy relative to the eye! The question is pertinent because, as shown in later chapters, local buoyancy in deep convection plays an important role in the dynamics of storm intensification.

Even if there are departures from hydrostatic and gradient wind balance in a tropical cyclone, it is clear that not all of the buoyancy force in clouds, based on reference profiles of density and pressure in the vortex environment, will be "available" for "driving" a secondary (or toroidal) circulation of the vortex that is necessary for vortex amplification. A part of the buoyancy will be associated with that in balance with the primary circulation of the vortex, itself. One way to estimate the local buoyancy of clouds is to define the perturbation pressure and perturbation density relative to some balanced vortex-scale pressure and density distributions. The simplest case is when the primary vortex is steady and axisymmetric. Then we may take reference distributions $p_{ref}(r, z)$ and $\rho_{ref}(r, z)$, respectively, that are in thermal wind balance with the tangential flow field $v(r, z)$. We saw how to calculate such reference distributions using the method of characteristics in Section 5.2. The idea is to use the local reference distributions $\rho_{ref}(r, z)$ and $p_{ref}(r, z)$ instead of the environmental soundings $\rho_{ref}(R, z)$ and $p_{ref}(R, z)$ to isolate the generalized buoyancy force of clouds, $\mathbf{b}_{loc}$, using Eq. (5.17). Here, we denote the buoyancy force relative to the environmental soundings as $\mathbf{b}$.

Clearly, the partition of force between perturbation pressure gradient force and buoyancy force using $\rho_{ref}(r, z)$ and $p_{ref}(r, z)$ will be different from that using the reference state characterized by $\rho_{ref}(R, z)$ and $p_{ref}(R, z)$ and interpretations of the dynamics involving buoyancy will be different also, albeit equivalent to those using the more conventional reference quantities that depend on height only. Even so, the net force given by the generalized buoyancy force plus the perturbation pressure gradient force will be independent of the choice of reference quantities, as in the simpler case discussed in Section 2.2. In Section 9.5, we examine the concept of "effective buoyancy", taking into account the static part of the perturbation pressure gradient force when a buoyancy force is distributed over a finite region of space.

5.12.1 Local buoyancy and system buoyancy

Axisymmetric balance models of tropical cyclone intensification appear to capture several important observed features of tropical cyclone behavior (see Chapter 8). However, for an axisymmetric balance vortex that evolves in strict thermal wind balance, $\mathbf{b}_{loc}(r, z, t) \equiv \mathbf{0}$ and $\nabla p' \equiv \mathbf{0}$, even though there may be heat sources or sinks present that create generalized buoyancy $\mathbf{b}$ as defined in this section. It is clear from the foregoing discussion that any diabatic heating or cooling rate in such models is incorporated directly into the balanced state, changing $\mathbf{b}(r, z, t)$, while $\mathbf{b}_{loc}(r, z, t)$ remains identically zero by definition.

If the vortex evolution is not strictly balanced, nonzero values of $\mathbf{b}_{loc}$ serve to characterize the *unbalanced component of buoyancy* provided that an appropriate reference state such as that defined above has been selected for the definition of buoyancy at any given time. It may be helpful in this case to think of $\mathbf{b} - \mathbf{b}_{loc}$ as characterizing the *system buoyancy* of the vortex and $\mathbf{b}_{loc}$ as characterizing the *local buoyancy* that typically would reside in convective clouds.

In the more general case, when the vortex-scale flow has marked asymmetries, it may be appropriate to base the reference state on the thermal field in balance with the azimuthally-averaged tangential wind field. Other alternatives are discussed in Smith et al. (2005), Section 5.

5.13 Enrichment topics

5.13.1 Toroidal vorticity equation

The tangential component of vorticity, or

toroidal vorticity is

$$\eta = \frac{\partial u}{\partial z} - \frac{\partial w}{\partial r} \qquad (5.78)$$

The tendency equation for η is derived as follows. Consider

$$\frac{\partial \eta}{\partial t} = \frac{\partial}{\partial t}\left(\frac{\partial u}{\partial z} - \frac{\partial w}{\partial r}\right) = \frac{\partial}{\partial z}\left(\frac{\partial u}{\partial t}\right) - \frac{\partial}{\partial r}\left(\frac{\partial w}{\partial t}\right).$$

Using the axisymmetric versions of the radial and vertical momentum equations (Eqs. (5.1), (5.3)), this expression

may be written

$$\frac{\partial \eta}{\partial t} = \frac{\partial}{\partial z}\left(-\mathbf{u} \cdot \nabla u + C - \frac{1}{\rho}\frac{\partial p}{\partial r} + F_u\right) -$$
$$\frac{\partial}{\partial r}\left(-\mathbf{u} \cdot \nabla w - \frac{1}{\rho}\frac{\partial p}{\partial z} + F_w\right), \qquad (5.79)$$

where here, $\mathbf{u}$ denotes the transverse velocity field (u, w), ∇ denotes the gradient operator in the transverse plane and the frictional stress divergences in the radial and vertical directions, respectively, are F_u and F_w. This equation reduces to

$$\frac{\partial \eta}{\partial t} = \frac{\partial C}{\partial z} + \frac{\partial}{\partial r}\left(\frac{1}{\rho}\right)\frac{\partial p}{\partial z} - \frac{\partial}{\partial z}\left(\frac{1}{\rho}\right)\frac{\partial p}{\partial r}$$
$$+ \frac{\partial}{\partial r}(\mathbf{u} \cdot \nabla w) - \frac{\partial}{\partial z}(\mathbf{u} \cdot \nabla u) + \frac{\partial F_u}{\partial z} - \frac{\partial F_w}{\partial r},$$

or

$$\frac{\partial \eta}{\partial t} + \mathbf{u} \cdot \nabla \eta = \frac{\partial C}{\partial z} + \frac{1}{\rho^2}\left(\frac{\partial \rho}{\partial z}\frac{\partial p}{\partial r} - \frac{\partial \rho}{\partial r}\frac{\partial p}{\partial z}\right) +$$
$$\frac{\partial \mathbf{u}}{\partial r} \cdot \nabla w - \frac{\partial \mathbf{u}}{\partial z} \cdot \nabla u + \frac{\partial F_u}{\partial z} - \frac{\partial F_w}{\partial r}.$$

Now

$$\ln \theta = (1 - \kappa) \ln p - \ln \rho + \kappa \ln p^* - \ln R = -\ln \chi,$$

so that

$$(1 - \kappa)\frac{dp}{p} - \frac{d\rho}{\rho} = -\frac{d\chi}{\chi}.$$

Then

$$\frac{1}{\rho^2}\left(\frac{\partial \rho}{\partial z}\frac{\partial p}{\partial r} - \frac{\partial \rho}{\partial r}\frac{\partial p}{\partial z}\right) = \frac{1}{\rho\chi}\left(\frac{\partial \chi}{\partial z}\frac{\partial p}{\partial r} - \frac{\partial \chi}{\partial r}\frac{\partial p}{\partial z}\right).$$

Again

$$\frac{\partial \mathbf{u}}{\partial r} \cdot \nabla w - \frac{\partial \mathbf{u}}{\partial z} \cdot \nabla u = \left(\frac{\partial u}{\partial r} + \frac{\partial w}{\partial z}\right)\left(\frac{\partial w}{\partial r} - \frac{\partial u}{\partial z}\right),$$

but the continuity equation now gives

$$\frac{\partial u}{\partial r} + \frac{\partial w}{\partial z} = -\frac{u}{r} - \frac{1}{\rho}\left(\frac{\partial \rho}{\partial t} + u\frac{\partial \rho}{\partial r} + w\frac{\partial \rho}{\partial z}\right)$$
$$= -\frac{u}{r} + \rho\frac{D}{Dt}\left(\frac{1}{\rho}\right),$$

where $D/Dt \equiv \partial/\partial t + \mathbf{u} \cdot \nabla$. Thus the toroidal vorticity equation is

$$\frac{\partial \eta}{\partial t} + \mathbf{u} \cdot \nabla \eta = \frac{\partial C}{\partial z} + \frac{1}{\rho\chi}\left(\frac{\partial \chi}{\partial z}\frac{\partial p}{\partial r} - \frac{\partial \chi}{\partial r}\frac{\partial p}{\partial z}\right) +$$
$$\left[\frac{u}{r} - \rho\frac{D}{Dt}\left(\frac{1}{\rho}\right)\right]\eta + \frac{\partial F_u}{\partial z} - \frac{\partial F_w}{\partial r},$$

or

$$r\frac{D}{Dt}\left(\frac{\eta}{r\rho}\right) = \frac{1}{\rho}\frac{\partial C}{\partial z} + \frac{1}{\rho^2\chi}\left(\frac{\partial\chi}{\partial z}\frac{\partial p}{\partial r} - \frac{\partial\chi}{\partial r}\frac{\partial p}{\partial z}\right) + \frac{1}{\rho}\left(\frac{\partial F_u}{\partial z} - \frac{\partial F_w}{\partial r}\right),$$

$$(5.80)$$

where $\eta/(r\rho)$ is the *potential toroidal vorticity*. In this formula, r is the radius of a toroidal vortex ring, analogous to a "depth" in an incompressible shallow water formulation (Holton, 2004, page 97).

5.13.2 Eliassen equation and toroidal vorticity equation

The Eliassen equation is related to the local time derivative of the potential toroidal vorticity equation (5.80) when the frictional terms are omitted. In this case, when thermal wind balance exists, the right-hand-side of this equation may be written as

$$\frac{1}{\rho\chi}\left(g\frac{\partial\chi}{\partial r} + \frac{\partial}{\partial z}(C\chi)\right).$$

Then the time derivative of Eq. (5.80) (with F_u and F_w both zero) is

$$\frac{\partial}{\partial t}\left[r\frac{D}{Dt}\left(\frac{\eta}{r\rho}\right)\right] = \frac{\partial}{\partial t}\left[\frac{1}{\rho\chi}\left(g\frac{\partial\chi}{\partial r} + \frac{\partial}{\partial z}(C\chi)\right)\right].$$

$$(5.81)$$

The right-hand-side of Eq. (5.81) gives the Eliassen equation when the thermal wind equation (5.19) is satisfied for all time. Then consistency requires that the left-hand-side is identically zero. In other words, the Eliassen equation is satisfied if the local time derivative of the material rate-of-change of the potential toroidal vorticity is zero.

5.13.3 Geopotential tendency equation

In Section 5.9 we outlined an approach for understanding and calculating the evolution of a tropical cyclone based on the Eliassen equation for the balanced overturning circulation forced by heat and tangential momentum sources. An alternative (and equivalent) approach is to use the geopotential tendency equation. The latter approach for describing the balanced evolution of a vortex has some advantages over the use of the Eliassen equation. To begin with, unlike the Eliassen equation, the derivation is not degenerate for the steady state. A mathematical advantage of using the geopotential tendency equation is that it avoids the need to first invert for the overturning circulation, then advect the tangential wind component by the radial and vertical flow, and finally to link the changes in tangential wind to changes in the mass field by solving the thermal wind equation. In fact,

as we shall show, the geopotential tendency equation gives a direct link between the heat and momentum forcing to the changes in the mass (height) field of the vortex.

To simplify the mathematical presentation, we adopt an inviscid, Boussinesq-like formulation of the axisymmetric balance dynamics in cylindrical, pseudo-height coordinates (r, λ, Z), where Z, as noted in Chapter 2, is defined by $Z = (c_{pd}\theta_0/g)[1 - (p/p_s)^\kappa]$, $\kappa = R_d/c_{pd}$, θ_0 is a reference potential temperature (300 K) at reference pressure p_s, p is the pressure, g is the acceleration due to gravity, c_{pd} is the specific heat of dry air at constant pressure and R_d is the gas constant for dry air. A useful property of this coordinate system is that Z is nearly equal to height z in the troposphere. For this reason, we will not distinguish between Z and z in the following discussion.[12]

In the case of axisymmetric flow, the Boussinesq balance equations comprising gradient wind and hydrostatic balance are, respectively,

$$fv + \frac{v^2}{r} = \frac{\partial\phi}{\partial r} \qquad (5.82)$$

$$g\frac{\theta}{\theta_0} = \frac{\partial\phi}{\partial z} \qquad (5.83)$$

where v, ϕ, θ are the tangential velocity, geopotential height and potential temperature fields, respectively. The pertinent tendency equations are the axisymmetric tangential momentum and potential temperature equations

$$\frac{\partial v}{\partial t} + u\eta + w\frac{\partial v}{\partial z} = F_\lambda \qquad (5.84)$$

$$\frac{\partial\theta}{\partial t} + u\frac{\partial\theta}{\partial r} + w\frac{\partial\theta}{\partial z} = \dot\theta \qquad (5.85)$$

where $\eta = f + (1/r)\partial(rv)/\partial r$ is here the absolute vertical vorticity of the axisymmetric vortex, F_λ is the tangential momentum forcing per unit mass and $\dot\theta = D\theta/Dt$ is the material rate of change of potential temperature following the transverse (overturning) circulation associated with condensation/evaporation and/or radiative warming/cooling.

A closed differential equation for the geopotential tendency

$$\phi_t = \partial\phi/\partial t \qquad (5.86)$$

can be obtained as follows. First, Eqs. (5.84) and (5.85) are solved for the radial and vertical velocity:

$$u = \frac{N^2\left(F_\lambda - \frac{1}{\xi}\frac{\partial\phi_t}{\partial z}\right) - \left(Q - \frac{\partial\phi_t}{\partial z}\right)}{P} \qquad (5.87)$$

12. Recall the fact that the hydrostatic Boussinesq equations are formally the same as the un-approximated quasi-static equations of motion in pressure coordinates. Therefore, we expect the solutions to the former will closely resemble those of the latter and, as a result, no generality is lost in using the Boussinesq model for obtaining basic understanding.

$$w = \frac{\eta\left(Q - \frac{\partial\phi_t}{\partial z}\right) - \left(F_\lambda - \frac{1}{\xi}\frac{\partial\phi_t}{\partial r}\right)\frac{\partial^2\phi}{\partial z \partial r}}{P} \tag{5.88}$$

where $N^2 = (g/\theta_0)\partial\theta/\partial z$ is the static stability of the mean vortex and environment, $Q = g\dot\theta/\theta_0$ is here the normalized diabatic heating rate, $\xi = f + 2v/r$ is twice the absolute angular velocity at radius r and

$$P = \eta N^2 - \xi\left(\frac{\partial v}{\partial z}\right)^2 \tag{5.89}$$

is the potential vorticity (PV) of the axisymmetric base-state vortex in the Boussinesq approximation. In obtaining Eqs. (5.87) and (5.88), we have used Eq. (5.83) to express θ in terms of ϕ and have used Eq. (5.82) to express $\partial v/\partial t$ in terms of ϕ_t via

$$\frac{\partial v}{\partial t} = \frac{1}{\xi}\frac{\partial\phi_t}{\partial r}. \tag{5.90}$$

To ensure that the meridional circulation satisfies continuity, one next substitutes the foregoing expressions for u and w into the continuity equation

$$\frac{\partial ru}{\partial r} + \frac{\partial rw}{\partial z} = 0 \tag{5.91}$$

to yield, after some rearrangement, a closed partial differential equation for ϕ_t in the meridional coordinates (r,z) forced by the sources of heating and tangential momentum:

$$\frac{1}{r}\frac{\partial}{\partial r}\left[\frac{rN^2}{P\xi}\frac{\partial\phi_t}{\partial r}\right] + \frac{\partial}{\partial z}\left[\frac{\eta}{P}\frac{\partial\phi_t}{\partial z}\right]$$
$$-\frac{1}{r}\frac{\partial}{\partial r}\left[\frac{r}{P}\frac{\partial v}{\partial z}\frac{\partial\phi_t}{\partial z}\right] - \frac{\partial}{\partial z}\left[\frac{1}{P}\frac{\partial v}{\partial z}\frac{\partial\phi_t}{\partial r}\right]$$
$$=\frac{\partial}{\partial z}\left(\frac{\eta}{P}Q\right) - \frac{1}{r}\frac{\partial}{\partial r}\left(\frac{r}{P}\frac{\partial v}{\partial z}Q\right) + \frac{\partial}{\partial r}\left(\frac{rN^2F_\lambda}{P}\right)$$
$$-\frac{\partial}{\partial z}\left(\frac{r\xi}{P}\frac{\partial v}{\partial z}F_\lambda\right). \tag{5.92}$$

The forcing terms on the right-hand-side of Eq. (5.92) can be rewritten to yield the final tendency equation

$$\mathcal{L}(\phi_t) = \boldsymbol{\eta}\cdot\nabla\left(\frac{Q}{P}\right) + \frac{g}{r\theta_0}J\left(\frac{rF_\lambda}{P},\theta\right) \equiv S, \tag{5.93}$$

where $\mathcal{L}$ is the r-z differential operator in Eq. (5.92) that acts on ϕ_t,

$$\boldsymbol{\eta} = \left(-\frac{\partial v}{\partial z}, f + \frac{1}{r}\frac{\partial rv}{\partial r}\right), \tag{5.94}$$

is the absolute vorticity vector of the axisymmetric base state vortex and

$$J(a,b) = \frac{\partial a}{\partial r}\frac{\partial b}{\partial z} - \frac{\partial b}{\partial r}\frac{\partial a}{\partial z}, \tag{5.95}$$

is the Jacobian operator in the meridional plane.

We call the function S the *spin-up function* since this quantity controls the geopotential tendency and hence the tendency of the gradient wind via Eq. (5.90). In obtaining the form of the forcing terms in Eq. (5.93), use has been made of the thermal wind equation

$$\frac{g}{\theta_0}\frac{\partial\theta}{\partial r} = \xi\frac{\partial v}{\partial z}, \tag{5.96}$$

which is a consequence of assuming that the base state vortex remains in gradient and hydrostatic balance (5.82) and (5.83).[13]

It is straightforward to show[14] that the geopotential tendency equation (5.92) is equivalent to the (balanced) potential vorticity equation

$$\frac{\partial P}{\partial t} + u\frac{\partial P}{\partial r} + w\frac{\partial P}{\partial z} = \boldsymbol{\eta}\cdot\nabla Q + \frac{g}{\theta_0}\nabla\theta\cdot\nabla\wedge\mathbf{F}. \tag{5.97}$$

Here, $\mathbf{F} = F_\lambda\hat\lambda$ is the tangential momentum forcing per unit mass and $\hat{\mathbf{j}}$ is a unit vector in the tangential direction. Whereas the radial and vertical advection terms are explicit in the PV equation, the radial and vertical advection of P is implicit in the geopotential tendency formulation (5.92).[15] That is to say, solving the geopotential tendency equation (5.92) is equivalent to advecting, modifying and inverting the PV as discussed in Section 2.13.

Given the equivalence between the PV equation (5.97) and the geopotential tendency equation (5.92), the astute reader might be wondering what the relationship is between the Eliassen equation discussed in the main text to the geopotential tendency equation (5.92)? It can be shown that solving the geopotential tendency equation is equivalent to solving the Eliassen equation for the transverse circulation (e.g., as given by Exercise 5.21 using the Boussinesq formulation) (Shapiro and Montgomery, 1993; Möller and Shapiro, 2002, their Appendix B). That is, solving the Eliassen equation, together with the tendency equation (5.84) for tangential velocity and related thermal wind equation (5.96), is equivalent to advancing the PV according to Eq. (5.97) and inverting the updated PV for the balanced geopotential, its tendency and the inferred balanced wind and temperature fields at the new time step. Thus, solving the Eliassen problem is equivalent to executing PV inversion in the sense described in Section 2.13.

---

13. In the forthcoming section, we present an extended Eliassen model that permits one to relax the assumption of thermal wind balance in the fluid, though the formulation still requires that the forced vortex evolves suitably slowly to be defined precisely below.
14. See, e.g., Shapiro and Montgomery (1993, their Eq. (4.10)).
15. Apart from the Boussinesq approximation in pseudo-height coordinates, the tendency equation (5.92) is analogous to the tendency equation derived by Vigh and Schubert (2009, their Eqs. (2.20) and (2.21).).

5.13.4 Deductions from the spin-up function

Taking the ocean surface and tropopause to be approximately isothermal boundaries, the boundary conditions on the top and bottom of the vortex require a vanishing vertical derivative of the geopotential tendency, i.e., $\partial\phi_t/\partial z = 0$. Symmetry at the axis of rotation implies that $\partial\phi_t/\partial r = 0$ at $r = 0$. Boundedness of the solution requires that $\phi_t \to 0$ at large r. If the discriminant of the mean vortex is everywhere positive (i.e., $\xi P > 0$), and if the spin-up function S vanishes throughout the domain, we conclude from the geopotential tendency equation, together with these boundary conditions, that $\phi_t = 0$ everywhere and the flow is in a steady state. This is the *non-acceleration theorem* for axisymmetric balance vortex dynamics.[16]

For illustrative purposes, let us neglect frictional effects for simplicity so that $F_\lambda = 0$. In this case, it is evident from Eq. (5.93) that there will be no PV generation in the vortex interior by condensation heating/cooling (and hence no change of the mass field of the bulk vortex) unless the gradient of the heating rate divided by the vortex PV projects non-trivially onto the absolute vorticity vector in the meridional plane. If, for the sake of argument, during the early stage of intensification, the vortex is approximately barotropic, the largest term in the spin-up function $S = \eta \cdot \nabla(Q/P)$ results from the vertical gradient of Q. In this situation, assuming that deep convection has a simple heating profile with a maximum in the middle troposphere (e.g., Gill, 1982), the magnitude of the vertical gradient of the heating rate (and its location with respect to the maximum vertical vorticity) controls principally the intensification rate.[17] However, as the hurricane reaches maturity and the baroclinicity of the mean vortex becomes significant, the absolute vorticity vector η rotates clockwise in the r-z plane and starts to become locally perpendicular to the vector gradient of (Q/P). The end result of this evolution is that the spin-up function approaches zero and the balanced flow tends towards a steady state. Further discussion on the issue of the steady state vortex is deferred until Chapter 14.

Based on the foregoing discussion, we see that in general the intensification rate in the axisymmetric balance model is controlled by the structure of the spin-up function S and its radial distribution relative to the vortex's vorticity and stratification.

---

16. This result is distinct from the non-acceleration theorem for non-axisymmetric eddy motions in a barotropic Rankine vortex, as illustrated using the toy vortex-wave model discussed at the end of Chapter 4.
17. See also Vigh and Schubert (2009), their Eq. (2.25), and accompanying discussion.

5.13.5 The linear approximation, the Eliassen equation and extension to include unbalanced forcing

In this section, we explore the connection between the linearized dynamics of an axisymmetric vortex, the Eliassen balance model of Section 5.9, and an extended Eliassen formulation that accounts for slowly evolving, yet unbalanced processes. We explore this connection because interesting questions have been raised in the literature about an alleged equivalence between the long-time ($t \to \infty$) limit of the linear solutions and the Eliassen model. Suffice it to say, although a correspondence between the linear and Eliassen systems can and will be derived, important caveats exist concerning the validity of the linear solutions at long times as well as the validity of the Eliassen model in tropical cyclones. An in-depth analysis of these issues is presented in Montgomery and Smith (2017b).

Here, we articulate the proper link between the linear and Eliassen models. The linear model (defined precisely below) presumes the presence of perturbation forcing terms associated with nonconservative processes as well as nonlinear mean and eddy terms, the latter of which can be estimated from the linear solutions. To keep the mathematics as simple as possible, we adopt the Boussinesq approximation detailed in Section 2.1 and use regular height coordinates. Here will use the simplest form of the Boussinesq approximation, namely, ρ_o = constant everywhere except when coupled with the buoyancy force. To attain maximum simplicity, we will employ regular height coordinates as opposed to pseudo-height coordinates of the prior subsection.

We consider then the linearized axisymmetric equations about an initially axisymmetric vortex in gradient and hydrostatic balance ($\bar{v}, \bar{p}$):

$$\frac{\bar{v}^2}{r} + f\bar{v} = \frac{1}{\rho_o}\frac{\partial\bar{p}}{\partial r}, \tag{5.98}$$

$$g\frac{\bar{\theta}}{\theta_o} = \frac{1}{\rho_o}\frac{\partial\bar{p}}{\partial z}. \tag{5.99}$$

Next, let $\theta = \bar{\theta} + \theta'$, $p = \bar{p} + p'$, $v = \bar{v} + v'$, $u = u'$, $w = w'$, where the prime indicates a perturbation from the assumed axisymmetric basic state vortex. The linearized equations of motion are:

$$\frac{\partial u'}{\partial t} - \bar{\xi}v' = -\frac{1}{\rho_o}\frac{\partial p'}{\partial r} + F_r', \tag{5.100}$$

$$\frac{\partial v'}{\partial t} + \bar{\eta}u' + w'\frac{\partial\bar{v}}{\partial z} = F_\lambda', \tag{5.101}$$

$$\frac{\partial w'}{\partial t} = -\frac{1}{\rho_o}\frac{\partial p'}{\partial z} + b' + F_z', \tag{5.102}$$

$$\frac{\partial\theta'}{\partial t} + u'\frac{\partial\bar{\theta}}{\partial r} + w'\frac{d\bar{\theta}}{dz} = \dot{\theta}, \tag{5.103}$$

$$\frac{1}{r}\frac{\partial r u'}{\partial r} + \frac{\partial w'}{\partial z} = 0, \qquad (5.104)$$

where $b' = g\theta'/\theta_o$ is the perturbation buoyancy force per unit mass (defined relative to the axisymmetric vortex in thermal wind balance). Multiplying Eq. (5.103) by g/θ_o gives an equation for the perturbation buoyancy force

$$\frac{\partial b'}{\partial t} + u'\frac{\partial \bar{b}}{\partial r} + w'N^2 = Q, \qquad (5.105)$$

where $N^2 = \partial \bar{b}/\partial z$ is the static stability, $\bar{b} = g\bar{\theta}/\theta_o$ is the mean (system) buoyancy, $\bar{\xi} = f + 2\bar{v}/r$ is twice the absolute vertical rotation rate at radius r, $\bar{\eta} = f + \bar{\zeta}$ is the basic state absolute vertical vorticity at radius r, and $Q = g\dot{\theta}/\theta_o$ is the normalized heating rate as defined previously in Section 5.13.3. The F quantities represent the component forcing terms in the perturbation momentum equations associated with non-conservative processes, such as surface friction and related sub grid scale parameterization of the vortex boundary layer as well as second-order mean and eddy advective accelerations (see Montgomery and Smith (2017a), Section 2 for details). The second-order terms implicit in the F and $\dot{\theta}$ quantities will be referred to below as "the N terms". Other variables in the foregoing equations take on their usual meaning from this chapter.

If the total vortex evolution were *strictly balanced*, we would have

$$\frac{\partial}{\partial z}\left(\frac{v^2}{r} + fv\right) = \frac{\partial}{\partial z}\left(\frac{1}{\rho_o}\frac{\partial p}{\partial r}\right),$$

or

$$\xi\frac{\partial v}{\partial z} = \frac{1}{\rho_o}\frac{\partial^2 p}{\partial r \partial z} = \frac{g}{\theta_o}\frac{\partial \theta}{\partial r}, \qquad (5.106)$$

after making use of Eq. (5.99), with $\xi = f + 2v/r$. This is a thermal wind balance for the total tangential wind, pressure and potential temperature fields. The linear approximation to the full thermal wind balance equation, after subtracting the base state balance, gives

$$\bar{\xi}\frac{\partial v'}{\partial z} + \frac{2v'}{r}\frac{\partial \bar{v}}{\partial z} = \frac{g}{\theta_o}\frac{\partial \theta'}{\partial r},$$

or

$$\bar{\xi}\frac{\partial v'}{\partial z} + \frac{2v'}{r}\frac{\partial \bar{v}}{\partial z} = \frac{\partial b'}{\partial r}. \qquad (5.107)$$

Eq. (5.107) is a perturbation thermal wind equation expressing a balance between the vertical shear of the azimuthal flow and the radial gradient of the perturbation buoyancy force. For the less restrictive unbalanced case considered here, one must account for processes that give rise to thermal wind imbalance, i.e., processes that contribute to a nonzero difference between the left-hand-side and right-hand-side of Eq. (5.107). Bearing this in mind, we return

now to the goal of finding the connection between the perturbation solutions and the Eliassen model.

Motivated by Eq. (5.107), we form first an equation for the time rate-of-change of $\partial b'/\partial r$:

$$\frac{\partial}{\partial t}\left(\frac{g}{\theta_o}\frac{\partial \theta'}{\partial r}\right) \quad \text{or} \quad \frac{\partial}{\partial t}\left(\frac{\partial b'}{\partial r}\right).$$

Eq. (5.105) gives

$$\frac{\partial}{\partial t}\left(\frac{\partial b'}{\partial r}\right) = -\frac{\partial}{\partial r}\left(u'\frac{\partial \bar{b}}{\partial r}\right) - \frac{\partial}{\partial r}\left(w'N^2\right) + \frac{\partial Q}{\partial r}. \qquad (5.108)$$

Next, form the equation

$$\frac{\partial}{\partial z}\left(\bar{\xi}\frac{\partial v'}{\partial t}\right) = \frac{2}{r}\frac{\partial \bar{v}}{\partial z}\frac{\partial v'}{\partial t} + \frac{\partial}{\partial t}\left(\bar{\xi}\frac{\partial v'}{\partial z}\right) = \frac{\partial}{\partial t}\left(\bar{\xi}\frac{\partial v'}{\partial z} + \frac{2v'}{r}\frac{\partial \bar{v}}{\partial z}\right),$$

assuming, as is usual in linear theory, that the basic state does not change, i.e., $\partial \bar{v}/\partial t = 0$ at this order of approximation.[18] Next we form $\partial/\partial z[\bar{\xi} \times (5.101)]$ and use the foregoing relationship to obtain

$$\frac{\partial}{\partial t}\left(\bar{\xi}\frac{\partial v'}{\partial z} + \frac{2v'}{r}\frac{\partial \bar{v}}{\partial z}\right) = -\frac{\partial}{\partial z}\left[u'\bar{\xi}\bar{\eta}\right]$$
$$-\frac{\partial}{\partial z}\left(w'\bar{\xi}\frac{\partial \bar{v}}{\partial z}\right) + \frac{\partial}{\partial z}\left(\bar{\xi}F'_\lambda\right). \qquad (5.109)$$

Subtracting Eq. (5.108) from Eq. (5.109) then gives an equation for the time rate-of-change of thermal wind imbalance:

$$\frac{\partial}{\partial t}\underbrace{\left(\bar{\xi}\frac{\partial v'}{\partial z} + \frac{2v'}{r}\frac{\partial \bar{v}}{\partial z} - \frac{\partial b'}{\partial r}\right)}_{\text{thermal wind imbalance}} = -\frac{\partial}{\partial z}\left(u'\bar{\xi}\bar{\eta}\right)$$
$$-\frac{\partial}{\partial z}\left(w'\bar{\xi}\frac{\partial \bar{v}}{\partial z}\right) + \frac{\partial}{\partial r}\left(u'\frac{\partial \bar{b}}{\partial r}\right)$$
$$+\frac{\partial}{\partial r}\left(w'N^2\right) + \frac{\partial}{\partial z}\left(\bar{\xi}F'_\lambda\right) - \frac{\partial Q}{\partial r}. \qquad (5.110)$$

The relation between thermal wind imbalance and the toroidal (azimuthal) vorticity equation emerges on forming the toroidal vorticity equation:

$$\frac{\partial}{\partial z}(5.100) \rightarrow \frac{\partial}{\partial t}\left(\frac{\partial u'}{\partial z}\right) = +\frac{\partial}{\partial z}'\left(\bar{\xi}v'\right)$$
$$-\frac{1}{\rho_o}\frac{\partial^2 p'}{\partial r \partial z} + \frac{\partial F'_r}{\partial z}, \qquad (5.111)$$

---

18. Changes to the basic state vortex can be estimated using the linear solutions to evaluate the tendencies in the basic state tangential velocity and buoyancy, similar to the quasi-linear methodology presented near the end of Chapter 4.

$$\frac{\partial}{\partial r}(5.102) \rightarrow \frac{\partial}{\partial t}\left(\frac{\partial w'}{\partial r}\right) = -\frac{\partial b'}{\partial r}$$
$$-\frac{1}{\rho_o}\frac{\partial^2 p'}{\partial z \partial r} + \frac{\partial F_z'}{\partial r}. \qquad (5.112)$$

Subtracting Eq. (5.112) from Eq. (5.111) gives:

$$\frac{\partial}{\partial t}\left(\frac{\partial u'}{\partial z} - \frac{\partial w'}{\partial r}\right) = \underbrace{\bar{\xi}\frac{\partial v'}{\partial z} + \frac{2v'}{r}\frac{\partial \bar{v}}{\partial z} - \frac{\partial b'}{\partial r}}_{\text{thermal wind imbalance}}$$
$$+ \frac{\partial F_r'}{\partial z} - \frac{\partial F_z'}{\partial r}. \qquad (5.113)$$

Now, take $\partial/\partial t$ of Eq. (5.113) and use the thermal wind imbalance equation (5.110) to obtain a closed evolution equation for the transverse circulation, assuming the forcing terms $(F_r', F_\lambda', F_z', Q)$ are known:

$$\frac{\partial^2}{\partial t^2}\left(\frac{\partial u'}{\partial z} - \frac{\partial w'}{\partial r}\right) = -\frac{\partial}{\partial z}\left(u'\bar{\xi}\bar{\eta}\right) - \frac{\partial}{\partial z}\left(w'\bar{\xi}\frac{\partial \bar{v}}{\partial z}\right)$$
$$+ \frac{\partial}{\partial r}\left(u'\frac{\partial \bar{b}}{\partial r}\right) + \frac{\partial}{\partial r}\left(w'N^2\right) + \frac{\partial}{\partial z}\left(\bar{\xi}F_\lambda'\right)$$
$$- \frac{\partial Q}{\partial r} + \frac{\partial^2 F_r'}{\partial z \partial t} - \frac{\partial^2 F_z'}{\partial r \partial t}. \qquad (5.114)$$

Rearranging terms yields the following:

$$\frac{\partial}{\partial z}\left[\left(\frac{\partial^2}{\partial t^2} + \bar{\xi}\bar{\eta}\right)u'\right] - \frac{\partial}{\partial r}\left[\left(\frac{\partial^2}{\partial t^2} + N^2\right)w'\right]$$
$$+ \frac{\partial}{\partial z}\left(\bar{\xi}\frac{\partial \bar{v}}{\partial z}w'\right) - \frac{\partial}{\partial r}\left(u'\frac{\partial \bar{b}}{\partial r}\right) =$$
$$\frac{\partial}{\partial z}\left(\bar{\xi}F_\lambda'\right) - \frac{\partial Q}{\partial r} + \frac{\partial^2 F_r'}{\partial z \partial t} - \frac{\partial^2 F_z'}{\partial r \partial t}. \qquad (5.115)$$

When a streamfunction ψ is introduced for the transverse circulation such that Eq. (5.104) is satisfied, i.e., $u' = -(1/r)\partial\psi/\partial z$ and $w' = (1/r)\partial\psi/\partial r$, then a single second-order partial differential equation results for ψ. Notice that if F_r', F_λ', F_z', and Q' and the response described by Eq. (5.115) are all assumed to *vary sufficiently slowly*, then we may neglect all the time derivatives in Eq. (5.115), giving us the standard Eliassen equation, in which strict thermal wind balance of the mean vortex (Eq. (5.106)) allows us to replace $\partial\bar{b}/\partial r$ by $\bar{\xi}\partial\bar{v}/\partial z$.

The result of the *ultra-slow* Eliassen balance approximation is then

$$\frac{\partial}{\partial r}\left[\frac{N^2}{r}\frac{\partial\psi}{\partial r} - \bar{\xi}\frac{\partial\bar{v}}{\partial z}\frac{1}{r}\frac{\partial\psi}{\partial z}\right] + \frac{\partial}{\partial z}\left[\frac{\bar{\xi}\bar{\eta}}{r}\frac{\partial\psi}{\partial z} - \frac{\bar{\xi}}{r}\frac{\partial\bar{v}}{\partial z}\frac{\partial\psi}{\partial r}\right]$$
$$= \frac{\partial Q}{\partial r} - \frac{\partial}{\partial z}\left(\bar{\xi}F_\lambda'\right), \qquad (5.116)$$

which is the Boussinesq analog of the anelastic Eliassen equation (5.67) derived in Section 5.9.[19]

Eq. (5.116) is elliptic if the vortex is symmetrically stable (i.e. if the inertial stability on isentropic surfaces is greater than zero); and symmetric stability is assured when $\bar{\eta}\bar{\xi}N^2 - (\bar{\xi}(\partial\bar{v}/\partial z))^2 > 0$, assuming that we can use the thermal wind balance relation $\partial\bar{b}/\partial r = \bar{\xi}\partial\bar{v}/\partial z$ for this purpose.

The generalized formulation associated with Eq. (5.115) permits the derivation of an extended Eliassen model that includes the forcing terms (F_r', F_z') that drive small, but persistent departures from gradient wind balance and hydrostatic balance, respectively. In this case we consider a less restrictive and more subtle distinguished limit of Eq. (5.115) by consistently keeping the first time derivatives on the right of Eq. (5.115), while neglecting the second time derivatives on the left, if rates of variation are "fairly slow" in the following sense. To neglect the second time derivatives we need $\partial^2/\partial t^2$ to be small in comparison with $\bar{\xi}\bar{\eta}$ and N^2, the squares of the local inertial frequency and buoyancy frequency, respectively. Nevertheless, there is a range of conditions for which the first derivatives can remain significant, especially in the third term on the right hand side of Eq. (5.115), whose magnitude in comparison with the first term on the same right hand side could be significant. On the reasonable assumption that $F_r' \sim F_\lambda'$, we need to compare $\partial/\partial t$ with $\bar{\xi}$, which is more like the *square root* of the *local* inertial frequency ($\sqrt{\bar{\eta}\bar{\xi}}$). In our judgment, such "fairly slow" evolution is often relevant to tropical cyclone evolution (Shapiro and Montgomery, 1993).

In the less restrictive *slow approximation*,

$$\frac{\partial}{\partial z}\left(\bar{\xi}\bar{\eta}u'\right) - \frac{\partial}{\partial r}\left(N^2 w'\right) + \frac{\partial}{\partial z}\left(\bar{\xi}\frac{\partial\bar{v}}{\partial z}w'\right) -$$
$$\frac{\partial}{\partial r}\left(u'\frac{\partial\bar{b}}{\partial r}\right) = \frac{\partial}{\partial z}\left(\bar{\xi}F_\lambda'\right) - \frac{\partial Q}{\partial r} + \frac{\partial^2 F_r'}{\partial z \partial t} - \frac{\partial^2 F_z'}{\partial r \partial t}. \quad (5.117)$$

Introducing the transverse streamfunction as defined above, and rearranging terms yields

---

19. E.g., Exercise 5.21, with N^2 denoting here the static stability of the environment plus base state vortex in the Boussinesq approximation. Eq. (5.116) is thus formally identical to Eq. (5.72).

$$\frac{\partial}{\partial r}\left[\frac{N^2}{r}\frac{\partial\psi}{\partial r}-\frac{\partial\bar{b}}{\partial r}\frac{1}{r}\frac{\partial\psi}{\partial z}\right]+\frac{\partial}{\partial z}\left[\frac{\bar{\xi}\bar{\eta}}{r}\frac{\partial\psi}{\partial z}-\frac{\bar{\xi}}{r}\frac{\partial\bar{v}}{\partial z}\frac{\partial\psi}{\partial r}\right]$$
$$=\frac{\partial Q}{\partial r}-\frac{\partial}{\partial z}(\bar{\xi}F'_\lambda)-\frac{\partial^2 F'_r}{\partial z\partial t}+\frac{\partial^2 F'_z}{\partial r\partial t}.$$
$$(5.118)$$

This equation has a diagnostic form, like the foregoing Eliassen equation (5.116) (and anelastic counterpart, Eq. (5.67)), if the right-hand side can be regarded as known. *We refer to this equation as the extended Eliassen equation.* It should be remembered that the nonlinear N terms on the right then need to be treated iteratively. It needs to be checked also whether the partial differential operator on the left is elliptic. Ellipticity can fail on account of the nonlinearity of the inner-core boundary layer (e.g., Montgomery and Persing, 2020).

The extended Eliassen equation (5.118) is elliptic if the vortex is symmetrically stable (i.e., if the inertial stability on isentropic surfaces is positive); and symmetric stability is assured when $\bar{\eta}\bar{\xi}(\partial\bar{b}/\partial z)-(\bar{\xi}(\partial\bar{v}/\partial z))^2 > 0$, assuming, again, that we can use the thermal wind balance relation $\partial\bar{b}/\partial r = \bar{\xi}\partial\bar{v}/\partial z$ for this purpose.[20]

As an illustration of the potential usefulness of the extended Eliassen equation (5.118), this equation can be used to diagnose the slowly evolving effects of nonlinear boundary-layer spin up (discussed later in Section 6.6) via iterative correction. That is, the N terms included on the right (as part of the forcing terms noted already) can be evaluated from a linear solution, then the equation solved again, the N terms evaluated more accurately, and so on. More rapidly time-varying effects, such as the emission of vertically and radially propagating inertia-buoyancy waves, where the inflowing air is rapidly decelerated and turns upward and outward at the base of the eyewall, require retention of the second time derivative terms in Eq. (5.115). Whereas a detailed analysis of these processes lies outside the scope of the present chapter, the virtue of this extended formulation is that it provides an overarching framework for quantifying the influence of thermal wind imbalance and the related forcing terms (F'_r, F'_z) associated with the departure from gradient and hydrostatic balance, effects not included in the conventional Eliassen balance model, cf. Eq. (5.116).

A special case of the extended Eliassen model occurs in the limit of slowly evolving forcing terms, wherein the time derivative terms on the right-hand-side of Eq. (5.118) are negligible compared to the forcing terms involving spatial derivatives of (F'_λ, Q). This limiting approximation corresponds to a "pseudo-balance" formulation because the coefficient terms on the left-hand-side of the differential equation do not strictly satisfy thermal wind balance, i.e., $\partial\bar{b}/\partial r - \bar{\xi}\partial\bar{v}/\partial z \neq 0$. This approximation has been used in several studies, but has been mistakenly referred to as "a balance model" (see Montgomery and Smith, 2018, for details).[21]

In the case of an intensifying tropical cyclone, the "pseudo-balance" solutions have been shown to be an improvement over the traditional Eliassen solutions in the vortex boundary layer where the traditional solutions significantly underestimate the mean inflow there. However, the tentative nature of neglecting the agradient forcing term $(\partial^2 F'_r/\partial t\partial z)$ in the vortex boundary layer appears to render the pseudo-balance approximation unreliable when applied to a finite time interval during vortex intensification (Montgomery and Persing, 2020).

---

20. See below for more when the thermal wind balance constraint is relaxed.

21. If strict thermal wind balance is not satisfied, the ellipticity condition needs to be modified to: $4N^2\bar{\eta}\bar{\xi} - (\partial\bar{b}/\partial r + \bar{\xi}\partial\bar{v}/\partial z)^2 > 0$.

Chapter 6

Frictional effects

In this chapter we consider the dynamical consequences of frictional effects and related turbulent momentum transfer on the structure and evolution of a tropical cyclone. Two important questions to be addressed are the reason for the calm eye near the storm center and the reason why the maximum tangential wind speed is found at low levels where frictional effects are non-negligible. The important thermodynamical effects of the near-surface friction layer will be addressed in subsequent chapters. Many of the topics to be discussed here carry over to other concentrated vortices subject to surface friction at the lower boundary such as tornadoes, waterspouts and dust devils.

To illustrate some of the key effects of friction in a rotating flow, we begin this chapter by considering the classical problem relating to the frictional spin down of a vortex. Then we derive approximate equations for the near-surface friction layer, or boundary layer, based on a scale analysis of the equations with the friction terms included. After that we show some solutions for approximations to the boundary layer equations and go on to discuss the limitations of these approximations. Finally, we highlight the importance of the boundary layer in tropical cyclone evolution and provide an explanation for the counter-intuitive observation shown in Chapter 1, and referred to above, that the maximum tangential wind occurs in the boundary-layer itself.

6.1 Vortex spin down

The classical spin down problem for a vortex considers the evolution of an axisymmetric vortex above a rigid boundary normal to the axis of rotation. We will show that the spin down is primarily a result of the generalized Coriolis force acting on the secondary circulation induced by friction above the boundary layer. The direct effect of the frictional diffusion of tangential momentum to the surface and the frictional drag at the surface are of secondary importance to the spin down in the parameter regimes relevant to tropical cyclones.

The boundary layer of a tropical cyclone is a shallow layer of air near the surface, typically 500-1000 m deep, in which frictional stresses are important. These stresses are linked to the frictional stress at the ocean surface and their effect at most radii[1] is to reduce the tangential wind speed in

the boundary layer and thereby the centrifugal and Coriolis forces in that layer. However, as the scale analysis will show, the inward-directed pressure gradient force in the boundary layer is approximately the same as that at the top of the boundary layer. Thus, in contrast to the approximate gradient wind balance in the vortex above the boundary layer, there is a net inward force within the boundary layer. This unbalanced force, or *agradient force*, drives air parcels inwards in this layer (see Section 6.4.1).

FIGURE 6.1 The beaker experiment showing the effects of frictionally-driven inflow near the bottom after the water has been stirred to produce rotation. This inflow carries tea leaves to form a neat pile near the axis of rotation.

One can demonstrate the frictionally-driven inflow simply by placing tea leaves in a beaker of water and vigorously stirring the water to set it in rotation. After a short time, the tea leaves congregate near the bottom of the beaker near the axis as shown in Fig. 6.1. They are swept there by the inflow in the friction layer. Slowly the rotation in the beaker declines because the inflow towards the rotation axis in the friction layer has to be accompanied by radially-outward motion in the vortex above this layer in order to satisfy mass continuity.

The depth of the friction layer depends on the viscosity of the water and the rotation rate and is typically only on the order of a millimeter or two in this experiment. Because the water is rotating about the vertical axis, it possesses angular momentum about this axis. Here angular momentum is defined as the product of the tangential flow speed and the radius.

---

1. The reasons for this qualification are discussed in Section 6.6.

Tropical Cyclones. https://doi.org/10.1016/B978-0-44-313449-4.00014-X

TABLE 6.1 Scaling of the terms in the vertical momentum equation (5.3) with frictional terms added. The eddy diffusivity, K, is assumed to be principally a function of altitude z. The quantities Δp_r and Δp_z are scales for the variation of perturbation pressure on the radial and vertical scales R and Z, respectively; $A = Z/R$ is an aspect ratio, $S_r = V/U$ is the swirl ratio, and $R_e = VZ/K^*$ is the Reynolds number based on the vertical eddy momentum diffusivity K^*.

Vertical momentum						
$\dfrac{\partial w}{\partial t}$	$+u\dfrac{\partial w}{\partial r}$	$+w\dfrac{\partial w}{\partial z}$	$=-\dfrac{1}{\rho}\dfrac{\partial p'}{\partial z}$	$+K\nabla_h^2 w$	$+\dfrac{\partial}{\partial z}\left(K\dfrac{\partial w}{\partial z}\right)$	(1a)
$\dfrac{W}{T}$	$U\dfrac{W}{R}$	$W\dfrac{W}{Z}$	$\dfrac{\Delta p_z}{\rho Z}$	$K_*\dfrac{W}{R^2}$	$K_*\dfrac{W}{Z^2}$	(1b)
A^2	A^2	A^2	$\dfrac{\Delta p_z}{\rho U^2}$	$A^3 S_r R_e^{-1}$	$A S_r R_e^{-1}$	(1c)

As water particles move outwards above the friction layer, they conserve their angular momentum and as they move to larger radii, they spin more slowly (see Section 5.6). The same process would lead to the decay of a tropical cyclone if the frictionally-induced outflow were to occur just above the friction layer, as in the beaker experiment. The spin down problem for laminar flow goes back to the classical study of Greenspan and Howard (1963). Illustrations of the spin down process for a hurricane-like vortex without deep convection are shown in Section 8.1.2 and with a neutrally-stratified atmosphere in Montgomery et al. (2001).

What then prevents the hurricane from spinning down, or, for that matter, what enables it to spin up in the first place? Clearly, if it is to intensify, there must be a mechanism capable of drawing air inwards above the friction layer, and of course, this air must be rotating about the vertical axis and possess angular momentum so that, as it converges towards the axis, it spins faster. The only conceivable mechanism for producing inflow above the friction layer is the collective effect of entrainment into the updrafts of deep convective clouds. In turn, these updrafts are driven ultimately by buoyancy forces arising from latent heat release (see Chapter 9).

6.2 Scale analysis of the equations with friction

We repeat now the scale analysis of the dynamical equations (5.1)-(5.3) with friction terms added to the right-hand sides. We assume the flow to be homogeneous with density ρ and axisymmetric ($\partial/\partial\lambda = 0$). We assume also a particularly simple form for friction incorporating an eddy diffusivity K, which is taken to be principally a function of z and to have a typical scale K_*. A more sophisticated scale analysis, treating horizontal diffusion of momentum differently from vertical diffusion, is sketched out by Rotunno and Bryan (2012) and we refer to their analysis in Subsection 6.7.2.

With the foregoing simplifying assumptions, the equations of motion with friction included are:

$$\frac{\partial u}{\partial t}+u\frac{\partial u}{\partial r}+w\frac{\partial u}{\partial z}-\frac{v^2}{r}-fv=-\frac{1}{\rho}\frac{\partial p'}{\partial r}$$
$$+K\left(\nabla_h^2 u-\frac{u}{r^2}\right)+\frac{\partial}{\partial z}\left(K\frac{\partial u}{\partial z}\right) \quad (6.1)$$

$$\frac{\partial v}{\partial t}+u\frac{\partial v}{\partial r}+w\frac{\partial v}{\partial z}+\frac{uv}{r}+fu=$$
$$+K\left(\nabla_h^2 v-\frac{v}{r^2}\right)+\frac{\partial}{\partial z}\left(K\frac{\partial v}{\partial z}\right) \quad (6.2)$$

$$\frac{\partial w}{\partial t}+u\frac{\partial w}{\partial r}+w\frac{\partial w}{\partial z}=-\frac{1}{\rho}\frac{\partial p'}{\partial z}$$
$$+K\nabla_h^2 w+\frac{\partial}{\partial z}\left(K\frac{\partial w}{\partial z}\right) \quad (6.3)$$

where $\nabla_h^2 = (1/r)(\partial/\partial r)(r\partial/\partial r)$.

As before, we define velocity scales (U, V, W), length scales (R, Z), where Z is now a scale for the boundary layer depth, and an advective time scale $T = R/U$. The time-independent form of the continuity equation, (5.4), yields the same relation between the vertical and radial velocity as before, namely, $W/Z \sim U/R$. Let Δp_r and Δp_z be scales for change in perturbation pressure, p', on the radial scale R and vertical scale Z, respectively. We define a *Reynolds number*, $R_e = VZ/K_*$, which is a nondimensional parameter that characterizes the importance of the inertial (or advective) terms in comparison with the frictional terms, and an *aspect ratio* $A = Z/R$, that measures the ratio of the boundary-layer depth to the radial scale. For reasons that will become apparent, we begin with a scale analysis of the vertical momentum equation (6.3).

6.2.1 *w*-momentum equation

The terms in Eq. (6.3) have the nondimensional scales shown in Table 6.1. The tropical cyclone boundary layer is

TABLE 6.2 Scaling of the terms in Eqs. (6.5) and (6.6). Here Λ_g and Ξ are scales for the vertical component of absolute vorticity ($\zeta_{ag} = dv_g/dr + v_g/r + f$) and twice the absolute angular velocity ($\xi_g = 2v/r + f$), $Ro_{v'} = V^*/(R\Lambda_g)$ and $Ro_{v'} = V^*/(R\Xi_g)$ are the perturbation Rossby numbers based on Λ_g and Ξ_g, $S^* = U/V^*$ is an inverse swirl ratio based on V^* and $E_u = K_*/(\Xi Z^2)$, $E_{v'} = K_*/(\Lambda Z^2)$ are two local Ekman numbers. In the derivation, line (2b) is divided by $\Xi_g V^*$ to obtain line (2c) and line (3b) is divided by $\Xi_g U$ to obtain line (3c).

Radial momentum							
$\dfrac{\partial u}{\partial t}$	$+u\dfrac{\partial u}{\partial r}$	$+w\dfrac{\partial u}{\partial z}$	$-\dfrac{v'^2}{r}$	$-\left(\dfrac{2v_g}{r}+f\right)v' \quad =$	$K\left(\nabla_h^2 u - \dfrac{u}{r^2}\right)$	$+\dfrac{\partial}{\partial z}\left(K\dfrac{\partial u}{\partial z}\right)$	(2a)
$\dfrac{U}{T}$	$\dfrac{U^2}{R}$	$W\dfrac{U}{Z}$	$\dfrac{V^{*2}}{R}$	$\Xi_g V^*$	$K_*\dfrac{U}{R^2}$	$K_*\dfrac{U}{Z^2}$	(2b)
$S_*^2 Ro_{v'}^*$	$S_*^2 Ro_{v'}^*$	$S_*^2 Ro_{v'}^*$	$Ro_{v'}^*$	1	$A^2 E_u$	E_u	(2c)
Perturbation tangential momentum							
$\dfrac{\partial v'}{\partial t}$	$+u\dfrac{\partial v'}{\partial r}$	$+w\dfrac{\partial v'}{\partial z}$	$+\dfrac{uv'}{r}$	$+\left(\dfrac{dv_g}{dr}+\dfrac{v_g}{r}+f\right)u \quad =$	$K\left(\nabla_h^2 v' - \dfrac{v'}{r^2}\right)$	$+\dfrac{\partial}{\partial z}\left(K\dfrac{\partial v'}{\partial z}\right)$	(3a)
$\dfrac{V^*}{T}$	$U\dfrac{V^*}{R}$	$W\dfrac{V^*}{Z}$	$\dfrac{UV^*}{R}$	$\Lambda_g U$	$K_*\dfrac{V^*}{R^2}$	$K_*\dfrac{V^*}{Z^2}$	(3b)
$Ro_{v'}$	$Ro_{v'}$	$Ro_{v'}$	$Ro_{v'}$	1	$A^2 E_{v'}$	$E_{v'}$	(3c)

relatively thin, not more than 500 m to 1000 m in depth, and the aspect ratio A is small compared with unity. Moreover, typical values of K_* are on the order of 50 m^2 s^{-2}, although, as shown later in Fig. 6.2, there is scatter in the observational data and a dependence also of K_* on mean wind speed. Therefore, taking $V = 50$ m s^{-1}, $R = 200$ km, and $Z = 1000$ m, one obtains $R_e = 1 \times 10^3$ and $A = 5 \times 10^{-3}$. It follows from line (1c) in Table 6.1 that

$$\frac{\Delta p_z}{\rho U^2} \sim \max\left(A^2, AS_r R_e^{-1}\right). \tag{6.4}$$

The maximum radial velocity in the tropical cyclone boundary layer is found to be a substantial fraction of the maximum tangential velocity (see e.g., Fig. 6.4), so we take $U = 20$ m s^{-1} giving $S_r = 50/20 = 2.5$. Then $\max\left(A^2, AS_r R_e^{-1}\right) = \max(2.5, 1.5) \times 10^{-5} = 2.5 \times 10^{-5}$. Taking $\rho = 1$ kg m^{-3} it follows that $\Delta p_z \leq 2.5 \times 10^{-5} \times 20^2 = 10^{-2}$ Pa $= 10^{-4}$ mb. Thus the vertical variation of p' across the boundary layer is only a tiny fraction of the radial variation of p' above the boundary layer, typically 50 mb and to a very good approximation we can safely assume that $\partial p'/\partial z = 0$ in the boundary layer. In other words, to a close approximation, the radial pressure gradient within the boundary layer is the same as that above the boundary layer. *The result that the radial pressure gradient force is transmitted essentially uniformly across the boundary layer is an extremely important result and was first articulated by L. Prandtl* (see e.g., Anderson, 2005; Kundu and Cohen, 2010).

6.2.2 u- and v-momentum equations

The scale analysis of the w-momentum equation in the previous subsection shows that the horizontal pressure gradient in the boundary layer is essentially the same as that above it. Thus we may replace the radial pressure gradient in Eq. (6.1) by the gradient wind, $v_g(r)$, at the top of the boundary layer (i.e., by $v_g^2/r + f v_g$), assuming for simplicity that the radial flow above the boundary layer is zero or small compared to the gradient wind. Then, with the substitution $v = v_g(r) + v'$, the radial and tangential momentum equations may be written as:

$$\frac{\partial u}{\partial t} + u\frac{\partial u}{\partial r} + w\frac{\partial u}{\partial z} - \frac{v'^2}{r} - \left(\frac{2v_g}{r}+f\right)v' =$$
$$+K\left(\nabla_h^2 u - \frac{u}{r^2}\right) + \frac{\partial}{\partial z}\left(K\frac{\partial u}{\partial z}\right) \tag{6.5}$$

$$\frac{\partial v'}{\partial t} + u\frac{\partial v'}{\partial r} + w\frac{\partial v'}{\partial z} + \left(\frac{dv_g}{dr}+\frac{v_g}{r}+f\right)u =$$
$$+K\left(\nabla_h^2 v' - \frac{v'}{r^2}\right) + \frac{\partial}{\partial z}\left(K\frac{\partial v'}{\partial z}\right) \tag{6.6}$$

The scale analysis of these equations is summarized in Table 6.2. Here we have explicitly omitted the time dependence of the gradient wind. This time dependence is accounted for implicitly through the time evolution of the radial pressure gradient at the top of the friction layer. Rows (2a) and (3a) show the terms in the radial and tangential momentum equations and rows (2b) and (3b) show the scales of each term in these equations. Rows (2c) and (3c) show the non-dimensional ratio of terms with each term in row

(2b) divided by $\Xi_g V^*$ to give the corresponding term in row (2c) and each term in row (3b) divided by $\Lambda_g U$ to give the corresponding term in row (3c). Here V^* is a scale for v', and Λ_g and Ξ are scales for the absolute vertical vorticity $\zeta_{ag} = dv_g/dr + v_g/r + f$ and twice the absolute angular velocity $\xi_g = 2v_g/r + f$ of the gradient wind above the boundary layer.

Altogether, the non-dimensionalization introduces five new nondimensional parameters:

- $Ro_{v'} = V^*/(R\Lambda_g)$ a local Rossby number in the tangential momentum equation based on the departure of the tangential wind from the gradient wind, V^*, and on ζ_{ag}, which has scale Λ_g;
- $Ro_{v'}^* = V^*/(R\Xi_g)$ a local Rossby number in the radial momentum equation, again based on V^*, but on ξ_g with scale Ξ instead of Λ_g;
- $S_* = U/V^*$, an inverse swirl ratio;
- $E_u = K/(\Lambda Z^2)$, an Ekman number based on Λ;
- $E_{v'} = K/(\Xi Z^2)$, an Ekman number based on Ξ

The quantities ζ_{ag} and ξ_g vary with radius by several orders of magnitude across a tropical cyclone with typical values in the inner-core region, within the radius of maximum tangential wind speed, being on the order of 10^{-3} s^{-1}. For example, using a radial scale of 50 km for the inner core, $Ro_{v'}$ and $Ro_{v'}^*$ have typical magnitudes $20/(5 \times 10^4 \times 10^{-3}) = 0.4$ so that there is little scale separation between the terms on the left-hand-sides of the radial and tangential momentum equations for the boundary-layer. For a realistic tropical cyclone boundary layer flow, U may often exceed V^*, hence $S_* > O(1)$. In these situations $S_*^2 Ro_{v'}^* \sim O(1)$ and the nonlinear terms in the radial momentum equation must be retained. On the right-hand sides, however, the smallness of the aspect ratio A compared with unity implies that only the friction terms involving the vertical derivatives need to be retained.

6.2.3 Boundary layer depth scale

An estimate for the boundary layer depth may be obtained from the scaling in Table 6.2. The requirement for the friction terms to be important in Eqs. (2a) and (3a) is that $\Xi_g V^* = K_* U/Z^2$ and $\Lambda_g U = K_* V^*/Z^2$, from which it follows that $\Xi_g \Lambda_g = K_*^2/Z^4$ and therefore $Z = \sqrt{K/I}$, where $I^2 = \xi_g \zeta_{ag}$.

6.2.4 Boundary layer equations

With the approximations indicated by the scale analysis above, Eqs. (6.5) and (6.6) become:

> **Boundary layer equations**
>
> $$\frac{\partial u}{\partial t} + u\frac{\partial u}{\partial r} + w\frac{\partial u}{\partial z} - \frac{v'^2}{r} - \xi_g v' =$$
> $$\frac{\partial}{\partial z}\left(K\frac{\partial u'}{\partial z}\right), \quad (6.7)$$
> $$\frac{\partial v'}{\partial t} + u\frac{\partial v'}{\partial r} + w\frac{\partial v'}{\partial z} + \frac{uv'}{r} + \zeta_{ag}u =$$
> $$\frac{\partial}{\partial z}\left(K\frac{\partial v'}{\partial z}\right). \quad (6.8)$$

The vertical velocity in the boundary layer is determined by the continuity equation. Because the boundary layer is typically thin, density variations across it may be neglected and the continuity equation takes the approximate form:

$$\frac{\partial ru}{\partial r} + \frac{\partial rw}{\partial z} = 0. \quad (6.9)$$

To summarize, the principal simplifications of the boundary layer approximations are that the horizontal pressure gradient of the flow above the boundary layer is transmitted unchanged through the boundary layer to the surface and that the friction terms are dominated by the vertical diffusion contributions. In the case of a steady boundary layer, the latter assumption changes the character of the flow equations from elliptic to parabolic, which has consequences for the direction of information flow (see Section 10.14.1).

With the background gradient wind balance removed, there is less of a scale separation between the remaining terms in the boundary-layer equations, but further simplifications are possible for certain parameter values if the flow is steady.

6.3 The Ekman layer

If the radius of curvature of the isobars is small so that the gradient wind can be approximated by the geostrophic wind, i.e., if the overall Rossby number $Ro = V/(Rf) \ll 1$, then $\zeta_{ag} \approx f$, $\xi_g \approx f$. Further, if the perturbation velocities in the boundary layer are u and v' are much smaller in magnitude than the geostrophic wind, then $Ro_{v'}$ and $Ro_{v'}^*$ are both small compared with unity and (6.7) and (6.8) reduce to the

> **Ekman layer equations**
>
> $$0 = fv' + \frac{\partial}{\partial z}\left(K\frac{\partial u}{\partial z}\right), \quad (6.10)$$
> $$0 = -fu + \frac{\partial}{\partial z}\left(K\frac{\partial v'}{\partial z}\right), \quad (6.11)$$

Here the Coriolis acceleration terms have been moved to the right-hand sides of the equations, where they are then interpreted as Coriolis forces per unit mass. This is the classical

Ekman layer approximation, which for the case of constant K is straightforward to solve. The details are left as a guided exercise (see Exercise 6.1).

6.4 The linear approximation

The classical Ekman formulation may be readily generalized to the case of circular flow provided that $Ro_{v'} \ll 1$, $Ro_{v'}^* \ll 1$ and $S_*^2 Ro_{v'}^* \ll 1$. In this case the nonlinear acceleration terms in Eqs. (6.7) and (6.8) may be ignored and the approximate equations obtained constitute the

---

linear boundary layer approximation

$$0 = \xi_g v' + \frac{\partial}{\partial z}\left(K\frac{\partial u}{\partial z}\right), \qquad (6.12)$$

$$0 = -\zeta_{ag} u + \frac{\partial}{\partial z}\left(K\frac{\partial v'}{\partial z}\right). \qquad (6.13)$$

Again the linear acceleration terms have been moved to the right-hand sides of the equations, where they are then interpreted as *generalized* Coriolis forces per unit mass. In this form, the linear equations are seen to express everywhere a local force balance.

The condition $Ro_{v'} \ll 1$ requires that $V^*/V \ll 1$, a condition that, strictly speaking, is only satisfied beyond the strong wind region of a tropical cyclone. However, as remarked above, the condition $S_*^2 Ro_{v'}^* \ll 1$ will not be satisfied when U becomes large compared to V^* in the inner-core boundary layer. Thus, only when $U \lesssim V^*$ can one safely neglect the nonlinear acceleration terms in the radial momentum equation. The integrity of the linear approximation is investigated further in Section 6.4.11.

6.4.1 Physical interpretation

The generalized Coriolis force in the radial momentum equation (6.12), $\xi_g v'$, is the residual of the radial pressure gradient and centrifugal and Coriolis forces per unit mass and is often referred to as the *agradient force*. This is because $\xi_g v'$ is a measure of the degree of gradient wind imbalance. Eqs. (6.12) and (6.13) express a local balance between the generalized Coriolis forces and the corresponding frictional forces.

In a similar way, the Ekman approximation is an expression of balance in the radial direction between the agradient force, fv', and the radial component of frictional force and in the tangential direction between the Coriolis force, $-fu$, and the tangential component of frictional force.

Note that the balance referred to here involves both agradient and frictional forces and is therefore fundamentally different in nature from gradient wind balance discussed in Chapter 5. Because of the similarities in the equations for Ekman balance and linear balance, the latter is sometimes referred to as *generalized Ekman balance*.

6.4.2 Mathematical solution

When K is treated as constant, Eqs. (6.12) and (6.13) may be readily solved by taking, for example, the second vertical derivative of Eq. (6.13) and using Eq. (6.12) to eliminate $\partial^2 u/\partial z^2$ leaving a single fourth order differential equation for v', i.e.,

$$\frac{\partial^4 v'}{\partial z^4} + \frac{I^2}{K^2} v' = 0, \qquad (6.14)$$

where, $I^2(r) = \xi_g \zeta_{ag}$ is the inertial stability of the gradient wind at the top of the boundary layer. It may be verified that the solution of Eq. (6.14) that is bounded as $z \to \infty$ is[2]

$$v'(z) = v_g(r)e^{-z/\delta}[a_1 \cos(z/\delta) + a_2 \sin(z/\delta)], \quad (6.15)$$

---

where $\delta = (2K/I)^{1/2}$ is a refined boundary-layer scale depth and a_1 and a_2 are real constants. The corresponding solution for u is obtained simply by substituting for v' in Eq. (6.13), i.e.,

---

$$u(z) = -\chi v_g(r)e^{-z/\delta}[a_2 \cos(z/\delta) - a_1 \sin(z/\delta)], \qquad (6.16)$$

---

where $\chi = (\xi_g/\zeta_{ag})^{1/2}$.

For a turbulent boundary layer like that in a tropical cyclone, an appropriate boundary condition at the surface is to prescribe the surface stress, τ_s, as a function of the near-surface wind speed, normally taken to be the wind speed at a height of 10 m, and a drag coefficient, C_D. The condition takes the form

---

$$\frac{\tau_s}{\rho} = K\frac{\partial \mathbf{u_s}}{\partial z} = C_D |\mathbf{u_s}|\mathbf{u_s}, \qquad (6.17)$$

---

where $\mathbf{u_s} = (u, v_g(r) + v')_s$ is the wind vector at a height of 10 m. We apply a linearized form of this condition at $z = 0$, appropriate for the linearized form of the equations, to determine the constants a_1 and a_2 in Eqs. (6.15) and (6.16). The derivation is as follows. The substitution of Eqs. (6.15) and (6.16) into the boundary condition (6.17) leads before linearization to the following pair of algebraic equations for a_1 and a_2:

$$a_2 + a_1 = -va_2\sqrt{X}, \qquad (6.18)$$

$$a_2 - a_1 = v(1 + a_1)\sqrt{X}, \qquad (6.19)$$

where $X = (1 + a_1)^2 + \chi^2 a_2^2$, $v = C_D R_e$, and $R_e = v_g\delta/K$. Here, R_e is a Reynolds' number for the boundary layer. When a_1 and a_2 are small compared with unity, consistent with the linear theory, the expression for X can be linearized

---

2. Note that Eq. (6.14) has solutions of the form $\exp(\alpha z)$, where $\alpha^4 = -(K^2/I^2)$, i.e., $(K^2/I^2)\exp(\pi i + 2n\pi i)$, $i = \sqrt{-1}$ and n is an integer. Then possible values of α are proportional to $\pm\exp(i\pi/4)$ and $\pm\exp(3i\pi/4)$, or $\pm(1 \pm i)/\sqrt{2}$. The two values that lead to bounded solutions as $z \to \infty$ are proportional to $-(1 \pm i)/\sqrt{2}$.

to give $X \approx 1 + 2a_1$, whereupon $\sqrt{X} \approx 1 + a_1$. Then the linearized form of Eqs. (6.18) and (6.19) is

$$a_2 + a_1 = -\nu a_2, \tag{6.20}$$

$$a_2 - a_1 = \nu(1 + 2a_1), \tag{6.21}$$

which have the unique solution

$$a_1 = -\frac{\nu(\nu + 1)}{2\nu^2 + 3\nu + 2}, \quad a_2 = \frac{\nu}{2\nu^2 + 3\nu + 2}. \tag{6.22}$$

The vertical velocity, $w(r, z)$ is obtained by integrating the continuity equation with respect to z (see Exercise 6.3). The result is:

$$w(r, z) = \frac{1}{r} \frac{\partial}{\partial r} \left[\frac{K r v_g}{\zeta_{ag} \delta} \left\{ (a_2 - a_1) \times \right. \right.$$
$$\left. \left. \left(1 - e^{-z/\delta} \cos \frac{z}{\delta} \right) + e^{-z/\delta} (a_1 + a_2) \sin \frac{z}{\delta} \right\} \right]. \tag{6.23}$$

Taking the limit of Eq. (6.23) as $z \to \infty$ gives the vertical velocity at large heights,

$$w(r, \infty) = \frac{1}{r} \frac{\partial}{\partial r} \left[\frac{r K v_g}{\zeta_{ag} \delta} (a_2 - a_1) \right], \tag{6.24}$$

i.e., the vertical velocity at the top of the boundary layer.

In the special case, with the Ekman approximation ($\zeta_{ag} = f$, $\xi_g = f$), constant drag coefficient and constant K, δ, a_1, and a_2 are independent of radius, whereupon

$$w(r, \infty) = \frac{K \zeta_g}{f \delta} (a_2 - a_1). \tag{6.25}$$

Since $a_1 < 0$, $a_2 > 0$, $w(r, \infty)$ is linearly proportional to the vertical vorticity, ζ_g, above the boundary layer.

Given a radial profile of $v_g(r)$ such as one of those shown in Fig. 6.2, together with values for K, C_D, and f, it is possible to calculate the full boundary-layer solution $(u(r, z), v(r, z), w(r, z))$ on the basis of Eqs. (6.15), (6.16), and (6.23). For illustration purposes, we choose typical values of the foregoing parameters: $K = 50$ m^2 s^{-1}, $C_D = 2.0 \times 10^{-3}$, and $f = 5 \times 10^{-5}$ s^{-1}. The former two values are chosen based on the observations described in Chapter 7, where a refined representation of C_D in terms of near-surface wind speed is presented.

Fig. 6.2 shows reference wind profiles for a Rankine ($v \propto 1/r$) and a modified Rankine ($v \propto 1/r^{0.5}$) vortex outside the radius of maximum winds for comparison. While the narrow vortex profile decays more rapidly than the modified Rankine vortex outside a radius of 100 km, the decay is still slower than that for the Rankine vortex out to 400 km radius and hence the profile is inertially (centrifugally) stable for any latitude in this radial span (see Section 5.7.1 and Exercise 5.10).

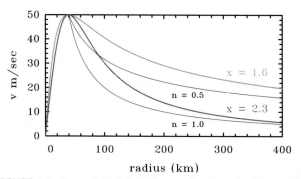

FIGURE 6.2 Tangential wind profiles as a function of radius used in the calculations later in this chapter. This profile has the form $v_g(r) = v_1 s/(1 + s^x)$, where $s = s_m r/r_m$, r is the radius, $r_m = 50$ km, and s_m and v_1 are constants chosen to make $v = v_{gm}$, the maximum tangential wind speed, when $r = r_m$. Red (mid gray in print version) curve has $x = 1.6$, blue (dark gray in print version) curve has $x = 2.3$. The thin black reference curves have the form $v = v_{gm}(r_m/r)^n$, where the exponent n equals either 0.5 or 1. These curves are discussed in the text. Adapted from Smith and Montgomery (2020). Republished with permission of Wiley.

For the foregoing parameter values, the boundary layer depth scale, $\delta = 447$ m and the surface values of u and v are $-0.24 v_g$ and $0.76 v_g$, respectively. The radial variation of the quantities v, I^2, a_1, and a_2 are shown in Appendix 1 for the two profiles of $v_g(r)$ in Fig. 6.2.

6.4.3 Vertical structure of the solution

The vertical profiles of $u(z)$ and $v(z)$ at a radius of 100 km, using the wind profile with $x = 1.6$, are shown in Fig. 6.3a and the hodograph thereof is shown in Fig. 6.3b. At this radius v_g is approximately 41 m s^{-1}. The profiles are qualitatively similar to those of the classical Ekman layer and the hodograph is similar to the spiral wind structure of the Ekman layer. The surface wind velocity, $\mathbf{v_s}$, makes an angle of about 30 deg to the left of the gradient wind and its magnitude, $|\mathbf{v_s}|$, is about 70% of that of the gradient wind, v_g. The wind vector $\mathbf{v}$ turns clockwise with height and the tangential component slightly exceeds the gradient wind in the altitude range from about 300 m to 1 km as it asymptotes back to the gradient wind at larger heights. We say that the tangential wind is supergradient where it exceeds the local gradient wind. The radial wind component at the surface is about 30% of v_g. On account of the nonlinear boundary condition, the ratio of the magnitude of both surface wind components to the gradient wind has a weak dependence on the magnitude of the gradient wind, itself.

6.4.4 Observed wind structure

Fig. 6.4 shows vertical profiles of storm-relative tangential (v) and radial (u) wind components in the eyewall composites for Atlantic Hurricane Isabel (2003) on three consecutive days of observations (12-14 September) as well as composites for many hurricanes. Despite the fact that the

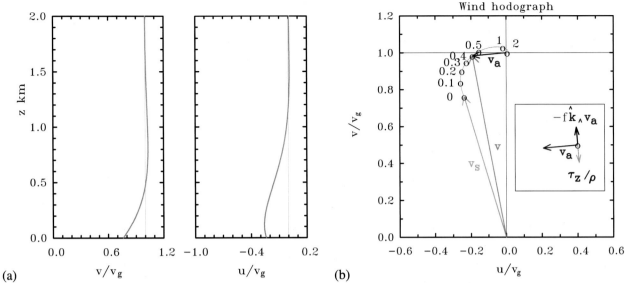

FIGURE 6.3 (a) Vertical profiles of tangential and radial wind components for the linear vortex boundary layer at a radius of 100 km using the radial profile of tangential wind profile with $x = 1.6$ shown in Fig. 6.2. (b) Hodograph of the two solutions showing the spiral of the wind vector. The surface wind vector $\mathbf{v_s}$ is shown in green (gray in print version), the wind vector at a height of 400 m, $\mathbf{v}$, is shown in red (mid gray in print version), and the ageostrophic wind vector, $\mathbf{v_a}$, in blue (dark gray in print version). Small circles indicate the wind vectors at heights of 100 m, 200 m, 300 m, 400 m, 500 m, 1 km, and 2 km. The inset shows the agradient wind vector at 400 m, the Coriolis force vector acting on an air parcel at this level, $-f\hat{\mathbf{k}} \wedge \mathbf{v_a}$, together with the frictional stress gradient at this level, $(1/\rho)\partial\boldsymbol{\tau}/\partial z$, denoted in the figure by $\boldsymbol{\tau}_z/\rho$.

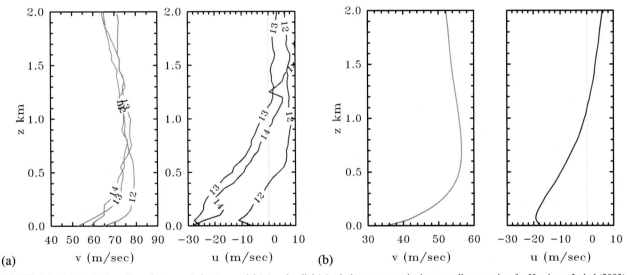

FIGURE 6.4 (a) Vertical profiles of storm-relative tangential (v) and radial (u) wind components in the eyewall composites for Hurricane Isabel (2003) on three consecutive days of observations (12-14 September). Numbers on curves denote the date. These data have a vertical resolution of 50 m. Data courtesy of M. A. Bell. From Smith and Montgomery (2014). Republished with permission of Wiley. (b) Vertical profiles of v and u in the eyewall composites of many hurricanes. Data courtesy of J. A. Zhang.

linear approximation cannot be justified in the inner core region of hurricanes, the profiles in Fig. 6.3 are somewhat similar to those observed, except that, because of the warm core structure of a hurricane, the tangential wind speed has a maximum at low levels. In fact, this maximum tends to be within the layer of strong inflow. An explanation for this behavior is given later in this chapter.

6.4.5 Radial-vertical structure

Fig. 6.5 shows radius-height cross-sections of the isotachs of u, v, and w below a height of 2 km for the two tangential wind profiles shown in Fig. 6.2. It shows also the radial variation of the boundary layer depth scale, δ. Note that δ decreases markedly with decreasing radius, while the inflow increases. The decrease of δ is simply related to the

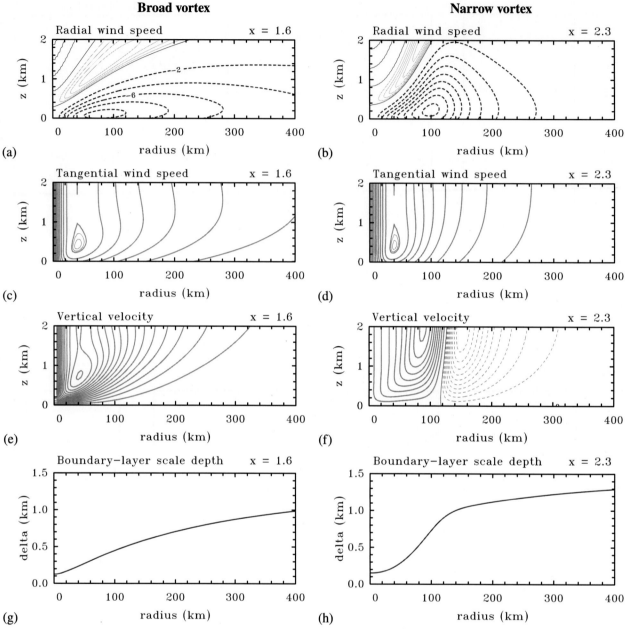

FIGURE 6.5 Isotachs of (a), (b) radial velocity u, (c), (d) tangential velocity v, and (e), (f) vertical velocity w in the r-z plane based on the solutions for the linear boundary layer given by Eqs. (6.15), (6.16) and (6.23) with the broad and narrow tangential wind profiles shown in Fig. 6.2. Left columns for the wind profile with $x = 1.6$, right columns with that for $x = 2.3$. Contour intervals: for u, 2 m s^{-1} for negative values (blue (dark gray in print version) contours), 0.1 m s^{-1} for positive values (thin red (mid gray in print version) contours); for v, 5 m s^{-1} for values < 50 m s^{-1} (red (mid gray in print version) contours), 0.5 m s^{-1} for values > 50 m s^{-1} (thin blue (dark gray in print version) contours); for w, when $x = 1.5$, 0.01 m s^{-1} for positive values (red (mid gray in print version) contours), 0.005 m s^{-1} for negative values (thin blue (dark gray in print version) contours), when $x = 2.3$, 0.05 m s^{-1} for positive values (red (mid gray in print version) contours), 0.01 m s^{-1} for negative values (thin blue (dark gray in print version) contours). (f), (g) Corresponding radial variation of boundary-layer scale depth, $\delta(r)$ in km. From Smith and Montgomery (2020). Panels (e) and (f) from Smith and Montgomery (2021a). Republished with permission of Wiley.

radial increase in the inertial stability parameter with decreasing radius, but it ignores the likely increase in eddy diffusivity as the wind speed (and hence vertical wind shear) increase.

With the broad wind profile ($x = 1.6$), the maximum inflow occurs at a radius of about 85 km, 45 km beyond the radius, r_m of maximum gradient wind speed (panel (a)). There is a region of weak outflow above the inflow layer with the maximum occurring at a similar radius to that of the maximum inflow. The tangential flow is slightly super-gradient in a region near the radius of maximum gradient wind r_m (panel (c)).

For $x = 1.6$, the vertical velocity (panel (e)) is positive at all radii and heights, but the maximum occurs at a radius close to r_m (panel (c)), and at an altitude of about 750 m, which is near the top of the inflow layer at this radius (panel (a)). The vertical velocity at a height of 2 km peaks also at a radius close to r_m and progressively declines beyond this radius. The boundary layer depth scale increases from just over 100 m near the rotation axis to about 1 km at $r = 400$ km (panel (g)).

With the narrower wind profile ($x = 2.3$), the radial inflow is markedly stronger and somewhat deeper, but the maximum inflow occurs further out near a radius of 100 km (panel (b)). The tangential flow is again slightly supergradient in a region near the radius r_m (panel (d)). The vertical velocity is considerably stronger (note the larger contour spacing in panel (f) compared with that in panel (e)) and the maximum vertical velocity occurs significantly further outwards, beyond 80 km radius, and at a significantly larger altitude, above 2 km. The slope of the region of ascent has greatly increased. Now there is subsidence beyond a certain radius and the strongest subsidence is confined radially and is relatively close to the region of maximum ascent.

With $x = 2.3$, the boundary layer depth scale increases more sharply with radius inside a radius of 120 km than for the broad profile, starting at 150 m near the rotation axis and reaching 1.3 km at $r = 400$ km (panel (h)).

The fact that the maximum vertical velocity peaks away from the axis for both tangential wind profiles suggests that the boundary layer plays a role in determining the radial location of the eyewall clouds, thereby providing a plausible explanation for the occurrence in a tropical cyclone of a central region free of deep clouds.

6.4.6 Interpretation, torque balance

Recall from Section 6.4.1 that the linear approximation is an expression of a local force balance in a situation where the material acceleration of air parcels is negligibly small. Therefore, one cannot appeal to the material acceleration terms in Newton's second law to "explain" the differences in behavior for the broad and narrow profiles. Any interpretations of the flow behavior must hinge on the assumption of force balance, which, of course, is reflected in the structure of the solution for the velocity components in Eqs. (6.15), (6.16), and (6.23).

Using the analysis in Section 2.2, the frictional stresses per unit area in the tangential and radial directions, $\tau_{\lambda z}$ and τ_{rz}, are given by the formulae

$$\tau_{\lambda z} = \rho K \frac{\partial v'}{\partial z}, \quad \tau_{rz} = \rho K \frac{\partial u}{\partial z}, \quad (6.26)$$

where v' and u are given by the formulae (6.15) and (6.16), respectively. The profiles of these stresses are shown in Fig. 6.6. Also from Section 2.2, the components of force per unit mass are just the vertical gradients of these stresses divided by density and, to a reasonable first approximation density can be treated as constant across the boundary layer. In the generalized Ekman layer, these are just the forces required to balance the generalized Coriolis forces as expressed, of course, by Eqs. (6.12) and (6.13).

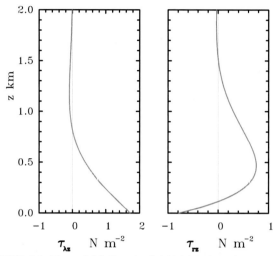

FIGURE 6.6 Tangential (left) and radial (right) horizontal stresses, $\tau_{\lambda z}$ and τ_{rz} as functions of height corresponding to the boundary layer wind profiles in Fig. 6.2.

Clearly, each stress component changes sign where the corresponding velocity component has a maximum or minimum (cf. Fig. 6.3a). One might argue that the boundary layer extends to some height at which the stress is small compared with the surface stress and has begun to asymptote to zero. For the tangential stress component, one might take this height to be about 800 m (or 1.8δ, where $\delta = 447 m$ is defined in Subsection 6.4.2), but for the radial component it is somewhat higher, say 1500 m (or 3.4δ) (see also Fig. 6.6). *With this criterion, the boundary layer depth in the radial direction would be about twice as large as that in the tangential direction.* This difference in depths is seen also in the observed wind profiles below the eyewall in Category 5 Hurricane Isabel (2003) (Fig. 6.4a) and in the eyewall composites of many hurricanes (Fig. 6.4b).

6.4.7 Factors determining the inflow and vertical motion

In the tangential wind direction, the force balance expressed by Eq. (6.13) is between the generalized Coriolis force, $-\zeta_{ag}u$ (minus the generalized Coriolis acceleration), which is positive, and the downward diffusion of tangential momentum, which is negative. This balance is sometimes referred to as *torque balance* when the equation is multiplied by the radius. It follows that, *in the linear theory*, the radial flow is determined by the tangential (*sic*) momentum equation. As discussed later, this generalized Ekman balance has led a number of authors[3] to erroneously argue that *the inflow in the nonlinear problem* is determined also by torque balance, when, in fact, *the inflow is determined by integrating the nonlinear radial acceleration*, $u\partial u/\partial r + w\partial u/\partial z - v'^2/r$, along the air parcel trajectories. Using Eq. (6.7), the radial acceleration is equal to the generalized Coriolis force, $\xi_g v'$, plus the frictional force. Because $\xi_g v'$ is a leading-order measure of the degree of gradient wind imbalance, we refer to it here as *the agradient force*.

At a given radius, the only parameter in the linear solution that contains information about *the local radial variation* of the flow is the absolute vorticity, $\zeta_{ag}(r)$. In conjunction with ξ_g, ζ_{ag} enters in determining the amplitude of the radial velocity in Eq. (6.16) through the factor $\chi v_g(r)$, where $\chi = (\xi_g/\zeta_{ag})^{1/2}$. Moreover, ζ_{ag} and ξ_g determine the inertial stability I^2, which, in turn, is a parameter involved in determining the boundary layer depth scale δ. It is only indirectly through the dependence of δ on I^2 that the inertial stability appears in the solution for the radial flow.

Since the coefficients $a_1(r)$ and $a_2(r)$ depend on δ and therefore on I^2 through their dependence on $v(r)$, it is difficult to discern the precise mathematical dependence of the radial inflow on I^2 because of the height dependence of $a_2 \cos(z/\delta) - a_1 \sin(z/\delta)$ in the formula for u in Eq. (6.16). Notwithstanding this dependence, the radial profiles of χ and $\chi v_g(r)$ for the two vortex profiles shown in Fig. 6.2 help to provide an understanding of the different structure of the radial flow seen in panels (a) and (b) of Fig. 6.5.

The profiles of χ and $\chi v_g(r)$ are compared in panels (a) and (b) of Fig. 6.7, respectively. First note that as $r \to 0$, ξ_g is dominated by $2v_g/r$ and ζ_{ag} is dominated by ζ_g so that $\chi \approx 1$. In contrast, as $r \to \infty$, both quantities are dominated by f, so that, again, $\chi \to 1$. In both vortex profiles, χ exceeds unity for all other radii, but whereas for the broad vortex profile with $x = 1.6$, the radial profile of χ is comparatively flat, for the narrower profile with $x = 2.3$, it has a sharp peak near a radius of 120 km. This peak is close to the radius of minimum ζ_{ag}, the magnitude of this minimum being smaller for the sharper profile on account of the smaller minimum of ζ_{ag}. The radial profiles of $\chi v_g(r)$ in panel (b) show a sharper peak for the narrower vortex profile also, the peak being located at a radius of 100 km, compared with only 60 km for the broader peak of the broader vortex profile.

The linear solution (Fig. 6.5) shows that the maximum radial inflow is stronger for the narrower vortex profile (a little over 16 m s^{-1} compared with a little over 10 m s^{-1}). This property is succinctly captured by the pre-factor, $\chi v_g(r)$, plotted in Fig. 6.7b. The stronger radial inflow at radii beyond r_{vmax} for the broader vortex profile (cf. Fig. 6.5) is

3. See Smith and Montgomery (2020).

(a)

(b)

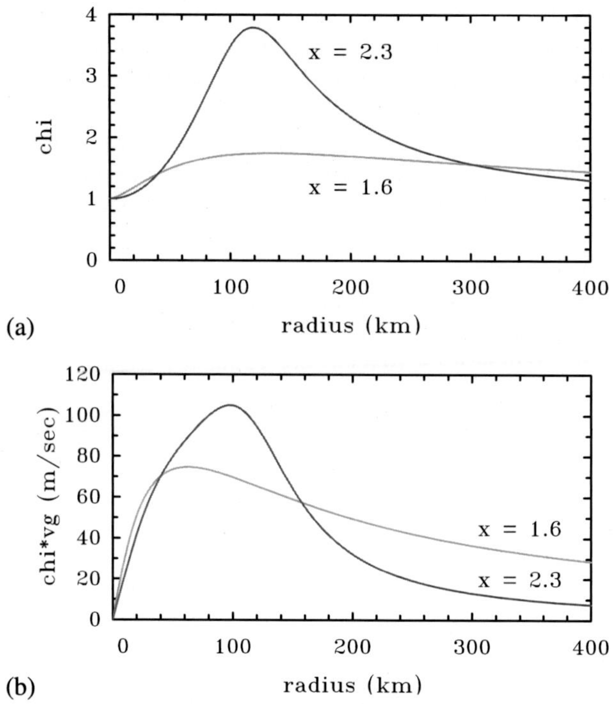

FIGURE 6.7 Radial variation of (a) χ and (b) $\chi v_g(r)$ (units: m s^{-1}) for the two vortex profiles in Fig. 6.2. Curves for profile exponent $x = 1.6$ in red (mid gray in print version), for $x = 2.3$ in blue (dark gray in print version). From Smith and Montgomery (2020). Republished with permission of Wiley.

captured also by the pre-factor $\chi v_g(r)$. If the boundary layer inflow was controlled primarily by the inertial stability, the radial inflow would be weaker for the broader profile. Precisely the opposite behavior is found!

6.4.8 Dependence on vortex size

At first sight, it might be tempting to attribute the more radially-confined pattern of vertical flow for the narrow vortex seen in Fig. 6.5f to a larger inertial stability. However, the narrower wind profile has a mostly smaller inertial stability than the broader profile, a fact that is reflected in a mostly deeper vertical depth scale in this case (compare panels (h) and (g) of Fig. 6.5). Thus, the larger radial confinement seen in panel (f) compared with panel (e) cannot be attributed to the differences in inertial stability. Rather, they are shown above to be due largely to the pre-factor χv_g, but partly also to the radial variation of δ.

To quantify further the dependencies on vortex size, Fig. 6.8 summarizes the changes in various metrics of the linear boundary layer solution as the imposed vortex profile becomes narrower, i.e., as the exponent x increases. Panel (a) shows that as x increases, both the maximum inflow u_{min} and maximum ascent w_{max} increase. The increase is slow at first, but begins to increase rapidly as x exceeds about 2 in the case of w_{max} and about 2.2 in the case of

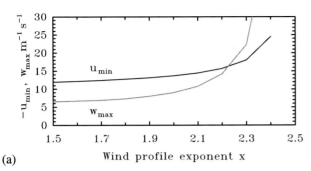

(a)

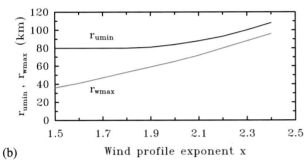

(b)

FIGURE 6.8 Variation of (a) u_{min} (units: m s^{-1}) and w_{max} (units: cm s^{-1}), and (b) r_{umin} and r_{wmax} in km with wind profile exponent x. Curves for u_{min} and r_{umin} in blue (dark gray in print version), w_{max} and r_{wmax} in red (mid gray in print version). From Smith and Montgomery (2020). Republished with permission of Wiley.

u_{min}. The radius of maximum inflow r_{umin} remains almost constant with a value of 80 km for $x \le 1.9$ after which it begins to increase reaching a value of 108 km for $x = 2.4$, shortly before the vortex profile becomes inertially unstable for a latitude of 20°. In comparison, the radius of maximum ascent, r_{wmax}, steadily increases with x at a rate that is approximately linear from a value of only 36 km for $x = 1.5$ to a value of 96 km for $x = 2.4$.

6.4.9 Supergradient winds in the linear solution

In Subsection 6.4.5 we showed that the tangential flow is slightly supergradient (i.e., $v' > 0$) in a region near the radius of maximum gradient wind r_m. In the linear boundary layer solution, in regions of supergradient winds, the agradient force in the radial momentum equation is radially outwards, i.e., $\xi_g v' > 0$, and this force is exactly balanced by the upward diffusion of negative radial momentum, i.e., $\partial \tau_{rz}/\partial z < 0$, where $\tau_{rz} = K \partial u/\partial z$ is the radial stress at height z (see Eq. (6.12)). In turn, the generalized Coriolis force associated with the negative radial momentum u is balanced by the downward diffusion of tangential momentum as represented by the second term on the right-hand side of Eq. (6.13).

6.4.10 Exercises

Exercise 6.1. The scale analysis of the u- and v-momentum equations in Table 6.1 shows that for small Rossby number ($Ro \ll 1$), there is an approximate balance between the net Coriolis force and the vertical diffusion of momentum. If K is assumed to be a constant, this balance is expressed by the equations:

$$f(v_g - v) = K \frac{\partial^2 u}{\partial z^2} \tag{6.27}$$

and

$$fu = K \frac{\partial^2 v}{\partial z^2}. \tag{6.28}$$

Show that:

(a) With the substitution $V_E = v + iu$, where $i = \sqrt{(-1)}$, Eqs. (6.27) and (6.28) reduce to the single differential equation

$$K \frac{d^2 V_E}{dz^2} - if V_E = -if V_{Eg}, \tag{6.29}$$

where V_{Eg} is the gradient wind above the boundary layer.

(b) Eq. (6.29) has the particular integral $V = V_{Eg}$ and two complementary functions proportional to $\exp(\pm(1 - i)z/\delta)$, where $\delta = \sqrt{(2K/f)}$.

(c) The solution of Eq. (6.29) that remains bounded as $z \to \infty$ has the form

$$V_E = V_{Eg}[1 - A \exp(-(1 - i)z/\delta),] \tag{6.30}$$

where A is a complex constant determined by a suitable boundary condition at $z = 0$.

(d) For a laminar viscous flow, the no-slip boundary condition at $z = 0$ gives $A = 1$. This assumption leads to the classical *Ekman solution*.

Exercise 6.2. Verify the expression for $u(z)$ in Eq. (6.16).

Exercise 6.3. Verify Eq. (6.23) for the vertical velocity at height z in the steady linear boundary layer solution.

Exercise 6.4. Show that in the classical Ekman layer solution, the magnitude of the maximum radial inflow and the vertical velocity at large heights both decrease with increasing latitude. Interpret these results physically.

Exercise 6.5. Show that the substitution of (6.15) and (6.16) into the boundary condition (6.17) leads to the following pair of algebraic equations for a_1 and a_2:

$$a_2 + a_1 = -va_2\sqrt{X},$$
$$a_2 - a_1 = v(1 + a_1)\sqrt{X}, \tag{6.31}$$

where $X = (1 + a_1)^2 + \chi^2 a_2^2$, $v = C_D R_e$, $\chi = I/\zeta_{ag}$, and $R_e = V\delta/K$. Here, R_e is a Reynolds' number for the boundary layer.

Show further that the constants a_1 and a_2 lie on a circle in the (a_1, a_2) with center at $(-\frac{1}{2}, -\frac{1}{2})$ and radius $1/\sqrt{2}$. [Hint: divide the two components of Eq. (6.31).]

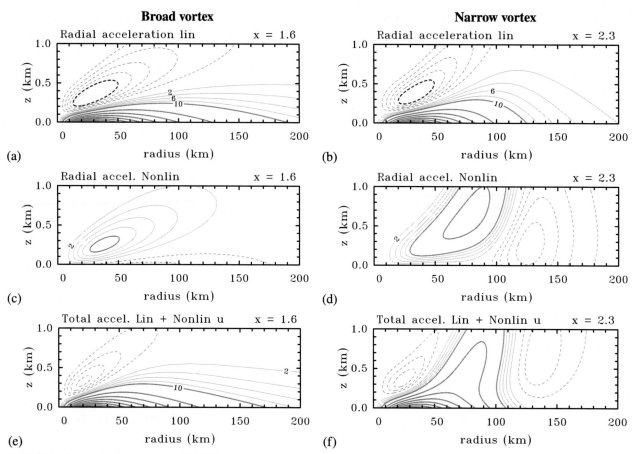

FIGURE 6.9 Isotachs in the $r - z$ plane of (a)(b) linear radial acceleration $-\xi_g v'$ in Eq. (6.7), (c), (d) nonlinear radial acceleration $u \partial u / \partial r + w \partial u / \partial z - v'^2 / r$, with the nonlinear term calculated from the linear solution to Eqs. (6.12) and (6.13), (e), (f) total radial acceleration (linear + nonlinear) $u \partial u / \partial r + w \partial u / \partial z - v'^2 / r - \xi_g v'$, continued overleaf (g), (h) linear tangential acceleration $\zeta_{ag} u$ in Eq. (6.8), and (i), (j) nonlinear tangential acceleration $u \partial v' / \partial r + w \partial v' / \partial z + u v' / r$ with the nonlinear term calculated from the linear solution to Eqs. (6.12) and (6.13), (k), (l) total tangential acceleration (linear + nonlinear) $u \partial v' / \partial r + w \partial v' / \partial z + u v' / r + \zeta_{ag} u$. Left columns for the tangential wind profile in Fig. 6.2 with $x = 1.6$, right column with the profile for $x = 2.3$. Contour intervals for radial terms: 5 m s^{-1} h^{-1} (thick contours), 1 m s^{-1} h^{-1} (thin contours); for tangential terms: 2 m s^{-1} h^{-1} (thick contours), 0.5 m s^{-1} h^{-1} (thin contours). Positive values red (mid gray in print version), solid; negative values blue, dashed. From Smith and Montgomery (2021a). Republished with permission of Wiley.

By setting $(a_1, a_2) = (\frac{1}{\sqrt{2}} \cos \theta - \frac{1}{2}, \frac{1}{\sqrt{2}} \sin \theta - \frac{1}{2})$, show that θ satisfies the equation:

$$F(\theta) = [(\sin \theta - c)^2 - (\cos \theta - c)^2] + v^2 \left[(\cos \theta + c)^2 + \chi^2 (\sin \theta - c)^2 \right] \times (\cos \theta + c)(\sin \theta - c) = 0, \quad (6.32)$$

where $\chi = I / \zeta_{ag}$. [Hint: multiply the two components of Eq. (6.31).]

Note that solving the foregoing equation for the angle θ provides a way of solving Eqs. (6.15) and (6.16) with the full nonlinear boundary condition (6.17).

6.4.11 Limitations of linear theory

According to Zhang et al. (2023), the foregoing linear solution offers useful zero-order insights into the kinematic structure of observed boundary layers for broad and narrow cyclones. By construction, the linear solution should provide a useful quantitative approximation in the outer part of the vortex where the nonlinear terms are small compared to the linear terms. While providing a qualitatively correct picture of the frictionally-induced convergence in the boundary layer, some reflection on the scale analysis of Section 6.2 would suggest that the linear approximation may become poor quantitatively in the inner-core region of a tropical cyclone strength vortex because the nonlinear acceleration terms may not be ignored: indeed, they may even dominate the linear terms.

Some effects of the nonlinear acceleration terms may be illustrated by estimating these terms from the linear solution and comparing their magnitude and structure with the linear terms, themselves.

Fig. 6.9 shows the linear and the sum of the nonlinear acceleration terms in the radial and tangential momentum

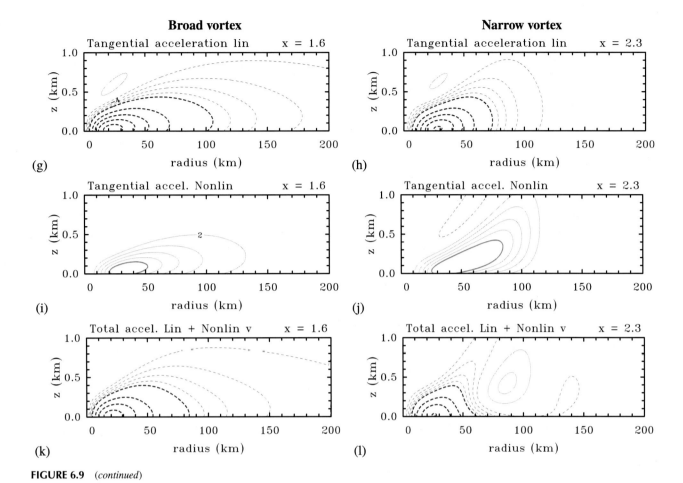

FIGURE 6.9 (*continued*)

equations, (6.7) and (6.8) respectively, for the broad and narrow gradient wind profiles shown in Fig. 6.2. The nonlinear acceleration terms on the left-hand side of each of these equations calculated from the linear solutions [Eqs. (6.15), (6.16), and (6.23)].

The linear radial acceleration, $-\xi_g v'$, in panels (a) and (b) shows a radially-outward acceleration at low levels with maximum values at the surface at radii between 20 and 40 km. This positive acceleration in the inertial frame is equivalent to an agradient force in the rotating frame that is of the opposite sign and in the linear solution is exactly balanced by the radial frictional force, which, for example, in Fig. 6.6 (right) is positive below approximately 450 m at 100 km. Therefore, in this solution, the positive acceleration cannot be interpreted as decelerating the boundary-layer inflow (Subsection 6.4.6, cf. also Subsections 5.8.2.3). At larger heights, the radial acceleration is negative with a minimum value between 20 and 40 km radius and an altitude between 200 and 400 m.

The nonlinear radial acceleration, $u\partial u/\partial r + w\partial u/\partial z - v'^2/r$, in panels (c) and (d) is positive in the innermost region and negative beyond. For the broad gradient wind profile ($x = 1.6$) this term is somewhat smaller in magni-

tude than the linear acceleration term, but for the narrow profile ($x = 2.3$), its magnitude is much larger in the region around 100 km radius and 500 m height and it makes a non-negligible contribution to the total radial acceleration in panel (f).

The linear tangential acceleration, $\zeta_{ag}u$, in panels (g) and (h) is mostly negative, but in each case there are small positive values aloft. These are most noticeable above 400 m in height and inside a radius of 50 km. As noted in Section 6.4.6, in the linear solution, the negative acceleration corresponds to a positive generalized Coriolis force that, in turn, is balanced by a negative frictional drag.

The nonlinear tangential acceleration, $u\partial v'/\partial r + w\partial v'/\partial z + uv'/r$, in panels (i) and (j) is positive for both the broad and narrow vortex profiles and it is comparatively larger in the case of the narrow vortex. In neither case is its magnitude negligibly small compared with the linear solution and the nonlinear acceleration makes a measurable contribution to the total acceleration in panels (k) and (l).

Significantly, the nonlinear acceleration terms in both the radial and tangential directions are positive compared with the corresponding linear acceleration. In the radial direction, the nonlinear acceleration adds to the linear ac-

celeration at small radii and reduces it at larger radii, while in the tangential direction, the positive nonlinear acceleration reduces the frictional deceleration of tangential winds in the lower part of the boundary layer. For these reasons one might expect the linear solution to produce a weaker and broader inflow than a corresponding nonlinear solution and thereby a weaker and less concentrated region of ascent at inner radii. Not surprisingly, the main features of the radial and tangential acceleration terms are more radially confined for the narrow vortex profile.

Since the foregoing estimates of the nonlinear terms are based on the linear solution, they can be expected to underestimate the actual contribution in a nonlinear boundary layer model due to the quadratic nature of the nonlinearity (see e.g., Figs. 2-4 in Smith and Montgomery, 2008).

6.5 A nonlinear slab boundary layer model

In the previous section we showed that the linear boundary layer model is not a consistent approximation for the inner core region of a tropical cyclone vortex because the magnitude of nonlinear terms that it neglects are not small compared with the linear terms, themselves. The simplest model that includes the nonlinear terms is the so-called "slab model" in which the vertical structure of the boundary layer is averaged.

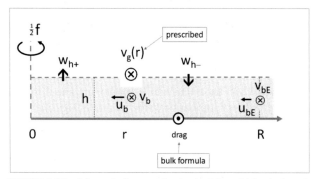

FIGURE 6.10 Schematic diagram of a simple slab model for the boundary layer of a stationary vortex. See text for discussion.

The structure of this slab model is shown schematically in Fig. 6.10. The vortex is assumed to be stationary on an f-plane and the flow at the top of the layer is assumed to be steady, axisymmetric and in gradient wind balance with tangential wind speed $v_g(r)$ and zero radial wind speed. The boundary layer is taken to be homogeneous *and to have uniform depth h*. The vertically-averaged radial and tangential wind speed components in the boundary layer are denoted by u_b and v_b. It is straightforward to include a *prescribed* variation of boundary layer depth as well as a formulation of thermodynamics,[4] but we focus here solely on the dy-

namical aspects. As in Section 6.4, a bulk formula is used to represent the effects of surface drag.

With the foregoing assumptions, we integrate the boundary layer equations with respect to depth to obtain a set of coupled ordinary differential equations governing the radial variation of u_b and v_b, together with an algebraic equation for w_h. These equations may be integrated inwards from some large radius, R, at which $v_g(R)$ is sufficiently small that the boundary layer can be approximated by the slab form of the linear solution, or even the slab form of the Ekman layer. In the former case, it is assumed that, at radius R, the nonlinear acceleration terms may be neglected and, in the latter case, that the centripetal acceleration may be neglected also. In either case, at radius R, the slab boundary layer equations may be easily solved analytically for u_b and v_b. The details of the full nonlinear solution are given in the next subsections. The solutions for the linear and Ekman approximations are left as exercises.

6.5.1 The boundary layer equations

For the model configuration described above, the steady form of the boundary layer equations (Eqs. (6.7) and (6.8)) may be written in flux form as:

$$\frac{1}{r}\frac{\partial}{\partial r}(ru^2) + \frac{\partial}{\partial z}(uw) + \frac{v_g^2 - v^2}{r} + f(v_g - v) = -\frac{\partial}{\partial z}\overline{(u'w')}, \quad (6.33)$$

$$\frac{1}{r^2}\frac{\partial}{\partial r}(r^2 uv) + \frac{\partial}{\partial z}(vw) + fu = -\frac{\partial}{\partial z}\overline{(v'w')}, \quad (6.34)$$

$$\frac{\partial}{\partial r}(ru) + \frac{\partial}{\partial z}(rw) = 0, \quad (6.35)$$

where (u, v, w) is the azimuthal mean velocity vector in a stationary cylindrical coordinate system (r, λ, z) and prime quantities here denote turbulent fluctuations about this mean (see Section 2.6 for guidance in the derivation of Eqs. (6.33) and (6.34)). Multiplying Eqs. (6.33)-(6.35) by a suitable power of r and then taking the integral with respect to z from $z = 0$ to the top of the boundary layer, $z = h$, and assuming that h is a constant, we obtain:

$$\frac{d}{dr}\left(r\int_0^h u^2 dz\right) + [ruw]|_{z=h} + \int_0^h (v_g^2 - v^2)dz +$$

$$rf\int_0^h (v_g - v)dz = -r\overline{(u'w')}|_{z=h} + r\overline{(u'w')}|_{z=0}, \quad (6.36)$$

$$\frac{d}{dr}\left(r^2\int_0^h uv dz\right) + [r^2 vw]|_{z=h} + fr^2\int_0^h u dz =$$

$$-r^2\overline{(v'w')}|_{z=h} + r^2\overline{(v'w')}|_{z=0}, \quad (6.37)$$

$$\frac{d}{dr}\int_0^h ru dz + [rw]|_{z=h} = 0. \quad (6.38)$$

---

4. See Smith (2003).

Now

$$[ruw]|_{z=h} = ru_b w_{h_+} + rU w_{h_-},$$

where U is the radial component of flow above the boundary layer, taken here to be zero, $w_{h_+} = \frac{1}{2}(w_h + |w_h|)$, and $w_{h_-} = \frac{1}{2}(w_h - |w_h|)$, and h_+ and h_- refer to values just above and just below the top of the boundary layer. In the slab formulation, it is assumed that the horizontal velocity at the top of the boundary layer differs from that in the boundary layer. Note that w_{h_+} is equal to w_h if the latter is positive and zero otherwise, while w_{h_-} is equal to w_h if the latter is negative and zero otherwise. Carrying out the integrals in Eqs. (6.36)-(6.38)[5] and dividing by h gives the radial and tangential momentum equations in flux form

$$\frac{d}{dr}(ru_b^2) = -\frac{w_{h_+}}{h}ru_b - (v_g^2 - v_b^2)-$$

$$rf(v_g - v_b) - r\frac{\overline{(u'w')}|_{z=h}}{h} + r\frac{\overline{(u'w')}|_{z=0}}{h}, \quad (6.39)$$

$$\frac{d}{dr}(ru_b rv_b) - -\frac{w_{h_+}}{h}r^2 v_b - \frac{w_{h_-}}{h}r^2 v_g -$$

$$r^2 fu_b - r^2\frac{\overline{(v'w')}|_{z=h}}{h} + r^2\frac{\overline{(v'w')}|_{z=0}}{h}, \quad (6.40)$$

and the continuity equation

$$\frac{d}{dr}(ru_b) = -r\frac{w_h}{h}. \quad (6.41)$$

The latter equation may be written as

$$\frac{du_b}{dr} = -\frac{w_h}{h} - \frac{u_b}{r}. \quad (6.42)$$

In these equations, a subscript 'b' denotes a mean value through the depth of the boundary layer. Before we can proceed further, we must find a way to represent the turbulent flux terms on the right-hand-side of Eqs. (6.39)-(6.40). We examine this problem in the next subsection.

6.5.2 Representation of surface and top fluxes

The surface momentum flux terms in (6.39)-(6.40) are represented by an empirical bulk drag law

$$-\overline{(u', v')w'}|_{z=0} = C_D|\mathbf{u_s}|\mathbf{u_s}, \quad (6.43)$$

where again C_D is a drag coefficient and $\mathbf{u_s} = (u_s, v_s)$ is the wind vector at a height of 10 m. This boundary condition is the same as that in Eq. (6.17) and represents the rate of which momentum is lost to the surface, which equals the frictional drag imposed by the surface on the air above. As

---

5. In this procedure, vertical integrals of products of flow variables are replaced by the corresponding product of the boundary velocities multiplied by the uniform depth.

in Section 6.4.2, we assume here for simplicity that C_D has the constant value 2.0×10^{-3}, although, in reality it does vary with surface wind speed (see Chapter 7).

Implementation of this boundary condition requires the surface wind vector $\mathbf{u_s}$ to be expressed in terms of the mean boundary later wind vector $\mathbf{u_b}$. For example, one might assume that $(u_s, v_s) = (\alpha_u u_b, \alpha_v v_b)$, where α_u and α_v are typical values suggested by an examination of the velocity profiles from the linear solution in Fig. 6.3, from observed profiles like those shown in Fig. 6.4, or from numerical model solutions such as those discussed in Chapters 10 and 11. While the tangential wind component increases with height through the boundary layer, $\alpha_v < 1$, the magnitude of the radial wind component has a maximum near the surface and subsequently decreases with height so that $\alpha_u > 1$. Typical values of these surface wind modification factors are around 0.7 for α_v and around 1.7 for α_u.

For simplicity, in what follows we will take α_u and α_v to be equal to unity and ignore the turbulent momentum fluxes at the top of the boundary layer, even though, in reality, these may be a non-negligible fraction of those at the surface. More refined calculations are possible, but we do not explore these here.

6.5.3 The final equations

For any dependent variable η

$$\frac{d}{dr}(ru_b\eta) = ru_b\frac{d\eta}{dr} + \eta\frac{d}{dr}(ru_b) = ru_b\frac{d\eta}{dr} - \frac{w_h}{h}r\eta,$$

where η is either u_b or rv_b. Then the radial and tangential momentum equations, (6.39) and (6.40) become the

> **slab boundary layer momentum equations**
>
> $$u_b\frac{du_b}{dr} = \frac{u_b w_{h_-}}{h} - \frac{v_g^2 - v_b^2}{r} - f(v_g - v_b)$$
> $$- \frac{C_D}{h}(u_b^2 + v_b^2)^{\frac{1}{2}}u_b, \quad (6.44)$$
>
> $$u_b\frac{dv_b}{dr} = \frac{w_{h_-}}{h}(v_b - v_g) - u_b\left(\frac{v_b}{r} + f\right)$$
> $$- \frac{C_D}{h}(u_b^2 + v_b^2)^{\frac{1}{2}}v_b. \quad (6.45)$$

Dividing each of these equations by u_b, Eqs. (6.44) and (6.45) become the effective boundary layer equations that will be integrated radially in this section.

$$\frac{du_b}{dr} = \frac{w_{h_-}}{h} - \frac{(v_g^2 - v_b^2)}{ru_b} - \frac{f(v_g - v_b)}{u_b}$$

$$- \frac{C_D}{h}(u_b^2 + v_b^2)^{\frac{1}{2}}, \quad (6.46)$$

$$\frac{dv_b}{dr} = \frac{w_{h-}}{h} \frac{(v_b - v_g)}{u_b} - \left(\frac{v_b}{r} + f\right)$$
$$- \frac{C_D}{h}(u_b^2 + v_b^2)^{\frac{1}{2}} \frac{v_b}{u_b}. \qquad (6.47)$$

Eqs. (6.46)-(6.47) form a pair of coupled ordinary differential equations that may be integrated radially inwards to find u_b and v_b as functions of r, given values of these quantities at some large radius $r = R$, and assuming that $v_g(r)$ and w_h are known. The quantity w_h is obtained by making sure that du_b/dr predicted by Eq. (6.46) is consistent with that which enters the continuity equation, Eq. (6.42).[6] Then

$$w_h = \frac{h}{1+\alpha} \left[\frac{1}{u_b} \left\{ \frac{v_g^2 - v_b^2}{r} + f(v_g - v_b) + \frac{C_D}{h}(u_b^2 + v_b^2)^{\frac{1}{2}} u_b \right\} - \frac{u_b}{r} \right], \qquad (6.48)$$

where α is zero if the expression in square brackets is negative and unity if this expression is positive.

6.5.4 Starting conditions at large radius

We assume that at $r = R$, far from the axis of rotation, the boundary layer is in generalized Ekman balance so that u_b and v_b satisfy equations analogous to those for the linear boundary layer, i.e.,

$$\left(\frac{2v_g}{r} + f\right)(v_g - v_b) = -\frac{C_D v_g}{h} u_b, \qquad (6.49)$$

$$\left(\frac{v_g}{r} + f\right) u_b = -\frac{C_D v_g}{h} v_b. \qquad (6.50)$$

In essence, we have assumed that $v_g - v_b \ll v_g$, that the total wind in the friction term can be approximated by the gradient wind, and that the downward transport of mean horizontal momentum from above the boundary layer can be neglected. Eqs. (6.49) and (6.50) are linear algebraic equations and can be readily solved for u_b and v_b (see Exercise 6.6).

With the starting values for u_b and v_b determined by Eqs. (6.49) and (6.50), Eqs. (6.46)-(6.47) may be integrated inwards numerically, given the radial profile of tangential velocity at the top of the boundary layer, $v_g(r)$. In the solutions presented below, the integration is accomplished using a Runge-Kutta method (see e.g., Press et al. (2007), Section 17.1).

---

6. Eq. (6.46) is written in the form

$$u_b \frac{w_{h-}}{h} = u_b \frac{du_b}{dr} + \{\ldots\} \quad \text{and} \quad \frac{du_b}{dr}$$

is eliminated from this expression using (6.42). Note that if $w_h < 0$, $w_h = w_{h-}$, in which case $\alpha = 1$. If $w_h > 0$, $w_{h-} = 0$, in which case $\alpha = 0$.

6.5.5 Exercise

Exercise 6.6. Show that the linearized form of

$$\frac{v_g^2 - v_b^2}{r} + f(v_g - v_b) \quad \text{is} \quad \xi(v_g - v_b),$$

where $\xi = 2v_g/r + f$. Hence show that the linear form of the steady slab boundary layer equations, analogous to Eqs. (6.12) and (6.13), is

$$-\xi(v_g - v_b) = \frac{C_D}{h} v_g u_b,$$

$$-\zeta_{ag} u_b = \frac{C_D}{h} v_g v_b.$$

Show that these linearized slab boundary layer equations have the solution

$$(u_b, v_b) = \frac{C_D}{h} \frac{v_g}{1 + v^2}(\mu, 1),$$

where $\mu = C_D v_g/(h\zeta_{ag})$, $v = C_D v_g/(hI)$ and $I^2 = \xi \zeta_{ag}$.

6.5.6 Slab boundary layer solutions

Fig. 6.11 shows examples of solutions for the slab boundary layer using the broad and narrow tangential wind profiles shown in Fig. 6.2. The boundary layer depth is taken to be 1000 m, the latitude 20° and the starting radius $R = 500$ km. The solutions for the broad vortex profile (with $x = 1.6$) are shown in the left column while those for the narrow profile (with $x = 2.3$) are shown in the right column. The top panels show radial profiles of wind components in the boundary layer and at the top of it and the middle panels show vertical velocity at the top of the layer.

The general characteristics of both solutions are that, beyond a certain radius, about 90 km for the broad vortex and about 40 km for the narrow vortex, the tangential velocity, v_b, is subgradient, i.e., less than that at the top of the boundary layer, v_g. In addition, the radial velocity is inwards and appreciably less than v_g. Ultimately, as the radius decreases, the tangential velocity becomes supergradient ($v_b > v_g$). When this happens, the radial flow decelerates rapidly and, in both cases, u_b becomes zero at some finite radius. At this radius, the boundary layer equations are singular and the solution has to be terminated.

In the broad vortex case (panel (a)), the maximum inflow attains a value of nearly 20 m s^{-1} at a radius near 100 km, 60 km beyond the radius of maximum v_g, while, in the narrow vortex case (panel (b)), the maximum inflow slightly exceeds 30 m s^{-1}, close to the radius of maximum v_g. The development of supergradient winds in the boundary layer is an important feature that we discuss further in Section 6.6.

In the broad vortex case (Fig. 6.11c), the mean vertical motion at the top of the boundary layer, w_h, is positive at all

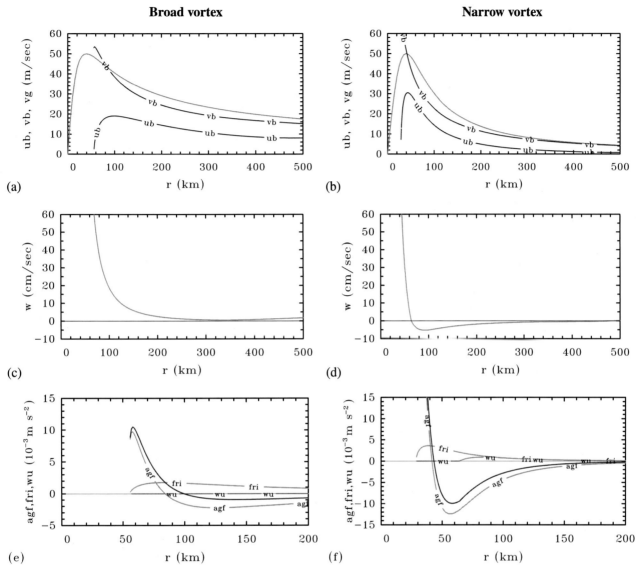

FIGURE 6.11 Radial profiles of radial (u_b) and tangential (v_b) velocity components in the boundary layer as well as the tangential wind speed above the boundary layer (v_g), in the slab boundary layer calculations for the (a) broad and (b) narrow vortex profiles shown in Fig. 6.2 [Units m s^{-1}]. (c) and (d) show the corresponding profiles of vertical velocity (w_h) at the top of the boundary layer [Units cm s^{-1}]. (e) and (f) show the force terms on the right-hand-side of the radial momentum equation for the two vortex profiles. Curves labeled 'pgf', 'fri', and 'wu' represent the net radial pressure gradient, frictional force and transport of radial momentum through the top of the boundary layer. The solid curve is the net total force. *Note the different scales on the abscissa in panels (e) and (f) compared with those in (a)-(d) and the different scales on the ordinates of panels (e) and (f).*

radii, increasing ever more rapidly as the radius decreases. In the narrow vortex case (Fig. 6.11d), w_h is negative, but mostly small beyond a radius of about 65 km, but it increases sharply as the radius decreases inside this radius. The behavior of w_h is similar in both cases to that in the linear boundary layer calculation in Section 6.4.

6.5.7 Physical interpretation

The behavior of the boundary layer in the slab model depends in a delicate way on the relative importance of various force terms in the radial and tangential momentum equa-

tions. The radial profiles of forces in the radial momentum equation (6.44) for the foregoing calculations are shown in the lower panels of Fig. 6.11. The first term in this equation, $u_b w_{h-}/h$, represents the dilution of radial momentum in the boundary layer because the radial flow above the boundary layer is zero and the last term represents effects of the frictional stress at the surface per unit depth. Generally, both of these terms, denoted in Figs. 6.11e, f by wu and fri, respectively, are positive and act to reduce the radial inflow. However, wu is nonzero only where there is subsidence into the boundary layer. When comparing the last two panels of

Fig. 6.11 with the first four, keep in mind the difference in scale on the abscissa.

The only term that can cause a radially-inward acceleration is the agradient force per unit mass, $(v_b^2 - v_g^2)/r + f(v_b - v_g)$ (denoted by *agf* in Figs. 6.11e, f), *which is negative only if the flow is subgradient, i.e.,* $v_b < v_g$*. If the flow is supergradient* $(v_b > v_g)$*, the agradient force is positive, reinforcing the frictional term*[7] *and leading to a rapid decline in* u_b *inside the radius of maximum inflow.*

For the typical tangential wind profiles used for the calculations here, v_b increases at first more slowly than v_g as r decreases from the starting radius so that the inward agradient force increases in magnitude. However, inside a certain radius (about 190 km for the broad vortex and about 70 km for the narrow vortex), v_b increases more rapidly than v_g when the *agf* decreases in magnitude and eventually reverses in sign the tangential winds become supergradient.

The radial frictional force *fri* is always positive and increases with decreasing radius until the radial inflow begins to decelerate strongly. For the broad vortex, $w_{h-} = 0$ as w_h is positive at all radii, but for the narrow vortex there is subsidence into the boundary layer beyond a radius of about 65 km. At radii beyond about 75 km, *wu* is comparable in magnitude with *fri*. The positive values of *wu* arise because air with zero radial momentum subsides into the boundary layer, reducing the inward momentum in the boundary layer, itself. In other words, the dilution of radial momentum by subsidence into the boundary layer is equivalent to a radially outward force.

For both vortices, the *net* radial force *agf − wu − fri* is negative at outer radii, but ultimately reverses sign as the radius decreases. The change occurs at a radius of about 100 km for the broad vortex and a little under 50 km for the narrow vortex. At this transition radius there is ascent out of the boundary layer in both vortices so that the term *wu* is zero.

When the net radial force becomes positive, the radial inflow decelerates, but v_b continues to increase as air parcels continue to move inwards. In fact, v_b increases inwards more rapidly than v_g, eventually becoming larger than v_g. At this point, which is a little inside the radius where *agf − wu − fri* becomes positive, *agf* reverses sign and increases rapidly with decreasing radius, becoming the dominant effect in decelerating the inflow. As a result of this rapid deceleration, the vertical velocity at the top of the boundary layer increases sharply.

In general, there are competing effects in the radial and tangential directions that determine the boundary layer behavior. First, the net inward pressure gradient force increases with the degree to which the tangential flow in the boundary layer is subgradient (i.e., with the magnitude of $v_g - v_b$). Thus the inward force increases as the effective frictional drag in the tangential direction increases to reduce v_b. Since the surface stress is distributed over the boundary layer depth, the frictional drag is inversely proportional to this depth. It follows that, *in the outer part of the vortex, shallower boundary layers favor lower tangential wind speeds because of the increased drag, but larger radial wind speeds, because they lead to a larger net inward pressure gradient force compared with the radial component of friction.*

The competing effect is that *large radial wind speeds favor large tangential wind speeds because then air parcels may penetrate rapidly to small radii, with minimal reduction of* v_b *by the frictional drag.* From a Lagrangian viewpoint, individual air parcels are spiraling inwards in the boundary layer. Stronger inflow means that fewer spirals are required for an air parcel to move a unit radial distance and therefore less drag is exerted on the air parcel in the tangential direction. This effect is most pronounced at inner radii because the length of spirals decreases with radius and the radial inflow is comparatively large. In mathematical terms, the rate at which the tangential wind, v_b is reduced by the frictional drag is represented by the last term in Eq. (6.47), which is inversely proportional to u_b. The effect explains why, inside a certain radius, v_b increases more rapidly than v_g, leading eventually to the development of supergradient winds. Conversely, the slower that air parcels spiral inwards, i.e., the smaller their u_b, the longer tracks they have along which friction can act to reduce v_b.

6.6 The boundary-layer spin up enhancement mechanism

The occurrence of the maximum tangential wind speed in the frictional boundary layer found in the nonlinear slab model (Section 6.5.6) may seem surprising at first sight. This is because friction would be expected to reduce the wind speed in that layer. As we recall from Subsection 6.4.9, even the linear theory produced a small amount of supergradient wind near the top of the strong inflow layer. However, the reason for the supergradient winds in the nonlinear theory is quite different and, under suitable circumstances, the supergradient winds can be quantitatively much more significant than those found in the linear theory. Indeed, as indicated in the previous subsection, the reasons for this occurrence are subtle and involve competing effects on account of the nonlinearity. An alternative way of understanding the occurrence of supergradient winds is as follows.

In the absence of frictional stresses, converging rings of air would conserve their absolute angular momentum, M, and spin faster (Section 5.6). Recall that $v = M/r - \frac{1}{2}fr$, so that for an air parcel spiraling inwards in the boundary layer, if the relative rate at which M is reduced by friction is less than the relative rate at which the parcel's radius decreases, then v will increase.

---

7. Note that regions of supergradient flow coincide with regions of ascent out of the boundary layer so that uw_{h-} is zero.

As explained above, it follows that the development of supergradient winds in the boundary layer requires a sufficiently rapid radial displacement of air parcels in the boundary layer, which in turn, requires sufficiently strong radial inflow. To determine whether or not supergradient winds occur requires one to perform a nonlinear boundary layer calculation. We showed that for the slab boundary layer calculation in Section 6.5.6, this is indeed the case.

We refer to the mechanism for amplifying the tangential wind in the boundary layer (relative to the gradient wind) as *the boundary-layer spin up mechanism*, since it explains how air parcels can acquire a tangential velocity in the boundary layer that is larger than the gradient wind at the same radius. However, this mechanism *does not* explain how the maximum tangential wind speed in the boundary layer increases with time. A spin up of the tangential wind above the boundary layer by the classical mechanism (Section 5.6) is generally required for the maximum tangential wind in the boundary layer to increase with time. In this sense, a more accurate description might be to refer to a boundary-layer spin up enhancement mechanism.

The boundary-layer spin up enhancement mechanism provides an explanation for the observations discussed in Chapter 1 that the maximum tangential wind occurs at low levels, near the top of the strong frictional inflow layer (e.g., Kepert, 2006b; Bell and Montgomery, 2008; Zhang et al., 2011b; Montgomery et al., 2014; Sanger et al., 2014), a feature found also in numerical modeling studies (see Chapters 10 and 11).

6.7 Limitations of the two boundary layer models

In this chapter we have examined two models for the boundary layer, a steady linear model that determines the vertical structure of the flow at each radius and a steady slab model that includes nonlinear terms, but represents only the vertically averaged properties of the boundary layer. Neither of these models is accurate in the sense that one would be comfortable to use them as part of an operational forecast model. However, they provide a useful framework for understanding important aspects of the boundary layer behavior in a rapidly-rotating vortex.

As discussed in Section 6.4.11, the linear solution is limited because it ignores the development of sharp radial gradients of all velocity components. Thus, the single radial scale of $R = 200$ km is no longer appropriate for determining the typical order of magnitude of the radial and tangential acceleration terms in the boundary layer in the scale analysis of Section 6.2. In particular, it was shown in Section 6.4.11 that, for narrow vortices, the linear boundary-layer solution becomes non-uniformly valid in the inner-core region and so it follows that the nonlinear terms cannot be ignored at high wind speeds.

A limitation of the slab model is the vertical averaging, which may introduce errors in estimating the nonlinear terms (Kepert, 2010a,b[8]). Moreover, when applying the surface boundary condition, one should use the wind at a height of 10 m and not the boundary-layer average wind speed.[9] Notwithstanding these limitations, both models illustrate important features of the boundary layer.

The scale analysis of Section 6.2 indicates that the boundary layer depth scales as $\sqrt{2K/I}$. Thus, if the diffusivity K does not depend on radius, the depth should decrease as the inertial stability based on the gradient wind, I^2, increases. While observations indicate that K increases with increasing wind speed (see Fig. 7.3), perhaps by a factor of 5-10 over wind speeds from 10 m s^{-1} to 100 m s^{-1}, I increases typically by two orders of magnitude, suggesting that the boundary layer depth should eventually decrease with decreasing radius. It is straightforward to modify Eqs. (6.46)-(6.49) to allow for a prescribed variation of boundary layer depth $h(r)$ (Smith and Vogl, 2008) and to assess the effect of a decrease in h with decreasing radius.

6.7.1 Advantages of the slab model

The slab boundary layer model illustrates one particularly important feature of the full steady boundary layer equations, namely their parabolic nature wherein information propagates radially inwards where there is inflow and radially outwards where there is outflow. The implication would be that if there is a region of deep convection removing air from low levels at some inner radii, the boundary layer beyond this radius, if governed by the boundary layer equations, cannot know about the removal. Thus boundary layer theory cannot be entirely valid in such regions. Further discussion of this issue will be presented in Section 10.14, where we introduce the important idea of boundary layer control.

Other attractive features of the slab boundary layer model are the relative ease with which thermodynamics quantities (Smith, 2003) and time dependence (Slocum et al., 2014) can be incorporated. Time-dependent numerical model simulations to be discussed in Chapters 10 and 11 show that there is an adjustment region where the flow exits the boundary layer and ascends into the eyewall. The adjustment has the form of a quasi-stationary centrifugal wave along the lower part of the eyewall.

The occurrence of supergradient winds in the inner core region is suggestive that boundary layer theory may have

---

8. We should say that we do not endorse all of the arguments in these two papers.
9. As noted previously, one could, for example, apply a reduction factor so that $v_s =$ (reduction factor) $\times v_b$, but this would be inappropriate for the radial velocity component at small radii where the maximum inflow is found to occur at or close to the surface as seen in Fig. 6.3.

problems in this region as the boundary layer would be expelling tangential momentum upwards that is not in balance with the pressure gradient above the boundary layer. However, boundary layer theory assumes that the horizontal wind above the boundary layer is prescribed and that it is in approximate gradient wind balance. These assumptions are too restrictive where there is ascent of supergradient tangential momentum out of the boundary layer. In essence, the top of the boundary layer must be regarded as an open boundary for the boundary layer, which means that the horizontal wind at this level should be determined by the boundary layer itself and not prescribed. This prescription would appear to be an over-constraint in the linear problem, but *is not necessary in the slab model* where the tangential velocity in the boundary layer can differ substantively from the gradient wind.

A hitherto un-appreciated property of the slab boundary layer model is that it provides a simple window into the nonlinear solutions that are possible under very realistic forcing from the overlying vortex. For example, recent work by Slocum et al. (2014) shows that the temporal slab equations possess a hyperbolic character and highly-localized (shock-like) solutions can readily emerge in the inner-core boundary layer. These solutions possess large radial gradients in radial inflow and tangential velocity associated with the development of supergradient winds near the radius of maximum wind of the overlying vortex.

In the simple case of an inviscid vortex, the nonlinear slab equations can be solved analytically and the solutions become singular in a finite time interval for both vorticity and convergence (hence vertical velocity) variables. When realistic surface friction and radial diffusion is added back in, the solutions are no longer singular, yet they retain near-singular characteristics of a vortex sheet and corresponding highly localized updraft maximum that would likely impart severe turbulence to reconnaissance aircraft that are unfortunate to be penetrating the storm within the boundary layer (see Slocum et al. (2014) for further details).

Such idealized predictions of the nonlinear slab model cannot be easily discounted. This is because low-level flight level observations, shown in Fig. 6.12, of the infamous flight of a NOAA Hurricane Hunter WP3D aircraft into rapidly intensifying Hurricane Hugo (1989) point to the existence of near singular structures in both the boundary layer vertical velocity and vertical vorticity (inferred from the horizontal wind speed profile).

The role of the nonlinear horizontal advection terms in the tropical cyclone boundary layer remains an important problem to understand more fully. Despite the limitations mentioned above, the slab model is arguably the simplest model for understanding the development of localized nonlinear structures in the inner-core region of a tropical cyclone vortex.

6.7.2 Limitations of boundary-layer theory in general

The region of the boundary layer inside the radius of maximum tangential wind speed, where there is a rapid deceleration of the radial inflow and strong ascent into the eyewall clouds, is one in which classical boundary-layer theory as presented in Section 6.2 becomes progressively invalid. For one thing, the relatively strong ascent out of the boundary layer requires an upward-directed perturbation pressure gradient force to drive it. As a result, the boundary-layer approximation that $\partial p'/\partial z = 0$ is no longer strictly valid. In fact, the presence of deep convection above the boundary layer will contribute also to the upward-directed perturbation pressure gradient force as a sort of "sucking effect", an effect that is not represented by the boundary-layer equations. Further discussion of this point is provided in Sections 11.5.3, 11.5.4, 16.1.6 and 16.2. For another thing, the effects of turbulent momentum transfer in the radial direction may become important also in this inner region.

In numerical models, in general, and in those for tropical cyclones in particular, horizontal diffusion of momentum is usually included, but with a much larger coefficient of turbulent diffusivity than that for vertical diffusion. This alone might make the radial diffusion terms that were neglected in the development of the boundary layer equations important. The much smaller radial scale of the eyewall updraft in comparison with the characteristic scale of flow variation in the outer region of the vortex would be an additional factor for elevating the importance of the radial diffusion terms.

In view of the complex flow structure where the inflow boundary layer terminates, the assumption of a constant diffusivity coefficient for characterizing horizontal turbulent momentum transport is unlikely to be appropriate. For this reason, the horizontal diffusivity of momentum in numerical models is generally related to the local mean straining motion. However, the appropriate magnitude is chosen without much observational guidance, but rather as an expedient to remove energy at scales close to the grid scale as a crude representation of the dissipation of turbulent kinetic energy at the viscous scale.

The effects of parameterized diffusion on the quasi-steady *axisymmetric* numerical simulations of tropical cyclones was examined in some detail by Rotunno and Bryan (2012). These authors carried out, in part, a scale analysis of the equations of motion with the horizontal diffusion terms included, confirming that these terms can play an important role in the region where the boundary layer is erupting into the eyewall. Their numerical calculations suggested that supergradient winds could be suppressed in the inner-core region of the vortex *if* the radial diffusion was sufficiently strong. These particular calculations do suppose that the radial diffusion is sufficiently strong in this region and assume also that the 'turbulent' eddies (asymmetric vortex waves,

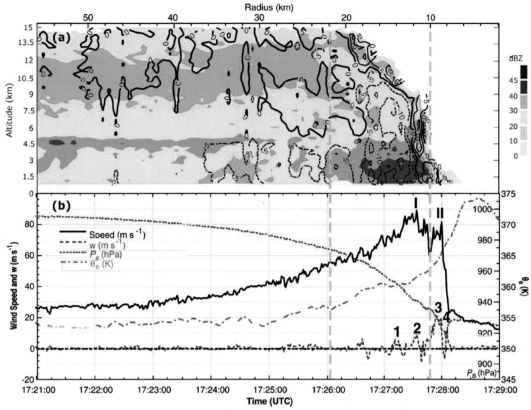

FIGURE 6.12 (a) Time-height cross section of vertical incidence tail radar reflectivity (dBZ) from a reconnaissance aircraft flying into Hurricane Hugo (1989) at an altitude of about 450 m during the period 1721-1728 UTC 15 September. Solid and dashed lines denote vertical velocity, w, and radar reflectivity is denoted by colors using the color scale on the right. (b) Time series plots of w, horizontal wind speed, local surface pressure P_s, and equivalent potential temperature θ_e for the same period UTC. Updrafts labeled 1, 2, 3, and 4 and wind speed peaks I and II are highlighted and described in detail in Marks et al. (2008). The thick vertical dashed lines in (b) approximately delineate the outer and inner radii of strong eyewall reflectivity maxima in the lower troposphere ($1 < z < 5$-km altitude). Distance along the abscissa in panel (a) is based on the aircraft position relative to the diagnosed center of circulation. From Marks et al. (2008). Republished with permission of the American Meteorological Society.

vortex structures, and small-scale turbulence) act diffusively and not counter-gradient.

We know from the idealized calculations shown in Chapter 4 that axisymmetrizing vorticity waves on a smooth vortex monopole can act in a counter-gradient manner. While the nondivergent barotropic model used there is a far cry from a real tropical cyclone vortex, we know from idealized three-dimensional simulations of an intensifying tropical cyclone that the resolved eddies possess counter-gradient structure in the inner-core region in association with the suction by convection, in and above the boundary layer, and attending axisymmetrizing vorticity structures in and around the eyewall updraft (Section 16.2, and Persing et al., 2013, their Section 6).

Accepting for the time being the down-gradient assumption,[10] first estimates of the horizontal diffusivity in major hurricanes using flight-level data from reconnaissance aircraft were obtained by Zhang and Montgomery (2012) (see Section 7.4).

6.7.3 Balanced boundary layer approximation

As noted in Section 6.4.1, the linear boundary layer expresses a balance between the generalized Coriolis force and frictional force. This generalized Ekman balance is a different type of balance to that discussed in Chapter 5. Nevertheless, the balance theory developed there provided the possibility to include a frictional force in the tangential momentum equation ($-\dot{V}$ in Eq. (5.62)), which appears as a forcing term $(\partial/\partial z)(\chi\xi\dot{V})$ on the right-hand-side of the Eliassen equation (5.67). This balance boundary-layer approximation has been used by many authors in the past (e.g., Ooyama, 1969; Emanuel, 1986), but it would seem to be at odds with the theory developed in the present chapter. For example, the reasons given for the existence of inflow in the surface boundary in Section 6.1 invoke the lack of balance to drive the inflow. Such an interpretation is supported by the non-vanishing agradient force $v'^2/r + \xi_g v'$ that would appear in the radial momentum equation, Eq. (6.5), if the corresponding acceleration term was placed on the right-hand-side of that equation.

---

10. We take up this issue again in Sections 11.5.3, 11.5.4 and 16.2.

It is not surprising that the assumption of *strict* gradient wind balance in the tropical cyclone boundary layer leads to results that are quantitatively poor (see e.g., Smith and Montgomery, 2008; Smith et al., 2008). Nevertheless, it turns out that the inclusion of a frictional force in the tangential momentum equation in balance theory leads to predictions that are qualitatively similar to those of more sophisticated treatments of the boundary layer and, while recognizing the intrinsic limitations of this approach, we will explore the role of friction in the context of a prognostic balance theory in Chapter 8.

6.8 Importance of the tropical cyclone boundary layer

The frictional boundary layer is a crucially important feature of a tropical cyclone as it provides a powerful coupling between the primary circulation and the secondary circulation. Moisture enters a hurricane from the sea surface and the radial distribution of moisture is strongly influenced by that of the boundary layer winds. Indeed, the boundary layer dynamics and thermodynamics determine the vertical transport of moisture and angular momentum out of the boundary layer and the radial distribution of these quantities on leaving the layer. In particular, the boundary layer exerts a constraint on the radial distribution of buoyancy to support deep moist convection in the developing/mature eyewall and in the rainbands exterior to the eyewall of the storm.

The theory presented so far in this chapter shows that, by itself, the boundary layer leads to vortex spin-down above the boundary layer. More precisely, if deep convection in the inner core region of a tropical cyclone is collectively too weak to remove all the mass that is converging in the boundary layer, there will be outflow just above the boundary layer and the tangential circulation at the top of the boundary layer will weaken. Clearly for vortex spin up to occur, deep convection needs to be strong enough not only to remove mass at the rate that is converging in the boundary layer, but to draw air inwards in the lower troposphere *above* the boundary layer as well. The role of deep convection in this regard will be addressed in several subsequent chapters. In particular, in Chapter 10 we introduce the notion of boundary layer control in the evolution of a tropical cyclone. Other aspects of boundary layer coupling with the flow above are discussed in Section 15.5.3.

6.9 Appendices

6.9.1 Appendix 1: radial variation of v, I^2, a_1, and a_2 in the linear boundary layer solution

Fig. 6.13 shows the radial variation of the quantities v, I^2, a_1, and a_2 that appear in the solution to the linear bound-

ary layer problem. These quantities are calculated with the parameter values given in Section 6.4.2 and the two vortex profiles shown in Fig. 6.2. It is seen that the parameter v, the drag coefficient multiplied by the boundary layer Reynolds number, lies within the range 0 to 1 (panel (a)), as do the profiles of a_1 and a_2 that appear in the solutions (6.15), (6.16) and (6.23) (panels (c) and (d)). The profiles of a_1 have the same qualitative behavior as those of v, but reversed in sign, starting with a value of zero at $r = 0$. For $x = 1.6$, the profile shows a broad minimum near $r = 220$ km, beyond which it slowly increases. For $x = 2.3$, the minimum of a_1 is more peaked, near $r = 120$ km, and it increases more rapidly as r increases. The a_2 profile is more peaked for $x = 2.3$ than for $x = 1.6$, but the maxima occur at similar radii. Since the coefficients a_1 and a_2 appear in the combination $a_2 \cos(z/\delta) - a_1 \sin(z/\delta)$ in the solution for $u(r, z)$ in Eq. (6.16), their individual contribution to $u(r, z)$ will vary with height. Note that a_1 and a_2 are both small in magnitude compared with unity (and consistent with the neglect of quadratic and higher order terms in the linear approximation of Section 6.4).

Fig. 6.13b shows the radial variation of the inertial stability, I^2, illustrating the rapid increase with decreasing radius inside the radius of maximum gradient wind and the fact that the broader vortex has a larger increase than the narrower vortex.

6.9.2 Appendix 2: what determines the vertical velocity in the linear boundary layer?

The formula for the vertical velocity (Eq. (6.24)) is sufficiently complex to obscure the main factors that determine the vertical velocity at the top of the boundary layer. The formula for $w(r, \infty)$ may be expanded out to give

$$w(r, \infty) = \underbrace{\frac{K \zeta_g}{\zeta_{ag} \delta}(a_2 - a_1)}_{w_1} + \underbrace{\frac{v_g}{\zeta_{ag} \delta}(a_2 - a_1)\frac{\partial K}{\partial r}}_{w_2}$$
$$- \underbrace{\frac{K v_g}{\zeta_{ag}^2 \delta}(a_2 - a_1)\frac{\partial \zeta_{ag}}{\partial r}}_{w_3} + \underbrace{\frac{K v_g}{\zeta_{ag}}\frac{d}{dr}\left(\frac{a_2 - a_1}{\delta}\right)}_{w_4}, \quad (6.51)$$

where ζ_g is the relative vorticity of the gradient wind. Here, for utmost generality, we have allowed the vertical turbulent diffusivity to be a function of r. The first term is familiar from Ekman theory itself (cf. Eq. (6.25)), in which ζ_{ag} is approximated by f, K is assumed to be constant and, for a no slip boundary condition, corresponding with the limit $C_d \to \infty$, $a_1 = -\frac{1}{2}$, and $a_2 = \frac{1}{2}$. In this limit, the terms w_2, w_3, and w_4 are all zero.

The second term on the right-hand-side of Eq. (6.51), w_2, is the contribution to $w(r, \infty)$ arising from the radial variation of turbulent diffusivity. This contribution is

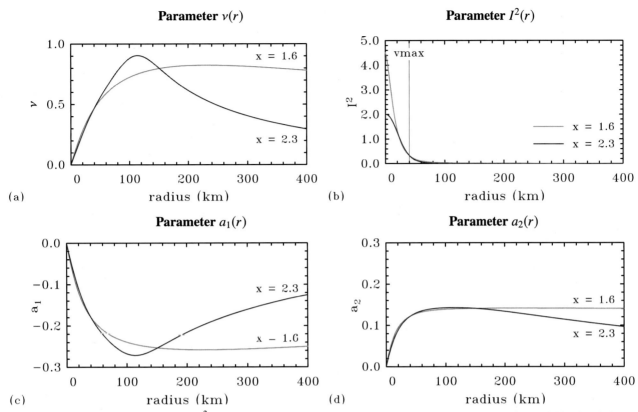

FIGURE 6.13 Radial variation of (a) v, (b) I^2, (c) a_1, and (d) a_2. Curves for $x = 1.6$ in red (mid gray in print version), for $x = 2.3$ in blue (dark gray in print version). The quantities v, a_1 and a_2 are dimensionless. One unit for I^2 corresponds to 1×10^{-5} s^{-2}. From Smith and Montgomery (2020). Republished with permission of Wiley.

zero for the calculations shown in Smith and Montgomery (2020) because K was taken to be constant for simplicity, but allowing K to vary radially might be quite important (see Section 6.9.2.1). The third term, w_3, is the contribution to $w(r, \infty)$ arising from the radial variation of relative vorticity of the gradient wind (note that, with f assumed constant, $\partial\zeta_{ag}/\partial r = \partial\zeta_g/\partial r$). The fourth term, w_4, contains the effects of radial changes in C_d through the dependence of a_1 and a_2 on C_d, which, in general depends on surface wind speed.[11] However, a_1 and a_2 depend also on v_g so that, even if C_d is taken to be constant as in the calculations in this chapter, w_4 depends on the radial gradient of v_g also.

Since the gradient wind profile for the broad vortex in Fig. 6.2 has positive relative vorticity ζ_g at all radii, Ekman theory would predict that $w(r, \infty) > 0$ at all radii also. This observation raises the question whether w_1 is the dominant term in the calculation of $w(r, \infty)$ for the generalized Ekman layer. To investigate this question we have computed the separate contributions to $w(r, \infty)$ from the terms w_1, w_3, and w_4 in the calculations relating to Fig. 6.2 (because K was assumed constant in these calculations, w_2 is zero).

These, together with $w(r, \infty)$, itself, are shown in the upper panels of Fig. 6.14.

It may be seen immediately that for both the narrow and broad radial profiles of gradient wind, the answer to the foregoing question is a resounding "no", at least in the inner-core region (specifically $r < 100$ km for the broad profile and $r < 150$ km for the narrow profile). In the case of the broad profile, w_3 and w_4 are opposite in sign at radii beyond about 30 km. At large radii, beyond about 300 km, they approximately cancel, whereupon $w(r, \infty)$ is well approximated by w_1. However, as the radius decreases, w_3 becomes larger in magnitude than w_4 and the sum $w_3 + w_4$ makes a significant contribution to $w(r, \infty)$, comparable in magnitude to w_1. The upshot is that the maximum in $w(r, \infty)$ is about twice as large as that of w_1 and it occurs at a somewhat larger radius, close to the radius of maximum gradient wind rather than well inside it. Note that the radius of maximum w_1 occurs well inside the radius of maximum gradient wind.

In contrast, for the narrow profile, w_3 and w_4 are both negative at radii beyond about 160 km and both are larger in magnitude than w_1 inside about 120 km.

---

11. The formulae for a_1 and a_2 are given in Eq. (6.22). The dependence of a_1 and a_2 on C_d arises from the dependence of v on C_d.

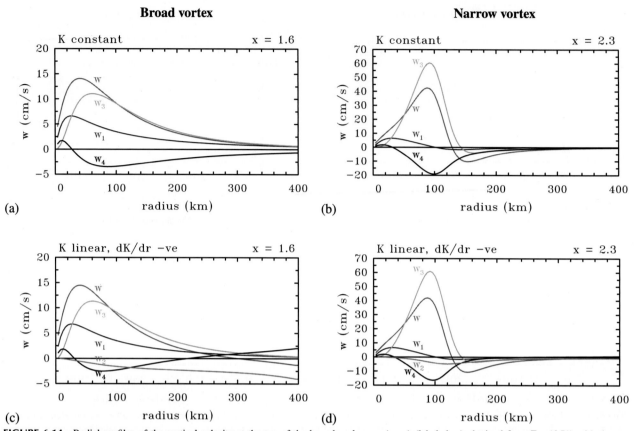

FIGURE 6.14 Radial profiles of the vertical velocity at the top of the boundary layer $w(r, \infty)$ (labeled w) obtained from Eq. (6.51) with the two tangential wind profiles shown in Fig. 6.1 together with the separate contributions from the terms w_1, w_3, and w_4. The values $x = 1.6$ and $x = 2.3$ refer to the broad and narrow gradient wind profiles, respectively. Panels (a) and (b) refer to the case where K is assumed a constant, while panels (c) and (d) refer to cases with a linear increase of turbulent diffusivity with decreasing radius as described in the text. From Smith and Montgomery (2021b). Republished with permission of Wiley.

6.9.2.1 Effects of radially-varying turbulent diffusivity

It is pertinent to ask how much the vertical velocity would be affected by the variations in the turbulent diffusivity with radius. These effects would lead to a nonzero contribution from w_2 in Eq. (6.51) as well as modifications to w_1, w_3, and w_4 through their dependence on K and δ, which depends also on K. While there is not much observational guidance in formulating such a variation, the observations that are available do point to a substantial increase in K with wind speed (see Chapter 7, Fig. 7.3). To illustrate the importance of such variations we show in the lower panels of Fig. 6.14 the vertical velocity at the top of the boundary layer and the contributions to it from the terms w_1-w_4 in two calculations where K increases linearly with decreasing radius. The slope is taken so that $K = 10 \ \text{m}^2 \ \text{s}^{-1}$ at radius 400 km and 50 $\text{m}^2 \ \text{s}^{-1}$ at radius 100 km. This postulated behavior is consistent with the observed dependence of K on wind speed shown in Fig. 7.3. The two calculations relate to the broad and narrow vortices used earlier. The boundary-layer

depth scale variations, $\delta(r)$, implied by the variation of K and ζ_{ag} are shown in Fig. 6.15.

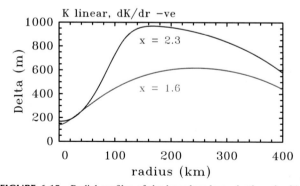

FIGURE 6.15 Radial profiles of the boundary layer depth scale, $\delta(r)$ with the two tangential wind profiles shown in Fig. 6.1 and a linear increase of turbulent diffusivity with decreasing radius as described in the text. The values $x = 1.6$ and $x = 2.3$ refer to the broad and narrow gradient wind profiles, respectively. From Smith and Montgomery (2020). Republished with permission of Wiley.

When K increases with decreasing radius (i.e., $dK/dr < 0$), so does the boundary layer depth scale, δ, at least at large radii (Fig. 6.15). Subsequently, as the rapid increase in ζ_{ag} in the denominator of the formula for δ begins to dominate the increase in K in the numerator, δ begins to decrease as r decreases. Since $a_1 < 0$ and $a_2 > 0$, all terms multiplying dK/dr in the expression for w_2 are positive implying that $w_2 < 0$. The effect is particularly noticeable for the broad profile (Fig. 6.14c) where it seen that w_2 is largest in magnitude at outer radii, but the large subsidence implied there is mitigated by the effect of w_4, which is now positive (recall that w_4 is mostly negative in the case of constant K: see Fig. 6.14a). Nevertheless, the effect of increasing K with decreasing radius does lead to subsidence at large radii for the broad profile, contrary to the situation for constant K.

The effect of increasing K with decreasing radius is much smaller in the case of the narrow vortex (compare Fig. 6.14d with Fig. 6.14b) and the relative magnitude of w_2 is somewhat smaller compared with the other contributions to w in Eq. (6.51).

6.9.2.2 An alternative form for $w(r, \infty)$

In a previous derivation, Kepert (2001) presented a formula for $w(r, \infty)$ that appears to have a rather different form to Eq. (6.51). In our notation, this formula, his Eq. (28), can be written

$$w(r, \infty) = \frac{1}{r} \frac{\partial}{\partial r} \left[\frac{C_d r v_g}{\zeta_{ag}} (v_g + 2v'(0)) \right]. \qquad (6.52)$$

Kepert remarked on the fact that this formula is nearly independent of the diffusivity K, stating that only the weak influence through $v'(0)$ remains. The same remark would not seem to be applicable to Eq. (6.51), where K appears explicitly as well as implicitly through the dependence of δ and the coefficients a_1 and a_2 on K. However, using Eq. (6.15) and Eq. (6.21), it follows that

$$v_g + 2v'(0) = v_g(1 + 2a_1) = v_g \frac{a_2 - a_1}{\nu}. \qquad (6.53)$$

Since $\nu = C_d v_g \delta / K$ it is easily verified that Eqs. (6.51) and (6.52) are identical. As shown in Section 6.9.2, there are several implicit dependencies in the formula for $w(r, \infty)$ that complicate interpretations.

In a more recent paper, Kepert (2013) expanded out his formula (6.52) to obtain, in our notation,

$$w(r, \infty) = -\frac{1}{\zeta_{ag}^2} \frac{\partial \zeta_{ag}}{\partial r} C_d v_g (v_g + 2v'(0))$$
$$+ \frac{1}{r\zeta_{ag}} \frac{\partial}{\partial r} [r C_d v_g (v_g + 2v'(0))]. \qquad (6.54)$$

Again, using Eq. (6.53), it follows that the first term on the right is w_3 in Eq. (6.51), while the second term is simply

the sum $w_1 + w_2 + w_4$. While Kepert's decomposition of $w(r, \infty)$ is simpler than that in Eq. (6.51), it does not lead to a simpler interpretation. Many effects are still implicit, including that involving the radial variation of K. In particular, for the broad and narrow profiles studied here, one cannot argue that either term is everywhere dominant over the other.

6.9.3 Appendix 3: the upper boundary condition

Kepert and Wang (2001) sought to implement an open boundary condition at the top of the boundary layer by choosing zero vertical gradient boundary conditions on the radial and tangential velocity components at the top of the boundary layer where air exits the layer. However, one can show that for a steady vortex boundary layer, the upper boundary condition that the vertical gradient of the horizontal velocity components is zero at the top of the boundary layer is equivalent to prescribing the radial wind component to be zero and the tangential component to be the gradient value.

For a steady boundary layer with $\partial/\partial t \equiv 0$ and assuming that frictional forces can be ignored at the top of the boundary layer (the usual assumption of boundary-layer theory), Eqs. (6.7) and (6.8) become

$$u \frac{\partial u}{\partial r} + w \frac{\partial u}{\partial z} - \frac{v'^2}{r} - \xi_g v' = 0, \qquad (6.55)$$

$$u \frac{\partial v'}{\partial r} + w \frac{\partial v'}{\partial z} + \frac{uv'}{r} + \zeta_g u = 0. \qquad (6.56)$$

Let us choose the upper boundary condition on u and v to be

$$\frac{\partial u}{\partial z} = 0 \quad \text{and} \quad \frac{\partial v}{\partial z} = 0 \quad \text{at} \quad z = h. \qquad (6.57)$$

The second of these conditions is equivalent, of course, to $\partial v'/\partial z = 0$. Then Eqs. (6.55) and (6.56) give

$$u \frac{\partial u}{\partial r} = \frac{v'^2}{r} + \xi_g v' \quad \text{at} \quad z = h, \qquad (6.58)$$

and

$$u(\zeta' + \zeta_g) = 0 \quad \text{at} \quad z = h, \qquad (6.59)$$

where $\zeta' = (1/r)(\partial r v'/\partial r)$ is the vertical component of the relative vorticity of the agradient flow. Since the vertical component of the relative vorticity $\zeta' + \zeta_g$ at $z = h$ is typically not zero, Eq. (6.59) requires that $u = 0$, whereupon from (6.58), $v' = 0$.

It follows from the analysis above that there is practically no difference between applying the zero-vertical-gradient condition (6.57) and applying the conventional boundary condition that the flow at the top of the boundary layer merges smoothly into the prescribed flow above, i.e., $u = 0$ and $v = v_g$ (or $v' = 0$) at $z = h$. Thus the zero-vertical-gradient condition does not provide a means of allowing a steady-state boundary-layer model to directly determine the flow in the vortex interior.

The foregoing result may be interpreted physically in terms of angular momentum. If frictional torques are negligible in an axisymmetric flow, the absolute angular momentum, $M = rv + \frac{1}{2}fr^2$, is materially conserved. The zero vertical gradient condition on the tangential velocity component at the upper boundary implies that the vertical advection of absolute angular momentum, $w(\partial M/\partial z)$, is zero. Then, if the flow is steady ($\partial M/\partial t = 0$), the radial advection of absolute angular momentum, $u(\partial M/\partial r)$, must vanish also. If $\partial M/\partial r > 0$, i.e., if the flow is centrifugally stable, it follows that u must be zero. Note that $u(\partial M/\partial r) = 0$ is equivalent to $u(\zeta' + \zeta_g) = 0$ (cf. Eq. (6.59)), which states that the radial flux of absolute vorticity is zero.

In view of the over-constraint imposed by the upper boundary condition in boundary layer models that account for vertical structure within the boundary layer, a constraint not shared by the slab boundary layer model, we think it is inappropriate to judge the integrity of the slab model based on a comparison with a model whose integrity is, itself, in question!

In addition to this over-constraint problem, boundary layer model formulations do not represent the pressure gradients in the boundary layer that would arise from the suction effect of deep convection above the boundary layer. This issue is discussed further in Section 16.1.6.

Chapter 7

Estimating boundary layer parameters

In this chapter we consider the specific problem of boundary layer measurements. As we have seen in Chapters 1 and 6, the boundary layer plays a very important role in constraining both the dynamics and energetics of a tropical cyclone. Being the ultimate energy source for the vortex, primarily through wind-enhanced evaporation of water at the sea surface, accurate forecasts of tropical cyclone intensity and structure require, in part, a quantification of the energy transfer rates between the ocean and the atmosphere at the sea surface. Specifically, the latent heat flux, which is determined by the rate of evaporation of sea water, and the sensible heat flux, which is just the rate of ordinary (turbulent) heat transfer, are the two energy inputs to the vortex system. The combined sea-to-air energy transfer is usually represented as a bulk aerodynamic enthalpy transfer process, with coefficient C_K. The forthcoming sections provide further details.

The main energy sink is that associated with surface friction and the related dissipation of mechanical energy by shear-generated turbulence associated with the frictional stress at the sea surface. This energy transfer rate generally varies with wind speed near the surface, but the environment in a mature tropical cyclone is not easily conducive to acquiring accurate measurements of these quantities.

Neglecting for the moment the loss of energy associated with electromagnetic radiation to space, one prerequisite for a quasi-steady vortex would be that the energy gain is balanced by the mechanical energy loss. Further discussion of the possible steady configuration will be deferred until later in Chapter 14. For now, however, it proves convenient to address the loss of mechanical energy by considering the frictional transfer of momentum between the atmosphere and ocean at the sea surface. The air-to-sea momentum transfer is characterized by a bulk aerodynamic drag coefficient, C_D. The momentum transfer problem requires also knowledge of the turbulent momentum diffusivities employed in Chapter 6. Again, the high wind and complex wave environment of a mature tropical cyclone makes direct measurements of these momentum quantities particularly challenging.

In the early days of research flights into hurricanes, some penetrations were made at low levels, typically no higher than 500 m, providing direct measurements of boundary layer structure. After an incident where a particular flight encountered severe turbulence at these levels (Marks et al., 2008), such flights were discontinued on safety grounds. Nowadays, the structure of the boundary layer is probed using dropsondes from aircraft flying at greater altitudes and by other remote sensing techniques (Hock and Franklin, 1999; Uhlhorn and Black, 2003). These methods are supplemented by more traditional laboratory measurements using a wind-wave tank (Curcic and Haus, 2020 and refs.). Another way to circumvent the dangers of low-level manned aircraft missions is to deploy unmanned drone aircraft released from NOAA P3 reconnaissance aircraft (Cione et al., 2020).

Besides providing information on boundary layer structure, an important requirement of observations is to yield data on parameters that are needed in both theoretical and numerical forecast models such as turbulent diffusivities and surface exchange coefficients. A few specific findings using both *in situ* measurements and dropsondes are reviewed below. A summary of some of the more recent developments will be provided at the end of this chapter.

7.1 Boundary layer structure, supergradient winds

Some observations of the low-level wind structure of tropical cyclones are discussed in Section 1.2.4 and some vertical profiles of boundary layer winds are shown in Section 6.4.4. Notable feature of those observations are that the maximum tangential wind speed occurs at low levels, typically in the layer of strong boundary layer inflow and that the radial wind component becomes a sizeable fraction of the tangential wind component in this region. This situation occurs in the high wind region of an intensifying and mature tropical cyclone vortex and results in tangential winds that exceed the gradient wind at the top of the boundary layer. The reasons for these surprising findings are discussed in Section 6.5.7.

Supergradient winds are observed to occur in real tropical cyclones. An illustration of the supergradient wind phenomenon in the inner-core region of a mature tropical cyclone, as well as in a cyclone undergoing rapid intensification, is presented in Figs. 7.1 and 7.2, respectively.

Fig. 7.1 is from Kepert (2006b), who conducted an analysis of the observed azimuthally-averaged tangential wind,

Tropical Cyclones. https://doi.org/10.1016/B978-0-44-313449-4.00015-1

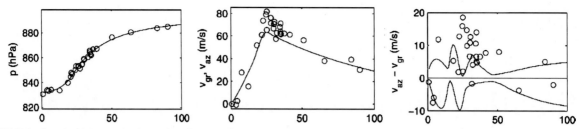

FIGURE 7.1 Sample of dropsonde observations from Hurricane Mitch (1998) at 500 m height as analyzed by Kepert (2006b). (left) Pressure data at 500 m height, together with a fitted profile. (middle) Gradient-wind at 500 m height calculated from the pressure analyses, together with observed storm-relative tangential wind. (right) Difference between storm-relative, tangential wind observations and the gradient wind at 500 m height, together with the 5th to 95th percentile confidence interval about the gradient wind. Adapted from Fig. 14 of Kepert (2006b). Republished with permission of the American Meteorological Society.

pressure and gradient wind derived from dropsonde data from Category five Hurricane Mitch (1998). The analysis shown corresponds to a time when the storm was not rapidly intensifying. The agradient wind V_a, also called "supergradient excess relative to the gradient wind", is obtained from the difference between the averaged tangential wind and the inferred gradient wind, $V_a = V_{az} - V_{gr}$. Fig. 7.1 shows a significant supergradient excess of approximately 23% near the radius of maximum gradient wind.

Fig. 7.2 is from Montgomery et al. (2014), who carried out a similar boundary layer analysis for both the rapid intensification and mature periods of Hurricane Earl (2010), with the latter period following a concentric eyewall cycle of the storm. Here the dropsonde analysis is presented at the height of the maximum tangential wind speed (typically at or below 500 m height). In this case, the average tangential wind is significantly greater than the corresponding gradient wind near the maximum wind radius. During the mature period (bottom row), the averaged tangential wind exceeds the *local* gradient wind by as much as 60%. During the mature period, we see also that the radius of maximum gradient wind is located significantly outside the radius of maximum tangential wind.

7.2 Subgrid-scale parameterizations

Because of the large Reynolds numbers found in a tropical cyclone and because of our inability to represent the small-scale motions in forecast and theoretical-idealized numerical models, numerical models of atmospheric flows and those of tropical cyclones, in particular, generally require the parameterization of subgrid-scale turbulent transfer of momentum, heat, and moisture. On account of the typical large disparity between the horizontal and vertical grid spacing employed in numerical weather prediction models, the parameterizations for horizontal and vertical transfer are treated separately.

While the formulation of turbulent transfer in the vertical is guided by observations, at least in the planetary boundary layer, the formulation of turbulent momentum and

heat transfer above the boundary layer and that for the horizontal transfer everywhere is more *ad hoc*. There have been some observationally-inferred estimates of vertical momentum diffusivity at both major and minimal tropical cyclone wind speeds (Zhang et al., 2011a; Zhang and Drennan, 2012). There have been some recent observational estimates also of the horizontal diffusivity in hurricanes by Zhang and Montgomery (2012). Here we summarize the efforts for the case of major hurricanes using flight-level data from research aircraft flown at altitudes around 500 m.

7.3 Vertical diffusivity in the boundary layer

Observational estimates of vertical diffusivity based on research aircraft flying at low levels, around 500 m altitude, are summarized in Fig. 7.3. The scientific basis underlying these vertical diffusivity estimates, along with corresponding error estimates, is detailed in Zhang et al. (2011a). In a nutshell, using flight-level data, in combination with surface wind speed data from the step-frequency-microwave-radiometer (SFMR) (available post 1988), three estimates of the vertical diffusivity may be inferred using the local vertical shear of the mean wind speed, the transverse momentum flux, and the turbulent kinetic energy spectrum. Typical values of K_v are on the order of 50 m^2 s^{-1}, although, as is evident in the figure, there is scatter in the observational data. An approximate monotonic increase of K_v with mean wind speed is evident also.

The foregoing observations of the vertical turbulent diffusivity have resulted in improvements in hurricane intensity forecasts. Supported by the NOAA Hurricane Forecast Improvement Project (HFIP), Zhang et al. (2015) applied the observational findings of Zhang et al. (2011b) to upgrade the boundary layer physics in the operational Hurricane Weather Research and Forecasting (HWRF) Model. Specifically, Zhang et al. (2015) showed that the incorporation of the observed estimates in the vertical momentum diffusivity yielded improvements in the simulated track, intensity, and structure of four hurricanes using retrospec-

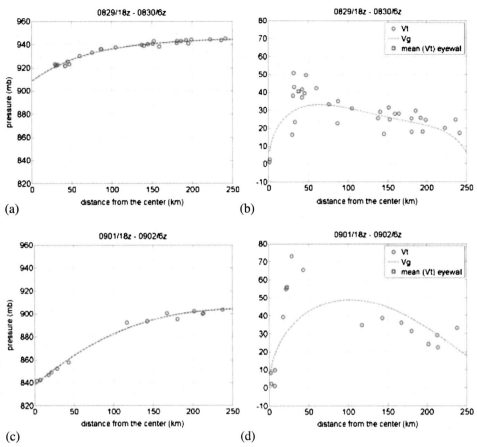

FIGURE 7.2 Sample of dropsonde observations from Hurricane Earl (2010) at the height of maximum tangential winds ($\approx$ 500 m) as analyzed by Montgomery et al. (2014). Azimuthally-averaged, storm-relative, tangential wind V_t and gradient wind (V_g) at the height of maximum tangential wind speed for periods 2 & 3 (0829/18z - 0830/6z; 0901/18z - 0902/6z) comprising a portion of Earl's rapid intensification and mature periods, respectively. Calculations are displayed in pairs with dropsonde-observed pressure (blue (dark gray in print version) circle) at the left and dropsonde-observed V_t (blue (dark gray in print version) circle) at the right of each plotted pair. For each pair, the left panel shows dropsonde pressure observations (blue (dark gray in print version)) as a function of radius with the fitted curve (red (mid gray in print version)) based on least-squares regression. Right panels show dropsonde observed V_t (blue (dark gray in print version)) and gradient wind V_g (green (gray in print version)) as a function of radius. V_g is calculated using the fitted pressure gradient by solving the gradient balance equation: $f V_g + V_g{}^2/r = (1/\rho)\partial p/\partial r$, where ρ is moist air density and r denotes the radius from storm center. The red (mid gray in print version) square in the V_t plot is the arithmetic average of V_t at the eyewall region within 5 km on either side of the radius of maximum tangential wind as inferred from the Doppler radar data at 2 km altitude. Adapted from Fig. 11 of Montgomery et al. (2014). Republished with permission of Wiley.

tive HWRF forecasts. Incorporation of these observations demonstrated also

> "substantial improvements in the simulated storm size, surface inflow angle, near-surface wind profile, and kinematic boundary layer heights in simulations with the improved physics, while only minor improvements are found in the thermodynamic boundary layer height, eyewall slope, and the distributions of vertical velocities in the eyewall."

Zhang et al. provided reasons for the improved vortex structure. Their work is a reminder of the power of fusing basic and applied research to improve hurricane intensity forecasts. Quoting from Zhang et al. (2015):

> "This work emphasizes the importance of aircraft observations in model diagnostics and development, endorsing a developmental framework for improving physical parameterizations in hurricane models."

7.4 Horizontal diffusivity in the boundary layer

The flight level data obtained through early aircraft penetrations of several Category 4 and 5 hurricanes have been utilized also to estimate the horizontal eddy diffusivity as a function of the mean flight-level wind speed, using a higher frequency sampling method. These storms include Hurricanes Allen (1980), David (1979), Hugo (1989), and

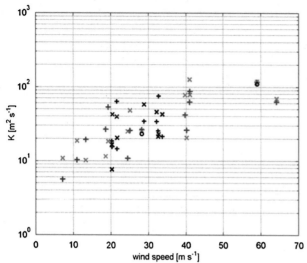

FIGURE 7.3 Vertical eddy diffusivity K as a function of mean wind speed at aircraft flight level for all the good runs below an altitude of 1 km in Hurricanes Allen (Category 4, 1980), Hugo (Category 5, 1989), and Frances (2004). The diffusivities are estimated using three methods: the direct method (○), Hanna's method (×), and the turbulent kinetic energy closure method (+). The Frances data are in blue (dark gray in print version). The Hugo and Allen data are in the other colors, depending on the method. From Fig. 10 of Zhang et al. (2011b). Republished with permission of the American Meteorological Society.

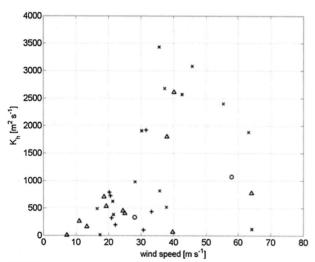

FIGURE 7.4 Calculated horizontal eddy diffusivity K_h as a function of mean wind speed at aircraft flight level for all flux runs in Hurricanes Allen (△), David (×), Hugo (○), and Frances (+). From Fig. 6 of Zhang and Montgomery (2012). Republished with permission of the American Meteorological Society.

Frances (2004). Fig. 7.4 shows that, in the high wind eyewall region, the average horizontal eddy diffusivity is approximately 1500 m² s⁻¹. Overall, it is found that the eddy diffusivity tends to increase with increasing mean wind speed at flight-level. Of course, it is evident from the figure that there is again a large scatter in the data and the sample size is small.

Clearly, further data are needed to narrow down uncertainties associated with the horizontal diffusivity and horizontal mixing length (not shown, see Zhang and Montgomery, 2012 for details). Recognizing the challenges associated with the collection of such valuable data in the dangerous environment of the hurricane boundary layer, Zhang and Montgomery (2012) noted that

"it is unlikely that we may see such data in the near future because of safety constraints for manned aircraft to be flown at such low altitudes again in the boundary layer. ... To more completely quantify turbulence characteristics in the intense eyewall and their impact on our understanding and prediction of hurricane intensification and maximum intensity, a focused field program is recommended ..., possibly with unmanned platforms employing advanced turbulent sensors on board or using advanced remote sensing techniques."

Acknowledging these realities, we must continue to make as much use as possible of this unique low-level dropsonde and *in situ* data and consider also new ways to probe this complex, but important region of the hurricane on the basis of observations (e.g., Cione et al., 2020).

7.5 Air-sea interaction, drag coefficient, enthalpy coefficient

The air-sea state below 50 m height is generally quite complex (e.g., Fig. 1.22). Surface wave heights in the high wind region of the vortex often exceed 10 m, waves often break, generating spray and spume, and contribute to forming an emulsive layer at very high wind speeds.

Until early this century, quantitative knowledge of the turbulent eddy momentum, heat, and moisture fluxes at the air-sea interface at surface wind speeds above about 30 m s⁻¹ was absent and one relied upon extrapolation of observational results and/or theoretical formulae for low wind speeds (≤ 20 m s⁻¹). Here we summarize briefly the latest efforts to quantify these surface exchange processes in major hurricane conditions (≥ 50 m s⁻¹), equivalent to Category 3 and above on the Saffir-Simpson hurricane intensity scale.

Because of the high Reynolds numbers involved, a wind-speed dependent, bulk aerodynamic formulation is normally implemented to represent these surface fluxes in terms of the exchange coefficients for momentum C_D, moisture C_E and dry enthalpy C_T. Such a formulation may be regarded as a turbulent closure model whose flow parameters are to be determined by experiment. The formulae used to represent these fluxes at the surface (at standard anemometer level of

10 m, say) are as follows:

$$\tau_{z\lambda} = -\rho\overline{w'v'} = \rho C_D|\mathbf{u_h}|v, \qquad (7.1)$$

$$\tau_{zr} = -\rho\overline{w'u'} = \rho C_D|\mathbf{u_h}|u, \qquad (7.2)$$

where $\tau_{z\lambda}$ and τ_{zr} denotes the tangential and radial components of the near-surface (10 m) vertical eddy stress, respectively, the overbar denotes a Reynolds (running time) average for turbulent flow with primes denoting a departure from the Reynolds average, $|\mathbf{u_h}|$ denotes the mean near-surface wind speed, (u, v) denotes the mean near-surface radial and tangential velocity components, respectively, and ρ denotes the mean near-surface density. This formulation for the surface momentum stress is aimed at a highly turbulent geophysical boundary layer - as opposed to a laminar boundary layer. Such a formulation is often referred to as a *semi-slip boundary formulation.*

As noted above, observations of the atmospheric boundary layer below 50 m altitude raise legitimate questions about the practical usefulness and physical validity of an alternative no-slip formulation in this extreme environment. It is unclear whether such an approach would yield additional insight on the fluid dynamics of hurricanes.[1] As noted in the Introduction, aspects pertaining to the lowest levels of the marine atmospheric boundary layer involving waves, breaking waves, spume, etc., lie beyond the scope of this review. The reader is referred to Sullivan et al. (2014) for a contemporary view and modeling approach of the marine atmospheric boundary layer.

The vertical turbulent heat and moisture eddy fluxes from the ocean into the atmosphere are represented similarly,

$$F_{zT} = \rho c_{pd}\overline{w'T'} = \rho c_{pd}C_T|\mathbf{u_h}|(T_s - T_{10}), \qquad (7.3)$$

$$F_{zr_v} = \rho\overline{w'r_v'} = \rho C_E|\mathbf{u_h}|(r_{vs} - r_{v10}). \qquad (7.4)$$

In Eq. (7.3), F_{zT} denotes the vertical, turbulent, dry enthalpy flux at the air-sea interface (commonly called the *sensible heat flux*), c_{pd} is the specific heat of dry air at constant pressure, T_s is the sea surface temperature, and T_{10} is

the air temperature at the 10 m level. Similarly, in Eq. (7.4), F_{zr_v} denotes the vertical, turbulent, moisture flux (expressed in terms of the water vapor mixing ratio r_v as defined in Section 2.5), r_{vs} is the saturation water vapor mixing ratio evaluated at the sea surface temperature and (local) surface air pressure, and r_{v10} is the water vapor mixing ratio of air at the 10 m level.

It is implicit herein that as water is removed from the ocean through a turbulent wave-breaking and/or emulsion process, the water droplets that do not fall back to the sea ultimately evaporate to form the vapor assumed in Eq. (7.4). The energy required to evaporate the suspended droplets is assumed to come from the ambient air.

From the foregoing formulae, it is clear that the turbulent fluxes transfer heat and moisture into the surface layer of the tropical cyclone if both $T_s > T_{10}$ and $r_{vs} > r_{v10}$, which is ordinarily the case in the maritime tropics where there is a thermodynamic disequilibrium between the atmosphere and underlying ocean. For the hurricane problem, what is pertinent is the moist enthalpy transfer from the ocean to the atmosphere.

The vertical turbulent moist enthalpy flux F_{zk} near the air-sea interface may then be represented by

$$F_{zk} = \rho c_{pd}\overline{w'T'} + \rho L_v\overline{w'r_v'} = C_K\rho|\mathbf{u_h}|(k_s - k_{10}), \quad (7.5)$$

where L_v denotes the latent heat of evaporation/condensation defined in Chapter 2, C_K is the moist enthalpy exchange coefficient, k_s is the saturation moist (static) enthalpy at the sea surface ($k_s = c_{pd}T_s + L_vr_{vs}$) and k_{10} is the corresponding moist enthalpy at standard anemometer level ($k_{10} = c_pT_{10} + L_vr_{v10}$).[2] This formulation assumes that the transfer coefficients for moisture and dry enthalpy are the same (i.e., $C_E = C_T$, see Emanuel, 1995).

It goes without saying that parameterizing the air-sea interaction at these wind speeds with 10-m bulk exchange coefficients is a simplification of a complex process. However, accurate estimates of these quantities continue to be a priority due to the established practical reliance on these coefficients in numerical forecast models, wind damage and wind surge predictions, and also theoretical studies. Moreover, improvements in the representation of air-sea interaction in numerical forecast models for major tropical cyclones are held to contribute to improved intensity forecasts issued by the Joint Typhoon Warning Center and the National Hurricane Center (NHC) (Rappaport et al., 2009).

Because direct turbulent flux measurements of heat, moisture and momentum in high wind speed maritime environments (Category 3 and above) are not easy to obtain,

---

1. Whereas the "semi-slip" formulation is physically defensible and routinely used in the hurricane community, it might be still argued that a "no-slip" formulation is more fundamental and less subject to model tuning. A no-slip formulation at the sea surface would entail matching the horizontal air velocity (and stress) with the corresponding ocean current velocity (and stress) at the air-sea interface (e.g., Gill, 1982). Observations of Franklin et al. (2003); Montgomery et al. (2006b); Kepert (2006a,b), and Zhang et al. (2011a), using the high-resolution Global Positioning System (GPS) dropwindsonde released from reconnaissance aircraft into hurricanes, show that the tangential and radial velocity are still relatively large at and below 50 m above the sea surface. Considering the fact that ocean currents are considerably smaller than their atmospheric counterparts, and given the fact that relatively large horizontal velocities persist at heights of 50 m (and often down to 10 m, see e.g., Powell et al., 2003), implicates an excessive vertical resolution to implement the no-slip boundary condition. In short, a no-slip formulation appears neither practical nor theoretically advantageous.

2. Some authors choose to define the above moisture flux and corresponding moist static enthalpy using the specific humidity, q_v, defined in Section 2.5, as well as the specific internal energy of water vapor (e.g., Jeong et al., 2012). However, because r_v is at most 30 g kg^{-1} and because the difference between r_v and q_v is at most a few percent and barely measurable, the difference between the moisture flux and moist static enthalpy quantities calculated using r_v or q_v is neglected here.

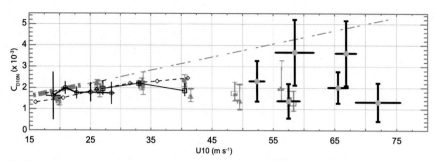

FIGURE 7.5 Wind-speed dependence of drag coefficient C_D deduced by Bell et al. (2012a) using an energy/angular momentum budget method (green (gray in print version) circles) compared with prior studies. Black symbols denote data adapted from French et al. (2007) and blue (dark gray in print version) symbols denote data adapted from Vickery et al. (2009). Red (mid gray in print version) line indicates measured (thick) and extrapolated (thin) Large and Pond (1981) drag coefficient. Figure and caption adapted from Bell et al. (2012a). Republished with permission of the American Meteorological Society.

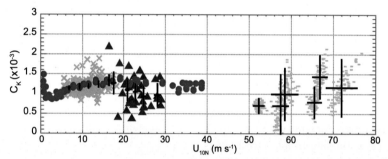

FIGURE 7.6 Wind-speed dependence of moist enthalpy coefficient C_K deduced by Bell et al. (2012a) using energy/angular momentum budget method (green squares) compared with previous studies. Laboratory results (blue circles) and CBLAST field campaign measurements (red triangles) shown with HEXOS results (gray crosses) adapted from Haus et al. (2010). The mean and 95% confidence intervals are shown in black. From Bell et al. (2012a). Republished with permission of the American Meteorological Society.

an alternative methodology was developed by Bell et al. (2012a) to infer the coefficients C_D and C_K. There, the surface momentum and moist enthalpy fluxes were deduced via a residual method using an annular control volume analysis on the budgets of absolute angular momentum, M, and total energy, E, around the eyewall region of the storm in a composite, azimuthally-averaged sense. The method is based on the idea that given observations of the lateral fluxes of M and E into and out of the annular volume, and given the fluxes of M and E out of the top of the volume, one can, in a quasi-steady-state configuration, infer the surface fluxes of M and E by residual and thereby obtain direct estimates of the bulk transfer coefficients C_D and C_K.

The new estimates are based on a coordinated field program during the 2003 Atlantic hurricane season. The details of the new variational methodology, its historical roots, and a detailed error analysis is presented by Bell et al. (2012a).

The field data were collected during six intensive observing periods conducted in Hurricane Fabian from 2 to 4 September and in Hurricane Isabel from 12 to 14 September 2003 as part of the collaborative Office of Naval Research's Coupled Boundary Layer Air-Sea Transfer (CBLAST) and National Oceanic and Atmospheric Administration (NOAA)/National Environmental Satellite, Data, and Infor-

mation Service (NESDIS) OceanWinds experiments (Black et al., 2007). Only the key findings are summarized here.

The Bell et al. (2012a) results for C_D and C_K are shown in Figs. 7.5 and 7.6, respectively. The data are plotted together with prior estimates from other field experiments and from wave-tank experiments with wind speeds up to approximately 40 m s^{-1} (Haus et al. (2010), Jeong et al. (2012) and refs.). Although there is certainly scatter in the data, there is no discernible systematic variation of C_D and C_K with wind speed for speeds above approximately 30 m s^{-1}. Moreover, the results provide support for the similarity hypothesis that C_D and C_K become constants at high wind speeds (Emanuel, 2003).

In the high wind speed regime (> 50 m s^{-1}), Bell et al. (2012a) obtained a mean C_D estimate of 2.4×10^{-3} with a standard deviation of 1.1×10^{-3}; they obtained similarly a mean C_K estimate of 1.0×10^{-3} with a standard deviation of 0.4×10^{-3}. Fig. 7.7 shows the ratio of exchange coefficients C_K/C_D as inferred from the residual methodology at high wind speeds, together with prior estimates.

The results of Bell et al. (2012a) concerning C_D and C_K are broadly consistent with the independent work of Powell et al. (2003), Holthuijsen et al. (2012), and Richter and Stern (2010) using high resolution GPS dropwindsondes released from reconnaissance aircraft flown into hurricanes.

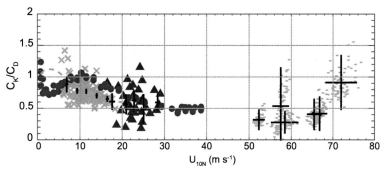

FIGURE 7.7 Wind-speed dependence of ratio C_K/C_D deduced by Bell et al. (2012a) using energy/angular momentum budget method (green squares) compared with previous studies. Symbols are as in Fig. 7.6. From Bell et al. (2012a). Republished with permission of the American Meteorological Society.

These studies obtained estimates of C_D using a methodology based on determining a least-squares fit of mean dropsonde wind speed profiles to an assumed log-layer structure for horizontal wind speed: $U(z) = (U_*/k) \ln(z/z_o)$, where $U(z)$ is the wind speed at height z, U_* is the friction velocity, k is here the von Kármán constant (≈ 0.4), and z_0 is the roughness length. By linearly extrapolating the derived mean profile to zero on a logarithmic-linear graph, and by determining the slope of this line, one obtains values for z_o and U_*. From these values, one then obtains the surface stress $\tau_s = \rho U_*^2 = \rho C_D U_{10}^2$ and hence the 10 m drag coefficient C_D.

Consistent with Bell et al.'s findings at high wind speeds (> 50 m s^{-1}), Powell et al. (2003) and subsequent coworkers suggested a roughly invariant C_D with wind speed between approximately 28 m s^{-1} and 40 m s^{-1} (Fig. 3 of Powell et al., 2003), but suggested also that C_D decreases somewhat with wind speed above 40 m s^{-1}.

The Powell et al. work was pioneering because it provided the first plausible quantitative estimates of C_D in hurricane-force conditions and because it suggested that C_D was significantly less than prior estimates and/or extrapolation procedures from other field experiments operating under considerably weaker atmospheric wind speeds. However, as discussed by Smith and Montgomery (2014), the validity of assuming a constant-stress layer with *logarithmically increasing tangential and radial wind components over an appreciable span of the hurricane inner-core boundary layer* is on questionable theoretical grounds in the hurricane boundary layer. Therefore, it is unclear whether to ascribe much physical significance to the reported decline of C_D above wind speeds of 40 m s^{-1} using such similarity-fitting methodology.

In contrast, the Bell et al. (2012a) methodology makes no log-layer assumption for the low-level boundary layer. A complementary and insightful study by Andreas et al. (2012) advocates the extrapolation of an alternative formulation of the friction velocity to major hurricane wind speeds. Their deduced drag coefficient for major hurricane

wind speeds (see their Fig. 10) is approximately constant with wind speed and only slightly larger than the *average drag coefficient* ($C_D = 2.4 \times 10^{-3}$) obtained by Bell et al. (2012a).

Finally, a re-analysis of prior wave-tank experiments by Curcic and Haus (2020) suggests that, on account of the discovery of a subtle coding error, the updated drag coefficient at hurricane wind speeds is some 30% higher than prior published estimates using the same source data (Fig. 7.8). Their re-analysis finds also that the updated value of C_D levels off at near major hurricane wind speeds to an approximately invariant value of $C_D = 3.1 \times 10^{-3}$.

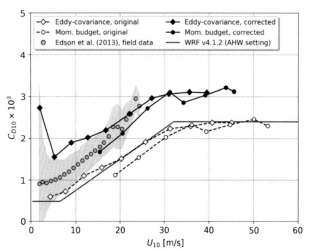

FIGURE 7.8 Corrected drag coefficient data from wind-wave tank experiment. Original (black, dashed) and corrected (black, solid) drag coefficient data by Donelan et al. (2004). Field data (mean $\pm$ 1 standard deviation) from Edson et al. (2013) are shown in yellow (light gray in print version). Surface drag parameterization in the Weather Research & Forecasting Model (WRF) (Advanced Hurricane WRF setting) is shown in blue (dark gray in print version). From Curcic and Haus (2020). Information on the WRF model may be found at: https://www.mmm.ucar.edu/models/wrf. Republished with permission of the American Geophysical Union.

In summary, the observational findings on the heat and momentum exchange coefficients in major hurricane conditions have important ramifications in the determination of the maximum possible hurricane intensity in a given environment. Both field data and laboratory data yield consistent results on the enthalpy coefficient C_K at high wind speeds. Although the mean drag coefficient is roughly similar between field data and the wind-wave tank data, the relatively large scatter in the drag coefficient at high wind speeds deduced using the residual method raises some question about the representativeness of the mean drag coefficient at major hurricane wind speeds. In recent re-analyses of the wind-wave tank data, a subtle coding error has been discovered in prior C_D estimates. The corrected data indicates an approximate 30% increase of C_D at high wind speeds. Despite the limited fetch of the wind-wave tank, the tank estimates of C_D are arguably the most thorough. Indeed, state-of-the-art numerical weather prediction models are presently tuned to be consistent with the prior C_D data from the wind-wave tank. On the basis of the evidence presented in this chapter, we consider it important to update the drag coefficient in hurricane intensity forecast models to be consistent with the corrected C_D data from the wind-wave tank at high wind speeds. We will explore the physical implications of an increased drag coefficient in more detail in Chapter 13.

Chapter 8

A prognostic balance theory for vortex evolution

The derivation of the Eliassen equation in Chapter 5 is an important step in formulating a balance theory for the evolution of an axisymmetric vortex. The assumption that the flow is everywhere in thermal wind balance paves the way for a method to solve an initial-value problem for the slow evolution of an axisymmetric vortex forced by sources of heat and tangential momentum. Given an initial tangential wind profile $v_i(r, z)$ and an environmental density sounding $\rho_o(z)$, one can proceed using the following steps:

(1) solve Eq. (5.16) for the initial balanced density and pressure fields corresponding to v_i;

(2) solve the Eliassen equation (5.67) for the streamfunction of the secondary circulation, ψ, that keeps v and θ in thermal-wind balance for a short time interval, Δt;

(3) solve for the velocity components u and w of the secondary circulation using Eq. (5.61);

(4) predict the new tangential wind field[1] using Eq. (5.62) at time Δt;

(5) Repeat the sequence of steps from item (1) to obtain the solution at time $n \Delta t$, where n is a positive integer.

The method is analogous to the vorticity-streamfunction method for solving for the evolution of barotropic flows described in Section 2.10 and is relatively straightforward to implement numerically. However, certain issues arise when the method is applied in the tropical cyclone context.

The first issue is how to formulate the forcing terms in the Eliassen equation, i.e., the diabatic heating rate, $\dot{\theta}$, and the near-surface frictional force, $\dot{V}$. The representation of diabatic heating in terms of the evolving flow constitutes the cumulus parameterization problem. Typically, this representation requires the inclusion of a prediction equation for water vapor and, in more sophisticated representations, for various species of water condensate. The representation of friction was examined in Chapter 6 and some effects of friction will be discussed below. In this chapter, diabatic heating will be represented by a prescribed heating distribution for simplicity. As explained in Section 16.1.1, a judicious choice of vertical structure of the heating leads to a pattern

of low-level horizontal convergence and upper-level horizontal divergence characteristic of active regions of deep convection. More sophisticated explicit representations of heating associated with deep convection are discussed in Chapter 10.

The second issue is that, as tropical cyclones intensify, the boundary layer and upper-tropospheric outflow generally develop regions of zero or negative discriminant ($D < 0$), where D is defined in Eq. (5.68). The discriminant is proportional to the quantity Δ defined in Eq. (5.48). A negative discriminant implies the existence of local symmetric instability (cf. Section 5.7.3) and the Eliassen equation is no longer globally elliptic. Nevertheless, if the regions of symmetric instability remain localized and do not extend throughout the mean vortex, it may be reasonable to apply some kind of regularization procedure to keep the Eliassen equation elliptic and thus solvable for a longer period of time.[2] Some solutions of such a model are shown in this chapter.

8.1 Solutions for the evolution of a balanced vortex

Examples of calculations for the evolution of a balanced vortex are described by Smith et al. (2018b). Similar examples are discussed below.[3] We begin with a case where there is diabatic heating only and go on to examine a case with only surface friction (Section 8.1.2) and finally one with both diabatic heating and surface friction (Section 8.1.3). In Section 8.1.3, we explore the interplay between heating and friction. The various simulations are listed in Table 8.1.

---

1. In principle it would appear equally acceptable to predict the potential temperature via Eq. (5.9). However, this method has the disadvantage that a strict balance solution cannot be guaranteed to exist everywhere for the corresponding gradient wind.

2. Possible strategies for regularizing the Eliassen equation are discussed by Smith et al. (2018b) and Wang and Smith (2019).

3. In Smith et al. (2018b) and in Smith and Wang (2018), the latitude was given as 20°N, but it was subsequently discovered that the latitude therein was only 10°N. This error has been corrected and, for this reason, the figures shown here differ slightly in detail from those in Smith et al. (2018b) and Smith and Wang (2018). However, the general conclusions of these studies are not affected.

Tropical Cyclones. https://doi.org/10.1016/B978-0-44-313449-4.00016-3

TABLE 8.1 Summary of the three simulations referred to in this section.

Simulation	Heating	Friction	Section
Ex-HO	Yes	No	8.1.1
Ex-FO	No	Yes	8.1.2
Ex-HF	Yes	Yes	8.1.3

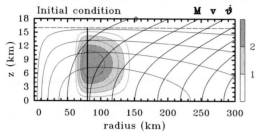

FIGURE 8.1 Initial tangential wind structure v, the corresponding M-surfaces, and heating rate distribution $\dot{\theta}$ for the balance calculation described below. Contour intervals are: for v, 2 m s^{-1}; for M, 5 $\times 10^5$ m^2 s^{-1}; for $\dot{\theta}$, 0.5 K h^{-1}. The thin dashed line indicates the tropopause level.

8.1.1 Diabatic heating, no friction

Fig. 8.1 shows the initial conditions for this calculation including the tangential velocity distribution, the corresponding structure of the M-surfaces and the prescribed heating rate function $\dot{\theta}$. In this example, the initial warm-cored vortex has a maximum tangential wind speed of 10 m s^{-1} at the surface at a radius of 100 km. The heating rate varies sinusoidally with height and has a maximum amplitude of 3 K/hour at an altitude of 8 km, consistent with heating rates diagnosed from observations in deep convection (Zagrodnik and Jiang, 2014, see their Fig. 11). The function has a skewed bell-shape in radius (inner radius 20 km, outer radius 70 km scale) and the maximum is centered on the location of a particular M surface at a height of 1 km. This M-surface is located initially inside the radius of maximum tangential wind speed and moves inwards in the lower troposphere as the vortex evolves. There is no representation of friction (i.e., $\dot{V} = 0$). The time step for the calculation is 5 minutes. The solutions are summarized in Figs. 8.2-8.6.

8.1.1.1 Metrics of vortex evolution

Fig. 8.2 shows the time evolution of the maximum tangential velocity component, V_{max}, the maximum radial velocity component, U_{max}, and the magnitude of the (negative) minimum radial velocity component, U_{min}, for the foregoing calculation. In the calculation, V_{max} increases essentially linearly with time to about 10 hours at a rate of approximately 0.7 m s^{-1} h^{-1}, shortly before the time that the discriminant in the Eliassen equation, D, becomes negative. This time is indicated by a vertical line in Fig. 8.2. Beyond this time, a regularization procedure is applied to remove the initially small regions where the flow would be symmetrically unstable. The regularization procedure applied here is

Simulation with diabatic heating, but no friction: Ex-HO

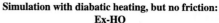

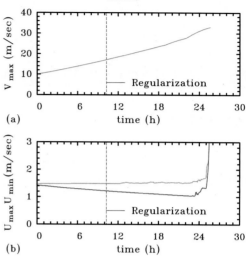

FIGURE 8.2 Time evolution of (a) $V_{max}(t)$, (b) $U_{max}(t)$ (red (mid gray in print version)), $U_{min}(t)$ (blue (dark gray in print version)), for the balanced vortex. The vertical dashed line indicates the first time at which the discriminant in the Eliassen equation first becomes negative.

the same as that in Smith et al. (2018b) and is described in an appendix (Section 8.5).

After this time, the rate at which V_{max} increases grows slowly until about 22 hours, beyond which it grows more rapidly. At 24 hours, one hour and 40 minutes before the integration terminates, V_{max} has attained a value of about 31 m s^{-1}. The reasons for the termination of the calculation are discussed in Subsection 8.1.1.4.

At $t = 0$ hours, $U_{max} = 1.47$ m s^{-1} and $U_{min} = 1.42$ m s^{-1}. Thereafter, U_{min} declines slowly and approximately linearly to 1.05 m s^{-1} until 23 hours, when it begins to increase rapidly. In contrast, U_{max} increases marginally reaching 1.54 m s^{-1} at 23 hours, before beginning to increase rapidly also. The rapid increases of U_{min} and U_{max} near the end of the simulation are associated with the physics of breakdown. Before this breakdown begins, V_{max} and U_{min} are found to occur at or just above the surface. Typically V_{max} ascends slowly from the surface during the first hour to 1 km at 24 hours, while U_{min} is at 200 m until about 15 hours and jumps to 400 m until 23 hours (see Exercise 8.1). In contrast, U_{max} occurs in the upper troposphere, initially at a height of about 12 km, but increasing to 13.4 km at 24 hours (see Figs. 8.3c, 8.3d, and 8.6b).

8.1.1.2 Radius-height cross sections

Figs. 8.3-8.5 show radius-height cross sections of various flow fields in relation to the M-surfaces at the initial time and at 10 hours, 15 minutes before the regularization procedure is first needed. Panels (a) and (b) show the distribution of heating rate, $\dot{\theta}$, as well as the streamfunction of the secondary circulation, ψ. As noted above, the $\dot{\theta}$ distribution is

Simulation with diabatic heating, but no friction: Ex-HO

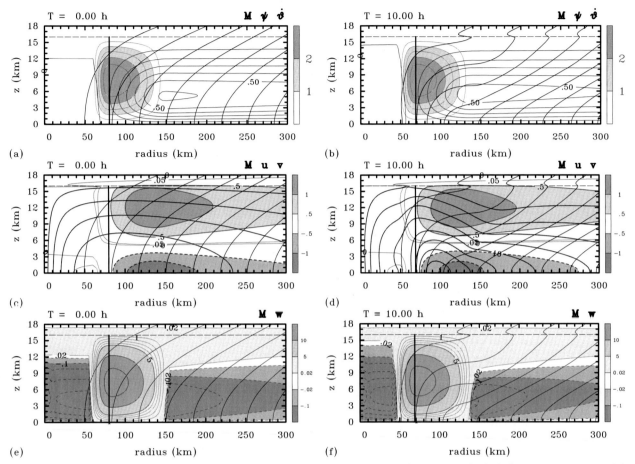

FIGURE 8.3 Radial-height cross sections of M-surfaces superimposed on various quantities for the simulation including: (a, b) streamlines of the secondary circulation, ψ, and diabatic heating rate $\dot{\theta}$ (shaded); (c, d) contours of tangential velocity, v, and radial velocity u (shaded); (e, f) contours of vertical velocity, w (shaded); (continued overleaf) (g, h) contours of potential temperature perturbation, $d\theta = \theta(r, z) - \theta(R, z)$ (shaded), and (i, j) the tangential wind tendency, $\partial v/\partial t$ (shaded). The left panels are at the initial time and the right panels are at 9 hours. Contour intervals are: for M, 5×10^5 m^2 s^{-1}; for ψ, 1×10^8 kg s^{-1}; for $\dot{\theta}$, 0.5 K h^{-1}; for u, 0.5 m s^{-1}, 0.05 m s^{-1} (very thin solid contour), 1 m s^{-1} for negative values dashed contours; for v, 2 m s^{-1}; for w, positive values: 5 cm s^{-1} (thick contours), 1 cm s^{-1} to 4 cm s^{-1} (thin contours), and 0.2 mm s^{-1} (very thin contour), negative values 2 mm s^{-1} (thin contours), 1 mm s^{-1} and 0.2 mm s^{-1} (very thin contours); for θ', 0.5 K; and for $\partial v/\partial t$, 0.5 m s^{-1} (6 h)$^{-1}$ (thick contours). For all fields: positive values (solid), negative values (dashed).

tied to a prescribed M-surface at a height of 1 km. This M-surface moves inwards from about 80 km at $t = 0$ hours to about 69.5 km at $t = 10$ hours. Radially outside the heating region there is inflow in the lower troposphere, below about 5 km, and outflow above this level. Most of the streamflow passes through the region of heating.

Figs. 8.3c and 8.3d show the radial and tangential velocity components in relation to the M-surfaces. The maximum tangential wind speed occurs at the surface, where the inflow is a maximum. This is to be expected as U_{min}, itself, occurs at the surface so that the largest inward advection of the M-surfaces occurs at this level. The asymmetry in the depths of inflow and outflow is a consequence of mass continuity and the fact that density decreases approximately exponentially with height.

Figs. 8.3e and 8.3f show the vertical velocity w in relation to the M-surfaces. At both times shown, there is strong ascent in the heating region and weak ascent in the upper troposphere at all radii, except interior to the heating region, where there is weak subsidence below about 12 km at the initial time and 14 km at 10 hours. In the lower and middle troposphere, there is subsidence both interior and exterior to the heating region.

At the initial time, the maximum vertical velocity, W_{max}, is about 17 cm s^{-1} and it occurs at a height just above 7 km. At 10 hours, W_{max} is only marginally stronger, but still less than 18 cm s^{-1} and marginally higher (by 200 m).

The subsidence is strongest at low levels interior to the heating region. At the initial time, the maximum subsidence, W_{min}, is 7 mm s^{-1} at a radius of 55 km and a height

Simulation with diabatic heating, but no friction: Ex-HO

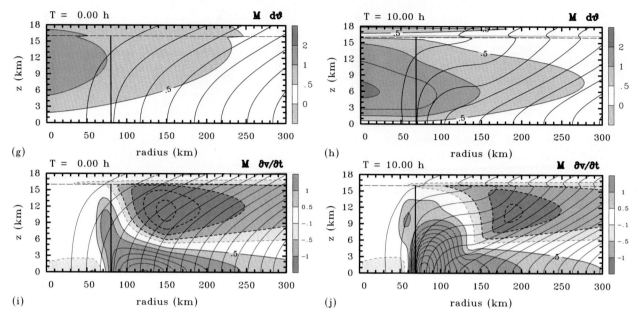

FIGURE 8.3 (continued)

of 3.8 km. As time proceeds, W_{min} increases somewhat in strength as the radius of the heating region contracts. At 10 hours, W_{min} is 9 mm s^{-1} at a radius of 40 km and a height of 4.6 km. The subsidence outside the heating region increases only marginally in strength with time and is a maximum at low levels just outside the heating region.

Figs. 8.3g and 8.3h show the perturbation potential temperature $d\theta = \theta(r, z) - \theta(R, z)$ relative to the ambient potential temperature profile at the outer boundary (1000 km, not shown). At the initial time, the warm anomaly in balance with the tangential wind field has a maximum of 1.4 K located on the rotation axis at an altitude of 10.8 km. By 10 hours, the warm anomaly has strengthened throughout the vortex core and the maximum, now 2.1 K, has shifted downwards to 6.2 km, again on the vortex axis.

At 10 hours, a shallow layer with negative values of $d\theta$ has developed just above the tropopause (Fig. 8.3h). The minimum value is -0.4 K and occurs at a radius of 85 km and a height of 16.2 km. This feature is seen to coincide with a region in which the M-surfaces fold sharply westwards with height, implying a positive vertical gradient of tangential wind speed as seen in Fig. 8.5d below. It is therefore a natural consequence of the differential horizontal advection of M in conjunction with the assumed thermal wind balance.

Figs. 8.3i and 8.3j show the local tangential wind tendency $\partial v/\partial t$, obtained by evaluating the terms on the right-hand-side of the tangential momentum equation (5.62) with

$\dot{V} = 0$:

$$\frac{\partial v}{\partial t} = -u\zeta_a - w\frac{\partial v}{\partial z}. \qquad (8.1)$$

At both the initial time and at 10 hours, there is a strong positive tendency in the lower troposphere, within and outside the region of heating, and a tongue of positive tendency extending to the high troposphere about the axis of heating. The positive tendency is a maximum at the surface. Elsewhere, the tendency is mostly negative in the upper troposphere, above an altitude of about 5 km, and this negative tendency is strongest in the region of strong outflow (compare panels (i) and (j) of Fig. 8.3 with panels (c) and (d), respectively). Near the surface and inside the axis of maximum heating there is a small spin down tendency due to the weak outflow under the main updraft (see panels (c) and (d)).

Figs. 8.4a and 8.4b show isentropes of potential temperature, θ, in relation to the M-surfaces at the initial time and at 10 hours. They show also the isopleths of the discriminant D at these times. Notably, the θ-surfaces are close to horizontal at both times, even though they dip down slightly in the inner region reflecting the warm core structure of the vortex. At neither times are there regions of static instability. At the initial time, D decreases monotonically as a function of radius and, like θ, has a sharp positive vertical gradient at the tropopause. At 10 hours there is a small closed region of low values of D in the middle troposphere at a radius of about 135 km and centered at a height of about 7 km (panel (b)).

Simulation with diabatic heating, but no friction: Ex-HO

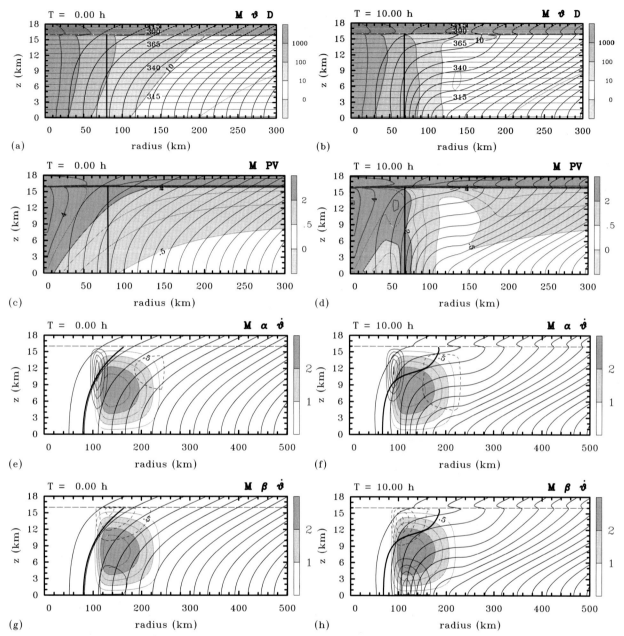

FIGURE 8.4 Radius-height cross sections of M-surfaces superimposed on various quantities for the simulation. These quantities include (a), (b) isentropes and the discriminant Δ of the Eliassen equation (shaded); (c), (d) PV (shaded); (e), (f) the forcing term α in the PV-tendency equation [Eq. (10)] superimposed on $\dot{\theta}$ (shaded); and (g), (h) the forcing term β in the PV tendency equation superimposed on $\dot{\theta}$ (shaded). The left panels are at the initial time and the right panels are at 9 hours. Contour intervals are as follows: for M, 5×10^5 m^2 s^{-1}; for D, 1×10^{-28} unit (very thin contour), 1×10^{-27} and 1×10^{-26} units (thin contour), 1×10^{-25} and 1×10^{-24} units (thick contours), where one unit has dimensions of m^4 s^{-4} kg^{-2} K^{-1}; for PV, 1 PV unit (PVU = 1×10^{-6} m^2 s^{-1} K kg^{-1}); for θ, 5 K; for $\dot{\theta}$, 0.5 K h^{-1}; for α and β, 1×10^{-10} m^2 s^{-2} K kg^{-1}.

8.1.1.3 Potential vorticity, development of inertially-unstable regions

To gain insight into the factors responsible for the evolution of the discriminant D, and in particular its eventual change from positive to negative values beyond the region of maximum heating, it proves useful to recall the dynamics

of dry Ertel potential vorticity (PV or P). For an axisymmetric flow expressed in cylindrical-polar coordinates, P is given by the formula

$$P = \frac{1}{\rho}\left(-\frac{\partial v}{\partial z}\frac{\partial \theta}{\partial r} + \zeta_a \frac{\partial \theta}{\partial z}\right), \qquad (8.2)$$

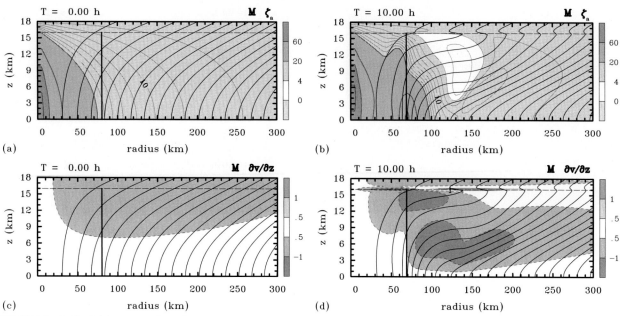

FIGURE 8.5 Radius-height cross sections of M-surfaces superimposed on various quantities for the simulation. These quantities include: (a, b) contours of absolute vorticity ζ_a (shaded); (c, d) contours of the vertical shear $\partial v/\partial z$. The left panels are at the initial time and the right panels are at 9 hours. Contour intervals are: for M, 5×10^5 m^2 s^{-1}; for ζ_a, 10^{-5} s^{-1}; for $\partial v/\partial z$, 5×10^{-5} s^{-1} (thin red (mid gray in print version) contours 1×10^{-5} s^{-1} and 5×10^{-5} s^{-1}).

where ζ_a denotes the absolute vertical vorticity (see Exercise 2.19). The relationship between D and P follows easily from Eqs. (5.58) and (5.68) and can be written as:

$$D = 4\gamma^2 \rho g \chi \xi P. \tag{8.3}$$

From this equation it is evident that the evolution of D is intimately tied to that of P and the parameter, ξ. For all examples considered here, the low-level tangential wind is everywhere cyclonic (i.e., $\xi > 0$) so that the only way for D to become negative is for the PV to become negative.

Figs. 8.4c and 8.4d show radial-height cross sections of the PV distribution at the initial time and 10 hours. At the initial time, the PV is a monotonically decreasing function of radius in the troposphere and has large values in the stratosphere on account of the large static stability there. However, at 10 hours, an extensive region of low PV values has formed in the mid-to-upper troposphere in an annulus extending between about 110 km and 160 km. This region is approximately centered on the region of low values of D seen in panel (b).

To understand how anomalously low (and eventually negative) PV can arise from an initially everywhere positive PV, we recall the equation for the material rate of change of PV:

$$\frac{DP}{Dt} = \frac{1}{\rho}\left(-\frac{\partial v}{\partial z}\frac{\partial \dot{\theta}}{\partial r} + \zeta_a \frac{\partial \dot{\theta}}{\partial z}\right), \tag{8.4}$$

where $D/Dt = \partial/\partial t + u\partial/\partial r + w\partial/\partial z$ is the material derivative operator in axisymmetric flow (cf. Eq. (2.63)). Figs. 8.4e-h show the two contributions on the right hand side of Eq. (8.4) to the material rate of change of PV at the initial time and 10 hours. The first term $\alpha = (-1/\rho)(\partial v/\partial z)(\partial \dot{\theta}/\partial r)$ represents the local generation or destruction of PV by the radial vorticity component times the radial gradient of diabatic heating rate, because, for an axisymmetric flow, $\partial v/\partial z$ is equal to minus the radial component of relative vorticity. The second term $\beta = (\zeta_a/\rho)(\partial \dot{\theta}/\partial z)$ represents the concentration or dilution of pre-existing PV by the vertical gradient of diabatic heating. At both times shown, the structure of α exhibits a radial dipole-like structure with a positive PV tendency just inside the axis of maximum heating in the mid to upper troposphere, and a weaker, but broader negative PV tendency beyond the heating axis at roughly similar altitudes. The structure of β shows initially a weak positive PV tendency at low-levels and a weak negative PV tendency at upper-levels centered along the axis of maximum heating. This structure resembles a vertical dipole and reflects the positive vertical gradient of diabatic heating at low levels and the negative vertical gradient at upper levels above the heating axis. The combined effect of the β and α terms is to produce an annulus of elevated (cyclonic) PV in the lower part of the heating region and reduced values of PV on the outer side of this annulus.

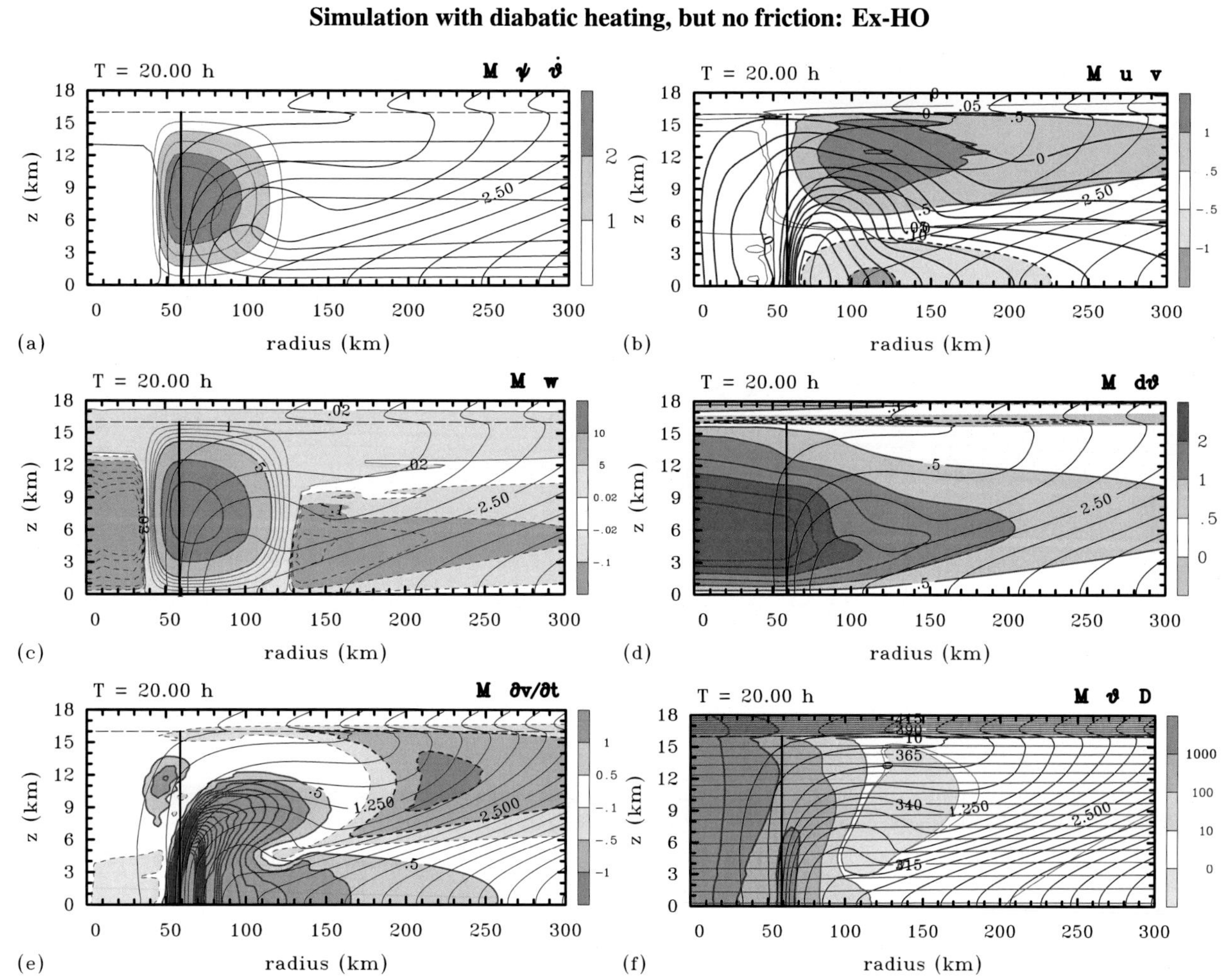

Simulation with diabatic heating, but no friction: Ex-HO

FIGURE 8.6 Legend as for Figs. 8.3 and 8.5, but at 20 hours: continued overleaf.

An important ingredient of both D and P is the vertical component of absolute vorticity, ζ_a, which characterizes the vortical structure of the horizontal wind field. For this reason it is of interest to examine vertical cross sections of this quantity.

Figs. 8.5(a) and (b) show cross sections of ζ_a in relation to the M-surfaces at the initial time and at 10 hours. At the initial time, ζ_a is everywhere positive and, like the PV, has a broad monopole structure (i.e., it decreases monotonically with radius). However, at 10 hours, the initially cyclonic ζ_a has become amplified in an annular region centered on the axis of diabatic heating and a region of low values has developed beyond the axis of maximum heating rate, mostly above 3 km. This region is clearly related to that of low values of D in Fig. 8.4b and PV in Fig. 8.4d.

Figs. 8.5c and 8.5d show the contours of vertical shear of the tangential wind component, $\partial v/\partial z$. At the initial time, the vertical shear is negative and the region of significant shear lies in the upper troposphere (panel (c)), whereas, by 10 hours (panel (d)), regions of stronger negative shear have

developed, the most prominent ones being in the upper part of the heating region with a thin layer extending near the tropopause inside the heating region, and in the low to mid troposphere, extending beyond the axis of maximum heating rate to a radius of about 200 km. The latter region is associated with the differential horizontal advection of the M-surfaces by the heating-induced inflow, which is a maximum near the surface.

Also at 10 hours, a thin layer of enhanced positive shear has developed beyond the heating region along the tropopause. As discussed in Subsection 8.1.1.2, this positive shear is associated with the differential horizontal advection of the M-surfaces near the tropopause.

Fig. 8.6 shows radius-height cross sections of selected quantities similar to those in Figs. 8.3-8.5, but at 20 hours, almost 12 hours after the Eliassen equation requires regularization. Even at this time, the fields are generally smooth. The axis of heating has moved further inwards to just under 60 km radius and some of the M-surfaces have folded in the mid and upper troposphere to form a "well-like" structure

Simulation with diabatic heating, but no friction: Ex-HO

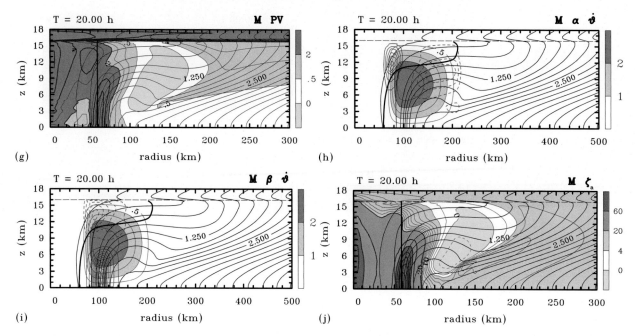

FIGURE 8.6 *(continued)*

beyond the heating region, partly on account of subsidence, but also because of the vertical advection of the M-surfaces in the ascending branch of the secondary circulation. Some distance inwards from the lowest point of the well, the radial gradient of M is negative, implying negative absolute vorticity and thereby inertial (centrifugal) instability. It is interesting that, even at this stage of evolution, the M-surfaces are generally not congruent with the streamlines in the upper troposphere as they would have to be if the flow were close to a steady state (Exercise 8.2).

Notably, the subsidence at radii inside the heating region has strengthened (panel (c)) and the eye has warmed further (panel (d)), θ' now being 3.4 K at a height of 7.5 km compared with 2.1 K at 7.8 km at 10 hours. The tangential wind tendency has increased further and remains positive at the location of V_{max} (compare panels (e) and (b)) consistent with the continued increase in V_{max} seen in Fig. 8.2a.

At 20 hours, the regions with low values of D, P, and ζ_a seen in Fig. 8.4 and Fig. 8.5 have become more pronounced and regions where these quantities are negative have formed (Figs. 8.6f, g, j). As noted above, the region of negative ζ_a coincides with that in which the M-surfaces dip down with increasing radius and because the static stability does not reverse sign anywhere (panel (f)), the region of inertial instability is approximately one also of symmetric instability in which both D and P are both negative. Panels (h) and (i) of Fig. 8.6 show similar patterns of PV generation as in Fig. 8.4e, f, g, h, respectively, with β now considerably strengthened in the upper and lower troposphere. Because, in the absence of heating, PV is materially conserved, the occurrence of regions of negative PV must be a result of the generation of negative PV in the upper troposphere, which is another way of viewing the formation of a region of symmetric instability.

8.1.1.4 Ultimate breakdown of solution

The breakdown of the regularized solution after 25 hours is brought about by the appearance and growth of small-scale features in the secondary circulation in the upper troposphere, near the edge of where $D < 0$ (not shown). The small-scale features arise from spatial irregularities introduced by the *ad-hoc* regularization procedure described in the appendix. This regularization procedure is necessarily arbitrary and only removes the effects of inertial instability in the Eliassen equation: it does not remove negative values of ζ_a (equivalent to a negative radial gradient of M) in the equation for the tangential wind tendency. As a result, regions of inertial instability are still seen by this tendency equation and a form of inertial instability can still manifest itself during the flow evolution. Typically, regions of static instability occur only in the presence of friction and heating as discussed in later sections.

Eventually, a time is reached when M becomes negative at some point, presumably on account of numerical issues. At this point, the solution is programmed to terminate. While it may be possible to extend the solution beyond this point for some time interval by refining the numerical algorithm (and in part smoothing the coefficients in the SE equation after each regularization step) or devising an alternate method for confronting the inertial instability (e.g., Wirth and Dunkerton, 2006), one of the aims of this chapter is to understand how the axisymmetric balance solution itself breaks down and not to devise necessarily *ad hoc* ways to extend the solution.

If, for the sake of argument, such a continuation method could be developed, we would expect to see continued sharpening of the radial gradient of M at the base of the updraft as air is drawn into the updraft from both sides near the surface. In essence, the flow there is trying to form a discontinuity in M (and a corresponding vortex sheet) by a process akin to frontogenesis (Hoskins and Bretherton, 1972; Emanuel, 1997).

Even if one could prolong the period in which the regularized solution could be obtained, then in the presence of nonaxisymmetric perturbations, one would expect that the annular vortex sheet would soon become baroclinically unstable on account of the reversal in sign of the radial and vertical gradients of the axisymmetric PV (Montgomery and Shapiro, 1995; Schubert et al., 1999; Naylor and Schecter, 2014). These unstable solutions with attending coherent vortex structures and vorticity mixing phenomenology could potentially dominate the inner-core evolution of the vortex so that the axisymmetric balance solution would cease to remain meaningful in a three-dimensional configuration. Such a setting is arguably the most relevant for tropical cyclone and weather forecasters on the bench.

An introduction to the topic of vortex shear instability was developed briefly in Chapter 4 within the context of horizontally nondivergent, two-dimensional vortex dynamics (Exercise 4.3). In that context, for example, one of the necessary conditions for shear instability was a reversal in sign of the axisymmetric radial vorticity gradient (the so-called generalized Rayleigh's theorem for shear instability). There we learned that this condition ensures the existence of counter propagating VR waves relative to the local angular velocity of the azimuthally-averaged vortex. In the example discussed here we see a natural tendency for the formation of an analogous vortex structure with the development of an annular region of axisymmetric PV (see, for example, Fig. 8.6g near 50 km radius).

In summary, the continuation issue has two components. The first is the continuation of the strictly axisymmetric solution to (and possibly beyond) a point of surface frontogenesis in the M field. The second concerns the growth and nonlinear lifecycle of asymmetric instabilities that are supported on account of the change in sign of the radial and vertical gradients of axisymmetric PV. Either component represents real deviations from the strictly axisymmetric balance solution. Questions concerning these and other nonaxisymmetric instabilities of the vorticity annulus, as well as longer-term evolution issues (as discussed in Chapter 15), lie beyond the scope of this chapter. These issues will be revisited in a planned sequel to this book. In the meantime, the reader is referred to Naylor and Schecter (2014), Menalou et al. (2016) and references cited therein for the latest quantitative results concerning vortex shear instabilities in simulated and idealized hurricanes.

8.1.2 Friction, no heating

It was shown in Section 6.7.3 that it is possible to include crudely the effects of frictional forcing in a balance context and it turns out to be instructive to do so. This is despite the fact that, in reality, radial inflow in the frictional boundary layer is fundamentally a result of *gradient wind imbalance* and not, for example, a manifestation of torque balance (Subsection 6.4.6). However, the balance formulation does lead to a low-level inflow that is qualitatively similar, albeit quantitatively weaker than a more complete boundary layer formulation (Smith and Montgomery, 2008).

In this subsection we examine the balanced frictional spin down of a vortex in the absence of diabatic heating. The vortex is the same as that used in the previous subsection. The effects of surface friction are represented by a body force corresponding with the surface frictional stress distributed through a layer with depth H. The body force has the spatial form

$$\dot{V}(r, z) = \frac{C_d}{H} |\mathbf{v_s}| v_s \left(1 - \frac{z}{H_1}\right)^2, (0 \le z \le H_1) \quad (8.5)$$

where C_d is the surface drag coefficient (assumed here to be a constant) and $\mathbf{v_s}(r, 0, t) = \sqrt{[u(r, 0, t)^2 + v(r, 0, t)^2]}$ is the surface wind speed at time t and v_s is the tangential component of $\mathbf{v_s}$. Here we choose $H = 500$ m, $H_1 = 2000$ m and set $C_d = 2 \times 10^{-3}$. These parameters are tuned so that the inflow occurs within a surface-based layer about 1 km in depth. This simple formulation is in the spirit of that proposed by Shapiro and Willoughby (1982), but it is different from that for a classical boundary layer discussed in Chapter 6 in that it assumes strict gradient wind balance and does not have a radial pressure gradient force that is uniform through the depth of the layer.

Fig. 8.7a shows time series of the maximum tangential wind component $V_{max}(z)$ at the surface ($z = 0$) and at heights z of 1 and 2 km in this simulation. At no time was the regularization procedure required. It is seen that there is a comparatively rapid spin down of the surface tangential wind component from 10 m s^{-1} at the initial time to slightly less than 3 m s^{-1} at 30 h. However, the rate of spin down becomes progressively slower as the height increases.

Fig. 8.7b shows the radial and tangential wind components as well as the M-surfaces at the initial time. The imposition of surface friction leads to a shallow layer of inflow near the surface with a weaker and somewhat deeper outward return flow above the inflow. The maximum return flow occurs at an altitude of about 2 km. As shown

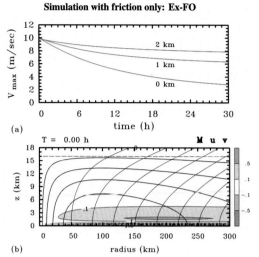

(a)

(b)

FIGURE 8.7 (a) Time series of maximum tangential wind component (here $V_{max}(z)$) at the surface and at heights z of 1 and 2 km. (b) Radius-height cross sections of M-surfaces superimposed on the radial and tangential wind components, u and v, respectively. Contour intervals: for M, 5×10^5 m^2 s^{-1}; for u (shaded), 0.2 m s^{-1} down to -1 m s^{-1}, 1 m s^{-1} for lower values (negative contours dashed); for v, 2 m s^{-1}.

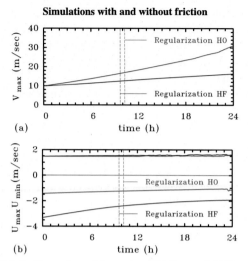

(a)

(b)

FIGURE 8.8 Comparison of time series of (a) V_{max} and (b) U_{max}, U_{min} in the calculations with heating and friction, Ex-HF (blue curves) and the calculation with diabatic heating only, Ex-HO (red curves). The dashed vertical lines indicate the time at which the discriminant in the Eliassen equation first becomes negative.

in Chapter 6, the spin down of the tangential wind above the surface inflow layer is largely a result of the generalized Coriolis force $-(v/r + f)u$ and this is confined to the height range where the radial outflow is appreciable. The shallowness of the return flow is a reflection of the stable stratification. Above the return flow, the spin down of the tangential circulation is minimal.

In the simple representation of friction considered here, the tangential component of frictional stress is reduced as the near-surface tangential flow weakens and so, therefore, does the secondary circulation induced by the friction (blue dashed curve in Fig. 8.8). Clearly, spin up of the vortex would require diabatic heating to induce inflow of sufficient strength to negate the frictionally induced outflow above the inflow layer. We investigate the combined effects of friction and diabatic heating in the next subsection.

8.1.3 Diabatic heating and friction

We examine next the effects of surface friction on the balanced vortex evolution in the presence of diabatic heating as formulated in Section 8.1.1. Fig. 8.8 compares time series of V_{max}, U_{max}, and U_{min} for this calculation with similar time series for the diabatic heating only calculation.

In the calculation with diabatic heating and friction, V_{max} increases approximately linearly from 10 m s^{-1} at the initial time to a little over 16 m s^{-1} at 24 hours, about two hours before the solution breaks down. The solution requires regularization to be applied a little before 10 hours, less than an hour earlier than in the calculation without friction, but it breaks down about an hour later than in the latter

calculation. Notably, in the presence of near-surface friction, U_{min} is generally at least more than twice in magnitude as large as in the no friction calculation: at the initial time its magnitude is 3.3 m s^{-1} compared with 1.4 m s^{-1} in the case without friction. In both cases, U_{min} declines progressively with time, while U_{max}, which occurs in the upper troposphere, changes little during the calculation and has practically the same value with or without friction.

Fig. 8.9 shows a selection of radius-height cross sections for the calculation with diabatic heating and friction at selected times, similar to those in Figs. 8.3c and 8.4c. Perhaps of most significance in comparing the radial motion between this simulation and that without friction is the fact that the near-surface inflow is much stronger at most radii. The stronger inflow is a result of the additional frictionally-induced inflow, a point that we return to in the next section. The shallow surface-based outflow seen at small radii inside the axis of heating in Figs. 8.3c, d and 8.4b no longer occurs and with friction, there is now weak inflow in this region (Figs. 8.9a, b).

Unlike the case without friction, where V_{max} remains near the surface at all times, in the calculation with diabatic heating and friction, at 10 hours V_{max} occurs above 2 km and its altitude continues to increase with time. This height is still within a region of significant radial inflow ($U_{min} > 1$ m s^{-1}), albeit above the layer of strongest inflow. We showed earlier that this is not a realistic feature of observed tropical cyclones, where the largest tangential wind speeds are found to occur within, but near the top of the frictional boundary layer (see Section 6.6).

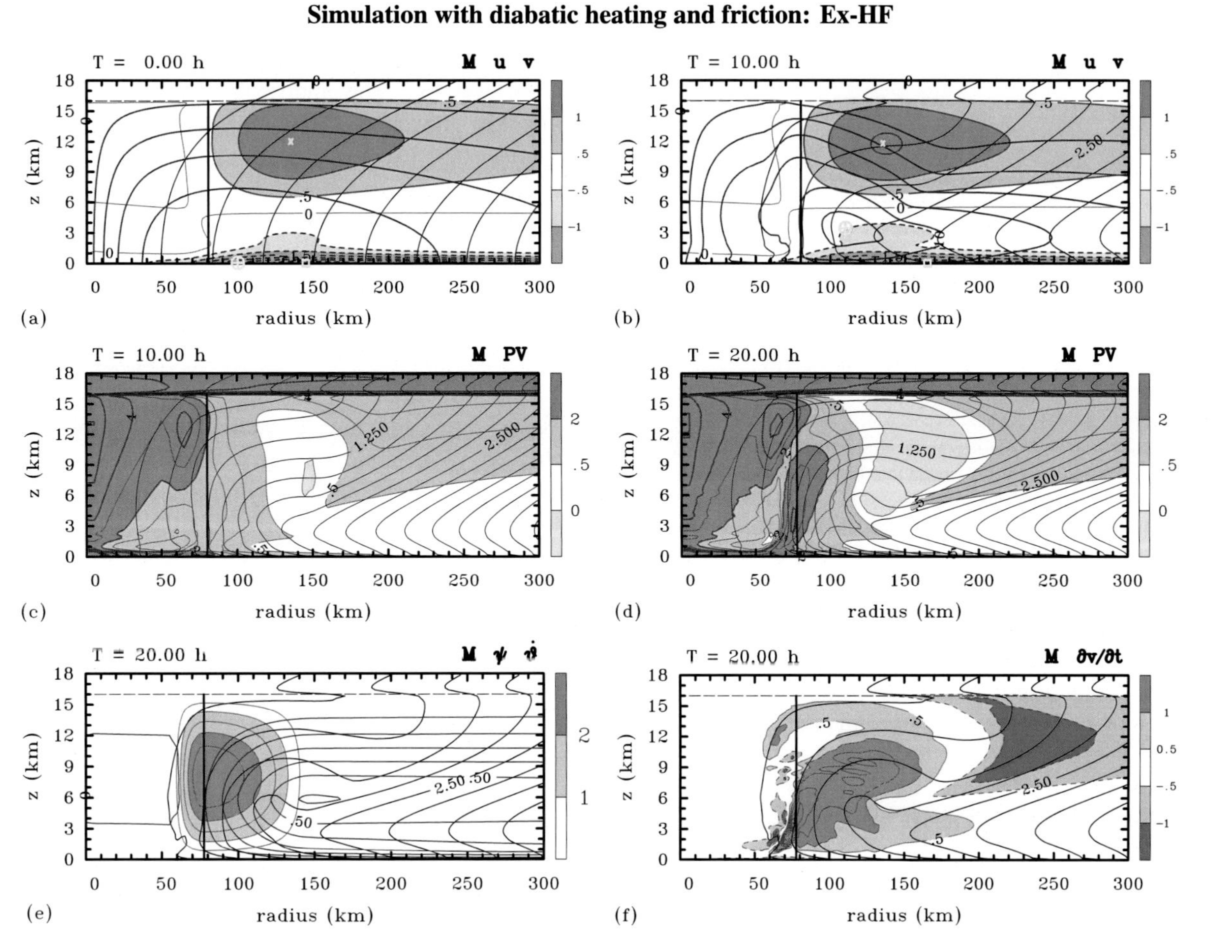

Simulation with diabatic heating and friction: Ex-HF

FIGURE 8.9 Results for the calculation with diabatic heating and friction. (a) and (b) as for panel (c) and (d) in Fig. 8.3. (c) and (d) as for panel (c) in Fig. 8.5, but at 10 hours and 20 hours, respectively. Panels (e) and (f) show, *inter alia*, the streamfunction and tangential wind tendency fields at 20 h, similar to Figs. 8.6a and 8.6e, respectively. The yellow symbols in (a) and (b) indicate the locations of maximum v, ($\oplus$), maximum u, ($\times$) and minimum u, ($\square$).

Figs. 8.9c, d show the PV fields at 10 hours and 20 hours for the calculation with diabatic heating and friction. These should be compared with those for the case without friction shown in Figs. 8.4d and 8.6g, respectively. It is seen that the differences are confined primarily to the lower troposphere (below about 5 km) and that the regions of negative PV are practically the same.

Figs. 8.9e, f show, *inter alia*, the streamfunction and tangential wind tendency fields at 20 hours, similar to Figs. 8.6a, e. As in the case without friction (Fig. 8.6), much of the low level inflow passes through the heated region and there is a positive tangential wind tendency through much of the heated region. Again, in the upper troposphere, the positive tendency occurs in a region of radial outflow so that the vertical advection of M must exceed the negative radial advection of M.

8.2 Interpretation: the classical spin up mechanism

The evolution of vortex structure in the foregoing balance solutions illustrates some of the individual processes discussed in Chapters 5 and 6. These processes form a basis for the classical spin up mechanism articulated by Ooyama (1969), which is illustrated schematically in Fig. 8.10. The idea is that the collective effects of deep convection in some inner region of the vortex circulation are to carry the air into the upper tropospheric outflow layer and at the same time to draw air inwards in the lower troposphere, both in and above the boundary layer.

Above the boundary layer, rings of air moving radially and/or vertically approximately conserve their individual value of M (Eq. 5.8). Those moving inwards in the lower troposphere spin faster, while those moving outwards in the upper troposphere spin more slowly. Eventually, at radii beyond $\sqrt{2M/f}$, $v < 0$, the outward-moving air parcels begin

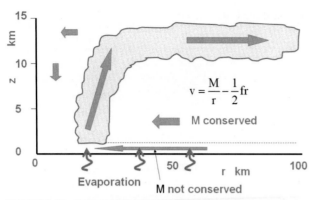

FIGURE 8.10 Schematic illustration of the azimuthally-averaged view of vortex spin-up in the thought experiment for tropical cyclone genesis and intensification. Blue (dark gray in print version) arrows denote the sense of transverse (secondary) circulation. Light blue (light dark gray in print version) shading denotes cloudy areas. The size of the arrows is not proportional to the flow speed. The curvy dark blue (dark gray in print version) arrows indicate evaporation from the sea. Adapted from Smith and Montgomery (2016). Republished with permission of Wiley.

to spiral anticyclonically. The spin up in the lower troposphere above the frictional boundary layer as a result of the conservation of M constitutes the *classical spin-up mechanism* mentioned in Section 5.6.

At large radii within the boundary layer, the tangential wind is less than that just above the layer and, as shown in Chapter 6, there is an inward agradient force that drives inflow, an effect that dominates the "suction effect" of deep convection except close to the convection, itself (Kilroy et al., 2016a).[4] The spiraling inflow in the boundary layer picks up moisture from the sea surface providing a source of moisture to replenish convective instability. This replenishment sustains inner core deep convection in support of the intensification process. Of course, the convectively-induced inflow above the boundary layer has to be strong enough to negate the tendency of frictionally-induced outflow there. From an alternative perspective, for intensification to occur, the convective mass flux in the deep convective clouds has to be *more than sufficient* to *ventilate* the air that is converging in the boundary layer at a particular time and carry this air into the upper troposphere (for more on this point, see Sections 8.4 and 15.5.4).

The latent heat release in the clouds tends to warm the inner core of the vortex, making the vortex less unstable to subsequent deep convection. In addition, precipitation-cooled downdrafts produced by deep convection further stabilize the atmosphere to deep convection, although this ef-

fect becomes progressively weaker as the mid-tropospheric relative humidity increases on account of moistening by prior convection. This moistening serves to reduce the strength of downdrafts. To compensate for these stabilization effects, the latent heat fluxes at the ocean surface serve to restore the convective instability. Ooyama (1969) showed in the context of a simple model that without such fluxes, a vortex would not intensify.

Of course, the foregoing summary of the classical spin up mechanism ignores the complications associated with potential asymmetric shear and other instabilities, including their accompanying nonlinear life cycle and related coherent vortex structures as well as vorticity mixing phenomenology, mentioned in this chapter and introduced briefly in Chapter 4. These asymmetric deviations to the axisymmetric cartoon summarized herein are generally nontrivial perturbations to the classical model, at least in the vortex core region possessing the largest vorticity (PV) gradients. However, it is our experience based on examining both observations and a number of hurricane simulations in different modeling platforms that the present cartoon still serves to summarize the azimuthally-averaged features of the classical spin up mechanism.

8.2.1 Exercises

Exercise 8.1. Explain why the location of V_{max} might be expected to progressively rise as the vortex evolves in the heating only simulation described in Section 8.1.1.

Exercise 8.2. Recall from Exercise 5.6 that in a steady, inviscid, axisymmetric flow, the M-surfaces and streamlines must be congruent. If the flow is unsteady, show that a local increase of M requires the flow to have a component in the direction of decreasing M.

Exercise 8.3. (a) Show that for a given tangential velocity distribution, a doubling of the diabatic heating strength in the Eliassen equation would lead to a doubling of the strength of the induced secondary circulation, but not to a change in the radial or vertical scales of the strengthened circulation.

(b) Speculate on the implications of these results for an intensifying vortex in the context of the balance vortex model presented in this chapter. Do these results suggest a brake on the final intensity and size of the emergent vortex under the assumption of a fixed diabatic heating rate structure? Hint: Consider the implication of the inertial stability parameter in the secondary circulation as the vortex intensifies under constant heating rate amplitude and vertical structure.

---

4. Note that in the present calculations, the linearity of the Eliassen equation implies that the frictional effects of the boundary layer in producing inflow are additive to those of the convective heating. However, this is not true in general because the boundary layer in the inner core region of a tropical cyclone is intrinsically nonlinear except, perhaps, at low maximum tangential wind speeds of a few m s^{-1} (see the scale analysis in Section 6.2).

8.3 Rotational stiffness, latitude dependence and vortex size evolution

The conservation of absolute angular momentum, M, above the boundary layer indicates that the amplification of the tangential velocity of an inward moving air parcel depends on the parcel's value of M at some earlier time and on the radial distance by which it is has moved from this time. Further, the basic principles developed in Chapter 5 suggest that the ability of deep convection to draw air parcels inwards above the boundary layer is constrained by the rotational stiffness of the vortex, which, ignoring radial density variations, is quantified by the inertial stability I^2. This quantity has the axisymmetric forms $(1/r^3)\partial M^2/\partial r$ or $\zeta_a(f+2v/r)$ (see Chapter 5).

For a cyclonic vortex, I^2 increases with latitude through its dependence on the Coriolis parameter, f, and for an air parcel at a given radial distance in a particular radial profile of tangential wind, the value of M will increase with latitude also. Thus, considering radial displacements alone, the amplification of the tangential velocity for an inward displaced air parcel from a given initial radius at a particular latitude depends on two competing effects: the radial displacement that can be achieved, which, for a given radial driving force decreases with increasing latitude, and the initial value of M, which increases with latitude. Clearly, one has to do the calculation to determine the net effect on the amplification possible.

These considerations are relevant to understanding the controls on the size of the inner vortex core, as measured by the radius of maximum tangential wind speed, or the outer size, as measured, for example, by the radius of gale force winds ($17\ \mathrm{m\ s^{-1}}$). At any radius, the tangential winds will continue to increase as long as M-surfaces are being drawn inwards at that radius (assuming that the vortex is inertially stable) and the vortex will expand in size. The increase in the radius of gale-force winds is an important observed feature of storms as they intensify and mature. An example is illustrated in Fig. 8.11 for Typhoon Chan-Hom (2015). This figure shows JTWC postseason-analyzed 34-kt ($17\ \mathrm{m\ s^{-1}}$) wind radii estimates from an objective, satellite-based technique to determine the extent of gales (see e.g., Sampson et al., 2017 and refs). It can be seen that the radius of gales grows about three-fold over a period of about five days.

The question is then, what factors control the growth in the radius of gales, or more specifically, what controls the inward movement of the M surfaces? The basic principles governing either measures of size, i.e., the radius of maximum tangential wind speed or the radius of gales, would appear to be encapsulated in the classical spin up mechanism articulated above and may be explored in the context of an axisymmetric model.

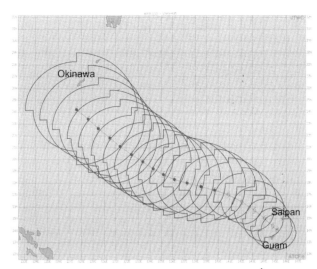

FIGURE 8.11 JTWC postseason-analyzed 34-kt ($17\ \mathrm{m\ s^{-1}}$) wind radii estimates from an objective technique for the 5 days leading up to Typhoon Chan-Hom (2015), which passed south of Okinawa. (Courtesy Buck Sampson. Adapted from Sampson et al., 2017).

8.3.1 A laboratory experiment

Perhaps the simplest answer to the question of size goes back to ideas developed in the context of tall, thin vortices as exemplified by a laboratory experiment by Turner and Lilly (1963), in which a vortex was produced in water contained in a rotating cylinder by releasing bubbles from a thin tube along the upper part of the rotation axis. A sketch of the flow configuration is shown in Fig. 8.12.

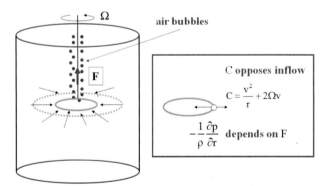

FIGURE 8.12 Schematic of flow configuration in the Turner and Lilly experiment. The right side indicates the principal forces per unit mass acting on an air parcel in the radial direction: an inward-directed pressure gradient force, $-(1/\rho)(\partial p/\partial r)$, and an outward force C, which is the sum of the centrifugal force (v^2/r) and Coriolis force ($2\Omega v$).

In general, there are two fundamental requirements for vortex amplification: a source of rotation and some forcing mechanism to concentrate the rotation. In the laboratory experiment sketched in Fig. 8.12, the drag on the fluid exerted by the ascending bubbles generates a secondary circulation in the water, producing radial convergence in the region below the source of bubbles and radial divergence near the top of the container. Except in a shallow boundary layer near the lower boundary, converging rings of fluid conserve their absolute angular momentum and spin faster. The ultimate degree of amplification of the angular velocity depends on how far rings of fluid can be drawn inwards by the overturning circulation produced by the rising bubbles, which, in turn, for a given bubbling rate, depends on the background rotation rate.

How far rings of fluid can be drawn inwards depends, inter alia, on the forcing strength, i.e., on the bubbling rate. If the forcing is sufficiently large for a given rotation rate, rings of fluid may be drawn in to relatively small radii before the centrifugal force (v^2/r) and Coriolis force ($2\Omega v$) opposing the inward motion balance the radial pressure gradient induced by the bubbles. It is this pressure gradient that drives the rings of fluid inwards (right of Fig. 8.12). Here v is the azimuthal velocity component, r is the radius from the rotation axis and Ω is the background rotation rate.

If the forcing is comparatively weak, or if the rotation rate is sufficiently strong, this balance may be achieved before the radial displacement of a fluid parcel is very large, so that a significant amplification of the background rotation will not be achieved. Of course, if there is no background rotation, there will be no amplification, and if the background rotation is very weak, the centrifugal and/or Coriolis forces never become large enough to balance the radial pressure gradient, except possibly at large radii from the source of bubbles. *Note that these arguments emphasize the unbalanced wind adjustment from the instant that the bubbling is turned on.*

The foregoing considerations point to the existence of an optimum forcing strength to produce the maximum amplification of the tangential velocity for a given strength of background rotation, or an optimum background rotation rate for a given forcing strength. These ideas were eloquently articulated by Morton (1966) and demonstrated in related numerical experiments by Smith and Leslie (1976, 1978).

The dust devil-like vortices investigated in Smith and Leslie (1976) are driven by the buoyancy arising from thermal heating of the surface. It turns out that the high-rotation-rate regime produces warm-cored, two-cell vortices that are an observed feature of some dust devils (Sinclair, 1969) and reminiscent of tropical cyclones.

Given the utility of the foregoing ideas to tornadoes, waterspouts and dust devils it is natural to ask to what extent the above ideas apply to tropical cyclones? Also, how must these ideas be modified to apply in a balance framework? The latter question is perhaps more appropriate when constructing elementary arguments to explain the behavior of tropical cyclones.

8.3.2 Balance considerations

The reason why the foregoing ideas need to be modified for a balance framework is that, within this framework, one can-not talk about radial or vertical force imbalances driving the flow as these imbalances are not present by definition (see Subsection 5.8.2.3). The question then is, does increasing inertial stability suppress radial motion in a balance framework? An answer to this question in the affirmative is provided by a Green function solution for unit jumps and point sources of diabatic heating by Eliassen (1951) (see e.g., Fig. 5.15). In the next section we demonstrate this result for the distributed heat source used earlier in this chapter.

Another feature of the balance framework presented in Section 5.10.2, is the effect of the generalized inertial stability I_g^2 on the radial scale of the induced secondary circulation. Referring to Exercise 8.3, the radial scale of the secondary circulation forced by the diabatic heating rate in the Eliassen equation depends on the radial distribution of I_g^2, which, in turn, depends on the radial profile of tangential velocity. This scale does not depend on the strength of the heating rate, even though the strength of the induced overturning circulation increases with the strength of the heating rate.

As the tangential winds strengthen in the inflow branch of the circulation, I_g^2 will generally increase, reducing the radial scale of the overturning circulation and thereby the radial range of M values that the developing vortex is able to access for further intensification. Further, the reduced inflow accompanying the increased inertial instability (as demonstrated later in Fig. 8.13) will reduce the rate at which M values are advected inwards.

In the foregoing thought experiment for balance evolution, the radial pressure gradient force would seem to have no limit if one were to allow fluid parcels to become arbitrarily close to the axis of rotation. However, in a balance vortex the flow must be in strict hydrostatic balance also, so the magnitude of the radial pressure gradient must be consistent with the temperature perturbations forced by the heating rate. Therefore, for a balanced vortex forced by diabatic heating, the net pressure change is constrained by both the radius at which the inflow vanishes and the column integrated warming. (Note that the kinetic energy of an inward-moving air parcel at radius r in a rotating flow increases as $1/r^2$ while conserving M, which points to an additional constraint on inward moving air parcels in a system with finite energy. The kinetic energy requirement suggests also why there must be a radius at which the inflow vanishes.)

In contrast, in the upper-level outflow region, the flow becomes centrifugally unstable ($I_g^2 < 0$). This instability occurs when either the absolute vorticity becomes negative (i.e., $\zeta_a < 0$) or the tangential velocity becomes sufficiently anticyclonic (i.e., $f + 2V/r < 0$).

We explore the foregoing issues further in the next subsection, where we present some simulations using the prognostic balance model described in Section 8.1.1.

Simulation with diabatic heating only: latitude dependence

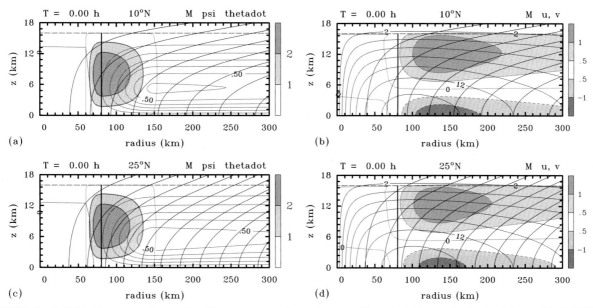

FIGURE 8.13 Radial-height cross sections of M-surfaces superimposed on various quantities at the initial time for the simulations at (a), (b) 10°N and (c), (d) 25°N. The left panels show streamlines of the secondary circulation, ψ, and diabatic heating rate $\dot\theta$ (shaded) and the right panels show contours of tangential velocity, v, and radial velocity u (shaded). Contour intervals are: for M, 5×10^5 m^2 s^{-1}; for ψ, 1×10^8 kg s^{-1}; for $\dot\theta$, 0.5 K h^{-1}; for u, 0.5 m s^{-1}, 0.05 m s^{-1} (very thin contour), 1 m s^{-1} for negative values; for v, 2 m s^{-1}. For all fields: positive values (solid), negative values (dashed).

Simulation with diabatic heating only: latitude dependence

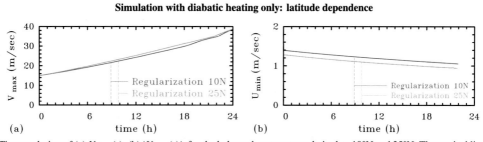

FIGURE 8.14 Time evolution of (a) $V_{max}(t)$, (b) $|U_{min}(t)|$, for the balanced vortex at two latitudes: 10°N and 25°N. The vertical line indicates the first time at which the discriminant in the Eliassen equation first becomes negative. The curves for $|U_{min}(t)|$ are terminated at a point where the location of $|U_{min}(t)|$ jumps several km in height, heralding the imminent breakdown of the solution a few hours later.

8.3.3 Idealized balance simulations

The prognostic balance model developed in this chapter offers the opportunity to explore the effect of latitude on vortex evolution and, in particular, the effects on vortex size. The calculations to be presented are essentially the same as that for the no-friction case described in Section 8.1.1, but they start with a stronger initial vortex (15 m s^{-1} instead of 10 m s^{-1}). The magnitude and structure of the diabatic heating rate are the same, as is the initial location of maximum heating. The only difference between the two new simulations is the assumed latitude, which has values of 10°N or 25°N.

The left panels of Fig. 8.13 shows the radius-height structure of the diabatic heating rate and contours of the streamfunction for the overturning circulation ψ at the initial time for these two latitudes, while the right panels show

corresponding contours of the radial and tangential velocity components. Contours of M are shown in all panels. Comparing panels (a) and (c) and panels (b) and (d) shows that the initial overturning circulation is a little stronger at 10°N than at 25°N. Nevertheless, the maximum inflow, which occurs close to the surface as in the calculation in Section 8.1.1, is only marginally stronger at 10°N (Fig. 8.14) and, despite the stronger inflow, the maximum tangential velocity is marginally smaller. It follows, then, that in this idealized flow configuration with the same prescribed heating and no friction,[5] the effect of increasing M with latitude slightly outweighs the effect of smaller radial displacements of air parcels, implied by slightly smaller radial velocities, in producing the maximum tangential wind speed.

---

5. One could of course question the realism of this assumption, an issue we return to in Section 16.3.

Simulation with diabatic heating only: latitude dependence

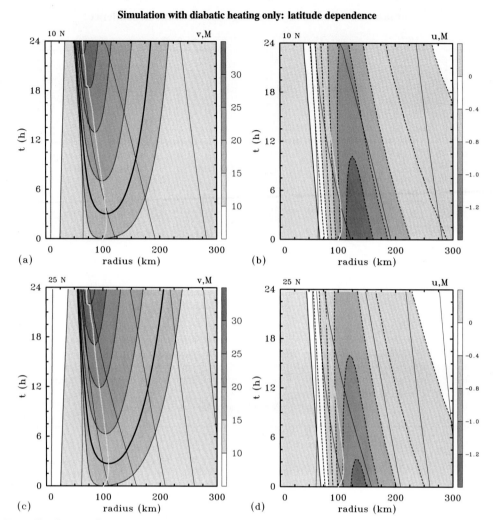

FIGURE 8.15   Hovmöller diagrams of the tangential (v, left column) and radial (u, right column) velocity components at the surface for the calculations with diabatic heating only at 10°N and 25°N. Contour interval: 5 m s^{-1} for v; 0.2 m s^{-1} for u. The thick solid black contour is the 17 m s^{-1} contour corresponding with the radius of gale-force tangential winds and the yellow (light gray in print version) curve shows the radial location of maximum tangential wind speed. Superimposed in each panel are contours of absolute angular momentum, M (thin black contours), contour interval 1×10^6 m^2 s^{-1}.

The rate of intensification differs only slightly between the two latitudes. Between 6 hours and 18 hours, the rate of increase per day is marginally larger at 25°N, 22.9 m s^{-1} per day compared with 21.2 m s^{-1} per day at 10°N.

Fig. 8.15 shows Hovmöller diagrams of the tangential and radial velocity components at the surface for the calculations at 10°N and 25°N. Comparing the tangential velocity components in panels (a) and (c), it is seen that there is little perceptible difference in the radius of maximum velocity (indicated by the yellow (light gray in print version) line), although the outer radius of gale force winds (solid black curve) is generally larger at 25°N, reflecting the ability of the vortex to access the larger values of M at outer radii.

Note that at early times at both latitudes, the outer radius of gales increases rapidly at first, but the rate of increase progressively declines as the vortex strengthens. This de-

cline is as would be expected from the increasing inertial stability and the accompanying local reduction of the inflow as the radial scale of response to the heating becomes smaller. Of course, in reality, frictional effects will be important near the surface and the insights from the heating only simulation may need to be modified. We examine this issue in the next subsection using the simple balance model with friction included described in Subsection 8.1.3.

8.3.4   Effects of friction on vortex size growth

To examine the additional effects of friction on the growth in vortex size, a pair of simulations similar to those in Section 8.1.3 were carried out, but starting with a stronger vortex, the same as in Section 8.3.3. Because of ventilation issues to be discussed in Section 8.4 below, the stronger vor-

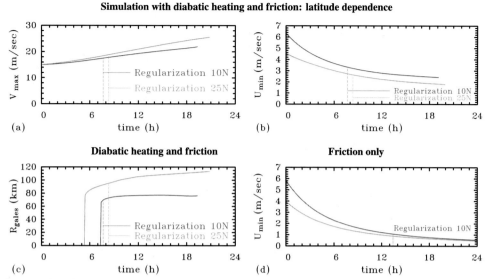

FIGURE 8.16 Time evolution of (a) $V_{max}(t)$, (b) $|U_{min}(t)|$, (c) R_{gales} for the balanced vortex with heating and friction at the two latitudes 10°N (blue (dark gray in print version) curves) and 25°N (red (mid gray in print version) curves). The vertical line indicates the first time at which the discriminant in the Eliassen equation first becomes negative. (d) Time evolution of $|U_{min}(t)|$ in the two friction only calculations. In these, only the calculation at 10°N requires regularization.

tex necessitated an increase in the maximum magnitude of diabatic heating rate and an inwards relocation of the axis of heating distribution to obtain solutions that intensify by more than just a few meters per second before the solution breaks down. The axis of heating is moved inwards from 80 km to 50 km in these calculations and the maximum strength of the heating rate is increased modestly to 4 K/hour.

8.3.5 Vortex intensity and size metrics

Figs. 8.16a-c summarize various metrics of vortex evolution for the two simulations with friction. These metrics include: the maximum tangential wind speed at an altitude of 2 km, V_{max}; the outer radius of gale-force winds at this altitude, R_{gales}; and the maximum radial inflow, $|U_{min}|$, which is typically within the lowest kilometer. In the presence of friction, the maximum tangential wind speeds are generally smaller than in the case without friction and the location of the maximum becomes elevated above the nominal depth of the friction layer (1 km) after about 4 hours, reaching 2 km by about 12 hours and exceeding 3 km after about 17 hours. The two simulations shown are relatively smooth until after 24 hours, but break down 2-3 hours after this time. The rate of intensification is a little less in these cases than in the cases without friction: between 6 hours and 18 hours, the rate is 18.1 m s^{-1} per day at 10°N and 20.1 m s^{-1} per day at 25°N, but, of course, the near surface winds are much reduced in the presence of friction.

Although the maximum inflow is stronger at 10°N, the maximum tangential winds at a height of 2 km are slightly higher at 25°N. This behavior is the same as that in the

simulations without friction and, again, *the effect of increasing M with latitude outweighs the effect of smaller radial displacements of air parcels in producing the maximum tangential wind speed.*

As in the cases without friction, R_{gales} is generally a little larger at the higher latitude, the difference here being the altitude at which the outer radius of maximum gales occurs is around 2 km in the presence of friction, while it is much closer to the surface when there is no friction.

8.3.6 Dependence of frictionally-driven inflow on latitude

In Section 8.1.3, we showed that, in the case with diabatic heating and friction, the frictionally-driven inflow approximately doubles the near-surface inflow induced by the heating rate. At the same time, friction reduces the tangential wind speed near the surface. The fact that the diabatic heating rate distribution is the same in the two foregoing simulations, the stronger maximum inflow at 10°N points to a dependence of frictionally-driven inflow on latitude. This feature is confirmed by Fig. 8.16d, which compares time-series of U_{min} in two analogous simulations at 10°N and 25°N *without diabatic heating*, similar to the simulation in Section 8.1.2. At the initial time, U_{min} is about 5.8 m s^{-1} at 10°N compared with only about 4 m s^{-1} at 25°N, but because the frictionally-induced overturning circulation is thereby stronger at 10°N, the vortex at this latitude decays more rapidly so that the two time series converge over time.

It turns out that, for the calculations with friction only, regularization is required only in the simulation at 10°N, a feature that can be traced to the higher vertical shear of the

Simulation with friction only: latitude dependence

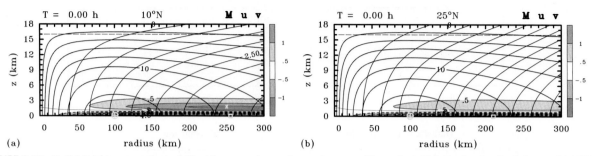

FIGURE 8.17 Radial-height cross sections of M-surfaces superimposed on various quantities at the initial time for the simulations at (a) 10°N and (b) 25°N. The left panels show streamlines of the secondary circulation, ψ, and diabatic heating rate $\dot\theta$ (shaded) and the right panels show contours of tangential velocity, v, and radial velocity u (shaded). Contour intervals are: for M, 5×10^5 m^2 s^{-1}; for ψ, 1×10^8 kg s^{-1}; for $\dot\theta$, 0.5 K h^{-1}; for u, 0.5 m s^{-1}, 0.05 m s^{-1} (very thin contour), 1 m s^{-1} for negative values; for v, 2 m s^{-1}. For all fields: positive values (solid), negative values (dashed). The yellow (light gray in print version) symbols indicate the locations of maximum v, ($\oplus$), maximum u, ($\times$) and minimum u, ($\square$).

near-surface tangential wind that develops at this latitude. In fact, the discriminant of the Eliassen equation only becomes negative for a finite strip of locations at the surface.

The stronger overturning circulation induced by friction as the latitude is decreased is further highlighted in Fig. 8.17, which shows isotachs of the radial component of flow in the two friction only simulations at 10°N and 25°N at the initial time. The latitudinal dependence of the inflow is a robust feature of the Ekman boundary layer (Chapter 6, see Exercise 6.4) as well as more sophisticated representations of the boundary layer and it has important implications for vortex structure and evolution.

The M-surfaces at the initial time are shown also in Fig. 8.17 and, as in the simulation shown in Section 8.1.2 (see Fig. 8.6), these have not yet been affected by friction. As discussed later in Section 8.4.4, this is a pathological feature of the boundary layer formulation in balance models.

8.3.7 Summary

The foregoing results illustrating aspects of the latitudinal dependence of the balance solutions provide a starting point for understanding the controls on vortex size in more realistic models where the structure of diabatic heating rate is not held fixed, but is determined as part of the solution. The latitudinal dependence of intensification rate and size in such models is examined in Section 16.3.

8.4 Interplay between diabatic heating and friction

The simulations with diabatic heating and friction at different latitudes indicate that the ensuing flow depends on the relative importance of these forcing processes, since, while both lead to inflow near the surface, friction would, by itself, lead to outflow above the frictional boundary layer. As discussed in Section 8.2, vortex intensification by the classical mechanism requires that the diabatically-forced inflow

above the boundary layer be strong enough to offset the frictionally-induced outflow there.

It is reasonable to surmise that, in reality, the ability of deep convection to ventilate the mass of air converging in the boundary layer depends, among other things, on the degree of convective instability and hence on the moisture content of the ascending air (for more on this idea see Section 9.9.1). However, the rate at which air converges in the boundary layer depends to a significant degree, through boundary-layer dynamics, on the radial structure of the gradient wind at the top of the boundary layer.

In Section 15.5.4, the relative importance of ventilation by deep convection is shown to vary systematically during the life cycle of a tropical cyclone, with ventilation having the upper hand during early intensification stages and the influx of mass in the boundary layer having the upper hand through much of the mature and decay stages. It is shown also that there is a general mismatch between ventilation and boundary layer mass influx on shorter, deep convective time scales, which accounts for short-term fluctuations in vortex intensity. *The possibility of such a mismatch appears to have gone unnoticed in many past theoretical studies, including the pioneering studies of Ooyama (1968, 1969) and Emanuel (1986, 1995, 1997), where it has been traditional to assume that all of the air that converges in the boundary layer ascends in deep convection to the upper troposphere.*

In this section we explore one aspect of the ventilation problem by comparing five simulations of the prognostic balance model in which the strength of diabatic heating is varied while keeping the formulation of surface friction the same. In particular, we examine the dependence of vortex intensification on the diabatic heating rate. As summarized in Table 8.2, these simulations differ only in the maximum magnitude of the heating distribution, the structure of which remains constant in time, but the axis of the heating is allowed to move with a prescribed M-surface at a height of 1 km as before. The first simulation, Ex-HF, is the same as the calculation with heating and friction in Section 8.1.3 and

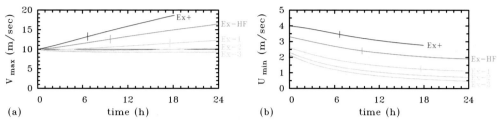

FIGURE 8.18 Time series of (a) V_{max} and (b) U_{min} for the simulations listed in Table 8.2. The short dashed vertical lines indicate the time at which the regularization is first required for Ex+, Ex-HF and Ex-1. It is not required for Ex-3 and Ex-3, at least during the 24 hours of integration.

serves as a control calculation. The other simulations start with the same initial warm-cored vortex (maximum tangential wind speed 10 m s^{-1} at the surface at 100 km radius) and share the same latitude (20°N).

TABLE 8.2 Summary of the five simulations referred to in this section.

Simulation	Maximum heating amplitude
Ex-HF	3 K h^{-1} (Control calculation)
Ex+	4.5 K h^{-1} (1.5 × Control)
Ex-1	1.5 K h^{-1} (0.5 × Control)
Ex-2	0.75 K h^{-1} (0.25 × Control)
Ex-3	0.3 K h^{-1} (0.1 × Control)

Fig. 8.18a shows the evolution of the maximum tangential velocity component in the five simulations and the times at which the Eliassen equation requires regularization, which depends on the strength of the heating rate and the secondary circulation it gives rise to. As expected, the rate of intensification and its sign depend also on the strength of the heating. In experiments Ex+, Ex-HF, Ex-1, and Ex-2, the rate of intensification is positive and increases with the strength of the heating, whereas in experiments Ex-3, the vortex decays slowly with time, a reflection of the complete dominance of friction over heating wherein the overturning circulation produced by the heating is too weak to reverse the comparatively strong outflow above the boundary layer induced by the friction.

The time at which the Eliassen equation requires regularization decreases as the maximum heating rate increases, as does the time of eventual breakdown of the solution. For example, the solution in Ex+, which has the strongest heating rate, breaks down just after 18 hours and the Eliassen equation requires regularization from $6\frac{2}{3}$ h. On the other hand, all the other simulations run for 24 hours and in the simulations with the two weakest heating rates, the Eliassen equation does not require regularization over this period.

Fig. 8.18b shows the evolution of the minimum radial velocity component (with the sign reversed), $|U_{min}|$, in the five simulations. In all simulations, $|U_{min}|$ decreases in magnitude with time, as does the rate of decrease. However,

at any given time, $|U_{min}|$ increases with the magnitude of the diabatic heating rate.

8.4.1 Flow structure at the initial time

Reasons for some of the foregoing behavior are highlighted by Fig. 8.19, which shows radial-height cross sections of M-surfaces superimposed on the radial and tangential flow components in Ex+ and Ex-1, Ex-2 and Ex-3 at the initial time (the corresponding cross section for Ex-HF is shown in Fig. 8.9a, albeit with slightly different contour shading). In Ex+ there is a deep layer of inflow throughout the lower troposphere, below a height of 4-5 km, while in Ex-1, this inflow layer becomes elevated beyond about 150 km radius and it does not extend beyond 240 km radius. Indeed, there is enhanced outflow just above the frictionally-driven inflow layer beyond 150 km radius. In Ex-2 and Ex-3, there is outflow on the axis of heating at all heights above the boundary layer, except at upper levels inside the heating. In experiments Ex-1, Ex-2, and Ex-3, there is a weak outward component of flow across the M-surfaces along the axis of heating at a height of 1 km. Thus, the M-surface to which the heating axis is tied will drift slowly outwards.

8.4.2 Flow structure at later times

Fig. 8.20 shows radius-height cross sections of the radial and tangential flow components at 15 hours in Ex+ and Ex-1 to Ex-3. In Ex+, this time is beyond that for which regularization is required. The panels in Fig. 8.20 should be compared with the corresponding fields in Fig. 8.19.

At 15 hours in Ex+ and Ex-1, the M-surfaces have moved inwards in the lower troposphere, although near the surface the inward movement is impeded by the tangential frictional force. Thus, the M-surfaces develop a nose-like structure in the lower troposphere. As a result, V_{max} is located at a height of 5 km in Ex+ and 3.4 km in Ex-1 at this time. In the heated region, the M-surfaces have moved upwards, presumably a result, in part, of the vertical advection of low values of M from the boundary layer, where M is reduced by the frictional force. Beyond this region, throughout much of the troposphere, the M-surfaces have

Simulations with diabatic heating and friction: heating rate dependence

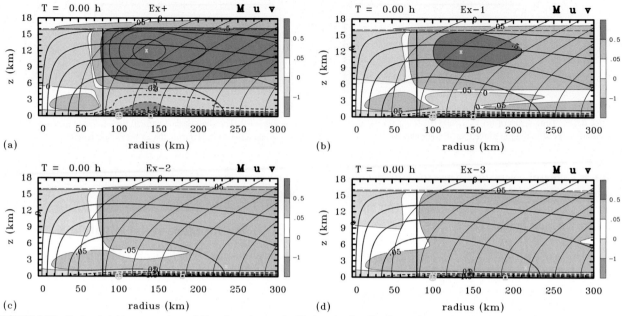

FIGURE 8.19 Radius-height cross sections of M-surfaces (composite black and yellow lines) superimposed on the contours of tangential velocity, v, (solid blue contours) and radial velocity u (shaded) at the initial time. Contour interval for M, 5×10^5 m^2 s^{-1}; for u, 0.5 m s^{-1} (shown also is the $u = 0.25$ m s^{-1} contour); for v, 2 m s^{-1}. Positive values (solid), negative values (dashed). The black vertical line shows the location of the maximum diabatic heating rate. The yellow symbols indicate the locations of maximum v, ($\oplus$), maximum u, ($\times$) and minimum u, ($\square$).

Simulations with diabatic heating and friction: heating rate dependence

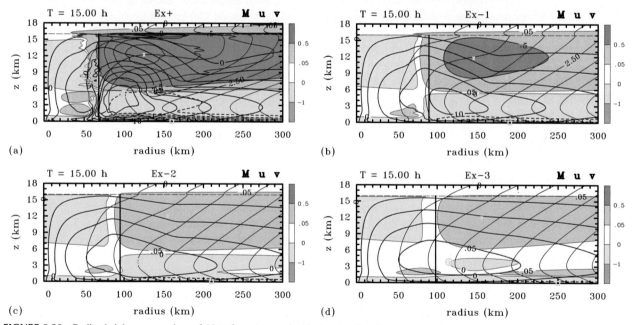

FIGURE 8.20 Radius-height cross sections of M-surfaces (composite black and yellow lines) superimposed on the contours of tangential velocity, v (solid blue contours), and radial velocity u (shaded) at 18 h. Contour interval for M, 5×10^5 m^2 s^{-1}; for u, 0.5 m s^{-1} (shown also is the $u = 0.25$ m s^{-1} contour); for v, 2 m s^{-1}. Positive values (solid), negative values (dashed). The black vertical line shows the location of the maximum diabatic heating rate. The yellow symbols indicate the locations of maximum v, ($\oplus$), maximum u, ($\times$) and minimum u, ($\square$).

descended. As explained in Subsection 8.1.1.3, the descent is associated with subsidence beyond the region of heating. This subsidence is particularly strong just outside the region of heating to the extent that it leads to a negative radial gradient of M and thereby to a region of inertial instability (see Subsection 8.1.1.2). Note that the axis of heating has moved slightly inwards in Ex+, but outwards in Ex-1. The behavior in Ex-HF is similar to that in Ex+ (not shown).

In Ex+, the radial velocity fields have started to exhibit small-scale irregularities inside the heating axis and in the upper troposphere, a sign of the imminent breakdown of the solution with the emergence of inertial instability. In Ex-1, radial inflow has become established throughout the lower troposphere and it is slightly negative at a height of 1 km on the axis of heating, even though over the 15 hour period, the heating axis has moved outwards by about 10 km. In Ex-2 and Ex-3, the radial flow remains outwards at 1 km on the axis of heating at 15 hours and in these simulations, the axis of heating has moved outwards to 95 km in Ex-2 and just under 100 km in Ex-3. In Ex-2, the radial outflow in the lower troposphere at the initial time has mostly reversed to weak inflow beyond the axis of heating, but in Ex-3, this reversal is confined to an elevated lens region between about 115 km and 205 km radius.

As in Ex+ and Ex-1, the M surfaces at 15 hours in Ex-2 and Ex-3 both exhibit the inward-pointing nose like structure in the lower troposphere on account of the tangential frictional force and again, there is some elevation of these surfaces in the region of heating. Of course, the elevation of the M surfaces is less pronounced when the heating is weaker as the vertical advection of M will be weaker in this case.

8.4.3 Summary: the issue of convective ventilation

Despite the obvious shortcomings of these five balance calculations with a prescribed heating, which are discussed below in Section 8.4.5, the calculations illustrate clearly the consequences of the deep convective mass flux being uncoupled from the mass flux of air converging in the boundary layer. If the convective mass flux is strong enough to more than ventilate the mass flux of air converging in the boundary layer, air will be drawn inwards above the boundary layer and lead to spin up by the classical mechanism. On the other hand, if the convective mass flux is too weak, the fraction of air converging in the boundary layer that cannot be ventilated will flow radially outwards in a shallow layer above the boundary layer causing the tangential wind in this layer to spin down as it conserves its absolute angular momentum. *Calculations using more sophisticated models discussed in later chapters suggest that the general mismatch between the rate at which air converges in the frictional boundary layer and the rate at which this air can be ventilated by deep convection is an intrinsic feature of tropical cyclone evolution.* A way of quantifying this mismatch is discussed in Section 15.5.4.

8.4.4 Pathological nature of the balanced boundary layer

The foregoing solutions highlight the somewhat pathological nature of the balanced boundary layer. Returning to Fig. 8.19, it will be noticed that, although at the initial instant, a fully developed layer of inflow is present on account of the specified boundary layer friction, no such boundary layer is evident in the tangential velocity field at this time. In particular, the M-surfaces intersect the surface at right angles, consistent with the prescribed zero vertical gradient of tangential velocity at the initial time. Since the tangential velocity experiences the initially-imposed frictional force, F_λ, through the tendency equation

$$\frac{\partial v}{\partial t} = -u\zeta_a - w\frac{\partial v}{\partial z} + F_\lambda, \qquad (8.6)$$

it takes time for the near surface tangential velocity to be decelerated. In contrast, in the balanced boundary layer formulation, the radial velocity component is determined by the need to keep the tangential flow in gradient wind and hydrostatic balance in the presence of diabatic heating and friction that act to destroy these balances. As a result, the inflow boundary layer is fully developed, even at the initial instant. Note that in this boundary layer formulation, there is no component of frictional force in the radial direction. Thus, while the balance formulation produces a vertical structure of inflow that is qualitatively similar to a true boundary layer, even at the initial instant, the same cannot be said for the tangential component, which, as the foregoing solutions show, takes a significant time (i.e. many hours) to develop.

If judged by the height of the nose of the M-surfaces at 12 hours (Fig. 8.20), the boundary layer would appear to be considerably deeper ($\approx$ 3 km) in the tangential direction than in the radial direction ($\approx$ 1 km). However, since the frictional force is confined primarily below 1 km in altitude, the elevated nose is due in part to the vertical advection of low values of M from the boundary layer as discussed in Section 8.1.3. In comparison, in a realistic boundary layer, which a scale analysis shows to be not in gradient wind balance, the process that determines the near-surface inflow, at least at radii somewhat beyond the radius of maximum gradient wind is the frictional retardation of the tangential flow near the surface (see e.g., Smith and Vogl (2008) and refs.). This retardation makes the near-surface flow subgradient, leaving a net inward pressure gradient force to drive the inflow.

8.4.5 Utility and limitations of the prognostic balance model

Although we believe the prognostic balance theory has considerable pedagogical value, in illustrating the classical spin up mechanism and in helping to understand some of the controls on vortex size, it does have certain limitations as itemized below.

1. The assumption of axial symmetry imposes a number of limitations since it constrains any representation of deep convection to be in concentric rings. Except, possibly, for the eyewall of a mature tropical cyclone, such a representation is rather unrealistic. Some of the consequences of assuming axial symmetry are discussed in Section 16.2 in a comparison of two similar numerical simulations, one that assumes axial symmetry and the other which does not.

2. The assumption of gradient wind balance in the boundary layer is problematic, since, as shown in Chapter 6, the inflow in a vortex boundary layer occurs primarily because of an imbalance of forces in the boundary layer.

3. In a prognostic theory, there is a natural tendency for the balanced flow to develop regions where the discriminant of the equation for the streamfunction of the overturning circulation becomes negative. In these regions, which include regions of inertial instability, it is necessary to regularize the streamfunction equation as discussed earlier. However, procedures to accomplish this regularization are necessarily *ad hoc* and the flow in and near to the regions of regularization can be sensitive to the procedure employed (Wang and Smith, 2019).

4. The prescription of the diabatic heating rate is a further limitation. In general, the spatial distribution of diabatic heating should be determined as part of the solution as the flow evolves. Recognition of this issue raises the question: what is an appropriate way to represent different types of convection in models? A prerequisite for answering this question is a basic understanding of convective clouds, themselves, which is the topic of the next chapter.

5. Last but not least, the limitations of the strictly balance model solutions are their continuation to (and possibly beyond) a point where the M field develops a discontinuity at or near to the surface. A second issue is the growth of asymmetric instabilities and related coherent vortex structures as well as vorticity mixing phenomenology supported by the formation of the axisymmetric annulus of PV. These issues are discussed in more detail in Sections 8.1.1.4 and 8.2.

8.5 Appendix

As noted in Section 5.9, the Eliassen equation (5.67) can be solved as a diagnostic equation for the streamfunction ψ

only if the equation is elliptic globally. Regularization is an *ad hoc* procedure to adjust the coefficients of the equation in localized regions where the ellipticity condition is not satisfied (i.e., regions in which $D < 0$), the "hope" being that the existence of such regions will not appreciably affect the solution for the streamfunction outside these regions. From Eq. (5.68), the discriminant of the Eliassen equation, D, is given by $4\gamma^2\Delta$, where $\Delta = N^2 I_g^2 - B^2$. Further, the equation is locally *elliptic* if $\Delta > 0$, locally *hyperbolic* if $\Delta < 0$ and locally *parabolic* if $\Delta = 0$. In Section 5.9 it was shown also that regions where the Eliassen equation is hyperbolic correspond with regions of negative PV, equivalent to the flow being symmetrically unstable. Regions where $\Delta < 0$ are ones in which the flow is inertially unstable ($I_g^2 < 0$), statically unstable ($N^2 < 0$) or where the baroclinicity B, a measure of the vertical wind shear, is large, specifically $\bar{B}^2 > \bar{A}\bar{C}$.

Various procedures have been proposed to regularize the Eliassen equation in regions where $\Delta \leq 0$. These are reviewed by Wang and Smith (2019): see their Section 5. The scheme used here is the one used by Wang et al. (2020). The procedure is as follows. At any grid point where the discriminant $\Delta = I_g^2 N^2 - B^2 < 0$ we examine whether $I_g^2 < 0$, $N^2 < 0$ or $B^2 > I_g^2 N^2$. Negative values of $I_g^2(r, z)$ are removed by adding the local value $|1.001 I_g^2(r, z)|$ and negative values of N^2 are removed by setting N^2 to a small positive value. Making adjustments to the magnitude of I_g^2 or N^2 generally requires a reduction in the magnitude of B to keep $\Delta > 0$. Here, at each grid point where Δ remains negative after modifying I_g^2 and/or N^2, $\bar{B}^2$ is replaced with $0.99 I_g^2 N^2$. In numerical models of tropical cyclones, regions of azimuthally averaged inertial instability are generally the most extensive, while regions of static instability are typically small in areal extent.

In regions of large vertical shear, typically in the frictional boundary layer, but possible elsewhere as well, B^2 may exceed $I_g^2 N^2$ when both these quantities are positive. Again, in this case we set $\bar{B}^2 = 0.99 I_g^2 N^2$ to keep $\Delta > 0$. The study by Wang and Smith (2019) recommends caution in making interpretations of flow structures in regions which require regularization.

Chapter 9

Moist convection

As discussed in Chapter 1, tropical cyclones originate from tropical disturbances with embedded convective cloud systems. A distinguishing feature of the mature tropical cyclone is the ring of deep convective clouds that form the eyewall encircling the largely cloud-free eye. In this chapter we review a few basic concepts necessary to understand the role of moist convection in tropical cyclone evolution and structure. Then, in Chapter 10 we explore the role of moist convection in the dynamics of tropical cyclone genesis and intensification. We begin here with a brief review of convective instability and go on to examine three types of penetrative convection: shallow convection, which as the name suggests has a limited vertical extent; deep convection, which extends through much or all of the troposphere; and intermediate convection, where the cloud tops remain either below the freezing level, about 5.5 km high in the Tropics, or do not extend high enough to fully glaciate. Glaciation occurs spontaneously when cloud temperatures fall below $-40\,°C$.

9.1 Convective instability

The occurrence of moist convection in the atmosphere is frequently explained in terms of the behavior of an ascending air parcel that does not mix with its environment. When an unsaturated air parcel is lifted adiabatically, it expands and cools, but its water vapor mixing ratio remains constant. Since the saturation water vapor pressure decreases rapidly with temperature, the air parcel eventually reaches a level at which it becomes saturated. This level is called the *lifting condensation level*, or LCL. If the parcel is lifted above this level, the expansion and cooling continue and water progressively condenses. However, the accompanying release of latent heat reduces the rate at which the parcel cools.

Typically, when an air parcel rises above the LCL, it is at first still denser than its environment and therefore negatively buoyant. However, if the lifting continues, the reduced rate of cooling may enable the air parcel to reach a level above which it becomes lighter than its environment and therefore positively buoyant. This level is called the *level of free convection*, or LFC.

Above the LFC, the air parcel can accelerate under its own buoyancy force until at some further level its temperature becomes less than that of its environment. This level is called the *level of neutral buoyancy*, or LNB. Above the

LNB, the parcel becomes negatively buoyant and decelerates, often rapidly if the LNB is near or just above the tropopause.

The various levels described above are usually estimated on the basis of an aerological diagram, which is a judiciously transformed thermodynamic diagram. In its most basic form, a thermodynamic diagram is one in which the pressure, p, and specific volume, α, are plotted on the ordinate and abscissa, respectively. Then, according to the equation of state (2.11), the isotherms are rectangular hyperbolae. The state of a parcel of dry air can be represented by a point in such a diagram and the change in state as the pressure and specific volume change can be represented as a curve. For example, an isothermal change is represented by a rectangular hyperbola, $p\alpha = R_d T$, while a dry adiabatic change is represented by a curve of constant potential temperature, $\theta = T(p_{**}/p)^\kappa = p_{**}\alpha(p_{**}/p)^{\kappa-1}/R_d$.

9.2 Aerological diagrams

The p-α diagram is of limited practical use in meteorology for a number of reasons, one being that the specific volume is not a quantity that is measured. An aerological diagram is a transformed version of a p-α diagram in which more relevant meteorological variables are displayed and lines representing state changes associated with important atmospheric processes are plotted.

There are various kinds of aerological diagram, one being the tephigram and another the skew T-log p diagram. These diagrams are broadly similar. In both, the isobars are straight horizontal lines and are plotted using a logarithmic scale with pressure decreasing upwards so that the ordinate is approximately equal to height. In addition, the isotherms are straight lines that run upwards to the right at an angle of 45 degrees. Fig. 9.1 shows a skew T-log p diagram with a mean tropical sounding plotted. The sounding is the Dunion moist tropical sounding, a mean for the Western Atlantic Ocean and Caribbean Sea during the hurricane season (Dunion, 2011).

While the state of a sample of dry air is represented by a single point (p, T) in the diagram, that of a moist sample requires two points, one corresponding to the pressure and temperature of the parcel and the other corresponding to the pressure and dew-point temperature, T_d. The state of the atmosphere obtained at a given location and time, for

Tropical Cyclones. https://doi.org/10.1016/B978-0-44-313449-4.00017-5

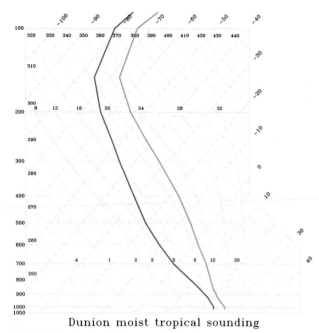

Dunion moist tropical sounding

FIGURE 9.1   A skew T-log p aerological diagram with a mean sounding for the Atlantic Ocean and Caribbean Sea for the hurricane season plotted. The thick red (right) curve is the temperature and the thick blue (left) curve the dew-point temperature.

example from a radiosonde sounding, can be plotted as pair of curves (p, T) and (p, T_d). The mean sounding for the Atlantic hurricane season is plotted as two thick curves in Fig. 9.1, the red (right) curve for (p, T) and the blue (left) curve for (p, T_d).

The skew T-log p diagram can be used to investigate changes of state of an air parcel as its pressure changes. For this purpose, several other sets of curves are plotted in the diagram. These include: the *dry adiabats*, curves along which the potential temperature, θ, is constant; the *pseudo-moist adiabats*, curves along which the *saturation pseudo-equivalent potential temperature*, θ_e^*, is constant; and curves of constant *saturation mixing ratio*, r^*. The isopleths of the various quantities are shown in Fig. 9.2.

The pseudo-equivalent potential temperature, θ_e is defined in Subsection 2.5.4. The quantities, θ_e^* and r^*, are defined respectively as the equivalent potential temperature and mixing ratio which air would possess if it were saturated with water vapor at the same temperature and pressure. Like θ, these quantities are both functions of state, depending only on p and T, while θ_e is a function of mixing ratio also and cannot be plotted on such a diagram.

The water vapor mixing ratio of an air parcel is given by the saturation mixing ratio line that intersects the point (p, T_d), since the *dew-point temperature* is the temperature at which an air parcel becomes saturated when cooled at constant pressure.

One empirical formula for the saturation vapor pressure, $e^*(T)$ is provided by Teten's formula, i.e.,

$$e^*(T) = e^*(0) \exp\left(\frac{17.67T}{T + 243.5}\right), \qquad (9.1)$$

where $e^*(0) = 6.112$ mb and the temperature T in this formula is in °C. The saturation mixing ratio can be obtained from the formula

$$r^*(p, T) = \frac{\epsilon e^*(T)}{p - e^*(T)}, \qquad (9.2)$$

where $\epsilon = R_d/R_v \approx 0.621$ (see Section 2.5).

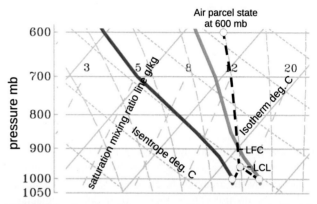

FIGURE 9.2   A portion of the same skew T-log p diagram as in Fig. 9.1 showing the location of the LCL and LFC for an air parcel initially at 1000 mb. The thick red (right) curve is the temperature, the thick blue (left) curve the dew-point temperature and the black dashed curve shows the state of the lifted air parcel.

If an unsaturated air parcel is lifted adiabatically (i.e., without mixing with its environment), its potential temperature and mixing ratio will remain the same. Thus, its state at a given pressure level is determined in the diagram by following its temperature along a dry adiabat through its initial point, say (p_i, T_i) and its dew-point temperature along the saturation mixing ratio through its initial point (p_i, T_{di}). Fig. 9.2 shows the lower right portion of Fig. 9.1 to illustrate this process. The points (p_i, T_{di}) and (p_i, T_i) are indicated by two small yellow (light gray in print version) circles, where, in this case, p_i is taken to be a pressure of 1000 mb. The dry adiabat through the point (p_i, T_i) and the saturation mixing ratio line through the point (p_i, T_{di}) are shown as approximately straight lines. When the dry adiabat meets the saturation mixing ratio curve, the air parcel will be saturated and the pressure at this intersection represents the LCL. The pressure and temperature at this intersection, say (p_{LCL}, T_{LCL}), are sometimes referred to as the saturation point of the air parcel. Typically, T_{LCL} is less than that of environmental air at that pressure level, i.e., the air parcel has negative buoyancy.

If the air parcel is lifted further, condensation will take place. In general, some of the condensed water in the form of tiny cloud droplets will remain within the rising air parcel and the larger droplets that develop will fall out as rain. Condensation is accompanied by latent heat release so that the potential temperature will no longer be conserved. To estimate the air parcel state as it rises above the LCL using the aerological diagram, we make the *pseudo-adiabatic assumption* that all condensate falls out of the parcel as soon as it forms. The heat lost from the moist air parcel from the falling condensate is assumed to be negligible in comparison with the heat liberated by the condensation of water vapor (Exercise 9.1). With this assumption, the air parcel will conserve the value of θ_e^* it had when it first became saturated, i.e. at the LCL, which is essentially cloud base. Since θ and r are conserved while the parcel is rising below cloud base, θ_e^* is just the value of θ_e at the air parcel's starting level.

As shown in Fig. 9.2, the LFC can be found using the aerological diagram by following the pseudo-adiabat or constant θ_e^* curve through the saturation point until it intersects the temperature curve of the sounding. In a similar way, the LNB is where this pseudo-adiabat again intersects the temperature curve of the sounding (Fig. 9.3).

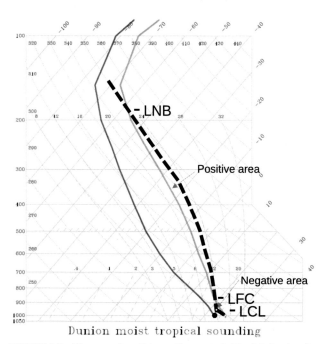

Dunion moist tropical sounding

FIGURE 9.3 The same skew T-log p diagram as in Fig. 9.1 showing the location of the LCL, LFC, and LNB for an air parcel initially at 1000 mb. The thick red (right) curve is the temperature of the sounding, the thick blue (left) curve the dew-point temperature of the sounding and the black dashed curve shows the state of the lifted air parcel.

9.2.1 CAPE and CIN

The potential energy of an air parcel rising from a particular level z_i to its LNB is called the convective available

potential energy $(CAPE)$[1] and may be written as

$$CAPE = \int_{z_i}^{LNB} \mathbf{b} \cdot \mathbf{dl} \qquad (9.3)$$

where $\mathbf{b}$ is the generalized buoyancy force per unit mass that acts on the parcel and $\mathbf{dl}$ is the vector displacement along the path of the displacement. If the parcel rises vertically, this integral may be written

$$CAPE = \int_{z_i}^{LNB} b\,dz, \qquad (9.4)$$

where z measures height (normal to geopotential surfaces) and b is the vertical component of the buoyancy force. Using (2.9), and assuming that the environment is in hydrostatic equilibrium, we can show that (see Exercise 9.2)

$$CAPE = \int_{p_{LNB}}^{p_i} R_d \left(T_{\rho p} - T_{\rho a} \right) d \ln p, \qquad (9.5)$$

where $T_{\rho p}$ and $T_{\rho a}$ are the density temperatures of the lifted parcel, defined in Section 2.5.1, and of the environment at the same pressure, respectively, p_i is the initial pressure of the air parcel and p_{LNB} is the pressure at the LNB. If the environment is unsaturated, $T_{\rho a}$ is just the virtual temperature of the environment.

The amount of work that must be expended to lift the air parcel to its LFC is called the *convective inhibition* (CIN) and is given by the integral

$$CIN = \int_{z_i}^{LFC} \mathbf{b} \cdot \mathbf{dl} \qquad (9.6)$$

Again, if the parcel is lifted vertically, this integral may be written

$$CIN = \int_{p_{LFC}}^{p_i} R_d \left(T_{\rho p} - T_{\rho a} \right) d \ln p, \qquad (9.7)$$

where p_{LFC} is the pressure at the LFC.

On an aerological diagram such as the skew T-log p diagram, where the coordinates are linear in temperature and $\ln p$, $CAPE$ is just the area enclosed by the temperatures $T_{\rho p}$ and $T_{\rho a}$, from p_i to p_{LNB}, and CIN is the area from p_i to p_{LFC}. Because $T_{\rho p} < T_{\rho a}$ between p_i and p_{LFC}, CIN is negative and its magnitude is referred to as the *negative area*, NA. The area between the two temperature curves from p_{LFC} to p_{LNB} is called the *positive area*, PA. Clearly, $CAPE = PA - NA$ (recall that some authors define $CAPE = PA$).

---

1. There are differences in the literature because some authors define $CAPE$ as the potential energy of the air parcel when it commences from its LFC. Here we follow Emanuel (1994).

In general, $CAPE$ and CIN depend on the initial pressure, temperature, and dew-point temperature of an air parcel and on the thermodynamic process assumed in lifting the parcel. Two extremes are the *pseudo-adiabatic process*, where all liquid condensate is assumed to fall out of the parcel immediately, or the *reversible process* in which all condensate is carried along with the parcel. These are two extremes because, in reality, tiny cloud drops are likely to be carried along with the air parcel, while the much larger rain drops fall out. The pseudo-adiabatic assumption may be a better approximation for precipitating clouds while the reversible assumption is more appropriate for shallow clouds that do not precipitate or produce only drizzle.

The values of $CAPE$ and CIN as well as those of the LCL, LFC, and LNB for the sounding in Fig. 9.1 are shown in Table 9.1 for air parcels lifted from the surface and from altitudes at intervals of 100 m to 500 m. For this mean sounding, the LCL is a monotonically increasing function of z_i, whereas the lowest LFC, the largest LNB and largest value of $CAPE$ are for an air parcel lifted from an altitude of 100 m. Above this level, the LFC increases with z_i, while the LNB and $CAPE$ decrease with z_i. The sensitivity of $CAPE$ to z_i is appreciable, with the value for $z_i = 500$ m being only about 60% of the value at $z_i = 100$ m. Absolute values of CIN show much less sensitivity to z_i and they are very much less than the values of $CAPE$.

TABLE 9.1 LCL, LFC, LNB, $CAPE$, and CIN for air parcels lifted from the surface and from altitudes at intervals of 100 m to 500 m for the Dunion moist tropical sounding in Fig. 9.1. The last row shows the arithmetic mean of the $CAPE$ and CIN values for parcels lifted from 100 m to 500 m.

z_i	LCL	LFC	LNB	$CAPE$	CIN
m	km	km	km	J/kg	J/kg
0	0.5	1.0 km	13.8	1996	−58
100	0.6	0.9 km	13.9	2076	−49
200	0.7	1.0 km	13.8	1972	−44
300	0.8	1.1 km	13.6	1747	−43
400	0.9	1.3 km	13.3	1466	−44
500	1.0	1.4 km	13.0	1263	−44
Mean				1704	−44

9.2.2 Height-temperature-difference diagram

An alternative way to illustrate the dependence of the buoyancy of a lifted parcel as a function of the initial height of the parcel is shown in Fig. 9.4. Rather than the buoyancy per unit mass, itself, the diagram shows the difference in virtual temperature between that of an air parcel lifted pseudo-adiabatically from a height, say z_i, to a height z and that of its environment at the same height z. The calculation is constructed for the Dunion moist tropical sounding with

values of z_i starting at the surface and going to a height of 500 m in intervals of 50 m. The lifted height is indicated by the lowest black contour. The second zero contour going up is the LFC and the one at, or above, 13 km height is the LNB.

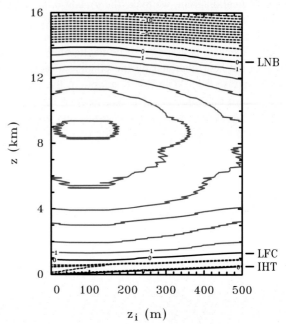

FIGURE 9.4 Contours of the virtual temperature difference between the virtual temperature of a lifted air parcel and that of its environment at the same height, based on the Dunion moist sounding. Values of the initial height, z_i are on the abscissa. Contour interval 1 °C. Positive values solid red, negative values dashed blue. Zero contours are black. The line labeled IHT refers to the initial height of the lifted air parcel and the curves labeled LFC and LNB indicate the level of free convection and level of neutral buoyancy as a function of z_i.

The largest temperature differences are found between 8.5 and 9 km height for air parcels lifted from between 50 m and 150 m height. In line with this result, for air parcels lifted from 200 m and above, the LFC increases, the LNB decreases and the maximum temperature difference decreases.

9.2.3 More on aerological diagrams

In practical applications of aerological diagrams, $T_{\rho p}$ and $T_{\rho a}$ are usually approximated by the actual temperatures: $T_{\rho p}$ by the temperature along the dry adiabat from the initial pressure to the LCL and along the pseudo-adiabat to the LNB, and $T_{\rho a}$ by the temperature of the environment.

Recall that in a non-cloudy atmosphere, however, the density of a moist air parcel is inversely proportional to the virtual temperature (Section 2.5.1), i.e., $\rho = p/R_d T_v$. The fact that the buoyancy of an air parcel involves the difference between the density of the moist air parcel and that of the environment at the same pressure is a reminder of the importance of calculating $CAPE$ using the virtual temperature. The density temperature is a generalization of the virtual temperature concept to include cloud water and ice in a cloudy region. In a moist column of air, the contribution from the moisture content is non-negligible and can make a significant difference when assessing whether the tropical atmosphere is unstable or stable. In practical applications of aerological diagrams, however, the difference between the density temperature or virtual temperature and the actual temperature is usually ignored.

The negative area may be interpreted as a potential barrier to convection, preventing it from occurring spontaneously. This barrier has no analogy in dry convection. Under certain meteorological conditions, for example when there is a strong inversion layer at the top of the moist air layer, the existence of this barrier allows $CAPE$ to accumulate. In such circumstances, the instability may be released explosively at some later time. In deep convective regions over the warm tropical oceans, values of CIN are often relatively small.

Assuming that the only vertical force acting on a rising air parcel is the buoyancy force, Newton's second law can be applied to show that

$$\frac{1}{2}w_{LNB}^2 - \frac{1}{2}w_{LFC}^2 = CAPE - CIN \qquad (9.8)$$

where w_{LNB} and w_{LFC} are the vertical velocity at the LNB and LFC, respectively (Exercise 9.3). To this level of approximation, the increase in kinetic energy per unit mass as an air parcel ascends from the LFC to the LNB is simply the positive area on the aerological diagram as defined above. Of course, the assumption that the only vertical force acting on a rising air parcel is the buoyancy force is not appropriate in the upper troposphere, where perturbation pressure gradient forces contribute to the vertical deceleration of air parcels and their radial acceleration outwards (see e.g., Section 9.5). As a result, maximum vertical velocities in convective updrafts usually fall well short of estimates obtained from Eq. (9.8).

9.2.4 The use of θ_e for assessing convective instability

An alternative way to investigate the degree of conditional instability is to plot vertical profiles of θ_e and its saturation value, θ_e^*. An example is shown in Fig. 9.5, which shows vertical profiles of θ_e and θ_e^* obtained from the Dunion moist tropical sounding in Fig. 9.1.

For reference, Fig. 9.5 shows also the profile of *virtual potential temperature* θ_v. In the tropics, this quantity is typically a few degrees warmer than the potential temperature, θ, in the lowest kilometer or so, where there are relatively high values of water vapor mixing ratio, but it rapidly asymptotes to θ above a height of a few kilometers.

Strictly, it is the vertical gradient of θ_v that characterizes the stability of the atmosphere to unsaturated vertical displacements of an air parcel. The θ_v in the mean sounding shows a progressive increase with height at all levels indicating that the atmosphere is everywhere stable to vertical displacements of air parcels that are unsaturated.

Prominent features of the mean sounding of θ_e are the relatively high values near the surface, the decrease with height to about 4 km and the increase above that level. The decrease at lower levels is a reflection of the strong decrease of water vapor mixing ratio with height, while the subsequent increase with height is because θ_e approaches θ (and θ_v) as the mixing ratio becomes small.

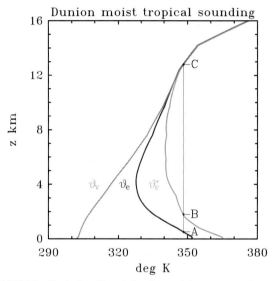

FIGURE 9.5 Vertical profiles of virtual potential temperature, θ_v, equivalent potential temperature, θ_e, and saturation equivalent potential temperature, θ_e^*, for the same mean tropical sounding as in Fig. 9.1. The saturation equivalent potential temperature is the equivalent potential temperature of a hypothetically saturated air with the same temperature and pressure.

Now, suppose that an air parcel is lifted pseudo-adiabatically from the boundary layer, say at an altitude of 500 m (point A in Fig. 9.5). Then, because θ_e is materially conserved for the air parcel, the path of changes in the parcel state in the figure would be a straight vertical line starting from point A. As seen in Fig. 9.2, condensation of water vapor begins before this path intersects the θ_e^* curve because the LCL is lower than the LFC. As soon as the parcel becomes saturated, the difference between the air parcel's θ_e and the θ_e^* of the environment is an approximate measure of the parcel buoyancy. This follows from the fact that, at a given pressure, both θ_e^* and the saturation mixing ratio are both monotonically increasing functions of temperature (Exercise 9.4). The reason why this difference is only an approximate measure of the parcel buoyancy is discussed below.

Above the intersection of the vertical θ_e line through point A with the θ_e^* curve of the sounding, at point B in

Fig. 9.5, the air parcel will become warmer than its environment and will continue to be accelerated by its buoyancy as it rises towards the tropopause. Of course, in reality, pseudo-adiabatic ascent of a parcel without mixing is unlikely to occur and a more realistic path of an air parcel that mixes with ambient air as it rises will be closer to the θ_e^* curve than to the vertical straight line.

If an air parcel is lifted pseudo-adiabatically from a point in the free atmosphere above about 1.5 km, its equivalent potential temperature will follow a vertical path which does not intersect the θ_e^* curve and this parcel will never acquire positive buoyancy, even though condensation may occur. Therefore, *the possibility of unstable moist convection in this atmosphere depends critically on the high values of θ_e in the boundary layer.*

The reason why the LFC is only approximately where the vertical line through the air parcel's initial θ_e first intersects the θ_e^* curve of the environment is as follows. Neglecting the weight of condensate, the *sign* of the air parcel buoyancy is proportional to

$$T_{vp} - T_{va} = T_p(1 + \epsilon r_p) - T_a(1 + \epsilon r_a), \qquad (9.9)$$

where T_p, and T_a are the parcel and environmental temperatures at some pressure level p, T_{vp}, and T_{va} are the corresponding virtual temperatures, and r_p and r_a are the corresponding mixing ratios. If the air parcel is saturated, $r_p = r_{ps}(p, T)$, and

$$\begin{aligned} T_{vp} - T_{va} = \big[&T_p(1 + \epsilon r^*(p, T_p)) - T_a(1 + \epsilon r^*(p, T_a)) \\ &+ \epsilon T_a(r^*(p, T_a) - r_a) \big]. \end{aligned}$$
$$(9.10)$$

Because $r^*(p, T_a) - r_a > 0$, the virtual temperature difference $\big[T_p(1 + \epsilon r^*(p, T_p)) - T_a(1 + \epsilon r^*(p, T_a)) \big]$ is an underestimate for the buoyancy so that the level of zero buoyancy, in this case the LFC, must be lower than that where $T_{vp}^* = T_{va}^*$.

Since T_v is a monotonic function of θ_e at constant p, altitudes where the parcel buoyancy changes sign such as the LFC or LNB are approximately where $\theta_{ep}^* = \theta_{ea}^*$, but slightly lower.

It follows that the LFC is somewhat lower than the intercept indicated in Fig. 9.5 if the environment is unsaturated. In fact, for the sounding plotted, the actual LFC is at a height of 1350 m, 425 m below that in Fig. 9.5. At the LNB, the difference in height is negligible as mixing ratio values are tiny.

A more conventional diagram illustrating the virtual temperature difference between the lifted air parcel and its environment along the line A-C in Fig. 9.5 and beyond is shown in Fig. 9.6. Note the shallow layer of negative difference below the LFC corresponding to the negative area in Fig. 9.2 and the layer of increasing negative difference above the LNB.

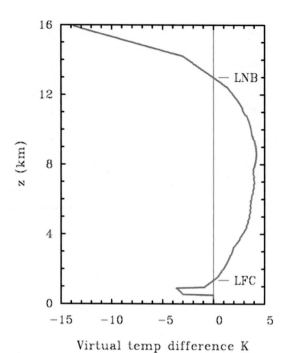

FIGURE 9.6 Vertical profiles of virtual temperature difference, $T_v - T_{va}$, for the same mean tropical sounding as in Figs. 9.1 and 9.5.

The local temperature and moisture stratification in a tropical cyclone is not necessarily the same as the mean state shown in Figs. 9.1 and 9.5. Persistent deep convection in the developing storm carries air with large near-surface values of θ_e aloft, thereby elevating the mean values of θ_e in the mid troposphere by the detrainment of moisture there. As shown in Section 9.3.4, the removal of these low values of θ_e has implications for the strength of convective downdrafts, which act to stabilize the atmosphere to further deep convection. In the next chapter, we show that tropical cyclone formation and maintenance requires organized convective activity to last for a period much longer than the time scale of individual convective clouds. In order to support such activity, it is important that there is continuing and adequate supply of moisture from the warm ocean to maintain convective instability against the stabilizing effect of downdrafts.

9.3 Types of penetrative convection

It is sometimes useful to divide tropical convection into three types: shallow convection, intermediate convection and deep convection (Johnson et al., 1999). Figs. 9.7 and 9.8 show some photographs of these three types of convection.

9.3.1 Shallow convection

Shallow convection refers to convective clouds that have a limited vertical extent, with tops perhaps no higher than

FIGURE 9.7 Some photographs of shallow trade-wind convection in the Tropics. In (a) one tower has managed to penetrate to a much larger altitude than the other clouds, but is not precipitating. Such long thin towers typically mix with surrounding air and evaporate.

one or two kilometers. In the Tropics they are often referred to as "trade-wind cumuli". Theoretical treatments of shallow convection assume that the clouds do not precipitate, although the tallest ones may do so. Fields of shallow convection are important for the larger scale because the convection facilitates mixing, leading to a vertical transport of heat and moisture into the cloud layer and the transport of drier air through intra-cloud subsidence into the subcloud layer. In this way, shallow convection counteracts the drying and warming effects of large-scale subsidence within the cloud layer and helps to maintain the trade inversion layer. Examples of shallow convection are shown in Fig. 9.7.

9.3.2 Intermediate convection

Intermediate convection refers to convective clouds that produce heavy rain showers, but which may not extend far enough in the vertical to contain much ice. In the Tropics, the height of the freezing level is around 5.5 km, but, depending on what aerosols are present, appreciable freezing may not occur for some distance above this level. Glaciation

occurs when temperatures fall below $-40\,°C$, the temperature below which all supercooled liquid water freezes spontaneously. In the tropical sounding shown in Fig. 9.1, the $-40\,°C$ isotherm occurs at a height of about 9 km. In precipitating cumulus clouds with tops below the freezing level, rain drops are produced only by the collision and coalescence of smaller cloud drops. We refer to the rain produced in such clouds as *warm rain*. Clouds in the category of intermediate convection are commonly referred to as cumulus congestus clouds. The World Meteorological Organization defines cumulus congestus clouds as "strongly sprouting cumulus with generally sharp outlines and often great vertical extent. The bulging upper part of cumulus congestus frequently resembles a cauliflower". Examples of congestus clouds are shown in Fig. 9.8.

9.3.3 Deep convection

Deep convection refers to precipitating clouds that extend vertically through a significant depth of the troposphere, typically well above the freezing level. Besides producing

FIGURE 9.8 Some photographs of intermediate and deep convection in the Tropics. The two towers seen in the center of (a) are precipitating, but even the tallest is wholly below the freezing level. These towers would be classified as cumulus congestus. Panel (b) shows some patches of intermediate convection with shallow convection on their flanks, an indication that the moisture structure of the atmosphere is horizontally inhomogeneous. Panel (c) shows two clusters of congestus convection in different stages of evolution. The one on the left looks to be in the early stages of development, while the one on the right seems to be battling to survive with the uppermost towers evaporating. The cluster in the background had managed to develop into a deep convective system. It has already begun to produce a spreading anvil. Panel (d) shows a common sight in the tropics with a deep layer of cirrus cloud in the upper troposphere, presumably a remnant of deep convection produced some hours earlier in the day, a largely cloud-free middle troposphere and patches of shallow convection at low levels.

heavy local precipitation, they may give rise to thunder and lightning. An important characteristic of precipitating convection is the production of downdrafts when precipitation evaporates in subsaturated air, or when it melts at the freezing level. An example of deep convection is the cumulonimbus cloud with a well-formed anvil shown in Fig. 9.8c. Deep convective clouds usually produce extensive anvils of ice particles that often outlive the convective cells that generate them (Fig. 9.8d). These anvil clouds take on a dynamics of themselves with slow ascent above the freezing level and subsidence below, the latter generated by moderate to weak precipitation. There may be significant convergence near the freezing level induced by the sharp increase in the fall speed of hydrometeors as snow melts to form raindrops. Tropical cumulonimbus clouds are often organized into larger-scale cloud systems such as squall lines or mesoscale convective systems.

Deep convective cloud systems produce not only updrafts, but also two types of downdraft, one on the scale of the convective updraft associated with heavy precipitation and one on the mesoscale associated with the melting of ice and evaporation of precipitation in unsaturated air below thick anvil clouds (Zipser, 1969, 1977). These downdrafts can have an important stabilizing effect on the subcloud and near surface air, making it cooler and drier (Section 9.3.4). Moreover, the lifting of warmer air along the boundaries of the low-level cold outflow can trigger new convection (Mapes, 1993). In this manner, convective systems have a propensity to generate new cells that subsequently aggregate. Over the warm tropical oceans, surface fluxes of heat and moisture lead to a warming and moistening of the cooler downdraft air on time scales on the order of half a day, thereby restoring convective instability. Where surface wind speeds are tropical storm strength and higher, the time scale for restoration is reduced to a few hours (see, e.g., Sections 10.4 and 10.11).

The warming of the atmosphere in regions of convection is a result not only of the direct release of latent heat in clouds, but also by the subsidence in the environment of clouds brought about by horizontally-propagating gravity waves that the growing clouds generate (Mapes, 1993; Bretherton and Smolarkiewicz, 1989). In this way, a whole region including clouds and their environments can undergo warming.

Typical updrafts speeds in oceanic deep convection in the Tropics are in the range 4-9 m s^{-1}, considerably weaker than typical speeds in deep convection over land in the middle latitudes (Lemone and Zipser, 1980; Zipser and Lemone, 1980; Jorgensen et al., 1985; Lucas et al., 1994). However, tropical cumulonimbus clouds are generally deeper than their middle latitude counterparts as the tropopause is several kilometers higher in the tropics. Maximum downdraft speeds are typically fractionally smaller than maximum updrafts speeds.

The foregoing summary of typical updraft speeds in oceanic deep convection does not cover updraft speeds in intense deep convective systems observed during the early stages of tropical cyclone development. Fig. 1.34 suggests the existence of "extreme convection" within the mesoscale convective systems in the parent tropical wave. A landmark observational analysis is that of Houze et al. (2009) who detailed one such updraft complex in a tropical depression that became Hurricane Ophelia (2005). The specific updraft documented was 10 km wide and had vertical velocities 10–25 m s^{-1} in the upper portion of the updraft. The peak vertical velocity within this updraft exceeded 30 m s^{-1}. Chapters 10 and 11 will show that such updrafts are dynamically significant, in part, because they generate enhanced vertical vorticity by one to two orders of magnitude larger than the background values.

9.3.4 Convective downdrafts

As discussed in Section 9.3.3, deep convective cloud systems produce not only updrafts, but also downdrafts. These downdrafts are driven by the negative buoyancy associated with precipitation. This negative buoyancy is caused by the drag exerted by the precipitation and the cooling that results from its evaporation below cloud base.

The strength of downdrafts is related in part to the presence of dry air in the middle troposphere. Typically, the drier the air, the greater is the evaporation of precipitation below cloud base and the cooler and stronger are the downdrafts. As the cool downdraft air spreads out near the surface as a so-called "cold pool", it reduces the degree of convective instability, even though the leading edge of the spreading cold air in the form of a gravity current may trigger further convection in air that has not been affected recently by convection.

In regions of disturbed weather in the tropics, the middle troposphere will be progressively moistened by the evaporation of moderate and deep convection, provided that the broadscale flow patterns are favorable, e.g., weak vertical shear and pouch-like regions of recirculating winds as described in Chapter 1. The progressive moistening acts to reduce the strength of downdrafts produced by successive bouts of convection, making it easier for convection to be sustained for long enough to form a tropical cyclone. In a part of the next chapter, we examine the downdrafts produced during tropical cyclogenesis in terms of the surface cold anomalies of θ_e that they produce.

9.3.4.1 The wet-bulb temperature and wet-bulb potential temperature

To discuss the thermodynamics of downdrafts it is necessary to introduce the *wet-bulb temperature*, T_w, which is defined as the equilibrium temperature of air that is cooled adiabatically at constant pressure when water is evaporated into it. In this process, the latent heat consumed by the evaporation is supplied by the air, itself. For an isobaric process ($dp = 0$), Eq. (2.16) reduces to $c_{pd}dT = dQ$ and, if dQ is associated with the latent heat consumed by the evaporation of water, $dQ = -L_v dr_v$, where dr_v is the increase in water vapor mixing ratio. Neglecting the weak temperature dependence of L_v, it follows that the quantity $c_{pd}T + L_v r_v$ is conserved in an isobaric process. This quantity is called the *specific moist enthalpy*.[2] Then, T_w is determined by the equation

$$c_{pd}T_w + L_v r_v^*(T_w) = c_{pd}T + L_v r_v, \qquad (9.11)$$

where $r_v^*(T_w)$ is the saturation mixing ratio at the wet-bulb temperature, and T and r_v are the temperature and mixing ratio of the air before evaporation takes place. This is an implicit equation for T_w and, given a formula for the saturation vapor pressure as a function of temperature, it can be solved by iteration (Exercise 9.5).

The wet-bulb temperature can be determined from an aerological diagram as illustrated in Fig. 9.9, which shows the lower portion of the mean tropical sounding in Fig. 9.1. Consider an air parcel with temperature T, dew-point temperature T_d and pressure p, which in the example is at 700 mb. The evaporation of water into the air parcel will raise its water content and hence its dew-point temperature, but will decrease the air temperature. The wet-bulb temperature is the point at which the new temperature and dew-point temperatures meet. Clearly, $T_d \leq T_w \leq T$.

It turns out that, as shown in the figure, T_w lies at the intersection of the isobar p and the pseudo-moist adiabat corresponding with the θ_e of the air parcel in question. This

---

2. A more accurate formula that includes heat changes of water substance is derived by Emanuel (1994): see his Eq. (4.5.4). This leads to a more accurate formula for T_w, see his Eq. (4.6.4).

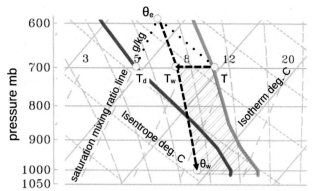

FIGURE 9.9 A portion of the same skew T-log p diagram as in Fig. 9.1 showing the location of the LCL for an air parcel initially at 700 mb with temperature T and dew-point temperature T_d. If this air parcel is saturated by the evaporation of precipitation falling into it, it will cool to the wet-bulb temperature T_w. This wet-bulb temperature lies on the pseudo-adiabat that intersects with the LCL of the air parcel and is shown as a black dashed curve descending to the surface (at 1015 mb). If the evaporation of rain continues to keep the descending air parcel saturated and if there is no mixing with ambient air during the descent, the state of the air parcel will follow the pseudo-adiabat to the surface. The $DCAPE$ is represented by the hatched area between the pseudo-adiabat and the environmental temperature curve.

pseudo-adiabat is the one that passes through the saturation point (p_{LCL}, T_{LCL}) of the air parcel. This construction for finding T_w is called *Normand's rule*. It follows that the evaporation of moisture and the associated cooling does not change the θ_e of the air parcel.

Note that the loss in heat content per unit mass of the air parcel in cooling isobarically from T to T_w is $c_{pd}(T - T_w)$, which, according to Eq. (9.11), is exactly balanced by the gain of latent heat associated with the moistening of the air parcel, $L_v(r_v^*(T_w, p) - r_v)$. If the now saturated air parcel ascends pseudo-adiabatically to the saturation point, condensation will occur and the same amount of latent heat will be released since the mixing ratio at the saturation point is r_v, the same as that for the original air parcel before evaporation occurs.

The *wet-bulb potential temperature*, θ_w, is just a different label for a pseudo-moist adiabat, being the temperature of a pseudo adiabat at a pressure of 1000 mb.

9.3.4.2 Downdraft thermodynamics

The possible strength of downdrafts can be examined also on an aerological diagram. The basic thought process is as follows. In the mean tropical sounding shown in Fig. 9.5, we saw that θ_e has a minimum at a height of about 4 km and that this minimum is associated in part with relatively dry air. This dryness is reflected in the relatively large separation between the θ_e and θ_e^* curves near this height as well as between the temperature and dew-point temperature curves in Fig. 9.1.

Using Fig. 9.9 as an illustrative example, imagine that rain falls into unsaturated environmental air at the 700 mb level. As noted earlier, the rain has two effects on the air: it exerts a drag on the air and some of the rain evaporates to cool the air. Cooled air parcels will be negatively buoyant and will begin to sink, aided by the downward drag of the rain, and a downdraft will form. If the subsiding air does not saturate, its temperature will rise at the dry adiabatic lapse rate, an effect that would tend to mitigate the cooling due to evaporation. At the other extreme, if the evaporation rate is sufficient to saturate the air, the air temperature will cool to its wet-bulb temperature and as long as the saturation is maintained, the temperature of descending air will follow along the pseudo-adiabat, assuming, of course, that there is no mixing with environmental air. In this case, the air would reach the surface with the same θ_e at the level from which it began its descent. However, as can be seen from the position of the pseudo-adiabat at the surface, the downdraft air (with temperature and dew point both equal to T_w) is both cooler and drier than environmental air at the same level.

Using the skew T-log p diagram like that in Fig. 9.1, it is relatively straightforward to show that a saturated downdraft will be stronger (i.e., cooler and drier) if the relative humidity of the environmental air is lower, i.e., the dew-point depression, $T - T_d$ is greater (Exercise 9.7, Fig. 9.9).

In reality, evaporation rates are rarely sufficient to produce saturated downdrafts over an appreciable depth and mixing with ambient air may be important also as is the case with entrainment into updrafts. Nevertheless, precipitation downdrafts do still bring air with low values of θ_e to the surface, which is a characteristic of cold pools to be discussed later (see Section 10.4).

9.3.4.3 Downdraft convective available potential energy

Just as it is possible to quantify the potential energy that can be converted to kinetic energy in an updraft (Section 9.2.3), it is possible to estimate the potential energy of a saturated downdraft, the so-called *downdraft convective available potential energy (DCAPE)*. In analogy with the formula for *CAPE*,

$$DCAPE = \int_{p_i}^{p_s} R_d\left(T_{\rho a} - T_{\rho p}\right) d\ln p, \qquad (9.12)$$

where p_i is the pressure at which the air parcel begins its descent and p_s is the pressure at the surface, or at a level of neutral buoyancy if this were to occur above the surface. Again, on a skew T-log p diagram (Fig. 9.9), $T_{\rho a}$ is approximated by the temperature of the environment and now $T_{\rho p}$ is approximated by the wet-bulb temperature of the descending air parcel. The $DCAPE$ is represented by the area between the two curves and the two pressure levels.

Clearly, the lower the environmental wet-bulb temperature, i.e., the drier the environmental air, the greater will be the $DCAPE$ (cf. Exercise 9.7).

Recalling that precipitation-driven downdrafts are usually unsaturated, $DCAPE$ represents an upper bound on the available downdraft energy associated with the cooling.

9.4 Understanding the effects of deep convection on the tropical circulation

An important consideration in understanding the interaction between deep convection and the broadscale flow in tropical disturbances is to realize that deep convection occurs as a result of convective instability. First, some air parcels must be brought to their LFC, either as a result of boundary layer turbulence or as a result of mechanical lifting, for example at the spreading outflow boundary produced by previous convection. Above the LFC, air parcels can rise through the depth of the troposphere in favorable conditions and they may even overshoot their LNB into the lower stratosphere.

It turns out that usually, only air parcels below a few 100 m to 1 km have any $CAPE$ so that only these can rise through a deep layer. Thus deep convection typically peels off a layer of warm moist air near the surface and expels (or detrains) it into the upper troposphere near the level of neutral buoyancy. As air parcels rise they entrain air from their environment. The entrained air is generally unsaturated and slightly cooler than the air parcel so that entrainment tends to reduce the amount of buoyancy in the updraft below that which would be predicted by an aerological diagram. As a result, entrainment tends to lower the LNB.

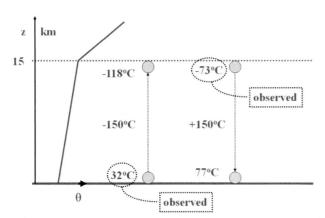

FIGURE 9.10 Schematic of air parcels undergoing hypothetical deep adiabatic ascent: see text for discussion.

We have seen that, when an air parcel ascends in a stably stratified environment, it expands and cools (see Fig. 9.10). As long as it remains unsaturated, it cools at the dry adiabatic lapse rate ($\approx 10\,°C\ km^{-1}$) and if a parcel could be lifted dry adiabatically from the surface in the tropics

(say with a temperature of $32\,°C$) to the tropopause (say at a height of 15 km), it would have a temperature when it reached the tropopause of $-118\,°C$. Since temperatures near the tropical tropopause are typically around $-73\,°C$, the air parcel would have a huge negative buoyancy when it got there and the lifting would require an unrealistically large amount of work to be done to lift the parcel. Since there is no conceivable mechanism in the atmosphere that could supply the needed work, we must conclude that *the air parcel must rise in cloud*, where the cooling can be offset by latent heat release, first as condensation occurs and later as freezing occurs.

A similar argument can be used to demonstrate that an air parcel cannot descend from the tropopause level to the surface unless it cools to offset the $150\,°C$ temperature rise that would be produced as a result of adiabatic compression (see Fig. 9.10). Typically, the subsiding air parcel would have to cool radiatively, the normal rate in clear sky conditions being on the order of $2\,°C$ per day. Thus, in order to arrive at the surface with a temperature of $30\,°C$, the parcel would have to cool by $45\,°C$, which could be accomplished only after more than 20 days. Since an air parcel can rise from the surface to the tropical tropopause in about 20 minutes, mass conservation would require that the area of descent be much larger than the area of ascent.

There is one way in which air can subside over an appreciable depth (although not the depth of the troposphere) on a much shorter time scale than the radiative time scale and that is within a precipitation driven downdraft in a deep convective system. The processes associated with downdrafts were discussed in Section 9.3.4.

9.5 Buoyancy in a finite horizontal domain

The analysis of convective instability using aerological diagrams and the concepts of $CAPE$ and CIN discussed earlier in this chapter are based on displaced parcel arguments. Air parcels are assumed to have infinitesimally small dimensions and when they are displaced vertically, they the are assumed to acquire the local pressure of their environment. Then, the buoyancy force acting upon them can be calculated using the local density of their environment. Together, these assumptions make the buoyancy force uniquely defined and imply that it is the only *static* force acting on the air parcel (see below).

As explained earlier in this chapter, as well as in Chapter 5, displaced parcel arguments are useful conceptually for investigating the stability of a region of fluid, but they are less useful for investigating fluid motion. In Section 2.2, we saw that, in general, buoyancy is not a unique force as it depends on the arbitrary definition of a reference density. Nevertheless, the total force in the vertical, the sum of the vertical perturbation pressure gradient force and the buoyancy force, i.e., $-(1/\rho)\partial p'/\partial z + b$, *is* a unique force.

We noted in Sections 2.1 and 2.7 that, in an incompressible flow, or in a situation where the anelastic approximation is valid, the perturbation pressure p' cannot be prescribed. In fact, as shown below, it has to be determined in a way that the local acceleration field leads to a velocity field that satisfies the continuity equation.

In Section 7.2 of his book on cloud dynamics, Houze (1993, 2014) explains how a region of buoyant air with finite size induces a static perturbation pressure, p'_b, and how, in general, the vertical component of the static perturbation pressure gradient force, $-(1/\rho)\partial p'_b/\partial z$, increases with the horizontal scale of the buoyant region in a way that opposes the buoyancy force. Thus, the force $b_e = -(1/\rho)\partial p'_b/\partial z + b$ may be thought of as an *effective buoyancy*.

When the buoyant region has infinitely small dimensions, i.e., the case of an air parcel, $-(1/\rho)\partial p'_b/\partial z$ is zero and $b_e = b$, while in the limit of infinite width, $-(1/\rho)\partial p'_b/\partial z$ exactly balances the buoyancy force, i.e., $b_e = 0$. If there were no motion, the flow would be in strict hydrostatic balance, but, in general, there will be a dynamic contribution to the perturbation pressure, say p'_d, which is required to ensure that the flow satisfies the mass continuity equation also. Houze's analysis is sketched below and provides the basis for a quantification of the relative contributions of $-(1/\rho)\partial p'_b/\partial z$ and b to b_e in a tropical context.

In a non-rotating, inviscid fluid, the three-dimensional momentum equation is

$$\frac{\partial \mathbf{u}}{\partial t} + \mathbf{u} \cdot \nabla \mathbf{u} = -\frac{1}{\rho}\nabla p' + b\mathbf{k}, \tag{9.13}$$

where $\mathbf{u}$ is the velocity vector, $p' = p - p_{ref}(z)$ is the *perturbation pressure* and $b = -g(\rho - \rho_{ref})/\rho$ is the *buoyancy force*, both per unit mass, ρ_{ref} is a reference density and p_{ref} a reference pressure chosen to be in hydrostatic balance (i.e., $dp_{ref}/dz = -g\rho_{ref}$).

First, we make the *anelastic approximation* and approximate the density by its horizontal average, $\rho_o(z)$, except where it appears in the buoyancy force. Further, we define ρ_{ref} to be $\rho_o(z)$. Multiplying Eq. (9.13) by ρ_o and taking the three-dimensional divergence with the momentum advection term moved to the right-hand side gives

$$\frac{\partial}{\partial t}(\nabla \cdot \rho_o \mathbf{u}) = -\nabla^2 p' + \frac{\partial}{\partial z}(\rho_o b) - \nabla \cdot (\rho_o \mathbf{u} \cdot \nabla \mathbf{u}). \tag{9.14}$$

Using the anelastic form of the mass continuity equation, $\nabla \cdot (\rho_o \mathbf{u}) = 0$, Eq. (9.14) may be written

$$\nabla^2 p' = \frac{\partial}{\partial z}(\rho_o b) - \nabla \cdot (\rho_o \mathbf{u} \cdot \nabla \mathbf{u})$$
$$= F_b + F_d, \tag{9.15}$$

where

$$F_b = \frac{\partial}{\partial z}(\rho_o b), \tag{9.16}$$

and

$$F_d = -\nabla \cdot (\rho_o \mathbf{u} \cdot \nabla \mathbf{u}). \tag{9.17}$$

Eq. (9.15) is a Poisson-type, second-order, elliptic partial differential equation from which the perturbation pressure may be diagnosed, given the spatial distribution of F_b and F_d together with suitable boundary conditions on p'. The key step in the derivation of this equation is the use of the continuity equation. Physically, the equation constrains the pressure field in such a way that the total force (pressure gradient force and buoyancy force) produces a local flow acceleration at every point that is consistent with mass continuity together with conditions on the flow boundary (Section 2.1).

If the solution domain of Eq. (9.15) is bounded by impermeable boundaries at the ground ($z = 0$) and at the tropopause ($z = H$) and if the forcing distributions are localized in the horizontal direction, Neumann boundary conditions, $\partial p'/\partial n = 0$, are the appropriate boundary conditions to take, assuming that the side boundaries are sufficiently far from the forcing. Here, n refers to the direction normal to the particular boundary. These conditions would determine p' to within an arbitrary constant that could be taken to be zero at some point on one side boundary. From a dynamical perspective, it is only the perturbation pressure gradient that is of consequence. The condition that $\partial p'/\partial z = 0$ at $z = 0$ and $z = H$ is required for consistency with the vertical momentum equation when the vertical velocity is set to zero at these boundaries.

At this point, it is convenient to partition the perturbation pressure into its static and dynamic parts by writing $p' = p'_b + p'_d$. Then, the solution of $\nabla^2 p'_b = F_b$ gives the static perturbation pressure distribution, p'_b, associated with the buoyancy distribution over a finite domain while that of $\nabla^2 p'_d = F_d$ gives the dynamic perturbation pressure distribution, p'_d, associated with the instantaneous distribution of flow acceleration. Because the equation for p' is linear, there would be no loss of generality by choosing the same boundary conditions on p'_b and p'_d as on p', since these would ensure the correct conditions on p'.

Houze (2014) invoked an analogy with electrostatics to elucidate the qualitative behavior of the solutions to $\nabla^2 p' = F_b + F_d$, but equally one could use the membrane analogy discussed in the appendix to Chapter 2, albeit with appropriate consideration given to the Neumann boundary conditions.[3] In this analogy, p' may be thought of as the equilibrium displacement of a stretched membrane in response to a steady force distribution $-(F_b + F_d)$ applied normal to the membrane.

---

3. With Neumann boundary conditions, a slightly more appropriate analogy would be a soap film between two vertical slippery glass plates along which the soap film displacement would have zero normal gradient at the plate boundaries (M.E. McIntyre, unpublished lecture notes).

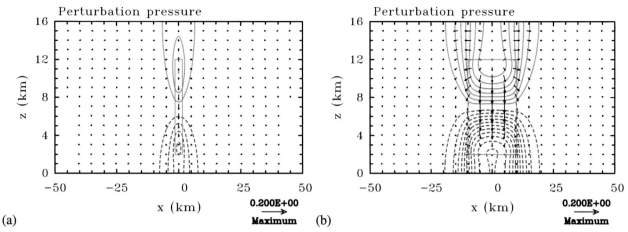

FIGURE 9.11 Static perturbation pressure field induced by a horizontally-uniform sinusoidal temperature perturbation in the vertical to an analytic approximation to the Dunion moist tropical sounding. This perturbation is applied in the region enclosed by the green (gray in print version) rectangle centered along the x-axis. It has a maximum amplitude of $+3\,°C$ and is zero at the upper and lower boundaries of the rectangle. Rectangle width in (a) 3 km, (b) 20 km. Contour interval for perturbation pressure 20 Pa. Shown also are the vectors of static perturbation pressure gradient force per unit mass. The scale for these vectors shown in the lower right is 0.2 m s^{-2}. From Smith and Montgomery (2022b). Republished with permission of Wiley.

In the next section we present some numerical solutions of Eq. (9.15) with $F_d = 0$ in a bounded domain for specific buoyancy distributions that relate to the tropical atmosphere.

9.6   Quantification of effective buoyancy

To elucidate the perturbation pressure distribution induced by a finite region of buoyant air in a stably-stratified environment, we examine two thought experiments. For simplicity, we assume a two-dimensional (slab symmetric) configuration in a domain that is 100 km wide, 16 km high and with rigid horizontal boundaries at top and bottom where $w = 0$.

First we consider the buoyancy distribution in a vertical column of finite width extending from an altitude of 2 km to 12 km. The background density within the domain, $\rho_o(z)$, is taken to be a piece-wise analytical approximation to the Dunion moist tropical sounding and the buoyancy in the column assumes a sinusoidal temperature perturbation over the column with a maximum amplitude of $+3\,°C$ above the sounding temperature. The temperature perturbation is uniform across the column and zero at the upper and lower edges of the column. Details of the solution method and boundary conditions on perturbation pressure are given in Appendix 9.11.2.

Fig. 9.11 shows the static perturbation pressure associated with columns widths of 3 km and 20 km. As anticipated by Houze (2014), the induced pressure perturbation increases as the width of the buoyant region increases. The buoyancy induces low perturbation pressure around the base of its region and high perturbation pressure around its top. This pattern would be consistent with the tendency for air to be drawn into the base of the buoyant region and to be forced upwards at the top, combined with a tendency for air

to rise within the region itself (see e.g. Smith et al. (2005), their figure 4 and related discussion, keeping in mind that a complete picture requires a consideration of the effects of F_d in Eq. (9.15) as well).

Fig. 9.12 shows the spatial distribution of perturbation pressure gradient force per unit mass as well as contours of the vertical component of this force corresponding to the calculations for the same columns shown in Fig. 9.11. Supplementary to these panels, Fig. 9.13 shows the vertical profiles of $-(1/\rho)\partial p_b'/\partial z$ and $-(1/\rho)\partial p_b'/\partial z + b$ along the axis of the buoyancy column for these calculations and for two additional ones with column widths of 5 km and 10 km. Panel (b) shows also the vertical profile of buoyancy, which corresponds with the total static force for a column of infinitesimal width. Again, as anticipated by Houze (2014), as the horizontal extent of the air parcel increases, the contribution of $-(1/\rho)\partial p_b'/\partial z$ increases in a way that opposes the buoyancy force, itself, with the expectation that as the horizontal scale becomes very large, the two forces almost cancel, leading to approximate hydrostatic balance between $-(1/\rho)\partial p_b'/\partial z$ and b.

The perceptive reader will notice that the total static force does not become negative below the top boundary, but the total force $-(1/\rho)\partial p'/\partial z + b$ must be negative since rising air parcels must decelerate before reaching this boundary. It follows that the deceleration effect must be contained in the solution of $\nabla^2 p_d' = F_d$, which gives the dynamical contribution to p'. To show that this is the case, we examine conditions along the vertical axis. Here, the lateral flow component is zero by symmetry and ignoring the decrease in density with height just below the top of the domain, F_d can be shown to be approximately equal to $-2\rho_o(\partial w/\partial z)^2$, which is negative (the derivation is sketched Appendix 9.11.3). According to the mem-

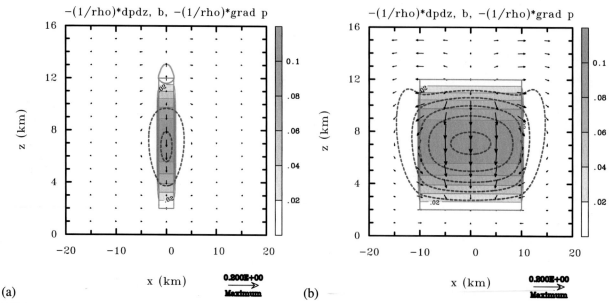

FIGURE 9.12 Buoyancy force per unit mass (green (gray in print version) contours, color shading) and static perturbation pressure gradient force per unit mass (arrows) corresponding to the calculations shown in Fig. 9.11. The actual calculations were carried out on the same domain as Fig. 9.11. Values on the color bar in m s^{-2}. The scale for the pressure gradient vectors, 0.2 m s^{-2}, is shown in the lower right. Red and blue (mid and dark gray in print version) contours show the vertical component of the static perturbation pressure gradient force per unit mass. Contour interval 0.02 m s^{-2}; solid red (mid gray in print version) contours positive, dashed blue (dark gray in print version) contours negative. From Smith and Montgomery (2022b). Republished with permission of Wiley.

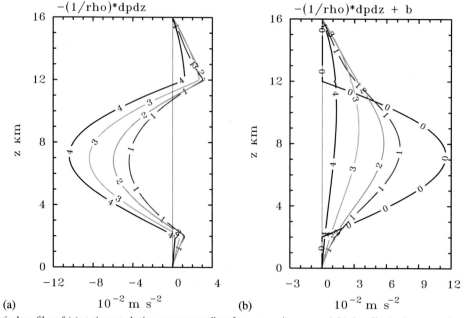

FIGURE 9.13 Vertical profiles of (a) static perturbation pressure gradient force per unit mass, and (b) the effective buoyancy force per unit mass, b_e, corresponding to the buoyancy distribution and sounding in Fig. 9.12. Curves 1-4 refer to the forces along the axis of vertical columns of widths 3 km, 5 km, 10 km and 20 km, respectively, while curve 0 refers to the buoyancy force alone corresponding with a column of infinitesimal width. From Smith and Montgomery (2022b). Republished with permission of Wiley.

brane analogy for solving the perturbation pressure equation (Chapter 2), a region of negative F_d would imply one of positive total perturbation pressure p' as would be required to decelerate the vertical flow induced by the buoyancy force near the domain top.

Assuming a judicious choice of reference density so that the difference between $1/\rho$ and $1/\rho_o$ can be ignored, the effective buoyancy force per unit mass would be essentially independent of reference density also. The question is then: what does the field of effective buoyancy force, say

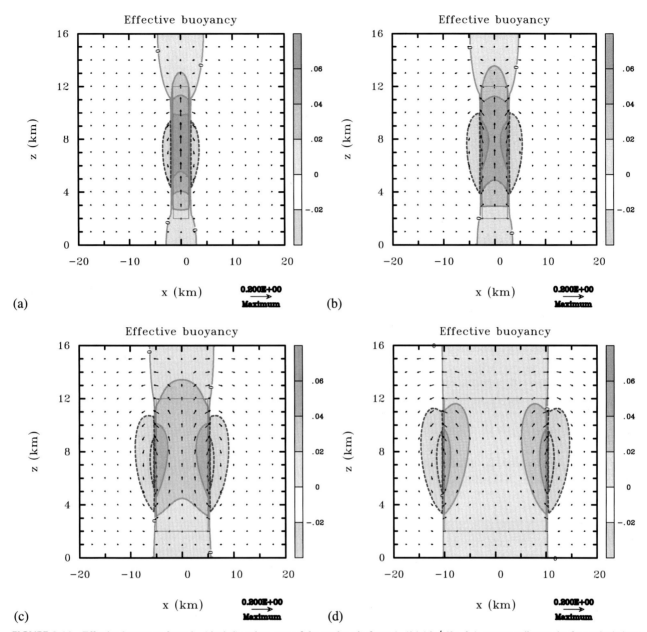

FIGURE 9.14 Effective buoyancy force b_e (shaded) and vectors of the total static force $(-(1/\rho)\partial p_b'/\partial x, b_e)$ corresponding to the four calculations discussed in Section 9.6. Width of buoyancy region in (a) 3 km, (b) 5 km, (c) 10 km, (d) 20 km. Contour interval for effective buoyancy force 0.02 m s^{-2}; solid red (mid gray in print version) contours positive, dashed blue (dark gray in print version) contours negative. Values on the color bar in m s^{-2}. The scale for the pressure gradient vectors, 0.2 m s^{-2}, is shown in the lower right. From Smith and Montgomery (2022b). Republished with permission of Wiley.

b_e, look like and how does its distribution vary with the width of the region of buoyancy? Answers to these questions are provided by Fig. 9.14, which shows contours of effective buoyancy force and vectors of the total static force $(-(1/\rho_o)\partial p'/\partial x, b_e)$ corresponding to the four calculations in Fig. 9.13.

For each width of the buoyancy column, the effective buoyancy force is positive, both above and below the column from the surface to the top boundary, and in the narrower columns the positive region is a little broader than the

buoyancy column, itself. Elsewhere, the effective buoyancy force is negative with a narrow sheath of strong negative buoyancy along the side of the buoyancy column. As the width of the buoyancy column is increased, the maximum effective buoyancy force moves away from the center of the column with the largest positive values along the inner edge of the column. At the same time, the effective buoyancy force becomes smaller in the central region of the column. In essence, the largest values of effective buoyancy force, both positive and negative, become concentrated along the sides of the buoyancy column. Note that, even though the buoyancy, itself, is horizontally uniform across the buoyancy column, the effective buoyancy force is not, a fact that will have consequences for calculating $CAPE$ in a region of buoyancy of finite width. The concept of *effective CAPE* is explored in the next section.

In contrast to our definition of effective buoyancy force per unit mass, Davies-Jones (2003) defines an *effective buoyancy force per unit volume*, which he shows to be completely independent of the reference density (see Appendix 9.11.1 herein).

9.7 Implications for $CAPE$

The definition of $CAPE$ given by Eq. (9.5) considers only the buoyancy force acting on a lifted air parcel of infinitesimal size. Although this force is made unique by taking the reference density along a vertical through the lifted air parcel, it is strictly not the only force acting on the air parcel if the air parcel belongs to a region of buoyancy with a finite size. As shown in the previous section, in this case, the static perturbation pressure gradient force p_b' should be taken into account also.

It is clear from the substantial dependence of total static vertical force on the width of the buoyant region shown in Fig. 9.13 that the conventional $CAPE$ is likely to be a significant overestimate of potential energy achievable by an ascending air parcel, at least an air parcel that rises along the axis of the buoyant column. By effective $CAPE$ we mean the $CAPE$ calculated with the effective buoyancy force per unit mass acting on the air parcel. To demonstrate this fact, we calculate first the buoyancy distribution and $CAPE$ in the usual way for an infinitesimally-small air parcel lifted from an altitude of 100 m in the Dunion moist tropical sounding to its LNB. The positive buoyancy distribution between the LFC (about 900 m) and LNB (about 13.8 km) is then used to calculate the total force on an air parcel within a buoyant column of finite width as in Section 9.6. Where the buoyancy is negative, below the LFC and above the LNB, we set the buoyancy equal to zero for simplicity.

Fig. 9.15 highlights conditions along the domain axis for four calculations corresponding to the same buoyancy columns in Fig. 9.13, but with the buoyancy distribution calculated as described in this section. The figure shows the vertical profile of effective buoyancy force, b_e, as in Fig. 9.13b. In these cases also, the vertical component of total static force is reduced significantly as the width of the buoyant column is increased on account of the increasing adverse perturbation pressure gradient force.

The effect of the decreasing total vertical static force on the effective $CAPE$ along the axis in these calculations is summarized in Table 9.2. The normal parcel $CAPE$ based on the buoyancy force alone, equivalent to a column width of 0 km is approximately 2130 J/kg, based on a vertical

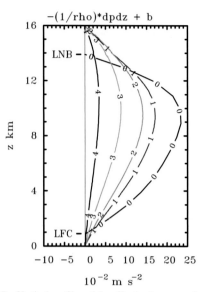

FIGURE 9.15 Vertical profiles of the effective buoyancy force for an air parcel lifted from a height of 100 m in the Dunion moist tropical sounding. Curves 1-4 refer to the specific forces along the axis of vertical columns of widths 3 km, 5 km, 10 km, and 20 km, respectively, while curve 0 refers to the buoyancy force in a column of zero width. The levels of free convection and neutral buoyancy for the lifted air parcel are denoted by LFC and LNB, respectively. From Smith and Montgomery (2022b). Republished with permission of Wiley.

interval for the integration of 100 m from the LFC to the LNB.[4] As the column width increases to 20 km, the effective $CAPE$ reduces to as little as 300 J/kg.

TABLE 9.2 Values of effective $CAPE$ for an infinitesimally small air parcel lifted along the axis of buoyant columns of finite width from a height of 100 m to the LNB in the Dunion moist tropical sounding. The calculations are based only on the "Positive Area", i.e., the total vertical force between the LFC and LNB.

Column width	Effective $CAPE$ J/kg
0 km	2130
3 km	1540
5 km	1280
10 km	800
20 km	300

It is worth noting that, like the effective buoyancy force, the effective $CAPE$ in the foregoing calculations is a minimum along the axis of the buoyancy column and increases towards edge of the column. Judging from Fig. 9.15, this effect should be most pronounced for the larger column widths. Fig. 9.16 highlights this feature for a lifted air parcel in the Dunion moist tropical sounding, showing the effective

---

4. The perceptive reader may notice that the value 2130 J/kg is a little larger than those given in Table 9.1. This is because, in the present calculation, the integration begins from the LFC and not the lifted parcel height, z_i.

buoyancy force and vectors of total force for the broadest buoyancy column with 20 km width. In the case of a buoyancy distribution without such an abrupt decline at its edge as here, it may be shown that the maximum effective $CAPE$ occurs where the horizontal Laplacian of buoyancy is a maximum (see Eq. (9.21) in Appendix 9.11.1).

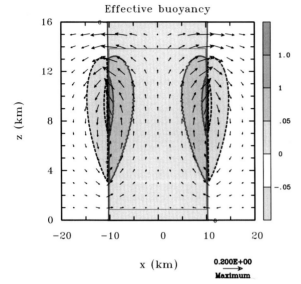

Effective buoyancy

FIGURE 9.16 Effective buoyancy force, b_e (shaded) and vectors of the total static force $(-(1/\rho)\partial p'_b/\partial x, b_e)$ corresponding to the simulation with a buoyant region of 20 km width discussed in Section 9.7. Contour interval for effective buoyancy force 0.02 m s^{-2}; solid red (mid gray in print version) contours positive, dashed blue (dark gray in print version) contours negative. Values on the color bar in m s^{-2}. The scale for the pressure gradient vectors, 0.2 m s^{-2}, is shown in the lower right.

9.8 More on $CAPE$

Many authors have cautioned against the indiscriminate use of parcel $CAPE$ as a measure of convective instability as well as the vigor of convective updrafts. Observations have shown that undilute ascent is rare in deep tropical convection, an indication that turbulent mixing must be important as air ascends in convective updrafts. Citing observational studies by Lucas et al. (1994) and Wei et al. (1998), Zipser (2003) points out that in the case of oceanic deep convection in the Tropics, large dilution of convective updrafts by entrainment is the norm. Nevertheless, he argues that despite large dilution, subcloud air can easily ascend within cumulonimbi to the tropopause if freezing of condensate is taken into account, supporting the results of an earlier study by Williams and Renno (1993).

In a recent review of atmospheric convection, Siebesma et al. (2020) point out that the actual $CAPE$ of a given atmospheric sounding is

"not a number without ambiguity and that if it is to be used quantitatively, the particular use of the concept must be made precise".

As shown in Section 9.2.1, $CAPE$ is a function of the state of the air parcel lifted, the environment in which it is lifted and the assumed thermodynamic process by which it is lifted. Even if the concept is made precise, comparison of $CAPE$ values from different studies is often difficult because of many differences in the way in which the initial thermodynamic state is determined, including the initial height of the lifted parcel and whether or not its properties are based on some layer average. Observations by Renno and Williams (1995) indicate that convective updrafts have their roots at low levels in the boundary layer and these authors recommended using lifted parcel heights between screen level and 100 m above the surface, a range similar to that suggested by Romps and Kuang (2011), see their section 1c. Notably, none of the authors mentioned in this section discuss the potential limitations of $CAPE$ because of the neglect of the vertical pressure gradient force, which has been investigated here.

9.9 Cloud structure in tropical cyclones

In a review article on tropical cyclone clouds, Houze (2010) writes:

"It is tempting to think of clouds in the inner regions of tropical cyclones as cumulonimbus that just happen to be located in a spinning vortex. However, this view is over simplified, as the clouds in a tropical cyclone are intrinsically connected with the dynamics of the cyclone itself."

One feature highlighted in the review is the tendency of deep convection to evolve from upright buoyant cells that readily develop into mesoscale convective systems in the formation stage of the tropical cyclone, to slantwise structures with little parcel buoyancy that form the eyewall cloud complex as the mature stage is approached. This feature coincides with the experience of pilots flying hurricane reconnaissance missions that the most vigorous local updrafts occur when the storm is in an early stage of development or rapidly intensifying. Nevertheless, Houze points out that, even in the mature stage, transient deep convective cells may develop in the eyewall complex that are more buoyant and more upright than the mean eyewall, itself. He goes on to cite the modeling study of Hurricane Bob (1991) by Braun (2002), who found that buoyant elements within the eyewall accounted for over 30% of the vertical mass flux in the eyewall. Other numerical studies have shown the existence of two maxima in vertical velocity in the eyewall, one at low levels near where air exits the boundary layer and a second one in the upper troposphere, which is presumably a result of a positive (effective) buoyancy (see e.g., Chapter 15, Figs. 15.6c and 15.6e).

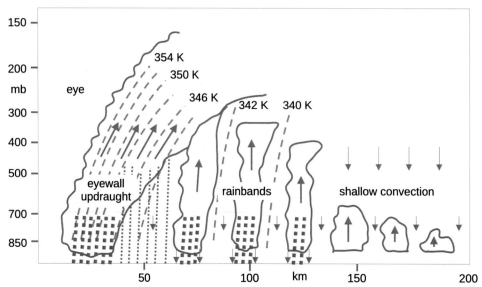

FIGURE 9.17 Schematic of the azimuthal mean cloud and θ_e structure of a mature tropical cyclone. Arrows show the main regions of ascent and subsidence and dotted lines indicate precipitation. Arrows beneath cloud base beyond the eyewall cloud indicate precipitation-driven downdrafts.

Fig. 9.17 shows a schematic of the azimuthal mean cloud structure and pseudo-equivalent potential temperature, θ_e, distribution in a mature tropical cyclone, inspired by Fig. 15.6 of Palmén and Newton (1969). Air parcels ascending in deep convective clouds approximately conserve their θ_e, at least until liquid water begins to freeze, so that the radial structure of θ_e aloft bears an imprint of that where the air enters the cloud at the top of the boundary layer (see Fig. 1.9d,f).

Since surface moisture fluxes tend to increase with wind speed until near-surface air becomes close to saturation with respect to the sea-surface temperature, near-surface values of θ_e typically increase with decreasing radius. Thus, because of deep convection, a negative radial gradient of θ_e develops throughout much of the troposphere. Also, because the air in the eyewall is mostly saturated, this negative radial gradient of θ_e implies a negative radial gradient of virtual temperature at a given pressure level and accounts for the warm core nature of the vortex beyond the eye.

The extension of the warm core into the eye, itself, is accounted for by subsidence that occurs within the eye as the vortex intensifies (see Section 5.5).

The cloud structures that occur in a tropical cyclone are related, in part, to the structure of the secondary circulation, which, in turn, they help to produce. The main regions of ascent occur in the eyewall clouds and in the rainbands with weaker ascent in the anvil region just beyond the eyewall.

Subsidence is generally present between convective-cloud updrafts and at large radii above the cloud layer where there are only shallow clouds. As explained in Section 9.4, prolonged subsidence is constrained by the need for descending air parcels to cool by radiation to space. Forced subsidence that is produced in the neighborhood of individ-

ual clouds leads to gravity waves or inertia-gravity waves and is essentially transient (see e.g. Mapes, 1993; Bretherton and Smolarkewicz, 1989). A succinct review of the effects of clouds on their environment is provided by Stevens (2005).

Within and below the shallow cloud layer, and indeed at all radii where the near-surface air is unsaturated, there is moistening on account of surface evaporation. Turbulent motions within the sub-cloud layer carry moist air upwards and the buoyant clouds carry the moisture higher, in part in the form of cloud water droplets that ultimately evaporate. Subsidence in the stably-stratified air between clouds, or in the free troposphere above the shallow cloud layer, has generally a drying effect.

In regions where the cloud field is reasonably homogeneous, there is an approximate balance between the net moistening and drying so that an areal average of the sounding would be approximately stationary. The drying above the cloud layer is responsible for the mimimum of θ_e at a height of about 4 km (about 600 mb).

If the near-surface wind speed increases and/or the upper-level subsidence weakens, the convection would be expected to become more vigorous and extend over a deeper layer, likely producing precipitation. As discussed in Section 9.3.4, precipitating clouds have the tendency to produce downdrafts at a later stage in their lifetime, in part, because of the drag exerted by water loading and, in part, because of cooling associated with the evaporation of rain below cloud base. However, these deeper clouds lead to moistening over a deeper layer, thereby elevating θ_e in their environment and collectively helping to remove the depressed values of θ_e where the precipitating downdrafts have their roots. As explained in Section 9.3.4.2, this moistening has the effect of reducing the strength of downdrafts.

As the eyewall is approached from the outside, the moistening effect of increasingly vigorous convection, often in the form of rainbands, is sufficient to largely remove the θ_e minimum in the local thermodynamic profile and downdrafts associated with precipitation become relatively weak or non-existent, allowing surface moisture fluxes to act unhindered to elevate the θ_e of air of the subcloud layer.

9.9.1   Ventilation by deep convection in tropical cyclones

In Chapter 6 we saw that near-surface friction effects in a vortex induce convergence in a shallow boundary layer, with ascent at the top of the boundary layer in the inner region. In a stably-stratified atmosphere that is stable also to moist convection (i.e. without any $CAPE$), the air that ascends out of the boundary layer will flow radially outwards in a relatively shallow layer. If there is some degree of convective instability, some of the ascending air parcels may reach their LFC and, without any mixing with their environment, they would rise under their own buoyancy to their LNB and perhaps a little above if they arrive at the LNB with a finite vertical velocity.

If the LNB is high enough, deep convection may occur and, based on the calculation in Section 9.2.3, one might expect the maximum vertical velocity that occurs in the convective updrafts to increase with the $CAPE$. However, the ability of the convection to remove the air that is ascending out of the boundary layer depends on the vertical mass flux in the convective towers, which depends on the area of the updrafts as well as the mean vertical velocity in the updrafts, themselves. Because the processes governing the vertical mass flux in the convective towers are different to those in producing boundary layer convergence, it would be fortuitous if, at any instant of time, the convective mass flux at the top of the boundary layer were exactly equal to the mass flux ascending out of the boundary layer.

If the vortex is relatively weak and the convective mass flux is relatively strong, the convection will be more than able to "ventilate" mass at the rate it converges in the boundary layer. In this case, the "suction effect" of the convection in its immediate locality will dominate the low-level convergence produced by the vortex boundary layer and it will lead to convergence above the boundary layer, itself. On the other hand, if the vortex is relatively strong, the convective mass flux may not be strong enough to "ventilate" the mass that is converging in the boundary layer. In this case, the fraction of air that cannot be ventilated will flow radially outwards in a shallow layer above the boundary layer (shallow because the air is typically stably stratified). These ventilation ideas were explored in an axisymmetric balance framework in Section 8.4, in which the strength of convec-

tion is related to spatial gradients of diabatic heating rate. The ideas are explored further in Chapter 15 in the context of a three-dimensional numerical model simulation of a tropical cyclone life cycle, where they are shown to be a key to understanding the long-term evolution of the vortex, including its decay phase (see Section 15.5.4).

Clearly, a steady state could be envisioned only when the boundary layer is supplying mass at exactly the rate that deep convection can ventilate it. As noted in Section 8.4, a deficiency of most models for tropical cyclone behavior to date, both time dependent or steady state, is the *assumption* that all the mass converging in the boundary layer ascends in deep convection to the upper troposphere. We discuss a popular steady-state model in Chapter 13 and question the realism of assuming a steady state in Chapter 14.

9.10   Exercises

Exercise 9.1. Show that c_L is small in calculating θ_e.

Exercise 9.2. Verify Eq. (9.5).

Exercise 9.3. Applying Newton's second law of motion to a rising air parcel rising solely under its buoyancy force in a steady updraft with vertical velocity $w(z)$, verify Eq. (9.8).

Exercise 9.4. Show that the virtual temperature of saturated air and the saturation pseudo-equivalent potential temperature given by the approximate formula Eq. (2.28) are functions of state, i.e., they depend only on pressure and temperature. Using Teten's formula for the saturation vapor pressure, Eq. (9.1), show that, at constant pressure, these quantities are monotonically increasing functions of temperature.

Exercise 9.5. The water vapor mixing ratio, r_v can be expressed in terms of the partial pressure of water vapor, e, by the formula $r_v = \epsilon e/(p - e)$, where $\epsilon = R_d/R_v = 0.6220$ and R_d and R_v are the specific gas constants for dry air and water vapor, respectively. Show how this formula can be used together with one for the saturation vapor pressure, $e^*(T)$ to obtain an implicit formula for T_w using Eq. (9.11).

Given the pressure, p, temperature T and dew-point temperature T_d, write an iterative code to determine T_w using the Newton-Rapheson method and assuming Teten's formula for $e^*(T)$.

Exercise 9.6. Modify Fig. 9.5 to illustrate the negative buoyancy of a saturated downdraft originating at a height of 3 km.

Exercise 9.7. Using the example illustrated in Fig. 9.9, show that if the environment is drier at 700 mb than in the mean sounding shown, the saturated downdraft will be cooler and drier than the one shown and conversely, if the environment is moister at 700 mb, the saturated downdraft will be warmer and moister when it reaches the surface.

9.11 Appendices

9.11.1 Appendix 1: effective buoyancy per unit volume

As an alternative approach to the derivation in Section 9.6, Davies-Jones (2003) partitions the static perturbation pressure p_b' into the sum of two parts p_{nh}' and p_h', where p_h' is chosen to satisfy the perturbation hydrostatic equation

$$\frac{1}{\rho_o}\frac{\partial p_h'}{\partial z} = b. \tag{9.18}$$

Then, the vertical component of the momentum equation has the form

$$\frac{Dw}{Dt} = \frac{1}{\rho_o}\left(\frac{\partial p_d'}{\partial z} + \frac{\partial p_{nh}'}{\partial z}\right), \tag{9.19}$$

where $Dw/Dt = \partial w/\partial t + \mathbf{u}\cdot\nabla w$ is the vertical acceleration. Note that the buoyancy b and vertical gradient of static perturbation pressure p_b' per unit mass have combined to give $(1/\rho_o)\partial p_{nh}'/\partial z$.

After a little algebra, the equation corresponding to Eq. (9.15) becomes one for p_{nh}, i.e.,

$$\nabla^2 p_{nh}' = F_h + F_d, \tag{9.20}$$

where $F_h = -\nabla_h^2 p_h'$, and F_d is as in Section 9.5. Davies-Jones (2003) takes $-\partial/\partial z$ of Eq. (9.20) and uses Eq. (9.18) to obtain

$$\nabla^2\left(-\frac{\partial p_{nh}'}{\partial z}\right) = -g\rho_o\nabla_h^2 b. \tag{9.21}$$

He notes also that if $w = 0$ at $z = 0$, the right-hand side of Eq. (9.19) must be zero at $z = 0$. Further, he goes on to assume that the two pressure gradient terms on the right-hand side of Eq. (9.19) are separately zero at $z = 0$, whereupon Eq. (9.21) may be solved, in principle, subject to Neumann boundary conditions.

Davies-Jones (2003) defines $-\partial p_{nh}'/\partial z$ as the *effective buoyancy per unit volume* and notes that, because the term $g\rho_o\nabla_h^2 b$ on the right-hand side of Eq. (9.21) is independent of the reference density, so must this effective buoyancy. Nevertheless, effective buoyancy per unit mass $-(1/\rho_o)\partial p_{nh}'/\partial z$, does depend on the choice of reference density, unless one knows the actual density ρ to use in place of ρ_o. Davies-Jones (2003) obtained a formal solution of Eq. (9.21) in terms of a Green function in the half space $z > 0$, but did not show specific solutions.

More recently, Peters (2016) attempted to solve the two-dimensional (x, z) form of Eq. (9.21) for a localized region of buoyant air with width δx and depth between the LFC to the LNB. However, this solution is limited by the fact that Neumann boundary conditions $\partial p_{nh}'/\partial z|_{LFC} = 0$, $\partial p_{nh}'/\partial z|_{LNB} = 0$, and $\partial p_{nh}'/\partial z|_{\pm\delta x} = 0$ are applied on the boundaries of the buoyant region rather than along the boundaries of the whole domain that encompasses the buoyant region and its environment.

9.11.2 Appendix 2: numerical solution of Eq. (9.15)

The Poisson-type second-order partial differential equation (9.15) is solved explicitly for the perturbation pressure p' on a rectangular grid using the solver helmholtz_ffacr.f90. The domain is 100 km wide and 16 km high with a grid spacing of 500 m in the horizontal direction and 100 m in the vertical, equivalent to 201×161 grid points. Neumann conditions are imposed on p' along each boundary and the solution is made unique by the requirement that $p' = 0$ at the lower left corner of the domain.

9.11.3 Appendix 3: forcing of p' by F_d in Eq. (9.15) on the upper domain axis

Taking $\mathbf{u} = (u, w)$, $F_d = -\nabla\cdot(\rho_o\mathbf{u}\cdot\nabla u, \rho_o\mathbf{u}\cdot\nabla w)$, or

$$F_d = -\frac{\partial}{\partial x}\left(\rho_o u\frac{\partial u}{\partial x} + \rho_o w\frac{\partial u}{\partial z}\right)$$
$$-\frac{\partial}{\partial z}\left(\rho_o u\frac{\partial w}{\partial x} + \rho_o w\frac{\partial w}{\partial z}\right), \tag{9.22}$$

while mass continuity, $\nabla\cdot(\rho_o\mathbf{u}) = 0$, gives

$$\frac{\partial u}{\partial x} = -\frac{1}{\rho_o}\frac{\partial}{\partial z}(\rho_o w). \tag{9.23}$$

Then Eq. (9.22) becomes

$$F_d = -\frac{1}{\rho_o}\left(\frac{\partial}{\partial z}(\rho_o w)\right)^2 + u\frac{\partial}{\partial x}\left(\frac{\partial}{\partial z}(\rho_o w)\right)$$
$$-\rho_o\frac{\partial w}{\partial x}\frac{\partial u}{\partial z} + \rho_o w\frac{\partial}{\partial z}\left(\frac{1}{\rho_o}\frac{\partial}{\partial z}(\rho_o w)\right)$$
$$-\frac{\partial}{\partial z}\left(\rho_o u\frac{\partial w}{\partial x} + \rho_o w\frac{\partial w}{\partial z}\right). \tag{9.24}$$

Along the axis $x = 0$, $u = 0$, and $\partial w/\partial x = 0$ by symmetry, whereupon

$$F_d = -\frac{1}{\rho_o}\left(\frac{\partial}{\partial z}(\rho_o w)\right)^2 - \frac{\partial}{\partial z}\left(\rho_o w\frac{\partial w}{\partial z}\right). \tag{9.25}$$

Assuming that near the domain top, the Boussinesq approximation is valid, i.e., ρ_o is approximately constant, it follows that

$$F_d \approx -2\rho_o\left(\frac{\partial w}{\partial z}\right)^2, \tag{9.26}$$

which is negative definite and, according to the membrane analogy (see Section 9.5), would lead to a positive perturbation pressure near the top of the domain, as expected.

Chapter 10

Tropical cyclone formation and intensification

Tropical cyclone formation or tropical cyclogenesis are somewhat fuzzy terms to define because of the multi-scale nature of the processes involved in building the mesoscale vortex from an initially weak parent disturbance. At a fundamental level, one needs to explain the natural tendency for the emergence of a mesoscale coherent convective-vortex structure from a much weaker initial disturbance with scattered cumulus convection that has no apparent organized structure.

Traditionally, the processes involved in tropical cyclone formation, i.e., tropical cyclogenesis, and those involved in the subsequent intensification of the cyclone have been treated separately. This may be, in part, because early models of the intensification process were axisymmetric, whereas it has long been recognized that the genesis process, itself, is far from axisymmetric. In fact, as discussed in Chapter 1, observations show that even mature tropical cyclones have only an appreciable degree of axial symmetry in their inner region. It will emerge in this chapter that there is no essential difference between the processes involved in tropical cyclone formation (or tropical cyclogenesis) and those involved in tropical cyclone intensification. As a result, one could question whether it is physically meaningful to define a genesis time.

While axisymmetric models have played an important role in understanding some of the basic processes of tropical cyclone intensification, including those described in Chapters 5, 6, and 8, they have one serious limitation. The limitation is that deep convection in an axisymmetric model can be represented explicitly only in the form of concentric rings, which is highly unrealistic except for the eyewall region of mature storms and possibly for secondary eyewalls. As an example, even when considering the intensification process starting from a relatively strong initial vortex possessing a maximum tangential wind of 15 m s^{-1}, axisymmetric solutions have been shown to have inherent limitations for representing the intensification process because of the omnipotence of noisy convective rings relative to analogous three-dimensional solutions (Persing et al., 2013). It should not be surprising then, that these models would be severely limited if applied to understand the genesis of storms. For this reason we will largely bypass axisymmetric models in this chapter.

To represent the complex three-dimensional processes involved in tropical cyclone dynamics, including deep cumulus convection and the frictional boundary layer, it is necessary to resort to numerical models, preferably ones that are able to represent deep convection explicitly. The use of numerical models to explore the basic features of tropical cyclone intensification has a long history. The models employed have ranged in sophistication from axisymmetric models with just two or three layers and a highly simplified representation of moist processes to three-dimensional models that include parameterizations of ice microphysical processes and electromagnetic radiation. Some of these models are coupled also to an ocean model. It would require a dedicated book to properly review all of these models and the findings that emerged from them. Here, we present a small hierarchy of simplified, but three-dimensional models that help to elucidate the basic physical processes involved in tropical cyclogenesis and intensification.

We assume that the reader is broadly familiar with the principal features of numerical models. In essence, the nonlinear partial differential equations comprising the three components of the vector momentum equation, the first law of thermodynamics, the moisture equation, the equation of state and the continuity equation are converted to a set of algebraic equations for the dependent variables of the problem. The conversion is accomplished by rewriting the partial derivatives in the equations by finite-differences on a mesh of grid points. The algebraic equations are written in a form in which the dependent variables can be advanced in time, starting from the known initial values (e.g., Richtmeyer and Morton, 1967). In a numerical weather prediction model, the initial condition might be an objective analysis of the state of the weather at the start of the forecast (e.g., Holton (2004), Chapter 13 entitled "Numerical Modeling and Prediction"). In an idealized problem such as the one described below, it may be an analytically prescribed basic state.

10.1 The prototype problem for genesis and intensification

The observation in Chapter 1 that tropical cyclogenesis takes place in a protected pouch-like region in which there

Tropical Cyclones. https://doi.org/10.1016/B978-0-44-313449-4.00018-7

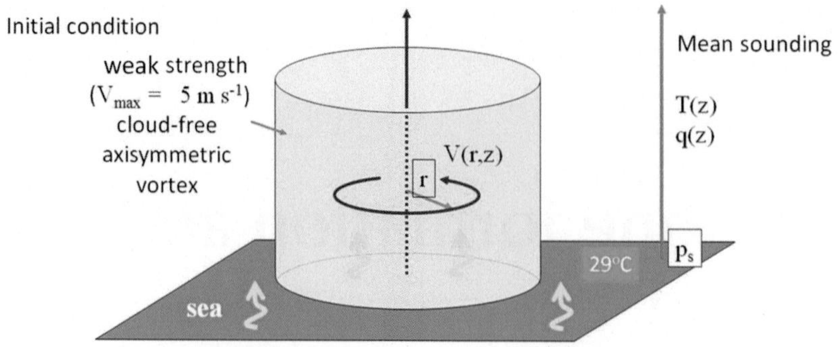

FIGURE 10.1 The prototype problem for understanding tropical-cyclone genesis and intensification. The yellow curvy arrows indicate surface enthalpy fluxes supplied mainly through the evaporation of sea water. These fluxes are necessary to support prolonged deep convective activity during the genesis and intensification process.

is a closed, but weak cyclonic circulation in a frame of reference moving with some incipient convective disturbance over the warm tropical ocean suggests the following prototype problem for understanding tropical cyclogenesis and subsequent intensification.

The problem considers the evolution of an initially cloud-free, axisymmetric, cyclonic vortex in an environment with no background flow on an f-plane with a suitable environmental thermodynamic sounding typical of the tropical-cyclone season and a prescribed sea surface temperature, T_s (Fig. 10.1). The initial vortex is assumed to be in thermal wind balance (Section 5.3). A prototype problem for genesis would begin with a relatively weak initial vortex with, say, a maximum tangential wind speed, V_{max}, of 5 m s^{-1} at a radius of, say 100 km or more, while one for intensification only might start with a stronger vortex, say $V_{max} = 15 \text{ m s}^{-1}$ at a similar radius.

Given the tangential wind structure of the initial vortex, $V(r, z, 0)$, the initial potential temperature field, $\theta(r, z, 0)$, in thermal wind balance with the vortex can be obtained using the method discussed in Section 5.2. To do this, one requires an environmental thermodynamic sounding to be specified in which the vertical profiles of temperature, dewpoint temperature and pressure are given. For example, one might choose a mean sounding for the tropical cyclone season over some ocean basin.

The plan is to investigate the evolution of such an initial vortex using a suitable numerical model, given an appropriate sea surface temperature.

10.2 A simplified numerical model experiment

The idealized three-dimensional numerical model used to explore the genesis and intensification of a vortex in the prototype problem is that described by Kilroy et al. (2017c). Echoing Ian James' sentiments discussed in the Preface, the formulation of the model is highly simplified, but represents the physical processes thought to be of leading order importance to tropical cyclogenesis and intensification. These

processes include a simple explicit representation of deep convection in which the clouds are allowed to produce rain and evaporatively-cooled downdrafts when rain evaporates below cloud base, but the freezing of cloud and rain drops is not represented. They include also a simple representation of the frictional boundary layer that incorporates a formulation of the surface stress and surface fluxes of heat and moisture (i.e., enthalpy fluxes).

This model should have the most important elements to be able to capture convective-vorticity organization and the ensuing intensification of a system-scale vortex. These elements are:

- the formation of a statistical population of convective clouds that collectively produce convergence in the low to mid troposphere, strong enough to oppose the natural tendency of the boundary layer to induce divergence above it; and

- an uninterrupted source of moisture at the sea surface to help maintain convectively-unstable conditions to support continued deep convection.

The numerical model (CM1 version 16) is a non-hydrostatic and fully compressible cloud model (Bryan and Fritsch, 2002). The calculations are carried out on a domain 3000×3000 km in size. The inner part of the domain is 300×300 km in size and has a uniform horizontal grid spacing of 500 m, while the outer part has a gradually expanding horizontal grid spacing reaching 10 km near the domain boundaries. There are 40 vertical levels extending to a height of 25 km. The vertical grid spacing expands gradually from 50 m near the surface to 1200 m at the top of the domain.

Subgrid-scale parameters and exchange coefficients for momentum, moisture, and heat are based on the latest observational estimates of vertical and horizontal turbulent diffusivities and air-sea exchange processes discussed in Chapter 7. In brief, the model has prediction equations for the three components of the velocity vector, specific humidity, suspended liquid, perturbation Exner function (see Section 2.3.3), and perturbation density potential tempera-

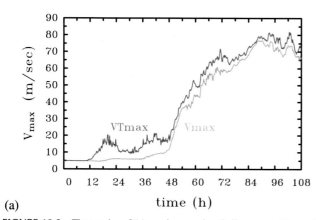
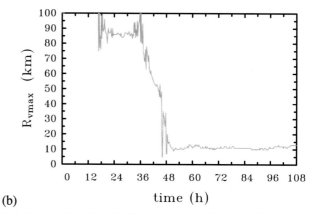

(a)   time (h)

(b)   time (h)

FIGURE 10.2 Time series of (a) maximum azimuthally-averaged tangential wind speed, V_{max} (in red (mid gray in print version)), and (b) the radius R_{vmax} at which V_{max} occurs in the genesis and intensification warm-rain simulation described in the text. Panel (a) shows also the maximum total wind speed, VT_{max} (in blue (dark gray in print version)).

ture, where perturbation quantities are defined relative to a prescribed basic state in hydrostatic balance.

In the simple warm-rain scheme for moist convection, rain has a fixed fall speed of 7 m s^{-1}. For simplicity, ice microphysical processes and dissipative heating are omitted. The lateral boundary conditions are taken to be open, that is, air is allowed to enter or exit the lateral boundary.[1] The genesis and intensification experiment with warm rain processes only will be referred to as *the warm-rain experiment*.

Later in this chapter, in Section 10.9, we summarize the key differences that arise when ice microphysical processes are included. The genesis and intensification experiment including ice microphysics will be referred to as *the ice experiment*.

The environmental sounding used is an averaged sounding taken from the PREDICT observations in the pre-Karl pouch region on 12 September, 2010 (Smith and Montgomery, 2012b). This sounding has a $CAPE$ of 2028 J kg^{-1}, a CIN of 47 J kg^{-1}, and a Total Precipitable Water of 61 kg m^{-2}. The calculation of $CAPE$ and CIN used here is described in Smith and Montgomery (2012b). The sea-surface temperature (SST) is taken to be constant (29 °C). This choice of parameters represents approximately those pertinent during the genesis of Hurricane Karl (2010): see Montgomery et al. (2012). The calculations are carried out on an f-plane with the Coriolis parameter $f = 2.53 \times 10^{-5}$ s^{-1}, corresponding with a latitude of 10°N. This value is lower than that of pre-Karl. It was chosen because spin-up occurs sooner at low latitudes (Smith et al., 2015a), requiring less computational time.

Following Rotunno and Emanuel (1987), their Experiment K, radiative effects are represented crudely by adopt-

ing a simple Newtonian cooling approximation in which the temperature is relaxed to the prescribed ambient temperature at the same height. The relaxation time is 12 hours and the cooling rate is capped at 2 K per day. The limitations of such a representation are discussed later in Section 15.1.

The integration is carried out for $4\frac{1}{2}$ days that span the period of vortex evolution from a weak, initial vortex, through a convective organization phase to a period of rapid intensification[2] (RI), and a mature stage. Here, we consider genesis to have occurred when RI begins. In more complex environments, RI might never occur, but the vortex may slowly intensify before finally weakening.

10.3 The numerical simulation

10.3.1 A summary of vortex evolution

A broad summary of vortex evolution is provided by Figs. 10.2 and 10.3. Fig. 10.2 shows time series of the maximum total horizontal wind speed, VT_{max}, the maximum azimuthally-averaged tangential wind speed, V_{max}, and the radius, R_{vmax}, at which V_{max} occurs. Fig. 10.3 shows two measures of outer core size: R_{gales} is the outer radius at which the azimuthally-averaged tangential wind speed at a height of 1 km is gale force (17 m s^{-1}) and R_{galesF} is the outer radius at which the azimuthally-averaged total horizontal wind speed interpolated from the lowest model level (25 m) to the surface is gale force.[3] In each case, the azimuthal average is carried out about some suitable vortex center (see Section 10.5).

The measure of size adopted by forecasters is the (outer) radius at which wind speed at a standard meteorological

---

1. This feature is the only difference from the calculation performed by Kilroy et al. (2017c), which used impervious boundary conditions. However, because of the relatively short integration time and the large domain size, the differences between that and the present calculation with open lateral boundary conditions are found to be minor.

2. The United States National Hurricane Center defines the rapid intensification of a tropical cyclone as occurring when the maximum sustained winds (based on a 1 minute average) increase by at least 30 knots (15 m s^{-1}) over a 24-hour period.

3. The reasons for choosing both these altitudes are discussed in Section 3.2 of Kilroy et al. (2016a).

height of 10 m is 17 m s^{-1}, which, in general, is a function of azimuth. In practice, the radius of gales is an important quantity to know for warning affected communities.

The evolution of the vortex begins with a period of about 10 hours during which both VT_{max} and V_{max} decrease slightly on account of friction. The imposition of surface friction at the initial time reduces the tangential wind speed near the surface, elevating the height of the maximum. As explained in Chapter 6, this reduction leads to an inward agradient force in the layer affected by friction. The agradient force drives inflow in the shallow layer affected by friction and outflow just above the inflow layer. At this early stage, there is no deep convection to ventilate any of the inflowing air.

During the early spin down phase, the boundary layer moistens because of evaporation from the underlying sea surface. This moistening and the lifting of air within the boundary layer due to the frictionally-induced convergence lead to a reduction of the CIN and an increase in the $CAPE$. These changes lead to convective instability and eventually to the initiation of isolated deep convective clouds. The rapid increase in VT_{max} a little after 12 hours is a result of the local overturning circulations accompanying these clouds. These local circulations break the axial flow symmetry of the initial vortex and account for the subsequent fluctuations of VT_{max}. Before convection begins, the vortex remains close to axisymmetric, but thereafter the flow becomes markedly asymmetric. It is not until about 36 hours that V_{max} begins a marked increase also.

From Fig. 10.2 we see that at about 45 hours, there is a turning point in the evolution when V_{max} begins to increase more rapidly. At this time, $V_{max} = 9.1$ m s^{-1} and $VT_{max} = 17.4$ m s^{-1}. At 48 hours, there is a sharp rate of increase in both V_{max} and VT_{max}. Following Kilroy et al. (2017c), we refer this time as the *intensification begin time* as it marks the onset of a period of RI. Significantly, at this time, $V_{max} = 12.0$ m s^{-1} which is appreciably less than gale force, while $VT_{max} = 20.5$ m s^{-1}, which exceeds *gale force*. Of major interest is the subtle, but spatially marked change in vortex structure as the intensification begin time is approached and passed, which is the topic of the next subsection.

It is worth noting that the intensity change between 48 and 72 hours is 47 m s^{-1}. This value exceeds the National Hurricane Center threshold for RI by a factor of more than three, showing that, despite the simplifications made, the model is easily able to capture RI. It is worth noting also that the time variation of the maximum intensity simulated here is not a result of vertical shear, or cooler sea surface temperatures, etc. Rather, this represents an internal mode of variability of intensity and structure change that, in our view, must be understood as a prerequisite to understanding observed or simulated tropical cyclone intensity change in more complex situations.

The onset of rapid intensification just after 48 hours is accompanied by a sharp contraction of R_{vmax} from about 54 km at 42 hours to 7.5 km at 48 hours. During the subsequent intensification period, beyond about 51 hours, R_{vmax} fluctuates, but remains approximately within the range of 10-13 km. The large fluctuations in R_{vmax} between 46 hours and 48 hours are a consequence of abrupt jumps in the diagnosed circulation center location during this period. They are not a reflection of location changes of the broader scale vortex circulation, itself, but rather of changes in local centers on the mesoscale within the broader scale circulation.

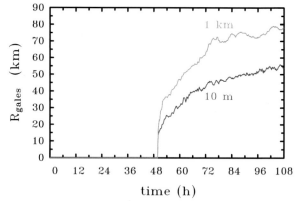

FIGURE 10.3 Time series of two metrics of vortex size as measured by the outer radius at which gale force winds occur in the warm-rain simulation described in the text. R_{gales} (red (mid gray in print version) curve labeled '1 km') is the outermost radius at which the azimuthally-averaged tangential wind speed at a height of 1 km reaches gale force, while R_{galesF} (blue (dark gray in print version) curve labeled '10 m') is the outermost radius at which the azimuthally-averaged near-surface *total wind* at the surface reaches gale force.

Fig. 10.3 shows time series of two metrics that characterize the overall size of the vortex circulation, namely the outermost radius of gale force winds. The first metric, R_{gales} is calculated at a height of 1 km and corresponds to the outer radius where the azimuthally-averaged tangential wind falls below gale force. The second metric, R_{galesF} is the outer radius where the azimuthally-averaged near-surface (nominally 10 m altitude) total wind falls below gale force. In either case, gale-force winds first develop at 50 hours and both metrics indicate a progressive broadening of the vortex with time as the vortex intensifies. As expected, because of the frictional stress near the surface, R_{galesF} is generally smaller than R_{gales}. The calculations show that at 60 hours, R_{gales} is 16 km larger than R_{galesF}, but this difference increases by up to 26 km at later times.

In order to place the evolution of the foregoing five metrics in context and to gain an understanding of the intensification of the simulated vortex, it is insightful to examine the structure of specific fields at particular times during the vortex evolution, focusing at first on the period between 42 and 48 hours, which is near the end of the genesis period.

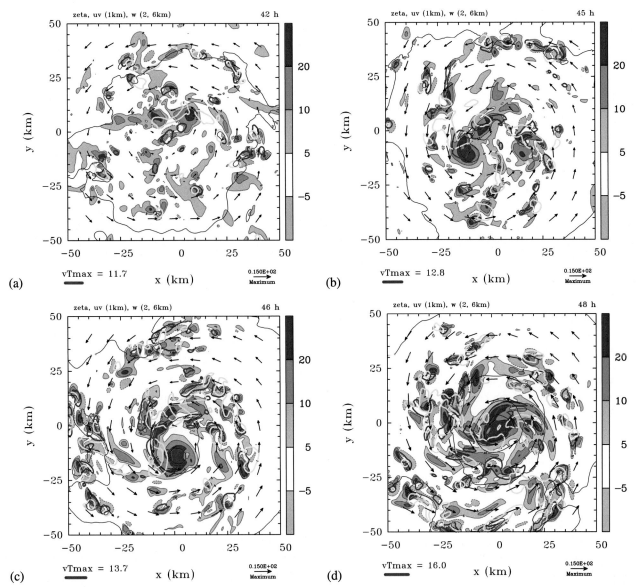

FIGURE 10.4 Horizontal cross sections of relative vertical vorticity and wind vectors at a height of 1 km for the warm-rain experiment, with contours of vertical velocity at heights of 2 km (aqua) and 6 km (yellow). Shown also are black contours of surface pressure, contoured every 1 mb. Values for the shading of vertical vorticity are given in the color bar, multiplied by 10^{-4} s^{-1}. Vertical velocity contour is 1 m s^{-1} for both heights. The wind vectors are in relation to the maximum reference vector at the bottom right, while on the bottom left the maximum total wind speed in the domain plotted is given in m s^{-1}. The times shown are: (a) 42 hours, (b) 45 hours, (c) 46 hours, (d) 48 hours.

10.3.2 Evolution of vorticity

Stokes' theorem discussed in Chapter 2 shows that the circulation about a fixed circuit in a flow is equal to the area mean of the vertical component of relative vorticity, ζ, enclosed by the circuit. It follows that the change in mean tangential wind speed about any circular contour in a vortical flow is equal to the change in mean vorticity enclosed by that circuit. For this reason it is of interest examine the evolution of vorticity as the vortex begins to intensify.

Fig. 10.4 shows horizontal cross sections of ζ in the central part of the fledgling vortex at selected times straddling the intensification begin time. Superimposed on each panel are the wind vectors at a height of 1 km, the isobars of surface pressure, and two 1 m s^{-1} contours of vertical velocity at different heights. The contours of vertical velocity are shown for heights of 2 km and 6 km and are chosen to indicate the location of strong updrafts at these levels. Where these contours overlap or lie in close proximity gives an indication of the location of vertically coherent or possibly sloping deep convective updrafts. Panel (a) shows the fields at 42 hours, six hours before the intensification begin time, while subsequent panels show the fields at 45 hours (panel (b)), 46 hours (panel (c)), and finally at 48 hours (panel (d)) when RI begins (cf. Fig. 10.2a).

At 42 hours (panel a), there are several irregular-shaped patches of locally-enhanced cyclonic vertical vorticity, the larger ones in area lie in a west-southwest[4] to east-northeast-oriented strip to the north of the vortex center and in a sector to the south west of the vortex center. These patches result from the stretching of ambient vortex vorticity by previous as well as current deep convective cells.

The circulation center at 42 hours, indicated by the velocity vectors, lies near the center of the computational grid, while the surface pressure field is quite diffuse at this time. In fact, the location of minimum pressure is not apparent at the 2 mb contour spacing shown. The value of V_{max} is 9.0 m s^{-1}, while that of R_{vmax} has reduced from 100 km at the initial time to 55 km (Fig. 10.2b).

There are a few patches of negative (anticyclonic) vertical vorticity also. These are associated with the tilting of horizontal vorticity into the vertical by convective updrafts and downdrafts, which tends to produce dipole structures. A few such dipoles are evident beneath strong updraft roots at 2 km height, for example near $(-20$ km, 35 km), (43 km, -2 km), and (34 km, -12 km).

At 45 hours (panel (b)), the center of circulation remains within about 10 km of the center of coordinates and the surface pressure field remains diffuse. The value of V_{max} has risen by only 0.2 m s^{-1}, but generally the winds have strengthened near the circulation center and there has been a marked aggregation of cyclonic relative vorticity. Individual patches of cyclonic vorticity have increased in size and have begun to consolidate around the vortex center. In particular, the area of patches with a magnitude exceeding $> 1 \times 10^{-3}$ s^{-1} has increased. There are many patches of negative vertical vorticity also, but again, these occur mostly on the periphery of the coherent region of cyclonic vorticity that surrounds the center of circulation. The regions with updraft speeds exceeding 1 m s^{-1} at a height of 6 km (the areas enclosed by yellow contours in the figure) have increased markedly in area over the three hours, indicating that deep convection has become more widespread.

By 46 hours (panel (c)), the cyclonic vorticity surrounding the center of circulation has consolidated further and there is a significant updraft region at 6 km height near the region with the largest area of elevated cyclonic vorticity. Although V_{max} at this time has increased by only 0.9 m s^{-1} in one hour, a small closed contour of surface pressure has formed near the center of circulation indicating that the pressure has started to fall within the central core of high cyclonic vorticity. At this time, the circulation center has moved about 12 km south of the center of coordinates.

Two hours later, at 48 hours (panel (d)), the monopole of high cyclonic vorticity near the center of circulation has grown in size and V_{max} has increased to 14.0 m s^{-1}. Similar plots at later times portray an acceleration of the intensification process with the appearance of more and more concentric surface isobars.

Despite the relatively simple configuration of the foregoing thought experiment, the flow evolution becomes rather complex. The complexity is largely because of the formation of deep convective clouds, which introduces a stochastic element to the evolution. Compounding the complexity of the processes involved, the moist thermodynamics of the transition to a self-sustaining, coherent, convective vortex entails subtle, but important changes in the θ_e structure of the flow near the emergent center of circulation. We examine these changes in the next subsection.

10.4 Moist instability and θ_e

In Chapter 9, it was shown that moist convective instability requires air with high, near-surface values of θ_e that exceed the saturation values of θ_e of the sounding, θ_{es}, at some range of altitudes. It was shown also that, for the mean tropical sounding in Fig. 9.1, the vertical profile of θ_e has a minimum at about 4 km altitude. In Sections 9.3 and 9.3.4 we showed that deep convection produces cool downdrafts, which bring air with lower values of θ_e to the surface, reducing near-surface values of θ_e and thereby depleting, or removing, local convective instability. However, we have argued that tropical cyclone formation requires deep convective activity to be sustained, which, in turn, would require downdrafts to be weak and the low values of near-surface θ_e to be quickly removed by the action of surface enthalpy fluxes, or to become displaced locally from active deep convection. It is therefore of interest to examine changes in surface values of θ_e during tropical cyclone formation.

Fig. 10.5 shows horizontal cross sections of the θ_e difference from its initial value in the warm-rain experiment at 46, 50, 54, and 62 hours. The first two times straddle the time at which RI begins (Fig. 10.2). At 46 hours, V_{max} is only 10 m s^{-1}, at 48 hours it is still only 12 m s^{-1}, but by 50 hours it has risen to 20 m s^{-1}. The next time, 54 hours, is well into the RI period with V_{max} already 36 m s^{-1} and 62 hours is later in the RI period with V_{max} nearly 43 m s^{-1}

At 46 hours (Fig. 10.5a), there are extensive patches of depressed θ_e values with little obvious pattern in relation to the vortex circulation, although the patches with the lowest depressed θ_e values lie some distance from the center of circulation and all are located in regions with active convective cells (with vertical velocity at 6 km altitude exceeding 1 m s^{-1}) and nearby regions of convective downdrafts. Of course, some patches may have been produced by earlier deep convection. Four hours later, at 50 hours (Fig. 10.5b), the θ_e pattern has become more structured with a positive

---

4. We refer to the coordinate directions using geographical descriptions, the horizontal coordinate in the figure pointing east and the vertical coordinate pointing north.

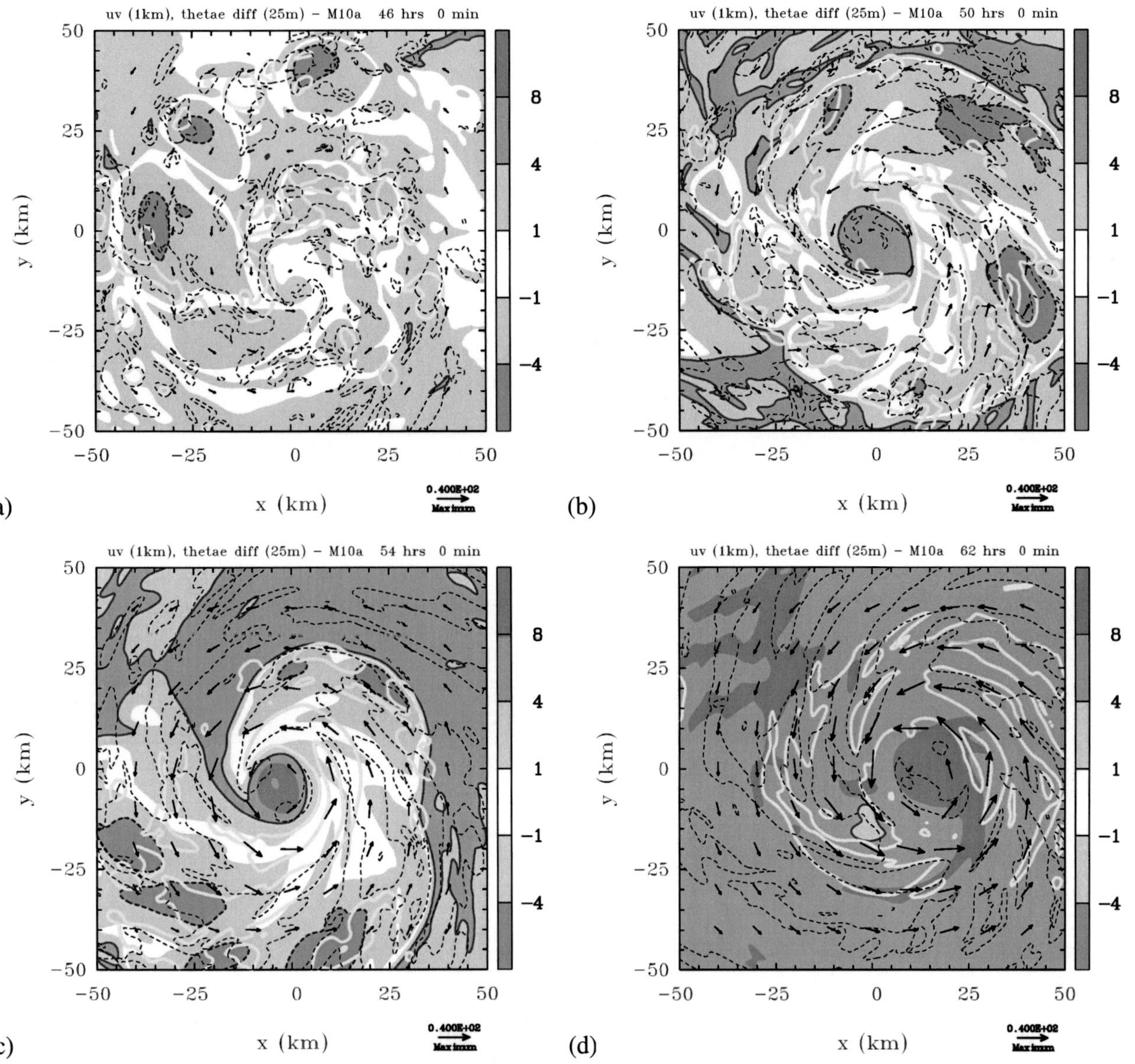

(a)

(b)

(c)

(d)

FIGURE 10.5   Horizontal cross sections of the difference between near-surface θ_e and its initial value in the warm rain simulation at (a) 46 hours, (b) 50 hours, (c) 54 hours, and (d) 62 hours. Shown also are horizontal wind vectors at a height of 1 km and two contours of vertical velocity: 1 m s^{-1} at a height of 6 km (yellow) and -10 cm s^{-1} at a height of 1 km (thin black dashed contours). Values for the shading of θ_e difference are given in the color bar, in K. The wind vectors are in relation to the maximum reference vector (40 m s^{-1}) at the bottom right.

anomaly exceeding 4 K and approximately 20 km in diameter straddling the center of circulation and an annular region with depressed values surrounding this anomaly. Notably, the region of depressed values is covered with areas of deep convection.

By 54 hours (Fig. 10.5c), deep convection has developed a more azimuthal wavenumber-one pattern with extensive regions to the south of the vortex center spiraling inwards to the east and north and accompanied by depressed values of near-surface θ_e. At larger radii beyond the convection, there are elevated values of θ_e on the eastern, northern and western sides of the center spiraling inwards to a near circular

region of elevated θ_e still straddling the center. Animations of the θ_e fields at 15 minute intervals show that as the vortex wind field strengthens, the areas of depressed θ_e do not survive for more than an hour and, as seen in Fig. 10.5d, by 62 hours, elevated values of θ_e cover the entire domain shown, despite the presence of deep convection in all sectors. The near circular region of elevated θ_e straddling the center remains a prominent feature, although at this stage, deep convection has not yet become organized to form a complete eyewall.

The evolution of the θ_e field during the genesis and early RI phase of the experiment, summarized in Fig. 10.5,

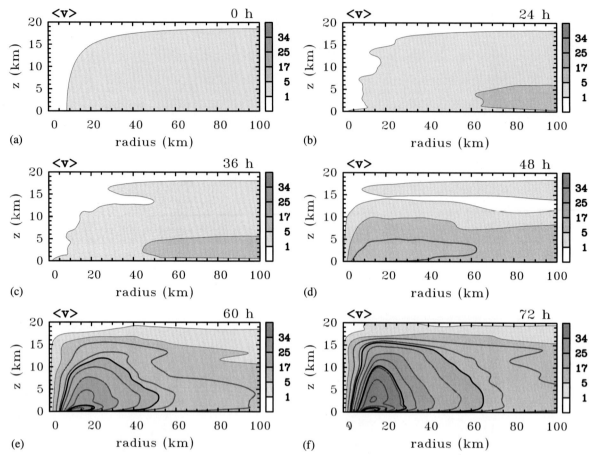

FIGURE 10.6 Vertical cross sections of the azimuthally-averaged, 3 hour time-averaged tangential velocity at various times in the warm-rain experiment. Contour interval: thin contours 1 m s^{-1} and 5 m s^{-1}. Thick contours every 5 m s^{-1} from 10 m s^{-1}. The first black contour encloses regions of gale force winds, greater than 17 m s^{-1}. Subsequent black contours are 34 m s^{-1} and 51 m s^{-1}. See label bar for shading.

highlights again the highly asymmetric nature of the genesis process. *Moreover, it cautions against explanations of the genesis process that invoke a casual chain of events tied primarily to the thermodynamics. As we can see in these results, the localized region of elevated θ_e anomaly encompassing the center of circulation emerges after the intensification process is already well underway, not before (compare Fig. 10.5d with Fig. 10.4c). It is the subtle change in the amalgamation of the otherwise patchy regions of enhanced cyclonic vorticity that seems to be a robust signature of completion of genesis and the transition to the RI phase.*

10.5 Azimuthal mean view of vortex evolution

To develop a conceptual framework for understanding the intensification of the vortex amid the complexities shown in the preceding subsections, it proves insightful to examine the azimuthally-averaged flow fields. To this end, Fig. 10.6 shows vertical cross sections of the azimuthally-averaged,

3 hour time-averaged[5] tangential velocity, $\langle v(r, z) \rangle$, at 12 hour intervals to 72 hours. The time averaging is based on 15 minute model output centered on the time shown. The vortex center used for azimuthal averaging is the location of the minimum pressure in a smoothed surface pressure field and is taken to be independent of height. The assumption of a vertical axis is defensible in the absence of background vertical shear. The smoothing prevents the center jumping between local pressure minima associated with deep convective clouds.

While the time series in Fig. 10.2 indicates little increase in V_{max} during the first 48 hours, with most of that increase occurring after 45 hours, the tangential wind fields in Fig. 10.6 show progressive changes in structure during the entire 72 hour period.

Between 48 hours and 60 hours there is a major change in vortex structure, with a significant contraction of the tangential wind field and the appearance of the maximum wind at a radial distance of only 10 km from the axis at an altitude

---

5. Here we use an angle bracket for this combined azimuthal and time average. In later chapters we use an overbar for an azimuthal average alone.

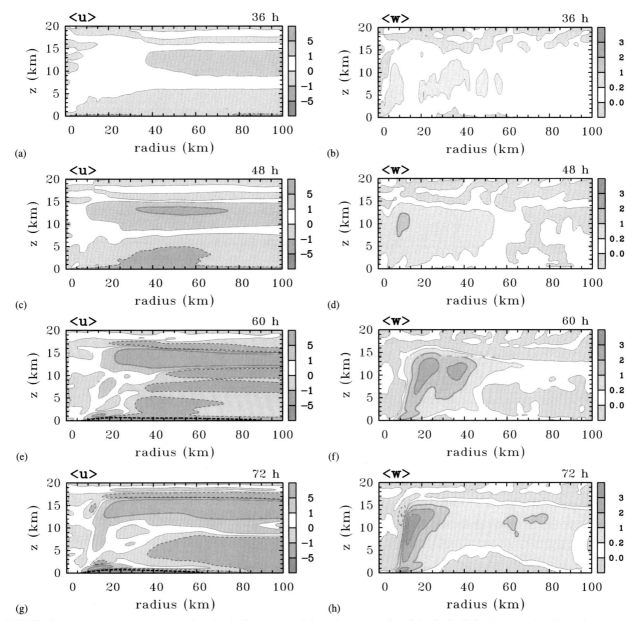

FIGURE 10.7 Vertical cross sections of the azimuthally-averaged, 3 hour time-averaged, radial velocity (left column) and vertical velocity (right column) at various times in the warm-rain experiment as indicated in the upper right of each panel. Contour interval: radial velocity 0, ± 1 and ± 5 m s^{-1} as indicated on the side bar, thick black dashed contour highlights inflow exceeding 5 m s^{-1}; positive vertical velocity 0, 0.2, 1, 2, 3 m s^{-1} as indicated on the side bar, negative vertical velocity, thin blue (dark gray in print version) dashed contours, interval 0.5 m s^{-1}. Solid contours positive, dashed contours negative.

of barely 200 m (Fig. 10.6e). By 72 hours (Fig. 10.6f), the vortex has intensified further, but with no additional contraction of the wind field. At this time there is a second tangential wind maximum at an altitude of about 3 km.

Fig. 10.7 shows vertical cross sections of the similarly averaged radial and vertical velocity fields, $\langle u(r, z)\rangle$ and $\langle w(r, z)\rangle$ respectively, at the last four times shown in Fig. 10.6, while Fig. 10.8 shows the corresponding absolute angular momentum field, $\langle M(r, z)\rangle$. The $\langle u \rangle$ and $\langle w \rangle$ fields at 36 hours depict the system-scale overturning circu-

lation generated by the collective effects of deep convection. The primary features are inflow in the lower troposphere, below about 7 km, with outflow in the upper troposphere (Fig. 10.7a). The maximum inflow above the frictional boundary layer occurs at about 3 km altitude, while the maximum outflow is found at an altitude of about 12.5 km.

Broadscale ascent with $\langle w \rangle \geq 0.1$ m s^{-1} occurs through much of the domain shown, with maximum ascent at a radius of about 12 km and an altitude of about 11 km at this time (Fig. 10.7b). The $\langle M \rangle$ surfaces are bowed inwards in the lower troposphere where the inflow is a maximum (Fig. 10.8a), reflecting the approximate material conservation of $\langle M \rangle$ above the boundary layer.

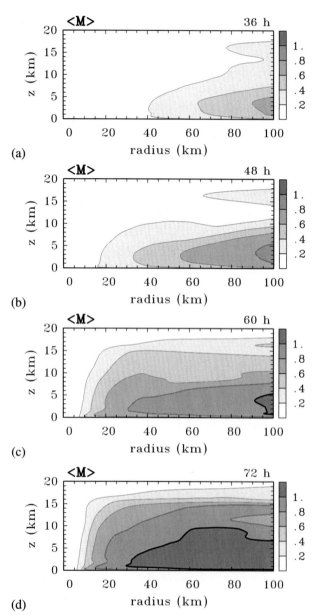

(a)

(b)

(c)

(d)

FIGURE 10.8 Vertical cross sections of the azimuthally-averaged, 3 hour time-averaged, absolute angular momentum, $\langle M \rangle$, at various times in the warm-rain simulation as indicated in the upper right of each panel. Shading on side bar. Contour interval: 0.2×10^6 m^2 s^{-1}. Thick black contour bounds regions where $\langle M \rangle$ is large ($> 1 \times 10^6$ m^2 s^{-1}).

As in the tangential wind field, significant structural changes occur in the radial and vertical velocity fields following the onset of RI (compare panels (e), (f) of Fig. 10.7 with (c), (d), respectively). Even as early as 60 hours, the vertical velocity shows a radially-narrow annular region of strong ascent ($\langle w \rangle \geq 2$ m s^{-1}) reminiscent of an eyewall

with a narrow region of subsidence along its inner edge (Fig. 10.7f). The inflow has strengthened markedly in a shallow frictional boundary layer near the surface and there is outflow in the eyewall (Fig. 10.7e), indicating an outward slope of the eyewall with height. There is strong outflow in a shallow layer just above where the boundary layer inflow terminates, indicative of the fact that the flow ascending out of the boundary layer into the eyewall is supergradient and the agradient force is positive (see forthcoming Figs. 11.9, 11.10 and related discussion).

Note that there is a low-level maximum in vertical velocity where the boundary layer inflow terminates, an indication of a strong vertical pressure gradient force that would be required to accelerate the ascending flow in this region. Again, the $\langle M \rangle$ surfaces have continued to move inwards in the lower troposphere and have become more erect in the eyewall (Fig. 10.8c). Significantly, there is radial outflow across a broad radial band outside the main eyewall updraft as defined above. Thus the classical mechanism for spin up, whereby $\langle M \rangle$ surfaces are advected inwards in the lower troposphere above the frictional boundary layer, does not operate to spin up the eyewall, nor does it act to spin up the maximum tangential wind speed, which occurs within the region of strong boundary layer inflow (compare Figs. 10.7e, g with Fig. 10.7c).

By 72 hours, the vortex has strengthened further and shows many features of a mature tropical cyclone. In particular, the boundary layer inflow has continued to strengthen, as has the region of outflow just above it where the inflow terminates and the air ascends into the eyewall. Now there is a prominent layer of enhanced inflow just above the foregoing region of outflow. These alternating inflow and outflow features, which extend up to about 4 km, are indicative of a standing inertial/centrifugal wave in the lower part of the eyewall as the eyewall updraft tries to adjust[6] towards gradient wind balance.

The region of moderate ascent outside the main eyewall updraft ($\langle w \rangle \geq 0.1$ m s^{-1}) has contracted radially inwards (Fig. 10.7h) and strengthened, although the inflow feeding the eyewall updraft has weakened a little, except in the frictional boundary layer. The weakening inflow above the boundary layer is presumably an indication that the increasing inertial stability[7] accompanying the strengthening vortex is acting to weaken the convectively-induced inflow. The strengthening boundary-layer inflow, itself, is consistent with a strengthening vortical flow above the boundary layer. With the continued inflow, above approximately 1.5 km height outside of the inner-core region, the M surfaces have moved further inwards in the lower troposphere, which

---

6. See Section 11.6.2 for more discussion about the dynamics of this process, where it is found that, in fact, most of the eyewall updraft during RI is supergradient.

7. Inertial stability is discussed in Chapter 5.

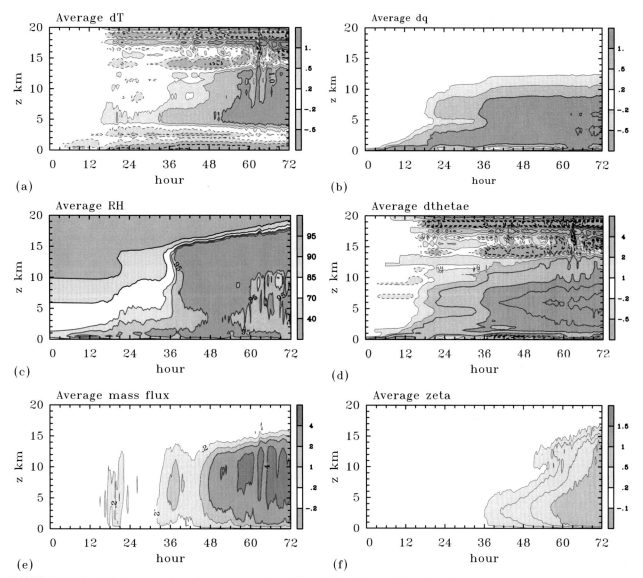

FIGURE 10.9 Time-height cross sections of system-averaged quantities within a 50 km ×50 km column for the warm-rain experiment, centered on the domain center. These include (a) the deviation of temperature from that at the start of the time series, (b) the deviation of water vapor mixing ratio from that at the start of the time series (in g kg^{-1}), (c) the relative humidity (in per cent), (d) the deviation of θ_e from that at the start of the time series, (e) the vertical mass flux per unit area (units kg m^{-2} s^{-1}), and (f) the relative vorticity (units s^{-1} multiplied by 10^{-4}).

accounts for the expansion of the outer tangential circulation (Fig. 10.6f).

In the next section we provide an interpretation of some of the main features of the azimuthally-averaged view of vortex spin up described above, deferring an interpretation of the three-dimensional aspects until the next chapter.

10.6 Modified view of spin up

In this azimuthally-averaged perspective, we recognize many elements of the classical paradigm for spin up discussed in Chapter 8. There is inflow in the lower troposphere, above the boundary layer. This inflow can only be

attributed to the collective effects of buoyancy generation by latent heat release in the inner-core deep convection. There is ascent in the heated region and outflow in the upper troposphere and because absolute angular momentum is approximately materially conserved in the outflow layer, the tangential velocity becomes anticyclonic beyond a certain radius (not shown in Fig. 10.6 due to the limited horizontal domain chosen for highlighting inner core structure.).

A feature of this solution that is not part of the classical paradigm is the occurrence of the maximum tangential wind speed in the layer of strong low-level inflow. Since the strong inflow is associated primarily with the reduction of the tangential wind by the frictional stress at the

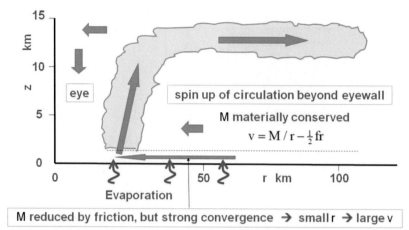

FIGURE 10.10 Modified schematic of the azimuthally-averaged view of vortex spin-up in the thought experiment for tropical cyclone genesis and intensification to include the boundary layer spin up enhancement mechanism. Dark blue arrows denote the sense of transverse (secondary) circulation. Light blue shading denotes cloudy areas. The size of the darker blue arrows is not proportional to the flow speed. The curvy dark blue arrows indicate evaporation from the sea. Adapted from Smith and Montgomery (2016).

ocean surface, we arrive at the conclusion that the maximum tangential wind occurs within the frictional boundary layer, a counter-intuitive result. As discussed in Section 6.6 and exemplified by Fig. 6.11, this feature is accounted for by the nonlinear boundary layer spin up mechanism, and leads to a modified cartoon of the spin up process from an azimuthally-averaged perspective. This modification is sketched in Fig. 10.10. Even so, since the flow is generally not axisymmetric, we do not expect the classical mechanism, including its extension to include the boundary layer spin up enhancement mechanism, to be complete, especially during the genesis and early intensification phase of evolution in which the flow is far from axisymmetric. A more complete framework for analyzing the three-dimensional nature of the vortex evolution, especially during the genesis and early intensification phases, is outlined in Chapter 11.

10.7 A system-averaged perspective

We examine now some important aspects of the system-averaged behavior. A system-averaged perspective of the vortex evolution is useful because it offers an aggregate view of the evolution that is not restricted to the strict azimuthally-averaged framework.

Fig. 10.9 shows time-height cross sections of system-averaged quantities within a column with horizontal cross section 50 km × 50 km, centered at the center of the circulation. These include the deviations of temperature, water vapor mixing ratio and equivalent potential temperature from their respective values at the start of the time series, the relative humidity, the vertical mass flux, and the vertical component of relative vorticity. It is seen that, like the vorticity fields discussed in Section 10.3.2, thermodynamic conditions in the column have evolved significantly

before the RI stage. In particular, the mid-to-upper troposphere warms while the lower troposphere cools.

Generally, the troposphere moistens in an absolute sense as evidenced by the increase of moisture throughout the troposphere, but the relative humidity, after first increasing at most levels, begins to develop a mid-tropospheric minimum after about 56 hours. This development is presumably because, as the vortex strengthens, it becomes narrower also so that the 50 km × 50 km column begins to sample part of the subsiding branch of the overturning circulation, including the eye. Significantly, the mid-tropospheric θ_e and the tropospheric relative humidity both increase[8] prior to RI. Since the increase in θ_e in this central region cannot be explained by radial advection on account of the general negative radial gradient of θ_e in the lower troposphere, it must originate from surface evaporation, which is necessary to maintain conditional convective instability.

It is evident that the moistening of the 50 km × 50 km column is accompanied by periods of enhanced positive vertical mass flux associated with deep convection. These periods correspond with a strengthening of the deep overturning circulation as shown in Figs. 10.7a-d. Recalling the discussion of the necessary conditions for genesis (Section 1.5.2), the moistening of the middle troposphere is considered to be an important aspect of the genesis process. Traditional reasoning would suggest that the moistening reduces the strength of cool downdrafts from deep convection that tend to stabilize the boundary layer to further convection until they are eliminated by surface enthalpy fluxes. However, it is worth noting the depression of θ_e values in the lowest kilometer between about 36 and 54 hours in Fig. 10.9d, which, as shown in Fig. 10.5, are the result of such downdrafts. Significantly, these depressed θ_e values are present

---

8. A similar result has been found in many earlier studies (e.g., Nolan, 2007 and refs.).

beyond the time that rapid intensification begins, underscoring the remarks made at the end of Section 10.4 cautioning against explanations of the genesis process that invoke a casual chain of events tied primarily to the thermodynamics. Clearly, traditional reasoning does not explain the sequence of events in this warm-rain experiment.

A more plausible explanation for the importance of moistening the low and middle troposphere is that it shields convective updrafts from adverse effects of the entrainment of lower θ_e air, which would not only produce cool downdrafts, but as shown by numerical studies of James and Markowski (2010) and Kilroy and Smith (2013), would reduce the strength of deep updrafts. The reduced strength of updrafts would have a concomitant effect of reducing the strength of the overturning circulation. This weakening of the overturning circulation is not seen in the present experiment. On the contrary, we have shown that the overturning circulation progressively strengthens. This strengthening enhances the influx of cyclonic vorticity above the boundary layer, leading to an amplification of the system-scale tangential wind field by the classical spin up mechanism.

Referring to panels (e) and (f) of Fig. 10.9 shows that deep convection in the column leads at first to an amplification of the vertical vorticity in the lower troposphere. From 48 hours onwards, the vertical extent of cyclonic vorticity increases rapidly during the period of RI.

10.8 Predictability issues

The association of the inner-core flow asymmetries with the development of deep cumulus convection suggests the real possibility that vortex-scale asymmetries, like the deep convection itself, may possess a limited interval of time over which they can be predicted accurately from initial conditions with a small, but nonzero, observational error. The reason is that the pattern of convection is strongly influenced by the low-level moisture field, which observations show has significant variability on small space scales and is not well sampled by observations, especially in tropical-depression and tropical cyclone environments. This fact points to the possibility of a limited ability to accurately forecast flow asymmetries in a tropical cyclone vortex.

The study of flow predictability, namely, the determination of the temporal range of meaningful weather prediction given initial conditions with suitably small error, was pioneered in the seminal studies of Thompson (1957) and Lorenz (1969). Although originally limited to the barotropic vorticity equation model for obtaining quantitative estimates, the underlying ideas have been subsequently applied to more realistic atmospheric and oceanic models spanning a vast range of scales covering the atmospheric boundary layer, deep convection, synoptic and global scale vortices and wave-like patterns (e.g., Judt, 2018; Zhang et al., 2019).

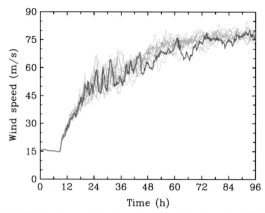

FIGURE 10.11 Time-series of maximum total wind speed at 900 hPa in the control experiment in Nguyen et al. (2008) (blue (dark gray in print version)) and in the 10 ensemble experiments in which a small (±0.5 g kg^{-1}) random moisture perturbation is added at every horizontal grid point in the innermost domain of the model below 900 mb (thin, red (mid gray in print version)). Republished with permission of Wiley.

The paradigm shift that emerged from the Lorenz study was the important finding that (p36):

> *"certain formally deterministic fluid systems possessing many scales of motion may be observationally indistinguishable from indeterministic systems (chaotic systems, our addition), in that they possess an intrinsic finite range of predictability which cannot be lengthened by reducing the error of observation to any value greater than zero."*

In the context of the tropical cyclone problem, this study suggests the possibility that uncertainties in the smallest scales of motion can progress upwards to the vortex and possibly even the synoptic scales, within a time interval comparable to the period over which tropical cyclone intensification forecasting is currently feasible. These ideas point to real challenges in obtaining accurate tropical cyclone intensity forecasts.

Although the study of upscale error growth is an important subject in itself,[9] beyond the scope of this book, we can illustrate the sensitivity of the flow asymmetries to low-level moisture in an intensifying model cyclone by performing an ensemble of simulations in which a small (e.g., between ±0.5 g kg^{-1}) random moisture perturbation is added at every horizontal grid point in the innermost domain of the model below a certain height, in our case 1 km. The results of such calculations are discussed below.

The perturbed moisture calculations shown below are taken from Nguyen et al. (2008) and relate to a calculation very similar to that in Section 10.3, but starting with

---

9. An insightful perspective on the strength and limitations of Lorenz's pioneering work, and its extensions to more geophysically apt formulations, is given in McWilliams (2019), Section 4.

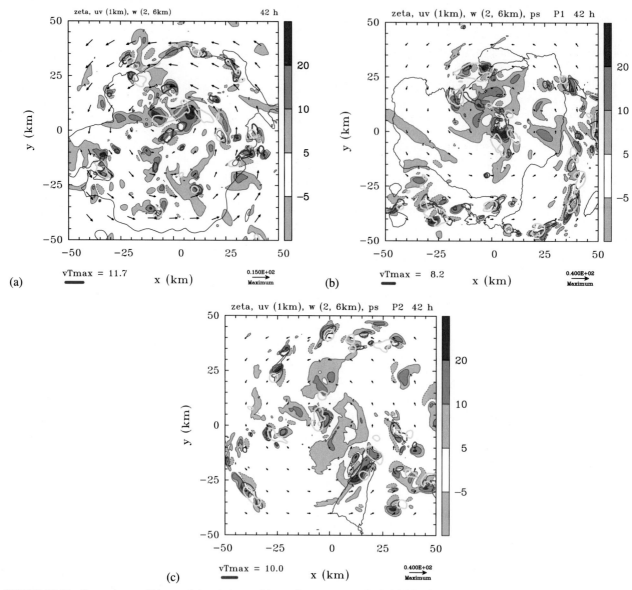

FIGURE 10.12 Comparison at 42 hours of the relative vorticity, surface pressure and wind fields of (a) the warm-rain experiment described in Section 10.3 with those of two similar experiments, but with random moisture perturbations added in the lowest 1 km (panel (a) is a repeat of Fig. 10.4a and the details are the same as there).

a stronger initial vortex ($V_{max} = 15$ m s^{-1}). Fig. 10.11 shows the evolution in the local intensity as characterized by VT_{max} at 900 mb in ten ensemble members. Note that there is a non-negligible spread in values, the maximum difference at any given time in the four-day simulation being as high as 20 m s^{-1}.

The pattern of evolution of the flow asymmetries is significantly different between ensemble members also. These differences are exemplified by the relative vorticity fields of two simulations like that described in Section 10.3, but with random moisture perturbations added in the same way as above. These fields are shown in Fig. 10.12 and should be compared with that in Fig. 10.4a (the calculation without

moisture perturbations). The pattern of the evolving flow (and its accompanying asymmetries) is significantly different between ensemble members. In general, inspection of the ζ and w fields of each ensemble member shows similar characteristics to those in the warm-rain simulation, the ζ and w fields being highly asymmetric and dominated by locally intense cyclonic updrafts, which punctuate the central region of the vortex. However, the detailed pattern of these updrafts is significantly different between the ensemble members.

Since the flow on the convective scales exhibits a degree of randomness, we conclude that *the convective-scale asymmetries are intrinsically chaotic and unpredictable*. By

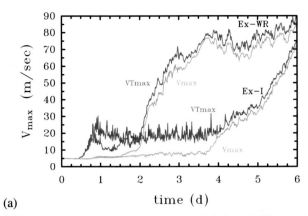
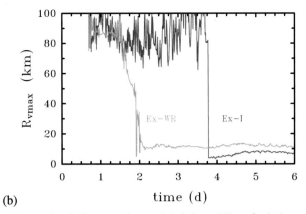

(a)   time (d)   (b)   time (d)

FIGURE 10.13 Time series of (a) maximum total wind speed VT_{max}, and maximum azimuthally-averaged tangential wind speed V_{max}, for the ice and warm-rain experiments. Panel (b) shows time series of the radius R_{vmax} at which V_{max} occurs.

chaotic is meant that small differences in the initial data lead to substantive qualitative and quantitative changes in the predicted flow at a later time (Lorenz, 1963). These calculations form a basis for the view that only the asymmetric features that survive in an ensemble average of many realizations can be regarded as robust features.

10.9   Inclusion of ice processes

A feature of the simulation described in Section 10.3 is that the vortex circulation builds first in the lower troposphere. This behavior is seen in both the tangential wind fields in Fig. 10.6 and the vorticity field in Fig. 10.9f. However, many previous studies of tropical cyclogenesis, both observational and numerical, had noted the existence of a mid-level vortex accompanying genesis. These studies are reviewed by Kilroy et al. (2018), except the most recent study from the PREDICT experiment by Bell and Montgomery (2019).

The question that emerges is: how is the vortex evolution described in Section 10.3 modified by the inclusion of a representation of ice microphysics? This question was addressed by Kilroy et al. (2018) and we summarize the results here. The only differences between the earlier simulation and the present one is the inclusion of ice microphysical processes using the Morrison double-moment ice microphysical scheme (Morrison et al., 2005) and the longer integration time (132 hours instead of 108 hours), because of the longer time required for the vortex to intensify.

10.10   Vortex evolution with and without ice

Fig. 10.13 compares the evolution of azimuthally-averaged quantities in the two experiments with and without ice, including the maximum tangential wind speed (V_{max}) and the radius at which this maximum occurs (R_{vmax}). Shown also is the maximum total horizontal wind speed (VT_{max}). A

similar comparison of the radius of total gale force winds (17 m s^{-1}) at the surface (R_{galesF}) and the radius of the gale-force tangential wind component at a height of 1 km (R_{gales}) is shown in Fig. 10.14.

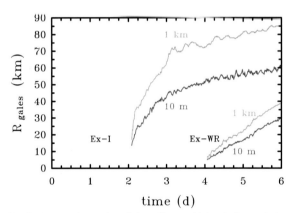

time (d)

FIGURE 10.14 Time series of the radius at which gale force winds occur (R_{gales}) in the ice (Ex-I) and warm-rain (Ex-WR) experiments, where R_{gales} is calculated at a height of 1 km, and corresponds to the radius of 17 m s^{-1} azimuthally-averaged tangential winds outside the eyewall. Shown also is the radius at which the azimuthally-averaged total near surface gale force winds outside the eyewall occur (labeled "10 m").

For the first 45 hours of the experiments, the V_{max} time series remain close to each other. Shortly thereafter, the vortex without ice begins to rapidly intensify. As explained in Section 10.3.1, VT_{max} increases before V_{max} as it incorporates localized wind speed fluctuations associated with deep convection that do not project significantly onto the azimuthal average. In the ice experiment, VT_{max} begins to exceed V_{max} at about the same time as in the warm-rain experiment indicating that deep convection begins to occur at about the same time in both experiments. However, the V_{max} and corresponding VT_{max} curves are further apart for a longer period of time than in the warm-rain experiment, a reflection of the fact that wind speed fluctuations associated with deep convection are larger when ice microphysical processes are included. The gestation period in the experiment with ice is more than double that in the experiment without ice.

While the time of RI (referred to earlier as the intensification begin time) is rather easily identified in the warm-rain experiment, it is less clear in that with ice, at least without an investigation of other metrics. On the basis of V_{max} one might judge it to be at 90 hours. In this experiment, V_{max} increases steadily from 5 m s^{-1} to 10 m s^{-1} in the first 95 hours and from 95 hours to 132 hours, it increases to almost 50 m s^{-1}. Note that the intensification rate in the ice experiment is barely half of that in the warm-rain experiment.

In both experiments, R_{vmax} begins to fluctuate as deep convection starts to occur, presumably as a result of radial wind fluctuations generated by the convection, which in an aggregate sense would advect the absolute angular momentum surfaces inwards or outwards. However, R_{vmax} shows a sharp decline at about 46 hours in the warm-rain experiment and at about 97 hours in the ice experiment, in both cases signifying the formation of a narrow vortex core at the respective times. Subsequently, this inner vortex has a time-mean R_{vmax} of about 7 km in the warm-rain experiment, compared with approximately 11 km in that with ice. While these radii are small, they are within the range of values observed.[10] In idealized numerical simulations, the mature inner core size of the simulated storm is related, *inter alia*, to the initial vortex size (see Section 16.5).

Coincidentally, in the warm-rain experiment, both R_{gales} and $R_{gales F}$ appear first at about 50 hours.[11] Thereafter, both increase steadily, with R_{gales} exceeding $R_{gales F}$. In contrast, in the ice experiment, gales occur only after about 115 hours.

The reasons why the period of genesis is prolonged in the ice experiment are subtle and are investigated in detail by Kilroy et al. (2017c). In brief, the reduction of inner-core CIN in the warm-rain experiment allows bouts of convection to occur near to the circulation center after about half a day. This reduction of CIN results predominantly from enhanced surface moisture fluxes and lifting accompanying frictional convergence in the boundary layer, but it is opposed to a degree by inner-core subsidence induced by convection at larger radii. In contrast, in the ice experiment, the induced subsidence is stronger at early times because deep convection at larger radii is stronger. As a result, the inner core region becomes drier and has higher CIN than in the ice experiment. When deep convection does occur in the inner core region in the ice experiment, the strong

downdrafts associated with ice processes flood the boundary layer with low θ_e air, which is detrimental to the persistence of convection. As time proceeds, this low θ_e air is steadily modified, predominantly by the persistent surface moisture fluxes, allowing subsequent bouts of convection to occur. Successive bouts of deep convection serve to moisten the middle troposphere in the inner core and eventually deep convection becomes persistent near the circulation center.

The recovery period following a bout of convection at 72 hours coincides with a period of little to no convective activity that lasts until about 80 hours. To illustrate the behavior at this and subsequent times, we show in Fig. 10.15 horizontal cross sections of vertical vorticity, wind vectors at a height of 1 km, and surface pressure, at 10 hour intervals from 80 hours to 110 hours. Contours of vertical velocity equal to 1 m s^{-1} at heights of 2 km and 6 km are superimposed to indicate the location of strong updrafts at these levels.

At 80 hours, the center of circulation as indicated by the pattern of the velocity vectors lies near the center of the computational grid. At this time there are no convective cells stronger than 1 m s^{-1} at heights of 2 km or 6 km, indicating that there is no deep convection occurring. Moreover, there are no patches of enhanced vertical vorticity exceeding 1×10^{-3} s^{-1} anywhere near the center of circulation.

After the inner core θ_e has recovered and the CIN has reduced to a value close to zero, deep convection is able to resume there. At 90 hours (Fig. 10.15b) there is a dramatic change in the vertical velocity and vertical vorticity fields near the circulation center from 10 hours earlier, with an approximate monopole of cyclonic vertical vorticity present near the circulation center and deep convection co-located over the monopole. There is what appears to be a spiral rainband feature also, with deep convection and vorticity dipoles extending out from the circulation center. These vorticity dipoles are the result of the tilting of background horizontal vorticity into the vertical. At 100 hours the rainband has all but decayed, although there remains a core of strong cyclonic vertical vorticity encompassing the circulation center. There is now persistent deep convection over the vorticity monopole, which stretches the cyclonic vorticity there further. At this time there is a lowering of the surface pressure in the vortex center, as indicated by the presence of black contours.

At around 100 hours the cycle of deep convection and recovery ends, and deep convection becomes continuous. At 110 hours there are significant differences evident in vortex structure from 10 hours before. While the central pressure has not fallen further, deep convection has become widespread. There are patches also of cyclonic vorticity just outside the inner core monopole that appear to be wrapping around the monopole. After this time V_{max} and VT_{max} begin to increase considerably in strength (see Fig. 10.13a).

10. For example, values of R_{vmax} in Australian region Tropical cyclones Ada (1970), Tracy (1974), and Kathy (1984) were 9 km, 7 km, and 10 km, respectively (Callaghan and Smith, 1998) and those in Atlantic Hurricanes Allen (1980), Charley (2004), and Ivan (2004) were 10 km, 13 km and 9 km, respectively (Lajoie and Walsh, 2008).

11. Recall that the definition of $R_{gales F}$ is based on the total near-surface wind speed whereas R_{gales} is based on the tangential wind component only.

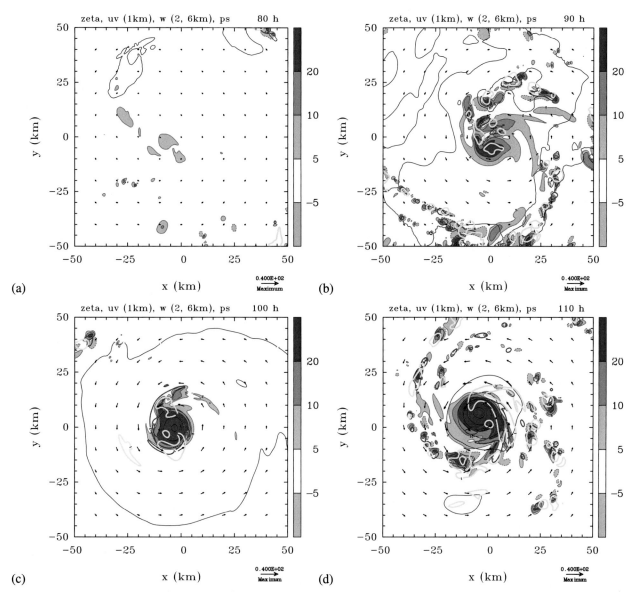

FIGURE 10.15 Horizontal cross sections of relative vertical vorticity and wind vector at (a) 80 hours and (b) 90 hours, (c) 100 hours and (d) 110 hours at 1 km altitude for the ice experiment. Shown also are contours of vertical velocity at heights of 2 km (aqua) and 6 km (yellow), and black contours of surface pressure, contoured every 2 mb. The minimum pressures are in: (a) 1005.9 mb, (b) 1004.6 mb, (c) 999.8, (d) 995.5 mb. Values for the shading of vertical vorticity are given in the color bar, multiplied by 10^{-4} s^{-1}. Vertical velocity contour is 1 m s^{-1} for both heights. The wind vectors are in relation to the maximum reference vector (40 m s^{-1}) at the bottom right.

10.11 Moist instability and θ_e

As in Section 10.4 for the warm-rain experiment, we investigate the evolution of near-surface θ_e during the early stages of RI. Fig. 10.16 shows horizontal cross sections of θ_e difference from its initial value in the ice experiment at 90, 96, 102, and 120 hours. The first time is around the time that RI begins (Fig. 10.13a) when V_{max} is only 9 m s^{-1}. By 96 hours, V_{max} has increased to nearly 18 m s^{-1} and by 102 hours it has reached 22 m s^{-1}. The final time, 120 hours is well into the RI period with V_{max} now just hurricane strength, 33 m s^{-1}.

At the start of RI (Fig. 10.16a) the θ_e difference is much like that at 46 hours, just before the start of RI in the warm-rain experiment (Fig. 10.5a). There are patches of depressed values of θ_e in the neighborhood of current deep convection and patches of elevated θ_e elsewhere with little organization. The most notable feature during the next six hours is the formation of a positive θ_e anomaly near the center of circulation, much as in the warm-rain experiment (cf. Figs. 10.5b and 10.5c). By 102 hours, all depressed θ_e anomalies have disappeared, but even though the vortex is rapidly intensifying, regions of depressed θ_e continue to form as evidenced by the situation at 120 hours (Fig. 10.16d). An animation of

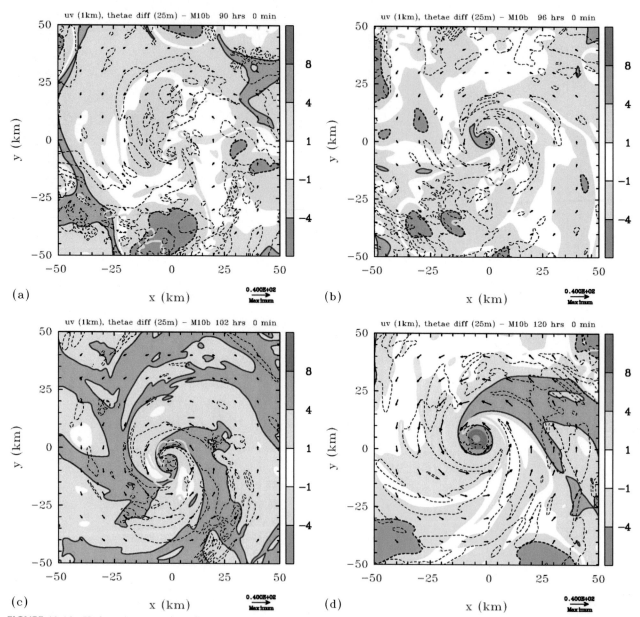

FIGURE 10.16 Horizontal cross sections of the difference between near-surface θ_e and its initial value in the ice simulation at (a) 90 hours, (b) 96 hours, (c) 102 hours, and (d) 120 hours. Shown also are horizontal wind vectors at a height of 1 km and two contours of vertical velocity: 1 m s^{-1} at a height of 6 km (yellow) and -10 cm s^{-1} at a height of 1 km (thin black dashed contours). Values for the shading of θ_e difference are given in the color bar, in K. The wind vectors are in relation to the maximum reference vector (40 m s^{-1}) at the bottom right.

the θ_e-field for this simulation shows that these depressed regions are found some distance from the circulation center and are removed on a time scale of an hour or two. Beyond about 130 hours, there are elevated values of θ_e at all times and the magnitude of the anomalies grows progressively (not shown).

As in the warm-rain experiment (see Fig. 10.5), the evolution of the θ_e field from the genesis to the early RI phase demonstrates again the highly asymmetric nature of the genesis process. Fig. 10.16 shows that the θ_e field is highly variable both spatially and temporally. Significantly, as the RI

phase is approached near 90 hours, there are a few isolated deep updraft cells near the center of circulation, but these cells are still producing weak downdrafts at 1 km height and there is a large region of weak negative θ_e anomaly to the west and north of the center. Despite this negative anomaly, the vorticity field has already become organized into a cyclonic monopole around the circulation center.

It is not until 96 hours that a marked positive anomaly in θ_e has developed and the large region of weak negative θ_e anomaly has disappeared. From this time, a positive θ_e anomaly becomes a persistent feature near the circulation

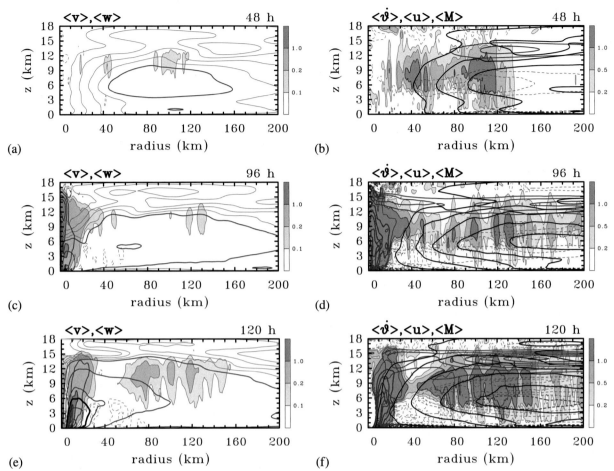

FIGURE 10.17 Vertical cross sections of the azimuthally-averaged, 3 hour time-averaged tangential and radial velocity components centered at selected times in the ice experiment. Superimposed on the tangential component (left panels) are contours of the averaged vertical velocity (shaded). Superimposed on the radial component (right panels) are the averaged diabatic heating rate (shaded) as well as selected contours of absolute angular momentum. Contour intervals are as follows. Tangential velocity: thin blue contours 1 m s^{-1} between 0 and 4 m s^{-1}, thick blue contours every 5 m s^{-1}. Vertical velocity: thin red contours 0.1 m s^{-1} and 0.2 m s^{-1}, thick red contour interval 0.5 m s^{-1}, thin dashed red contours indicate subsidence at intervals of 0.02 m s^{-1}. Radial velocity: thick blue contours 1 m s^{-1}, dashed negative, thin blue dashed contours every 0.2 m s^{-1} down to -0.8 m s^{-1}. Diabatic heating rate: thin red contours 0.2 and 0.5 K h^{-1}, medium thickness red contours 1 and 2 K h^{-1}, thick red contours every 5 K h^{-1}, dashed red contours negative with values -0.2 and -0.5 K h^{-1}. Shading as indicated on the side bar. Absolute angular momentum: thick black contours every 2×10^5 m^2 s^{-1}.

center. However, at this stage, RI has been taking place for six hours already and a small-scale vortex has already become established (Fig. 10.15).

The foregoing features are consistent with those found in the warm-rain experiment and, again, they call into question explanations of the genesis process that invoke a casual chain of events tied primarily to the thermodynamics. Our analyses do not show any evidence of a thermodynamic trigger for genesis in either the warm-rain experiment or the ice experiment.

10.12 An azimuthal mean view of vortex evolution

Again, as in the warm-rain experiment, the features of the genesis process in the ice experiment are intrinsi-

cally asymmetric. However, as before, it is insightful to consider an azimuthal-mean view of the process. To this end, Fig. 10.17 shows vertical cross sections of various azimuthally-averaged and 3 hour time-averaged quantities in the ice experiment at 48 hours, 96 hours, and 120 hours. These times span the gestation period prior to genesis and the intensification period. As in Section 10.5, the time averaging is based on 15 minute model output centered on the time indicated and averaged quantities are denoted angle brackets $\langle \rangle$. The quantities shown are mean tangential, $\langle v(r, z) \rangle$, radial, $\langle u(r, z) \rangle$, and vertical $\langle w(r, z) \rangle$, velocity components, the mean diabatic heating rate, $\langle \dot{\theta}(r, z) \rangle$ and the contours of mean absolute angular momentum $\langle M(r, z) \rangle$. The vertical velocity and diabatic heating rate provide a perspective on the evolution of the convectively-induced secondary circulation in relation to features in the horizontal velocity component fields.

Even though the time series in Fig. 10.13a indicates little change in V_{max} before about 90 hours in the ice experiment, there are considerable changes in the structure of $\langle v(r, z) \rangle$ during this time. In particular, in contrast to the vortex in the warm-rain experiment, the strongest tangential winds develop first at middle tropospheric levels rather than at low levels (e.g., left columns of Fig. 10.17) and by 120 hours there is a band of moderately strong (> 10 m s^{-1}) tangential winds at mid-levels extending out beyond a radius of 100 km in the ice experiment (Fig. 10.17e).

At 48 hours, the most prominent diabatic heating lies in the mid-to-upper troposphere within a radius of about 125 km (Fig. 10.17a). The effect of the heating is to produce inflow with a peak near 6 km in altitude beyond a radius of 85 km (Fig. 10.17b). The inflow is evident in the inward protruding "noses" of the $\langle M \rangle$ surfaces, a feature that becomes more prominent with time (right panels of Fig. 10.17).

Some deep convection, as characterized by either the distribution of vertical motion or that of diabatic heating rate, becomes progressively focused in a central region ($r \lesssim 30$ km) near the axis of rotation as the vortex intensifies. This is, of course, a favorable location for the convectively-induced inflow to continually draw in the $\langle M \rangle$ surfaces, thereby intensifying the tangential wind, at least above the frictional boundary layer. The focusing, itself, must be aided by the strengthening boundary-layer inflow, which is evident Fig. 10.17f.

At 96 hours (Fig. 10.17c), a second maximum in $\langle v(r, z) \rangle$ has formed near to the surface, at a radius of less than 10 km, and by 120 hours, this tangential wind maximum has grown in intensity to become the most prominent feature (Fig. 10.17e). This feature accounts for the sharp decrease of R_{max} to less than 10 km (Fig. 10.13b) and is similar to the behavior in the warm-rain experiment (cf. Fig. 10.6c, e) and, again, is explained by the boundary layer spin up enhancement mechanism.

Although V_{max} at 120 hours is less than 25 m s^{-1}, the vortex structure has the main features of a tropical cyclone, at least from an azimuthally-averaged perspective. These include the strong low-level inflow, a tangential wind field that decreases in strength with height, a strong annular core of ascent with the maximum updraft away from the axis of rotation and a strong outflow layer in the upper troposphere (Fig. 10.17f). The strengthening near-surface inflow as the vortex develops is seen in Figs. 10.17d, f. Since the flow in this figure is azimuthally-averaged, increasing near-surface inflow must be reflected in increasing convergence.

The foregoing pattern of evolution is similar in many respects to that in the warm-rain experiment where the genesis process is much more rapid and the formation of a low-level concentrated vortex is not preceded by the development of a broad tangential wind maximum in the middle troposphere

(the early tangential wind maximum seen in Figs. 10.6c, d is located generally below a height of about 3 km).

We have discussed already the reasons for the delayed genesis in the ice experiment and turn now to examine the reasons for the early formation of a mid-level vortex in the ice experiment.

10.13 Mid-level vortex development with ice microphysics

As discussed in Kilroy et al. (2018) and refs., the occurrence of a mid-level vortex in individual mesoscale convective systems has been attributed to the vertical structure of the diabatic heating rate that exists in the extensive stratiform regions of these systems (e.g., Chen and Frank, 2001; Rogers and Fritsch, 2001). In these regions, there is heating due to latent heat release by the condensation of water vapor and the progressive freezing of liquid water in the weakly ascending anvils, but cooling in mesoscale downdrafts due to sublimation, melting and evaporation beneath the anvils (see e.g., Houze, 2014, Section 9.6.3). This configuration of upper-tropospheric heating and lower-tropospheric cooling leads to a maximum vertical gradient of heating rate in the middle troposphere, rather than in the lower troposphere, associated primarily with vigorous deep convective updrafts. Because, as discussed in Section 5.8, the heating rate is approximately proportional to the vertical velocity, this pattern of heating rate is consistent through mass continuity with the occurrence of horizontal convergence in the middle troposphere rather than the lower troposphere.

Near the freezing level, the mid-level convergence should be reinforced by the vertical divergence of vertical stress generated by the melting of snow crystals just below the freezing level (typically 5.5 km in the Tropics). Since raindrops have fall speeds many times greater than snowflakes, they exert a much larger collective drag on air below the freezing level than snowflakes exert above the freezing level. The vertical stress divergence leads to a horizontal convergence of the flow near the freezing level. In either case, the convergence will lead to an amplification of any existing vortical circulation by the classical spin up mechanism.

10.13.1 Increasing influence of the boundary layer

The vertical cross sections in Fig. 10.17 show that, rather than the mid-tropospheric tangential wind maximum moving downwards, the tropical-cyclone vortex develops *in situ* by processes similar to those in the warm-rain experiment. In essence, the strengthening boundary-layer inflow seen in Figs. 10.17d, f is accompanied by a focusing and strengthening of the distribution of diabatic heating rate within a radius of less than 30 km and a progressive inward dis-

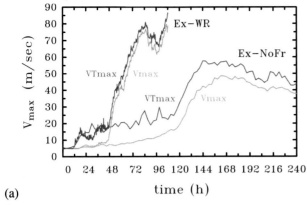
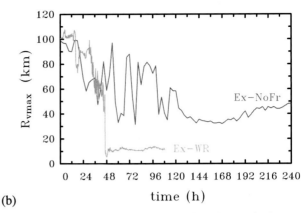

FIGURE 10.18 Time series of (a) maximum total wind speed VT_{max}, and maximum azimuthally-averaged tangential wind speed, V_{max}, for the warm-rain simulation and corresponding no friction simulation (Ex-NoFr). Panel (b) shows time series of the radius R_{vmax} at which V_{max} occurs.

placement of the $\langle M \rangle$ surfaces, at least above the shallow boundary layer. The inward displacement of the $\langle M \rangle$ surfaces at low levels seen in the right panels of Fig. 10.17 is accompanied by a spin up of the tangential wind field at low levels, as seen in the left panels of Fig. 10.17. The fact that the maximum tangential wind speed at 120 hours lies within the relatively strong inflow layer indicates that the boundary-layer spin up enhancement mechanism discussed in Section 6.6 has begun to operate in conjunction with the classical mechanism. The dynamics of formation of the low-level vortex broadly corroborate the results of earlier studies of the genesis problem by Montgomery et al. (2006b); Nolan (2007), and Nicholls and Montgomery (2013).

As a means to examine further the role of the boundary layer on tropical cyclogenesis, Kilroy et al. (2017a) compared the warm-rain experiment with a corresponding one in which the surface drag is set to zero. Fig. 10.18 compares the evolution of V_{max}, VT_{max}, and R_{vmax} for these two simulations. It is seen that, while spin up occurs in both simulations, the vortex in the one without surface drag takes over twice as long to reach its intensification begin time. In addition, when surface friction is not included, the inner core size of the simulated vortex is considerably larger and the subsequent vortex intensity is significantly weaker than in the case with friction.

Kilroy et al. found that, in the absence of surface drag, the convection eventually develops without any systematic organization and lies often outside the radius of maximum azimuthally-averaged tangential winds. *The results underscore the subtle, but crucial, role of surface friction in organizing deep convection in the inner core of the nascent vortex and point to the likelihood that the timing of tropical cyclogenesis in numerical models may have an important dependence on the boundary-layer parametrization scheme used in the model.*

10.13.2 Synthesis

When ice processes are included, the gestation time for genesis is more than twice as long as in the warm-rain experiment. The reasons for this longer gestation period are discussed by Kilroy et al. (2018). Further, this gestation period is characterized by the formation of a moderate strength system-scale vortex in the middle troposphere.

Kilroy et al. (2018) carried out diagnostic axisymmetric balance calculations using methods similar to those described in Chapters 5 and 8 to show that the spin-up of the mid-level vortex is related to the larger horizontal and vertical gradient of diabatic heating rate in the middle troposphere than in the warm-rain experiment. The elevated gradient of heating rate leads to a system-scale radial influx of absolute vorticity in the middle troposphere, or equivalently, from an azimuthally-averaged perspective, to an inward displacement of the $\langle M \rangle$ surfaces at these altitudes. In either perspective, the inflow leads to a spin-up of the tangential wind at middle levels.

The formation of the tropical-cyclone vortex in the ice experiment is similar to that in the warm-rain experiment, with the strengthening frictional boundary layer exerting a progressively important role in focusing inner-core deep convection. This vortex develops *in situ* on a much smaller scale than the mid-level vortex by processes that are similar to those in the warm-rain experiment, in which no such mid-level vortex forms. Indeed, there is no evidence that the low-level vortex forms as a result of the mid-level vortex being somehow carried downwards.

In Section 11.1, we will show that, from a vorticity perspective, the enhanced low-level vorticity near the circulation center can come about only by the *horizontal divergence* of a vorticity flux. Accordingly, there can be no net downward transport of vorticity from the middle-level vortex, ruling out the possibility of genesis being a *top-down process*.

Taken together, the results for the warm-rain and ice experiments suggest that the formation of the mid-level vortex is not a prerequisite for explaining tropical cyclogenesis in a favorable, moist, pouch-like environment with minimal vertical shear. This does not mean, however, that a mid-level vortex cannot be helpful in providing some protection from the injection of dry air from the environment (Bell and Montgomery, 2019).

10.14 Boundary layer control

In the early stages of tropical cyclone formation, while the primary circulation is still relatively weak, the inflow induced by deep convection dominates that which occurs in the boundary layer. Moreover, this inflow is distributed through the lower troposphere and not confined to a near-surface layer. As a result of the inflow above the boundary layer, the classical spin up mechanism (Section 8.2) acts to spin up the nascent vortex.

Even so, that is not to say that the boundary layer is unimportant at this stage. As explained in Section 10.3.1, the weak inflow in the boundary layer produces lifting in the inner region, which helps to remove any CIN present in the thermodynamic sounding and is therefore conducive to the initiation of the deep convection, itself.

As noted in the prior section, the weak but persistent frictional inflow helps also to foster convective organization near the nascent center of circulation in comparison to an identical simulation without surface drag (Persing et al., 2013; Kilroy et al., 2017a; see also Sections 16.2 and 16.10.1).

However, as the vortex strengthens, the boundary layer plays an increasingly important dynamical and thermodynamical role in the vortex intensification. This importance is highlighted by a 30-day numerical simulation by Kilroy et al. (2016a) in the context of the prototype problem for cyclone intensification discussed in Section 10.1. The simulation used an earlier model than CM1, the Pennsylvania State University/National Center for Atmospheric Research Mesoscale model (MM5).

After intensifying rapidly and reaching a mature intensity, the model vortex progressively decays. After a few days, both the inner-core size, characterized by the radius of the eyewall updraft, and the size of the outer circulation (measured for example by the radius of gale-force winds) progressively increase. This behavior is illustrated in Fig. 10.19, which shows a Hovmöller plot of tangential velocity at a height of 1 km near the top of the boundary layer from the MM5 simulation.

Kilroy et al. (2016a) pointed out that the tight coupling between the flow above the boundary layer and that within the boundary layer makes it impossible, in general, to present simple cause and effect arguments to *explain* the structural evolution of the vortex. The best one can do is to articulate the individual elements of the coupling, which

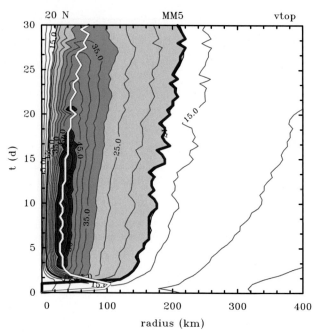

FIGURE 10.19 Azimuthally-averaged and temporally-smoothed Hovmöller plot of tangential velocity at a height of 1 km near the top of the boundary layer from MM5 output at 20°N. Contour interval: red contours every 5 m s^{-1}. Velocities larger than 50 m s^{-1} are highlighted with black contours. The interval from 17 m s^{-1} to 30 m s^{-1} is shaded light red, that from 30 m s^{-1} to 50 m s^{-1} by a slightly darker red shading, and values of 50 m s^{-1} and above are shaded dark red. The white curve represents the radius of the time- and azimuthally-averaged maximum tangential velocity at a height of 1 km, while the blue curve represents the time- and azimuthally-averaged radius of gales at this height. Adapted from Kilroy et al. (2016a). Republished with permission of the American Meteorological Society.

might be described as a set of coupled mechanisms encapsulating the idea of boundary layer control.

10.14.1 Boundary layer coupling in brief

Boundary layer dynamics exert a control on the radii at which air is forced to ascend to feed the inner-core convection. They exert a control also on the maximum tangential wind speed that can be attained because of how far $\langle M \rangle$-surfaces can be drawn inwards and on the strength of the inflow that helps to bring these $\langle M \rangle$-surfaces inwards.

It is for the above reason, for example, that as the cyclone matures, deep convection tends to form into an annular ring at some finite radius from the rotation axis - the so-called eyewall cloud. In turn, the ascent in this ring leads to subsidence inside the ring, the eye of the storm, which is free of deep cloud.

A result of boundary layer dynamics is that, as the tangential wind in the lower troposphere in the outer part of the vortex expands, so do the radii of strongest forced ascent at the top of the boundary layer (Subsection 6.5.6). Since the inner-core deep convection feeds on moisture supplied by the boundary layer inflow, and since the source of moisture

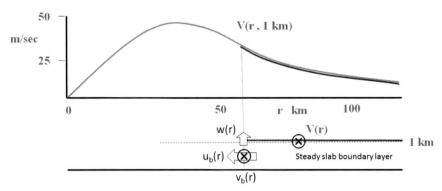

FIGURE 10.20 Schematic illustrating the idea behind the boundary layer control mechanism. See text for discussion. Adapted from Montgomery and Smith (2017b). Reproduced with permission from the Annual Review of Fluid Mechanics, Volume 49; copyright 2017 Annual Reviews.

lies at the ocean surface, boundary-layer thermodynamics exert an important control on the vortex evolution also. For one thing, the changes in the radial distribution of boundary-layer θ_e that accompany this expansion lead to a change in the radial and vertical distribution of diabatic heating rate. For another, the thermodynamic processes affect the ability of the deep convection to ventilate the air that is converging in the boundary layer as discussed in Section 9.9.1.

As a way to break into the chain of coupled mechanisms referred to above, Kilroy et al. (2016a) employed the simple, steady, slab boundary layer model described in Section 6.5. The assumption is that, because the boundary layer is relatively shallow, it adjusts rapidly to the flow above it. Since the partial differential equations from which the steady slab boundary layer model is derived are parabolic in the radially inward direction, the inflow and hence the ascent (or descent) at the top of the boundary layer at a given radius R know only about the tangential wind profile at radii $r > R$ (see Fig. 10.20). The inflow at radius R knows nothing directly about the vertical motion at the top of the boundary layer at radii $r < R$, including the pattern of ascent into the eyewall cloud associated with convection under the eyewall. In contrast, the numerical simulation does not solve the boundary layer equations separately, and it does not make any special boundary layer approximation. Thus, the ability of the slab boundary layer model to produce a radial distribution of radial, tangential and vertical motion similar in behavior to those in the time-dependent numerical simulation provides a useful measure of the degree of boundary layer control in the evolution of the vortex.

10.14.2 A demonstration of boundary layer coupling

As an example, we show in Fig. 10.21 a comparison of velocity fields from the slab boundary layer calculations compared with the corresponding azimuthally-averaged fields from the full numerical simulation at latitude 20°N ana-

lyzed by Kilroy et al. (2016a). The radial and tangential wind components from the numerical simulation are averaged over the lowest 1-km depth, corresponding to an average over the depth of the boundary layer, to provide a fair comparison with the slab boundary layer fields. The slab boundary layer calculations are performed every 12 hours using the azimuthally-averaged tangential wind profile extracted from the numerical simulation shown in Fig. 10.19.

Even though as noted in Section 6.5.6, the integration of the slab boundary layer equations breaks down at some inner radius, where the radial velocity tends to zero and the vertical velocity becomes large, the calculations capture many important features of the corresponding depth-averaged boundary layer fields from the numerical simulation. For example, they capture the broadening of the vortex core with time, i.e., the increase in the radii of maximum tangential wind speed and eyewall location, the latter characterized by the location of maximum vertical velocity. They capture also the broadening of the outer radial and tangential wind field. However, they overestimate the radial extent of the subsidence outside the eyewall (cf. Figs. 10.21e, f). For reasons articulated above that the slab boundary layer solution at some radius knows nothing explicitly about deep convection inside this radius, these results provide strong support for the existence of a dynamical control by the boundary layer on the evolution of the vortex.

Kilroy et al. (2016a) investigated further the thermodynamic control of the boundary layer and other aspects of the coupling discussed above. We return to these issues in Chapter 15, where we investigate a long term integration similar to that discussed above from their study.

10.15 Towards a conceptual model for tropical cyclogenesis

All of the vortex developments examined in this chapter share strong similarities in that the organization process involves a non-axisymmetric and upscale vorticity consol-

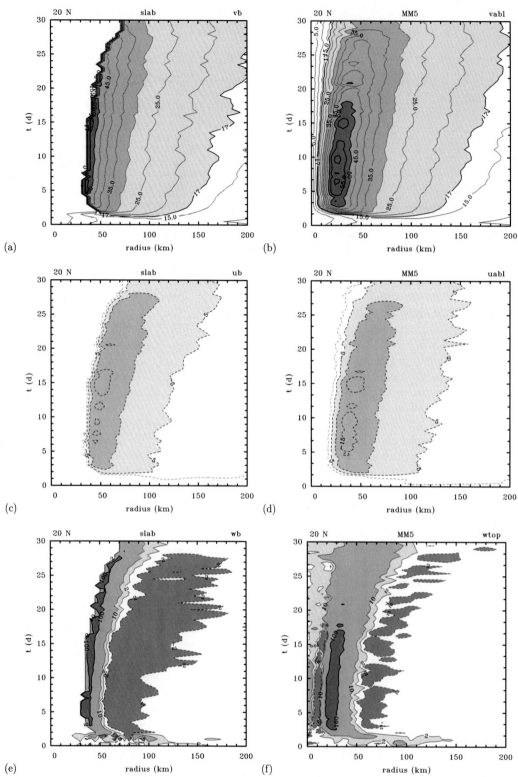

FIGURE 10.21 The left panels show azimuthally-averaged and temporally-smoothed Hovmöller plots of radial, tangential and vertical velocities at the top of the boundary layer (a height of 1 km) from the slab boundary layer model, with a constant depth of 1000 m for the experiment at 20°N. The right panels show the corresponding variables from the MM5 output. Contour interval: top panel contours every 5 m s^{-1} with the 17 m s^{-1} colored black. The interval from 17 m s^{-1} to 30 m s^{-1} is shaded light red (light gray in print version), that from 30 m s^{-1} to 50 m s^{-1} by a slightly darker red (light mid gray in print version) shading, and values of 50 m s^{-1} and above are shaded dark red (mid gray in print version). Middle panels: first contour 1 m s^{-1}, then in 5 m s^{-1} intervals. Light blue (light dark gray in print version) shading from −5 m s^{-1} to −10 m s^{-1}, darker shading below 10 m s^{-1}. Bottom panels: first (red (mid gray in print version)) contour is 2 cm s^{-1}, with light red (light gray in print version) shading up to 10 cm s^{-1}. Darker shading from 10 cm s^{-1} to 100 cm s^{-1}. The darkest (red (mid gray in print version)) shading is enclosed by a 100 cm s^{-1} black contour. Regions of downward motion are shaded blue (dark gray in print version) and enclosed by a 2 cm s^{-1} blue (dark gray in print version) contour. Solid (red (mid gray in print version)) contours positive, dashed (blue) contours cynegative. From Kilroy et al. (2016a). Republished with permission of the American Meteorological Society.

idation process in which deep convection and vortex-tube stretching is a key ingredient. Of course, the convective activity is maintained by modest sea-to-air fluxes of water vapor, which serves to elevate the near-surface equivalent potential temperature necessary to support convective instability as the stabilizing warm core develops. It would seem obvious to most readers that the totality of these processes cannot be described by existing axisymmetric paradigms for tropical cyclone intensification. To remedy this situation, a conceptual framework to understand the three-dimensional organization and intensification process is developed in the next chapter.

Chapter 11

The rotating-convection paradigm

In order to interpret and understand the results of any observational study, or any numerical model simulation such as those described in Chapter 10, one needs to have a suitable conceptual framework. The classical intensification paradigm discussed in Chapter 8 provides such a framework for understanding the behavior of the axisymmetric balance model simulations, while a modified version of that is required to interpret the evolution of the azimuthally-averaged structure of the simulations described in Chapter 10. However, as we have shown, the flow evolution in the simulations in Chapter 10 is intrinsically three dimensional, a major feature being the organization and aggregation of the convectively-enhanced relative vorticity field of the system-scale vortex. Clearly, an extended framework is required to interpret and understand how this organization comes about. It was the search for such a framework that led to the development of the rotating-convection paradigm.

The dynamics of the rotating-convection paradigm are encapsulated, in part, by the azimuthal-mean tangential momentum equation (Persing et al., 2013), or equivalently by the vertical vorticity equation in conjunction with Stokes' theorem (Montgomery and Smith, 2017b, see their section 4.1). Stokes' theorem (Section 2.11) equates the circulation about any fixed closed loop to the area integral of the vorticity within that loop. In the present application we take the area A and loop C that encloses the area to be in a horizontal plane. Then Stokes' theorem may be written in the form

$$\oint_C \mathbf{V} \cdot d\mathbf{s} = \iint_A \zeta \, dA, \qquad (11.1)$$

where $\mathbf{V}$ is the velocity vector, $d\mathbf{s}$ is a vector increment along the curve C, dA is an area element of A, and ζ is the vertical component of relative vorticity.

While the spin up tendency derived directly from the tangential momentum equation and that inferred indirectly from the vorticity equation in conjunction with Stokes' theorem are mathematically equivalent, the vorticity approach leads arguably to deeper insights, especially in relation to the generation of local rotation by convection. For this reason, we begin by presenting the vorticity approach.

A schematic diagram showing the essence of the rotating-convection paradigm is shown in Fig. 11.1. The diagram encapsulates some of the basic three-dimensional features of the numerical simulations of tropical cyclogenesis described in Chapter 10. Deep convection growing in an existing vortical environment enhances the vertical vorticity locally by vortex stretching to produce a field of initially-isolated vortex cores in the lower troposphere. The vorticity in these cores can be one to two orders of magnitude larger than the surrounding vorticity.

The calculations described in Chapter 10 show that during the period leading up to genesis, these areas of enhanced vorticity, with the same parity as the parent circulation, progressively aggregate and strengthen to form a monopole-like structure in the vorticity field. Two processes can be envisioned to accomplish this aggregation. We know from Chapter 4 that if two barotropic vortices are sufficiently close to one another, they will induce flow fields that cause them to merge. An alternative process would be the overturning circulation produced by the collective effects of deep convection, itself, and the influence of frictionally-driven inflow. Such collective effects lead to convergent flow in the lower troposphere, which would tend to concentrate the convectively-induced same parity vorticity. Calculations by Kilroy et al. (2017c) suggest that this is the dominant aggregation process.

In a warm-cored vortex such as a tropical cyclone, the tangential velocity first increases with height as the frictional effect of the surface stress lessens and then decreases with height as described in Section 5.3. These changes with height are associated with a radial component of vorticity, which, when tilted by convective updrafts will tend to produce vertical vorticity anomalies in the form of dipoles (see Section 11.1). The way in which the vorticity structures induced by deep convection contribute to vortex spin up are discussed in the following sections.

While the main focus of the rotating-convection paradigm is on the dynamical processes of spin up leading to the amplification of the maximum tangential velocity, the paradigm recognizes the need for a modest elevation of surface moisture fluxes. These moisture fluxes, in conjunction with reduced entrainment of comparatively dry air into the boundary layer as air parcels spiral in towards the developing eyewall, serve to elevate the boundary layer θ_e and thereby maintain deep convective instability in the inner core region of the vortex. Sustained deep convection in this region is required for continued intensification of the vortex (Montgomery and Smith, 2017b, p551).

Tropical Cyclones. https://doi.org/10.1016/B978-0-44-313449-4.00019-9

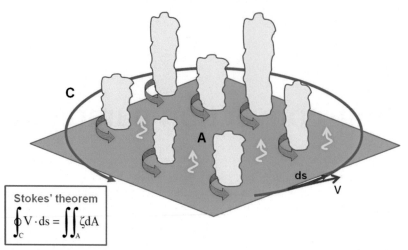

FIGURE 11.1 Schematic of a region of deep rotating updrafts, with a hypothetical horizontal circuit indicated by a circle. By Stokes' theorem, the relative circulation about the circuit is equal to the areal integral of the relative vorticity enclosed by the circuit. The yellow squiggly arrows indicate moisture fluxes from the ocean. To avoid clutter, the local and system-scale overturning circulations associated with the updrafts are not shown. See text for further discussion. Adapted from Smith and Montgomery (2016). Republished with permission of Wiley.

The rotating-convection paradigm represents the only conceptual framework we are aware of that applies to vortex evolution in three-dimensions, which is arguably the proper benchmark for interpreting observations and numerical model simulations of real-world and hypothetical future-world tropical cyclones.

Naturally, one can develop an azimuthally-averaged perspective of vortex evolution in three-dimensions. The mathematical framework to accomplish this is presented in Section 11.4 and the application of such an analysis is detailed in Section 11.5.

11.1   Flux form of the vorticity equation

In Section 2.15, we discussed the flux form of the vertical vorticity equation. In pressure coordinates, this equation has a particularly concise form in terms of a *horizontal* divergence of a *horizontal* flux, i.e.,

$$\frac{\partial \zeta_a}{\partial t} = -\nabla_h \cdot \mathbf{F}_{\zeta_a}, \qquad (11.2)$$

where $\mathbf{F}_{\zeta_a} = \mathbf{F}_{af} + \mathbf{F}_{naf}$, $\mathbf{F}_{af} = \zeta_a \mathbf{u_h}$ is the *advective flux*, and $\mathbf{F}_{naf} = \zeta_h \omega + \mathbf{k} \wedge \mathbf{F}_{fri}$ is the *non advective flux*. Here $\mathbf{u_h} = (u, v, 0)$ is the horizontal velocity vector, $\zeta_h = -\hat{\mathbf{k}} \wedge \partial \mathbf{u_h}/\partial p$ is the horizontal component of "vorticity",[1] ω is *here* the material derivative of pressure, Dp/Dt, and plays the role of *vertical velocity* in pressure coordinates, $\mathbf{F}_{fri}$ is

the horizontal force per unit mass due to molecular effects and subgrid-scale eddy momentum fluxes associated with unresolved turbulence, and $\mathbf{k}$ is a unit vector in the vertical. Eq. (11.2) applies equally to rectangular coordinates (x, y, p) or cylindrical coordinates (r, λ, p), in which r is the radius, λ is the azimuth, and u and v are the radial and tangential velocity components.

An important deduction about low-level vorticity organization follows immediately from Eq. (11.2): *The formation of a mesoscale region of intense cyclonic vorticity in the lower troposphere cannot arise by a downward flux of cyclonic vorticity from above. Rather, the generation of such a vortex can only occur by a sustained horizontal convergence of the flux* $\mathbf{F}_{\zeta_a}$. That is, the formation of a mesoscale region of intense cyclonic vorticity in the lower troposphere, as in the genesis simulations in Chapter 10 (see Section 10.13.2), can only arise through a horizontal concentration of pre-existing cyclonic vertical vorticity, in combination with a preferential generation of cyclonic vorticity (or preferential destruction of anticyclonic vorticity) associated with the divergence of the non-advective flux.

The physical mechanisms behind the three terms of the vorticity flux are illustrated in Fig. 11.2, which is borrowed from Raymond et al. (2014). The advective flux, $\mathbf{F}_{af}$, is illustrated in the left panel of the figure for the case when there is inflow in the presence of positive absolute vertical vorticity. The convergence of this flux then leads to a positive vorticity tendency. The convergence of $\mathbf{F}_{\zeta_a}$ incorporates the effects of horizontal vorticity advection and the amplification of vertical vorticity by stretching, effects that appear separately in the material form of the vorticity equation (cf. Eq. (2.46)).

Above the frictional boundary layer, where $\mathbf{F}_{fri}$ can be neglected, the *non-advective flux*, $\mathbf{F}_{naf}$, incorporates the ef-

---

1. In pressure coordinates with the hydrostatic approximation, the horizontal vorticity ω_h defined in Section 2.16 is $\hat{\mathbf{k}} \wedge \partial \mathbf{u_h}/\partial z = \rho g \zeta_h$, where here $\zeta_h = -\hat{\mathbf{k}} \wedge \partial \mathbf{u_h}/\partial p$ is a horizontal vorticity-like vector that points in the same direction as ω_h. Since within the hydrostatic approximation $\omega = -\rho g w$, it follows that, in the case of weak potential temperature deviation as defined in Section 2.16, the inviscid non-advective flux $-\omega_h w = \zeta_h \omega$.

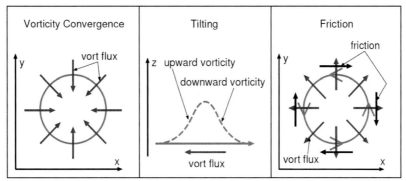

FIGURE 11.2 Illustration of the physical effect of each of the three terms in the vorticity flux equation. See text for further explanation. Adapted from Raymond et al. (2014). Republished with permission of the Australian Bureau of Meteorology.

fects of vertical vorticity advection and the tilting of horizontal vorticity into the vertical. Again, these terms appear separately in the material form of the vorticity equation (2.46). The occurrence of this flux is illustrated in the middle panel of Fig. 11.2. The red (mid gray in print version) horizontal arrow represents a vortex line with an initially zero vertical component. Localized ascent would tend to advect this vortex line as indicated by the red (mid gray in print version) dashed line, resulting in positive vertical vorticity on the left side of the maximum ascent and negative vertical vorticity on the right of the maximum ascent. The result is an equivalent flux of vertical vorticity from right to left and the creation of a vorticity tendency dipole, with the positive and negative components located on opposite sides of the updrafts. This orientation of the vorticity dipole is normal to the vertical shear of the horizontal wind represented by the vortex lines.

The contribution of the frictional term, $\mathbf{k} \wedge \mathbf{F_{fri}}$, to the non-advective flux is illustrated in the right panel of Fig. 11.2. This horizontal plan view shows the effect of a frictional force (the black arrows) opposing a counter-clockwise circulation indicated by the red (mid gray in print version) arrows and circle. Then $\mathbf{k} \wedge \mathbf{F_{fri}}$ represents a flux of vorticity normal to both $\mathbf{F_{fri}}$ and $\hat{\mathbf{k}}$ and is directed away from the center of the circulation as indicated by the blue (dark gray in print version) arrows. This vorticity flux expulsion leads to a dilution of vorticity within the circuit and hence a spin down tendency.

11.2 Axisymmetric flow

For an axisymmetric flow, $\mathbf{F_{af}}$ reduces to $u\zeta_a\hat{\mathbf{r}}$ and, above the frictional boundary layer, $\mathbf{F_{naf}}$ reduces to $\omega(\partial v/\partial p)\hat{\mathbf{r}}$, where $\hat{\mathbf{r}}$ is a unit vector in the radial direction. In this case, the vorticity equation has the flux form

$$\frac{\partial \zeta_a}{\partial t} = -\frac{1}{r}\frac{\partial}{\partial r}\left[r\left(u\zeta_a + \omega\frac{\partial v}{\partial p}\right)\right], \quad (11.3)$$

which is equivalent to the perhaps more familiar material form[2]:

$$\frac{\partial \zeta_a}{\partial t} + u\frac{\partial \zeta_a}{\partial r} + \omega\frac{\partial \zeta_a}{\partial p} = \zeta_a\frac{\partial \omega}{\partial p} - \frac{\partial v}{\partial p}\frac{\partial \omega}{\partial r}, \quad (11.4)$$

in which the left-hand-side is just the material derivative of ζ_a and the terms on the right-hand-side represent the generation of vorticity by the vertical stretching of existing ζ_a and the tilting of radial vorticity (proportional to $\partial v/\partial p$) by the radial gradient of ω (cf. Section 2.9).

As shown in Chapter 5, the tangential velocity above the boundary layer in a warm-cored vortex decreases with altitude so that the non-advective flux vector points inwards. In terms of vorticity, tilting by a convectively-driven, axisymmetric overturning circulation would lead to enhanced cyclonic vorticity at radii where the vertical velocity increases with radius, i.e., where $\partial \omega/\partial r < 0$, and reduced vorticity at larger radii where $\partial \omega/\partial r > 0$, the situation sketched in the middle panel of Fig. 11.2. Within the boundary layer, $\omega(\partial v/\partial p)\hat{\mathbf{r}}$ points outwards and the vorticity tendencies due to tilting are reversed in sign, but there one cannot, of course, ignore the contribution of $\mathbf{F_{fri}}$ to $\mathbf{F_{naf}}$, which is shown in the right panel of Fig. 11.2.

The corresponding axisymmetric, inviscid, isobaric, tangential momentum equation has the form:

$$\frac{\partial v}{\partial t} = -u\zeta_a - \omega\frac{\partial v}{\partial p}. \quad (11.5)$$

Since, as noted in Chapter 5, $\zeta_a = (1/r)\partial M/\partial r$, where $M = rv + \frac{1}{2}fr^2$ is the absolute angular momentum, the term $u\zeta_a$ is proportional to the radial advection of M and characterizes the local spin up or spin down that occurs as air parcels move radially while materially conserving their value of M. In essence, *this term represents the classical spin up mechanism* discussed in Section 8.2. The term $\omega(\partial v/\partial p)$ is proportional to the vertical advection of M. Note further that the two terms on the right-hand-side of

2. It is easy to verify that expansion of the radial divergence in Eq. (11.3) yields Eq. (11.4).

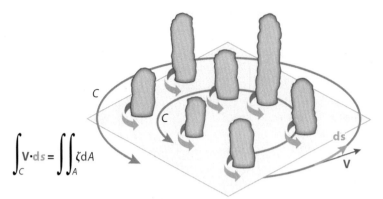

$$\int_C \mathbf{V} \cdot d\mathbf{s} = \iint_A \zeta \, dA$$

FIGURE 11.3 Schematic illustration of a region of deep rotating updrafts with two hypothetical circuits (gray circles). By Stokes' theorem, the circulation about either circle is equal to the areal integral of the vorticity enclosed by that circuit. Wide dark yellow (light gray in print version) arrows denote the local rotational flow associated with the rotating convective updrafts. To avoid clutter, the local and system-scale overturning circulations associated with the updrafts are not shown. See text for details. From Montgomery and Smith (2017b). Reproduced with permission from the Annual Review of Fluid Mechanics, Volume 49; copyright 2017 Annual Reviews.

(11.5) are just the advective and non-advective fluxes, respectively, that appear on the right-hand-side of Eq. (11.3). Taken together, these two terms constitute the generalized radial vorticity flux.

11.3 Non-axisymmetric flow

In a flow that is not axisymmetric, one can consider the change in circulation about a fixed circular loop at a given pressure level and radius. The circulation at a particular time is equal to the areally-integrated vorticity within the loop. Hence, the circulation about the loop will increase if there is a flux of positive absolute vorticity into the loop. Since the circulation about the loop is proportional to the mean tangential velocity around the loop, an increase in circulation implies an increase in the mean tangential wind speed at this radius.

The rotating-convection paradigm invokes the lower tropospheric branch of the overturning circulation produced by the collective effects of deep convection within the loop to provide the influx of vorticity required for vortex intensification. Imagine now a fixed loop surrounding a region of deep convection as sketched in Fig. 11.3. If the convection develops in an environment of cyclonic vertical vorticity, it will amplify this vorticity locally by the process of vortex tube stretching. However, because of mass continuity, vortex-tube stretching reduces the cross-sectional area of the enhanced vorticity and there is no *net* contribution from the stretching, by itself, to the circulation around the loop. What does contribute to this circulation is the horizontal flux of absolute vorticity into the loop by the convectively-induced inflow.

On the convective scale, where the flow is fully three-dimensional, the non-advective flux leads to localized vorticity dipoles (Montgomery et al., 2006b; Kilroy and Smith, 2015). If a particular convective element occurs entirely within some circular loop in a pressure surface, the associated vorticity dipole will not contribute to the circulation change about the loop. For this reason, only dipoles that are intersected by a particular loop would contribute to the circulation about that loop. We will show in Section 11.5.1 below that the eddy contribution to the non-advective flux is important for circular loops within the main eyewall updraft.

11.4 Azimuthally-averaged tangential and radial wind tendency

An alternative approach to investigating tropical cyclone spin up for the case of a non-axisymmetric vortex is to use an azimuthally-averaged form of the tangential momentum equation, which for inviscid flow in cylindrical polar coordinates is given by

$$\frac{\partial \bar{v}}{\partial t} = -\overline{u \zeta_a} - \overline{\omega \frac{\partial v}{\partial p}}, \tag{11.6}$$

where the overbar denotes an azimuthal average on a constant pressure surface at a fixed radius centered on an appropriate definition of the vortex center. For any dependent variable γ, the averaging operator is defined by

$$\bar{\gamma} = \frac{1}{2\pi} \int_0^{2\pi} \gamma \, d\lambda. \tag{11.7}$$

Then γ may be written as the sum of a mean part $\bar{\gamma}$, and an asymmetry (or "eddy part") γ', i.e., $\gamma = \bar{\gamma} + \gamma'$, where, by definition, $\overline{\gamma'} = 0$. Eq. (11.6) is a generalization of the axisymmetric form, Eq. (11.5).

As discussed in more detail in the next section, quantities on the right-hand-side of Eq. (11.6) can be separated into azimuthal mean terms and terms involving products of perturbations leading to so called co-variance or eddy contributions to spin up or spin down. The eddy terms complete the description of the spin up process by accounting for the effects of localized deep convection. As in the axisymmetric form (Eq. (11.5)) and as already noted above, comparison between (11.6) and (11.2) indicates that the two terms on the right-hand-side of (11.6) are just minus the advective and non-advective vorticity fluxes, respectively, that appear on the right-hand-side of Eq. (11.2).

As explained above, the main contribution to the non-advective flux is from the azimuthally-averaged vortex-scale dipole. This equivalent effect is contained in the second term on the right-hand-side of Eq. (11.6) as a vertical advection of tangential momentum. This effect is a particularly important element of eyewall spin up because the vorticity flux $-\overline{u\zeta_a}$ in the eyewall is typically outwards on account of the outward radial flow there (Persing et al., 2013; Schmidt and Smith, 2016).

The alternative articulation of the rotating-convection paradigm framed in terms of the azimuthally-averaged momentum equations is discussed in detail by Persing et al. (2013) and reviewed by Montgomery and Smith (2017b), Section 4.2. Its extension to encompass the effects of vertical shear associated with the ambient flow is discussed by Smith et al. (2017). In the next section we summarize the basic ideas behind these studies and provide some snippets of what can be learned from such analyses.

11.4.1 Characterizing eddy processes

In cylindrical polar coordinates defined relative to some center of circulation and using height z as the vertical coordinate, the azimuthally-averaged tangential momentum equation takes the form:

$$\frac{\partial \bar{v}}{\partial t} = \underbrace{-\bar{u}(\bar{\zeta}+f)}_{V_{m\zeta}} \underbrace{-\bar{w}\frac{\partial \bar{v}}{\partial z}}_{V_{mv}} \underbrace{-\overline{u'\zeta'}}_{V_{e\zeta}} \underbrace{-\overline{w'\frac{\partial v'}{\partial z}}}_{V_{ev}} + \underbrace{\bar{F}_\lambda}_{V_d},$$
(11.8)

where $\bar{F}_\lambda$ is the tangential component of the subgrid-scale eddy-momentum flux divergence. Here, we have neglected the azimuthally-averaged pressure gradient term involving perturbations of density in the azimuthal direction. As noted by Persing et al. (2013), this term (V_{ppg} in their Eq. (12)) is tiny compared with all other terms.

The five terms on the right-hand-side of Eq. (11.8) are interpreted as follows: $V_{m\zeta}$ is the mean radial influx of absolute vertical vorticity, V_{mv} is the mean vertical advection of mean tangential momentum, $V_{e\zeta}$ is the eddy radial vorticity flux, V_{ev} is the eddy vertical advection of eddy tangential momentum, and V_d is the combined mean horizontal and vertical diffusive tendency of tangential momentum, given

by:

$$\underbrace{\bar{F}_\lambda}_{V_d} = \underbrace{\frac{1}{r^2\bar{\rho}}\frac{\partial r^2 \overline{\bar{\rho}\,\overline{\tau_{r\lambda}}}}{\partial r}}_{V_{dr}} + \underbrace{\frac{1}{\bar{\rho}}\frac{\partial \overline{\bar{\rho}\,\overline{\tau_{\lambda z}}}}{\partial z}}_{V_{dz}},$$
(11.9)

where the stress tensors (e.g., Landau and Lifshitz, 1966, p51) generalized to account for anisotropic eddy momentum diffusivities for the subgrid-scale motions are:

$$\overline{\tau_{r\lambda}} = K_{m,h}\overline{\left(\frac{1}{r}\frac{\partial u}{\partial \lambda} + r\frac{\partial v/r}{\partial r}\right)},$$
(11.10)

$$\overline{\tau_{\lambda z}} = K_{m,v}\overline{\left(\frac{1}{r}\frac{\partial w}{\partial \lambda} + \frac{\partial v}{\partial z}\right)},$$
(11.11)

and $K_{m,h}$ and $K_{m,v}$ are the horizontal and vertical momentum diffusivities, respectively.

The azimuthally-averaged radial momentum equation can be written similarly as

$$\underbrace{\frac{\partial \bar{u}}{\partial t}}_{U_t} + \underbrace{\bar{u}\frac{\partial \bar{u}}{\partial r}}_{U_{mr}} + \underbrace{\overline{\left(u'\frac{\partial u'}{\partial r} + \frac{v'}{r}\frac{\partial u'}{\partial \lambda}\right)}}_{U_{eh}} = \underbrace{-\bar{w}\frac{\partial \bar{u}}{\partial z}}_{U_{mv}} \underbrace{-\overline{w'\frac{\partial u'}{\partial z}}}_{U_{ev}}$$

$$\underbrace{+\frac{\bar{v}^2}{r} + f\bar{v} - \frac{1}{\bar{\rho}}\frac{\partial \bar{p}}{\partial r}}_{U_{magf}} + \underbrace{\overline{\frac{v'^2}{r} - \frac{1}{\rho}\frac{\partial p'}{\partial r}}}_{U_{eagf}} + \underbrace{\bar{F}_r}_{U_d}$$

(11.12)

Here, $\bar{F}_r$ is the radial component of the subgrid-scale eddy-momentum flux divergence given by:

$$\underbrace{\bar{F}_r}_{U_d} = \underbrace{\frac{1}{r\bar{\rho}}\frac{\partial r\overline{\bar{\rho}\,\overline{\tau_{rr}}}}{\partial r} - \frac{\overline{\tau_{\lambda\lambda}}}{r}}_{U_{dh}} + \underbrace{\frac{1}{\bar{\rho}}\frac{\partial \overline{\bar{\rho}\,\overline{\tau_{rz}}}}{\partial z}}_{U_{dz}},$$
(11.13)

where the stress tensors (e.g., Landau and Lifshitz, 1966, p51) for the subgrid-scale motions[3] are:

$$\overline{\tau_{rr}} = 2K_{m,h}\overline{\left(\frac{\partial u}{\partial r}\right)},$$
(11.14)

$$\overline{\tau_{\lambda\lambda}} = 2K_{m,h}\overline{\left(\frac{1}{r}\frac{\partial v}{\partial \lambda} + \frac{u}{r}\right)},$$
(11.15)

$$\overline{\tau_{rz}} = K_{m,v}\overline{\left(\frac{\partial u}{\partial z} + \frac{\partial w}{\partial r}\right)}.$$
(11.16)

The individual terms on the left-hand-side of Eq. (11.12) represent the following: the local tendency of the mean radial velocity, U_t, the mean radial advection of radial momentum per unit mass, U_{mr}, and the mean horizontal advection of eddy radial momentum per unit mass, U_{eh}.

---

3. The expression for $\overline{\tau_{rz}}$ corrects the expression given in (Persing et al., 2013, their Eq. (20)) and (Montgomery and Smith, 2017b, their Eq. (15)). The difference is found to be negligible.

The terms on the right-hand-side of Eq. (11.12) are in order: U_{mv} is minus the mean vertical advection of mean radial momentum per unit mass and U_{ev} is minus the eddy vertical advection of eddy radial momentum per unit mass; U_{magf} and U_{eagf} are the mean and eddy agradient force per unit mass, respectively; and U_d is the combined mean radial and vertical diffusive tendency of radial momentum.

Note that we have chosen to write U_{magf} with $\bar{\rho}$ in the denominator of the radial pressure gradient force term. This choice requires that the azimuthal variation of ρ be retained in the definition of U_{eagf}, and assumes that $|\rho - \bar{\rho}| \ll \rho$, which is always well satisfied in the numerical simulation to be described.

In contrast to the tangential momentum equation, we have chosen this pseudo-Lagrangian form of the radial momentum equation in which the left-hand-side represents the material acceleration in the radial direction following the horizontal wind. Then, the sum of terms on the right-hand-side can be interpreted as forces that produce pseudo-material acceleration in the radial direction. This pseudo-Lagrangian form is preferred because it facilitates a layer-wise perspective on the forces generating horizontal acceleration within the boundary layer as well as on the formation of the upper-tropospheric inflow and outflow layers. Such a choice would seem less appropriate in the tangential momentum equation as one would then lose the neat form of the radial vorticity flux term.

11.4.2 Attributes of the mean-eddy flow partitioning

The partitioning of the flow into azimuthal mean and eddy contributions is a natural one for an isolated vortex, but one aspect should be borne in mind when interpreting the individual contributions in situations where strong and highly azimuthally-localized features punctuate the vortex flow in a particular annulus. This is because *such features project on both components of the partition*. For example, the mean strength, $\overline{w}$, of a localized updraft, w, will have a small positive value, while the eddy, w', will have a large positive value in its particular location, but *a small negative value elsewhere*. The latter is a consequence of the partitioning result that $\overline{w'} = 0$. The small negative value for the eddy will identically cancel the small positive value from the mean outside the region of the eddy. Despite this cancellation issue, the formulation in terms of mean and eddy components is generally useful for providing insight.

11.4.3 Eddy effects of an isolated deep convective cloud

In general, it is hard to anticipate the structure of the eddy terms and one has to do the calculations. However, a simple example presented by Kilroy and Smith (2016), see their

Section 6, provides some insight into what is involved. In this example, eddy momentum fluxes associated with velocity perturbations due to an isolated convective cloud spanning a small range of azimuths in an idealized vortex are examined and the effects of these fluxes on the mean tangential velocity tendency are determined. The focus is on a buoyant air parcel within the updraft as shown schematically in Fig. 11.4.

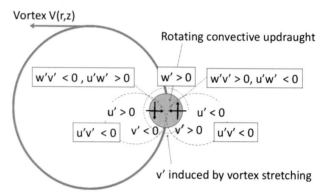

FIGURE 11.4 Horizontal cross-section schematic indicating the velocity perturbations induced by an isolated updraft at some radius from the circulation center of a cyclonic axisymmetric vortex. The perturbation momentum flux terms about the parent vortex along the center line of the updraft have signs as shown. See text for discussion. Adapted From Kilroy and Smith (2016). Republished with permission of Wiley.

Most of the flow perturbation induced by the initial thermal perturbation will appear in the perturbation velocity (u', v', w'), but the question here is: What is the effect of the flow perturbation on the magnitude and spatial distribution of the tendency of the azimuthal-mean tangential velocity, $\bar{v}$? More specifically: what is the structure of the associated eddy momentum fluxes: $\overline{u'v'}$, $\overline{u'w'}$, and $\overline{v'w'}$?

To be specific, consider the velocity perturbations induced by an isolated updraft at some radius from the circulation center of a cyclonic axisymmetric vortex, such as those shown in Figs. 5.5 or 8.1. At levels below that where the perturbation vertical mass flux, $\rho w'$, is a maximum, the vertical velocity perturbation w' will be accompanied through continuity by entrainment into the updraft characterized by a radial velocity perturbation u' as shown in Fig. 11.4. The w' perturbation will stretch the existing cyclonic vorticity to produce a positive vertical vorticity perturbation ζ' and an associated tangential wind perturbation v' about the vortical updraft as indicated. Above levels where the perturbation vertical mass flux, $\rho w'$, is a maximum, the perturbations in u' and v' will be reversed.

From these perturbations it is easy to infer the sign of the eddy covariance terms $u'v'$, $u'w'$, $v'w'$ as indicated in Fig. 11.4. If the updraft is axisymmetric about its axis, the maximum amplitude of the covariances will lie in a plane through the axes of the parent vortex and the updraft, itself. Of course, to obtain the amplitude structure of the azimuthal mean terms $\overline{u'v'}$, $\overline{u'w'}$, $\overline{v'w'}$ about the axis of the parent vortex, one has to do a full calculation. Such calculations are presented in detail by Kilroy and Smith (2016), see their Section 6. These authors calculated not only the covariance terms, but also their effect on the azimuthal wind tendency, $\partial\bar{v}/\partial t$ for a single cloud updraft at different times of the cloud life cycle.

During intensification, the multiple vortical updrafts excite vortex Rossby waves (e.g., Chapter 4) and inertia-buoyancy waves (as discussed, e.g., in Chen et al., 2003 and Reasor and Montgomery, 2015), which in turn contribute to the sign and structure of the eddy momentum fluxes. The intensification process generally comprises a turbulent system of rotating, deep moist convection and vortex waves. For these reasons, it is hard, in general, to anticipate the structure of the eddy terms and one has to do the calculations. The next section examines the effects of the eddy momentum fluxes in an intensifying tropical cyclone as they appear in the material rate-of-change of mean tangential and radial velocities. A diagnosis of the fluxes, themselves, is summarized later in Section 16.2.

11.5 Applications to a numerical model simulation

The numerical simulation used to illustrate the role of eddy momentum fluxes is slightly different from that used in the genesis simulation in Section 10.2 and is described in detail by Montgomery et al. (2020). The main differences are in the chosen latitude (here 20°N), SST (here 27 °C), strength of the initial vortex (here 15 m s^{-1}) and the horizontal grid spacing in the inner region (here 1 km). Because the simulation starts with a moderately strong vortex, we refer to it as the intensification simulation.

The details of vortex evolution are similar to that in the genesis simulations with deep convection developing within the initial radius of maximum tangential wind speed after several hours and the convectively amplified vorticity anomalies subsequently aggregating to form an approximate monopole of intense cyclonic vorticity. The main difference is that, because of the stronger initial vortex, the gestation period before the vortex begins to rapidly intensify is much shorter. This may be seen in Fig. 11.5, which shows the time evolution of maximum azimuthally-averaged tangential wind speed, V_{max}, for this simulation. In this case the vortex begins to rapidly intensify a little after 24 h, nearly a day earlier than that in the corresponding genesis simulation (cf. Fig. 10.2a).

Fig. 11.6 shows vertical cross sections of three-hour time-averaged and azimuthally-averaged tangential and radial velocity fields centered at 30 hours and 42 hours, both during the period of RI. At both times, the tangential wind maximum is below 1 km in height and the radius of tangential wind speed maximum shows a sharp increase with

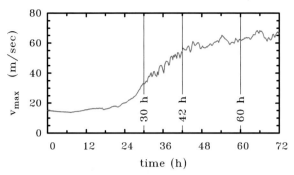

FIGURE 11.5 Time series of maximum azimuthally-averaged tangential wind speed, V_{max}, in the intensification simulation described in Section 11.5. The thin vertical line shows the time at which the tendency analyses in Sections 11.5.1 and 11.5.3 are presented.

height in the lowest 2 km. Above that height the maximum moves inwards slightly up to about 7 km and then outwards to a height of about 14 km. During the 12 hour period between the panels, the tangential wind speed strengthens and broadens throughout the entire troposphere. The tangential wind and inflow maximum move inwards also.

At 30 hours, the radial flow beyond the tangential wind maximum has a simple structure with inflow in the lower troposphere, outflow in the upper troposphere and inflow again just below the upper-level outflow layer and above the tropopause (about 16 km altitude in this simulation). In particular, there is strong inflow in a shallow surface-based layer and strong outflow above about 10 km to the tropopause. Significantly, there is a shallow region of outflow just above where the surface inflow terminates and ascends into the eyewall (marked approximately by the 0.5 m s^{-1} contour of vertical velocity). At 42 hours, the inflow in the frictional boundary layer has strengthened as has the outflow just above where this inflow terminates. The upper tropospheric outflow has strengthened also.

The behavior of both the tangential and radial wind components is similar to that of the corresponding wind components during the RI phase in the genesis simulation as exemplified by the fields at 60 hours and 72 hours shown in Figs. 10.6e, f and 10.7e, g, respectively.

11.5.1 Tangential velocity tendency analysis

We focus our detailed analysis of spin up at 30 hours, which is several hours into the period of rapid intensification (Fig. 11.5), but still before the vortex has developed an appreciable degree of axial symmetry. Fig. 11.7 shows radius-height plots of the three-hour time averaged terms in the azimuthally-averaged tangential velocity tendency equation at this time. The time average is constructed from model output stored at 15 min intervals.

The contributions to the tendency from the mean radial vorticity flux and vertical advection are shown in panels (a) and (b) and their sum is shown in panel (c). The sum of

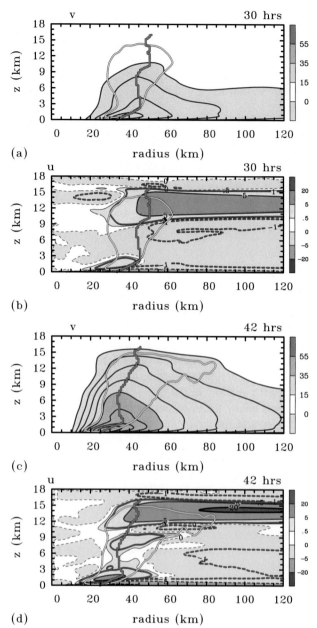

FIGURE 11.6 Radius-height plots of the three-hour time averaged and azimuthally-averaged tangential and radial velocity fields centered at 30 hours. (a), (c) tangential velocity, (b), (d) radial velocity. Upper panels for 30 hours, lower panels for 42 hours. Thick black contour highlights ± 20 m s^{-1} radial velocity. The yellow contour shows the 0.5 m s^{-1} vertical velocity, while the purple contours show ± 1 m s^{-1} radial velocity. The green curve shows the azimuthally-averaged R_{vmax}. Solid contours positive, dashed contours negative.

the corresponding eddy advection terms is shown in panel (d) and the sum of the two mean and two eddy advection terms is shown in panel (e). For reference, the location of the mean eyewall updraft, characterized by the 0.5 m s^{-1} isopleth of vertical velocity, is highlighted in each panel and regions of mean inflow and outflow are highlighted by the ± 1 m s^{-1} contours of radial velocity. Further, to provide

a spatial reference for the acceleration terms, the location of the azimuthally-averaged radius of maximum tangential velocity is shown also.

Because the absolute vorticity is mostly positive, the mean vorticity influx (panel (a)) has a similar pattern to that of the radial velocity shown in Fig. 11.6b: it is large and positive in a shallow layer near the surface, a result of the strong boundary-layer inflow, and large and negative in a broader sloping sheath just above where this layer terminates. The mean vorticity influx is mostly negative in the developing eyewall updraft, except in the layer from about 3-8 km in altitude. The region of negative vorticity flux extends into the upper-tropospheric outflow layer, a reflection of radial outflow in the updraft. There are a few localized regions of small positive vorticity influx, mostly in regions of inflow, but including a larger more coherent region in the upper-tropospheric outflow layer outside a radius of about 60 km.

The main features in the mean vertical advection tendency (panel (b)) are similar to those in panel (a), but are opposite in sign so that there is considerable cancellation in their sum as seen in panel (c). This cancellation is to be expected because, above the frictional boundary layer, absolute angular momentum is approximately materially conserved. Closer inspection of panel (c) shows that there is a sloping region between 1 and 3 km in altitude where the net mean tendency is to produce spin down. As shown below, this spin down tendency is more than negated by the vertical eddy momentum transport. Otherwise this low-level portion of the eyewall updraft would spin down rapidly.

The contribution to the advective tendency from the eddies is confined mainly to the developing eyewall updraft region (panel (d)). For reasons discussed in Section 11.4.2, the main features of the eddy terms are opposite in sign to those of the corresponding mean terms (not shown). Thus, the combined eddy tendency in panel (d) shows a pattern that is quite similar to the sum of the mean tendencies (panel (c)), but of opposite sign. Significantly, the combined eddy tendency in the developing eyewall region spanning approximately 1 to 3 km in height more than compensates the spin down effect from the combined mean terms (panel (e)). Since the eddy vorticity flux term is largely negative in this layer on account of the centrifugal effect (not shown), the positive combined eddy tendency indicates that the convective eddy structures are providing most of the vertical momentum transport in this layer. At 30 h, the net advective tendency in Fig. 11.7e is mostly positive through much of the troposphere beyond a radius of about 30 km.

The horizontal diffusive tendency shows a weak negative tendency in the eyewall (panel (f)), while the vertical diffusive tendency is strongly negative in a shallow layer near the surface with a small region of positive tendency below the base of the eyewall updraft (panel (g)). Elsewhere, the diffusive tendencies are relatively small. Thus, the sum

Tangential wind tendencies at 30 h

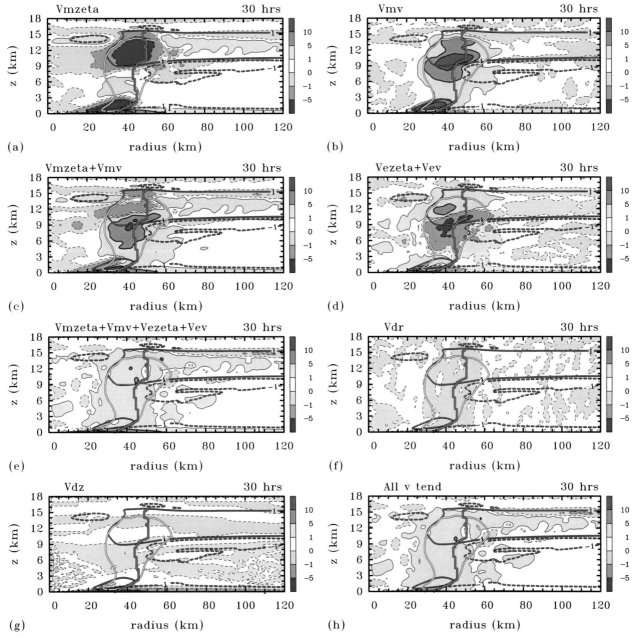

FIGURE 11.7 Radius-height plots of the three-hour time-averaged terms in the azimuthally-averaged tangential wind tendency equation centered at 30 h. (a) $V_{m\zeta}$, (b) V_{mv}, (c) $V_{m\zeta} + V_{mv}$, (d) $V_{e\zeta} + V_{ev}$, (e) $V_{m\zeta} + V_{mv} + V_{e\zeta} + V_{ev}$, (f) V_{dr}, (g) V_{dz}, (h) sum of all tendencies, including diffusive tendency terms. Shading as indicated on the side bar in m s^{-1} h^{-1}, with shaded regions enclosed by contours. The yellow contour shows the 0.5 m s^{-1} vertical velocity, while the purple contours show $\pm$ 1 m s^{-1} radial velocity. The green curve shows the azimuthally-averaged R_{vmax}. Solid contours positive, dashed contours negative.

of all tendency terms on the right-hand side of Eq. (11.8), shown in Fig. 11.7h, is similar to the net advective tendency in Fig. 11.7e.

A comparison of the radial and vertical eddy terms with the corresponding sub-grid-scale diffusion terms shows that the pattern of the eddy terms is generally quite different from that of the diffusion terms. The discrepancy in the pattern of tendencies implies that *the resolved eddy contributions cannot be regarded simply as a down-gradient diffusive process.*

Comparison of panels (a) and (g) in Fig. 11.7 shows that the horizontal influx of cyclonic vorticity is generally larger

Tangential wind tendencies at 30 h

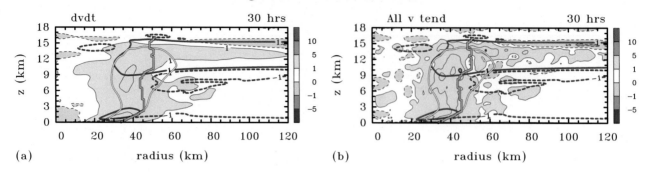

(a)

(b)

Tangential wind tendencies at 42 h

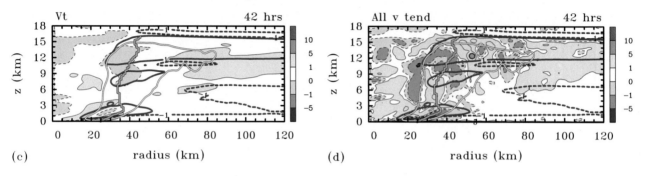

(c)

(d)

Tangential wind tendencies at 60 h

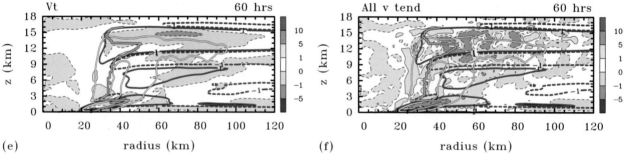

(e)

(f)

FIGURE 11.8 Comparison of radius-height plots of (a) the three-hour time-averaged tendency, Vt, on the left-hand side of the azimuthally-averaged tangential wind tendency equation, Eq. (11.12), centered at 30 h with (b) the corresponding sum of the terms on the right-hand side of this equation. The former is calculated as a difference in the tangential wind over the 3 hour interval divided by 3 hours. Panels (c) and (d) show similar fields for 42 hours and panels (e) and (f) for 60 hours. Shading as indicated on the side bar in m s^{-1} h^{-1}, with shaded regions enclosed by contours. The thick red contour is from 1 m s^{-1} h^{-1} to 4 m s^{-1} h^{-1} in intervals of 1 m s^{-1} h^{-1}. The yellow contour shows the 0.5 m s^{-1} vertical velocity, while the purple contours show $\pm$ 1 m s^{-1} radial velocity. The green curve shows the azimuthally-averaged R_{vmax}. Solid contours positive, dashed contours negative.

than the mean vertical diffusion tendency in the boundary layer. Even after adding the slight negative tendency from the combined eddy term near the surface (panel (d)), the net tendency of tangential wind is positive in the inner-core boundary layer (panel (h)). This result is confirmation that the nonlinear boundary layer spin up mechanism is operating to spin up the maximum tangential wind in the boundary layer. Another manifestation of this spin up mechanism is the outward sloping region of positive values below about 3 km in height in panel (d), which as noted above originate from the term V_{ev} (this is the eddy component of the non-advective flux discussed in Section 11.3). The positive values of v', which presumably contribute to the supergradient excess, are being lofted to give the positive values of V_{ev} and must be generated within the boundary layer. *These positive tendency values in panel (d) must be a result of the nonlinear boundary layer spin-up mechanism, which acts on the asymmetric component of flow also.*

The time-averaged tangential velocity tendency, the left-hand side of Eq. (11.8), is shown in Fig. 11.8a. This tendency is calculated directly from the model output as the difference in the azimuthally-averaged tangential velocity over the 3 hour period divided by the 3 hour time span. The panel shows clearly that there is spin up throughout much of the eyewall updraft, including the boundary layer beneath the eyewall and extending to approximately 60 km radius. Fig. 11.8b is a repeat of Fig. 11.7h, but with an additional contour ($3 \text{ m s}^{-1} \text{ h}^{-1}$) to facilitate a comparison with Fig. 11.8a. It shows the sum of all tendency terms on the right-hand side of Eq. (11.8). While the direct estimate in panel (a) is smoother, there is reasonable agreement between the two fields. This agreement provides strong support to the physical interpretations of the various tendency terms given above.

11.5.2 Spin up at later times

Radius-height plots similar to those in Fig. 11.7 have been constructed for 42 hours and 60 hours (not shown), which are near the end of the period of rapid intensification and in the mature stage, respectively (Fig. 11.5). Despite the fact that one might expect the vortex to become progressively more axisymmetric on account of vortex axisymmetrization as discussed in Chapter 4, the individual tendency terms have structures that are similar to their counterparts in Fig. 11.7, but are generally larger in magnitude.

In particular, there is still a strong mean tendency associated with the radial influx of vorticity in the boundary layer, which is partly opposed by the negative tendency due to friction. In addition, the strong negative mean tendency $V_{m\zeta} + V_{mv}$ where the air ascends as it exits the boundary layer is opposed by a positive tendency from the sum of the eddy terms $V_{e\zeta} + V_{ev}$. This implies, again, that the eddies are playing an important role in intensifying the tangential flow in the eyewall. In the upper part of the eyewall, there is much cancellation between the sums $V_{m\zeta} + V_{mv}$ and $V_{e\zeta} + V_{ev}$.

The middle and lower panels of Fig. 11.8 show similar comparison of terms on the left and right sides of Eq. (11.8) for 42 h and 60 h. While the quantitative agreement between the left-hand side and right-hand sides of Eq. (11.8) has become degraded somewhat relative to that at 30 h, the patterns of the terms are nonetheless similar in sign.

Unfortunately, there are several possible sources for the error in diagnosing the left and right sides of Eq. (11.8) from the azimuthally-averaged fields. These sources of error are discussed by Persing et al. (2013), p12318. In brief, they include: the coarse temporal sampling of output data (typically every 15 min); the diagnosis of parameterized internal diffusion and surface momentum fluxes using the interpolated data in cylindrical coordinates; and the use of centered spatial differences to calculate all advection terms. Note that, a 5th-order upstream advection scheme was used in the present simulation.

11.5.3 Radial velocity tendency analysis

We turn our attention now to the radial force fields that influence the radial flow dynamics during the simulated spin up process. Such an analysis is needed to understand the origin of supergradient winds as well as the upper-level inflow and outflow structure in simulated and observed tropical cyclones.

Fig. 11.9 shows radius-height plots of the three-hour time-averaged terms in the azimuthally-averaged, pseudo-Lagrangian radial momentum equation (11.12) centered at 30 hours. As in Fig. 11.7, the location of the mean eyewall updraft and regions of inflow and outflow are highlighted, as is the location of the azimuthally-averaged radius of maximum tangential velocity.

Panels (a), (b), and (c) show the mean and eddy agradient force fields, U_{magf} and U_{eagf}, and their sum, respectively. As expected, U_{magf} is strongly negative in a shallow surface-based layer at radii beyond about 30 km. This negative region coincides with the outer part of the frictional boundary layer and *it is primarily this negative force that drives the boundary layer inflow*. Kilroy et al. (2016a) showed that much of the low-level inflow can be attributed to boundary layer dynamics, whereupon the *suction effect* of deep convection plays a secondary role, except possibly beneath and near the eyewall updraft.

Near the surface, the magnitude of U_{magf} increases with decreasing radius to about 50 km and then declines rapidly. The corresponding tangential winds in this region of negative values are *subgradient* (i.e., $v < v_g$, where v_g satisfies the equation $v_g^2/r + f v_g = -(1/\rho)(\partial p/\partial r)$).[4] The mean radial inflow progressively increases with decreasing radius in this layer, reaching values of about 13 m s^{-1} (not shown), which is a significant fraction (about 40%) of V_{max} at this time.

Inside approximately 43 km radius and in a shallow layer that slopes upwards with radius, U_{magf} is strongly positive and serves to decelerate the inflow and even accelerate the flow outwards just above the boundary layer into the eyewall updraft. In this positive region, the tangential winds are supergradient (i.e., $v > v_g$). There is a region of strong negative values of U_{magf} between heights of about 2 and 4 km within the eyewall updraft. Above the negative region, U_{magf} is generally positive, being particularly strong in the inner portion of the eyewall updraft and in the outflow region to more than 70 km radius. This pattern is not surprising because air with high angular momentum is being transported vertically by deep convective cores in the developing eyewall.

The eddy agradient force field, U_{eagf}, in Fig. 11.9b indicates that the asymmetric part of the tangential flow

---

4. Of course, there might be localized regions where the radial pressure gradient is sufficiently negative that a real-valued solution for v_g is not possible.

Radial wind tendencies at 30 h

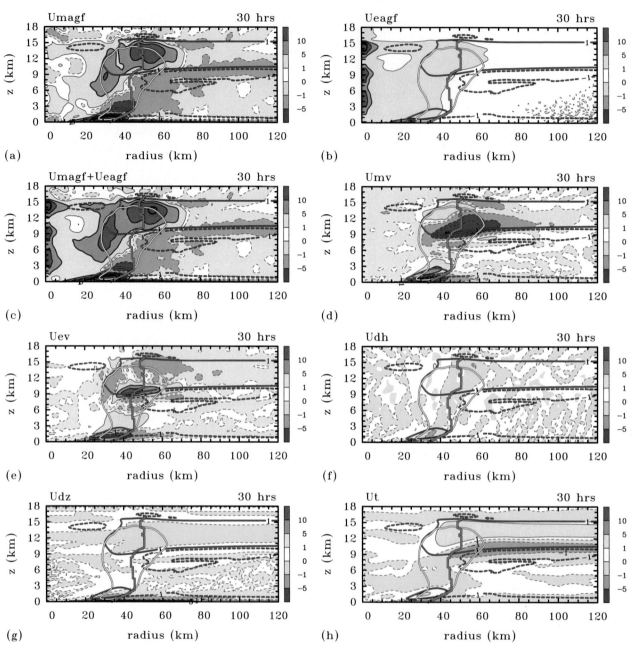

FIGURE 11.9 Radius-height plots of the three-hour time-averaged terms in the azimuthally-averaged radial wind tendency equation, Eq. (11.12), centered at 30 h. (a) U_{magf}, (b) U_{eagf}, (c) $U_{magf} + U_{eagf}$, (d) U_{mv}, (e) U_{ev}, (f) U_{dh}, (g) U_{dz}, (h) U_t. Shading as indicated on the side bar in m s^{-1} h^{-1}, with shaded regions enclosed by contours. Thick black contour in panels (a) and (c) is ± 20 m s^{-1} h^{-1}. The yellow contour shows the 0.5 m s^{-1} vertical velocity, while the purple contours show ± 1 m s^{-1} radial velocity. The green curve shows the azimuthally-averaged R_{vmax}. Solid contours positive, dashed contours negative.

is generally not in gradient wind balance. This eddy term contributes to a positive radial acceleration in excess of 1 m s^{-1} h^{-1} throughout the troposphere out to more than 60 km radius, where it represents an additional centrifuge effect to that of the mean term U_{magf}.

In particular, U_{eagf} shows strong positive values near the surface at radii between approximately 15 and 30 km near where the surface inflow terminates, reinforcing the deceleration of the boundary layer inflow. The relatively large values of U_{eagf} near the axis should not be taken too seriously and are presumably due to numerical inaccuracy.[5]

Fig. 11.9c shows the sum of the mean and eddy agradient force terms, which, beyond the outer boundary of the eyewall updraft, is similar to U_{magf}, itself. In the upper troposphere there is a region of positive $U_{magf} + U_{eagf}$ near the top of the developing eyewall, which is thickest near its top and decreases to zero at a radius just beyond 70 km. At larger radii in the upper troposphere and in much of the rest of the troposphere beyond the developing eyewall updraft, $U_{magf} + U_{eagf}$ has mostly negative values with magnitude not more than 5 m s^{-1} h^{-1}. *The subgradient tangential wind in this region acts to decelerate outflow and accelerate inflow.*

The radius-height structure of the mean vertical advection U_{mv} in Fig. 11.9d shows a series of layers in which U_{mv} has alternating sign in and near the eyewall region. In the lowest layer, near the base of the developing eyewall updraft, $U_{mv} < 0$, indicating a contribution to an increase of the radial inflow by the vertical advection of inward radial momentum from near the surface. In the sloping layer above, $U_{mv} > 0$, reflecting the mean vertical transport of positive radial momentum associated with the strong outflow of air just above the boundary layer. The layer of large negative U_{mv} between about 8.5 and 13 km in height is associated with the upward transport of mean radial inflow in the eyewall updraft from the lower troposphere (cf. Fig. 11.6b).

Fig. 11.9e shows the eddy vertical advection of eddy radial momentum per unit mass, U_{ev}. In the lower troposphere, the U_{ev} field broadly reinforces that of U_{mv}, which is not surprising because the deep convective cells, effectively the eddies, are projecting, in part, on the mean. *In particular, in the region underneath and just outside of the developing eyewall, both the mean and eddy advection terms contribute to increasing the low-level inflow. Presumably, this effect is associated, at least in part, with the buoyant deep convective cells drawing air inwards at the lowest levels. This suction effect acts in addition to the boundary layer spin up mechanism as discussed Section 6.6.* Above the strong inflow layer near the surface, the structure of the U_{ev} term in and near the developing eyewall region exhibits a series of layers in which U_{ev} has alternating sign. In the upper troposphere there is a degree of cancellation between U_{mv} and U_{ev}.

Figs. 11.9f and 11.9g show the time-averaged and azimuthally-averaged sub-grid-scale tendencies, U_{dh} and U_{dz}. The horizontal diffusion of radial momentum (U_{dh}) is relatively small, exceeding 1 m s^{-1} h^{-1} in magnitude only in two small parts in the developing eyewall region. The vertical diffusion of radial momentum (U_{dz}) shows a very shallow surface-based layer of strong positive tendency beyond a radius of about 20 km, which is a manifestation of surface friction slowing down the inflow. Above this layer lies a somewhat thicker layer of negative tendency, which is associated with the vertical diffusion of inward radial momentum through the inflow layer. This diffusion becomes particularly strong near where the boundary layer inflow terminates.

Fig. 11.9h shows the local radial wind tendency, U_t, averaged over the three hour period. The figure indicates that over this time interval there has been a strengthening of the overturning circulation (positive tendency where there is outflow, negative tendency where there is inflow). In particular, the low-level inflow is becoming stronger, the outflow just above the boundary layer where the boundary layer terminates is becoming stronger, the upper-level outflow is becoming stronger and the upper-level inflow is becoming markedly stronger. In addition, a region of inflow being drawn in by the developing eyewall updraft in a layer between 2 and 5 km in height is becoming stronger inside a radius of about 50-60 km, depending on the height.

11.5.4 Summary of radial velocity analysis at 30 h

The strengthening of the overturning circulation inferred from the local radial wind tendency field (Fig. 11.9h) can be explained only in terms of the radial force field and the response thereto. The essence of the radial momentum budget is encapsulated in Figs. 11.10a and 11.10b, which show the sum of the time-averaged and azimuthally-averaged tendencies on each side of Eq. (11.8). Panel (a) shows the time-averaged pseudo-Lagrangian radial acceleration, while panel (b) shows the corresponding net time-averaged net radial force leading to this acceleration. Generally, the principal features of these two fields match each other quite well, despite there being a few local discrepancies in detail that are presumably associated with interpolation errors and the like (see Section 11.2). These discrepancies are particularly evident inside the developing eyewall region.

Broadly speaking, there is a net inward force field through much of the low and mid troposphere, including the outer part of the developing eyewall region, itself. This inward force is particularly strong in the inner-core boundary layer near the surface and just below the eyewall updraft, reflecting in part the suction effect there of localized deep convective cells and in part by the inward agradient force produced in the boundary layer. Here, this time-averaged

---

5. Because the vortex center is not exactly stationary, there is a weak flow across the vortex axis, even in the problem studied here where the vortex environment is quiescent. Since there is no source or sink of mass at the axis, both $\bar{u}$ and $\bar{v}$ must vanish at the axis, the latter since the vorticity at the axis is finite. As a result, both mean tendency terms must be zero implying that the sums of terms on the right-hand-sides of Eqs. (11.8) and (11.12) must sum to zero at the axis. Because one of the terms in the expressions for U_{eh} and U_{eagf} involve v'/r, and v' may be finite at the axis, these terms must cancel. However, on a finite mesh, this cancellation may be susceptible to appreciable numerical discretization error.

Radial wind tendencies at 30 h

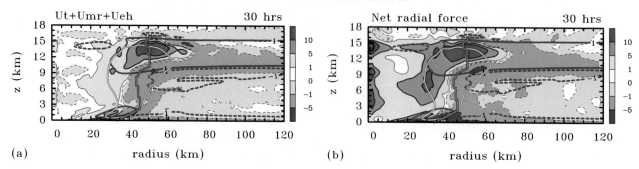

(a) radius (km)

(b) radius (km)

Radial wind tendencies at 42 h

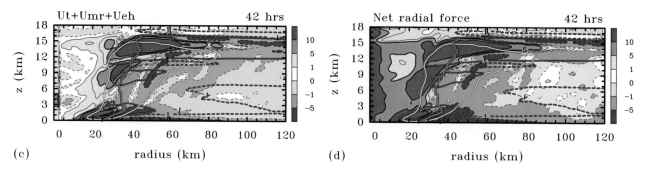

(c) radius (km)

(d) radius (km)

Radial wind tendencies at 60 h

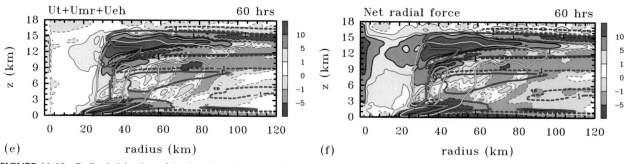

(e) radius (km)

(f) radius (km)

FIGURE 11.10 Radius-height plots of the three hour time-averaged terms in the azimuthally-averaged radial wind tendency equation centered at 30 hours. (a) $U_t + U_{mr} + U_{eh}$, the left-hand side of Eq. (11.12), (b) Net radial force, the right-hand side of Eq. (11.12). Panels (c) and (d) show similar fields for 42 hours and panels (e) and (f) for 60 hours. Shading as indicated on the side bar in m s^{-1} h^{-1}, with shaded regions enclosed by contours. The yellow contour shows the 0.5 m s^{-1} vertical velocity, while the purple contours show ± 1 m s^{-1} radial velocity. The green curve shows the azimuthally averaged R_{vmax}. Solid contours positive, dashed contours negative.

inward force exceeds 5 m s^{-1} h^{-1}. Inside the developing eyewall, near its inner edge, and throughout much of the upper troposphere out to a radius of about 100 km, there is strong positive radial acceleration, exceeding 10 m s^{-1} h^{-1} in places. The net radial force is associated primarily with the pattern of mean agradient force and the mean and eddy vertical advection terms (Fig. 11.9a, d, e).

In summary, the flow in much of the inner-core region is significantly unbalanced in the radial direction, especially at low levels and in the upper troposphere. Thus, while the assumption of gradient wind balance may be a reasonable zero-order approximation through much of the middle troposphere during spin up, the assumption cannot be justified

in the inner-core boundary layer, in the eyewall updraft, or in the upper troposphere in the region of strong outflow. The radial force fields demonstrate that even a generalized Ekman balance (i.e., a balance between the linearized form of U_{magf} and U_{dz}, see also Section 6.4.1) is strongly violated in the inner-core boundary layer, where the terms U_{mv}, U_{ev}, and U_{eagf} are comparable in magnitude to U_{magf} and cannot be neglected in a zero-order approximation.

11.5.4.1 Later times

Similar radius-height cross sections to those in Fig. 11.10a, b have been constructed at later times during the intensification process, 42 h and 60 h (Fig. 11.10c-f.). The tendencies in the individual panels have structures that are similar to their counterparts at 30 hours, but, as in the case of the tangential wind tendency equation, the flow field progressively strengthens and hence the individual tendency terms increase markedly in magnitude because of the quadratic nonlinearity of the equations. Again, there is good agreement between the mean pseudo-Lagrangian tendencies and the sum of the radial force contributions, except at radii inside the developing eyewall. Moreover, the net radial force increases considerably and the region of imbalance that it represents increases in areal extent. *These increases are particularly striking in both the lowest 2 km and in the upper troposphere between about 10 and 16 km. In the lowest 2 km, the increase reflects the combined effect of the increasing inward agradient force and the suction effect of localized deep convective cells, while in the upper troposphere it includes the expulsion of radial momentum by these cells.*

Another noteworthy feature at later times is the development of layers of strong inflow above and below the upper-level outflow layer. These inflow layers are illustrated and discussed in the next section.

11.6 Other features of the numerical simulation

11.6.1 Upper level inflow jets

A striking feature of the foregoing numerical simulation for tropical cyclone intensification, as well as the genesis and intensification simulation in Section 10.2 at later times is the development of a layer of strong inflow just below the upper-level outflow and a layer of weaker inflow just above it (Figs. 11.11a, b, e, f). The existence of such layers is seen in several previous studies going back to that of Rotunno and Emanuel (1987), but to our knowledge, only recently has a credible dynamical explanation for their existence been offered (Montgomery et al., 2020; Wang et al., 2020).

It would seem significant that, beyond a certain radius, the upper level outflow is pushing into a region where the agradient force is mostly inward. This feature is exemplified in Figs. 11.11c, d, g, h which show the combined agradient force, $U_{magf} + U_{eagf}$ corresponding with the radial velocity fields in panels (a), (b), (e), and (f).

A way to think about the upper tropospheric outflow layer is to consider it as an expanding jet of air emanating from a mass and radial momentum source where the eyewall convection terminates (Ooyama, 1987). The outward expansion is resisted, in part, by a radially-inward pressure gradient force. This is the only significant radial force that can resist this outflow because the centrifugal force is always positive and the Coriolis force in the radial direction is positive as long as the tangential flow remains cyclonic.

The inward pressure gradient force arises on account of the geometric requirement that the flow slows down as it moves outwards. Since the pressure field extends vertically beyond the outflow layer itself, one can expect a flow response laterally beyond the outflow layer as well. Where this inward agradient force persists, it will act to accelerate air parcels inwards. Of course, one should not expect a one-to-one relationship in sign between radial acceleration and radial velocity: where the radial flow is inwards, but the radial acceleration is outwards, the inflow will simply be slowed down and vice versa.

A more thorough investigation of these inflow jets is presented in a study by Wang et al. (2020) in which the vertical resolution is increased in the upper troposphere at the expense of that at low levels within the frictional boundary layer. In particular, Wang et al. show that the inflow jets have a significant degree of azimuthal asymmetry.

11.6.2 Centrifugal recoil effect

One prominent feature that develops at later times in the time and azimuthally-averaged flow fields in the genesis simulation of Section 10.2 is a secondary maximum of tangential wind at an altitude of 3 km in the eyewall updraft, together with a region of alternating positive and negative radial flow from the surface to a height of about 4 km in this updraft. These patterns develop after 66 h and are exemplified by the fields at 72 h shown in Figs. 11.12a, b.

The above features are a manifestation of a centrifugal recoil effect as the air rises out of the boundary layer and ascends in the eyewall (Montgomery and Smith, 2017b, and refs). A preliminary understanding of this effect is as follows. The air rising out of the boundary layer is supergradient, i.e., $U_{magf} > 0$ (Fig. 11.12c), even when the corresponding eddy term is accounted for (Fig. 11.12d). With the initially outward flow in the presence of cyclonic absolute vorticity, the mean vorticity flux, $V_{m\zeta}$, is negative (Fig. 11.12e), which, by itself, would contribute to a decline in tangential velocity and thereby to a decline in U_{magf}, ultimately to negative values. A negative U_{magf} would imply a negative radial acceleration and eventually to a reversal to inflowing air parcels. This inflowing air would then result in a positive mean vorticity influx and hence an increase in the tangential velocity in this region, once more leading to supergradient values and so on.

Of course, this argument focuses solely on the gradient wind imbalance and the $V_{m\zeta}$ terms. However, a more complete understanding requires a consideration of the total tangential wind tendency. In particular, the negative contribution of $V_{m\zeta}$ to the tangential wind tendency will be offset by the effects of vertical advection, V_{mv}, (Fig. 11.12f) and the eddy tendencies $V_{e\zeta} + V_{ev}$. The mean advective tendency, $V_{m\zeta} + V_{mv}$, is shown in Fig. 11.12g and the total advective tendency, $V_{m\zeta} + V_{mv} + V_{e\zeta} + V_{ev}$, including the

Upper-level inflow layers in the intensification simulation (Section 11.5)

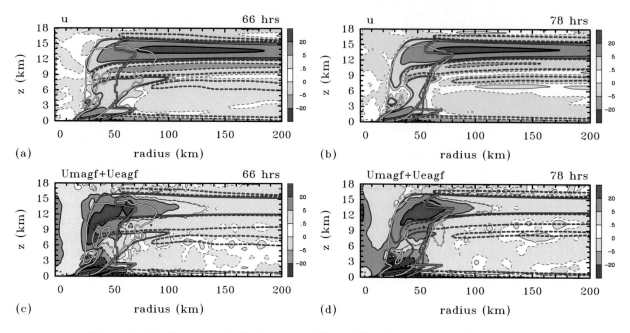

Upper-level inflow layers in the genesis and intensification simulation (Section 10.3)

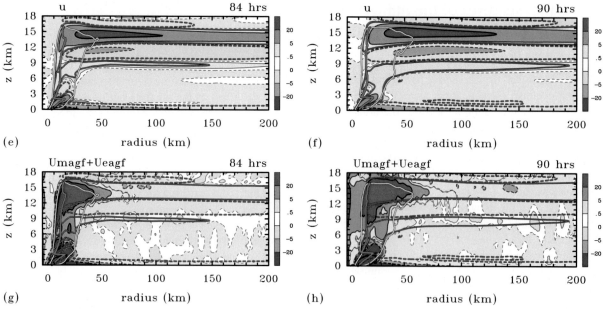

FIGURE 11.11 Radius-height cross sections of three-hour time-averaged and azimuthally-averaged radial velocity field, $\langle u \rangle$, from the numerical model simulation of Section 11.5 at (a) 66 hours, and (b) 78 hours. Panels (c) and (d) show the similarly averaged fields of $U_{magf} + U_{eagf}$ at these times. Panels (e), (f), (g) and (h) show a similar set of cross sections from the numerical model simulation of Section 10.3 at 84 hours and 90 hours, respectively. The quantity $\langle u \rangle$ is shaded with values indicated on the color bar. Red solid contours indicate positive values, blue dashed contours indicate negative values. Two additional magenta contours of $\langle u \rangle$ have values ± 1 m s^{-1}. These are included in panels (c), (d), (g) and (h) for reference. Contour interval for $U_{magf} + U_{eagf}$ as indicated on the color bar in m s^{-1} per hour. The yellow contour in each panel is the 0.5 m s^{-1} contour of similarly averaged vertical velocity.

eddy contribution, is shown in Fig. 11.12h. Although the pattern of total advective tendency differs in magnitude and sign from that of $V_{m\zeta}$ on account of the dominance of V_{mv}, it still shows an oscillatory structure.

The phenomenon that we have been describing is, in essence, a quasi-stationary, damped centrifugal wave that is adjusting the flow in the eyewall *towards a coherent supergradient regime in the mid to upper troposphere* as is

Centrifugal recoil effect

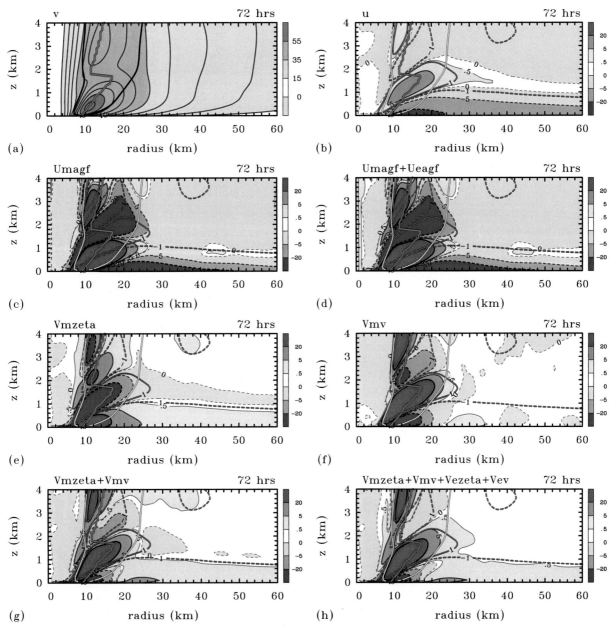

FIGURE 11.12 Radius-height plots of selected three-hour time- and azimuthally-averaged quantities centered at 72 hours in the numerical simulation. (a) tangential velocity, v, (b) radial velocity u, (b) mean advective tangential wind tendency, $V_{m\zeta} + V_{mv}$, (c) mean agradient force U_{magf}, (d) $U_{magf} + U_{eagf}$, (e) $V_{m\zeta}$, (f) V_{mv}, (g) $V_{m\zeta} + V_{mv}$, and (h) $V_{m\zeta} + V_{mv} + V_{e\zeta} + V_{ev}$. Shading as indicated on the side bar in m s^{-1} h^{-1}, with shaded regions enclosed by contours. Thick black contour ± 20 m s^{-1} h^{-1}. The yellow contour shows the 0.5 m s^{-1} vertical velocity, while the purple contours show ± 1 m s^{-1} radial velocity. The green curve shows the azimuthally averaged r_{vmax}. Solid contours positive, dashed contours negative.

evident, for example in Fig. 11.11c, d, g, h. This behavior indicates that the imposition of the gradient wind balance at the top of the boundary layer employed in some boundary layer calculations will not be valid in the strong ascent region that feeds the eyewall updraft (cf. Section 6.9.3). This feature appears to have been first described by Persing et al. (2013) and it has been documented since in observations by Stern et al. (2020).

11.7 Summary of the rotating-convection paradigm

A cartoon summarizing the rotating-convection paradigm for tropical cyclogenesis and intensification in a favorable environment is shown in Fig. 11.1. The depiction highlights the role of localized, rotating deep convection that grows in the cyclonic rotation-rich environment of the incipient

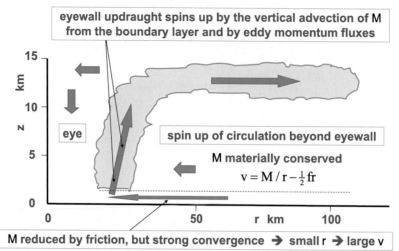

FIGURE 11.13   Schematic illustration of the revised view of system-scale spin-up in the rotating convection paradigm. See text for discussion. Gray gradient arrows denote the sense of transverse (secondary) circulation. Blue shading denotes cloudy areas. The size of the arrows is not proportional to the flow speed. To avoid clutter in this updated schematic, we have omitted to show the upper-level inflow layers and centrifugal recoil effect as explained in this chapter as well as the vertical arrows highlighting the surface moisture fluxes. Adapted from Smith and Montgomery (2016). Republished with permission of Wiley.

storm, structures that are intrinsically three dimensional. The convective updrafts greatly amplify the vertical vorticity locally by vortex-tube stretching, and the patches of enhanced cyclonic vorticity subsequently aggregate to form a central monolith of cyclonic vorticity, the aggregation being facilitated mainly by the axisymmetric-mean overturning circulation associated with the convection, itself. Convective instability is maintained principally by latent heat fluxes at the air-sea interface, which are enhanced as the near-surface wind speed increases. The stochastic nature of deep convection arising from the instability means that local asymmetric features of the developing vortex have limited predictability.

An azimuthally-averaged perspective of the paradigm is encapsulated in Fig. 10.10 and discussed in Section 10.6. This perspective is a modification to the classical axisymmetric spin up mechanism that goes back to Ooyama (1969) and which is discussed in Section 8.2. Recognition of the important role of eddy processes in tropical cyclone intensification calls for a further modification of the classical mechanism. A revised schematic is shown in Fig. 11.13. In addition to the nonlinear boundary layer spin-up mechanism discussed in Section 6.6, the revised schematic recognizes the role of eddy processes in the system-scale spin up of the vortex. In particular, eddy momentum transfer contributes to the spin up of the eyewall as well as an enhancement of the near-surface inflow by the suction effect of deep convection.

As far as we are aware, the rotating-convection paradigm is the only conceptual framework that has been advanced to understand the three-dimensional aspects of tropical cyclone behavior. So far, the paradigm has been applied mainly to understand the behavior in a quiescent environ-

ment, the exception being the extension by Smith et al. (2017) to interpret the behavior of Atlantic Hurricane Earl (2010), which intensified in an environment of moderate vertical wind shear. Three earlier paradigms that have been prominent in the literature, the so-called CISK paradigm proposed by Charney and Eliassen (1964), the classical paradigm of Ooyama and air-sea interaction (or so-called "WISHE") paradigm of Emanuel are all based on axisymmetric models. These paradigms are reviewed and compared by Montgomery and Smith (2014). The WISHE paradigm is discussed in the next chapter and is compared with the classical mechanism (Section 8.2) in Section 16.1.

One important question arises as to whether there are important differences between three-dimensional tropical cyclones and their purely axisymmetric counterparts. A study by Persing et al. (2013) compared tropical cyclone intensification in a three-dimensional model with that in an axisymmetric model that was formulated to be as close as possible to its three-dimensional counterpart. An important difference is the fact that in an axisymmetric configuration, deep convection is constrained to have the form of concentric rings. In general, such a constraint is unrealistic until a tropical cyclone reaches maturity and an eyewall cloud forms. Even then, it is only approximately realistic for the eyewall, itself. Persing et al. (2013) suggested that axisymmetric convection occurring in concentric rings may be overly efficient in generating buoyancy fluxes compared to three-dimensional convection in isolated thermals, leading to excessive condensation heating and an overly rapid spin-up. Important findings of the Persing et al. study are summarized in Section 16.2.

Chapter 12

Emanuel's intensification theories

This book would be incomplete without some substantive discussion of Kerry Emanuel's highly influential theories of tropical cyclone behavior, which have been a cornerstone of thinking about tropical cyclone structure and evolution for well over three decades. The wide acceptance of the theories is evidenced by their adoption as the basis for a theoretical articulation of tropical cyclone behavior in chapters of two modern textbooks in Dynamic Meteorology (Holton and Hakim, 2013, Section 9.7) and Cloud Dynamics (Houze, 2014, Chapter 10). The theories have become entrenched also to account for the intensification of tropical-cyclone-like lows that are occasionally observed over the Mediterranean Sea (Emanuel, 2005b; Carrió et al., 2017; Miglietta et al., 2020).

We should say at the outset that we do not consider these theories to be robust alternatives to the classical theory of Ooyama and its extensions embodied in the rotating-convection paradigm, despite certain limitations of the cumulus parameterization scheme in Ooyama's model discussed below. However, they have highlighted important aspects of tropical cyclone behavior.

The Emanuel theories can be split essentially into two groups:

- Those that address the intensification problem (e.g., Emanuel, 1989, 1995, 1997, 2012);
- Those that invoke a steady state (e.g., Emanuel, 1986, 1988; Tang and Emanuel, 2010; Emanuel and Rotunno, 2011).

In either group, the theories represent a novel view of the tropical cyclone as an Eady-like problem, arguably "the standard model" for representing the essential dynamics of middle-latitude synoptic weather systems. In the traditional Eady model, the flow dynamics are controlled essentially by a troposphere of uniform potential vorticity in conjunction with tendency equations for the perturbation potential temperature (effective potential vorticity anomalies) at the earth's surface and tropopause (e.g. Gill, 1982; Hoskins et al., 1985; Holton and Hakim, 2013).

In the tropical cyclone case, the potential vorticity in the Emanuel formulation is moist potential vorticity, which is based on the moist saturated equivalent potential temperature, θ_e^*, and is assumed to be everywhere zero. This assumption is equivalent to assuming that the surfaces of absolute angular momentum M and θ_e^* are congruent. The

upper- and lower-level boundary equations are equations for the pseudo-equivalent potential temperature, θ_e, constrained by radial advection at these levels and subsidence (including convective and mesoscale downdrafts) into the boundary layer and bulk-aerodynamic fluxes of moisture and heat at the sea surface (see below for more). In effect, the determination of the steady or intensifying tropical cyclone becomes[1] an inversion problem for zero moist potential vorticity subject to boundary conditions for θ_e at the ocean surface and near tropopause.

For either time-dependent or steady-state formulations, the flow dynamics are assumed to be strictly axisymmetric. That is, the flow is assumed to possess axial symmetry about the rotation axis and all asymmetric wave/eddy effects are neglected or are parameterized in terms of axisymmetric field variables. The early intensification theory is reviewed by Emanuel (2003) and the steady-state theory by Emanuel (2004) with interesting historical linkages.

The steady-state formulations provided a basis to develop a theory for the axisymmetric maximum tangential velocity that a tropical cyclone could achieve in a particular environment (the so-called maximum potential intensity, or "potential intensity" (PI), see Chapter 13). Over the years we have sought to appraise particular aspects of these theories and to relate them to other theories that have been proposed to explain the basic behavior of a tropical cyclone (Parks and Montgomery, 2001; Smith et al., 2008; Montgomery et al., 2009; Montgomery and Smith, 2014, 2017b; Montgomery et al., 2015; Montgomery and Smith, 2019; Smith et al., 2021).

In this chapter we review our understanding of these theories and how they relate to other theories developed in this textbook and elsewhere. In particular, we provide an appraisal of some of the underlying assumptions of the Emanuel formulations. This review will prove useful in the remaining chapters that re-examine aspects of the cyclone energetics and the steady-state assumption, together with the widely-accepted view that an indefinitely sustained hur-

---

1. However, recall the discussion near the end of Chapter 2 concerning the incorporation of explicit moist variables in the potential vorticity inversion problem. The analysis of Schubert et al. (2001) shows that in three-dimensional flows in which the air is not everywhere saturated, one no longer enjoys the material conservation of moist potential vorticity when the potential vorticity is based on θ_e. In this case the potential vorticity is not invertible.

Tropical Cyclones. https://doi.org/10.1016/B978-0-44-313449-4.00020-5

ricane is to be expected in a favorable environment on an f-plane.

We begin with reviewing the axisymmetric intensification theory formulated by Emanuel (1989), henceforth E89, and go on to examine the theories of Emanuel (1995), henceforth E95, Emanuel (1997), henceforth E97, and its later modification by Emanuel (2012), henceforth E12. These last two theories share many features with the companion steady-state problem. We have chosen to start with this time-dependent problem of vortex spin up because, for one thing, intensification chronologically precedes the attainment of maximum intensity, but there are issues with either approach.

12.1 The intensification theories

12.1.1 The Emanuel 1989 theory

The E89 study presented a new simplified axisymmetric balance model to investigate tropical cyclone intensification, similar in spirit to the three-layer model of Ooyama (1969), but with a vastly different formulation of moist convective processes. The model was developed to help explain, in simple terms, the finite-amplitude nature of the instability mechanism responsible for tropical cyclones in more sophisticated nonhydrostatic axisymmetric models, such as that of Rotunno and Emanuel (1987).

The study is monumental in a number of ways. First, the model is formulated in potential radius and pressure coordinates (R, p) in which angular momentum surfaces are vertical (see Exercise 5.23). The main thermodynamic variable is chosen to be the saturation moist entropy, s^*, which is the model's effective temperature variable. Second, the parameterization of deep convection is based on a simple model that invokes rapid convective adjustment along angular-momentum surfaces.

Unlike the formulation by Ooyama and that of many early models (e.g., Sundqvist, 1970a,b), the parameterization scheme does not depend on resolved-scale moisture convergence. Moreover, the representation of moist convection allows for low-precipitation-efficiency (LPE) clouds as well as for deep, high-precipitation-efficiency (HPE) clouds. Finally, the initial sounding is taken to be conditionally neutral to convection. The configuration of the model is illustrated in Fig. 12.1.

In contrast to Ooyama's model, the prediction equations for the tangential velocity are replaced by ones for the radius of a particular angular momentum surface at the top of the subcloud layer, r_b, and at the model top, r_t, a change that is necessitated by the use of potential radius coordinates. Moreover, the overturning circulation is characterized by a streamfunction, ψ, defined in the middle troposphere and a streamfunction, ψ_o, defined at the top of the subcloud layer. Aspects of the convection scheme are described in the caption to Fig. 12.1.

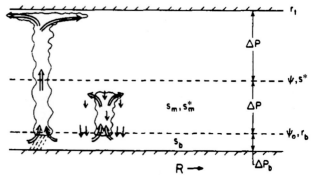

FIGURE 12.1 Vertical structure of Emanuel's (1989) numerical model. Streamfunction ψ is defined at the middle level and top of the subcloud layer, ψ_o and physical radius is calculated at the top of the model, r_t, and at the top of the subcloud layer, r_b. The temperature variable, s^*, represents a vertically averaged value, but is ascribed to the middle level. Entropy, s, is calculated in the subcloud layer, s_b, and within the lower-tropospheric layer, s_m. The subsidiary temperature variable, s_m^*, is used to predict the shallow cumulus activity. Deep clouds lift mass from the subcloud layer to the top layer while shallow clouds exchange entropy between the lower troposphere and the subcloud layer without producing a net mass flux or vertical momentum transfer. From Emanuel (1989). Republished with permission of the American Meteorological Society.

The streamfunction ψ is obtained from a diagnostic second-order ordinary differential equation, which, in essence, is an analog of the Eliassen equation derived in Section 5.9. As discussed there, this equation arises in a balance framework to determine the overturning circulation required to maintain thermal wind balance in the presence of processes that would otherwise destroy this balance. In the E89 model, these processes include the parameterized deep convective mass flux at the middle level, G, the radiative cooling, the subgrid-scale diffusion of both angular momentum and entropy, and forcing by the subcloud layer mass flux, ψ_o. The quantity G is determined as part of the convective parameterization. The equation for ψ_o is determined, in part, by assuming frictional torque balance in the boundary layer (see Section 6.4.6), and in part by the convective mass flux G, implying that the parameterized deep clouds do not entrain.[2] Nevertheless, the deep clouds do transfer azimuthal momentum vertically.

Emanuel's interpretation of the development process from a series of model simulations is summarized in the schematic in his Fig. 13, reproduced here as Fig. 12.2. He points out on p3454 that:

> *"The cyclone can only spin up if there is inflow above the boundary layer in this model",*

consistent with the classical spin-up mechanism (Section 8.2). However, he goes on to argue that:

---

2. This feature is in contrast to Ooyama's representation of deep convection, in which the entrainment is a key feature of the model.

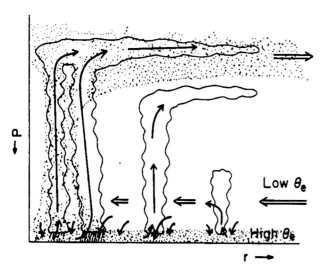

FIGURE 12.2 Schematic of purported air flow in a developing tropical cyclone according to E89. Dots show air with high θ_e. E89 writes: "In order for the cyclone to spin up, potentially cold air in the lower troposphere must flow inward. Were it to ascend, the vortex core would rapidly cool and the cyclone would decay. Instead, the potentially cold air descends within shallow clouds (sic), within precipitating downdrafts, and outside of clouds due to Ekman suction. These downdrafts reduce the entropy of the subcloud layer. Only if the surface fluxes are large enough to offset the drying effect of the downdrafts will the subcloud-layer entropy rise above ambient values, enabling the vortex core to become warmer than its environment. In the developing storm, individual air parcels flow inward in the lower troposphere, sink downward in downdrafts, receive entropy from the ocean, and then ascend in deep convective clouds". From Emanuel (1989). Republished with permission of the American Meteorological Society.

> "... the inflowing air is, in the model and in nature, potentially cold (i.e., it has low θ_e). If this air were to simply flow inward and upward, the vortex core would cool. What happens instead is that the lower-tropospheric air follows an indirect route to the tropopause. Before ascending, air above the subcloud layer first descends in convective downdrafts, is moistened by surface evaporation, and finally ascends in deep, HPE convective clouds. Only if fluxes from the ocean succeed in raising the entropy above that of the ambient subcloud layer can the vortex core warm and the cyclone amplify."

While these interpretations seem at first sight reasonable, they do raise a number of questions.

First, they point to a spin-up process more complex than the classical mechanism as the air entering the subcloud layer would be subject to a frictional torque opposing spin up. Indeed, if *all* the air ascending in deep convection does so from the subcloud layer (as indicated in Fig. 12.2), it is unclear (at least to us) how the tangential wind field in the convective eyewall region could increase with time. This is because, in physical coordinates portrayed in Fig. 12.2, there would be no mechanism other than radial diffusion to

draw the M surfaces into this region above the boundary layer to increase the tangential wind speed there. From the pattern of inflow portrayed in this figure, one could foresee the tendency for a sheet of cyclonic vorticity to form at the outer edge of the mean eyewall updraft in the lower troposphere as M-surfaces congregate there.

Curiously, in the schematic, the radial flow in the lower troposphere is portrayed as decreasing inward. This decrease would suggest significant vortex-scale subsidence into the subcloud layer, otherwise rings of inflowing air above the subcloud layer would approximately conserve their volume flux ru per unit azimuth. Then, decreasing radius r would imply an increase in the magnitude of the radial flow $|u|$ at a rate proportional to $1/r$. Of course, the calculations show that the vortex does spin up and that the M-surfaces do move inwards in the lower troposphere (i.e., r_b decreases with time). These facts raise question about the veracity of the schematic and the process of spin up in the model, especially in view of the fact that the HPE clouds do not entrain to produce vortex-scale inflow above the boundary layer.

12.1.2 The Emanuel 1995 theory

The E95 model is similar to the E89 model, but incorporates a different representation of deep convection based on the concept of sub-cloud layer quasi-equilibrium (Raymond, 1995). This concept holds that there is an approximate balance between surface enthalpy fluxes and the input of low-entropy air into the subcloud layer by convective downdrafts. This assumed balance determines an equilibrium cloud-base mass flux contribution to the total vertical velocity at the top of the subcloud layer associated with deep convection. Finally, the actual cloud-base mass flux is determined by a relaxation to the equilibrium mass flux on a prescribed time scale.

The model does not have a separate cloud model as in the E89 model. Rather, the cloud model is replaced by the simple assumption that the convective updraft and downdraft mass fluxes are proportional, with the proportionality related to a variable bulk precipitation efficiency. It appears that the effects of deep convection above the sub-cloud layer are represented explicitly as part of the total vertical velocity, constrained by the assumption that the surfaces of constant absolute angular momentum and constant saturation entropy are congruent (see the next section).

A primary objective of the E95 model was to demonstrate the efficacy of the sub-cloud layer quasi-equilibrium formulation in simulating important features of tropical cyclone evolution. Despite the claim that

> "... this type of convection scheme ... works very well in a simple model of tropical cyclones,"

the theory was soon superceded by the formulation described in the next subsection. Even so, the E95 model does form the basis for the Coupled Hurricane Intensity Prediction System (CHIPS) forecast tool that continues to be used for a variety of applications including both intensity forecasts and climate change assessments (Emanuel, 2017 and refs.).

12.1.3 The later theories

The main features of the flow configuration in the E97 and E12 theories have much in common with the E95 theory, except in the sub-cloud layer. These are sketched in Fig. 12.3 in radius-height coordinates (r, z). The many assumptions involved call for a range of important caveats to be noted. In an effort not to let these obscure the broad picture, we have taken the liberty to relegate many of the caveats to footnotes.

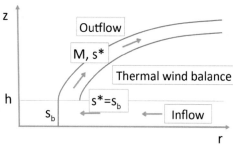

FIGURE 12.3 Geometric configuration of Emanuel's axisymmetric formulations (E95, E97, and E12) for an intensifying tropical cyclone in cylindrical polar coordinates (r, z). In the schematic, h denotes the depth of the layer influenced by friction, which is assumed constant, s_b is the specific entropy in the boundary layer, s^* is the saturation specific entropy above the boundary layer, and M is the absolute angular momentum. It is assumed that s_b is independent of height. The arrows indicate the secondary circulation with radial inflow in the friction layer and outflow above it. Adapted from Montgomery and Smith (2019).

As noted above, the flow is assumed to be axisymmetric and the tangential velocity is taken to be in thermal wind balance above a shallow frictional boundary layer and in approximate balance[3] within this layer. Air flows inwards in this boundary layer, which is assumed to have constant depth h, acquiring moisture from the sea surface as it does so. *As air parcels ascend out of this layer at inner radii, they are assumed to flow immediately upwards and radially outwards to exit the storm in the upper troposphere or lower stratosphere.*

A key assumption is that, along this ascent leg, the saturation specific entropy, s^*, and absolute angular momentum, M, are materially conserved.[4] This assumption is inherited

from the steady-state models, where it would seem to be perfectly reasonable since M and θ_e^* are approximately materially conserved as air parcels rise along streamlines in the eyewall updraft. However, the assumption is less easy to justify in a model for intensification because, in that case, the streamlines are not congruent with either the constant s^*-surfaces or the constant M-surfaces. Indeed, there must be a component of flow across the M surfaces towards low values of M in order for the vortex to spin up locally (Exercise 8.2). At best, the intensification theories would seem to be limited to stages of vortex evolution when the constant s^* and constant M-surfaces are nearly congruent, which axisymmetric numerical simulations show to be near the end of the intensification stage (Rotunno and Emanuel, 1987; Peng et al., 2018, 2019) - see e.g., Section 12.5. In other words, the theories are expected to apply only to a moderately well-developed vortex that is already thermodynamically saturated in its core.

A further assumption is that, as the constant M and s^* surfaces flare outwards with height, they do not fold over, forcing the flow to remain everywhere centrifugally- or inertially-stable.[5]

Time dependence in the theories appears explicitly only in the equation for the vertically-averaged specific entropy, s_b, within the frictional boundary layer, while *the tangential momentum equation,*[6] or alternatively the absolute angular momentum equation, *is used diagnostically to determine the vertically-averaged radial wind component, u_b, in the boundary layer.* With the assumption that $s^* = s_o$ at the top of the boundary layer, this equation provides one relationship between M and s^*. A second relationship between these quantities is obtained by integrating the thermal wind equation upwards along angular momentum surfaces. The derivations are a little involved and those relating to E12 are relegated to the Appendix in Section 12.8. It suffices here to explain the constraints leading to closure of both theories.

12.1.4 Specifics of the E97 theory

The derivation of the E97 theory is carried out effectively in angular momentum and pseudo-height coordinates (Sec-

---

3. In this class of models, there is torque balance in the azimuthal direction (Section 6.4.6) in M coordinates and the assumption of a well-mixed boundary layer in M and v to leading order implies tacitly that gradient wind balance prevails in the radial direction at this order of approximation.
4. These two quantities are defined in prior chapters with $s^* = c_{pd} \ln \theta_e^*$ and $M = rv + \frac{1}{2}fr^2$, where c_{pd} is the specific heat of dry air, θ_e^* is the

saturation equivalent potential temperature (assuming pseudo-adiabatic ascent in which all condensed water instantly precipitates), v is the tangential velocity component relative to an assumed invariant center of circulation, and f is the Coriolis parameter, assumed constant for simplicity.
5. The balance calculations of vortex evolution by Smith et al. (2018b) similar to those described in Chapter 8 would appear to question the validity of this assumption. In specific axisymmetric balance calculations in which the heating rate is specified along M surfaces centered around the radius of maximum tangential wind, which is qualitatively equivalent to assuming that s^* is constant along M surfaces, the M surfaces are found to turn over. Some hours after this overturning occurs, the balance calculations break down.
6. The assumption of balance precludes the use of the radial momentum equation for this purpose (as in the balance theory of Chapter 5), where the radial flow is obtained by solving the Eliassen equation.

tion 2.17), making the mathematics a little involved. However, this choice of coordinates simplifies the upward integration of the thermal wind equation along angular momentum surfaces. It is assumed that air parcels rising in the eyewall exit in the lower stratosphere in a region of approximately constant absolute temperature. This assumption provides the upper-level constraint between M and s^* referred to above. The lower-level constraint is obtained from the equation for s_b, *invoking the presumed crucial role of downdrafts outside of the main updraft region of the vortex.*

The overarching premise here is that convective downdrafts are necessary to maintain a strong negative radial gradient of moist equivalent potential temperature θ_e, or equivalently s_b, in the boundary layer.[7] This premise appears to stem from the conclusions of E89 reviewed in Section 12.1.1. In E97, the effects of these downdrafts are introduced through an *ad hoc* parameter β in the prognostic equation for s_b.

The main outcome of the E97 theory is an expression for the time rate of change of tangential wind at the radius of maximum tangential wind: Eq. (20) of E97. This equation equates the tangential wind tendency to the sum of three terms, two of which are negative definite. The third term is positive if the radial gradient of the parameter β is negative. Thus, the β-parameter turns out to be an essential element of intensification in the E97 theory, because the radial gradient of this β would have to be sufficiently negative to offset the other two terms for the vortex to spin up. In a subsequent review paper, Emanuel (2003) offered this simple, axisymmetric, moist-neutral balance model employing the β-formulation as the quintessential explanation for tropical cyclone intensification.

One weakness of the E97 theory is the lack of a clear physical basis for the *ad hoc* parameter β. A second weakness is the assumption of gradient wind balance in the boundary layer (Smith and Montgomery, 2008; Montgomery and Smith, 2014). In fact, the β-formulation was tested in two studies by Nguyen et al. (2008) and Montgomery et al. (2009) who, *inter alia*, conducted three-dimensional numerical experiments in which downdrafts were suppressed altogether (the case $\beta = 0$). It was found that, in comparison to experiments with downdrafts fully active, when downdrafts were eliminated the vortex intensified at an increased rate. These comparisons appear to refute the validity of the β-formulation, and therefore the E97 theory.

12.2 The air-sea interaction intensification theory, WISHE

The numerical simulations of Rotunno and Emanuel (1987), together with the time-dependent theories of E89, E95, and E97, led to the conceptualization of a so-called air-sea interaction instability/feedback mechanism. The proposed instability mechanism as discussed below is now commonly referred to as the Wind-Induced-Surface-Heat-Exchange (WISHE) feedback mechanism of tropical cyclone intensification. The acronym WISHE was first coined by Yano and Emanuel (1991) in the context of a model for the Madden-Julian oscillation. Following the review paper by Emanuel (2003), the physics of tropical cyclone intensification has been popularly explained by invoking the WISHE feedback mechanism, which is presented as a finite-amplitude instability of a moist vortex.

Emanuel (2003) specifically describes the WISHE intensification process as follows:

> *"Intensification proceeds through a feedback mechanism wherein increasing surface wind speeds produce increasing surface enthalpy flux ..., while the increased heat transfer leads to increasing storm winds."*

The increase of surface enthalpy flux with increasing wind speed is simply described by an empirical (aerodynamic) formula (Eq. (7.5) in Chapter 7), but the way in which the increased heat transfer leads to increasing storm winds is more involved and must invoke some chain of linkages for tropical cyclone behavior. A diagram attempting to illustrate the feedback loop is shown in Fig. 12.4, taken from a later study by Montgomery et al. (2009).[8] The mechanism outlined in the figure is described in the figure caption and is framed in part by the Rotunno and Emanuel (1987) study.

Two weaknesses in the proffered explanation of the WISHE feedback mechanism of Fig. 12.4 are:

(i) the invocation of the relationship between v_{max} and the radial gradient of θ_e at the top of the boundary layer; and

(ii) the silence about the explicit role of the overturning circulation in drawing surfaces of absolute angular momentum inwards in the lower troposphere.

The former relation between v_{max} and $\partial \theta_{eb}/\partial r$ is borrowed from the steady-state theory of Emanuel (1986) and is arguably a defensible relationship for slowly-intensifying, axisymmetric, storms under otherwise similar assumptions as invoked for the steady-state hurricane model (see Chapter 13). However, as discussed in Chapters 1, 10, and 11, intensification is usually a highly non-axisymmetric process

---

7. E97, p. 1019, noted the "... crucial presence of downdrafts by reducing the entropy tendency there (outside the radius of maximum tangential wind, our insertion) by a factor β."

8. The ensuing description by Emanuel (2003) is unclear because two of the key parameters α and β in Eq. (10) of the 2003 paper are undefined.

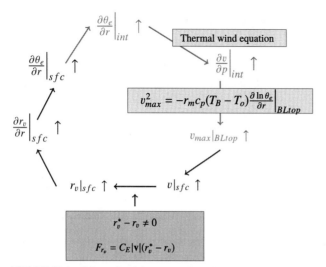

FIGURE 12.4 Schematic of the evaporation-wind intensification mechanism known as WISHE. The pertinent variables are defined in the text and are otherwise standard. Thermal wind balance refers to the axisymmetric thermal wind equation in pressure coordinates that relates the radial gradient of θ to the vertical shear of the mean tangential velocity v. The discussion assumes that for some reason the near-surface water vapor mixing ratio r_v is increased in the core region. This increase leads to an increase in the corresponding mean radial gradient and also an increase in the mean radial gradient of θ_e through the boundary layer. The secondary circulation is imagined to loft this increase in θ_e into the vortex interior thereby warming the core region of the vortex. By the thermal wind balance relation and its corollary (obtained upon applying a Maxwell thermodynamic relation and then integrating this equation along a surface of constant absolute angular momentum), this increased warmth is invoked to imply an increase in the mean tangential wind at the top of the vortex boundary layer. It is understood here that θ_e values at the boundary-layer top correspond to saturated values. Turbulent mixing processes in the boundary layer are assumed to communicate this wind speed increase to the surface, and thereby increase the potential for increasing the sea-to-air water vapor flux F_{r_v}. The saturation mixing ratio must increase approximately in step with the putative increase in near-surface r_v so as to maintain a thermodynamic disequilibrium. If this occurs then the near-surface specific humidity will increase, thereby leading to a further increase in the mean tangential wind, and so on. Adapted from Montgomery et al. (2009).

that can be quite rapid also and so we must consider this component of the explanation as tentative. Both of the foregoing concerns will be addressed in Chapter 16, where a revised explanation of the WISHE feedback mechanism is presented in the context of a family of minimal models of vortex intensification for the azimuthally-averaged dynamics.

Studies by Nguyen et al. (2008) and Montgomery et al. (2009) showed that the putative WISHE feedback mechanism for the maximum tangential velocity is not essential to explain tropical cyclone intensification and does not supercede the conventional spin up mechanism for the prototype intensification problem discussed in Chapters 5, 8, 10, and 11. This was done by carrying out three-dimensional numerical simulations similar to those in Chapter 10, but with the wind speed dependence of the surface enthalpy

fluxes capped at some modest tropical value. This procedure suppresses the possibility of the envisioned WISHE mechanism at wind speeds larger than the capping value.

When the wind speed cap is only 10 m s^{-1}, the rate of intensification and maximum horizontal wind speed are less than in the uncapped experiments, but the characteristics of the vortex evolution are qualitatively similar. Nevertheless, if the enthalpy fluxes are suppressed altogether, the vortex does not intensify, a finding that was obtained by Ooyama (1969) in the framework of a highly idealized, axisymmetric, three-layer model. This result is true whether or not the initial sounding has any degree of convective instability.

The role of the surface enthalpy fluxes is to replenish the low θ_e values that are transported into the boundary layer by frictional subsidence and by convective and mesoscale downdrafts (Chapter 9). These fluxes serve to replenish convective instability in the inner region of the circulation and maintain the deep convection in the presence of a developing warm core temperature field aloft. If the wind speed dependence of the enthalpy fluxes is capped, the boundary layer θ_e and convective instability will continue to be replenished, but on a longer time scale, provided, of course, the air near the sea surface remains unsaturated.

Montgomery et al. (2015) showed that there is much confusion in the literature about the WISHE mechanism and precisely what the feedback process is. To add to this confusion, Zhang and Emanuel (2016) appear to have redefined the "WISHE feedback process" as simply the formula relating the surface enthalpy flux to the surface wind speed and to the degree of thermodynamic disequilibrium near the surface. Just before their closing summary, they say

"Clearly in at least some real-world cases in which the tropical cyclone is influenced by external factors - for example, vertical shear of the environmental wind - the WISHE feedback is quantitatively important and may make the difference between growth and decay".

Unfortunately, nowhere in their paper do Zhang and Emanuel explain how the increased enthalpy fluxes lead to an increase in surface wind speed as, for example, in the cartoon in Fig. 12.4.

The fact that not capping the heat fluxes leads to stronger storms and more rapid intensification is not in question and Zhang and Emanuel's results do no more than confirm the calculations of Nguyen et al. (2008), Montgomery et al. (2009), and Montgomery et al. (2015). However, describing the formula relating surface heat fluxes to the surface wind speed as a "feedback" raises serious questions that remain unanswered.

In Section 2 of their paper, Zhang and Emanuel do discuss an equation system for a purported feedback process in which an air parcel rises under its buoyancy, b, in an *unstable* density stratified fluid. In this model, the parcel's motion is resisted by a quadratic drag proportional to the square of the parcel's vertical velocity, w. The dimensional form of the equation system is

$$\frac{dw}{dt} = b - \alpha|w|w, \qquad (12.1)$$

$$\frac{db}{dt} = -N^2 w, \qquad (12.2)$$

where α is some positive constant characterizing the strength of the drag and N^2 is the Brunt-Väisälä frequency squared, which is assumed to be a negative constant. Zhang and Emanuel offer an interpretation in which Eq. (12.1) says that "convection is driven by buoyancy" and Eq. (12.2) that "convective instability results from a feedback between vertical velocity and buoyancy". However, as pointed out by Kilroy et al. (2022), "convective instability" in this problem resides in the quantity N^2 and the fact that N^2 is taken to be negative.[9] The relevant energy equation for this system is

$$\frac{d}{dt}\left[\frac{1}{2}w^2 + \frac{1}{2}\frac{b^2}{N^2}\right] = -\alpha|w|w^2, \qquad (12.3)$$

where the first term on the left is the kinetic energy of the air parcel per unit mass and the second term is its available potential energy per unit mass. The term on the right of Eq. (12.3) is negative definite and the only "source term" for the total energy: there is no obvious (at least to us) source analogous to surface enthalpy fluxes in the tropical cyclone problem and the instability residing in the negative N^2 would seem more analogous to a situation where the instability is a direct result of the convective instability of the environment rather than some mechanism analogous to WISHE. While the intention of this section of Zhang and Emanuel's paper would seem to legitimize the use of the word "feedback" in the WISHE context, as noted above, these authors do not articulate the actual WISHE feedback to explain how the increase in surface enthalpy fluxes *feeds back* to increase the near-surface tangential winds.

12.3 The E12 theory

As a motivation for a revised intensification theory, E12 argued that the tenet of the E97 theory, wherein the air parcels rising in the eyewall exit in the lower stratosphere in a region of approximately constant absolute temperature, was questionable for the time-dependent problem. A further motivation was evidently to dispense with the assumed dependence on the *ad hoc* and radially variable parameter β, which, as explained above was introduced by E97 as a way of parameterizing the effects of mesoscale and convective-scale downdrafts into the boundary layer of an intensifying tropical cyclone.

The development of the E12 model shifted focus from the earlier-presumed crucial role of convective downdrafts into the boundary layer to that of small-scale Kelvin-Helmholtz instability and turbulent mixing in the storm's upper tropospheric outflow layer. Emanuel postulated that this mixing plays a crucial role in determining the spatial distribution of outflow temperature and, in particular, the vertical stratification there.

The new theory rapidly gained prominence in the scientific literature and it appears to underpin recent predictions of a new era of more frequent rapidly intensifying tropical cyclones on account of global warming (Hirji and Aldhous, 2020; Emanuel, 2017). However, the idea that turbulent mixing in the storm's outflow layer controls spin up of the system-scale circulation seemed quite implausible to us from a fluid dynamics standpoint. In particular, it is unclear physically how such mixing could facilitate the inward movement of the M-surfaces in the lower troposphere as would be necessary for a storm to intensify. Moreover, the study by Montgomery et al. (2019) investigating the impact of such mixing in a three-dimensional model simulation showed the mixing to be inconsequential to the spin-up process. Since the three-dimensional formulation is arguably the proper benchmark for understanding the dynamics and thermodynamics of vortex spin up, an inquiry into the basic physics of the E12 model and how it relates to other paradigms for vortex spin-up, including the rotating-convection paradigm discussed in the last two chapters, is clearly necessary.

In the next subsection we summarize the main elements of the E12 model in ordinary cylindrical-polar coordinates and we articulate our understanding of how spin up comes about in the model. These findings are compared against some key attributes of the spin up process deduced from solutions to the prototype genesis and intensification experiment presented in Chapter 10. The following summary and analysis of the E12 model is based on the study by Montgomery and Smith (2019), with supporting details provided in Appendix A of Montgomery et al. (2019).

12.3.1 Specifics of the E12 theory

12.3.1.1 A tendency equation for the maximum gradient wind

Fig. 12.5 shows the flow configuration as in Fig. 12.3, but with two additions specific to the E12 model. As noted in Section 12.1, the master prognostic equation is one for the vertical mean moist entropy, s_b, in the boundary layer, (Eq. (12.7) in the Appendix, Section 12.8), which is assumed to have constant depth[10] h. In turn, M and its time derivative are constrained by the following assumptions above the boundary layer:

---

9. This conclusion is not so apparent in the non-dimensionalization used by Zhang and Emanuel.

10. Curiously, Emanuel chose a value of 5 km for h! See page 993 in his paper.

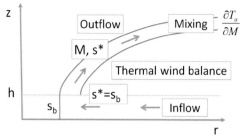

FIGURE 12.5 Updated configuration of the E12 axisymmetric formulation for an intensifying tropical cyclone. The only change from Figure 12.3 is the addition of mixing by shear-stratified turbulence in the upper-tropospheric outflow layer of a hypothetical developing cyclone. This mixing is assumed to determine the thermal stratification of the outflow ($\partial T_o/\partial M$, where T_o is the outflow temperature). From Montgomery and Smith (2019). Republished with permission of the American Meteorological Society.

(1) the flow there is in hydrostatic and gradient wind balance and therefore in thermal wind balance;

(2) the M- and s^*-surfaces are congruent;

(3) $s^* = s_b$ at $z = h$ and $M = M_b$ at $z = h$, where M_b is the vertical mean of M in the boundary layer; and

(4) a closure assumption in the upper-tropospheric outflow relating the partial derivative of outflow temperature T_o with respect to M, i.e., $\partial T_o/\partial M$, to the derivative of s^* with respect to M, i.e., $\partial s^*/\partial M$.

The closure assumption in (4), in essence a parameterization for $\partial T_o/\partial M$, is based on the premise that "the thermal stratification of the outflow ($\partial T_o/\partial M$, our insertion) is set by small-scale turbulence that limits the Richardson number" to a critical value (E12, p988).

The frictional boundary layer is assumed to be close to gradient wind balance so that mean radial inflow therein, which in physical coordinates is required to predict the time evolution of s_b, is determined by integrating vertically the tendency equation for M, across the layer. It is assumed that both horizontal velocity components, and therefore M, are essentially uniform through the depth of the layer. This formulation is represented by Eq. (A13) in Peng et al. (2018), which gives the inflow as a function of the temporal and radial derivatives of M_b together with the surface torque.

Because the equation for M_b is used to determine the radial flow, it is no longer available to determine the tendency of M_b in the friction layer. Rather, the tendency of M_b is determined by the tendency of s_b in conjunction with the assumed constraint that the M- and s^*-surfaces are congruent. *In essence, the M-surfaces are dragged in with the s_b surfaces on account of this assumed congruence.* It follows that the dynamics are slaved to the thermodynamics in the inner-core region where there is ascent out of the friction layer. It is unclear what is assumed at larger radii where there is subsidence into the friction layer and consequently the evolution of the outer part of the vortex is unknown.

The main mathematical outcome of the theory (Appendix, Section 12.8) is an equation for the intensification rate, $\partial v_m/\partial \tau$, of the maximum gradient wind, denoted here as v_m, given by

$$\frac{\partial v_m}{\partial \tau} = \underbrace{\frac{C_D}{2h}\frac{Ri_c}{r_t^2}M^2}_{\text{turbulent mixing term}} - \frac{C_K}{2h}v_m^2, \qquad (12.4)$$

and, with some further approximations, including the assumption that $v_m = 0$ at $\tau = 0$, an analytical solution for v_m of the form

$$v_m(\tau) = V_{max} \tanh\left(\frac{C_K v_{max}}{2h}\tau\right), \qquad (12.5)$$

where τ is the time,[11] C_D and C_K are the surface drag and enthalpy coefficients, both assumed constant with wind speed for simplicity, Ri_c is the critical Richardson number that determines the onset of turbulent mixing in the upper-tropospheric outflow, r_t is the radius where the gradient Richardson number first becomes critical, and V_{max} is a revised potential intensity determined by Eq. (18)[12] in E12 and summarized in some detail in Section 13.3.

12.3.1.2 Demystifying the E12 theory

It follows from Eq. (12.4) that spin up occurs only if the first term on the right-hand-side, which contains the effect of the parameterized turbulent mixing, is sufficiently large. This is because the second term on the right is negative definite. Invoking the classical paradigm for spin up, which refers to physical coordinates, spin up in the model can occur in the layer of friction only if the M_b surfaces therein move inwards at a sufficient rate that the radial advection of M_b exceeds the azimuthal torque per unit depth.

Since the radial flow above the friction layer is radially outwards, spin up of the flow above the friction layer has to occur by the vertical advection of M_b from the friction layer. It follows that, in physical coordinates, spin up of the maximum tangential wind in the E12 model must occur within, or at the top of, the friction layer.

While the mathematical constraints leading to vortex spin up in the E12 model as outlined above are reasonably clear, the physical processes they represent are not. Despite the fact that the mixing parameterization relating $\partial s^*/\partial M$

---

11. The time variable τ defined in M and p coordinates, wherein partial derivatives with respect to τ hold M and p constant.

12. This equation is a formula for V_{max} in terms of C_D, C_K, T_b, T_t, s_o, s_e^*, where T_b is the absolute temperature evaluated at the top of the inflow layer, T_t is equal to the outflow temperature T_o corresponding to the M-surface passing through the radius of maximum winds and is assumed equal to the initial tropopause temperature, s_o "is the environmental (constant) saturation entropy of air at sea surface temperature and ambient surface pressure", and s_e^* "is the value of s^* in the undisturbed environment" (E12, p991-2).

to $\partial T_o/\partial M$ in the upper-tropospheric outflow layer is a crucial element of the theory, without which the vortex will not spin up (Appendix, Section 12.8), *the physics of spin up brought about by this mixing are mysterious. In particular, it is unclear how spin up in the E12 model relates to the classical paradigm for tropical cyclone spin up* (Section 8.2).

The parameterization of turbulent mixing in the E12 model introduces the parameter r_t in the tendency equation for the maximum gradient wind. This radius is unknown *a priori*. The positive term in this tendency equation predicted by the theory is inversely proportional to r_t^2, but r_t is not determined by the theory: it must be prescribed. In this sense alone, *the theory is not a closed theory for intensification.*

Based on the foregoing analysis, the premise that spin up in the E12 model is controlled by mixing in the upper tropospheric outflow layer may be a red herring. As shown in the Appendix, Section 12.8, if one stops short of applying the mixing parameterization in the E12 theory, Eq. (12.4) would take the form

$$\frac{\partial v_m}{\partial \tau} = \frac{C_D v_m^2 M}{2h(T_b - T_t)}\frac{\partial T_o}{\partial M} - \frac{C_K}{2h}v_m^2, \qquad (12.6)$$

The symbols T_b and T_t are defined in the prior footnote. Comparing this equation with Eq. (12.4), it would seem that *any functional form for $\partial T_o/\partial M$ that leads to the M-surfaces being dragged inwards in the friction layer at a sufficient rate that the radial advection of M exceeds the azimuthal torque per unit depth would yield a model for vortex spin up.* This is true, whether or not the functional form has any physical meaning!

The foregoing deductions help explain why the tests of the E12 theory by Montgomery et al. (2019) using idealized, three-dimensional numerical model experiments showed that vertical mixing in the upper tropospheric outflow layer had no appreciable effect on vortex intensification. The deductions indicate further that any attempt to find a physical interpretation of spin up in the friction layer brought about by vertical mixing in the upper troposphere may be a fruitless exercise.

Where then does the main limitation of the E12 theory arise? In our view, the most serious issue is the assumption that the M- and s^*-surfaces are everywhere congruent, even during the intensification stage. *This assumption artificially forces the upper level mixing to determine the relationship between the values of M and s^*, where the flow ascends out of the inflow layer.* Several other issues we have with the E12 theory are discussed in Montgomery and Smith (2019). Since the assumed congruence of the M- and s^*-surfaces is a common feature of *all* the Emanuel theories, we examine this assumption more closely in the next section.

One curious feature of the analytical solution (12.5) is that the crucial effects of turbulent mixing represented by the first term on the right-hand-side of Eq. (12.4), without which intensification cannot occur, have totally disappeared

as a result of the further approximations made to derive Eq. (12.5) from Eq. (12.4). In particular the $1/r_t^2$ term has disappeared. This analytical solution raises additional concerns. Although, as noted above, the theory is only strictly valid when the vortex has attained some degree of maturity when the M and s^* surfaces have become approximately congruent at leading order, it is unclear how the vortex model can intensify from a quiescent state as there would be no surface enthalpy flux initially and no turbulent mixing to "initiate" intensification at this time (the first term in Eq. (12.4)). Indeed, with this initial condition, both terms on the right-hand-side of Eq. (12.4) would be zero,[13] implying that there would be no intensification from $\tau = 0$, in contradiction to the analytical solution (12.5), which intensifies from zero initial winds.

12.4   A boundary layer explanation for spin up

In a recent monograph reviewing the history of tropical cyclone research during the preceding 100 years, (Emanuel, 2018, p15, Section d. entitled Intensification physics) presents a new account of how storms intensify. Referring to the simple models of E97 and E12, Emanuel notes that,

> *"in either case, the boundary layer flow near the radius of maximum winds is strongly frontogenetical, with convergence of the Ekman boundary layer flow guaranteed by the large radially inward increase of inertial stability as the vorticity rapidly increases inward. This strongly converging flow increases the radial gradient of the boundary layer moist entropy, which further intensifies the vortex, leading to stronger Ekman convergence, and so on".*

Surprisingly, the purported role of upper-level turbulent mixing is no longer at the forefront of this explanation, despite the fact that, as explained in Section 12.3.1, the vortex in the E12 theory requires the mixing in order to intensify (see Eq. (12.4)). Not only is the concept of inertial stability problematic in the boundary layer (Smith and Montgomery, 2020), there is again no explanation as to how the increase in the radial gradient of the boundary layer moist entropy further intensifies the vortex (cf. Section 12.3.1.2).

12.5   Congruence of M and θ_e^* surfaces during spin up?

As noted above, one of the key tenets of the E95, E97, and E12 intensification models is that the M- and s^*-surfaces

---

13. In the absence of an initial vortex, the radius r_t where the gradient Richardson number first becomes critical is infinity, i.e., $r_t = \infty$, and hence the first term on the right-hand-side of Eq. (12.4) vanishes.

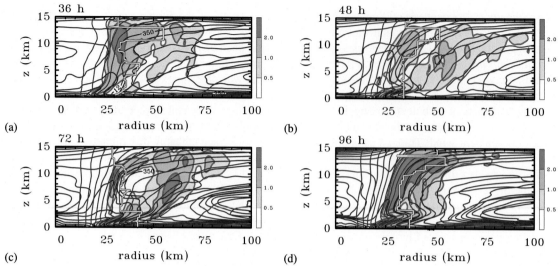

FIGURE 12.6 Radius-height cross sections of the three hour time-averaged and azimuthally-averaged surfaces of θ_e (red curves) and M (blue curves) at (a) 36 h, (b) 48 h, (c) 72 h, and (d) 96 h from the three-dimensional, warm-rain intensification experiment discussed in Chapter 11 (see Fig. 11.5 and related discussion). The time averaging is centered on the time shown. Contour intervals are: θ_e (red) contours every 2 K. Absolute angular momentum: (blue) contours every 2×10^5 m^2 s^{-1}. The solid yellow contour shows the contour of vertical velocity equal to 0.5 m s^{-1} and helps delineate the primary, azimuthally-averaged, cloud updraft. The green region depicts mean vertical updraft contours, ranging from 0 to more than 2 ms^{-1} as indicated by the color bar. Courtesy of G. Kilroy.

are congruent in and around the main convective updraft during the intensification process. Given the relationship between s^* and saturation equivalent potential temperature, i.e., $s^* = c_{pd} \ln \theta_e^*$, this is equivalent to assuming the congruency between M and θ_e^* surfaces during spin up. As noted in Section 12.1, the congruence assumption has its origins in the steady-state model of Emanuel (1986) and is based on the idea that deep convection leads to a state of moist neutrality in which M and θ_e^* are materially conserved as the moist air that is converged in the boundary layer rises immediately out of this layer and ascends to the upper troposphere. See Fig. 12.6.

Peng et al. (2018) sought to evaluate the assumptions of the E12 model, including the assumption that the M- and θ_e^* surfaces are congruent, by simulating the intensification of a tropical cyclone in an axisymmetric version of the CM1 model. Starting with an initial vortex with a maximum tangential wind speed of 22.5 m s^{-1} in an initially saturated environment, they found that this assumption was reasonable after a certain point in the intensification when the maximum tangential wind speed exceeded about 60 m s^{-1}, well into the period of rapid intensification. However, echoing a conclusion of Montgomery and Smith (2014, p60), since the proper benchmark to evaluate any theory of tropical cyclones is a simulation using a three-dimensional model, it is of interest to carry out a similar assessment of the congruence issue in the context of the rotating-convection paradigm developed in this textbook, using the idealized three-dimensional numerical simulation from Chapter 11.

The experiment to be analyzed here begins with a moderate strength initial vortex in a more realistic sub-saturated

atmosphere based on the composite moist Atlantic sounding of Dunion (2011). The initial vortex is axisymmetric and in thermal wind balance. The initial tangential wind speed has a maximum of 15 m s^{-1} at the surface and at a radius of 100 km (see Chapter 11 for other details).

Since the E12 theory and its predecessor appears to be the only one that has analytical solutions, for reasons given above it is of interest to examine at what stage, if any, the azimuthally-averaged M-surfaces and θ_e^* surfaces become appreciably congruent in a three-dimensional formulation that explicitly represents rotating deep convective clouds. Although this would seem to be a reasonable task to address, subtleties arise in such an endeavor. The air ascending in the developing eyewall of a simulated *axisymmetric vortex* is necessarily saturated and to a good first approximation, neglecting the effects of subgrid scale diffusion, θ_e^* is materially conserved.[14]

In contrast, in a simulated *three-dimensional vortex*, the air in the developing eyewall is not necessarily saturated at all azimuthal locations at a particular radius. For this reason, an azimuthal average of θ_e^* seems to be an inappropriate quantity to examine and this average may no longer be materially conserved to as good an approximation as in an axisymmetric configuration. While the latter may be true of θ_e, itself, we believe θ_e to be the rugged and appropriate invariant for the purpose of assessing the congruity question in three-dimensional hurricane flows. On the basis of these considerations, we have chosen here to examine the degree

---

14. Of course, if freezing were allowed to occur, θ_e^* would not be approximately materially conserved.

of congruence between the azimuthally-averaged M and θ_e surfaces.

Fig. 12.6 shows radius-height cross sections of the three hour time-averaged and azimuthally-averaged equivalent potential temperature θ_e and M-surfaces at 36 h, 48 h, 72 h, and 96 h. It is clear that, even at 36 h and 48 h, the latter being well into the period of vortex intensification, the θ_e and M-surfaces are nowhere congruent in the developing eyewall region. This non-congruity is especially evident in the region interior to the maximum mean updraft where the θ_e surfaces in the low- to mid-troposphere fold inward into the vortex eye while the M surfaces slant upwards and outwards.[15] Even at 72 h, which is well into the development of the mature stage (Fig. 12.5c), there remains a tendency for the two surfaces to cross each other except, perhaps, below a height of 3-4 km near the mean primary updraft, where the air exits the boundary layer and begins to ascend in the eyewall cloud.

Evidently, the assumption that the M- and θ_e surfaces are congruent does not hold to a plausible degree during intensification in the three-dimensional configuration.[16] In particular, the timing at which rapid intensification begins (just prior to 48 h, see Chapter 11 for details) bears no obvious relationship to this congruence condition. These findings do not support one of the cornerstone assumptions on which the time-dependent theory of E12 (or E95, E97) is based.

12.6 Appraisal of the Emanuel intensification theories

The many issues articulated above concerning the various Emanuel theories for tropical cyclone intensification, often referred to as WISHE theories, are reasons why *we* do not view them as robust alternatives to the classical theory of Ooyama and its extensions embodied in the rotating-convection paradigm. This is notwithstanding the fact that these theories are the only ones that allow for analytical treatment. However, we recognize that these theories have had a tremendous influence on the field and acknowledge that they have been a major stimulus for our own research.

Referring back to the animations of the near-surface θ_e fields discussed in Sections 10.4 and 10.11, it is hard to imagine that the dynamics of the intensification is slaved to the evolution of the azimuthally-averaged thermodynamic fields in the boundary layer as in the E95, E97, and E12 theories. This interpretation is not supported by either the warm-rain experiment, or that which includes ice, because in both cases, the RI phase has already commenced well before the near-surface θ_e field shows much organization. In these simulations, the patch of elevated near-surface θ_e is barely 5 km in diameter when RI begins.

12.7 Relevance to hurricanes in a warmer world?

Our choice to examine more deeply the basis of the E12 model enables us to make an informed appraisal of some recent predictions about the proffered effect of global warming on the rapid intensification of tropical cyclones.[17] The position is that on account of a slightly warmer climate,[18] we have entered a new era in which tropical cyclones are more prone to rapid intensification, making populated coastal communities in tropical regions of the world more vulnerable to catastrophic destruction and loss of life from the increased risk of rapid intensification near land.

Upon examining the source information, this prediction appears to stem from the E12 intensification model, which, as noted by Emanuel (2017) (and readily inferred from Eq. (12.5) above), predicts an intensification rate $\partial v_m / \partial \tau$, that is proportional to the square of the so-called potential intensity (PI) . As discussed in the next chapter, PI refers to the maximum gradient wind that is achievable given several environmental factors under a set of assumptions discussed in more detail later. The implication of the E12 theory is that, since the square of the potential intensity is a more sensitive metric than the potential intensity, itself, the increase in rapid intensification rate would be a more detectable cause of global warming than the potential intensity used previously in climate change assessments.

In our view, because of the generally limited spatio-temporal sampling of rapidly intensifying tropical cyclones by reconnaissance aircraft, it is questionable whether the historical tropical cyclone data base is currently adequate to directly test this prediction. Moreover, on the basis of the questions and inconsistencies raised in the foregoing subsections, we would argue that inferences drawn from the E12 theory to suggest that real-world tropical cyclones will intensify more rapidly in association with (natural or anthropogenic) global warming are questionable and should be viewed with skepticism.

---

15. The three hour time averaging serves to minimize the transient waviness one would expect in association with deep convection and the emission of inertia gravity, vortex Rossby waves and possible symmetric instabilities in and above the eyewall updraft. The state of non-congruity depicted herein is thus not an artifact of limited time sampling around a hypothetical slantwise moist-neutral configuration.

16. This finding transcends flow dimensionality. The axisymmetric numerical study by Rotunno and Emanuel (1987) showed similar results (their Figs. 7 and 11) in that congruity was well established only after the vortex attained a quasi-steady intensity.

---

17. Following an apparent heightened number of rapidly intensifying tropical cyclones in the North Atlantic basin in the 2020 hurricane season as reported in Hirji and Aldhous (2020).

18. as defined by the record of globally-averaged surface temperature anomalies (e.g., Lindzen and Christy, 2020).

12.8 Appendix: derivation of $\partial v_m / \partial \tau$ in the E12 theory, Eq. (12.4)

Here we detail the key approximations and assumptions underpinning the tendency equation in the intensification theory of E12 and stated in Eq. (12.4) in the main text. In particular, we derive the tendency equation for the gradient wind from the stated assumptions.[19] The derivation herein reveals the tangential force that is responsible for amplifying the maximum gradient wind at the top of the boundary layer in the new axisymmetric theory. Amidst the derivation, some additional questions arise about the theory and its proffered analytical solution.

The starting point of the revised intensification theory is the axisymmetric equation for the depth-averaged boundary layer moist entropy s_b in absolute angular momentum ($M = rv + \frac{1}{2}fr^2$) and pressure coordinates (Eq. (12) in E12):

$$h\frac{\partial s_b}{\partial \tau} - C_D r|\mathbf{v}|v\frac{\partial s_b}{\partial M} = C_K|\mathbf{v}|(s_o^* - s_b) + C_D\frac{|\mathbf{v}|^3}{T_s}, \quad (12.7)$$

where C_D and C_K are the surface exchange coefficients for momentum and enthalpy, $\mathbf{v}$ is the Reynolds-averaged velocity vector averaged across the boundary layer, s_o^* is the saturation moist entropy (defined below) at the sea surface temperature, T_s, and h is the boundary layer depth, assumed to be constant. Here τ is the dimensional time variable wherein partial derivatives with respect to τ hold M and p constant. The boundary layer depth $h = \Delta p_b/\rho g$, where Δp_b is the boundary layer depth (in pressure units), ρ is the density averaged over the boundary layer and g is the gravitational acceleration.

The two terms on the left hand side of the entropy equation are the local time tendency of s_b and the depth-averaged radial advection of s_b in M-coordinates. (The latter uses the fact that, in M-coordinates, the radial velocity u_M is defined by $u_M = DM/Dt$ and $hDM/Dt = -C_D r|\mathbf{v}|v$ from the angular momentum equation integrated over the boundary layer.) The right hand side of the entropy equation consists, respectively, of the bulk-aerodynamic parameterization of the vertical transfer of moist entropy between the underlying ocean and the moist air at anemometer level and a bulk representation of dissipative heating rate divided by surface temperature. By definition, the eddy radial and vertical entropy and angular momentum fluxes are zero in the axisymmetric theory.

---

19. Although we accept the basic assumptions of the E12 theory for the purposes of gaining an understanding of the theory, this should not be interpreted as our endorsement of these assumptions when applied to real or simulated storms in a three-dimensional configuration. In particular, we would step back from endorsing the assumption of axisymmetric moist neutral flow on the grounds that intensification is intrinsically non-axisymmetric and that M and θ_e surfaces only approach a state of congruence *after* the intensification process is already well under way (see Section 12.5 for a demonstration of this finding).

The saturation moist entropy is defined by

$$s^* = c_{pd}\ln T - R_d\ln p + \frac{L_v r_v^*}{T} \quad (12.8)$$

where c_{pd} is the specific heat of dry air at constant pressure, T is the temperature, R_d is the gas constant of dry air, p is the pressure, r_v^* is the saturation water vapor mixing ratio, and L_v is the latent heat of condensation.

The quantity s_o^* denotes the saturation entropy at the sea surface temperature

$$s_o^* = c_{pd}\ln T_s - R_d\ln p_o + \frac{L_v r_{v0}^*}{T_s} \quad (12.9)$$

where p_o is the surface pressure and r_{v0}^* is the surface saturated vapor mixing ratio. It will prove useful later to note that if we neglect the pressure dependence of s_o^* and use the environment values to evaluate this quantity, then the M derivative of s_o^* will vanish (see below for more).

As already described in the main text, the three principal approximations that are invoked to derive the tendency equation in the new theory are the following:

(a) neglect the pressure dependence of s_o^*;
(b) neglect of dissipative heating;
(c) approximate $|\mathbf{v}|$ and v by v_g, where v_g is the gradient wind.

In what follows v will refer to the gradient wind.

In the eyewall region, moist air rises out of the boundary layer, rapidly condenses, and ascends in cloud along M surfaces. Since, as discussed in Section 12.1, the M surfaces are assumed to be congruent to the saturated entropy surfaces, the rising moist air parcels are convectively neutral along these slantwise surfaces. Now, along an M surface, the saturated entropy above the boundary layer is assumed equal to the originating boundary layer entropy, $s^* = s_b$. Then it may be shown that

$$v^2 = -(T_b - T_o)M\frac{\partial s^*}{\partial M}, \quad (12.10)$$

where T_b is the boundary layer temperature and T_o is the outflow temperature. This equation is a form of the thermal wind equation (see Section 13.6.2.1 for derivation). The variation of the outflow temperature with M is then assumed to be controlled by the action of shear-stratified turbulence wherein the gradient Richardson number is bound to a near critical value

$$\frac{\partial T_o}{\partial M} = -\frac{Ri_c}{r_t^2}\left(\frac{\partial s^*}{\partial M}\right)^{-1} \quad (12.11)$$

which is a turbulence closure assumption on the upper-level outflow. In this equation, r_t denotes the radius where the gradient Richardson number first becomes critical in the outflow layer.

The moist entropy tendency equation at the top of the boundary layer is then

$$h\frac{\partial s^*}{\partial \tau} - C_D vM\frac{\partial s^*}{\partial M} = C_K v(s_o^* - s^*) \qquad (12.12)$$

Now $\partial/\partial M$ of Eq. (12.12) gives

$$h\frac{\partial}{\partial \tau}\frac{\partial s^*}{\partial M} - C_D\frac{\partial}{\partial M}\left(vM\frac{\partial s^*}{\partial M}\right) = C_K\frac{\partial}{\partial M}[v(s_o^* - s^*)] \qquad (12.13)$$

but Eq. (12.10) gives

$$\frac{\partial s^*}{\partial M} = \frac{-v^2}{(T_b - T_o)}\frac{1}{M}$$

whereupon, the first term on the left hand side of Eq. (12.13) becomes[20]

$$-\frac{h}{M}\frac{\partial}{\partial \tau}\left(\frac{v^2}{T_b - T_o}\right) \qquad (12.14)$$

and the second term on the left hand side becomes

$$-C_D\frac{\partial}{\partial M}\left(vM\frac{\partial s^*}{\partial M}\right)$$
$$= C_D\frac{\partial}{\partial M}\left(\frac{v^3}{T_b - T_o}\right)$$
$$= \frac{3C_D v^2}{T_b - T_o}\frac{\partial v}{\partial M} + C_D v^3\frac{\partial}{\partial M}\left(\frac{1}{T_b - T_o}\right). \qquad (12.15)$$

E12 makes the additional approximation

$$\frac{\partial}{\partial M}\left(\frac{1}{T_b - T_o}\right) \approx \frac{1}{(T_b - T_o)^2}\frac{\partial T_o}{\partial M}$$

This approximation follows from the assumption that T_b, the temperature after integrating across the boundary layer, is assumed to be a constant in radius (and hence M).

At this point the turbulence closure assumption, Eq. (12.11), is applied and $\partial s^*/\partial M$ is substituted using Eq. (12.10) so that

$$\frac{\partial}{\partial M}\left(\frac{1}{T_b - T_o}\right) = \frac{-1}{(T_b - T_o)^2}\frac{\text{Ri}_c}{r_t^2}\left(\frac{\partial s^*}{\partial M}\right)^{-1}$$
$$= \frac{-1}{(T_b - T_o)^2}\frac{\text{Ri}_c}{r_t^2} \times -\frac{M(T_b - T_o)}{v^2}$$
$$= \frac{M}{v^2(T_b - T_o)}\frac{\text{Ri}_c}{r_t^2} \qquad (12.16)$$

The term on the right of Eq. (12.13) becomes

$$C_K\frac{\partial}{\partial M}[v(s_o^* - s^*)]$$
$$= C_K(s_o^* - s^*)\frac{\partial v}{\partial M} + C_K v\frac{\partial}{\partial M}(s_o^* - s^*) \qquad (12.17)$$

---

20. Note that M is an independent variable so that $\partial M/\partial \tau = 0$.

and E12 writes the second term on the right hand side of this equation as

$$C_K v\frac{\partial}{\partial M}(s_o^* - s^*) \approx -C_K v\frac{\partial s^*}{\partial M} = \frac{C_K v^3}{M(T_b - T_o)}, \qquad (12.18)$$

using Eq. (12.10) and neglecting the pressure dependence of s_o^* (as foreshadowed above).

Collecting the terms in Eq. (12.13) together now gives (using (12.14)-(12.18)),

$$-\frac{h}{M}\frac{\partial}{\partial \tau}\left(\frac{v^2}{T_b - T_o}\right) = -\frac{3C_D v^2}{T_b - T_o}\frac{\partial v}{\partial M}$$
$$- \frac{C_D vM}{(T_b - T_o)}\frac{\text{Ri}_c}{r_t^2}$$
$$+ C_K(s_o^* - s^*)\frac{\partial v}{\partial M}$$
$$+ \frac{C_K v^3}{M(T_b - T_o)} \qquad (12.19)$$

or, cleaning up,

$$\frac{h}{M}\frac{\partial}{\partial \tau}\left(\frac{v^2}{T_b - T_o}\right) = \frac{\partial v}{\partial M}\left[\frac{3C_D v^2}{T_b - T_o} - C_K(s_o^* - s^*)\right]$$
$$+ \frac{C_D vM}{(T_b - T_o)}\frac{\text{Ri}_c}{r_t^2} - \frac{C_K v^3}{M(T_b - T_o)} \qquad (12.20)$$

Multiplying the last equation by $M(T_b - T_o)/(hv)$ gives Eq. (16) of E12:

$$\frac{T_b - T_o}{v}\frac{\partial}{\partial \tau}\left(\frac{v^2}{T_b - T_o}\right)$$
$$= \frac{M}{hv}\frac{\partial v}{\partial M}\left[3C_D v^2 - C_K(T_b - T_o)(s_o^* - s^*)\right]$$
$$+ \frac{C_D}{h}\frac{\text{Ri}_c}{r_t^2}M^2 - \frac{C_K}{h}v^2 \qquad (12.21)$$

At v_m (the maximum tangential wind), $\partial v/\partial M = 0$, whereupon Eq. (12.21) simplifies to

$$\frac{T_b - T_o}{v}\frac{\partial}{\partial \tau}\left(\frac{v^2}{T_b - T_o}\right) = \frac{C_D}{h}\frac{\text{Ri}_c}{r_t^2}M^2 - \frac{C_K}{h}v^2. \qquad (12.22)$$

E12 (p. 992) assumes that the outflow temperature at the radius of maximum gradient wind (defined henceforth as RMW) equals the tropopause temperature, i.e., $T_o = T_t$, with the latter assumed constant in time. This implies that the time derivative of T_o vanishes at the RMW. We can thus simplify the time derivative in the foregoing equation to obtain the tangential velocity tendency equation at the RMW:

$$\frac{\partial v_m}{\partial \tau} = \underbrace{\frac{C_D}{2h}\frac{\text{Ri}_c}{r_t^2}M^2}_{> 0} - \underbrace{\frac{C_K}{2h}v_m^2}_{> 0}, \qquad (12.23)$$

where v_m denotes the maximum tangential velocity at the top of the boundary layer.

The foregoing is a tendency equation for v_m forced by two terms on the right hand side. The first term is positive and denotes the tangential (generalized Coriolis) force per unit mass that increases v_m with time. This term is proportional to the drag coefficient and the critical gradient Richardson number. The second term is negative and arises in association with the depletion of tangential momentum in the boundary layer. Curiously, however, this second term is proportional to the enthalpy transfer coefficient C_K. One would ordinarily expect this term to be proportional to the drag coefficient C_D. E12 notes however that the equation is not yet closed and argues that it is possible that the global solution dependence on C_K may be different than would be apparent solely from an examination of this term at this stage in the derivation.

E12 proceeds to make an additional (and, in our view, unsubstantiated) assumption that the RMW always lies on the same M surface during vortex intensification. In fact, this assumption was falsified by Stern et al. (2015) (p1296) who found that:

"In this study, we have reexamined the relationship between the intensification of the maximum winds and the contraction of the radius of maximum winds (RMW) within tropical cyclones. From idealized simulations, we found that, in general, most contraction occurs prior to most intensification, and a quasi-steady size is often reached well before a quasi-steady intensity is achieved. In these simulations, the RMW first propagates inward across M surfaces, moving from higher to lower values of M during the initial period of rapid contraction. Once contraction of the RMW ceases, M surfaces propagate inward across the RMW, as intensification continues "in place"."

Acknowledging this challenge to the validity of the foregoing Emanuel assumption,[21] if one combines this assumption with an algebraic relation deduced from the revised steady-state theory of Emanuel and Rotunno (2011) (not written here), E12 presents an analytical (closed-form) solution for the evolution of v_m (his Eq. (19) and our Eq. (12.5) in Section 12.3.1.1).[22] The novelty of the result notwithstanding, the apparent elegance of the analytical solution conceals the essential role of the azimuthal force in amplifying v_m.

The critical role of the tangential force may be exposed by repeating the foregoing derivation while discarding the closure equation for the outflow temperature. In this case, one obtains at the RMW

$$\frac{\partial v_m}{\partial \tau} = \frac{C_D v_m^2 M}{2h(T_b - T_t)}\frac{\partial T_o}{\partial M} - \frac{C_K}{2h}v_m^2, \qquad (12.24)$$

where T_t denotes the tropopause temperature at the RMW (assumed equal to T_o and independent of time). While other outflow closures are conceivable that would, in turn, change the specification of $\partial T_o/\partial M$, if one employs the traditional Emanuel formulation of a constant outflow temperature (e.g. E97 with $\beta = 0$), then $\partial T_o/\partial M$ would be identically zero. The resulting tendency equation for v_m in this case consists only of the second term on the right hand side of Eq. (12.24), which is negative definite. Thus without the tangential force in the traditional formulation of the upper boundary condition, the vortex spins down! Notwithstanding this fact, we have a more fundamental issue with the physics encompassed by Eq. (12.24). We find it puzzling how, in reality, the stratification of the outflow layer ($\partial T_o/\partial M$) would act to move the M surfaces inwards in a way to amplify the tangential wind at the top of the boundary layer. A similar remark would apply to Eq. (12.23), in which $\partial T_o/\partial M$ has a specific parameterization. Other questions arising from the mathematical equations (12.23), (12.24) are discussed in Section 12.3.1.2.

---

21. While our own results (not shown) support the quoted findings of Stern et al. (2015), our citation of Stern et al. (2015) should not be interpreted as an endorsement of their other findings concerning the validity of a linearized model (the so-called 3DVPAS model) when applied to the rapid intensification of a tropical cyclone. In fact, Montgomery and Smith (2017a) concluded that the linearized 3DVPAS solutions "cannot be used (a) to dismiss the importance of the nonlinear boundary-layer spinup mechanism (discussed in Chapters 6 and 11, our addition), nor (b) to isolate the separate effects of diabatic heating from those of friction, within the nonlinear boundary layer at least. Such separation depends on the linear superposition principle, which fails whenever nonlinearity is important. Similar caveats apply to the use of another linear model, the traditional Eliassen balance model (discussed in Chapter 5, our addition). Its applicability is limited not only by linearity, but also by its assumption of strictly balanced motion. Both are incompatible with nonlinear spinup."

---

22. The closed-form solution strictly applies only for the case in which the initial tangential velocity is everywhere zero. The inconsistency of this latter assumption is discussed in Section 12.3.1.2.

Chapter 13

Emanuel's maximum intensity theory

Our understanding of hurricanes has been influenced strongly by the simple, axisymmetric, steady-state hurricane model described in a pioneering study by Emanuel (1986), henceforth E86. This model has served to underpin many ideas about how tropical cyclones function. It provided the foundation for the so-called "potential intensity (PI) theory" of tropical cyclones (Emanuel, 1988, 1995; Bister and Emanuel, 1998) and its time-dependent extension led to the formulation of intensification theories discussed in Chapter 12.

PI theory refers to a theory for the maximum possible intensity that a storm could achieve in a particular environment, based on the maximum possible tangential wind component in a strictly axisymmetric formulation of the problem (specifically the maximum azimuthal mean *gradient wind*).[1] That such an upper bound on intensity should exist follows from global energy considerations. Under normal circumstances, the energy dissipation associated with surface friction scales as the cube of the near-surface wind speed, while the energy input via moist entropy fluxes scales generally with the first power of the near-surface wind speed. It follows that the frictional dissipation will exceed the input of latent heat energy to the vortex from the underlying ocean at some point during the cyclone's intensification. There is a subtle caveat with this scaling argument because the linear dependence of the energy input on wind speed may be suppressed if the degree of moisture disequilibrium at the sea surface is reduced as the wind speed increases. Such a reduction would shift the intersection point of these two behavior curves leftward to a lower maximum tangential wind speed.[2]

Most storms never reach their PI (Merrill, 1988, Fig. 1; DeMaria and Kaplan, 1994, Fig. 1; Emanuel, 1999), a fact that is attributed to the deleterious effects of vertical wind shear, which tends to tilt and deform a developing vortex and open pathways for dry air intrusion (Riemer et al., 2010; Tang and Emanuel, 2010; Riemer and Montgomery, 2011; Riemer et al., 2013). These topics are of course important, but lie outside the main scope of this book.

13.1 The E86 steady-state model

Fig. 13.1 shows a schematic of the E86 steady-state hurricane model. The energetics of this model are often likened to that of a Carnot cycle in which the inflowing air in the boundary layer acquires sensible and latent heat (principally latent heat) while remaining approximately isothermal (cf. Chapter 1). The ascending air is assumed to be pseudomoist adiabatic[3] and the outflowing air at large radius is assumed to descend isothermally in the upper atmosphere. This isothermal subsidence would require the accompanying adiabatic compression to be exactly balanced by radiative cooling to space. The final leg in the cycle is assumed to follow a reversible moist adiabat. This leg is arguably the most implausible since for one thing, the time scale required for air parcels to descend back to the sea surface is on the order of one month (see Section 9.4), far longer than the typical life cycle of an individual hurricane in the real atmosphere. For another, in clear sky conditions, the radiative cooling acts throughout the troposphere and so θ_e cannot be materially conserved in the descending leg. See Exercise 13.7 and related footnote for our assessment of a recent attempt to circumvent these realities. Research by Bister et al. (2011) has pointed out, *inter alia*, that this hypothetical dissipative heat engine does no useful work on its environment.

The E86 model assumes hydrostatic balance and gradient wind balance above the boundary layer and uses a slab boundary layer model in which departures from gradient

---

1. The formulation of Emanuel (1988) for the maximum intensity of hurricanes characterized the intensity by the minimum surface pressure. Subsequent papers (Emanuel, 1995, 1997) revised the formulation for the minimum surface pressure and shifted focus to the maximum gradient wind.

2. It is thought presently that there is an exception to this argument when either the sea surface temperature is sufficiently warm or the upper tropospheric temperature is sufficiently cold, or some combination of the two prevails (Emanuel, 1986, 1988). Under such extreme conditions, the vortex is believed to be capable of generating enough latent heat energy via surface moisture fluxes to more than offset the dissipation of energy and a 'runaway' hurricane - the so-called 'hypercane regime' - is predicted. The hypercane has been implicated in global extinction scenarios associated with meteor impacts. Strictly speaking, however, these predictions have been formulated only in the context of axisymmetric theory and simulated using subsonic, axisymmetric flow codes. It remains an

open question whether hypercanes are dynamically realizable in a realistic, three-dimensional flow configuration.

3. Contrary to statements made in E86, the formulation assumes pseudo-adiabatic rather than reversible thermodynamics in which all condensate rains out instantaneously (Bryan and Rotunno, 2009a, p3044). It is not a true Carnot cycle, in part, because of the irreversible nature of the precipitation process in the eyewall region of the vortex.

Tropical Cyclones. https://doi.org/10.1016/B978-0-44-313449-4.00021-7

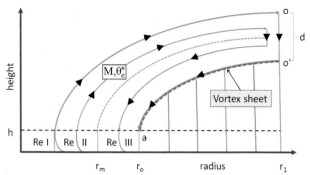

FIGURE 13.1 Schematic diagram of Emanuel's 1986 model for a steady-state mature hurricane. The arrows indicate the direction of the overturning circulation. See text for discussion. From Montgomery and Smith (2017b). Reproduced with permission from the Annual Review of Fluid Mechanics, Volume 49; copyright 2017 Annual Reviews.

wind balance are assumed negligibly small (cf. Chapter 6 and see below). The boundary layer is taken to have constant depth h in which the absolute angular momentum, M, and pseudo-equivalent potential temperature, θ_e, are well mixed at leading order. This layer is divided into three regions as shown in Fig. 13.1: the eye region (Re I), the eyewall region (Re II) and outside the eyewall region (Re III) where spiral rainbands and shallow convection are assumed to operate in the vortex above (not depicted). When moist air rises out of the boundary layer in the eyewall region, it is assumed to quickly saturate and materially conserve its saturation pseudo-equivalent potential temperature θ_e^* and M after the parcel leaves the boundary layer and immediately ascends in the eyewall cloud.

The precise values of M and θ_e at a particular radius are determined by the frictional boundary layer. The model assumes that the radius of maximum tangential wind speed, r_m, is located at the outer edge of the eyewall cloud (outer edge of Re II in Fig. 13.1), although subsequent observations show that r_m is closer to the inner edge (Marks et al., 2008, Fig. 3). The middle dashed curve emanating from r_m is the M-surface along which the vertical velocity is zero and demarcates the region of ascent in the eyewall region from that of large-scale descent outside the eyewall region. The outer dashed curve indicates the location of the vortex sheet as described in Smith et al. (2014). The flow segment between o and o' in the upper right corner of the figure represents the assumed isothermal leg noted above and is the location at which air parcels are assumed to steadily gain cyclonic relative angular momentum (RAM, rv) from the environment. The gain of RAM is needed in order to replace the frictional loss of angular momentum at the surface where the flow is cyclonic. This source of RAM is required for a steady state to exist and it is pertinent to inquire whether this source is physically plausible and whether a sustained hurricane vortex is, in fact, the likely outcome in an idealized configuration under strong boundary layer control. These interesting scientific topics are taken up in

Chapters 14 and 15 in a three-dimensional model framework. For the moment, we will set aside these issues and focus on the presumed steady solution.

In addition to the assumption of axial symmetry and the realism of the source of cyclonic RAM needed to replenish the loss of M at the ocean by the surface frictional stress, the E86 model suffers a range of issues as discussed by Parks and Montgomery (2001), Smith et al. (2008), Bryan and Rotunno (2009c), Emanuel (2012), and Montgomery and Smith (2014). The most pertinent issues are discussed in turn.

The original formulation for the PI rests, in part, on a simple slab boundary layer formulation, together with the development of an expression for the radial distribution of θ_e and pressure at the top of the boundary layer.[4] To keep the current presentation succinct, we focus here on the boundary layer aspect of the formulation. Near the radius of maximum tangential wind, the two limiting equations for an assumed well-mixed slab boundary layer express:

(i) a balance between the radial advection of moist entropy, s_b, and the evaporative gain of s_b by (primarily) surface moisture fluxes, and

(ii) a balance between the radial advection of M and the loss of M by frictional torque at the sea surface.

The limiting advection-source/sink equations describing these processes, respectively, are as follows:

$$hu_b\frac{\partial s_b}{\partial r} = C_K |\mathbf{v}|_0(s_0^* - s_b), \tag{13.1}$$

and

$$hu_b\frac{\partial M_b}{\partial r} = -C_D |\mathbf{v}|_0 r v_0, \tag{13.2}$$

where u_b (<0) denotes the slab-averaged radial velocity for the boundary layer, s_b denotes the (well-mixed) moist entropy of the boundary layer, s_0^* denotes the saturation moist entropy at the sea surface temperature[5] and pres-

---

4. See Appendix B, Section 13.7 and supporting Exercises 13.2-13.5 for details.

5. The sea surface temperature is assumed spatially constant for simplicity. Wind-driven upwelling of cold water from below the thermocline in the right-rear quadrant of the moving cyclone, and outside the maximum wind radius as discussed in Chapter 1, is neglected in this axisymmetric formulation. The near-surface air temperature is assumed constant with radius also. The latter is tantamount to an isothermal expansion for inflowing air parcels and is supported by observations showing an approximately invariant near-surface air temperature with decreasing radius (e.g., Malkus and Riehl, 1960; Cione et al., 2000). Deviations from the isothermal assumption do occur, however, and some storms exhibit a several degree Celsius cooling near the radius of maximum wind speed (e.g., Cione et al., 2000 and Montgomery et al., 2006a). Recent work of Makarieva et al. (2019) has developed an alternative formulation for the maximum tangential wind for the case in which the expansion in the boundary layer is not strictly isothermal.

sure, C_K denotes the coefficient of moist entropy transfer,[6] $M_b = rv_b + fr^2/2$ denotes the slab-averaged absolute angular momentum of the boundary layer, v_b denotes the slab-averaged tangential velocity, $|\vec{v}|_0$ and v_0 denote the wind speed and tangential velocity at anemometer level (10 m), and C_D is the drag coefficient at this level. When Eq. (13.1) is divided by Eq. (13.2), the slab-averaged radial velocity is eliminated, yielding

$$\frac{\partial s_b/\partial r}{\partial M_b/\partial r} = -\frac{C_K}{C_D}\frac{(s_0^* - s_b)}{rv_0}. \qquad (13.3)$$

The well-mixed assumption on M_b implies that $v_0 \approx v_b$ and to leading order $v_b \approx v_g$.[7] The well-mixed assumption implies also that the M_b and s_b surfaces are congruent in the boundary layer, i.e., $s_b = s_b(M_b)$. Then, according to the chain rule, the derivative quotient on the left side of Eq. (13.3) can be written as

$$\boxed{\frac{ds}{dM} = -\frac{C_K}{C_D}\frac{(s_0^* - s_b)}{rv_0} \quad \text{at} \quad (r,z) = (r_{gm}, z_{gm}).}$$
$$(13.4)$$

This expression is an entropy derivative with respect to angular momentum evaluated at the radius of maximum gradient wind r_{gm} and boundary layer top h (assumed equal to the height of maximum gradient wind, i.e., $h = z_{gm}$). Here, the subscripts on angular momentum and entropy have been dropped on the left hand side because the boundary layer values are assumed equal to their corresponding values at the top of the layer. In particular, the assumption that $v_b \approx v_g$ (noted above) tacitly implies that the tangential wind in the boundary layer is in gradient wind balance at leading order. As one might suspect, this logical deduction from the model formulation contradicts the basic property of a vortex boundary layer where, on account of surface friction, the reduction of tangential winds outside r_{gm} leads to an inward-directed agradient force ($AGF < 0$), where

$$AGF = -\alpha_d \frac{\partial p}{\partial r} + fv + \frac{v^2}{r}, \qquad (13.5)$$

and where α_d denotes the specific volume of dry air (neglecting the mass of water vapor in the horizontal momentum balance). When $AGF < 0$, this force accelerates air parcels radially inwards down the effective radial pressure gradient towards the vortex center (Section 6.1). Despite the retarding drag force on the swirling flow, as air parcels are accelerated inwards, the tangential velocity may increase relative to the gradient wind at the boundary layer top. A necessary condition for this circumstance to occur

is that the generalized Coriolis force, $-(f + v_b/r)u_b$, exceeds the frictional drag force, $-C_D|\mathbf{v}|_0v_0/h$. As shown in Sections 6.5.6, 6.5.7 and 6.6, under such circumstances, the tangential wind is expected to become supergradient and the agradient force will reverse sign accordingly ($AGF > 0$). Such a force reversal leads to an arrest of the inflowing air parcels just outside r_m. However, *in the E86 model the agradient force plays no role in the dynamics of the inflow layer.* Further discussion of the nonlinear boundary layer dynamics is given below and in Appendix A, Subsection 13.6.2.2.[8]

Eq. (13.4) for ds/dM together with the expression relating the radial distribution of θ_e and pressure at the boundary layer top are not yet sufficient to close the problem. This is because the moisture content of the boundary layer has yet to be specified. In general, the determination of the spatial distribution of moisture in the boundary layer is a challenging problem because the frictionally-driven vortex-scale subsidence into the boundary layer and the convective and mesoscale downdrafts outside r_{gm} (Re III in Fig. 13.1) together serve to bring relatively dry air with low values of θ_e into the boundary layer from above (Chapters 6, 9, and 11).

As illustrated in Sections 10.4 and 10.11, convective downdrafts in sub-saturated air act to dilute the boundary layer moisture. This dilution process will be opposed by the wind-forced evaporation of moisture from the sea surface. However, the outcome of this three-way competition is not known *a priori* and one must perform a complex calculation with a cloud-representing, mesoscale numerical model to determine the outcome. E86 tackled this issue by postulating a zero-order closure in which the relative humidity at the top of the surface layer is assumed to remain constant from some large radius inwards to r_{gm}.

As discussed by Parks and Montgomery (2001), pp1712-1713, Montgomery et al. (2006a), Fig. 4d, and Cione et al. (2000), the near-surface RH typically increases from approximately 80% in the environment to approximately 95% near r_{gm}. Meanwhile, the relatively small cooling of the sub-cloud layer air (noted above) tends to compensate the increasing RH, leaving the near-surface θ_e approximately unchanged relative to its constant RH counterpart. Clearly, the constant RH assumption is not ideal. Nonetheless, we judge it to be a crude, though adequate, zero-order closure outside r_{gm}.

The foregoing results, in conjunction with the constant RH closure up to r_{gm}, lead ultimately to a coupled system of equations for the Exner function π at the surface and v_{gmax}^2 at $z = h$, where v_g is the gradient wind and v_{gmax} is its maximum (assumed to occur at the boundary layer top, $z = h$). In the limit of $r_o^2 \gg r_{gm}^2$ and assuming that $T_B \approx T_s$, the solution for v_{gmax}^2 is approximated by (Exercises 13.4 and 13.5):

---

6. Equal also to the moist enthalpy transfer coefficient employed in Section 12.3, Exercise 13.1. Note that E86 uses the notation C_θ for this coefficient.

7. Bryan and Rotunno, 2009c, p3045, right column.

8. Additional discussion of the nonlinear dynamics of the inner-core "corner flow" region of the vortex is provided in Sections 11.5.3 and 11.5.4.

$$v_{gmax}^2 = \frac{C_K}{C_D} \epsilon L_v r_{va}^* (1 - RH_{as}) \times$$

$$\frac{1 - \dfrac{f^2 r_o^2}{4\beta R_d T_B}}{1 - \dfrac{\epsilon L_v r_{va}^* (1 - RH_{as}) C_K}{2\beta R_d T_s C_D}}, \qquad (13.6)$$

where L_v is the latent heat of condensation of water vapor, R_d is the specific gas constant for dry air, T_s is the sea surface temperature, $\epsilon = (T_B - T_0)/T_B$ is the thermodynamic efficiency factor, T_B is the temperature at the top of the boundary layer (assumed constant with radius also), T_0 is the average outflow temperature weighted with the saturated moist entropy of the outflow angular momentum surfaces (Eq. (19) of Emanuel, 1986),[9] r_{va}^* is the saturation mixing ratio at the top of the surface layer in the environment, RH_{as} is the ambient relative humidity at the top of the surface layer, $\beta = 1 - \epsilon(1 + L_v r_{va}^* RH_{as}/R_d T_s)$,[10] and r_o is the radial extent of the storm near sea level (nominally the radius at which $v = 0$).[11]

From Eq. (13.6), Emanuel constructed curves for v_{gmax} as a function of upper-level outflow temperature and sea surface temperature. As an example, for a sea surface temperature of 28 °C and an outflow temperature of −60 °C, the formula predicts a v_{gmax} of approximately 58 m s⁻¹ (see Fig. 13.2). In this calculation, it is assumed that $C_K/C_D = 1$, but the latest field observations and laboratory measurements synthesized in Bell et al. (2012a) and Curcic and Haus (2020) suggest a mean value of approximately $C_K/C_D \approx 0.3$ in the high wind speed range (Exercise 13.6). Although Bell et al. (2012a) acknowledge the scatter in the observational estimates of C_K/C_D in the high wind speed regime (see Chapter 7, where the recent findings of Curcic and Haus, 2020 are discussed also), these data still represent our best mean estimates at the time of writing. For this reduced ratio of mean exchange coefficients, and for the selected temperatures, Eq. (13.6) predicts a reduced v_{gmax} of approximately 35 m s⁻¹, significantly less than that obtained using $C_K/C_D = 1$. (Exercise 13.6 explores further the ramifications of this finding.)

One puzzling feature of both the E86 derivation and later extensions discussed below is that there seems to be no constraint that $\partial v_g/\partial r = 0$ at the radius of maximum gradient wind (r_{gm}). Moreover, all these derivations appear not to predict the radius of maximum tangential wind, at least in terms of *a priori* known quantities.

By the early 2000s, the E86 theory proved useful to the community because it offered a relatively simple framework for estimating a storm's maximum intensity that is well

---

9. This is the temperature at which air parcels are assumed to descend approximately isothermally in the upper atmosphere.
10. Not to be confused with the β parameter defined in Chapter 3 or discussed in Chapter 12 in the context of the E97 intensification model.
11. The mathematical definition for r_o is given by Eq. (20) of E86.

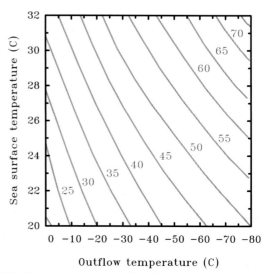

Outflow temperature (C)

FIGURE 13.2 Predicted v_{gmax} (blue (dark gray in print version) numbers adjacent to curves in m s⁻¹) from Eq. (13.6) as a function of sea surface temperature (T_s) and outflow temperature (T_o) from Emanuel's 1986 model for a mature steady-state hurricane. Temperature is in Celsius. The ratio of moist entropy to momentum transfer coefficients C_K/C_D is assumed to be unity. Calculations assume an ambient surface pressure of 1015 mb, an ambient near-surface relative humidity (RH) of 80 %, a Coriolis parameter f evaluated at 20 degrees latitude, and an outer radius r_0 equal to 500 km. See text for further details. Adapted from Emanuel (1986). Republished with permission from The American Meteorological Society.

within an order of magnitude of observations. Although there were competing PI theories that predicted intensities of a similar order of magnitude, those theories suffered physical inconsistencies. The E86 theory was arguably the most physically consistent of available theories (see Parks and Montgomery, 2001 for a review).

Despite the uncertainties within the theory to be discussed further below, PI theory has been used widely to estimate the impact of global climate change on tropical cyclone intensity and structure. As an example of its far-reaching influence, the E86 theory is still used as a basis for deriving updates to the *a priori* PI theory (e.g., Garner, 2015) as well as for estimating the impact of tropical cyclone intensity and structure due to global warming scenarios (Emanuel, 1988; Camargo et al., 2014).

Although, as previously mentioned, most tropical cyclones fail to reach their PI, some major hurricanes exhibit sustained tangential wind speeds that significantly exceed the value predicted by Eq. (13.6), even with an arguably liberal value $C_K/C_D = 1$. One such example is the case of Hurricane Isabel (2003) (Montgomery et al., 2006a and Bell and Montgomery, 2008), which was shown to significantly exceed its predicted PI for three consecutive days during an intensive observational campaign using multiple reconnaissance aircraft (see Figs. 13.3 and 13.4).

As an example, on 13 September, Isabel had an observed maximum, system-scale, tangential wind of approximately

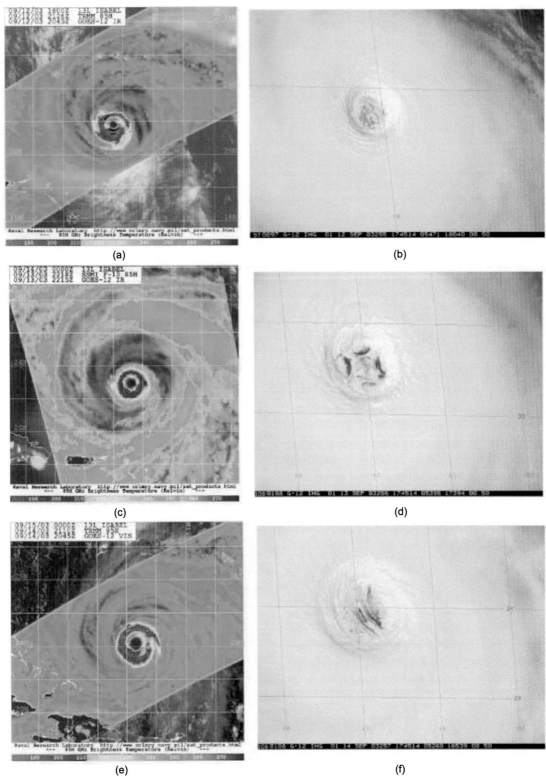

(a)   (b)

(c)   (d)

(e)   (f)

FIGURE 13.3   Satellite views of Hurricane Isabel (2003): (left) Satellite appearance of Hurricane Isabel at 85 GHz (courtesy NRL/Monterey) and (right) visible (courtesy CIRA/CSU) wavelengths during each intensive observational period. The 85-GHz imagery is from (a) TMI at 2126 UTC 12 Sep, (c) SSM/I at 2218 UTC 13 Sep, and (e) TMI at 2110 UTC 14 Sep. (b), (d), (f) Corresponding visible images are from GOES super-rapid-scan operations at 1745 UTC on each day. See Bell and Montgomery (2008) for further details about this data and other data collected as part of the intensive observations of this Category 5 hurricane. From Bell and Montgomery (2008). Republished with permission from The American Meteorological Society.

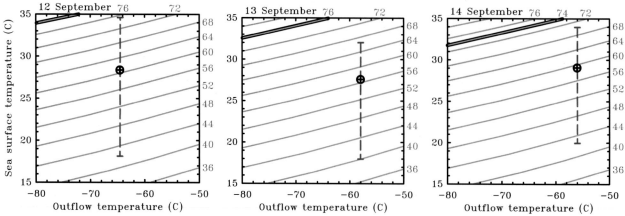

FIGURE 13.4 Theoretically predicted maximum mean gradient wind v_{gmax} for Hurricane Isabel (2003) at the nominal boundary layer top for varying outflow temperature and near core SST with a constant near-surface RH = 80%. The "X" indicates the primary estimate of the gradient wind potential intensity for the observed near environment around Isabel. The thick red (mid gray in print version) curve represents the average storm-relative tangential wind speed at the top of the boundary layer derived from the dropwindsonde measurements. To account for uncertainty in the extent of ocean cooling and dissipative heating effects, a range of *a priori* PI estimates is indicated also by the vertical dashed blue (dark gray in print version) line. The upper bound for the predicted v_{gmax} takes C_K/C_D equal to unity, assumes a full contribution from dissipative heating (as predicted by Bister and Emanuel, 1998 and discussed in the next section) and assumes no storm-induced ocean cooling. The lower bound takes C_K/C_D equal to 0.5, and assumes that dissipative heating is entirely offset by ocean cooling feedback. The primary estimate compromises these assumptions, with a bulk exchange coefficient ratio of unity, and offsetting dissipative heating and ocean cooling. See Bell and Montgomery (2008) for justification of assumptions leading to the upper and lower bound estimates. Adapted from Bell and Montgomery (2008). Republished with permission from The American Meteorological Society.

76 m s^{-1}, yet a best estimate based on Eq. (13.6) using in retrospect a liberal value of $C_K/C_D = 1$ gives only 56.6 m s^{-1} (see Fig. 13.4b). This main estimate takes into account the uncertainty of the exchange coefficients and the ocean cooling effect by turbulence-induced upwelling of cooler water from below the thermocline by assuming a ratio of exchange coefficients of unity and that dissipative heating offsets ocean cooling.[12] In this case, there is a discrepancy of approximately 20 m s^{-1} between the theory and the observations. The discrepancy spans at least two intensity categories on the Saffir-Simpson scale (Category 3 to Category 5).

As noted above, if we were to use the more likely ratio of $C_K/C_D = 0.3$, the discrepancy between theory as given by Eq. (13.6) and observations is increased considerably. The fact that some major tropical cyclones in regions of the world oceans can significantly exceed this theoretical predicted intensity for sea surface temperatures of 28 °C (or higher) is presumably a consequence of certain assumptions made in formulating the E86 model (Exercise 13.6 explores this issue further.).

13.1.1 Dissipative heating

The inclusion of dissipative heating (i.e., the increase in internal energy associated with the dissipation of kinetic energy by shear turbulence primarily in the atmospheric boundary layer) was first explored in the study by Bister and Emanuel (1998). The main finding was that the heat engine efficiency factor $\epsilon = (T_B - T_o)/T_B$ should be replaced by the quantity $\epsilon \times (T_B/T_o)$. That is, the effective efficiency is equivalent to using the cold outflow temperature T_o in the denominator of the efficiency instead of the warm inflow temperature T_B there. This change implies a boosting of the engine's efficiency by as much as 50% for a boundary layer temperature of approximately 300 K and a tropopause temperature of 200 K.

The inclusion of dissipative heating in PI theory has been controversial. Makarieva et al. (2011) critiqued the way in which Bister and Emanuel included this heating and suggested that

> "*such a dissipative heat engine violates the laws of thermodynamics*".

Bister et al. (2011) countered that their work does not violate fundamental physical laws and suggested that

> "*Makarieva et al. (2011) were misinterpreting expressions for wind speed as ones for work done on external objects.*"

---

12. See caption to Fig. 13.4 as well as Montgomery et al. (2006a) and Bell and Montgomery (2008) for further details and caveats, which include an extensive analysis of the other two days of the intensive observation campaign, as well as justification for using a sea surface temperature of 27.5 °C on 13 Sep. as Isabel's eyewall passed over the wake of prior hurricane Fabian (2003). See also discussion in Section 13.1.1 regarding an update on the matter of dissipative heating.

Notwithstanding the Bister et al. rebuttal and working along independent lines, Kieu (2015) has pointed to an inconsistency of the Bister and Emanuel formulation and the related assumption that all dissipative heating in the atmospheric surface layer can return to the atmosphere as an "additional" heat source, which acts to augment the maximum gradient wind.

Kieu (2015) recommends use of the original PI formulation of Emanuel (1986), since (conclusions)

> *"it can be reinterpreted as a rational estimation of the TC MPI (tropical cyclone maximum potential intensity - our insertion) even in the presence of the internal dissipative heating ...".*

Drawing upon the observational analyses of Zhang (2010), Kieu (2015) suggests that the dissipative heating formulation used by Bister and Emanuel (1998) and Bryan and Rotunno (2009a,c) overestimates the true dissipative heating, in part, due to an over-estimate of the viscous work term in the boundary layer and also, in part, to the radiation of energy out of the hurricane by wind-induced surface gravity waves at the ocean interface. This topic would appear to merit further study. Exercise 13.7 presents a novel and complementary analysis of dissipative heat engines as they relate to a hurricane and explores some of the implications for real-world storms.

13.1.2 High resolution tests of the E86 PI theory

High resolution, axisymmetric, numerical simulations were carried out to test the E86 PI theory in a controlled setting (Persing and Montgomery, 2003; Hausman et al., 2006; Bryan and Rotunno, 2009a). For horizontal mixing lengths consistent with recently observed estimates from flight-level data in major hurricanes ($L_h \approx 700$ m, Chapter 7), the numerical studies of Persing and Montgomery (2003) and Bryan and Rotunno (2009a) confirmed the tendency for solutions to significantly exceed the theoretically predicted PI given by Eq. (13.6) and its modifications summarized below,[13] even assuming that $C_D/C_K = 1$.

---

13. Not to belabor the point too much, but it was not until the work of Persing and Montgomery (2003) and Hausman et al. (2006) that compelling evidence was presented legitimately challenging the position that Emanuel's PI theory represented a rigorous upper bound for tropical cyclone intensity. The Isabel analyses provided a first detailed observational assessment of PI theory as it was understood at that time. The hypothesized "superintensity mechanism" for explaining Isabel's intensity put forward by Persing and Montgomery (2003) was shown by Bryan and Rotunno (2009b) to be too weak to explain the discrepancy between PI theory and experiment in a strictly axisymmetric model configuration. As discussed in Section 13.2, we understand better now the dynamical origin of "superintense storms" and the important distinction between r_{gm} and r_m. In particular, scientific questions remain concerning the impact of "high-octane" θ_e air generated at radii inside r_{gm} (including the low-level eye of the storm) in supporting

A point that should not be overlooked is that, by the early 2000s, the scientific conversation had advanced markedly beyond order-of-magnitude intensity estimates. Specifically, the PI theory given by Eq. (13.6) was widely regarded as a rigorous upper bound of the time and azimuthal-mean tangential wind intensity. Subsequent work, as summarized above, examining the dynamics and thermodynamics of so-called "superintense hurricanes" whose intensities exceeded the PI opened up new perspectives on the maximum intensity problem.[14] The next section summarizes these new perspectives and, as we shall see, this material builds heavily on the nonlinear boundary layer ideas presented in Chapter 6.

13.2 Unbalanced effects

An important limitation of the E86 theory is its neglect of nonlinear, unbalanced dynamics in the frictional boundary layer (Smith et al., 2008) and above the boundary layer (Bryan and Rotunno, 2009c). Bryan and Rotunno derived a modified formula for the maximum sustained tangential wind v_{max}, which accounts for unbalanced processes above

---

locally buoyant updrafts comprising a realistic, three-dimensional eyewall (e.g., Willoughby, 1998; Zhang et al., 2002; Cram et al., 2007; Persing et al., 2013; Houze, 2014, his Figs. 10.12b, 10.16, and 10.17 and accompanying discussions). A notable example is the recent study by Wadler et al. (2021a) who demonstrate (p3517) the quantitative importance of "low-level outflow from inside the eye and eye-eyewall mixing" in supporting relatively high θ_e air ascending the eyewall in their Category-5, hurricane simulation experiment. That study re-affirms the hypothesized non-negligible influence of eye-eyewall mixing in a three-dimensional, Category-5, hurricane. The relationship between the "interior" low-level θ_e structure and nonlinear boundary dynamics in a three-dimensional configuration is a scientifically interesting topic and part of the boundary layer control ideas discussed in Chapters 10 and 15.

14. Our emphasis here is on the upper limit to the time and azimuthal mean tangential wind speed. By definition, such considerations apply only to the system scale flow. Small-scale intense vorticity structures often pack higher and more transient winds locally. An example of an extreme horizontal wind speed of 107 m s^{-1} and collocated vertical updraft of approximately 25 m s^{-1} at about 1400 m above seal level was documented observationally in Hurricane Isabel (2003) by Aberson et al. (2006). This line of work has been extended and shows that locally intense wind structures are not uncommon in intense storms and are typically found within or just above the boundary layer region of the vortex just interior to the radius of maximum tangential wind along the eye-eyewall interface (Stern et al., 2016). A complementary experimental study was conducted by Montgomery et al. (2002). In retrospect, this finding is to be expected considering the nature of the nonlinear boundary layer spin up mechanism articulated in Chapter 6. Theoretical predictions for the *maximum local wind speed* and *updraft* have not yet been formulated, but such a program would necessarily involve considering the solution properties of the Navier-Stokes equations (e.g., Ladyzhenskaya, 1969) and the near singularities that these solutions support at large Reynolds number (Majda and Bertozzi, 2002, Ch. 5; Chorin, 1994, Ch. 5) after formulating the problem to incorporate inertial forcing from the system-scale hurricane vortex. Similar issues arise in the tornado problem and the extreme local damage that severe tornadoes are observed to inflict upon structures. These important topics lie outside the scope of this chapter and book.

the boundary layer. The extended formula for v_{max}^2 and its relationship to the E86 formalism is derived and discussed in Appendix B. The extended formula for v_{max}^2 is given by

$$v_{max}^2 = PI^2 + \gamma, \qquad (13.7)$$

where PI is nominally v_{gmax} as given by Eq. (13.6), $\gamma = \alpha_{diss} r_m \eta_b w_b$ and these latter terms are evaluated at the top of the boundary layer at the location of the maximum tangential velocity, (r_m, z_m). Here, $\eta = \partial u/\partial z - \partial w/\partial r$ is the azimuthal component of the vorticity vector, r_m and z_m are the radius and height of v_{max}, w_b is the vertical velocity at the top of the boundary layer at this same radius and $\alpha_{diss} = T_s/T_o$, where T_s is the SST and T_o is the outflow temperature. The α_{diss} term is associated with the inclusion of dissipative heating. When dissipative heating is omitted, $\alpha_{diss} = 1$. As noted above, Kieu (2015) argues that this is the more defensible formulation.

To gain some understanding of the link between the dynamics and the γ term in Eq. (13.7), we pause briefly to consider the nature of the γ term. At the top of the boundary layer, the steady-state inviscid radial momentum equation in cylindrical polar coordinates may be written as follows

$$u\frac{\partial u}{\partial r} + w\eta + w\frac{\partial w}{\partial r} - \frac{v^2}{r} - fv = -\alpha_d \frac{\partial p}{\partial r}, \qquad (13.8)$$

where all variables are evaluated at the boundary layer top (z_b), α_d is, again, the specific volume of dry air (neglecting the mass of water vapor in the horizontal momentum balance) and both radial and vertical turbulent diffusion of u are assumed vanishingly small compared to the retained terms. Near r_m, one finds that $u \approx 0$ and $\partial w_b/\partial r \approx 0$. The former follows from the fact that in a steady state, the influx of cyclonic vorticity must vanish at the top of the well mixed boundary layer where the vertical derivative of the tangential velocity and turbulent tangential momentum flux derivative are assumed to vanish. The latter follows from the fact that the maximum ascent out of the boundary layer occurs typically just inside the radius of v_{max}, so this ascent radius may be approximated as equal to r_m for the present purposes. The foregoing properties allow one to neglect the first and third terms on the left hand side of the radial momentum equation, leaving the limiting radial momentum balance:

$$w_b \eta_b \approx \frac{(v^2 - v_g^2)}{r} + f(v - v_g), \qquad (13.9)$$

where v is the total tangential velocity, v_g is the tangential velocity in gradient wind balance with the radial pressure gradient force per unit mass (i.e., $fv_g + v_g^2/r = \alpha_d \partial p/\partial r$) and both tangential velocities are evaluated at the top of the boundary layer. In the context of the modified formula for v_{max}^2 given by Eq. (13.7), it follows that in supergradient flow ($v > v_g$) near r_m, the right-hand-side of Eq. (13.9) is

positive, whereupon γ on the right-hand-side of Eq. (13.7) is a positive contribution to v_{max}.

In the limiting case of small horizontal mixing length, Bryan and Rotunno found that $\gamma \approx PI^2$ so that (p3055)

"... the effects of unbalanced flow contribute as much to maximum intensity as balanced flow for this case".

In the case of Hurricane Isabel, the new formula is found to be a significant improvement and despite uncertainties in the observations, Bryan and Rotunno (p3506) concluded that

"... unbalanced flow effects are not negligible in some tropical cyclones and that they contribute significantly to maximum intensity".

It is important to note that Bryan and Rotunno's extended analytical theory is not an *a priori* form of PI in the sense of using only environmental conditions as input: it requires knowledge also of η_b, r_m, and w_b. Thus one cannot make graphs of v_{max} as functions of SST and outflow temperature similar to Fig. 13.2. The theory continues to use the same boundary layer formulation as E86 (see their Section 2b and Appendix A, Subsection 13.6.2.2 for further discussion). In essence, the flow in the boundary layer is assumed to be in approximate gradient wind balance. For this reason, the new theory does not address the concerns raised by Smith et al. (2008) about the boundary layer in the original PI theory.

When η_b, r_m, and w_b are diagnosed from an axisymmetric numerical simulation, which has a nonlinear, unbalanced boundary layer, the new formula for v_{max} is shown to provide an improved estimate for the maximum intensity of numerically simulated vortices by Bryan and Rotunno (2009c) for a range of values of the horizontal mixing length (see Bryan and Rotunno, 2009a, their Fig. 12). Clearly, a more complete boundary layer formulation that includes a radial momentum equation would be required to determine η_b, r_m, and w_b, and thereby v_{max} from initial data. Indeed, it is these three quantities (η_b, r_m, and w_b) that characterize the effect of "boundary layer control" on the eyewall dynamics discussed in Chapters 10, 11, and 15. In this sense, Bryan and Rotunno's analysis provides further evidence that unbalanced effects in the boundary layer are responsible for those in the eyewall.

Notwithstanding the reasonably good agreement, there would appear to be two issues with the comparison of the extended PI theory with their numerical calculation, both acknowledged by Bryan and Rotunno (2009a). Firstly, the PI is calculated at the radius of v_{max} in the model rather than at the radius of maximum gradient wind of either the theory or the model. (The radius of maximum v_g typically lies some 10 km (or more) outside the radius of v_{max} for a strong hurricane (e.g., Bell and Montgomery, 2008, Fig. 9a; Bryan and Rotunno, 2009a, Fig. 10; Sanger et al., 2014, Fig. 11; Montgomery et al., 2014, Figs. 11 and 12). The second issue is the calculation of the gradient wind from the pressure field in the full model output, which incorporates substantial unbalanced effects as well as the balanced effects contained in the E86 model.

The work of Bryan and Rotunno (2009a) and subsequent work by Rotunno and Bryan (2012) and Bryan (2013) has emphasized the strong dependence of the simulated intensity in *axisymmetric* models to the horizontal mixing length (and related diffusivity) used to parameterize asymmetric mixing and small-scale turbulence. Appendix A, Subsection 13.6.2.2 discusses this topic further.

13.3 A revised theory

Emanuel and Rotunno (2011) and Emanuel (2012) argued that the assumption of the original model that the air parcels rising in the eyewall exit in the lower stratosphere in a region of approximately constant absolute temperature is questionable. In the second of these papers, Emanuel (2012) stated that

> "... Emanuel and Rotunno (2011, hereafter Part I) demonstrated that in numerically simulated tropical cyclones, the assumption of constant outflow temperature is poor and that, in the simulations, the outflow temperature increases rapidly with angular momentum."

To address these and the other issues discussed above in Section 12.1.4 concerning the β-parameterization of downdrafts outside the eyewall, a revised theory was proposed in which the absolute temperature stratification of the outflow is determined by small-scale turbulence that limits the gradient Richardson number to a critical value. Ordinarily, the Richardson number criterion demarcates the boundary between stratified shear stability and instability/turbulence. Here it seems that small-scale turbulence in the outflow layer is presumed to operate and limit the Richardson number to a critical value.

The new theory represents a major shift in the way that the storm is assumed to be influenced by its environment. In the previous version, it was assumed that the thermal structure of the lower stratosphere determined the (constant) outflow temperature. In the revised theory, the vertical structure of the outflow temperature is set internally within the vortex and, in principle, no longer matches the temperature structure of the environment.

For C_K/C_D near unity, the revised steady-state theory generally predicts a reduced intensity by a factor of approximately $1/\sqrt{2}$ compared with the original formula given above (see Emanuel and Rotunno, 2011, pp2246-2247).

However, it is difficult to assess the precise change in v_{max} between the two theories for the case of Hurricane Isabel or other observed storms because, *in the new theory, the effects of dissipative heating and the increase of the saturation specific humidity with decreasing pressure are excluded.*[15] In axisymmetric numerical model simulations used to test the new theory, (Emanuel, 2012, p. 994) used also large values of vertical mixing length to

> "... prevent the boundary flow from becoming appreciably supergradient",

thereby keeping the tests consistent with a key assumption of the theory. Because the axisymmetric numerical model includes the foregoing effects and provides good agreement with the theory, they argued that these effects must approximately cancel. This cancellation would imply that the revised theory is not an improvement to explain the discrepancy between the theory and observations on all three days of Hurricane Isabel (2003) as discussed above.

13.4 Three dimensional effects

As we have noted in Chapter 11, there are significant differences in behavior between tropical cyclone simulations in axisymmetric and three-dimensional models (Yang et al., 2007; Bryan et al., 2009; Persing et al., 2013). These studies have shown that three-dimensional models predict a significantly reduced intensity (15-20%) compared to their axisymmetric counterparts. In three-dimensional model simulations with parameter settings that are consistent with recent observations of turbulence in hurricanes, Montgomery et al. (2019) (their Fig. 4) confirmed the presence of subgrid-scale turbulent mixing in the upper-level outflow during rapid intensification and also during the quasi-steady vortex phase (not shown). However, the small-scale mixing in the upper level outflow was found to be essentially irrelevant to spin up (see their Fig. 1 and accompanying discussion, especially their Section 3.4). Given the significant questions raised in Chapter 12 regarding the E12 model, one is led naturally to question the importance of shear turbulent mixing in the mature stage for the three-dimensional vortex. In the context of the postulated importance of mixing in the upper-level outflow layer, it is important to remember that correlations found between the reduced mean gradient Richardson number in the outflow layer and the corresponding temperature stratification generally do not imply causality. In view of these considerations, it seems clear that, at a minimum, three-dimensional effects should be properly accounted for in a consistent formulation of the maximum intensity problem.

---

15. In addition, as per Section 12.3.1, the assumed boundary layer depth of 5 km in the companion intensification theory would seem incongruent with observed cyclones.

Bryan and Rotunno (2009c), p1771 pointed out that some of their own reported results

"might be specific to axisymmetric models and should someday be re-evaluated using three-dimensional simulations".

This remark would seem to apply not only to their results, but to many aspects of axisymmetric models (cf. Section 16.2).

13.5 Summary of Emanuel's steady-state PI theories

In this chapter we have reviewed Emanuel's pioneering work describing mature tropical cyclones as axisymmetric, dissipative heat engines. This formulation leads to a theory for the potential intensity of the gradient wind for these storms in an idealized environment without relative flow, vertical shear or cooling of the upper ocean. Several limitations to the basic theory exist and some have resulted in modifications of the original theory. Despite efforts to revise the theory, significant limitations of the revised theory have been noted. Certainly, the steady-state model has been influential in our thinking about the dynamics of a mature tropical cyclone. Subsequent work examining the nonlinear boundary layer dynamics of a mature cyclone provides a new perspective on the importance of the vortex boundary layer in the potential intensity problem. Forthcoming chapters will demonstrate the challenges that accompany the steady-state (or quasi-steady state) vortex model, particularly when the mature vortex is subject to strong boundary layer control.

13.6 Appendix A: Derivation of an extended PI model, Eq. (13.7)

This appendix details the derivation of the extended PI formulation for v_{max}^2 given by Eq. (13.7) in the main text. The derivation follows the semi-analytical model of Bryan and Rotunno (2009a) and provides some details not included by them. The derivation includes nonlinear, unbalanced effects in an axisymmetric, inviscid formulation of a mature tropical cyclone. In the limit of small deviations from gradient wind balance, the formulation reduces to the original formulation of E86 for the gradient wind. In the extended formulation, the inclusion of unbalanced flow effects applies strictly to the flow at the top of and above the boundary layer. The underlying boundary layer dynamics that, in reality, determine the radius of maximum tangential wind, and other terms required in the extended formulation, is represented by the crude slab formulation of the E86 model and related assumptions. Further discussion on the limitations of the formulation will be discussed in Subsection 13.6.2.2.

13.6.1 Formulation for the free troposphere

We begin by drawing two neighboring streamlines in the meridional $r - z$ plane (see Fig. 13.5). Each streamline is defined by $\psi =$ constant, where $\psi = \psi(r, z)$ is the meridional streamfunction. At low levels, the two streamlines depict inflowing moist air at two different levels in the boundary layer. At inner radii, the boundary layer inflow decelerates just outside the radius of maximum tangential wind and the air is then progressively forced to turn upwards out of the boundary layer. Immediately after leaving the boundary layer, moist air parcels are assumed to be saturated, with the value of saturation pseudo-adiabatic entropy, s^*, equal to the originating boundary layer value, i.e., $s^* = s_b$. All the air converging in the boundary layer is taken to rise in the eyewall updraft along a saturated pseudo-adiabat to the upper troposphere, where the air detrains near the tropopause and then flows outward approximately horizontally. The assumption is that deep convection is always able to ventilate any amount of air that converges (see Chapters 8 and 15 for a discussion on the limitations of this assumption).

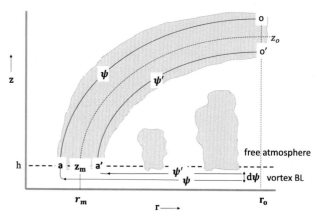

FIGURE 13.5 Radius-height schematic of the vortex boundary layer and eyewall updraft in an extended PI formulation, following the formulation of Bryan and Rotunno (2009a). The radius of maximum tangential velocity is denoted by r_m and approximately demarcates where the inflowing boundary layer air has decelerated and turned upwards to then rise out of the boundary layer at height h. The moist air is assumed to become saturated at height h and rise upwards and outwards pseudo-adiabatically to the tropopause, where the air then flows outwards approximately horizontally. The two highlighted curves denote two neighboring streamlines ψ and ψ'. By definition, the flow on these streamlines materially conserves the respective value of streamfunction. For axially symmetric motion, the specific absolute angular momentum M, saturation pseudo-adiabatic entropy s^* and total moist energy E are assumed to be materially conserved along streamlines above the boundary layer. The top of the boundary layer h is defined to be the height at which turbulent vertical momentum fluxes vanish or become constant. In this model, h is also taken to equal z_m, the height of maximum tangential velocity v_{max}. The horizontal segment of the eyewall updraft containing (r_m, z_m) is denoted $a - a'$. The vertical segment of the upper-tropospheric outflow containing the outflow maximum (r_o, z_o) is denoted $o - o'$. In the derivation, the limit $a \to a'$ and $o \to o'$ is eventually taken to furnish a relationship among the gradients of M, E, and s^* across the streamline passing through r_m, z_m in the free atmosphere above the boundary layer.

As they rise in the eyewall updraft along streamlines, air parcels are assumed to materially conserve s^* (instantaneously raining out liquid water as the vapor condenses) and to materially conserve M and total moist energy E (defined below) (Exercise 13.9). Since the flow is assumed to be steady, the cylindrical surfaces of ψ, M, s^*, and E must be congruent in the eyewall updraft (cf. Exercise 8.2). The two streamlines highlighted in Fig. 13.5 are assumed to flow just interior and exterior to the radius and height of maximum tangential velocity (r_m, z_m). Contrary to the original model of E86, r_m is assumed here to reside within the eyewall updraft at low levels and not at its outer edge (cf. Fig. 13.1).

A differential relationship among the conserved variables across streamlines may be derived by making use of the steady-state, primitive momentum, thermodynamic and continuity equations in cylindrical polar coordinates. Adopting the very reasonable approximation of neglecting the mass of water vapor in the horizontal momentum equations, the axisymmetric radial and vertical momentum equations are, respectively, as follows:

$$u\frac{\partial u}{\partial r} + w\frac{\partial u}{\partial z} - \frac{v^2}{r} - fv = -\alpha_d\frac{\partial p}{\partial r} + F_r \quad (13.10)$$

$$u\frac{\partial w}{\partial r} + w\frac{\partial w}{\partial z} = -\alpha_d\frac{\partial p}{\partial z} - g + F_z \quad (13.11)$$

where u denotes the radial velocity, w denotes the vertical velocity, p denotes the pressure, α_d the specific volume of dry air, g the gravitational acceleration, and (F_r, F_z) the divergence of the combined azimuthally-averaged eddy and Reynolds-averaged sub grid-scale momentum fluxes in the radial and vertical directions, respectively. If we restrict the foregoing equations to the region at the top of and above the boundary layer in the eyewall updraft (where the sub grid-scale turbulent fluxes are assumed to vanish or become constant), and if we neglect resolved eddy covariance processes, then we can set both of the F terms to zero.[16] Recalling that the azimuthal vorticity in the tangential direction is given by $\eta = \partial u/\partial z - \partial w/\partial r$, the transverse momentum equations may be recast into rotational-Bernoulli form as follows:

$$\frac{\partial}{\partial r}\frac{(u^2+w^2)}{2} + w\eta - \frac{v^2}{r} - fv = -\alpha_d\frac{\partial p}{\partial r}, \quad (13.12)$$

16. From Chapter 11, we know that this approximation is generally not justified in the presence of substantive eddy momentum and heat fluxes in the primary cumulus zone of the vortex in association with deep moist convection, vortex Rossby waves, inertia buoyancy waves and related nonlinear vorticity mixing - see also Schubert et al. (1999), Schecter and Montgomery (2007), and refs. However, one can plausibly argue that at the mature stage of a vortex life cycle such vorticity waves (e.g., Chapter 4) and nonlinear vorticity mixing processes act to weaken the vortex's maximum mean tangential wind, much like the eddies in the upper tropical troposphere act to weaken the subtropical jet relative to that obtained in the purely zonally-symmetric formulation of the Hadley circulation (Lindzen, 1990 and refs.).

$$\frac{\partial}{\partial z}\frac{(u^2+w^2)}{2} - u\eta = -\alpha_d\frac{\partial p}{\partial z} - g. \quad (13.13)$$

For small but otherwise arbitrary radial and vertical spatial increment (dr, dz) in the transverse plane, we multiply Eqs. (13.12) and (13.13) by $r\rho_d dr$ and $r\rho_d dz$, respectively, and add the resulting equations. Here, $\rho_d = 1/\alpha_d$ is the density of dry air. The result is

$$\rho_d r\left[\frac{\partial}{\partial r}\frac{(u^2+w^2)}{2}dr + \frac{\partial}{\partial z}\frac{(u^2+w^2)}{2}dz\right]$$
$$+ \rho_d r(w\eta dr - u\eta dz) = -r(\frac{\partial p}{\partial r}dr + \frac{\partial p}{\partial z}dz)$$
$$+ \rho_d(vrf dr + v^2 dr) - r\rho_d g dz. \quad (13.14)$$

This equation may be streamlined by defining the total differential of any smooth scalar quantity $F = F(r, z)$ as

$$dF = \frac{\partial F}{\partial r}dr + \frac{\partial F}{\partial z}dz, \quad (13.15)$$

and recalling that the differential of the streamfunction $d\psi$ defines the radial and vertical velocity fields in the anelastic formulation,

$$d\psi = \frac{\partial \psi}{\partial r}dr + \frac{\partial \psi}{\partial z}dz = -\rho_d rw dr + \rho_d ru dz. \quad (13.16)$$

Noting also that

$$\rho_d(vrf + v^2)dr = \rho_d\left(v dM - r d\frac{v^2}{2}\right), \quad (13.17)$$

enables Eq. (13.14) to be simplified:

$$\rho_d r d\left(\frac{u^2+w^2}{2}\right) - \eta d\psi =$$
$$-r dp + \rho_d v dM - \rho_d r d\left(\frac{v^2}{2} + gz\right). \quad (13.18)$$

Dividing this equation by $d\psi$, and taking the limit, gives

$$\rho_d r\frac{d}{d\psi}\left(\frac{u^2+w^2}{2}\right) - \eta =$$
$$-r\frac{dp}{d\psi} + \rho_d v\frac{dM}{d\psi} - \rho_d r\frac{d}{d\psi}\left(\frac{v^2}{2} + gz\right). \quad (13.19)$$

Now, from the first and second law of thermodynamics for a saturated air parcel, one can derive a differential equation for the saturation enthalpy:

$$T ds^* = dh - \alpha_d dp = c_{pd}dT + L_v dr_v^* - \alpha_d dp, \quad (13.20)$$

where $dh = c_{pd}dT + L_v dr_v^*$ is the moist saturation enthalpy increment per unit mass of dry air, ds^* is the approximate pseudo-adiabatic entropy increment per unit mass of a saturated air parcel, dr_v^* denotes the saturation water vapor mixing ratio increment, L_v is the latent heat of condensation (assumed constant $= L_{v0}$), and c_{pd} is the specific heat capacity at constant pressure for dry air. Recalling that for a moist air parcel rising pseudo-adiabatically, all liquid water that condenses is assumed to instantaneously rain out of the parcel; also, the heat lost by the removal of the condensate is small in comparison to the heat liberated by the phase transition from water vapor to liquid water (Exercise 13.8). *The foregoing equation (13.20) thus serves to define s^* as a state function.* Dividing Eq. (13.20) by the streamfunction increment $d\psi$ between neighboring streamlines, and taking the limit, furnishes a differential equation for the moist saturation entropy across streamlines:

$$T\frac{ds^*}{d\psi} = c_{pd}\frac{dT}{d\psi} + L_{v0}\frac{dr_v^*}{d\psi} - \alpha_d\frac{dp}{d\psi}. \quad (13.21)$$

Using this equation to solve for $-\alpha_d dp/d\psi$ allows one to eliminate the pressure derivative in Eq. (13.19):

$$\frac{d}{d\psi}\left(\frac{u^2 + w^2}{2}\right) - \frac{\eta}{\rho_d r} =$$
$$-\frac{d}{d\psi}\left(\frac{v^2}{2} + gz + c_{pd}T + L_{v0}r_v^*\right) + T\frac{ds^*}{d\psi} + \frac{v}{r}\frac{dM}{d\psi}. \quad (13.22)$$

Recalling that $v = M/r - \frac{1}{2}fr$, the last term in Eq. (13.22) may be rewritten as follows:

$$\frac{v}{r}\frac{dM}{d\psi} = \frac{1}{2r^2}\frac{dM^2}{d\psi} - \frac{f}{2}\frac{dM}{d\psi}. \quad (13.23)$$

Substituting Eq. (13.23) into Eq. (13.22) then yields a "Long's" equation (cf. Baines, 1995, Section 5.3; Long, 1953) as applied to a steady, inviscid, axisymmetric tropical cyclone:

$$\frac{d}{d\psi}\left(\frac{u^2 + w^2}{2}\right) - \frac{\eta}{\rho_d r} =$$
$$-\frac{d}{d\psi}\left(\frac{v^2}{2} + gz + c_{pd}T + L_{v0}r_v^* + \frac{fM}{2}\right)$$
$$+\frac{1}{2r^2}\frac{dM^2}{d\psi} + T\frac{dS^*}{d\psi}. \quad (13.24)$$

This Long's equation is valid within a steady-state, inviscid, axisymmetric, eyewall updraft at the top of and above the boundary layer.

Eq. (13.24) may be condensed further by defining

$$E = \frac{u^2 + v^2 + w^2}{2} + gz + c_{pd}T + L_{v0}r_v^*, \quad (13.25)$$

to be the total moist-saturated energy comprising kinetic, internal, latent heat and gravitational potential energies per

unit mass of dry air. For pseudo-adiabatic, inviscid, and axisymmetric motions in the eyewall updraft, E is materially conserved, i.e., $DE/Dt = 0$ following ascending air parcels in the updraft (Exercise 13.9, see also Exercise 2.4). The first term on the left side of Eq. (13.24) may be combined with the first term on the right side of Eq. (13.24) to give:

$$\frac{dE}{d\psi} - \frac{\eta}{\rho_d r} = -\frac{d}{d\psi}\left(\frac{fM}{2}\right) + \frac{1}{2r^2}\frac{dM^2}{d\psi} + T\frac{ds^*}{d\psi}. \quad (13.26)$$

Referring to the sketch in Fig. 13.3, we evaluate now Eq. (13.26) at the midpoint of segment $a - a'$ along the top of the boundary layer (the nominal radius of maximum tangential wind, r_m) and at the midpoint of segment $o - o'$ in the upper level outflow. The subscripts b and o will denote these values at the top of the boundary layer and in the upper-level outflow, respectively. Evaluating Eq. (13.26) at b gives

$$\frac{dE_b}{d\psi} - \frac{\eta_b}{\rho_d r_b} = -\frac{d}{d\psi}\left(\frac{fM_b}{2}\right) + \frac{1}{2r_b^2}\frac{dM_b^2}{d\psi} + T_b\frac{ds_b^*}{d\psi}, \quad (13.27)$$

while evaluation of Eq. (13.26) at o yields

$$\frac{dE_o}{d\psi} - \frac{\eta_o}{\rho_d r_o} = -\frac{d}{d\psi}\left(\frac{fM_o}{2}\right) + \frac{1}{2r_o^2}\frac{dM_o^2}{d\psi} + T_o\frac{ds_o^*}{d\psi}. \quad (13.28)$$

Since E, s^*, and M are materially conserved on streamlines for inviscid, axisymmetric pseudo-adiabatic flow in the eyewall updraft, it follows that

$$M_b = M_o, \qquad E_b = E_o, \qquad s_b^* = s_o^*, \quad (13.29)$$

and also that

$$\frac{dE_b}{d\psi} = \frac{dE_o}{d\psi}, \qquad \frac{dM_b}{d\psi} = \frac{dM_o}{d\psi}, \qquad \frac{ds_b^*}{d\psi} = \frac{ds_o^*}{d\psi}. \quad (13.30)$$

Subtracting Eq. (13.28) from Eq. (13.27), using Eqs. (13.29) and (13.30) to eliminate common terms, and taking the limit as the two streamlines become vanishingly close together, yields

$$-\frac{\eta_b}{\rho_d r_b} + \frac{\eta_o}{\rho_d r_o} =$$
$$\left(\frac{1}{2r_b^2} - \frac{1}{2r_o^2}\right)\frac{dM_b^2}{d\psi} + (T_b - T_o)\frac{ds_b^*}{d\psi}, \quad (13.31)$$

where T_b is the temperature at the top of the boundary layer and T_o is the outflow temperature. This equation serves to connect the azimuthal vorticity and various streamfunction derivatives at the top of the boundary layer with those in the upper-level outflow of the tropical cyclone.[17]

---

17. In their derivation of this equation, Bryan and Rotunno (2009a) suggested that an integration over the control volume defined by the two

In a well-developed tropical cyclone, the radius of maximum tangential wind is typically much smaller than the upper-level outflow radius, i.e., $r_m \ll r_o$. Consequently, provided η_o is not significantly larger than η_b, the foregoing Long's vortex equation may be approximated as

$$-\frac{\eta_b}{\rho_d r_b} = \frac{1}{2r_b^2}\frac{dM^2}{d\psi} + (T_b - T_o)\frac{ds^*}{d\psi}$$
$$\text{at } (r,z) = (r_m, z_m), \quad (13.32)$$

where z_m denotes the height of the maximum tangential wind. Eq. (13.32) may be simplified further by expanding

$$\frac{dM^2}{d\psi} = 2M\frac{dM}{d\psi} \quad (13.33)$$

and dividing both sides by $dM/d\psi$

$$\frac{M_b}{r_b^2} = -(T_b - T_o)\frac{\frac{ds^*}{d\psi}}{\frac{dM}{d\psi}} - \frac{\eta_b}{\rho_d r_b}\frac{d\psi}{dM}$$
$$\text{at } (r,z) = (r_m, z_m). \quad (13.34)$$

Because s^* and M must be congruent at the top of the boundary layer and above it, we have $s^* = s^*(M)$ at the base of the cloud updraft. Applying the chain rule along $z = z_m = h$ then gives

$$\frac{ds^*}{d\psi} = \frac{ds^*}{dM} \cdot \frac{dM}{d\psi}. \quad (13.35)$$

Upon multiplying both sides of Eq. (13.34) by M_b, and using the foregoing chain rule, Eq. (13.34) becomes

$$\frac{M_b^2}{r_b^2} = -M_b(T_b - T_o)\frac{ds^*}{dM} - \frac{\eta_b}{\rho_d r_b}M_b\frac{d\psi}{dM}$$
$$\text{at } (r,z) = (r_m, z_m). \quad (13.36)$$

In a well-developed tropical cyclone, the inner-core Rossby number $Ro = v_{max}/f r_m$ is large compared to unity. i.e., $Ro \gg 1$. Consequently, $M(r_m) = r_m v_{max} + f r_m^2/2 \approx r_m v_{max}$ and Eq. (13.36) simplifies to

$$v_{max}^2 = -M_b(T_b - T_o)\frac{ds^*}{dM} - \frac{\eta_b}{\rho_d r_b}M_b\frac{d\psi}{dM}$$
$$\text{at } (r,z) = (r_m, z_m). \quad (13.37)$$

Since the second term on the right side of Eq. (13.37) is not easily computed from observations or numerical mod-

els, a more convenient form of the second term may be obtained by applying the chain rule to $d\psi/dM$ along $z = z_m$:

$$\frac{d\psi}{dM} = \frac{\partial\psi}{\partial r}\frac{\partial r}{\partial M} = -\rho_b w_b r_b \frac{\partial r}{\partial M}$$
$$= -\frac{\rho_b w_b}{\zeta_b} = -\frac{\rho_b w_b r_b}{v_b}. \quad (13.38)$$

In this equation we have used the defining relation between streamfunction and vertical velocity, $-\rho r w = \partial\psi/\partial r$, together with the fact that at the radius of v_{max}, the vertical component of relative vorticity, $\zeta_b = v_b/r_b$, because $\partial v_b/\partial r = 0$ at this radius. (To avoid clutter, we have suppressed also the 'd' subscript on the dry air density.) The second term on the right-hand-side of Eq. (13.37) thus simplifies to

$$-\frac{\eta_b M_b}{\rho_d r_b}\frac{d\psi}{dM} = \frac{\eta_b M_b}{\rho_b r_b} \times \frac{\rho_b w_b r_b}{v_b} = \eta_b r_b w_b, \quad (13.39)$$

where, again, we have used the large Rossby number approximation $M \approx r v_b$ in the inner-core region. Collecting the results and cleaning up, gives an extended expression for v_{max}^2 that formally includes unbalanced effects at the top of the boundary layer:

$$v_{max}^2 = -M_b(T_b - T_o)\frac{ds_b^*}{dM} + \eta_b r_b w_b$$
$$\text{at } (r,z) = (r_m, z_m). \quad (13.40)$$

13.6.2 Boundary layer closure

The foregoing extended expression for v_{max}^2 is not a closed prediction equation since it is not yet linked with the boundary layer. To address this coupling problem we consider first the special case of near gradient wind balance dynamics. We then consider a more general case.

13.6.2.1 Gradient wind balance limit

If non-gradient wind effects are neglected based on (i) the smallness of the second term (in Eq. (13.40)) in comparison to the first term and (ii) the approximation $v_b \approx v_g$ in the boundary layer, the extended formulation of the prior subsection reduces to a more familiar equation for the maximum gradient wind:

$$v_{gmax}^2 = -M_b(T_b - T_o)\frac{ds^*}{dM}$$
$$\text{at } (r,z) = (r_{gm}, z_{gm}), \quad (13.41)$$

where, here, M_b is the absolute angular momentum using the gradient wind and it is understood that the right side is to be evaluated at the location of the maximum *gradient wind*, (r_{gm}, z_{gm}). Observations and high-resolution numerical simulations indicate that r_{gm} generally lies markedly

---

neighboring streamlines ψ and ψ' and $z > h$ was needed to obtain the result. However, this does not appear to be the case. As demonstrated here, this equation is a simple consequence of evaluating Eq. (13.26) at the two points 'b' and 'o' and subtracting the resulting equations.

outside r_m (e.g., Bell and Montgomery, 2008, Fig. 9a; Bryan and Rotunno, 2009a, Fig. 10; Sanger et al., 2014, Fig. 11; Montgomery et al., 2014, Figs. 11 & 12; Montgomery et al., 2020, Fig. 2). The chain rule formula connecting ds^*/dM to ds^*/dr at $z = z_m$, together with the large Rossby number approximation $M_g \approx r v_g$, allows Eq. (13.41) to be expressed as a thermal wind equation:

$$v_{gmax}^2 = -r_{gm}(T_b - T_o)\frac{\partial s^*}{\partial r}$$
$$\text{at } (r, z) = (r_{gm}, z_{gm}), \qquad (13.42)$$

which relates the maximum gradient wind to the radial moist saturated entropy gradient at the top of the boundary layer. This equation has proven useful in evaluating the self-consistency of the outflow temperature in estimating the PI for an observed category five hurricane (see Section 13.1 and Fig. 13.4).

When the E86 slab boundary layer (Eq. (13.4)) is invoked to connect ds^*/dM in Eq. (13.41) to the sea-to-air fluxes of moist entropy, one obtains (Exercise 13.10)

$$v_{gmax}^2 = \frac{C_K}{C_D} T_b \epsilon (s_o^* - s_b)$$
$$\text{at } (r, z) = (r_{gm}, z_{gm}), \qquad (13.43)$$

where, again, C_K is the moist enthalpy (and entropy) transfer coefficient, C_D is the surface drag coefficient, T_b is the temperature at the top of the boundary layer, $\epsilon = (T_b - T_o)/T_b$ is the thermodynamic efficiency factor, s_o^* is the saturation specific entropy at the sea surface temperature and s_b is the (unsaturated) moist entropy at anemometer level, with the latter two quantities evaluated at (r_{gm}, z_{gm}).

Bryan and Rotunno (2009c) associate the foregoing expressions for v_{gmax}^2 with the square of Emanuel's PI, or PI^2 for short, as used in Eq. (13.7) in the main text. Strictly speaking, however, *we would argue that one should avoid referring to these formulae as PI expressions since they are not closed expressions*. In particular, whereas Eq. (13.43) appears to suggest that the PI of an idealized hurricane boils down to a thermodynamic disequilibrium at one location, namely, at $r = r_{gm}$, the equation for v_{gmax}^2 is not closed in terms of initial quantities, such as the Coriolis parameter and the thermodynamic properties of the environment. In addition, although the theory is for the steady-state flow, the solution should, in principle, depend also on the structure of the initial vortex, as typified by its initial radius of maximum winds. Moreover, r_{gm} is not yet known. Even though the isothermal approximation allows one to hold the boundary layer air temperature constant as one moves inwards to the vortex center, the surface pressure at r_{gm} is not yet known, and as a result, the saturation entropy at the sea surface temperature and the boundary layer entropy at anemometer level are unknown at r_{gm} also.

In a hydrostatic and gradient balance vortex, the surface pressure at r_{gm} follows from vertically integrating the virtual temperature together with *a priori* knowledge of the pressure at some high altitude such as the tropopause, or by radially integrating the gradient wind equation inwards from large radius in conjunction with an ambient surface pressure (of 1015 mb, say). Both methods require knowledge of the radial distribution of the gradient wind beyond r_{gm}. Eq. (13.43), itself, does not provide such information. Incorporating the E86 closure assumptions and approximations, Appendix B, Section 13.7, details the solution to Eq. (13.43) that culminates in Eq. (13.6) of the main text.

13.6.2.2 General case

Bryan and Rotunno (2009c) invoked the same slab boundary layer model of E86 to connect ds^*/dM in Eq. (13.40) to the sea-to-air entropy fluxes. In this case, Eq. (13.40) may be updated using the E86 boundary layer closure given above:

$$v_{max}^2 = \frac{C_K}{C_D} T_b \epsilon (s_o^* - s_b) + \eta_b r_b w_b$$
$$\text{at } (r, z) = (r_m, z_m), \qquad (13.44)$$

where all terms in this equation have been defined above. When the first term on the right hand side of Eq. (13.44) is identified with the square of Emanuel's PI, this expression corresponds to Eq. (13.7) of the main text. Just as Eq. (13.43) does not predict the radius of maximum gradient wind (r_{gm}) or surface pressure at this radius from initial data, Eq. (13.44) suffers similar shortcomings and, indeed, requires also a theory to determine the azimuthal vorticity, η, and vertical velocity, w, at (r_m, z_m).

Once again, as in the gradient wind limit, Eq. (13.44) *should not be referred to as an extended PI formulation until a theory for these aforementioned quantities is developed*. It should be interpreted only as a diagnostic relationship that must be approximately satisfied at the maximum wind radius at the boundary layer top.

As in the E86 model (Exercise 13.2), the derivation of Eq. (13.44) assumes that the boundary layer is well mixed and that $s_b = s_b(M_b)$, i.e., that the moist entropy and absolute angular momentum surfaces are congruent in the boundary layer. While the well-mixed assumption for the scalar s_b would seem a plausible first approximation for a boundary layer dominated by shear turbulence, the congruity assumption between s_b and M_b in the boundary layer seems more tenuous. Indeed, the well mixed assumption on s_b together with the congruity assumption is inconsistent with the development of a "nose structure" in the M field (e.g., Rotunno and Bryan (2012), their Fig. 2c). Moreover, high resolution simulation experiments (e.g., Fig. 14 from Montgomery et al., 2014 and Fig. 12.6 herein) together with observational analyses summarized in Section 16.7 suggest that, in fact, θ_e is not well mixed in the boundary layer.

The development of supergradient winds in the upper portion of the boundary layer is tantamount to the development of such a nose structure in M. The net result is that invocation of the E86 boundary layer model effectively assumes that the tangential wind is close to gradient wind balance in the boundary layer.

Rotunno and Bryan (2012), p2296, conjectured that for practical purposes the E86 slab boundary layer model should serve as an adequate boundary layer closure in Eq. (13.44) since:

(i) the overshoot (the excess of the tangential wind over the gradient wind) is found to be relatively small in axisymmetric analog calculations, and

(ii) the E86 slab model represents a compromise between offsetting effects involving nonlinear boundary layer dynamics and the radial diffusion of horizontal momentum.

Their conjecture is based on the supposition that although nonlinear boundary layer dynamics act to concentrate angular momentum and moist entropy which, in turn, leads to enhanced tangential and radial wind speeds relative to the E86 theory for the gradient wind, horizontal diffusion acts to ameliorate the concentration of momentum and entropy in the boundary layer and the lofted values thereto in the vortex interior.

In Section 13.2 we argued that a more complete boundary layer formulation, which includes a radial momentum equation, is required to determine η_b, r_m, and w_b in Eq. (13.44), and thereby v_{max}. This position has been validated recently by Rotunno (2022). In Section 7.1 and Section 16.7, intensive observations of two major hurricanes analyzed by Montgomery et al. (2014) and Sanger et al. (2014) indicate that the local overshoot in tangential velocity can vary between 20% and 60%, arguably not small. Since the E86 slab model rests upon an inaccurate boundary layer approximation, this model can only provide a zero-order boundary layer closure in the original and extended axisymmetric models (cf. Montgomery et al. (2020), their Section 4.4 and refs.).

In realistic, three-dimensional, situations, there are also important issues related to the interaction of a tropical cyclone with vertical shear of the ambient horizontal flow, as well as the presence of resolved eddy processes around the maximum wind region, as noted above. For one thing, the results from Chapter 11 demonstrate that the resolved eddy processes do not always behave diffusively, contrary to what was assumed in Rotunno and Bryan (2012). For another, vertical shear is known to limit the intensity of the vortex relative to a strictly axisymmetric configuration (Riemer et al., 2010, their Figs. 1 and 2; Tang and Emanuel, 2010, their Figs. 1 and 2). The reason is because a vertically sheared cyclone opens up flow pathways for dry air ("*anti-fuel*") to negatively impact the convection in the vortex core region

(Riemer and Montgomery, 2011; Riemer et al., 2013). *Such intrusions of relatively dry air act ultimately to reduce the near-core boundary layer θ_e, thereby weakening the ability of the convection to ventilate the air leaving the boundary layer and counter the frictional outflow just above the boundary layer, which, by itself, spins down the vortex.*

On the basis of the foregoing considerations, we conclude that the influence of nonlinear boundary layer dynamics in supporting rapidly intensifying and intense storms cannot be generally dismissed in a boundary layer closure. Indeed, this conclusion is supported by the recent revision of the drag coefficient data at near major hurricane intensity conditions based on wind-wave tank experiments, which indicates an approximate 30% increase in the drag coefficient relative to prior estimates that contained a subtle coding error (Curcic and Haus, 2020). Exercise 13.6 explores the implications of these recent experimental findings further. On the basis of the above argument and recent scientific findings, the contribution of nonlinear boundary layer dynamics to the maximum intensity of the system-scale vortex in both quiescent and vertically shear environments would seem to be an important topic for further study.

13.7 Appendix B: Construction of E86 steady-state hurricane solution

This appendix sketches the necessary details to obtain Eq. (13.6) for the maximum gradient wind as presented in Section 13.1, which summarize, *inter alia*, the essential facets of the steady-state hurricane model of Emanuel (1986). As noted there, the formula for the maximum gradient wind requires two equations linking the radial variation of pressure and θ_e at the top of the vortex boundary layer. These two relationships are derived herein. The solution of these equations is shown to yield Eq. (13.43), together with closed expressions for the minimum pressure at the center of the storm, the surface pressure at the radius of maximum gradient wind, and some other properties of the flow field. The resulting solution will be shown also to support the interpretation that a steady, axisymmetric, tropical cyclone operates as a dissipative heat engine with a cycle coinciding with that of a Carnot cycle.

13.7.1 Conceptual overview

Before delving into the mathematics of the model, it is helpful to give a recap of the model's conceptual basis and key assumptions. As discussed in Section 13.1, the basis of the E86 model is encapsulated in Fig. 13.6, which is similar to Fig. 13.1. The model divides the vortex domain into three regions - I, II, and III. Regions I and II encompass the eye and eyewall regions, respectively, while Region III refers to that beyond the primary eyewall clouds. Region II is where the upward mass flux at the top of the boundary layer is large compared with the mass fluxes associated with shallow convection and precipitation-driven downdrafts. The boundary layer is taken to have uniform depth, h. The radius of maximum tangential wind, r_m, is assumed to lie at the outer edge of the eyewall cloud complex[18] and the maximum tangential wind is assumed to occur at the top of the boundary layer.

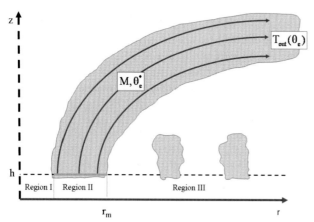

FIGURE 13.6 Schematic diagram of the secondary circulation of a mature tropical cyclone showing the eye and the eyewall clouds in the steady-state E86 model. The absolute angular momentum per unit mass, M, and saturation equivalent potential temperature, θ_e^* of an air parcel are conserved after the parcel leaves the boundary layer and ascends in the eyewall clouds. The precise values of these quantities depend on the radius at which the parcel exits the boundary layer. At radii beyond the eyewall cloud, shallow convection plays an important role in moistening and cooling the lower troposphere above the boundary layer and warming and drying the boundary layer as indicated. From Smith et al. (2008). Republished with permission from Wiley.

As discussed in the main text, the model assumes strict axial symmetry about the circulation center. The tendency of strong vortices to exhibit a high degree of axial symmetry, largely absent from strong asymmetric forcing or internal asymmetric instabilities, is an intrinsic property of coherent vortices, the essence of which is explained in Chapter 4. Therefore, the axisymmetric assumption is a reasonable zero-order approximation, at least within several hundred kilometers from the center. The model tacitly assumes also that the tangential wind is approximated everywhere, including the boundary layer, by the gradient wind. As a result, unbalanced effects as discussed in the main text (and Appendix A) are neglected. Consequently, in this model formulation, there is no distinction between the radius of maximum tangential wind speed and the radius of maximum gradient wind, r_{gm}. Above the boundary layer, the tangential wind is assumed to be in thermal wind balance and air parcels flowing upwards and outwards into the

18. As noted already in the main text, contrary to Emanuel's assumption in this figure, observations show that r_m is located very close to the inside edge of the low-level eyewall cloud (e.g. Marks et al., 2008, Fig. 3). The dynamical significance of this discrepancy is discussed in Section 13.1 and Subsection 13.6.2.2.

upper troposphere are assumed to materially conserve their absolute angular momentum per unit mass, M, and saturation moist entropy per unit mass, s^* (ostensibly calculated reversibly, but see pertinent footnote in Section 13.1). The corresponding surfaces of M and s^* are assumed to flare out monotonically in the upper troposphere and remain congruent *at all radii*. Further discussion on the limitations of the assumed congruity of M and reversible θ_e^* in Regions I and III of the steady vortex is presented in the main text. Finally, to address the competing effects of surface moisture fluxes, convective scale downdrafts and Ekman subsidence outside of the main eyewall updraft, the model invokes a zero-order moisture closure in which the near-surface relative humidity, RH, is assumed invariant, extending inwards to r_m.

13.7.2 Deductions from thermal wind balance and moist neutrality

As noted in the main text, s^*, is defined as

$$s^* = c_{pd} \ln \theta_e{}^*, \tag{13.45}$$

where θ_e^* is the (ostensibly) reversible saturation equivalent potential temperature and c_p denotes the specific heat at constant pressure of dry air. Because the saturation vapor pressure of moist air is a unique function of temperature, both s^* and θ_e^* are state variables.

In pressure coordinates, the gradient-wind and hydrostatic equations may be written, respectively, as:

$$g\left(\frac{\partial z}{\partial r}\right)_p = \frac{v^2}{r} + fv = \frac{M^2}{r^3} - \tfrac{1}{4}rf^2, \tag{13.46}$$

and

$$g\left(\frac{\partial z}{\partial p}\right)_r = -\alpha, \tag{13.47}$$

where

$$M = rv + \frac{1}{2}fr^2, \tag{13.48}$$

is the absolute angular momentum per unit mass, α is the specific volume, p is the pressure, z is the height of a pressure surface and g is the acceleration due to gravity. Eliminating the geopotential of the pressure surface, gz, gives an alternative form of the thermal wind equation:

$$\frac{1}{r^3}\left(\frac{\partial M^2}{\partial p}\right)_r = -\left(\frac{\partial \alpha}{\partial r}\right)_p. \tag{13.49}$$

Since s^* is a state variable, α can be regarded as a function of p and s^*. Applying the chain rule, the thermal wind equation (13.49) becomes

$$\frac{1}{r^3}\left(\frac{\partial M^2}{\partial p}\right)_r = -\left(\frac{\partial \alpha}{\partial s*}\right)_p\left(\frac{\partial s^*}{\partial r}\right)_p. \tag{13.50}$$

E86 invokes next one of Maxwell's thermodynamic relations for moist saturated air in the form

$$\left(\frac{\partial \alpha}{\partial s^*}\right)_p = \left(\frac{\partial T}{\partial p}\right)_{s*}, \qquad (13.51)$$

so that Eq. (13.50) becomes

$$\frac{1}{r^3}\left(\frac{\partial M^2}{\partial p}\right)_r = -\left(\frac{\partial T}{\partial p}\right)_{s*}\left(\frac{\partial s^*}{\partial r}\right)_p. \qquad (13.52)$$

With the assumption that M and s^* surfaces are everywhere congruent above the boundary layer, i.e., $M = M(s^*)$, Eq. (13.52) becomes

$$\frac{2M}{r^3}\left(\frac{\partial M}{\partial p}\right)_r = -\left(\frac{\partial T}{\partial p}\right)_{s*}\frac{ds^*}{dM}\left(\frac{\partial M}{\partial r}\right)_p. \qquad (13.53)$$

Note that $(\partial T/\partial p)_{s*}$ is just the temperature lapse rate as a function of pressure along a moist adiabat. Now, along an M surface,

$$\left(\frac{\partial M}{\partial r}\right)_p dr + \left(\frac{\partial M}{\partial p}\right)_r dp = 0, \qquad (13.54)$$

so that the slope of an M surface in (r, p) space is

$$\left(\frac{dr}{dp}\right)_M = -\left(\frac{\partial M}{\partial p}\right)_r \Big/ \left(\frac{\partial M}{\partial r}\right)_p. \qquad (13.55)$$

Combining Eqs. (13.53) and (13.55), the thermal wind equation (13.52) becomes

$$\frac{1}{2}\left(\frac{dr^{-2}}{dp}\right)_M = -\frac{1}{2M}\left(\frac{\partial T}{\partial p}\right)_{s*}\frac{ds^*}{dM}, \qquad (13.56)$$

which may be integrated upwards along the M (or s^*) surfaces, starting from the top of the boundary layer $z = h$ to an outer radius r_{out}. The result is

$$\frac{1}{r^2}\Big|_M - \frac{1}{r_{out}^2}\Big|_M = -\frac{1}{M}\frac{ds^*}{dM}[T - T_{out}(s^*, p_{out})], \qquad (13.57)$$

where the M subscript denotes evaluation along a particular M surface. Assuming that $r_{out} \gg r$, and using the chain rule, Eq. (13.57) becomes

$$-[T_B - T_{out}(s^*, p_{out})]\frac{\partial s^*}{\partial r} = \frac{1}{2r^2}\frac{\partial M^2}{\partial r} \quad \text{at } z = h, \qquad (13.58)$$

where T_B is the temperature at the top of the boundary layer and T_{out} is the outflow temperature along the M (or s^*) surface at r_{out}. Using the Exner function, $\pi = (p/p_o)^\kappa$, instead of pressure, the gradient wind equation (13.46) takes the form

$$M^2 = r^3\left[c_{pd}T_B\left(\frac{\partial \ln \pi}{\partial r}\right)_z + \frac{1}{4}rf^2\right]. \qquad (13.59)$$

In the expression for π, $\kappa = R_d/c_{pd}$, where R_d is the specific gas constant for dry air and p_o is a constant reference pressure, taken by E86 to be 1015 mb. Substituting Eq. (13.59) into Eq. (13.58) results in

$$-\frac{T_B - T_{out}(s^*, p_{out})}{T_B}\frac{\partial \ln \theta_e^*}{\partial r} = \frac{\partial \ln \pi}{\partial r}$$
$$+ \frac{\partial}{\partial r}\left(r\frac{\partial \ln \pi}{\partial r}\right) + \frac{1}{2}\frac{rf^2}{c_{pd}T_B} \quad \text{at } z = h, \qquad (13.60)$$

where it is assumed that $\theta_e = \theta_e^*$ at $z = h$. This equation may be integrated once more with respect to radius from r to some large radius $r = r_o$, where it is assumed that $\ln \pi/\pi_a$ and its radial derivative vanish, π_a being the value of π at $z = h$ and $r = r_o$. Remembering that T_B is assumed constant, the result is:

$$-\ln \theta_{ea}^* + \ln \theta_e^* + \frac{1}{T_B}\int_r^{r_o} T_{out}(s^*, p_{out})\frac{\partial \ln \theta_e}{\partial r}dr$$
$$= \ln \pi_a - \ln \pi + \frac{1}{2}\left(r\frac{\partial \ln \pi}{\partial r}\right)_a - \frac{1}{2}\left(r\frac{\partial \ln \pi}{\partial r}\right)$$
$$+ \frac{1}{4}\frac{f^2}{c_{pd}T_B}(r_o^2 - r^2) \quad \text{at } z = h. \qquad (13.61)$$

Following E86, a mean outflow temperature may be defined as follows:

$$\bar{T}_{out} = \frac{1}{\ln(\theta_e^*/\theta_{ea}^*)}\int_{\ln\theta_{ea}^*}^{\ln\theta_e^*} T_{out}\, d\ln\theta_e^*, \qquad (13.62)$$

which is an average outflow temperature weighted by the saturation moist entropy of the outflow angular momentum surfaces. Remembering that θ_e^* along angular momentum surfaces is taken equal to the equivalent potential temperature, θ_e, where these surfaces meet the top of the boundary layer, Eqs. (13.61) and (13.62) then furnish Eq. (18) of E86:

$$-\frac{T_B - \bar{T}_{out}}{T_B}\ln\left(\frac{\theta_e}{\theta_{ea}}\right) = \ln\left(\frac{\pi}{\pi_a}\right)$$
$$+ \frac{1}{2}\left(r\frac{\partial \ln \pi}{\partial r}\right) + \frac{1}{4}\frac{f^2}{c_{pd}T_B}(r^2 - r_o^2)$$
$$\text{at } z = h. \qquad (13.63)$$

This relationship at the boundary layer top ($z = h$) between $\pi(r)$ (pressure) and $\theta_e(r)$, which is valid at all radii where M- and s^*-surfaces coincide, exerts a powerful constraint on the structure of the mature, steady, axisymmetric tropical cyclone in the E86 model.

To complete the solution, a second relationship between π and θ_e is needed along the boundary layer top $z = h$. This relationship is obtained in the next subsection after considering first the constraints of the boundary layer flow in

Regions I, II, and III. The solution will be obtained for Regions I + II and III, separately, and then ultimately matched at the maximum wind radius, r_{gm}.

13.7.3 Boundary layer constraints

13.7.3.1 Model for Regions I + II

In Regions I + II we consider the dynamics and thermodynamics of the boundary layer, which imposes a further constraint on the steady solution as it determines the radial distribution of both M and θ_e exiting the top of the layer. Referring to Chapter 6, it is possible to define a streamfunction ψ for the flow in the boundary layer, given by:

$$\rho r u_b = -\frac{\partial \psi}{\partial z}, \quad \rho r w_b = \frac{\partial \psi}{\partial r}. \quad (13.64)$$

Then

$$u_b = -\frac{\psi(r, h)}{\rho r h}, \quad w_b = \frac{1}{\rho r} \frac{\partial \psi}{\partial r}. \quad (13.65)$$

Now, let φ_b be the absolute angular momentum M or the moist entropy, s. Then under steady state conditions φ_b satisfies

$$\psi(r, h)\frac{d\varphi_b}{dr} - r w_h(\varphi_h - \varphi_b) = -\frac{r}{\rho}\tau_\varphi(r, 0), \quad (13.66)$$

where τ_φ is the surface flux of φ. If the flow is out of the boundary layer, $(w > 0)$, then $\varphi_h = \varphi_b$ and neglecting shallow convection (see Chapter 9), φ_b satisfies

$$\psi(r, h)\frac{d\varphi_b}{dr} = -\frac{r}{\rho}\tau_\varphi(r, 0), \quad (13.67)$$

whereas if it is into the boundary layer $(w < 0)$,

$$\psi(r, h)\frac{d\varphi_b}{dr} = -\frac{r}{\rho}\tau_\varphi(r, 0) + r w_h(\varphi_h - \varphi_b). \quad (13.68)$$

Then, in Regions I + II, where the flow is out of the boundary layer and where convective downdrafts may be neglected,

$$\left.\frac{\partial s}{\partial M}\right|_{z=h} = \left.\frac{\tau_s}{\tau_M}\right|_{z=0}. \quad (13.69)$$

In the derivation of Eq. (13.69) it is assumed that the specific entropy, s, and the equivalent potential temperature, θ_e, are essentially vertically uniform across the boundary layer and that the air at the top of the sub-cloud layer is saturated so that $s_b = s^*$ and $\theta_e = \theta_e^*$ there.[19]

The standard aerodynamic formulae are used to quantify the surface fluxes of entropy and angular momentum:

$$\tau_s = -c_p C_K |\mathbf{v_s}|(\ln \theta_e - \ln \theta_{es}^*)$$
$$\tau_M = -C_D |\mathbf{v_s}| r \mathbf{v_s}, \quad (13.70)$$

where $|\mathbf{v_s}|$ is the magnitude of the surface horizontal velocity, v_s is the tangential component of the surface wind, C_K and C_D are exchange coefficients[20] for enthalpy and momentum, respectively, and θ_{es}^* is the saturation equivalent potential temperature at the sea surface temperature. If the relations (13.70) are inserted into Eq. (13.69) and $\partial s^*/\partial M$ is eliminated between Eqs. (13.69) and (13.58) one obtains at $z = h$

$$\ln \theta_e = \ln \theta_{es}^* - \mu \frac{C_D}{C_k} \frac{1}{c_{pd}(T_B - \bar{T}_{out})}\left(v^2 + \tfrac{1}{2}rfv\right) \quad (13.71)$$

where Eq. (13.48) has been used to express M in terms of v and μ is the ratio of v_s to v. Note that C_D and C_K do not enter separately, but only as a ratio. Since $v(r)$ is related to $p(r)$ via the gradient wind relation, we see that Eq. (13.71) provides an additional constraint relating θ_e and $p(r)$. The other constraint is Eq. (13.63).

As long as $rf \ll v$, the second term in parentheses on the right side of Eq. (13.71) can be neglected in comparison with v^2. Eq. (13.71) may then be simplified and rewritten as

$$\mu v_g^2 = \frac{C_K}{C_D}c_{pd}(T_B - \bar{T}_{out})(\ln \theta_{es}^* - \ln \theta_{es}) \text{ at } z = h, \quad (13.72)$$

where $\mu = v_s/v$ is a reduction factor for the tangential wind. In the E86 model, the tangential wind is assumed to be essentially uniform in the boundary layer[21] so that for consistency μ must be taken as unity.

In Regions I + II the tangential flow speed is relatively large and the radius of curvature is relatively small, so that $rf \ll v$. Here the gradient wind is well approximated by cyclostrophic balance:

$$v_g^2 \approx c_{pd}T_B r \frac{\partial}{\partial r}\ln \pi. \quad (13.73)$$

Eq. (13.72) states that in Regions I and II, v^2 is determined by the product of the temperature difference between the mean boundary layer and upper level outflow layer and

---

19. While this assumption is valid at cloud base in Region II, it does not represent the state of affairs inside the eye of the storm (Region I) where the air just above the boundary layer top is relatively dry and unsaturated. Also, the M and s^* surfaces are generally not congruent in the eye. The saturation and congruity assumptions are nevertheless imposed in the model to obtain a zero-order description of the vortex inside r_{gm}. A revision of this formulation inside the storm's r_{gm} is presented in Emanuel (1995).

20. Assumed constant for simplicity.
21. From E86 (p593): "Consider the conservation equation for any conservative variable c assumed to be well mixed in the vertical within a turbulent boundary layer of depth h. ... We take as two conservative variables angular momentum M (although it may not be well mixed in the vertical) and moist entropy, s." Nevertheless, E86 proceeds to treat M as well mixed in the vertical: "If M and s are used for c in (34) and the resulting two equations are divided into each other, the result is ..."

the thermodynamic disequilibrium between the air in the well-mixed portion of the boundary layer and the sea surface. Since from Eq. (13.73) the square of the cyclostrophic wind is proportional to the radial gradient of $\ln \pi$, v_g^2 may be eliminated in Eq. (13.72) so that Eqs. (13.63) and (13.72) comprise a two-by-two system for the logarithm of the Exner function and the logarithm of equivalent potential temperature as a function of radius in Regions I and II (at the top of the boundary layer (cf. E86, Eqs. (41) and (45)). If $\theta_e(z = h)$ $(= \theta_{es})$ is further eliminated between Eqs. (13.72) and (13.63), a single equation for $\ln \pi$ is obtained (Eq. (13.90) below). Once this equation is solved, the corresponding tangential velocity profile in Regions I + II may be deduced through either Eq. (13.73) or Eq. (13.72). Of course, the solution for the Exner function in Regions I + II must match the solution in Region III at the radius of maximum gradient wind r_{gm}. The details of this inner solution and its matching with the exterior solution are presented below in Section 13.7.5.

As a side note, remembering the thermodynamic efficiency factor, $\epsilon = (T_B - T_{out})/T_B$, and the definition of s^* given by Eq. (13.45), Eq. (13.72) evaluated at r_{gm} with $\mu = 1$ becomes Eq. (13.43) from Appendix A:

$$v_{gmax}^2 = \frac{C_K}{C_D} T_B \epsilon \, (s_0^* - s_b) \ \text{at} \ (r, z) = (r_{gm}, z_{gm}),$$

with v_{gmax} equal to $v(r = r_m)$ in the notation of this section and $r_m = r_{gm}$, etc.

It is worth re-iterating that Eq. (13.72) cannot be used to solve for the Exner function in Region III because the fluxes of moist entropy into the boundary layer from above it, which were neglected in deriving Eq. (13.69), make an essential contribution to the θ_e budget in real tropical cyclones. That is, outside of r_{gm}, entropy fluxes at the top of the boundary layer associated with convective downdrafts and Ekman suction in regions of anticyclonic relative vorticity act to transport relatively low θ_e air from above the boundary layer into the layer. Indeed, E86 inferred from Eq. (13.71) that without the incorporation of these fluxes, the boundary layer balance given by Eq. (13.71) (or Eq. (13.72)) yields absurd results for the radial distribution of θ_e outside of r_{gm}. The upshot is that an alternative solution method is needed in Region III.

13.7.3.2 Model for Region III

As noted in the foregoing discussion, in Region III we would have to use Eq. (13.68) rather than Eq. (13.67), but we do not have an expression for w_h. E86 circumvented this problem by assuming that the combined effect of boundary-layer-induced subsidence and turbulent fluxes at the top of the boundary layer is to keep the relative humidity (RH) of the boundary layer at a uniform level (typically about 80%). This step allows one to obtain the desired second relation-

ship between $\theta_e(r)$ and $p(r)$ at the boundary layer top in Region III.

The derivation of the second relationship starts from the approximate formula for θ_e:

$$\ln \theta_e = \ln T - \ln \pi + \frac{L_v r_v}{c_{pd} T}, \tag{13.74}$$

where r_v is the water vapor mixing ratio and L_v is the latent heat of condensation (assumed constant for simplicity). The model assumes an isothermal boundary layer in which θ_e is well mixed between the top of the surface layer ($z = z_s$) and the top of the boundary layer ($z = h$). The well-mixed assumption implies that $\theta_e(z = z_s) = \theta_e(z = h)$. In other words, $\theta_{es} = \theta_{eh}$. As employed above, a subscript 's' denotes evaluation of a quantity at the top of the surface layer and a superscript '*' denotes a saturation value at the indicated levels.

With the assumption that the air temperature at the sea surface does not vary with radius (isothermal approximation), and that θ_e is uniform through the depth of the boundary layer above the surface layer (well-mixed approximation), we obtain:

$$\ln \theta_e - \ln \theta_{ea} = -\ln \pi + \ln \pi_a + \frac{L_v}{c_{pd} T_s}(r_v - r_{va})$$
$$\text{at} \, z = h, \tag{13.75}$$

where a suffix "a" refers to ambient value at suitably large radius (r_o) and at height $z = h$. The hydrostatic equation may be written as $-d\ln \pi / dz = g/(c_{pd} T)$, whereupon

$$\ln \pi_{(z=h)} = \ln \pi_{(z=0)} - \int_0^h \frac{g \, dz'}{c_{pd} T(z')}. \tag{13.76}$$

If T does not vary with radius, it follows that $\ln(\pi_h/\pi_s) = \ln(\pi_{ha}/\pi_{sa})$, where the suffix "h" refers to values at $z = h$ and the suffix "s" refers to surface values. The foregoing relation implies that

$$\ln \frac{\pi_h}{\pi_{ha}} = \ln \frac{\pi_s}{\pi_{as}}. \tag{13.77}$$

Eq. (13.75) then becomes

$$\ln \frac{\theta_e}{\theta_{ea}} = -\ln \frac{\pi_s}{\pi_{as}} + \frac{L_v}{c_{pd} T_s}(r_{vs}^* RH_s - r_{va}^* RH_{as}), \tag{13.78}$$

where we have used the fact that in the well mixed layer, $r_v = r_s = r_{vs}^* RH_s$ and $r_{va} = r_{vas} = r_{va}^* RH_{as}$. Now, since r_v is essentially equal to the specific humidity, we have $r_v^* \approx \epsilon e^*(T_B)/p$, whereupon

$$r_{vs}^* = r_{va}^*(p_{as}/p_s) = r_{va}^*(\pi_{as}/\pi_s)^{1/\kappa}$$
$$= r_{va}^* \exp[-(1/\kappa) \ln(\pi_s/\pi_{as})]. \tag{13.79}$$

Here, $e^*(T)$ denotes the saturation vapor pressure as a function of temperature. Since $\ln(\pi_s/\pi_{as})$ is small compared to unity (even for a mature tropical cyclone), a Taylor expansion of the exponential function, retaining only the first two terms, gives

$$r_{va}^* \exp\left(-\frac{1}{\kappa}\ln\frac{\pi_s}{\pi_{as}}\right) \approx r_{va}^*\left(1 - \frac{1}{\kappa}\ln\frac{\pi_s}{\pi_{as}}\right). \quad (13.80)$$

Substituting Eq. (13.80) into Eq. (13.78) yields

$$\log\frac{\theta_e}{\theta_{ea}} \approx -\ln\frac{\pi_s}{\pi_{as}}\left[1 + \frac{L_v r_{va}^* R H_{as}}{R_d T_s}\right]$$
$$+ \frac{L r_{va}^*}{c_{pd} T_s}(RH_s - RH_{as}). \quad (13.81)$$

Eq. (13.81) connects the variation of the boundary layer θ_e with the boundary layer pressure and near-surface relative humidity.

We may now specialize Eq. (13.81) to Region III wherein, as noted above, the surface relative humidity is assumed constant with radius and equal to the ambient value, i.e., $RH_s(r) = RH_{as}$. Eq. (13.81) then simplifies to

$$\ln\frac{\theta_e}{\theta_{ea}} = -\ln\frac{\pi_s}{\pi_{as}}\left[1 + \frac{L_v r_{va}^* R H_{as}}{R_d T_s}\right]. \quad (13.82)$$

This is the desired second relationship between θ_e and $\ln\pi$ in Region III, but we note that it was obtained from thermodynamic considerations alone. This equation is the same as Eq. (25) of E86 if one assumes that the reference pressure in the definition of the Exner function is p_{as} ($= 1015$ mb) rather than 1000 mb as is usual. In this event, we have $\pi_{as} = 1$, $\ln\pi_{as} = 0$ and thus

$$\ln\frac{\theta_e}{\theta_{ea}} = -\ln\pi_s\left[1 + \frac{L_v r_{va}^* R H_{as}}{R_d T_s}\right]. \quad (13.83)$$

13.7.4 Solution for $\ln\pi$ in Region III

When the foregoing expression (13.83) is inserted into Eq. (13.63), and the mean outflow temperature (which depends on the yet unknown distribution of M and θ_e) is approximated by the *a priori* T_o, the differential equation (13.63) becomes

$$\frac{1}{2}\left(r\frac{\partial\ln\pi}{\partial r}\right) + \beta\ln\pi_s = \frac{1}{4}\frac{f^2}{c_{pd}T_B}(r_o^2 - r^2)$$
$$\text{at } z = h, \quad (13.84)$$

where

$$\beta = 1 - \epsilon\left(1 + \frac{L_v r_{va}^* R H_{as}}{R_d T_s}\right), \quad (13.85)$$

and $\epsilon = (T_B - T_o)/T_B$.

In Eq. (13.84), the previously derived relationship $\ln(\pi/\pi_a) = \ln\pi_s$ has been used to eliminate $\ln\pi_a$. For simplicity, in the differential equation (13.84) we approximate $\ln\pi_s$ as $\ln\pi$ so that the Exner functions are evaluated at the same pressure level thereby making the differential equation analytically solvable.

Since the differential equation (13.84) for $\ln\pi$ is linear and of the first order, the equation can be solved exactly. The solution is obtained by multiplying both sides of Eq. (13.84) by the integrating factor $r^{2\beta}$, integrating radially from r to r_o and invoking an outer boundary condition: Here, we follow E86 (p594) and assume $\ln\pi = 0$ at $r = r_o$.[22] The solution to Eq. (13.84) is then (Exercise 13.4):

$$\ln\pi(r, z = h) = \frac{f^2}{4c_{pd}T_B}\left[\frac{r_o^2}{\beta} - \frac{r^2}{(1+\beta)}\right.$$
$$\left. - \left(\frac{r_o}{r}\right)^{2\beta}\frac{r_o^2}{\beta(1+\beta)}\right]. \quad (13.86)$$

The tangential velocity (gradient wind) at the top of the boundary layer is obtained using Eq. (13.86) and solving the gradient wind equation (13.59) for v. It is worth pointing out that because the tangential wind at the top of Region III is obtained from thermodynamic considerations in the boundary layer, together with assumptions of thermal wind balance and congruity of M and reversible s^* surfaces above the boundary layer, the *dynamics* of the boundary layer in Region III have been ignored altogether. To us, this seems to be a limitation of the model that may become important when a deeper understanding of the structure of the outer circulation is sought. This topic will be considered in the sequel to this book.

13.7.5 Solution for $\ln\pi$ in Regions I + II

For the region inside r_{gm} (Regions I + II), the distribution of pressure is determined from Eqs. (13.72), (13.73), and (13.63), again with the mean upper-level outflow temperature approximated by the *a priori* outflow temperature estimate T_o. Remembering that for $r \leq r_{gm}$ the gradient wind may be approximated by the cyclostrophic wind, Eq. (13.73) may be substituted into Eq. (13.72) to eliminate v^2. Eq. (13.72) then becomes:

$$\epsilon\ln\theta_{es} \approx \epsilon\ln\theta_{es}^* - \frac{C_D}{C_K}r\frac{\partial\ln\pi}{\partial r} \quad (13.87)$$

If Eqs. (13.87) and (13.63) are added together we obtain

$$\epsilon\ln\frac{\theta_{ea}}{\theta_{esa}^*} = \epsilon\ln\frac{\theta_{es}^*}{\theta_{esa}^*} + \left(\frac{1}{2} - \frac{C_D}{C_K}\right)r\frac{\partial\ln\pi}{\partial r}$$

---

22. Strictly speaking, the reference pressure p_0 at $r = r_o$ is unequal to the ambient pressure at the boundary layer top. This inconsistency notwithstanding, this approximation seems a reasonable expedient for the far-field pressure and has the effect of constraining the Exner function to vary weakly in the vertical across the vortex.

$$+ \ln \frac{\pi}{\pi_a} + \frac{f^2}{4c_{pd}T_B}(r^2 - r_o^2) \qquad (13.88)$$

where for convenience $-\ln\theta^*_{esa}$ has been added to both sides of Eq. (13.88) and θ^*_{esa} denotes the logarithm of ambient saturation equivalent potential temperature at the sea surface temperature. We use next Eq. (13.83) to express $\ln(\theta^*_{es}/\theta^*_{esa})$ in terms of pressure and other parameters. That is,

$$\ln \frac{\theta^*_{es}}{\theta^*_{esa}} = -\ln\pi_s \left[1 + \frac{L_v r_{va}^*}{R_d T_s} \right]. \qquad (13.89)$$

Substituting Eq. (13.89) into Eq. (13.88) gives:

$$\left(\frac{1}{2} - \frac{C_D}{C_K} \right) r \frac{\partial \ln \pi}{\partial r} + \ln\pi_s \left[1 - \epsilon \left(1 + \frac{L_v r_{va}^*}{R_d T_s} \right) \right]$$
$$= -\epsilon \ln \frac{\theta^*_{esa}}{\theta_{ea}} + \frac{f^2}{4c_{pd}T_B}(r_o^2 - r^2). \qquad (13.90)$$

In the foregoing expression, the relation $\ln\pi_h/\pi_a = \ln\pi_s$ has been used to simplify the equation. Recalling the definitions of $\ln\theta^*_{esa}$ and $\ln\theta_{ea}$, the first term on the right side of Eq. (13.90) may be related to the ambient saturation mixing ratio at the top of the surface layer and sea surface temperature T_s (Exercise 13.3). We have

$$\ln\theta^*_{esa} = \ln T_s - \ln\pi_s + \frac{L_v r_{va}^*}{c_{pd}T_s}, \qquad (13.91)$$

and

$$\ln\theta_{ea} = \ln T_s - \ln\pi_s + \frac{L_v r_{va}}{c_{pd}T_s}. \qquad (13.92)$$

Therefore

$$\ln \frac{\theta^*_{esa}}{\theta_{ea}} = \frac{L_v}{c_{pd}T_s}(r_{va}^* - r_{va}) = \frac{L_v r_{va}^*}{c_{pd}T_s}(1 - RH_{as}), \quad (13.93)$$

where the foregoing relation $r_{va} \approx RH_{as}r_{va}^*$ has been used to obtain the final expression in Eq. (13.93). Inserting Eq. (13.93) into Eq. (13.90), and again approximating $\ln\pi_s$ as $\ln\pi$ in Eq. (13.90), the equation governing the radial variation of π on $z = h$ in Regions I + II becomes Eq. (40) of E86:

$$\left(\frac{1}{2} - \frac{C_D}{C_K} \right) r \frac{\partial \ln\pi}{\partial r} + \tilde{\beta}\ln\pi =$$
$$-\epsilon \frac{L_v r_{va}^*}{c_{pd}T_s}(1 - RH_{as}) + \frac{f^2}{4c_{pd}T_B}(r_o^2 - r^2) \quad (13.94)$$

where

$$\tilde{\beta} = 1 - \epsilon \left[1 + \frac{L_v r_{va}^*}{R_d T_s} \right] \qquad (13.95)$$

where the tilde on β is introduced to note that in this model the boundary layer air in Regions I + II (for $r < r_{gm}$) is assumed to be saturated.

The differential equation (13.94) (with Eq. (13.95)) may be solved exactly. The solution (Exercise 13.4) is:

$$\ln\pi(r, z = h) = Ar^\chi - \frac{\epsilon L_v r_{va}^*(1 - RH_{as})}{c_{pd}T_s\tilde{\beta}}$$
$$+ \frac{f^2 r_o^2}{4c_{pd}T_B\tilde{\beta}} + \frac{fr^2}{4c_{pd}T_B} \frac{1}{\left[2 - \epsilon(1 + \frac{L_v r_{va}^*}{R_d T_s}) - 2\frac{C_D}{C_K} \right]}, \qquad (13.96)$$

where

$$\chi = \frac{\tilde{\beta}}{\left(\frac{C_D}{C_K} - \frac{1}{2} \right)} \qquad (13.97)$$

is the exponent controlling the radial variation of the function Ar^χ and A is a constant of integration that must be determined through matching with the exterior solution in Region III.[23]

In obtaining this interior solution, only bounded solutions are admissible. That is, as $r \to 0$, bounded solutions to Eq. (13.94) exist provided[24] $C_D/C_K > 1/2$. The inequality $C_D/C_K > 1/2$ places a potentially nontrivial constraint on the ratio of drag and enthalpy coefficients for well-behaved solutions in this model. Recall from Chapter 7 that, while there is appreciable scatter in the observational data for C_D and C_K at major hurricane wind speeds, the mean values of $C_K \approx 1.0 \times 10^{-3}$ and $C_D \approx 2.4 \times 10^{-3}$ found by Bell et al. (2012a) imply a ratio of mean coefficients $C_D/C_K \approx 2.4$, easily satisfying the inequality. The validity of the foregoing inequality will be assumed hereafter.

From Eq. (13.96), we see that the first and fourth terms on the right hand side of Eq. (13.96) vanish at $r = 0$. This implies that $\ln\pi$ at the center of the storm must equal the sum of the second and third terms on the right hand side of Eq. (13.96), i.e.,

23. The matching is discussed further in Emanuel (1986). As noted above, the variation of pressure and tangential winds inside r_{gm} was reformulated by Emanuel (1995) who assumed that the mean tangential flow in the eye is in rigid body rotation. Although Chapter 4 briefly noted the deviation from solid body rotation in the eye, which, under certain circumstances helps support the formation of polygonal eyewalls, eyewall mesovortices and related vorticity mixing therein - as discussed in Schubert et al. (1999) and subsequent refs. - we will not dwell on the precise variation of the pressure and tangential wind within r_{gm} here. The main aim of this Appendix is to derive analytical expressions for the central surface pressure and the maximum gradient wind within the steady theory of E86.

24. It is readily verified that $\tilde{\beta} > 0$ for a wide range of thermodynamic parameters relevant to the tropical atmosphere and ocean.

$$\ln \pi_c = \frac{-\dfrac{\epsilon L_v r_{va}^*}{c_{pd}T_s}(1 - RH_{as}) + \dfrac{f^2 r_o^2}{4c_{pd}T_B}}{1 - \epsilon\left(1 + \dfrac{L_v r_{va}^*}{R_d T_s}\right)}. \tag{13.98}$$

Alternatively, upon evaluating Eq. (13.63) and Eq. (13.81) at $r = 0$ and using the relationship $\ln \pi_h/\pi_a = \ln \pi_s$ there, we see that $\ln \pi$ at the surface center must equal the right side of Eq. (13.98) also. That is, at $r = 0$, the model equations (13.63) and (13.81), together with the relation $\ln \pi_h/\pi_a = \ln \pi_s$, imply that

$$\ln \pi_{cs} = \frac{-\dfrac{\epsilon L r_{va}^*}{c_{pd}T_s}(1 - RH_{as}) + \dfrac{f^2 r_o^2}{4c_{pd}T_B}}{1 - \epsilon\left(1 + \dfrac{L r_{va}^*}{R_d T_s}\right)}, \tag{13.99}$$

where π_{cs} denotes Exner function at the storm's surface center. In other words, when Eq. (13.96) is evaluated at $r = 0$, the steady model implies $\ln \pi_c = \ln \pi_{cs}$, i.e., $\ln \pi$ at *the boundary layer top equals* $\ln \pi$ *at the surface!* When Eq. (13.98) is incorporated into Eq. (13.96), the resulting expression matches exactly Eq. (41) of E86. The ostensibly minor inconsistency involving $\ln \pi_{cs}$ and $\ln \pi_c$ is possibly one reason why the model of the hurricane eye (Region I) was revised in a subsequent paper by Emanuel (1995).

13.7.6 Solution for v_{gmax}^2 and $\ln \pi_s$ at $r = r_{gm}$

In this subsection we show that, to a first approximation, the maximum gradient wind v_m is given by Eq. (13.6) from the main text. An expression for the logarithm of the surface Exner function ($\ln \pi_s$) at $r = r_{gm}$ will be obtained also that coincides approximately with Eq. (44) of E86.

The solution for the maximum gradient wind squared v_{gmax}^2 at the boundary layer top and the logarithm of the surface pressure $\ln \pi_s$ at $r = r_{gm}$ is obtained as follows. We first evaluate Eq. (13.94) at $r = r_{gm}$ and substitute the cyclostrophic balance equation (13.73) into this equation to obtain the following:

$$\left(\frac{1}{2} - \frac{C_D}{C_K}\right)\frac{v_{gmax}^2}{c_{pd}T_B} + \tilde{\beta}\ln \pi_s =$$
$$-\epsilon\frac{L_v r_{va}^*}{c_{pd}T_s}(1 - RH_{as}) + \frac{f^2}{4c_{pd}T_B}(r_o^2 - r_{gm}^2), \tag{13.100}$$

where $\tilde{\beta}$ is given by Eq. (13.95). We use next Eq. (13.63), which, in this model, is valid on $z = h$ at all radii, but we evaluate the equation outside r_{gm} and subsequently take the limit as $r \to r_{gm}^+$. To close the expression, the cyclostrophic balance equation (13.73) is used at r_{gm} together with the thermodynamic equation (13.83) as $r \to r_{gm}^+$. These substi-

tutions and limit give:

$$-\frac{v_{gmax}^2}{2c_{pd}T_B} - \beta \ln \pi_s = \frac{f^2}{4c_{pd}T_B}(r_{gm}^2 - r_o^2), \tag{13.101}$$

where

$$\beta = 1 - \epsilon\left[1 + \frac{L_v r_{va}^* RH_{as}}{R_d T_s}\right]. \tag{13.102}$$

The foregoing equations (13.100) and (13.101) comprise a two-by-two system for v_{gmax}^2 and $\ln \pi_s$, both quantities being evaluated at $r = r_{gm}$.

The solution can be found e.g., using Cramer's rule. Since we anticipate that $r_o^2 \gg r_{gm}^2$, we may neglect in a first approximation the explicit dependence of the solution for v_{gmax}^2 and $\ln \pi_s$ on r_{gm}. Of course, the validity of this approximation needs to be checked *a posteriori*, but since we lack a means of determining r_{gm} independently from r_o in the steady state problem, this approximation seems necessary at this stage. E86 derived an estimate of r_{gm} in terms of r_o from the limiting solution (see his Eq. (46) and corresponding Fig. 7). As discussed in the main text, the relationship between r_{gm} and r_o is *not* equivalent to determining r_{gm} from the initial vortex data, which includes the maximum tangential wind and the radius of maximum wind of the initial vortex, *etc.*

Under the assumption $r_o^2 \gg r_{gm}^2$, the limiting 2-by-2 system is given by:

$$\left(\frac{1}{2} - \frac{C_D}{C_K}\right)\frac{v_{gmax}^2}{c_{pd}T_B} + \tilde{\beta}\ln \pi_s =$$
$$-\epsilon\frac{L_v r_{va}^*}{c_{pd}T_s}(1 - RH_{as}) + \frac{f^2 r_o^2}{4c_{pd}T_B}, \tag{13.103}$$

$$-\frac{v_{gmax}^2}{2c_{pd}T_B} - \beta \ln \pi_s = -\frac{f^2 r_o^2}{4c_{pd}T_B}, \tag{13.104}$$

where $\tilde{\beta}$, β and other parameters in these two equations are as defined above. Cramer's rule applied to Eqs. (13.103) and (13.104), together with the approximation $T_B \approx T_s$, then gives the solution for v_{gmax}^2 (Exercise 13.6):

$$v_{gmax}^2 \approx \frac{C_K}{C_D}\epsilon L_v r_{va}^*(1 - RH_{as})\times$$
$$\frac{1 - \dfrac{f^2 r_o^2}{4\beta R_d T_B}}{1 - \dfrac{1}{2}\dfrac{C_K}{C_D}\dfrac{\epsilon L_v r_{va}^*}{\beta R_d T_s}(1 - RH_{as})}, \tag{13.105}$$

which coincides exactly with Eq. (13.6) and Eq. (43) of Emanuel (1986). The solution for $\ln \pi_s$ at $r = r_{gm}$ may be deduced by inserting the solution for v_m^2 into Eq. (13.104) and solving for $\ln \pi_s$. Explicitly, Eq. (13.104) gives:

$$\ln \pi_s = \frac{1}{\beta}\left(\frac{f^2 r_o^2}{4 c_{pd} T_B} - \frac{v_{gmax}^2}{2 c_{pd} T_B}\right). \qquad (13.106)$$

Substituting Eq. (13.105) into Eq. (13.106) yields:

$$\ln \pi_s = \frac{1}{\beta}\left(\frac{f^2 r_o^2}{4 c_p T_B} \right.$$

$$\left. \frac{C_K}{C_D}\epsilon \frac{L_v r_{va}^*}{2 c_{pd} T_B} \frac{(1 - RH_{as})\left(1 - \frac{f^2 r_o^2}{4\beta R_d T_B}\right)}{1 - \frac{1}{2}\frac{C_K}{C_D}\epsilon \frac{L_v r_{va}^*}{\beta R_d T_s}(1 - RH_{as})}\right), \qquad (13.107)$$

which may be rewritten as a single ratio

$$\ln \pi_s = \frac{\text{Num}}{\text{Den}}, \qquad (13.108)$$

where Num (Numerator) and Den (Denominator) are given by

$$\text{Num} = \frac{f^2 r_o^2}{2}\left[1 - \frac{1}{2}\frac{C_K}{C_D}\epsilon \frac{L_v r_{va}^*}{\beta R_d T_s}(1 - RH_{as})\right]$$
$$- \frac{C_K}{C_D}\epsilon L r_{va}^*(1 - RH_{as})\left(1 - \frac{f^2 r_o^2}{4\beta R_d T_B}\right) \qquad (13.109)$$

and

$$\text{Den} = 2 c_{pd} T_B\left[\beta - \frac{1}{2}\frac{C_K}{C_D}\epsilon \frac{L r_{va}^*}{R_d T_s}(1 - RH_{as})\right]. \qquad (13.110)$$

The relative contribution of terms in the numerator (13.109) may be assessed by applying typical tropical values near 20 N latitude during Boreal summer conditions: $RH_{as} = 0.80$, SST = 300 K, $L_v = 2.5 \times 10^6$ J kg^{-1}, $r_{va}^* \lesssim 20$ g kg^{-1}, $c_{pd} = 1005$ J K^{-1} kg^{-1}, $R_d = 287$ J K^{-1} kg^{-1}, $f = 5 \times 10^{-5}$ s^{-1}, $T_B \approx 300$ K, $C_K/C_D = 0.5$. Anticipating $T_o \approx 200$ K and $r_o = 500$ km for the outflow temperature and outer radius, respectively, we find $\epsilon = 1/3$, $L r_{va}^*/R_d T_s = 0.58$, $L r_{va}^*/c_{pd} T_s = 0.16$, $\beta \approx \tilde{\beta} = 0.5$ and $f^2 r_o^2/(4 c_{pd} T_B) = 4.9 \times 10^{-4}$. We determine also:

$$\frac{f^2 r_o^2}{4\beta R_d T_B} \approx 3 \times 10^{-3} \ll 1,$$
$$\frac{1}{2}\frac{C_K}{C_D}\epsilon \frac{L r_{va}^*(1 - RH_{as})}{\beta R_d T_s} \approx 0.01 \ll 1. \qquad (13.111)$$

The smallness of the latter two dimensionless quantities permits the neglect of corresponding terms in the numerator (13.109) and hence a simplified formula for $\ln \pi_s$ at r_{gm}:

$$\boxed{\ln \pi_s \approx \frac{-\frac{1}{2}\frac{C_K}{C_D}\epsilon \frac{L r_{va}^*}{c_{pd} T_B}(1 - RH_{as}) + \frac{f^2 r_o^2}{4 c_{pd} T_B}}{\beta - \frac{1}{2}\frac{C_K}{C_D}\epsilon \frac{L r_{va}^*}{R_d T_s}(1 - RH_{as})},}$$
$$(13.112)$$

which is identical to Eq. (44) of Emanuel (1986).[25]

13.7.7 The complete solution

The steady flow field above the boundary layer may be obtained by evaluating quantities along absolute angular momentum M surfaces whose shape is given by Eq. (13.57), which may be written as

$$\left.\frac{1}{r^2}\right|_M = \frac{1}{M}\frac{ds^*}{dM}[T - T_{out}(s^*, p_{out})], \qquad (13.113)$$

on the assumption that $r \ll r_{out}$. (The reader is referred to Emanuel (1986) for details of the calculations.) E86 shows an example of a calculation for the following parameter values: $T_s = 27\,°$C, $T_B = 22\,°$C, $T_{out} = -67\,°$C, f evaluated at $20°$ latitude, $p_o = 1015$ mb, $r_o = 400$ km, $C_\theta = C_D$, $RH_a = 80\%$, and $\gamma = 2$, corresponding to a Brunt-Väisälä frequency of 1.5×10^{-2} s^{-1}. Under these conditions the central pressure is 941 mb, the maximum tangential wind speed is 58 m s^{-1}, the radius of maximum winds is 36 km, and the ambient boundary layer θ_e is 349 K. The distributions of M, θ_e^*, v, and the temperature perturbation from the far environment at the same altitude are shown in Fig. 13.7.

The solution captures the main observed features of a mature tropical cyclone including the warm core at high altitude, the outward-sloping velocity maximum, and the strong radial gradient of θ_e^* near and inside the radius of maximum tangential wind speed. As discussed in the Section 13.2 and Appendix A, this solution does not capture the supergradient winds generated in the nonlinear boundary layer inside the radius of maximum gradient wind. In favorable environments, and assuming only a minor weakening by asymmetric vortex Rossby waves and vorticity mixing in the region bordering the eye/eyewall, this supergradient excess can lead to a significant increase of tangential wind speed and heightened damage to property relative to the gradient wind theory.

E86 estimated the streamfunction at the top of the boundary layer assuming that Eq. (13.67) gives the correct momentum balance in the boundary layer without considering turbulent fluxes at the top of the layer, even if the neglect of such fluxes yields an incorrect heat budget. Setting $\varphi_b = M$ in Eq. (13.67) and using Eq. (13.67) we can solve for the boundary-layer streamfunction, from which we can obtain the vertical velocity using $\rho r w_h = \partial \psi/\partial r$.

---

25. Allowance being made for the minor discrepancy between $\ln \pi_s$ and $\ln \pi$ in Eq. (13.90) as discussed above.

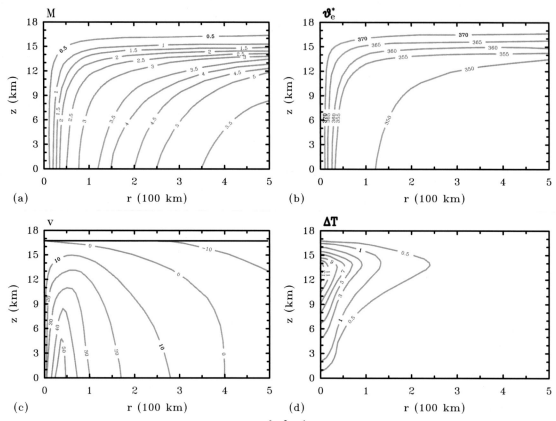

FIGURE 13.7 Distributions of: (a) absolute angular momentum, M (10^6 m^2 s^{-1}), (b) saturation equivalent potential temperature, θ_e^* (K), (c) gradient wind v, (m s^{-1}), and (d) temperature departure, ΔT (°C) from the far environment at the same altitude, for the vortex discussed above. Adapted from Emanuel (1986). Republished with permission of the American Meteorological Society.

The mean radial velocity in the boundary layer is given by $r\bar{u} = -\psi/(\bar{\rho}h)$, where h is the nominal depth of the layer and $\bar{\rho}$ is the mean density. The radial distributions of w_h and $\bar{u}$ for the vortex described above are shown in Fig. 13.8. These calculations are based on the assumptions that $\bar{\rho}$ and h are constants, with $h = 1$ km, and $C_D = 2 \times 10^{-3}$.

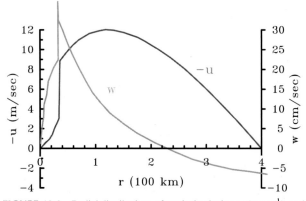

FIGURE 13.8 Radial distributions of vertical velocity, w (cm s^{-1}), and mean radial velocity, u (m s^{-1}), within the boundary layer for the vortex discussed above. Adapted from Emanuel (1986). Republished with permission of the American Meteorological Society.

The vertical velocity profile in Fig. 13.8 shows a sharp peak at the radius of maximum tangential wind (36 km), but the radial velocity reaches its maximum *at a much larger radius of about 120 km*. The streamfunction has a discontinuity at r_{gm} as a consequence of matching two separate boundary layers there, since the radial gradient of angular momentum is discontinuous. This results in a jump in u and a delta function spike of vertical velocity at r_{gm}. According to E86, these unrealistic features would not be present had a single boundary layer representation been applied throughout the vortex. However, again, as noted in the main text and Appendix A, Subsection 13.6.2.2, usage of a consistent nonlinear boundary layer formulation exposes nontrivial limitations of the E86 boundary layer model and the accompanying gradient wind theory as demonstrated in Smith et al. (2008), Smith and Montgomery (2008), and Bryan and Rotunno (2009a). Note also that within the context of the E86 slab boundary layer model, w_h becomes negative beyond a radius of about 220 km and has maximum subsidence around 400 km.

An interesting property of a mature hurricane is the relatively slow decay of the tangential velocity with radius. Typically, for a mature hurricane, the tangential velocity decays approximately as $r^{-0.5}$, the decay commencing just outside the radius of maximum tangential wind (Mallen et al. (2005), and refs.). This decay is much slower than the r^{-1} decay observed in an inviscid Rankine vortex outside its radius of maximum velocity. The radial decay of v outside of r_{gm} can be estimated in the E86 model by substituting Eq. (13.86) into the gradient wind balance equation (13.46) and deducing the asymptotic behavior outside r_{gm}. For r not too distant from r_{gm} such that the Rossby number is still greater than unity (i.e., $v > fr$), the cyclostrophic balance approximation (13.73) may be used for this purpose. Radially differentiating Eq. (13.86), and combining with Eq. (13.70), gives

$$v_g^2(r) = \frac{f^2}{4} r \left[\frac{2r}{1+\beta} + \frac{2\beta}{r_o} \left(\frac{r_o}{r} \right)^{2\beta+1} \cdot \frac{r_o^2}{\beta(\beta+1)} \right]. \quad (13.114)$$

For r much smaller than the outer radius r_o, it may be verified that the second term in brackets dominates the first term in brackets. In this case, the square of the gradient wind is approximated by

$$v_g^2(r) \approx \frac{f^2}{2} \frac{rr_o}{1+\beta} \left(\frac{r_o}{r} \right)^{2\beta+1} \sim O(r^{-2\beta}). \quad (13.115)$$

Since $\beta \approx 0.5$, it follows that $v_g \sim r^{-0.5}$ for r near and outside r_{gm}, in agreement with the observed decay with radius cited above, at least for radii in which the Rossby number is greater than unity and cyclostrophic balance is a valid first approximation.

It is noteworthy that the decay of the tangential velocity in this model is controlled by boundary layer thermodynamics together with thermal wind balance and congruence between M and s^* surfaces above the boundary layer. However, the solution in the outer region never makes use of the full tangential or radial momentum equations in the boundary layer. Given recent understanding that has emerged on the importance of the vortex boundary layer in controlling the evolution of the near-core wind field (Kilroy et al., 2016a; Smith et al., 2021), we must be open to the possibility that the dynamics of the boundary layer will enter in a nontrivial manner to influence the evolution of the near-core and outer-core wind field corresponding to the slow outward expansion of the primary eyewall or the formation of an outer (secondary) eyewall as summarized in Chapter 1 and cited refs. The role of boundary layer dynamics in determining the structure of the wind field beyond r_{gm} lies outside the scope of this Appendix, but this topic is considered in Chapter 15 and the proposed sequel to this book.

13.7.8 The tropical cyclone as a Carnot-like heat engine

Emanuel demonstrated that the E86 steady tropical cyclone operates as a Carnot-like, heat engine in which air flowing inwards in the boundary layer acquires heat energy (mostly in latent form[26]) from the sea surface, ascends, and ultimately gives off heat at the much lower temperature of the upper troposphere or lower stratosphere. A schematic of this heat engine is shown in Fig. 13.9. Air begins to flow inwards at constant temperature along the lower boundary at radius r_o and acquires an incremental amount of heat

$$\Delta Q_1 = \int_{\theta_{ea}}^{\theta_e} c_{pd} T_B d \ln \theta_e = c_{pd} T_B \ln \left(\frac{\theta_e}{\theta_{ea}} \right), \quad (13.116)$$

where θ_{ea} is the equivalent potential temperature at r_o. The air ascends at constant entropy along an M surface and flows out to large radius.

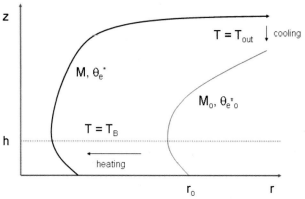

FIGURE 13.9 The tropical cyclone as a Carnot-like heat engine. Adapted from Emanuel (1986). Republished with permission of the American Meteorological Society.

To complete the circuit, the air eventually loses enough total heat through radiative cooling to return to its ambient θ_e so that

$$\Delta Q_2 = \int_{\theta_e}^{\theta_{ea}} c_{pd} \bar{T}_{out} d \ln \theta_e = -c_{pd} \bar{T}_{out} \ln \left(\frac{\theta_e}{\theta_{ea}} \right), \quad (13.117)$$

where $\bar{T}_{out}$ is given by Eq. (13.62). The total heating, from Eq. (13.116) and Eq. (13.117), is therefore

$$\Delta Q = \Delta Q_1 + \Delta Q_2 = c_{pd} T_B \epsilon \ln \left(\frac{\theta_e}{\theta_{ea}} \right), \quad (13.118)$$

---

26. Recall that when water evaporates from the ocean, it takes heat out of the ocean and this energy then resides in the water vapor content of the air. Because it does not immediately increase the temperature of the air, it is called latent heat. Ultimately, when the water vapor condenses inside clouds, the latent heat is converted to sensible heat and the temperature then actually increases somewhere in the system.

where $\varepsilon = (T_B - \bar{T}_{out})/T_B$ is the thermodynamic efficiency. This net heating is used to do work against frictional dissipation in the steady tropical cyclone. Referring to Fig. 13.9, it is seen that work is done against friction in the inflowing boundary-layer air and also to restore the angular momentum back to its ambient value at large radii in the outflow. Kinetic energy is also dissipated by turbulence within cumulus clouds; however, Emanuel argues that this sink primarily balances kinetic energy generated by release of the ambient convective available potential energy as is probably the case in the unperturbed tropical atmosphere. This is simply a statement that convective clouds in tropical cyclones are locally similar to those away from such disturbances. The balance between total heating and frictional dissipation in the inflow and outflow may be written symbolically as

$$\Delta Q = W_{PBL} + W_o, \qquad (13.119)$$

where W_{PBL}, and W_o are the work done in the boundary layer and outflow, respectively. The latter is proportional to the change in kinetic energy needed to restore the angular momentum of the outflow, M, back to its ambient value M_o:

$$\begin{aligned}
W_o &= \frac{1}{2}\Delta(v^2) \\
&= \frac{1}{2}\left[\left(\frac{M}{r_1} - \frac{1}{2}fr_1\right)^2 - \left(\frac{M_o}{r_1} - \frac{1}{2}fr_1\right)^2\right] \\
&= \frac{1}{2}\left[\frac{M^2 - M_o^2}{r_1^2} + f(M_o - M)\right], \qquad (13.120)
\end{aligned}$$

where we have related the azimuthal velocity to angular momentum using Eq. (1.1) from Chapter 1 and r_1 is some large radius at which the exchange is envisioned to take place. In the limit of large r_1:

$$\lim_{r_1 \to \infty} W_o = \frac{1}{2}f(M_o - M) = \frac{1}{4}f^2(r_o^2 - r^2) - \frac{1}{2}frv. \qquad (13.121)$$

Using Eq. (13.121) we infer the (irreversible) work done in the boundary layer:

$$W_{PBL} = c_{pd}T_B\varepsilon \ln\frac{\theta_e}{\theta_{ea}} + \frac{1}{2}frv - \frac{1}{4}f^2(r_o^2 - r^2). \qquad (13.122)$$

Finally, knowledge of the work done against dissipation in the boundary layer allows an evaluation of the pressure distribution in the boundary layer through the use of Bernoulli's equation evaluated along a horizontal parcel trajectory that spirals inward along the top of the surface layer. The latter, when integrated inward from r_0 at constant temperature, may be written

$$\frac{1}{2}v^2 + c_{pd}T_B\ln\pi_s + W_{PBL} = 0 \quad \text{at} \quad z \approx 0, \qquad (13.123)$$

where the well-mixed approximation, as discussed in Section 13.1, has been used to replace the tangential wind at the top of the surface layer with the gradient wind v at the top of the boundary layer. When Eq. (13.122) is substituted into this relationship and the gradient wind balance equation is used for the sum $v^2 + frv$, the result is

$$\ln\pi_s + \frac{1}{2}r\frac{\partial\ln\pi}{\partial r} + \varepsilon\ln\frac{\theta_e}{\theta_{ea}} \qquad (13.124)$$

$$-\frac{1}{4}\frac{f^2}{c_{pd}T_B}(r_o^2 - r^2) = 0 \quad \text{at} \quad z \approx 0, \qquad (13.125)$$

which is identical to Eq. (13.63).

The foregoing confirms the interpretation of the results of the previous section in terms of a Carnot-like heat engine.

13.8 Exercises

Exercise 13.1. Show that the bulk aerodynamic coefficient of moist entropy transfer in a tropical cyclone boundary layer is equal to the bulk aerodynamic coefficient of moist enthalpy transfer.

Exercise 13.2. Step through the formulation developed in Section 13.7.2 of Appendix B and derive Eq. (13.63). This relationship links the radial variation of Exner function π (hence pressure) and θ_e at the top of the boundary layer in the E86 steady-state hurricane model.

Exercise 13.3. Step through the boundary layer formulation and solution for Regions I + II detailed in Subsections 13.7.3.1 and 13.7.5 of Appendix B, respectively. The first step requires the derivation of Eq. (13.72). The second step requires solving Eq. (13.94) to yield Eq. (13.96). Hint: The first relationship links the square of the boundary layer tangential velocity to the thermodynamic disequilibrium at the air-sea interface in the high wind region of the vortex for $r \leq r_{gm}$. Since the tangential velocity is assumed to be essentially well-mixed in the boundary layer (and thus equal to the gradient wind at the boundary layer top at leading order), one may use the cyclostrophic balance approximation at $z = h$ (Eq. (13.73)) to eliminate the cyclostrophic wind so that Eqs. (13.63) and (13.72) comprise a two-by-two system for $\ln\pi$ and $\ln\theta_e$ as a function of radius in Regions I and II (along $z = h$, cf. E86, Eqs. (41) and (45)).

Exercise 13.4. Step through the boundary layer formulation and solution for Region III detailed in Subsections 13.7.3.2 and 13.7.4 of Appendix B, respectively. Hint: The first step requires the derivation of Eqs. (13.83) and (13.86). The first Eq. (13.83) links the θ_e surfeit (under the closure assumption of constant near-surface RH in Region III) to the surface Exner function $\ln \pi_s$ (and hence surface pressure, p_s). This relationship, in conjunction with Eq. (13.63) of Exercise 13.2, comprises a two-by-two system for $\ln \pi$ and $\ln \theta_e$ as a function of radius in Region III along $z = h$. The second Eq. (13.86) is obtained by inserting Eq. (13.83) into Eq. (13.63) and solving the resulting linear differential equation. For simplicity, assume that the outflow temperature may be approximated by the *a priori* temperature of the undisturbed tropopause, T_o.

Exercise 13.5. In Subsection 13.7.6 of Appendix B, step through the derivation of the coupled system of Eqs. (13.100) and (13.101) for v_{gmax}^2 and $\ln \pi_s$ at $r = r_{gm}$. Next, assume that $r_o^2 \gg r_{gm}^2$ and $T_B \approx T_s$ in the ratio T_B / T_s to obtain the limiting two-by-two system of equations for v_{gmax}^2 and $\ln \pi_s$, given by Eqs. (13.103) and (13.104). Deduce the solution given by Eqs. (13.105) and (13.112). Physically interpret the solution. The usefulness of this solution in estimating the maximum intensity for intense cyclones in the real tropical atmosphere is assessed in Exercise 13.6.

Exercise 13.6. In reference to the E86 PI theory for v_{gmax} derived in Exercise 13.5, consider the PI as plotted in Fig. 13.2. Use Eq. (13.6) to verify the estimated PI for the Hurricane Isabel (2003) case as discussed in the main text. Next, choose a couple of other reasonable values for sea surface temperature and outflow temperature so as to verify the nature of the displayed sensitivity of v_{gmax} to these two temperature variables.

The revised data at near major hurricane wind speeds in a wind-wave tank for the surface drag coefficient by Curcic and Haus (2020) suggests that C_D saturates at $\approx 3.1 \times 10^{-3}$ at major cyclone wind speeds and above. This increased value of C_D, together with the observationally inferred mean value of the moist enthalpy coefficient $C_K = 1 \times 10^{-3}$ analyzed by Bell et al. (2012a) (see Chapter 7), suggests that the mean value of C_K / C_D is more likely around 0.32 for major storms. Use this revised ratio of C_K / C_D to estimate the PI for the Isabel (2003) case and for your chosen sea surface and outflow temperatures used above.

Consider next the finding by Pauluis and Zhang (2017) that the thermodynamic efficiency for deep convection in the eyewall region of a three-dimensional tropical cyclone is reduced to 70% of the theoretical value, ϵ, as given by the E86 theory summarized in Section 13.1. Use this result, together with the above calculation incorporating the increased drag coefficient, to assess the utility of the E86 theory for realistic situations.

Finally, based on your calculations, together with the discussion of the nonlinear boundary layer dynamics as presented in the main text, speculate on the implications of your findings for intense cyclones observed in the real tropical atmosphere.

Exercise 13.7. This exercise develops the model of a mature hurricane as a dissipative heat engine, which draws its source heat from a high temperature reservoir at temperature $T_1 = T_B$ (mean temperature of boundary layer top) and dumps its waste heat to a low temperature reservoir in the lower stratosphere/upper troposphere at temperature $T_2 = T_o$. Fig. 13.10 sketches the set up of the proposed dissipative heat engine.

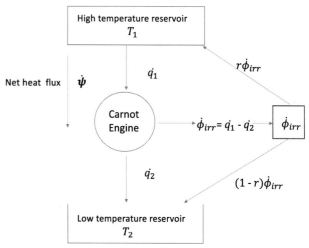

FIGURE 13.10 Schematic of the hurricane modeled as a dissipative heat engine. This schematic focuses only on the bulk aspects of the engine and not the individual legs of the cycle, which have been already discussed in Section 13.1. Adapted from Fig. 2.5 of Goody (1995), which was formulated originally for the atmospheric general circulation.

Consistent with the hypothesis that the overall heat engine does *no useful work* on the environment, the irreversible work ($\dot{\phi}_{irr}$) generated by the engine is assumed to be proportionately dumped back to the source and waste heat reservoirs, respectively, as sketched. The parameter r reflects the portion of the irreversible power (heating rate) generated by the engine that is returned to the high temperature reservoir, while $1 - r$ reflects the portion of the engine's irreversible power that is dumped to the low temperature reservoir.

As indicated by the cartoon, the internal heat engine is assumed to be that of a Carnot Engine, which operates between the foregoing warm and cold temperature reservoirs, respectively. By definition, the internal engine undergoes a Carnot cycle whose detailed features have been described already in Section 13.1. For convenience to the reader, we recap the main elements.

The first leg consists of an isothermal expansion in which near-surface air parcels flow inwards toward the circulation center, maintaining an approximately constant temperature. The second leg is that of a moist adiabatic expansion reflecting the approximation that moist parcels ascend the eyewall updraft and subsequently flow outward in the upper troposphere/lower stratosphere, all while conserving their moist equivalent potential temperature, θ_e^*. Strictly speaking, for the internal engine to be a true Carnot engine, all of the thermodynamic transformations *must be reversible!* In particular, the entropy in the eyewall updraft should be the reversible moist entropy and corresponding reversible equivalent potential temperature, both of which retain the parcel's liquid water (and ice) as the parcel ascends and cools in the eyewall updraft and flows outward into the lower stratosphere/upper troposphere.

The third leg consists of isothermal cooling (and compression) in the lower stratosphere and upper troposphere in the outer portion of the hurricane. The fourth leg is that of moist adiabatic warming (and compression) that hypothetically returns the air to its starting value of ambient near-surface pressure, 1015 mb, relative humidity, RH_{as}, and boundary layer temperature T_B.[27]

For the set up as sketched in Fig. 13.10, show first that the net heat flux $\dot{\psi}$ flowing from the high temperature reservoir to the cold reservoir $\dot{\psi} = \dot{q}_1 - r\dot{\phi}_{irr}$. Show next that the irreversible power generated by the dissipative engine is given by $\dot{\phi}_{irr} = \dot{q}_1 - \dot{q}_2$. Carnot's theorem for the steady-state Carnot engine implies that

$$\frac{\dot{q}_1}{T_1} = \frac{\dot{q}_2}{T_2}. \tag{13.126}$$

The overall engine's efficiency ϵ is given by the ratio of the power generated by the engine $\dot{\phi}_{irr}$ to the net heat flux flowing from the high temperature reservoir to the cold reservoir, $\dot{\psi}$, i.e.,

$$\epsilon = \frac{\dot{\phi}_{irr}}{\dot{\psi}}. \tag{13.127}$$

Use Eq. (13.126) in combination with Eq. (13.127) to show that

$$\epsilon = \frac{T_1 - T_2}{T_1 - r(T_1 - T_2)} \tag{13.128}$$

Now, interpret the result (13.128) physically. In particular, consider the asymptotic limit $r \to 1$, i.e., all of the irreversible work (heat) is dumped solely to the warm temperature reservoir. In this limit, we see that $\epsilon \to (T_1 - T_2)/T_2$, i.e., the ordinary Carnot efficiency ϵ is increased by the multiplicative factor T_1/T_2 - essentially using the cold reservoir temperature T_2 in place of T_1 in the denominator of ϵ.

For the present tropical atmosphere, the limit $r \to 1$ effectively boosts the efficiency of the engine by approximately 50% and increases the theoretical potential inten-

sity of the gradient wind by approximately 22.4% (cf. Section 13.1.1; see also Exercise 13.10). It is worth recalling that this asymptotic limit neglects the radiation of surface gravity wave energy out of the hurricane at the ocean surface and assumes that the portion of the engine's power being returned to the source reservoir goes solely into heating the atmospheric boundary layer and not the ocean. The foregoing formulation represents an alternative and direct derivation of the dissipative heating formulation of Bister and Emanuel (1998) and implicitly addresses the main objection raised by Makarieva et al. (2011).

Using the foregoing formulation, proceed to estimate the reduction of the engine's efficiency relative to the $r = 1$ limit if the energy dissipation associated with mixing and non-axisymmetric shear turbulence in the upper-level outflow is 50% of the engine's irreversible power. What would be the implications of such a result on the estimated PI of the hurricane vortex? Is this hypothetical situation a plausible possibility in real-world storms? Discuss your answer.

Exercise 13.8. For the given set up of an axisymmetric, steady-state hurricane vortex, show that for a moist air parcel rising in pseudo-adiabatic ascent, the heat lost by the removal of the condensate is small in comparison to the heat liberated by the phase transition from water vapor to liquid water.

Exercise 13.9. For the given set up of an axisymmetric, steady-state, hurricane vortex, show that the total moist energy (E) defined in Appendix A is materially conserved for a moist air parcel that ascends the eyewall cloud and upper-tropospheric outflow without momentum or heat diffusion.

Exercise 13.10. For the given set up of an axisymmetric, steady-state hurricane vortex, show that when the E86 slab boundary layer detailed in Section 13.1 is invoked to connect ds^*/dM in Eq. (13.41) to the sea-to-air fluxes of moist entropy, one obtains

$$v_{gmax}^2 = \frac{C_K}{C_D} T_b \epsilon (s_0^* - s_b) \text{ at } (r, z) = (r_{gm}, z_{gm}), \tag{13.129}$$

where all variables above have been defined in the main text. This expression is the limiting *boundary layer balance* in the strict E86 PI formulation for the maximum square of the gradient wind.

---

27. This last leg is arguably implausible since, in reality, the time scale required for air parcels to descend back to the sea surface is on the order of one month, far longer than the typical lifecycle of an individual hurricane in the real tropical atmosphere. An attempt to circumvent the strict requirements of these last two legs has been proposed using the device of a "differential Carnot cycle" that *eliminates* the latter two legs of the traditional Carnot cycle. However, serious scientific concerns have been raised about this approach (Montgomery and Smith, 2020), which, as of the time of this writing, remain unanswered (see also Makarieva et al., 2020).

Chapter 14

Global budgets and steady state considerations

In Chapters 10 and 11 we examined the dynamical and thermodynamical processes involved in tropical cyclone formation and intensification. We turn now to examine the sources of moisture, energy and angular momentum required to support the intensification of these storms. In particular, we examine the water budget, the energy budget and the absolute angular momentum budget of a vortex in the prototype problem for intensification, similar to that discussed in Chapter 11. We go on then to examine the theoretical requirements for storms to exist in a globally-steady state. As a prelude, we discuss briefly the numerical model used to calculate the budgets.

14.1 The numerical simulation

The numerical model simulation used for the budget calculations is essentially the same as that used in Chapter 11, but the horizontal grid spacing is 3 km instead of 1 km and the SST is one degree higher at $28\,°C$.[1] The calculations are carried out for a period of 4 days with data output every 15 min.

Fig. 14.1 shows time series of the maximum azimuthally-averaged tangential wind speed, V_{max}, in the simulation. As before, V_{max} is located no more than a few hundred meters above the surface. The evolution is broadly similar to that described in Section 11.5. After a gestation period lasting about a day, during which deep convection becomes established inside the radius of maximum tangential wind, R_{vmax}, the vortex undergoes an RI phase lasting about 36 hours, before V_{max} reaches a quasi-steady state. Initially R_{vmax} is located at a radius of 100 km, but contracts to a little more than 20 km after about $2\frac{1}{4}$ days (not shown). The most rapid contraction occurs during the RI phase as absolute angular momentum surfaces are drawn briskly inwards within and above the boundary layer.

Fig. 14.2 shows vertical cross sections of the azimuthally-averaged, 3 hour time-averaged, radial, and tangential velocity components, the vertical velocity component, and the M-surfaces during the intensification phase of the vortex. The time averages are centered on 36 hours during the period of rapid intensification and at 60 hours near the end of

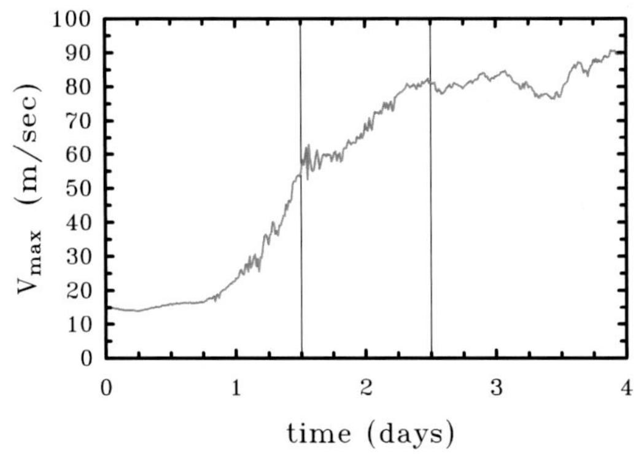

FIGURE 14.1 Time series of (a) maximum azimuthally-averaged tangential wind speed (V_{max}) in the simulation described in Section 14.1. The two vertical lines indicate the times of the vertical cross sections shown in Fig. 14.2 and related analyses. Adapted from Smith et al. (2018c). Republished with permission of Wiley.

this period. These two times are indicated by vertical lines in Fig. 14.1. The basic features of the flow are qualitatively similar at both times, but all three velocity components strengthen over the period, the absolute angular momentum surfaces moving inwards in the lower troposphere and outwards in the upper troposphere. The flow structure is similar to that in Fig. 11.6 with a layer of strong shallow inflow marking the frictional boundary layer, a layer of weaker inflow in the lower troposphere above the boundary layer, a region of strong outflow in the upper troposphere and a layer of enhanced inflow below this outflow. The maximum tangential wind speed occurs within, but near the top of the frictional boundary layer. Much of the ascent occurs in an annular region on the order of 50-60 km in radius. The region inside this annulus shows mostly subsidence.

14.2 Budget calculations

The budget of a particular quantity γ can be obtained by integrating the tendency of the relevant quantity over a cylindrical volume of space with radius R and height H centered on the storm[2] using the boundary conditions that $u = 0$ at $r = 0$, and $w = 0$ at $z = 0$ and $z = H$. We denote an integral

---

1. The calculations described here are an expansion of those presented by Smith et al. (2018c).

2. The vortex center is obtained by the method described at the start of Section 10.5.

Tropical Cyclones. https://doi.org/10.1016/B978-0-44-313449-4.00022-9

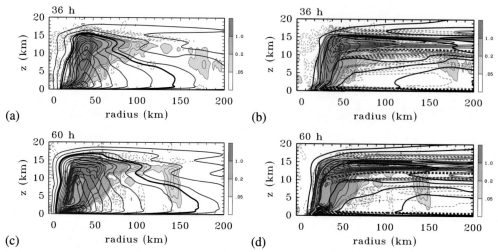

FIGURE 14.2 Left panels: Vertical cross sections of the azimuthally-averaged, 3-hour time averaged tangential velocity component (blue contours) centered at 36 hours and 60 hours in the simulation described in Section 14.1. Superimposed are contours and shading of the averaged vertical velocity. Contour intervals are as follows. Tangential velocity: blue contours every 5 m s^{-1}, with a thick black contour highlighting the 17 m s^{-1} contour. Vertical velocity: thin red contours every 0.05 m s^{-1} to 0.2 m s^{-1}, thick red contour interval 0.5 m s^{-1}, thin dashed red contours indicate subsidence at intervals of 0.02 m s^{-1}. Right panels: Vertical cross sections of the azimuthally-averaged, 3 hour time-averaged radial velocity component together with the averaged vertical velocity centered at the same times. Contour intervals are as follows. Radial velocity: thick blue contours 4 m s^{-1}, dashed negative, thin blue dashed contours every 0.5 m s^{-1} down to -3.5 m s^{-1}. Absolute angular momentum: thick black contours every 2×10^5 m^2 s^{-1}, with the 6×10^5 m^2 s^{-1} contour highlighted in yellow. From Smith et al. (2018c). Republished with permission of Wiley.

of γ over this volume by

$$\overline{[\gamma]} = \int_0^R r\,dr \int_0^{2\pi} d\lambda \int_0^H \gamma\,dz. \qquad (14.1)$$

14.2.1 Water budget

Using Eq. (2.71) (see Exercise 2.1) with γ equal to the total water mixing ratio, r_T, gives an equation for $\partial r_T/\partial t$. Integrating this equation over the cylindrical volume gives an equation for the total water budget

$$\frac{d}{dt}\overline{[\rho r_T]} = F_{r_T} + F_E - P_{rain}, \qquad (14.2)$$

where

$$F_{r_T} = R \int_0^{2\pi} d\lambda \int_0^H [\rho r_T u]_{r=R}\,dz \qquad (14.3)$$

is the net flux of water substance (principally water vapor) through the side boundary $r = R$,

$$F_E = \int_0^{2\pi} d\lambda \int_0^R (\text{surface evaporation rate})r\,dr \qquad (14.4)$$

is the rate of evaporation of water vapor from the sea surface, and

$$P_{rain} = \int_0^{2\pi} d\lambda \int_0^R (\text{surface precipitation rate})r\,dr \qquad (14.5)$$

is the rate of loss of water by precipitation reaching the surface. As shown in Chapter 7, the surface evaporation rate may be calculated using the empirical formula $\rho C_E \mathbf{u_{10}}(r_{vs}^* - r_{v10})$, where C_E is a coefficient of evaporation, $\mathbf{u_{10}}$ is the wind speed at a height of 10 m, r_{vs}^* is the saturation mixing ratio at the SST and r_{v10} is the mixing ratio at a height of 10 m. Typical values of C_E are similar to those of C_K, about 1.1-1.2 $\times 10^{-3}$ for wind speeds < 18 m s^{-1} (Chapter 7).

The vertical integral of the quantity ρr_T that appears on the left-hand-side of Eq. (14.2) is the total water substance (vapor + liquid + ice) per unit horizontal area in a column. We refer to this integral as the *total column water*, TW. Likewise, the vertical integral of the vapor mixing ratio, r_v, is often referred to as the *total precipitable water*, or TPW. This quantity is the theoretical maximum amount of precipitation that could be obtained if all the water vapor in the column were to precipitate out.

Fig. 14.3 shows the radial distributions of TW and TPW for the simulation of Section 14.1 as a function of time. In this simulation, there is no ice so that the total water is made up of vapor and liquid water only. Both quantities have a structure that reflects the mean overturning circulation as the vortex intensifies.

The largest increase in total column water starts a little after half a day at radii between about 10 km and 100 km (Fig. 14.3a). This increase is a result of the developing eyewall convection. There is a weaker increase that develops first to a radius of about 200 km, but which subsequently narrows to 120 km towards the end of the simulation. After

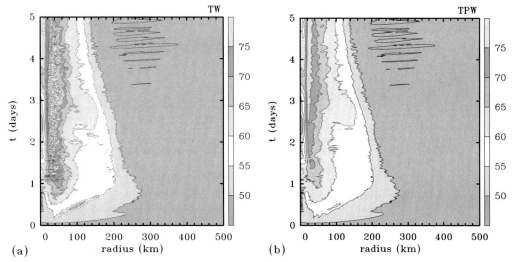

FIGURE 14.3 Hovmöller time-radius plots of (a) the total column water, TW, and (b) total precipitable water, TPW, for a cylinder of radius $R = 500$ km in the simulation described in Section 14.1. Values indicated on the side bar in kg m^{-2}. Contours at intervals of 5 kg m^{-2} starting at 45 kg m^{-2}.

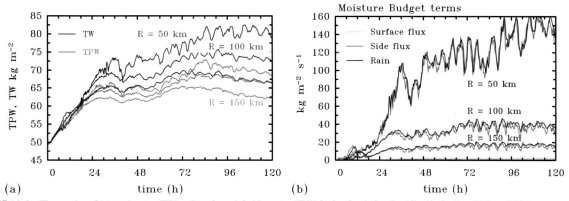

FIGURE 14.4 Time series of (a) total water, TW, and total precipitable water, TPW, in the simulation described in Section 14.1, and (b) the corresponding budgets in the water budget equation (14.2) for cylinders radius $R = 50$, 100, and 150 km. In (a), not all curves are labeled, but at a given time, larger R implies a smaller TPW or TW. In (b), "side flux" refers to net flux of water substance through the side boundary, F_{r_T}; "rain" refers to the rate of loss of water by precipitation, P_{rain}; "surface flux" refers to the rate of evaporation of water from the ocean surface. In panel (b) all quantities have been normalized by the area of the respective cylinder.

about a day, the lower values of TW migrate radially inwards beyond a radius of about 150 km, a feature that can be attributed to the subsidence of drier air from aloft. This subsidence is associated, in part by the balanced overturning circulation and in part by the cooling required by relaxing the local temperature to that of the ambient sounding at the same height. The evolution of TPW is qualitatively similar to that of TW, with values somewhat less, by definition, in the strongly precipitating eyewall region (Fig. 14.3b).

Fig. 14.4 shows the water budget expressed by Eq. (14.2) for averaging cylinders of radius 50 km, 100 km, and 150 km in the simulation of Section 14.1. In each case, the total water $\overline{[\rho r_T]}$ increases from its initial value (49.5 kg m^{-2})[3] to a maximum value of nearly 66 kg m^{-2} after about $2\frac{1}{2}$

days. This increase must be supported by a net flux through the side boundary, F_{r_T}, and from surface evaporation F_E minus the water lost by precipitation, P_{rain}. The time series for TPW is similar, but TPW values are less than those for total column water.

It is seen that the main balance for these cylinder radii is that between the boundary moisture flux and precipitation, with the surface vapor flux being small in comparison. This result goes back to a finding of a seminal paper by Malkus and Riehl (1960), which was confirmed later in a comprehensive study of the moisture budget of a numerically-simulated hurricane by Braun (2006). However, as pointed out in both these studies, the finding does not mean that the surface flux is unimportant. For one thing, of course, the boundary moisture flux between 100 km and 150 km contributes to the increased side boundary flux at 100 km radius. For another, as discussed in Section 14.2.5 below,

---

3. This mean value is slightly less than for the Dunion moist tropical sounding, 51.5 kg m^{-2} because the warm core of the initial vortex leads to a reduced density locally.

the surface moisture flux plays a subtle, but important role in the energetics of a tropical cyclone by elevating the near-surface θ_e in the eye and eyewall to values exceeding those in the near environment beyond the eyewall.

It turns out that, because of the near cancellation in the total water budget between the side flux and the rain, and the relative smallness of the surface evaporative flux, the time tendency calculated numerically from the right-hand-side of Eq. (14.2) is rather noisy compared with that calculated directly and for this reason is not shown.

14.2.2 Kinetic energy budget (Gill form)

The kinetic energy budget for a simulated storm in the prototype problem for intensification can be obtained by taking the scalar product of the momentum equation with the velocity vector, as in Section 2.8, and integrating the result over the cylindrical volume with the appropriate boundary conditions discussed above. Thus, one possibility would be to integrate Eq. (2.37) directly to obtain

> **Gill form**
> $$\frac{d}{dt}\overline{\left[\tfrac{1}{2}\rho \mathbf{u}^2\right]} = \overline{[-\rho g w]} + \overline{[p\nabla \cdot \mathbf{u}]} - F_{KEG} - D, \tag{14.6}$$

where

$$F_{KEG} = \int_0^{2\pi} d\lambda \int_0^H u\left(p + \tfrac{1}{2}\rho \mathbf{u}^2\right) dz, \tag{14.7}$$

is the flux of mechanical energy through the side boundary $r = R$, and $D = \overline{[\rho \epsilon_D]}$ is the total frictional dissipation. Our focus in this subsection is on the generation of kinetic energy and so for simplicity, in Eq. (14.7) we have omitted the small horizontal diffusion of kinetic energy. We have omitted also the loss of kinetic energy associated with the diffusion of kinetic energy at the surface, which is not small. Both contributions are accounted for in Section 14.2.5. Note that the Coriolis acceleration does not appear in the energy equation because it is orthogonal to $\mathbf{u}$. The first term on the right-hand-side of Eq. (14.6) is the rate of working by the gravitational force per unit volume, ρg. Since $\nabla \cdot \mathbf{u}$ is the fractional change in volume of an air parcel, the second term is the cumulative effect of the kinetic energy generated locally when an air parcel expands or contracts in volume.

For a Newtonian fluid with dynamic viscosity coefficient μ,

$$D = \overline{[\mu \Phi_v]}, \tag{14.8}$$

where, in cylindrical coordinates,

$$\Phi_v = 2\left[\left(\frac{\partial u}{\partial r}\right)^2 + \left(\frac{1}{r}\frac{\partial v}{\partial \lambda} + \frac{u}{r}\right)^2 + \left(\frac{\partial w}{\partial z}\right)^2\right]$$
$$+ \left[r\frac{\partial}{\partial r}\left(\frac{v}{r}\right) + \frac{1}{r}\frac{\partial u}{\partial \lambda}\right]^2 + \left[\frac{1}{r}\frac{\partial w}{\partial \lambda} + \frac{\partial v}{\partial z}\right]^2$$
$$+ \left[\frac{\partial u}{\partial z} + \frac{\partial w}{\partial r}\right]^2 - \frac{2}{3}(\nabla \cdot \mathbf{u})^2 \tag{14.9}$$

14.2.3 Kinetic energy budget (Anthes form)

An alternative equation for the kinetic energy budget can be obtained by taking the scalar product of $\mathbf{u}$ with the momentum equation in the form of Eq. (2.10) and using the identity $\mathbf{u} \cdot \nabla \mathbf{u} = \nabla(\tfrac{1}{2}\mathbf{u}^2) + \boldsymbol{\omega} \wedge \mathbf{u}$ to obtain

$$\frac{\partial}{\partial t}(\tfrac{1}{2}\mathbf{u}^2) + \mathbf{u} \cdot \nabla(\tfrac{1}{2}\mathbf{u}^2) = -\frac{1}{\rho}\mathbf{u_h} \cdot \nabla_h p' + Bw - \mathbf{u} \cdot \mathbf{F}, \tag{14.10}$$

where $\mathbf{u_h}$ is the horizontal velocity vector, ∇_h is the horizontal gradient operator, and $B = -(1/\rho)(\partial p'/\partial z) + b$ is the net vertical force per unit mass. Despite the explicit appearance of p' in the first term on the right-hand-side, all the terms in Eq. (14.10) are independent of the reference pressure $p_{ref}(z)$, since, in particular, $\mathbf{u_h} \cdot \nabla_h p_{ref}(z) = 0$. The first term on the right-hand-side of Eq. (14.10) represents the kinetic energy generated by the rate of working by the horizontal component of the perturbation pressure gradient, while the second term represents the rate of kinetic energy production by air rising in the presence of a positive vertical force ($B > 0$) and air sinking in the presence of a negative vertical force ($B < 0$).

Multiplying Eq. (14.10) by ρ and using Eq. (2.6) gives the flux form

$$\frac{\partial}{\partial t}(\tfrac{1}{2}\rho \mathbf{u}^2) + \nabla \cdot (\tfrac{1}{2}\rho \mathbf{u}^2)\mathbf{u} = -\mathbf{u_h} \cdot \nabla_h p'$$
$$+ \rho Bw - \rho \mathbf{u} \cdot \mathbf{F}. \tag{14.11}$$

Integrating over the cylindrical region and using the appropriate boundary conditions discussed above, Eq. (14.11) becomes

> **Anthes form**
> $$\frac{d}{dt}\overline{\left[\tfrac{1}{2}\rho \mathbf{u}^2\right]} = -\overline{[\mathbf{u_h} \cdot \nabla_h p']}$$
> $$+ \overline{[\rho Bw]} - F_{KEA} - D, \tag{14.12}$$

where

$$F_{KEA} = \int_0^{2\pi} d\lambda \int_0^H \left[u(\tfrac{1}{2}\rho \mathbf{u}^2)\right]_{r=R} dz. \tag{14.13}$$

Again, for reasons given above, we have omitted the lateral diffusive flux and vertical flux at the surface. We refer to

Eq. (14.12) as the Anthes form of the equation as it forms the basis for the seminal study of Anthes (1974).

In contrast to Eq. (14.6), the pressure-work term, $-\overline{[\mathbf{u_h} \cdot \nabla_h p']}$, now appears explicitly in the global form of the kinetic energy equation. For an axisymmetric flow, this term is simply $\overline{[-u \partial p'/\partial r]}$. At first sight, one might question the prominence of the pressure-work term as a source of kinetic energy, since, for example, $\partial p'/\partial r$ is *not* the only radial force acting on fluid parcels en route to the storm core. Above the frictional boundary layer, the radial pressure gradient is closely balanced by the sum of the centrifugal force and the radial component of the Coriolis force. Moreover, this source term does not appear in Eq. (14.6), although it is replaced by the term $-\overline{[p' \nabla_h \cdot \mathbf{u_h}]}$. Further, the boundary flux terms are different. Even so, one should bear in mind that even in the axisymmetric case, $\overline{[-u \partial p'/\partial r]}$ generates *not only a radial contribution to the kinetic energy, but also an azimuthal contribution* through the action of the generalized Coriolis force $-(f + v/r)u$. The generation of this azimuthal contribution is implicit in the kinetic energy equation as the generalized Coriolis force does no work, but it does convert radial momentum to tangential momentum.

14.2.4 Kinetic energy budget calculations

We examine now the generation terms in Anthes form of the kinetic energy equation for the case of the idealized tropical cyclone simulation detailed in Section 14.1. Both forms are discussed in the study by Smith et al. (2018c), where it was argued that the Anthes form is more intuitive in the case of a tropical cyclone.

14.2.4.1 Kinetic energy evolution

Fig. 14.5 shows time series of the domain-averaged kinetic energy per unit volume, $\overline{\left[\frac{1}{2}\rho \mathbf{u^2}\right]}$, for domain radii 300 km and 500 km. As anticipated by Anthes (1974), this quantity is dominated by the horizontal velocity components: in fact, the curves for $\overline{\left[\frac{1}{2}\rho \mathbf{u^2}\right]}$ and $\overline{\left[\frac{1}{2}\rho \mathbf{u_h^2}\right]}$ essentially overlap. It follows that the contribution of the vertical velocity to the global kinetic energy is negligible. Notable features of the curves for both domain sizes are the slight decrease during the first 12 hours on account of surface friction, followed by a rapid increase as the vortex intensifies. As time proceeds, the rate of increase progressively declines.

14.2.4.2 Kinetic energy generation

Fig. 14.6 shows time series of the principal terms in the generalized Anthes formulation (the right-hand-side of Eq. (14.12)), excluding only the global dissipation term since the focus of the paper is on kinetic energy generation. For both domain radii, 300 km (Fig. 14.6a) and 500 km (Fig. 14.6b), both the terms $\overline{[-\mathbf{u_h} \cdot \nabla_h p']}$ and $\overline{[\rho Bw]}$

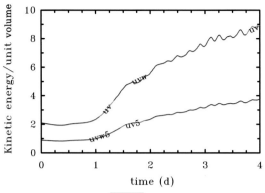

FIGURE 14.5 Time series of $\overline{\left[\frac{1}{2}\rho \mathbf{u^2}\right]}$ (curves labeled uvw) compared to $\overline{\left[\frac{1}{2}\rho \mathbf{u_h^2}\right]}$ (curves labeled uv) for cylinders of 300 km and 500 km in the simulation described in Section 14.1. All quantities have been normalized by the total volume of the respective cylinders. The curves for the 500 km domain are labeled with a "5". The curves for each cylinder size lie essentially on top of each other so that only a single curve is evident. Units on the ordinate 10^{-3} J m^{-3}. From Smith et al. (2018c). Republished with permission of Wiley.

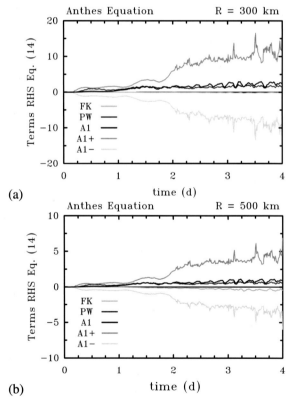

FIGURE 14.6 Time series of the kinetic energy tendency terms on the right-hand-side of Eq. (14.12), the Anthes' formulation, averaged over a cylinder of size (a) 300 km and (b) 500 km in the simulation described in Section 14.1. Units on the ordinate are $10^4 \times$ W m^{-3}. The dissipation term is not shown. A1 stands for $\overline{[-\mathbf{u_h} \cdot \nabla_h p']}$, FK for F_{KEA} and PW for $\overline{[\rho Bw]}$. A1+ and A1- stand for the contributions to A1 from regions where the argument $-\mathbf{u_h} \cdot \nabla_h p'$ is positive and negative, respectively. From Smith et al. (2018c). Republished with permission of Wiley.

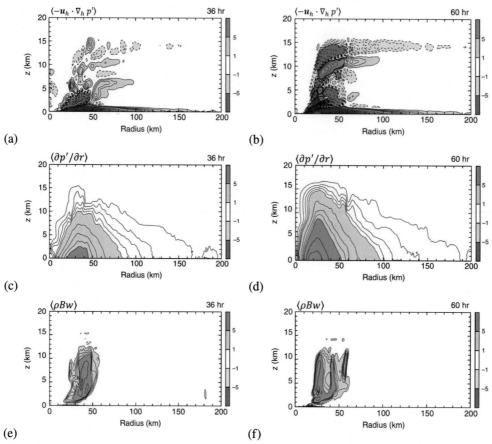

FIGURE 14.7 Radius-height cross sections of azimuthally-averaged quantities in Eq. (14.12), before performing the cylindrical average: $\langle -\mathbf{u_h} \cdot \nabla_h p' \rangle$ (panels (a), (b)); and $\langle \rho B w \rangle$ (panels (e), (f)), at 36 hours (left panels) and 60 hours (right panels) in the simulation described in Section 14.1. Panels (c) and (d) show similar cross sections of $\langle \partial p'/\partial r \rangle$ at these times. Contour intervals are as follows. Panels (a), (b), (e), and (f): thick contours 5×10^2 W m^{-3}: thin contours 1×10^2 W m^{-3}. Solid red contours positive, dashed blue contours negative. Shading as indicated on the side bar. Panels (c) and (d): thin contours 0.2×10^2 Pa m^{-1} to 0.8×10^2 Pa m^{-1}; medium thick contours 1.0×10^2 Pa m^{-1} to 5.0×10^2 Pa m^{-1}; thick contours every 5.0×10^2 Pa m^{-1}. Shading as indicated on the side bar. Adapted from Smith et al. (2018c). Republished with permission of Wiley.

are positive, but, perhaps surprisingly, the former term is not appreciably larger than the latter, even beyond 2 days when the differences are largest. The boundary flux term F_{KEA} is virtually zero throughout the calculation.

For the larger domain size (R = 500 km), the temporal behavior of the various terms is similar, but, as expected, the magnitudes of the respective terms are appreciably smaller (Fig. 14.6b), since the largest contributions to the averages are from well inside a 300 km radius (note the different scales on the ordinate in Figs. 14.6a, b).

The finding that the two terms $\overline{[-\mathbf{u_h} \cdot \nabla_h p']}$ and $\overline{[\rho B w]}$ are not appreciably different in magnitude is at first sight surprising since, as shown in Fig. 14.5, the contribution of the vertical velocity to the total kinetic energy is negligible. Moreover, the $\overline{[\rho B w]}$ term does not appear in Anthes' original formulation, where it was considered to be unimportant.

An explanation of this surprising result emerges from an examination of the radial-height structure of the azimuthally-averaged generation term before completing the cylindri-

cal average, i.e., $\langle -\mathbf{u_h} \cdot \nabla_h p' \rangle$, where the angle brackets here denote an azimuthal average. The structure of this average together with those of the other generation term, $\langle \rho B w \rangle$, at 36 hours and 60 hours, is shown in Fig. 14.7. At both times, the Anthes generation term $\langle -\mathbf{u_h} \cdot \nabla_h p' \rangle$ shows coherent regions of large kinetic energy generation and of large kinetic energy destruction. The main region of generation is at low levels, below about 2 km, where the strongest inflow occurs and where the inward directed radial pressure gradient force is particularly strong (panels (a) and (b) of Fig. 14.7). There is a second region of generation in an annular column, mostly on the outer side of the eyewall updraft below about 9 km at 36 hours and below about 12 km at 60 hours.

Since the radial pressure gradient is positive at all heights (see panels (c) and (d) of Fig. 14.7), these generation regions must be ones in which there is generally inflow.[4] For the same reason, where there is outflow, there

---

4. Note that eddy effects are included in all generation terms.

is kinetic energy removal as seen in the two principal coherent regions in panels (a) and (b) where $\langle -\mathbf{u_h} \cdot \nabla_h p' \rangle < 0$. It follows that the computed value of $\overline{[-\mathbf{u_h} \cdot \nabla_h p']}$ is the remainder resulting from the cancellation of two comparatively large contributions from $\langle \mathbf{u_h} \cdot \nabla_h p' \rangle$ of opposite sign, namely $\langle -\mathbf{u_h} \cdot \nabla_h p' \rangle_+$ and $\langle -\mathbf{u_h} \cdot \nabla_h p' \rangle_-$, the former being the sum of all positive values of $-\mathbf{u_h} \cdot \nabla_h p'$ and the latter $\langle -\mathbf{u_h} \cdot \nabla_h p' \rangle_-$ to be the sum of all negative values. This large cancellation is evident in the time series shown in Fig. 14.6.

In summary, a substantial fraction of the kinetic energy that is generated is removed in regions where there is outflow and the residual is relatively small. Indeed, it is comparable with the kinetic energy generated by the rate-of-working of the net vertical perturbation pressure gradient force (buoyancy plus perturbation pressure gradient), principally in the region of diabatically-forced ascent.

The structure of the net vertical perturbation pressure gradient force at 36 hours and 60 hours is shown in panels (e) and (f) of Fig. 14.7. As expected, this force is concentrated in an annular region overlapping the region of diabatic heating.

14.2.4.3 Discussion

There are some subtle issues concerning the generation of kinetic energy in tropical cyclones. Long ago, Anthes (1974) wrote that

> "the important source of kinetic energy production in the hurricane is the radial flow toward lower pressure in the inflow layer, represented by $-u\partial p/\partial r$".

This statement may seem at first sight problematic because, above the boundary layer, the radial pressure gradient is very closely in balance with sum of centrifugal and Coriolis forces. However, the kinetic energy equation doesn't recognize the balance constraint. In this equation, the radial pressure gradient acts to generate not only kinetic energy of radial motion, but also that of tangential motion through the action of the generalized Coriolis force $-(f + v/r)u$, a term that appears in the tangential momentum equation in cylindrical coordinates. Since the generalized Coriolis force does not appear explicitly in the kinetic energy equation, the conversion of kinetic energy from one component to another is transparent to the equation.

Anthes recognized that much of the inflow into the storm is

> "... a result of surface friction, which reduces the tangential wind speed and thereby destroys the gradient balance, so that the inward pressure gradient force exceeds the Coriolis and centripetal[5] forces"

and he pointed out that

> "In the warm core low the maximum pressure gradient $(-\partial p/\partial r < 0$ [sign corrected: our insertion]) occurs at the lowest level, at which the inflow ($u < 0$) is maximum. In the outflow layer, where the radial flow is reversed, the pressure gradient is much weaker. The result is a net production of kinetic energy, dominated by the contribution from the inflow region."

While the foregoing view is broadly supported by the calculations presented herein, these calculations provide a sharper view of the *net* production of kinetic energy. They indicate a region of significant kinetic energy generation accompanying inflow *throughout* the lower troposphere above the boundary layer as well as significant regions where kinetic energy is consumed as air flows outwards, against the radial pressure gradient, above the boundary layer.

The generation above the boundary layer is a manifestation of spin up by the classical mechanism articulated in Section 8.2, while the generation within the boundary layer, highlighted by Anthes, is a manifestation of the nonlinear boundary layer spin up enhancement mechanism articulated in Sections 6.6 and 11.5.

Anthes argued that the boundary layer

> "... must be responsible for a net gain of kinetic energy"

even though

> "a substantial dissipation of kinetic energy in the hurricane occurs in the boundary layer through turbulent diffusion and ultimate loss of energy to the sea surface".

As a result, he was led to the paradox that

> "surface friction is responsible for a net increase in kinetic energy and without friction the hurricane could not exist."

The resolution of this paradox would appear to be Anthes' de-emphasis of the role of the classical mechanism for spin up in the kinetic energy budget.

The results here, especially the noted cancellation of relatively large generation and consumption contributions to the term $\overline{[-\mathbf{u_h} \cdot \nabla_h p']}$, point to limitations in the utility of a global kinetic energy budget in revealing the underlying dynamics of tropical cyclone intensification.

An alternative approach to examining kinetic energy production in tropical cyclones is to examine the energetics of individual air parcels as they move around some hypothetical circuit (Chapter 13; see also Emanuel (2004) and

---

5. Presumably, Anthes means the centrifugal force.

references). However, this approach relies on assumptions about the circuits traversed, circuits that may or may not be realizable in practice.

The foregoing discussion highlights some of the limitations of a global kinetic energy budget in revealing the underlying dynamics of tropical cyclones.

14.2.5 Total energy budget

In Chapter 2 we derived a total energy equation, Eq. (2.39), in the form

$$\frac{\partial}{\partial t}\left[\rho\left(E+\Phi+\frac{1}{2}\mathbf{u}^2\right)\right]+\nabla\cdot\mathbf{F}_{\text{tot}}=\rho\dot{Q}, \qquad (14.14)$$

where

$$\mathbf{F}_{\text{tot}}=\rho\mathbf{u}\left(E+\frac{p}{\rho}+\Phi+\frac{1}{2}\mathbf{u}^2\right)-\mu\nabla\left(\frac{1}{2}\mathbf{u}^2\right). \qquad (14.15)$$

In applications to a turbulent atmosphere, the molecular dynamic viscosity $\mu=\rho\nu$ should be replaced by a turbulent dynamic viscosity, $\mu_{turb}=\rho K_M$, with K_M a turbulent, or eddy, momentum diffusivity, taken here as constant for simplicity. In Eq. (14.14), $\dot{Q}$ represents the material heating rate per unit mass associated with the sum of the divergence of the conductive heat flux and the electromagnetic radiation heat flux, the release of latent heat of condensation or fusion or sublimation, the cooling associated with the evaporation of liquid water or ice particles or the melting of ice.

The total energy equation may be written alternatively with the turbulent heat and radiation heat flux terms added to the total heat flux (following Gill, 1982). In this form, two additional terms representing the radiative heat flux, $\mathbf{F}_{\text{rad}}$, and the turbulent heat flux as represented by simple K-theory, $-K_H\nabla T$, where K_H is the turbulent conductivity of heat, are added to $\mathbf{F}_{\text{tot}}$. In this case Eq. (14.15) is rewritten as

$$\mathbf{F}_{\text{tot}}=\rho\mathbf{u}\left(c_p T+\Phi+\frac{1}{2}\mathbf{u}^2\right)+$$
$$\mathbf{F}_{\text{rad}}-K_H\nabla T-\rho K_M\nabla\left(\frac{1}{2}\mathbf{u}^2\right) \qquad (14.16)$$

and Eq. (2.41) is rewritten as

$$\rho\frac{D}{Dt}\left(c_{pd}T+\Phi+\frac{1}{2}\mathbf{u}^2\right)+$$
$$\nabla\cdot\left[\mathbf{F}_{\text{rad}}-K_H\nabla T-\rho K_M\nabla\left(\frac{1}{2}\mathbf{u}^2\right)\right]$$
$$=\rho\dot{Q}+\frac{\partial p}{\partial t}, \qquad (14.17)$$

where $\dot{Q}$ now represents only the material heating/cooling associated with phase changes of water. For a saturated

pseudo-adiabatic process with $\dot{Q}$ given by Eq. (2.26) and the diffusion of water vapor neglected, the foregoing total energy equation becomes

$$\rho\frac{D}{Dt}\left(c_{pd}T+\Phi+L_v r_v^*+\frac{1}{2}\mathbf{u}^2\right)+$$
$$\nabla\cdot\left[\mathbf{F}_{\text{rad}}-K_H\nabla T-\rho K_M\nabla\left(\frac{1}{2}\mathbf{u}^2\right)\right]=\frac{\partial p}{\partial t}. \qquad (14.18)$$

For unsaturated motions (such as within the boundary layer of a tropical cyclone), the turbulent diffusion of water vapor cannot be neglected, especially near the sea surface where the wind-forced evaporation of water occurs. In this case the equation for the water vapor mixing ratio obeys an advection diffusion equation, much like that for salinity in the ocean (e.g., Eq. (4.4.3) of Gill, 1982):

$$\rho\left(\frac{\partial r_v}{\partial t}+\mathbf{u}\cdot\nabla r_v\right)=\nabla\cdot(\rho K_r\nabla r_v), \qquad (14.19)$$

where K_r is the turbulent eddy diffusivity for water vapor. When this equation is multiplied by L_v (assumed constant for simplicity) and added to Eq. (2.41) (with $\dot{Q}=0$) one obtains, after a little manipulation,

$$\rho\frac{D}{Dt}\left(h_m+\frac{1}{2}\mathbf{u}^2\right)+\nabla\cdot\left[\mathbf{F}_{\text{rad}}-K_H\nabla T\right.$$
$$\left.-L_v\rho K_r\nabla r_v-\rho K_M\nabla\left(\frac{1}{2}\mathbf{u}^2\right)\right]=\frac{\partial p}{\partial t}, \qquad (14.20)$$

where, again, $h_m=c_{pd}T+L_v r_v+\Phi$ is the moist static energy per unit mass of dry air as defined in Section 2.5.4. Of course, Eq. (14.20) may be put in an equivalent flux form:

Total energy equation - flux form

$$\frac{\partial}{\partial t}\left[\rho\left(h_m+\frac{1}{2}\mathbf{u}^2\right)\right]+\nabla\cdot\tilde{\mathbf{F}}_{\text{tot}}=\frac{\partial p}{\partial t}, \qquad (14.21)$$

---

where

Total energy flux

$$\tilde{\mathbf{F}}_{\text{tot}}=\rho\mathbf{u}\left(h_m+\frac{1}{2}\mathbf{u}^2\right)+\mathbf{F}_{\text{rad}}$$
$$-K_H\nabla T-L_v\rho K_r\nabla r-\rho K_M\nabla\left(\frac{1}{2}\mathbf{u}^2\right). \qquad (14.22)$$

This flux equation proves useful when Eq. (14.21) is integrated over a cylindrical volume containing the ocean surface where the transfer of sensible and latent heat occurs by turbulent eddies near the surface. In this case it is useful to remember that the dry turbulent (sensible) heat flux $\mathbf{Q_T}$ and moist turbulent (latent) heat flux $\mathbf{Q_r}$ are represented as

$$\mathbf{Q_T}=-K_H\nabla T, \quad \mathbf{Q_r}=-L_v\rho K_r\nabla r_v, \qquad (14.23)$$

respectively.

The total energy budget for a simulated storm in the prototype problem for intensification can be obtained by integrating Eq. (14.22) over the cylindrical region defined at the beginning of Section 14.2. Then, upon neglecting horizontal radiation flux processes (i.e., scattering), one obtains

---

Total energy equation over a cylinder

$$\frac{d}{dt}\left[\rho\left(c_{pd}T + \Phi + L_v r_V + \frac{1}{2}\mathbf{u}^2\right)\right]$$

$$= \left[\frac{\partial p}{\partial t}\right] + F_R + F_{s_b} + F_{s_t},$$

(14.24)

---

where

Energy flux at side boundary of cylinder

$$F_R = -\int_0^{2\pi} d\lambda \int_0^H r\left[\rho u\left(h_m + \frac{1}{2}\mathbf{u}^2\right)\right.$$

$$\left. - (K_H\nabla T + L_v\rho K_r\nabla r_v - \rho K_M\nabla(\frac{1}{2}\mathbf{u}^2)\,)\cdot\hat{\mathbf{r}}\right]dz$$

(14.25)

---

is the advective flux of moist static energy and kinetic energy, plus the diffusive flux of heat, latent heat and kinetic energy through the side boundary $r = R$, $\hat{\mathbf{r}}$ is the radial unit vector, and

---

Energy fluxes at base of cylinder

$$F_{s_b} = \int_0^{2\pi} d\lambda \int_0^R \left[(Q_T + Q_r) + \hat{\mathbf{k}}\cdot\mathbf{F_{rad}}\right.$$

$$\left. -\rho C_D|\mathbf{u_h}|^3 - \text{Precip}\right]_{z_b} r\,dr$$

(14.26)

---

and

Radiative flux at top of cylinder

$$F_{s_t} = -\int_0^{2\pi} d\lambda \int_0^R \hat{\mathbf{k}}\cdot\mathbf{F_{rad}}|_{z_t} r\,dr$$

(14.27)

In the foregoing surface integrals along the top and bottom of the cylinder, the term $\hat{\mathbf{k}}\cdot\mathbf{F_{rad}}$ represents the vertical radiative flux on the respective surfaces. At the base of the cylinder, the term $-\rho C_D|\mathbf{u_h}|^3$ represents a definite loss of energy in association with the surface drag times the surface wind, where $\mathbf{u_h}$ denotes the horizontal velocity vector. The cubic dependence of this term on the wind speed emerges after vertically integrating the divergence of the turbulent flux of kinetic energy and recalling from Chapter 7 the formula equating the turbulent stress with the bulk aerodynamic formula for the stress at $z = z_b = 10$ m (anemometer level). The term $-\text{Precip}$ represents the loss of internal energy associated with rainwater exiting the cylindrical volume at near

surface temperature.[6] Finally, the sensible and latent heat fluxes, Q_T and Q_r are the vertical components of $\mathbf{Q_T}$ and $\mathbf{Q_r}$, respectively, and, again, are evaluated near the sea surface ($z = z_b$). Of course, the sum of these two fluxes is just the local moist enthalpy flux at the sea surface.

As discussed in Chapter 7, the local surface enthalpy flux may be calculated using the empirical formula $C_K|\mathbf{u_{10}}|[(c_{pd}(T_s - T_{10}) + L_v(r_{vs} - r_{v10})]$, where C_K is the specific enthalpy transfer coefficient, $\mathbf{u_{10}}$ is the wind speed at a height of 10 m, r_{vs} is the saturation mixing ratio[7] at the sea surface temperature T_s, and T_{10} and r_{v10} are the air temperature and mixing ratio, respectively, at a height of 10 m. Typical values of C_K are about 1.0×10^{-3} for wind speeds between 50 and 70 m s^{-1} (as summarized in Chapter 7).

While it seems worthwhile for completeness to write down these budget equations for the total energy, we are unaware of any studies that have estimated systematically the various terms.[8] We had hoped to carry out and include such calculations in the book, but at the time of writing we have not been able to accomplish this task. Since, at this stage, it is unclear when we will have the resources to do so, we must leave this task for future work.

To complete the discussion of budgets, we need to examine also the absolute angular momentum budget. This will be done in Section 14.4. First, in the next section, we examine in more depth the role of surface enthalpy fluxes in supporting tropical cyclone intensification.

14.3 Role of surface enthalpy fluxes

In Section 14.2.1 we showed that the surface moisture flux represents only a small fraction of the total water budget. However, global budgets provide limited information about processes taking place within the cylindrical averaging domain. For one thing, the energy equation is just one combination of the several governing equations discussed in Chapter 5. One should not expect to combine all the governing equations into a single equation while retaining all the information that the separate governing equations represent.

---

6. Although the precipitation term was formally neglected in the differential equation (14.21), it has been retained in the integral expression because the magnitude of this loss term is comparable to the turbulent flux of kinetic energy out of the cylinder. In Eq. (14.26), the precipitation term, $\text{Precip} = \rho r_l c_{pl} w_l$, where ρ is air density, r_l is liquid water mixing ratio at the surface, c_{pl} is specific heat of liquid water at constant pressure and w_l is the precipitation fall velocity just above the surface (J. Persing, personal communication).

7. Recall from Section 2.5 that one can use the specific humidity and mixing ratio interchangeably as there is negligible difference between these quantities in practice.

8. A simple scale analysis reveals that the dominant terms in this total energy equation for a circular cylinder are the moist static enthalpy, the dissipation of kinetic energy at the lower surface in association with surface drag and the precipitation leaving the volume.

In the case of the surface enthalpy flux, it is not so much the magnitude of F_{s_b} compared with the other terms in the right-hand-side of Eq. (14.14) that is important, rather it is the radial distribution of the local surface enthalpy flux that is of consequence. The reason is that this flux is responsible for restoring convective instability in the presence of convective downdrafts that have a stabilizing effect on an atmospheric column.

As explained in Chapter 9, the degree of convective instability is characterized in part by the magnitude of near-surface θ_e (Section 9.2.4) and is influenced by convective downdrafts, which typically bring air with low values of θ_e to the surface (Section 9.3.4). When an unsaturated air parcel ascends below cloud base into deep convection without appreciable turbulent mixing with its surroundings, θ_e is approximately materially conserved. After this air parcel becomes saturated, its virtual temperature, T_v, at any given pressure is a monotonically increasing function of θ_e (Exercise 14.1). Ignoring water loading, higher values of T_v imply lower values of air density compared with the air parcel's environment at the same pressure and a higher degree of buoyancy.

The importance of surface enthalpy fluxes and their role in tropical cyclone energetics has a long history, which goes back to seminal papers by Kleinschmidt (1951), Riehl (1967), Malkus and Riehl (1960), Riehl and Malkus (1961), and Riehl (1963). A comprehensive review of these papers is given by Emanuel (2004). The surface enthalpy fluxes are seen as a means of elevating values of near-surface θ_e in the eyewall region of mature storms beyond typical tropical values so that, when the air becomes saturated as it ascends in the eyewall, the negative radial gradient of near-surface θ_e is reflected in a negative radial gradient of θ_e in the eyewall, which implies a negative radial gradient of θ_v in the eyewall.

As an illustration, Malkus and Riehl (1960) assumed that the ascent in the eyewall is close to vertical and by making the assumption that the pressure at some level above the storm had a negligible radial gradient and that the ascending flow is in approximate hydrostatic balance,[9] they showed that the difference in surface pressure, δp_s between the interior of the eyewall and its environment is linearly proportional to the difference of near-surface θ_e over the same radial range, specifically $\delta p_s = -2.5\delta\theta_e$. Emanuel (1986) relaxed the idea that the eyewall was upright and obtained a similar formula (his Eq. (22)). The foregoing ideas relate specifically to axisymmetric aspects of the problem and there was no explicit mention of the elevated values of θ_e being required to maintain local convective instability.

As reviewed in Chapters 12 and 13, the importance of surface moist enthalpy fluxes was emphasized also in

the Emanuel models of tropical cyclone intensification and steady-state maximum intensity. In the steady-state model of Emanuel (1986), the enthalpy fluxes outside the maximum wind radius were invoked to offset the adverse influence of downdrafts and boundary-layer-induced subsidence, which together bring low values of θ_e into the boundary layer. Of course, at radii inside r_{gm}, the downdrafts are assumed to essentially disappear and θ_e rapidly increases with decreasing radius, providing the needed boost to maintain convective instability, despite the developing warm core temperature anomaly in the upper troposphere.

14.3.1 Contributions to θ_e changes

It is commonly assumed that the increase in boundary-layer θ_e, and hence in θ_e^* above the boundary layer, with decreasing radius is dominated by high surface moisture fluxes (e.g., Rotunno and Emanuel (1987), their Section 4b). Using the approximate formula for θ_e derived in Chapter 2, i.e., $\theta_e = \theta \exp[L_v r_v/(c_{pd}T)]$, the radial variation in near-surface θ_e can be written approximately as

$$\Delta\theta_e = \Delta\theta + \frac{L_v}{c_{pd}\pi}\Delta r_v, \qquad (14.28)$$

where T is strictly the temperature of an air parcel lifted adiabatically to its lifting condensation level, $\pi = (p/p_{**})^{\kappa}$ is the Exner function and other quantities are defined in Chapter 2. Here, the symbol Δ represents the increase in the indicated quantity between a given radius and the environment and since $L_v/(c_{pd}T)$ is $O(1)$ and $r_v \ll 1$, the exponential term has been linearized.

If there were no heat or moisture sources, an inward-moving air parcel would conserve its θ and r_v, and therefore its θ_e, but the temperature would decrease with decreasing pressure. Observations (including those to be presented) indicate that the low-level inflow into a tropical cyclone is often approximately isothermal (see Chapter 13 for further discussion and references), which implies that there must be a flux of sensible heat from the ocean. It is this flux that elevates θ through the first term in Eq. (14.28).

Because the saturation mixing ratio, r_v^*, increases with decreasing pressure, *isothermal* expansion would lead to a reduction in the relative humidity in the absence of surface moisture fluxes. This is true whichever definition one uses for relative humidity in Section 2.5 (see Exercise 14.2). In reality, of course, the moisture flux is considerable and the second term on the right-hand-side of Eq. (14.28) is not only positive, it may considerably exceed the first term.

At this stage it is insightful to write $r_v = RHr_v^*$, where RH is the alternative definition of relative humidity (see Section 2.5). Then, Eq. (14.28) becomes

$$\Delta\theta_e = \Delta\theta + \frac{L_v}{c_{pd}\pi}RH \times \Delta r_v^* + \frac{L_v r_v^*}{c_{pd}\pi} \times \Delta RH. \quad (14.29)$$

---

9. We have evidence from some of our own calculations that the neglect of the radial pressure gradient force at an altitude of 20 km is a tolerable assumption (Kilroy, personal communication), although it is by no means obvious why this should be the case and, if made, the assumption needs to be justified. See Smith and Montgomery (2015).

TABLE 14.1 Mean surface data in the eye and eyewall and outside the eyewall in Hurricane Earl on 1 and 2 September 2010. From Smith and Montgomery (2012a). Republished with permission of Wiley.

Date	location of mean	p mb	T °C	q g kg^{-1}	RH %	θ K	θ_v K	θ_e K
Sep 1	outside eyewall	1003.1	27.6	18.7	79	300.5	303.9	355.5
	in eyewall	938.3	27.2	24.7	99	305.9	310.5	380.8
Sep 2	outside eyewall	997.6	27.4	20.4	87	300.7	304.5	361.1
	in eyewall	948.0	27.6	24.9	99	305.4	310.0	380.8

We refer to the contributions from the three terms on the right-hand-side of this equation as $\Delta\theta_{e1}$, $\Delta\theta_{e2}$, and $\Delta\theta_{e3}$, respectively. One can envisage a situation in which the surface moisture flux is just sufficient to keep the relative humidity constant. Then $\Delta\theta_{e3} = 0$ and $\Delta\theta_{e2}$ represents the increase in θ_e from the moisture flux in this situation. It follows then, in the general case, that $\Delta\theta_{e3}$ must be positive in order to raise the relative humidity of inflowing air.

The premise of the air-sea interaction model of Malkus and Riehl (1960) and Emanuel (1986) (and later refinements referenced in Chapter 13) is that isothermal expansion, by itself (i.e., $\Delta\theta_{e3} = 0$), cannot provide a sufficient increment in θ_e to support a strong hurricane. In other words, latent heat transfer over and above that required to keep the relative humidity constant in the presence of isothermal expansion is assumed to be crucial for storm maintenance. This view was supported by the numerical model calculations of (Rotunno and Emanuel, 1987, their Section 4b) who concluded that "... latent heat transfer beyond that due to isothermal expansion is responsible for more than half the inward increase in θ_e."

The study by Montgomery et al. (2009) questioned the need for greatly augmented latent heat fluxes and, in particular, the need to allow surface fluxes to increase with wind speed beyond some nominal Trade-wind value, say 10 m s^{-1}, showing that a vortex in both a three-dimensional and axisymmetric model simulation still intensifies to a mature vortex, but at a somewhat reduced rate. The mean intensity in the simulation experiments with a 10 m s^{-1} cap in the wind speed for the sensible and latent heat fluxes was found to be only modestly less than that in the simulations where the heat fluxes were not constrained (see Section 12.6 and related work for details).

These studies motivate a fundamental question, framed in the context of Eq. (14.29): what is the relative contribution of the increase in eyewall θ_e arising from isothermal expansion and the elevation of the boundary layer relative humidity? The data presented in the next subsection provide an opportunity to estimate the relative contribution of the various terms in this equation from high-density observations of a major hurricane.

14.3.2 Some observations

An observational study, based on dropwindsonde observations from the high-flying NASA DC-8 aircraft in Hurricane Earl on 1 and 2 September 2010 serve to illustrate some of the basic ideas discussed above. Recalling Fig. 1.12 showing vertical profiles of θ_v and θ_e for the 25 soundings released on 1 September and 29 soundings released on 2 September. On each day, one observes a natural division of the soundings into two bins: those in the eyewall or eye and those at larger radii. Those in the eyewall have significantly higher values of θ_e and are distinctly warmer than the latter in terms of θ_v. The eyewall profiles of θ_e can be distinguished from those in the eye as they are almost vertical, a feature that is suggestive of moist adiabatic ascent up to flight level. Of course, the eyewall tends to flare outwards with height above this level. The soundings at larger radii were made within a radius of about 250 km from the storm center.

Taking the subdivision of soundings suggested by Fig. 1.12, one can construct 'bin-means' of various quantities in the two 'sounding bins': eye/eyewall soundings and soundings beyond the eyewall. Table 14.1 compares differences in 'bin-mean' values of various thermodynamic quantities at the surface. Note the consistency in the various quantities on the two successive days of observation. In particular, the surface temperature on 1 September decreases by 0.4 °C between the outer region and eyewall region and on 2 September it increases very slightly by 0.2 °C. These data affirm the approximate isothermal nature of the expansion of inflowing air parcels. In the absence of sensible heat transfer, adiabatic cooling would result in a temperature decrease of about 5-6 °C.

The corresponding increase in surface mixing ratio is 6 g kg^{-1} on 1 September and 4.5 g kg^{-1} on 2 September and the relative humidity increases from 79% to 99% on 1 September and from 87% to 99% on 2 September. The surface pressure reduction is on the order of 60 mb on 1 September and 50 mb on 2 September. These data are used to estimate the terms in Eqs. (14.28) and (14.29).

Table 14.2 shows estimates of $\Delta\theta_e$ in Eq. (14.29) and the three contributions thereto: $\Delta\theta_{e1}$, $\Delta\theta_{e2}$, and $\Delta\theta_{e3}$ de-

TABLE 14.2 Estimates of $\Delta\theta_e$ in Eq. (14.29) and the three contributions thereto: $\Delta\theta_{e1}$, $\Delta\theta_{e2}$, and $\Delta\theta_{e3}$, using the values of relevant quantities in Table 14.1 for the observations on 1 and 2 September. Listed also is the isothermal contribution, $\Delta\theta_{e\,iso} = \Delta\theta_{e1} + \Delta\theta_{e2}$, the ratio of this to the total contribution $\Delta\theta_{e\,iso}/\Delta\theta_e$ expressed as a percentage. The two right columns give the observed change, $\Delta\theta_{e\,obs1}$, calculated directly using the approximate formula for $\Delta\theta_e$ in Eq. (14.29), and $\Delta\theta_{e\,obs2}$, calculated using the more accurate formula of Bolton (1980). From Smith and Montgomery (2012a). Republished with permission of Wiley.

Date	$\Delta\theta_{e1}$ K	$\Delta\theta_{e2}$ K	$\Delta\theta_{e3}$ K	$\Delta\theta_e$ K	$\Delta\theta_{e\,iso}$ K	$\Delta\theta_{e\,iso}/\Delta\theta_e$ %	$\Delta\theta_{e\,obs1}$ K	$\Delta\theta_{e\,obs2}$ K
Sep 1	5.4	3.8	12.1	21.2	9.2	43	24.5	25.3
Sep 2	4.7	3.0	7.3	15.0	7.7	51	19.3	19.7

fined above, using the values of relevant quantities[10] in Table 14.1. It lists also the isothermal contribution to the total change, $\Delta\theta_{e\,iso} = \Delta\theta_{e1} + \Delta\theta_{e2}$, and the fractional contribution of this term as a percentage. These estimates are based on the use of the linear approximation for θ_e. The total change is compared with those computed directly from the observations, $\Delta\theta_{e\,obs1}$, which uses the linear approximation for θ_e, and $\Delta\theta_{e\,obs2}$, which uses the more accurate Bolton's formula (Bolton, 1980).

The increase in θ_e with decreasing radius on account of sensible heat input during the isothermal expansion of air parcels ($\Delta\theta_{e1}$) is about 5-6 K, while the contribution by latent heat input through surface evaporation to maintain the relative humidity ($\Delta\theta_{e2}$) is slightly less, about 3-4 K. The increase in θ_e associated with the moisture contribution that boosts the relative humidity ($\Delta\theta_{e3}$) is about 10-12 K at the surface, but only 6-7 at a height of 200 m (Smith and Montgomery, 2012a, Table 2). The total isothermal contribution ($\Delta\theta_{e\,iso}$) accounts for between 40% and 60% of the total change.

At the surface, the total increments in θ_e in Eq. (14.29) are about 3-5 K smaller than those determined directly from the data using Bolton's formula, consistent with the underestimate in θ_e provided by the linear formula in the range 8-12 K for the range of values found in the boundary layer of a hurricane.

Note that if one uses the subdivision of terms represented in Eq. (14.28), the contribution to the elevation of θ_e by evaporation, effectively $\Delta\theta_{e2} + \Delta\theta_{e3}$, is substantially larger than the contribution from the sensible heat flux, $\Delta\theta_{e1}$, being in the range 64%-75% of the total change, $\Delta\theta_e$, in the data presented in Table 14.1.

While the values in Table 14.1 are approximately similar on the two days of observation, there is likely to be some variation with storm intensity and from storm to storm. For example, in their control calculations, Rotunno and Emanuel (1987), page 557, reported a value for $\Delta\theta_e$ of 11.6 K in going from a radius of 150 km to 20 km in a mature storm with a maximum *tangential* wind speed of about 45 m

s^{-1}, with corresponding values $\Delta\theta_{e1} = 2.8$ K, $\Delta\theta_{e2} = 2.1$ K, and $\Delta\theta_{e3} = 6.7$ K. Then $\Delta\theta_{e\,iso} = 4.9$ K, or 42% of $\Delta\theta_e$, which is at the lower end of values found in Hurricane Earl. Correspondingly, $\Delta\theta_{e3}$ is 58% of $\Delta\theta_e$, which is at the upper end of the values obtained for Hurricane Earl.

Finally, it should be noted that the calculated differences in $\Delta\theta_e$ are a little higher when calculated directly using the approximate formula given just above Eq. (14.28) than with the linear approximation thereto, but they are considerably lower when using the more accurate formula of Bolton as shown by a comparison of columns 5, 8, and 9 in Table 14.2.

14.4 Absolute angular momentum budget

Consider a vertically-aligned vortex on an f-plane expressed in cylindrical coordinates (r, λ, z) with corresponding azimuthal-mean velocity components $(\bar{u}, \bar{v}, \bar{w})$ and departures therefrom denoted by primes. Using the notation of Section 11.4, the equation for the azimuthal-mean absolute angular momentum, $\bar{M} = r\bar{v} + \frac{1}{2}fr^2$, on an f-plane[11] may be written as

$$\frac{\partial \bar{M}}{\partial t} + \bar{u}\frac{\partial \bar{M}}{\partial r} + \bar{w}\frac{\partial \bar{M}}{\partial z} = E_1 + \bar{D}_M, \qquad (14.30)$$

where

$$E_1 = -\overline{u'\frac{\partial M'}{\partial r}} - \overline{w'\frac{\partial M'}{\partial z}}, \qquad (14.31)$$

represents the momentum fluxes associated with eddy processes and

$$\bar{D}_M = \frac{1}{r\rho}\frac{\partial}{\partial r}\left[\rho r^3 K_r \frac{\partial}{\partial r}\left(\frac{\bar{v}}{r}\right)\right] + \frac{1}{\rho}\frac{\partial}{\partial z}\left[\rho K_z \frac{\partial \bar{M}}{\partial z}\right], \qquad (14.32)$$

represents the unresolved horizontal and vertical (eddy) diffusive processes. Here K_r and K_z are horizontal and vertical eddy diffusivities, respectively, the azimuthal variations of which have been ignored.

---

10. For the purpose of computing the saturation mixing ratio in the expression for $\Delta\theta_{e2}$, we used the mean temperature of the eye/eyewall region and the region outside.

11. On a β-plane, there would be an additional term $-\beta r^2 \overline{u' \sin\lambda}$ on the right-hand-side of this equation.

With the reasonable anelastic assumption that $\rho = \rho(z)$, the flux form of Eq. (14.30) is (Exercise 14.3)

$$\frac{\partial \bar{M}}{\partial t} + \frac{1}{r}\frac{\partial r\bar{u}\bar{M}}{\partial r} + \frac{1}{\rho}\frac{\partial \rho \bar{w}\bar{M}}{\partial z} = E_2 + \bar{D}_M, \qquad (14.33)$$

where

$$E_2 = -\frac{1}{r}\frac{\overline{\partial ru'M'}}{\partial r} - \frac{1}{\rho}\frac{\overline{\partial \rho w'M'}}{\partial z}. \qquad (14.34)$$

Integrating (14.33) over the control volume defined in Section 14.2.1 and rearranging gives

$$\frac{d}{dt}\overline{[\rho\bar{M}]} = -\int_0^H \left(\rho r\bar{u}\bar{M} + \overline{\rho ru'M'}\right)_{r=R} dz$$
$$- \int_0^R \left(\rho K_z \frac{\partial \bar{M}}{\partial z}\right)_{z=0} r\,dr$$
$$+ \int_0^H \left(\rho K_r r^3 \frac{\partial}{\partial r}\left(\frac{\bar{v}}{r}\right)\right)_{r=R} dz. \qquad (14.35)$$

The term on the left of this equation is just the tendency of the mass-weighted integral of $\bar{M}$, while the terms on the right are the boundary source terms. The first represents the net influx of $\bar{M}$ through the side boundary, the second represents the sink of $\bar{M}$ on account of the surface frictional torque (although it would be a source if the surface flow is anticyclonic), and the third represents the net rate of gain of $\bar{M}$ associated with the turbulent diffusion of angular velocity at the side boundary.[12]

Eq. (14.35) has important consequences for vortex spin up. It shows that, for a cyclonic vortex to intensify globally, it requires not only an energy source, but a source of angular momentum also to replace the frictional sink of angular momentum at the surface. Like the kinetic energy budget constraint expressed by Eq. (14.12), Eq. (14.35) does not rule out a local increase in energy that might arise, for example, by a local amplification of the broad scale vortical circulation with energy derived from this circulation. However, it does rule out a global increase in angular momentum without a commensurate supply of angular momentum at the domain boundary. In particular, it has implications for the existence for a steady state vortex that are discussed in the next section.

14.5 Exercises

Exercise 14.1. Using the full continuity Eq. (5.1) together with Eq. (14.17), verify Eq. (14.20).

Exercise 14.2. Compare the relative humidity calculated from two definitions for this quantity in Section 2.5 for an air parcel at a pressure of 1000 mb, temperature 28 °C and dew-point temperature 23 °C.

Exercise 14.3. Derive the flux form of Eq. (14.30), i.e., Eq. (14.33).

14.6 Global steady-state requirements

Tropical cyclones undergo a life cycle that may be subdivided into the genesis phase, an intensification phase, a mature phase and a decay phase, although in some cases there may be sub-stages with shorter periods of intensification and decay. Many previous theories and many numerical simulations consider the mature stage either as a strict steady state or at best a quasi-steady state. In numerical simulations, the steady state is often taken to be a period when some metric like the maximum near-surface wind speed or minimum surface pressure become quasi-steady. The notion of a steady state using these criteria can be misleading as other metrics, such as the maximum anticyclonic wind speed in the upper-level outflow, may still be far from steady or quasi-steady (see e.g., Smith et al., 2014; Persing et al., 2019). In fact, the existence of a realistic globally steady state, even in theory, is highly questionable.[13]

In this section we explore reasons to question the existence of a realistic globally steady state and in Chapter 15 we examine an idealized simulation of a full tropical cyclone life cycle, focusing especially on the mature and decay phases.

While it is easy to imagine the existence of a global steady state on the basis of energy arguments wherein the latent heat energy supplied to the storm through evaporation of sea water balances the frictional dissipation of kinetic energy within the cyclone and the net energy exchange with the environment, there would be additional constraints on the system imposed by the need to satisfy global budgets such as the water budget, the total energy budget and the angular momentum budget in a way that the left-hand-sides of Eqs. (14.2), (14.14) and (14.35) are all zero. The last condition tells us that, since cyclonic angular momentum is lost

---

12. Here we have neglected the source of $\bar{M}$ associated with the Rayleigh damping of tangential velocity that is inserted to damp vertically-propagating inertia-buoyancy waves that flux $\bar{M}$ out of the cylindrical domain. See Persing et al. (2019) for an assessment of this effect, which is generally not negligible for long-time simulations with large outer domains and zero lateral damping there.

13. For the ostensibly more general case of a quasi steady-state, there must be no statistical trend in all dependent variables on time scales longer than the averaging interval. Such a theory would be constructed by time-averaging all of the equations of motion to obtain a system of equations for the time mean of the quasi-steady vortex. This procedure is analogous to Reynolds averaging in a turbulent flow. Because of the intrinsic nonlinearity of the field equations, such a theory for a quasi-steady hurricane vortex would necessarily include covariance terms contributing to the time-mean of all dependent variables. Without a temporal eddy closure for these terms, the equations do not close and the system is unsolvable in general. Thus without a defensible closure theory, talk of a quasi-steady theory is vapid. It is worthwhile at this point to remind the reader that the E86 theory reviewed in Chapter 13 is a strict steady-state theory and not a quasi-steady theory. For the foregoing reasons, we will limit our discussion to the requirements for a globally steady theory.

by the system at the sea surface because of the frictional torque acting there, a globally steady state would require a *steady* supply of cyclonic angular momentum to maintain it.

In general, a globally steady state would require first that the convective instability in the core region of the storm be maintained by fluxes of moisture at the underlying sea-surface and/or radiative cooling of the upper atmosphere. The convective heating in the inner-core region and the radiative cooling largely outside of the main convective region comprising the eyewall and spiral bands provide a pattern of net diabatic heating that forces an overturning circulation. Above the frictional boundary layer, this steady-state circulation must be along absolute angular momentum (or M-) surfaces (see Exercise 8.2).

In the upward branch of the secondary circulation, the diabatic heating rate associated with moist convection must be consistent with the approximate material conservation of pseudo-equivalent potential temperature, θ_e. In the clear-sky part of the descending branch of the circulation, where radiative cooling is mainly operative, θ_e is not materially conserved. However, the radiative cooling must exactly balance the adiabatic warming of air parcels as they descend. Above the boundary layer, these conditions are encapsulated by the need for the spin-up function, S, to vanish as discussed in Chapter 5 (see Section 5.13.3).

The severity of the global constraint associated with $S \equiv 0$, alone, led Persing et al. (2013), Smith et al. (2014), and Persing et al. (2019) to question the theoretical existence of a globally-steady tropical cyclone. However, there are other issues, since the foregoing simple picture assumes that the upward branch of the secondary circulation takes place in the eyewall and that elsewhere there is subsidence. This subsidence would bring extremely dry air from the upper troposphere to the surface.

Another layer of complexity is the maintenance of the moisture structure of the ambient thermodynamic sounding. This maintenance relies on the existence of a population of convective clouds with different detrainment levels to moisten the dry air descending into the boundary layer. This convection would need to be a part of the steady-state model. In addition, a global steady state would require just the fluxes of sensible and latent heat necessary to maintain the foregoing thermodynamic cycle.

Chapter 15

Tropical cyclone life cycle

Observations show that tropical cyclones undergo a life cycle that includes a genesis phase, an intensification phase, a mature phase, possibly accompanied by intensity fluctuations, and a decay phase. Idealized simulations of the genesis and intensification phases were the subject of Chapters 10 and 11, respectively. In this chapter we present an idealized simulation of an entire life cycle, focusing on the mature phase and decay phase. Understanding the decay phase, which has been less well studied than the other phases is a pertinent problem for forecasters, especially if rapid decay occurs just prior to landfall. An example is the rapid decay of Category 5 Hurricane Patricia discussed in Chapter 1, shortly before it made landfall over Mexico.

The life-cycle study is based on that of Smith et al. (2021), which uses the same warm rain model as in the genesis simulation of Chapter 10, the only differences being a broader initial vortex, a coarser horizontal grid spacing and a longer relaxation time scale for Newtonian cooling. The simulation here starts with an initial vortex with a maximum tangential wind speed of 5 m s^{-1} at a radius of 200 km instead of 100 km as before. The horizontal grid spacing is 1 km and the size of the inner fine grid is 600 km square. This larger domain size is more appropriate for the broader vortex and the larger grid spacing is a compromise to keep the computational demands within bounds. The model domain is 3010 × 3010 km, almost the same as before. Even so, the much longer integration time raises issues concerning the use of fixed environmental profiles of temperature and humidity and the use of a Newtonian cooling formulation in lieu of implementing a more complex radiation scheme. These issues are important enough to merit an extended discussion before analyzing the life-cycle simulation, itself.

15.1 Newtonian cooling

A question that arises in formulating any long term integration of a tropical cyclone model is how to represent the physical processes (radiation and convection) that determine, *inter alia*, the vertical profiles of thermodynamic quantities in the environment. If radiation is modeled directly in a simulation, the environment profiles must evolve with time until some quasi-radiative-convective equilibrium is achieved in the outer part of the domain. A realistic simulation of radiative-convective equilibrium in the tropics

would require a domain size appreciably larger than the domain size normally used to simulate tropical cyclones, extending perhaps out to 10,000 km. Such a large domain is not computationally feasible if one seeks to have a small enough grid spacing to adequately resolve deep convection (see e.g. Jakob et al., 2019). Even so, such an approach would be of questionable realism because in the real atmosphere, the processes determining the atmospheric sounding do not involve persistent deep convection at all latitudes.

In order to formulate a clean thought experiment, it has become common to apply Newtonian relaxation to the temperature (or equivalently the potential temperature) field, relaxing the temperature in and around the storm to that of the environment on a prescribed time scale (e.g. Rotunno and Emanuel, 1987 and many others). While certainly crude, this procedure is no more than an expedient to obtain a closed formulation that does not allow an excessive warming of the storm environment in the upper troposphere. In reality, such cooling would be accomplished by infrared radiation to space.

If the relaxation formulation is implemented, the question is: what is the appropriate relaxation time scale? Rotunno and Emanuel chose a 12 hour time scale on the basis that this value

"yields cooling rates of approximately 2 K d^{-1} in the outer regions and is approximately enough to balance the gentle, but persistent subsidence in the outer regions of the vortex".

However, they noted that this choice

"produces unrealistically large cooling rates in the inner core",

which has undesirable consequences for the parcel buoyancies produced in the eyewall (see their Section 3e). For these reasons they carried out tests in which the magnitude of local cooling was not allowed to exceed 2 K day^{-1} and one in which there was no cooling. They found, *inter alia*, that the simulation without Newtonian cooling

"produces a colder outflow (and a more intense vortex)".

Tropical Cyclones. https://doi.org/10.1016/B978-0-44-313449-4.00023-0

In a different context, Mapes and Zuidema (1996) carried out a study of dry layers that are frequently observed in atmospheric soundings from the climatologically humid western Pacific warm pool region. In this study they examined the effects of radiation on thermal and moisture perturbations typical of those generated through gravity waves produced in the environment of tropical mesoscale convective systems. These waves lead to ascent or subsidence with downward displacements of air parcels causing warm temperature and dry humidity anomalies in the upper troposphere, above the 500-mb level, and upward displacements causing cool temperature and moist humidity anomalies below (Mapes, 1993; Mapes and Houze, 1995).

Mapes and Zuidema applied the above kind of vertical displacement profile to the mean tropical sounding from their study and computed clear-air radiative cooling rates using the modified profiles for two cases: one in which the displacement affects the temperature only and one in which the displacement acts upon both the temperature and humidity fields. The radiative cooling rates calculated from the temperature perturbation alone were found to be close to those based on a Newtonian cooling formulation with a relaxation time scale of 10 days rather than 12 hours. However, Mapes and Zuidema pointed out that humidity variations are somewhat more important in their radiative cooling calculations than just the temperature variations alone, a finding that serves to reinforce the acknowledged crudity of the physics embodied in the Newtonian relaxation procedure. This procedure does not consider the moisture field.

Without obvious viable alternatives to employing a Newtonian relaxation scheme in the life-cycle simulation to be described, Kilroy (personal communication) carried out three preliminary idealized, 16-day tropical cyclone simulations that by-passed the genesis stage by starting with a moderately strong vortex, but included an RI phase, a mature phase and a decay phase (not shown). In the first simulation, the default 12-hour relaxation time scale in the CM1 model was retained (following Rotunno and Emanuel, 1987). In the second simulation, the 10-day time scale suggested by Mapes and Zuidema (1996) (their p631) was used, while in the third simulation, there was no thermal relaxation to the background state.

In the simulation with the 10-day time scale, warming occurs progressively over the entire domain. At mid to upper levels, the warming is greater than 4 K over the 16 days and the system-scale vortex decays markedly during this time.

The vortex in the simulation with no relaxation evolves in a similar way to that with the 10-day time scale, but the warming is about 1 K larger. With a 12-hour time scale, the temperature of the outer domain remains close to the initial sounding and the vortex takes considerably longer to decay.

It transpires that, as in the study by Rotunno and Emanuel (1987), the simulation without any cooling pro-

duces a slightly more intense vortex. This is presumably because, in this case, the eyewall updraft is stronger and leads to a stronger overturning circulation.

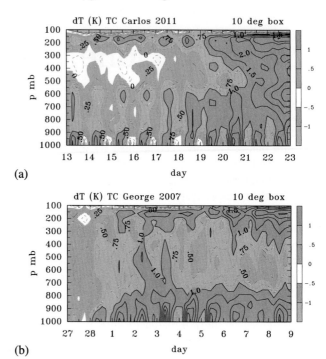

FIGURE 15.1 Time-height cross section of areally-averaged temperature deviation (in K) from that at the start of the time series within a column 10×10 deg lat, centered on the location of the minimum geopotential at 850 mb for the development of Tropical Cyclones (a) Georges (2007; Jan 27 - Feb 9) and (b) Carlos (2011, Feb 13 - 23). Shading as indicated by the label bar on the right. Additional contours every 0.5 K. This figure was constructed from the ECMWF operational analysis, which has a horizontal grid spacing of 0.125 deg and a temporal output of 6 hours. The evolution of Tropical Cyclone Georges is described in more detail in Kilroy et al. (2017a). From Smith et al. (2021). Republished with permission of Wiley.

The mid- and upper-level warming of the environment that occurs in the simulation with the 10-day relaxation time scale is not necessarily unrealistic. In nature there is often a warming of the upper troposphere in the environment surrounding tropical cyclones on the order of about 2-4 K. To support this claim, Fig. 15.1 shows time-height cross sections of the areally-averaged temperature difference on an isobaric surface in a 10×10 deg column following two tropical cyclones *Georges* and *Carlos*, which occurred in the Australian region[1] (Kilroy et al., 2017b). These cross sections are based on data from ECMWF operational analyses for the two cyclones. The ECMWF analyses are constrained by observations in a system that includes a sophisticated representation of radiation (Hogan and Bozzo, 2018).

The diagnosed warming of a few degrees suggests that the default time scale for relaxation to the initial sounding in

---

1. In addition to the upper tropospheric warming, these storms reveal a lower tropospheric warming with a prominent diurnal cycle associated with the land track of these systems.

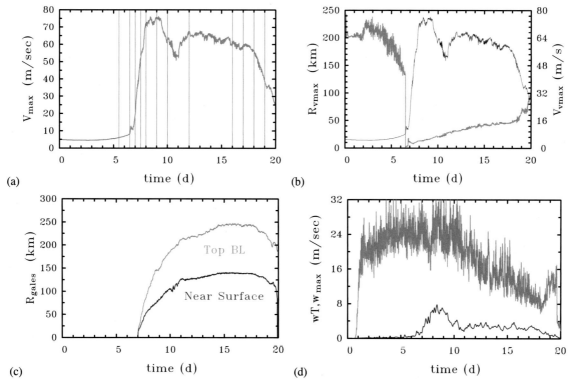

FIGURE 15.2 Life cycle simulation time series of (a) maximum azimuthally-averaged tangential wind speed (V_{max}), (b) the radius R_{vmax} at which V_{max} occurs, (c) the outermost radius of gale force (17 m s^{-1}) tangential winds, (R_{gales}), at a height of 1 km and the outermost radius $R_{gales F}$ where the total surface wind speed falls below gale force, and (d) the maximum azimuthally-averaged vertical velocity (W_{max}) (blue (dark gray in print version) curve) and maximum local vertical velocity (w_{max}) (red (mid gray in print version) curve). The thin vertical lines show times at which horizontal or vertical cross sections are shown later in the text. From Smith et al. (2021). Republished with permission of Wiley.

CM1 is too strong in the case of a mature tropical cyclone. For the foregoing reasons, we use a 10-day relaxation time-scale for the life-cycle simulation to be described.

15.2 Life cycle metrics

Returning now to the life-cycle simulation, we summarize in Fig. 15.2 the evolution of the vortex in terms of azimuthally-averaged metrics similar to those in Section 10.3.1: these include the maximum tangential wind speed, V_{max}, the radius at which this maximum occurs, R_{vmax}, two measures of outer vortex size R_{gales} and $R_{gales F}$, and the maximum vertical velocity, W_{max}. Shown also is the maximum local vertical velocity, w_{max}. Here, as in Section 10.3.1, R_{gales} refers to the outer radius at which the azimuthally-averaged tangential wind component equals gale force (17 m s^{-1}) and $R_{gales F}$ refers to the outer radius at which the azimuthally-averaged near-surface total wind speed falls to gale force.

The evolution to a mature vortex in the life-cycle simulation is similar to that of the genesis simulation described in Section 10.3.1. After a gestation period lasting a little more than $6\frac{1}{2}$ days, the vortex undergoes a period of RI (Fig. 15.2a), which lasts about a day and a half. This gestation period is more than three times as long as for the

smaller vortex in the genesis simulation of Section 10.3.1, but the increase in intensity between day 7 and day 8 is 51 m s^{-1}, slightly larger than that between day 2 and day 3 for the smaller vortex.[2] The maximum intensity, with $V_{max} = 75$ m s^{-1}, is reached at about $8\frac{1}{2}$ days.

The first mature phase, the period between 8 and $9\frac{1}{2}$ days, is relatively short-lived and is followed by a period of about $1\frac{1}{2}$ days during which V_{max} declines to just over 50 m s^{-1}. This marked decline is followed by a second period of RI in which V_{max} increases to about 65 m s^{-1} at about 12 days.

After a second mature phase that lasts about a day, the time-mean of V_{max} begins to decline, slowly at first, but from 18 days the rate of decline increases with V_{max} falling well below hurricane strength (32 m s^{-1}) by 20 days. As shown later in Section 15.4 and as in the genesis simulations in Chapter 10, V_{max} is always below a height of 1 km and within the frictionally-driven inflow boundary layer.

A little before the onset of the first period of RI, there is a sharp decline in R_{vmax} (Fig. 15.2b). This decline is accompanied by the formation of a strong, but localized, transient cyclonic vortex induced by a vigorous convective

---

2. The effects of initial vortex size on the gestation period are examined in Section 16.5.

cell near the center of the broader scale circulation. This vortex, which has an R_{vmax} of only 2 km and a V_{max} on the order of 10 m s^{-1}, has a strong vorticity signature near the domain center in Fig. 15.3b. This vortex is a precursor to the formation of a mesoscale convective vortex. At 6.5 days, R_{vmax} jumps from 4 km to 17 km and V_{max} increases slightly from 10.5 m s^{-1} to 11.2 m s^{-1}. Subsequently, after a brief decline, R_{vmax} steadily increases.

Both measures of size (R_{gales} and $R_{gales F}$) show a sharp increase as RI begins, but the rate of increase declines while remaining positive until about 18 days. Thereafter, both measures begin to decline.

The quantity W_{max} is one metric for characterizing the strength of the overturning circulation. This quantity fluctuates on the convective time scale, but its time mean over a few hours increases steadily during the period of RI and is largest during the mature phase (Fig. 15.2d). From about 9 days onwards, the time mean of W_{max} declines steadily as the vortex decays. Evidently, the increase in near-surface pseudo-equivalent potential temperature, θ_e, near the base of the eyewall increases sufficiently to outweigh the upper-level warming that accompanies the vortex intensification so that, presumably, the maximum $CAPE$ continues to increase through to maturity before declining (see Section 9.2.1).

Fig. 15.2d shows also the time evolution of the maximum local updraft velocity w_{max}. This quantity increases rapidly on the first day of the gestation period, reflecting the formation of the first deep convective clouds locally. It remains large until about 11 days, after which its time mean over a few hours progressively declines. This decline is a further indication of decreasing convective instability associated with the upper-level warming. Of course, w_{max} is much larger than W_{max}, since deep convective clouds are generally localized and thereby highly asymmetric relative to the nominal circulation center. During most of the cyclone's life cycle, deep convective clouds make the most important contribution to the mean transverse overturning circulation. The simulated maximum local updraft velocities are of similar order of magnitude to those estimated from airborne Doppler radar observations, typically 20–25 m s^{-1} (Houze et al., 2009; Guimond et al., 2010).

15.3 Vortex asymmetries

Like the genesis simulations in Chapter 10, the evolution of the tropical cyclone life cycle in three dimensions is somewhat complex, but it is necessary to highlight the main features of the evolution before attempting to explain these features. Thus, the next two sections are largely descriptive. As in Chapter 10, it proves insightful to examine the structure of specific fields during different phases of the vortex evolution. Again, we begin by documenting the intrinsically asymmetric nature of the various phases of evolution, showing diagrams similar to those in Section 10.3.2.

To complement these descriptions of flow evolution mainly in the lower troposphere, important features of the radial and vertical structure of selected azimuthally-averaged flow fields during the same phases of vortex evolution are discussed in Section 15.4.

Figs. 15.3, 15.4, and 15.5 show horizontal cross sections at a height of 1 km of the vertical component of relative vorticity and wind vectors at selected times. These figures show also the location of strong, deep convective updrafts as characterized by the 1 m s^{-1} contours of vertical velocity at a height of 6 km. Fig. 15.3 spans the 2 day period beginning at $5\frac{1}{2}$ days during which genesis occurs to the point at which the RI phase is two-thirds complete, while Fig. 15.4 shows selected times from 8 days to 12 days, during the first mature phase and the subsequent period of intensity decline and recovery. Fig. 15.5 shows similar daily fields from 16 to 19 days during the decay phase. The times at which these figures are shown are indicated by thin vertical lines in Fig. 15.2, at and beyond $5\frac{1}{2}$ days.

15.3.1 Genesis and RI phases

The vortex evolution during these phases is similar to that shown in Section 10.3.2, but is summarized here for completeness. At $5\frac{1}{2}$ days (Fig. 15.3a) during the later part of the gestation period, there are a few scattered clusters of deep convection within the weak broadscale vortex circulation. Individual deep convective cells are indicated by the yellow contours enclosing regions where the vertical velocity at 6 km altitude exceeds 1 m s^{-1}. Seen also are isolated patches of enhanced cyclonic relative vorticity, some of which are remnants from earlier bouts of convection, but like the current patches of deep convection, their distribution is quite asymmetric and has a noticeable lack of organization. Small areas of anticyclonic vorticity are found also, mostly as part of dipole structures which are a result of the tilting of horizontal vorticity into the vertical by local cloud updrafts (Sections 10.3.2, 10.10, and 11.1).

Fig. 15.3b shows the same fields one day later, at $6\frac{1}{2}$ days, just about the start of RI. During the intervening time, the clusters of deep convection and the patches of enhanced cyclonic vorticity have grown in size and become more numerous, with a tendency to organize into banded structures around the circulation center. As in the genesis simulation (see Section 10.3.2), this organization occurs during a period when V_{max} is less than 10 m s^{-1} (cf. Fig. 15.2a). At $6\frac{1}{2}$ days there are some deep convective cells within 20 km of the circulation center, one of which is the precursor to the formation of a mesoscale convective vortex referred to in Section 15.2.

The convection and vorticity continue to organize during the next day (panels (c)-(d)) and V_{max} increases sharply.

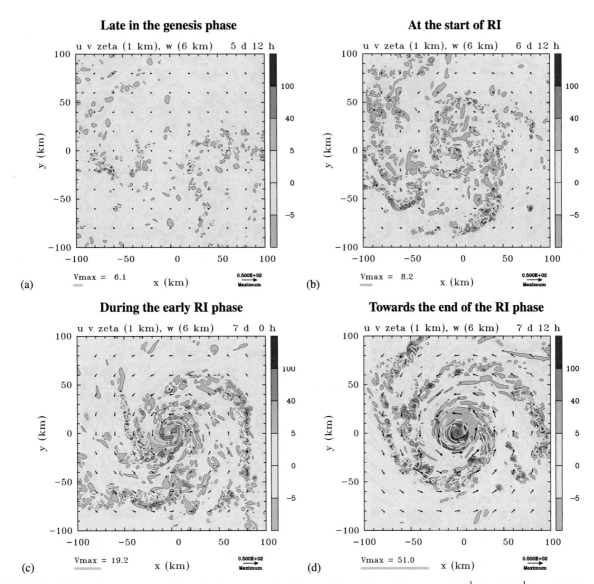

FIGURE 15.3 Horizontal cross sections of the vertical component of relative vorticity and the wind vectors at (a) $5\frac{1}{2}$ days, (b) $6\frac{1}{2}$ days, (c) 7 days, and (d) $7\frac{1}{2}$ days at 1 km altitude in the life cycle simulation. Shown also in each panel are contours of vertical velocity equal to 1 m s^{-1} at a height of 6 km (yellow). Values for the shading of vertical vorticity are given in the color bar, multiplied by 10^{-4} s^{-1}. Beyond 5×10^{-4} s^{-1}, the red solid contours are relative vorticity at intervals of 2×10^{-3} and below -5×10^{-4} s^{-1}, the blue dashed contours have the same interval. The wind vectors are in relation to the maximum reference vector (50 m s^{-1}) at the bottom right, while at the bottom left, the value of V_{max}, indicated with a thick light blue line with units of m s^{-1}, is here the maximum speed of the vectors plotted. From Smith et al. (2021). Republished with permission of Wiley.

By 7 days (Fig. 15.3c), early in the RI phase and before V_{max} has reached hurricane strength, the region within a radius of 20-30 km about the circulation center is largely covered with enhanced cyclonic vorticity ($> 5 \times 10^{-4}$ s^{-1}). This region is surrounded by an incomplete ring of 1 m s^{-1} updrafts at a height of 6 km. Inside this ring are patches of particularly large vorticity ($> 4 \times 10^{-3}$ s^{-1} in magnitude), which partially surround the circulation center and lie within about 10 km of the center.

As the vortex continues to rapidly intensify, typified by the fields at $7\frac{1}{2}$ days, the core of enhanced cyclonic vor-

ticity expands in radius with continuous bands of strong vertical velocity forming around it. There are now partial rings of vorticity inside these convective bands with values $> 1 \times 10^{-2}$ s^{-1}. At larger radii, convection has organized into banded structures also and patches of anticyclonic vorticity are found mostly in these outer bands.[3] Notably, de-

---

3. The foregoing vorticity distributions are consistent with the results of idealized numerical model simulations with a single deep cloud initiated in a realistic tropical-cyclone-like vortex (Kilroy and Smith, 2016). These authors showed that deep convection occurring close to the vortex axis generates predominantly cyclonic vertical vorticity, whereas near and beyond the radius of maximum winds it generates dipoles of vertical vorticity.

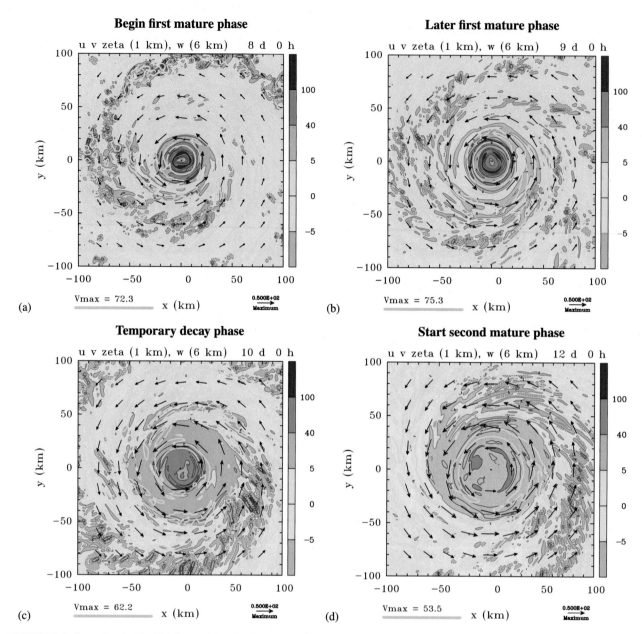

FIGURE 15.4 Legend as for Fig. 15.3, but (a) 8 days, (b) 9 days, (c) 10 days, and (d) 12 days.

spite the intensity of the vortex at this time (55 m s^{-1} in Fig. 15.2a), the patterns of deep convection and low-level vorticity continue to show a significant degree of asymmetry.

Kilroy and Smith showed also that the closer the convection is located to the vortex axis, the stronger the cyclonic vorticity anomaly produced, even though convection tends to be weaker near the axis. They showed further that the structure of the vorticity anomaly becomes more complex with increasing radius, changing from a monopole of cyclonic vorticity on the axis to dipoles that change in sign with height at larger radii. Explanations for the more complex vorticity structure at larger radii are given by Kilroy and Smith (2016).

In spite of the tendency for the flow to axisymmetrize during the RI phase, the pattern of convection retains an appreciable degree of asymmetry, even during the mature phase. Such behavior is found in previous simulations with coarser resolution (e.g. Nguyen et al., 2008; Shin and Smith, 2008; Fang and Zhang, 2011; Wang, 2002).

Although intensity fluctuations in terms of V_{max} are small during the mature phase (Fig. 15.2a), the vortex core steadily expands in terms of R_{vmax} (Fig. 15.2b). The expansion is manifest in Fig. 15.4b for 9 days as a radial expansion of the core of enhanced cyclonic vorticity compared with that in Fig. 15.4a for 8 days. As at 8 days, there remains an approximately annular region of deep convection within the region of enhanced vorticity, but its mean radius has increased compared with that at 8 days, consistent with the increase in R_{vmax}.

There is a noticeable difference in the spiral rainband structure between 8 and 9 days with a large increase in deep convection in the sector between the southwest and east-northeast, outside the inner convective ring. Further, deep convection in the outermost band at 8 days has decayed, although vorticity anomalies associated with this convection still remain.

15.3.2 First mature phase

The vorticity and vertical velocity fields at 10 days and 12 days are shown in Figs. 15.4c and 15.4d. The first time is roughly midway in the decline of V_{max} after the mature phase and the second time is near the start of a second mature phase that follows a further intensification of V_{max} (Fig. 15.2a). In both figures, a prominent feature is the continued expansion of the core of enhanced vorticity. By Stokes' theorem, this expansion is consistent with that of the tangential wind field in the outer region of the vortex (see Chapter 11, Eq. (11.1)).

The decline in V_{max} beyond 10 days is accompanied by fragmentation of the ring of "eyewall convection", which continues for about a day. A convective rain band complex develops beyond the ring and remains a dominant feature as the vortex reintensifies after 11 days (not shown). After the vortex has re-intensified at 12 days (Fig. 15.4d), the eyewall of deep convective cloud has almost re-consolidated into an annular ring, but with a significantly larger mean radius than before. By this time, the convective band complex has weakened a little and is no longer present in all sectors. This convective rain band is shown in Section 15.4.3 to be an important feature during the weakening and subsequent re-intensification of the vortex between $9\frac{1}{2}$ and 12 days.

15.3.3 Decay phase

The vorticity and vertical velocity fields at times during the decay phase are shown in Fig. 15.5. This phase begins after about 13 days. The upper panels show the fields at 16 days and 17 days, times when the decay rate is still relatively modest. At each of these times, the structure of both fields is similar with a large core of enhanced cyclonic relative vorticity, the strength of which declines rapidly beyond a radius of between 60 and 70 km.

The eyewall updraft at 6 km height at 16 days has become rather large, about 50 km in mean radius, and there have developed fine scale structures in the relative vorticity field on the inner edge of this updraft (Fig. 15.5a) primarily along the southwestern semicircle. These structures point to the presence of horizontal shear instability on the inner edge of the eyewall. The occurrence of small patches of

negative vorticity within these structures indicates that horizontal vorticity is being tilted into the vertical. It may be significant that the eyewall updraft is broken into localized structures at lower altitudes with a similar horizontal scale to that of the vorticity structures (not shown).

The eyewall updraft at 6 km height at 17 days has become even larger, but remains ring-like (Fig. 15.5b). The small-scale vorticity structures remain, but are now largest in the semicircle centered to the southeast.

The eyewall updraft at 6 km height at 18 days has begun to break up (Fig. 15.5c). The small-scale vorticity structures remain along the inner eyewall, but are now largest in the semicircle centered to the east. By 19 days (Fig. 15.5d), the deep convective updrafts stronger than 1 m s^{-1} and the small-scale vorticity features have virtually disappeared. All that remains is a large and almost circular region of enhanced relative vorticity with a radius on the order of 80 km and with values exceeding 5×10^{-4} s^{-1}.

The foregoing analysis shows that the decay phase has a strong asymmetric component involving some form of shear instability on the inner edge of the eyewall. The effects of this instability would be a challenge to properly represent in an axisymmetric model. Dynamical and thermodynamical aspects of the decay process are discussed in Section 15.5.

15.4 Azimuthally-averaged view of vortex evolution

Although, as shown above, the evolution of the flow is intrinsically three-dimensional, it proves insightful to examine the evolution of the azimuthally-averaged flow fields as in Chapters 10 and 11. We confine the discussion here to the period spanning the two mature phases and the decay phase, since, as shown in Section 15.3.1, the three-dimensional flow evolution prior to genesis and during the RI phase is quite similar to that described in Section 10.3.2 and the azimuthally-averaged view of those phases was discussed in Chapter 10 also.

Figs. 15.6 and 15.7 show radius-height cross sections of the azimuthally-averaged and 3 hour time-averaged tangential and radial velocity components at various times during the mature and decay phases.[4] As in Section 10.5, this type of average is denoted by ⟨⟩. Superimposed in each panel is the similarly averaged vertical velocity field, ⟨w⟩. The corresponding averaged ⟨M⟩-surfaces are shown along with the averaged radial velocity fields, ⟨u⟩.

15.4.1 Mature phase

Radius-height cross sections of the velocity fields at 9 days are shown in Figs. 15.6a and 15.6b. This time is well within

---

4. Note that because of the time averaging present in Figs. 15.6 and 15.7, the values of V_{max} deduced from these figures do not correspond exactly with those in Fig. 15.2, which are not based on a time average.

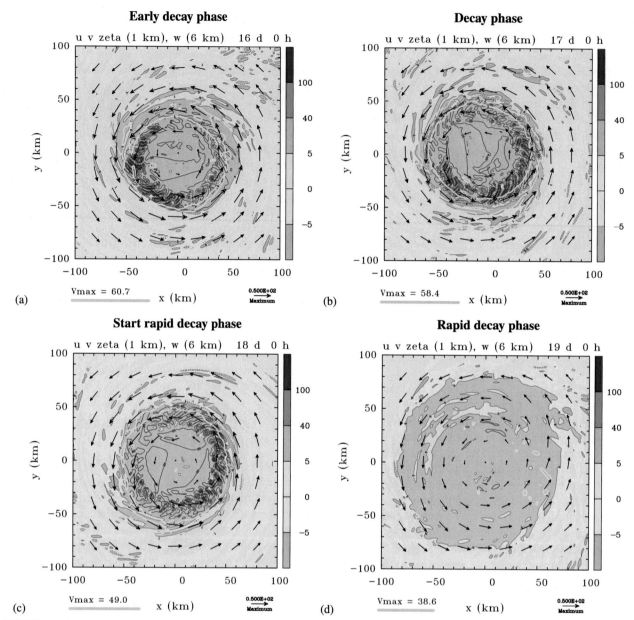

FIGURE 15.5 Legend as for Fig. 15.3, but at (a) 16 days, (b) 17 days, (c) 18 days, and (d) 19 days. From Smith et al. (2021). Republished with permission of Wiley.

the first mature phase (Fig. 15.2a). The mean updraft at this time is comparatively narrow and the strongest part ($\langle w \rangle$ exceeding 1 m s^{-1}) is centered at a radius of 20 km. The maximum mean tangential velocity, about 75 m s^{-1}, occurs at a radius of 16 km and a height of 300 m (Fig. 15.6a). As in many previous studies, there is a shallow surface-based layer of strong frictionally-driven inflow (Fig. 15.6b). Although hard to see in this figure, the maximum inflow speed, 35 m s^{-1}, occurs very close to the surface at a radius of 19 km.

Another feature of Fig. 15.6b to note is a shallow layer of strong outflow with maximum speed 16 m s^{-1} just above

where the boundary layer inflow terminates. Above this layer is a shallow layer of inflow with outflow above it. This outflow has a local maximum just above the inflow layer. These features are a result of the centrifugal recoil effect described in Section 11.6.2. The centrifugal wave appears first during the RI period (not shown) and it tends to be a persistent feature of the first mature phase, accounting for the second maximum of tangential wind at an altitude near 3 km in Fig. 15.5a.

The shallow tongue of outflow just above the surface-based inflow radial velocity exceeding 1 m s^{-1} extends radially to about 90 km. If, as in the classical model for vor-

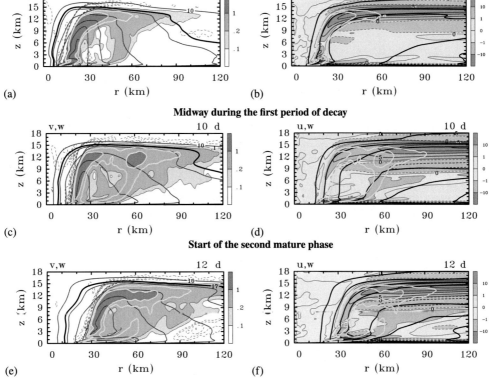

FIGURE 15.6 Vertical cross sections of the azimuthally-averaged, 3 h time-averaged tangential (left column) and radial (right column) velocity components centered at various times: (a, b) 9 days, (c, d) 10 days, (e, f) 12 days in the life cycle simulation. Superimposed in the left panels are contours and shading of the averaged vertical velocity. Contour intervals are as follows. Tangential velocity: blue contours every 5 m s^{-1}. The thick black contour is the radius of gale-force tangential wind (17 m s^{-1}), the outer portion of which provides a suitable measure of vortex size at different levels. Vertical velocity: thin red contours 0.1 m s^{-1} and 0.2 m s^{-1}, thick yellow contours are 0.5 m s^{-1} and 1 m s^{-1}, thin dashed red contours indicate subsidence at intervals of 0.02 m s^{-1}. Shading as indicated on the side bar. The right panels show contours and shading of the averaged radial velocity with contours of absolute angular momentum (black contours) and the two yellow contours of the averaged vertical velocity superimposed. Contour intervals are as follows. Radial velocity: 5 m s^{-1}, solid red contours positive, dashed blue contours negative, thin contours are 1 m s^{-1} as well as the zero contour. Shading as indicated on the side bar. Absolute angular momentum: thick black contours every 5 × 10^5 m^2 s^{-1}. From Smith et al. (2021). Republished with permission of Wiley.

tex spin up, the mean absolute angular momentum, $\langle M \rangle$, is approximately materially conserved in this tongue, the tangential wind speed at the top of the boundary layer would be spinning down as $\langle M \rangle$ surfaces are advected outwards. This spin down has consequences for the boundary layer inflow and is an element of the boundary layer control mechanism described in Section 10.14. Elsewhere, above the boundary layer and below about 9 km, there is mostly weak inflow ($|u| < 1$ m s^{-1}).

The mean eyewall updraft is highlighted in each panel of Fig. 15.6 by the two yellow contours of mean vertical velocity, $\langle w \rangle$, equal to 0.5 and 1 m s^{-1}. At 9 days, the updraft is quite narrow in the low- to mid-troposphere, less than 10 km wide, and it has only a small outward tilt up to about 6 km height. Above this height, the tilt increases on account of the progressively increasing strength of the radial outflow in the eyewall updraft.

There is a kink at a height of about 1.5 km in both $\langle w \rangle$ contours shown. Below this height, at $z = 1.2$ km, there is a localized maximum of strong ascent (2.3 m s^{-1}) that is a consequence of dynamically-forced ascent where the boundary-layer inflow terminates. The maximum value of $\langle w \rangle$, 6.1 m s^{-1} at an altitude of 8.8 km, is a reflection of buoyantly-forced ascent in the main eyewall updraft, itself.

The main outflow layer of the storm occurs at altitudes between about 12 km and nearly 16 km, the latter being the approximate height of the tropopause in the ambient thermodynamic sounding. The maximum outflow velocity exceeds 18 m s^{-1} at the edge of the domain shown. Wang et al. (2020) showed that beyond this domain edge, the outflow velocity declines with increasing radius as the outflow spreads out. They showed further that the decline is brought about by an inward-directed mean agradient force in the outflow layer.

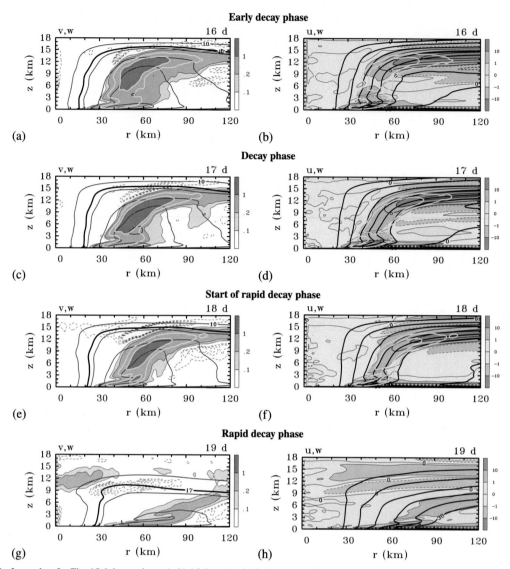

FIGURE 15.7 Legend as for Fig. 15.6, but at times: (a, b) 16 days, (c, d) 17 days, (e, f) 18 days, (g, h) 19 days. From Smith et al. (2021). Republished with permission of Wiley.

Other prominent features in Fig. 15.6 are the layers of inflow sandwiching the upper-tropospheric outflow layer at all three times. The strongest one lies beneath the main out-flow in the height range of about 9-12 km. At 10 and 12 days, the maximum inflow in this layer is around 3 m s^{-1} at a height of about 10.3 km. Such inflow layers are discussed in Section 11.6.1.

A characteristics of the $\langle M \rangle$-surfaces shown in Fig. 15.6 is the inward-pointing nose near the top of the boundary layer inflow, a feature that may be understood in terms of the approximate material conservation of $\langle M \rangle$ and the fact that, in the frictional boundary layer, $\langle M \rangle$ has a large positive vertical gradient on account of the frictional stress there.

15.4.2 Temporary decay and reintensification phase

The middle panels of Fig. 15.6 show the azimuthally-averaged structure at 10 days, which is roughly midway during the first period of vortex decay, and the lower panels show the fields at 12 days, which is near the start of the second mature phase.

There are three notable changes in the fields between 9 and 10 days. The first change is the increase in the ra-dial extent of ascent. In particular, the weaker secondary updraft near a radius of 50 km in Fig. 15.6c, seen at 9 days in Fig. 15.6a has strengthened. An examination of Fig. 15.4c shows that the secondary updraft in the azimuthally-averaged field in Fig. 15.6c is a result of several convective bands that have built up outside the ring of eyewall convection, but by 10 days, this ring has become fragmented (at least judged by the 1 m s^{-1} contour of vertical velocity in this figure). The maximum mean updraft in Fig. 15.6c is much weaker than that in Fig. 15.6a (only 3.1 m s^{-1} compared with 6.1 m s^{-1}), but it occurs 1 km higher at 9.8 km. The maximum mean tangential velocity at 10 days is 62 m s^{-1} and the radius at which it occurs has moved 5 km outwards since 9 days.

The second change is a substantial reduction of the radial inflow in the low-mid troposphere. At 10 days, this inflow is largely confined to radii beyond 60 km and below 5 km height. The maximum low-level inflow has fallen from 35 m s^{-1} at 9 days to 24 m s^{-1}, but the radius at which the maximum occurs has increased from 19 km to 24 km. At smaller radii, there is mostly outflow, at least above the boundary layer (Fig. 15.6d).

The third change is the strengthening of the upper-level outflow as well as the inflow layer just below this outflow. A foundation for interpreting the three changes can be obtained by recognizing the link between the strength of the boundary layer inflow and the proportion of this air that cannot be ventilated by convection in the primary eyewall. This link is investigated in detail in Section 15.5.

15.4.3 A new pathway to inner-core rainband formation

The mean updraft seen in Figs. 15.6c-15.6f is part of a transient spiral rainband complex that forms around this time. This connection is seen by comparing these figures with Fig. 15.4c and Fig. 15.4d. The inner-core rainband structures evolve in ways that resemble vortex Rossby waves (e.g., retrograde azimuthal propagation relative to local tangential wind and slow outward propagation that tends to stagnate in time). Vortex Rossby waves are discussed in Chapter 4.

The formation of these inner-core rainbands (and the corresponding mean updraft) is different from that of a classical secondary eyewall discussed in Chapter 1. Rather, its formation is linked to the low-level radial outflow that results from the un-ventilated part of the primary eyewall. In other words, the primary eyewall ceases to be able to evacuate the mass that is converging in the boundary layer, whereupon the excess mass flows radially outwards to trigger convection at a larger radius. The question as to why a classical secondary eyewall does not form in this simulation remains to be answered.

15.4.4 Decay phase

Radius-height cross sections like those in Fig. 15.6 are shown in Fig. 15.7 for the decay phase. A highlight of these is the further weakening and radial spread of the inner-core convection during the decay phase. In particular, in all panels, the $\langle M \rangle$-surfaces are moving outwards just above the boundary layer because of the positive radial wind component there. In turn, the outflow must be a consequence of the inability of deep convection to ventilate mass at the rate that it is converging in the boundary layer (Kilroy et al., 2016a; Smith and Wang, 2018). The reasons for the progressive outward expansion of the radii of forced ascent and the subsequent weakening of the vortex core are discussed below in Section 15.5.

The eyewall becomes progressively more tilted during the decay phase and a region of strong subsidence develops in the upper troposphere just above the eyewall updraft leading to a local overturning circulation there (cf. Figs. 15.7a, c, and e).

15.5 Interpretations of the life cycle

In general, as noted in Section 10.14, it is generally not possible to present simple cause and effect arguments to explain tropical cyclone behavior because of the tight coupling between the flow above the boundary layer and that within the boundary layer. The most one can hope to do is to articulate the individual elements of the coupling. Here, we extend the ideas discussed in Section 10.14 to interpret the vortex behavior described in Sections 15.2 to 15.4.

15.5.1 Important kinematical features

A series of Hovmöller diagrams showing the evolution of certain low-level velocity fields together with the vertical velocity at a height of 6 km provides a context for an interpretation of the vortex life-cycle in this chapter.

15.5.1.1 Tangential velocity evolution

A Hovmöller diagram of the azimuthally-averaged tangential velocity component at a height of 1 km is shown in Fig. 15.8. This height is nominally at the top of the frictional boundary layer. Notable features are the sharp contraction of the velocity maximum at about $6\frac{1}{2}$ days and the subsequent expansion beyond 7 days for the remainder of the life-cycle. This behavior is similar to that of V_{max} and R_{Vmax} shown in Figs. 15.2a, b. Seen also is the rapid expansion of R_{gales} (the contour with tangential wind speed equal to 17 m s^{-1}, cf. Fig. 15.2c) as the vortex rapidly intensifies. After a few days of intensification, the rate of expansion of R_{gales} slows down and after about 14.5 days stabilizes at about 240 km. After about 18 days, R_{gales} begins to contract, accompanied by the rapid demise of the vortex intensity.

The expansion of the vortex is as found in observations (e.g., Fig. 8.12) and, as shown below, may be attributed to the classical mechanism for intensification discussed in Section 8.2. To demonstrate this, it is necessary to examine the radial motion at the top of the boundary layer, where

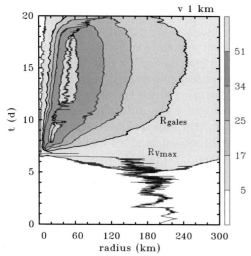

FIGURE 15.8 Hovmöller plots of the azimuthally-averaged tangential velocity component at a height of 1 km out to 20 days in the life cycle simulation. Shown also is the radius of the azimuthally-averaged maximum tangential wind, R_{Vmax} (blue curve) and radius of gale-force winds, R_{gales} (black curve). Shading as indicated on the side bar. Contours at 5, 17, 25, 34, 51 m s^{-1}. Units are m s^{-1}. Adapted from Smith et al. (2021). Republished with permission of Wiley.

frictional effects are small. According to the classical mechanism, the tangential wind will continue to increase locally as long as the radial flow there is inwards and the vertical advection of absolute angular momentum can be ignored.

15.5.1.2 Radial velocity evolution

Hovmöller diagrams of the azimuthally-averaged radial velocity component at heights of 1 km, near the boundary-layer top, and 500 m, within the boundary layer, are shown in Figs. 15.9a, b. The behavior of all fields is similar to that shown in the MM5 simulation described in Section 10.14 (see Fig. 10.19), suggesting that such behavior is generic, although the latter simulation does not have a rapid decay phase.

There is a narrow strip of outflow at the top of the boundary layer in Fig. 15.9a. This outflow is co-located approximately with the updraft shown in Fig. 15.9c. There is generally inflow beyond this strip, except for brief periods after about 10 days when there is outflow, even beyond 100 km radius.

During the main intensification period and the first period of maturity (approximately from 7 to 10 days) when the outer radius of gale-force winds increases most rapidly, there is relatively strong inflow at all outer radii. This feature is an indication that the classical mechanism for intensification is responsible for the expansion of the outer storm size. In the later period, as the rate of expansion measured by R_{gales} decreases, the inflow at outer radii is punctuated by brief periods of outflow so that the time-integrated effect of the inflow is reduced and eventually terminated. Thus,

R_{gales} stops increasing. As the vortex rapidly decays beyond 18 days, there is outflow at most if not all radii and R_{gales} steadily declines.

The flow within the boundary layer (Fig. 15.9b) is mostly inwards, but there is a narrow strip at small radii where it is outwards. This outflow is due, presumably, to a localized *suction effect* by the eyewall convection.

For much of the time, the locations of maximum inflow and maximum outflow as well as the dividing line separating inflow and outflow move radially outwards with time. This outward movement is clearest during the first period of vortex weakening between about $9\frac{1}{2}$ and 11 days (see Fig. 15.2a as well as 15.9a) and as the vortex rapidly decays beyond 18 days. Both of these periods are ones during which the inflow at a height of 1 km weakens markedly.

15.5.1.3 Vertical velocity evolution

Hovmöller diagrams of the azimuthally-averaged vertical velocity component at heights of 1 km and 6 km are shown in Figs. 15.9c, d. At 1 km height, a narrow, but intense mean updraft emerges from the boundary layer at inner radii. This feature marks the root of the mean eyewall updraft at this level. Beyond about 7 days, the radial location of the maximum vertical velocity increases progressively with time. At larger radii, there is weak subsidence into the boundary layer. At 6 km height (Fig. 15.9d), the pattern of vertical motion is similar, but the eyewall updraft is a little broader and the maximum is at a larger radius, a reflection of the outward slope of the eyewall.

15.5.2 Boundary layer dynamics

The low-level inflow that feeds into the developing inner-core deep convection is driven partly by the convection itself, essentially a suction effect, and partly by the frictionally-induced convergence associated with boundary layer dynamics. The intrinsic nonlinearity of the boundary layer dynamics in the inner-core region precludes the possibility of directly isolating the separate effects of the two processes that lead to low-level inflow (see Sections 6.7.2, 11.5.3-11.5.4, and 16.1.6).

An indirect method to isolate and estimate the boundary layer contribution to the inflow and thereby the degree of control exerted by the boundary layer on the vortex evolution was explored by Kilroy et al. (2016a). The idea is to compare the azimuthally-averaged velocity fields derived from the three-dimensional numerical model simulation with those of a simple, steady,[5] axisymmetric, slab boundary layer model driven by the time dependent, azimuthally-averaged tangential wind field at a height of 1 km obtained from the numerical model.

---

5. The justification for using a steady boundary layer model is that the boundary layer is a relatively shallow layer that can be expected to respond rapidly to the flow above it.

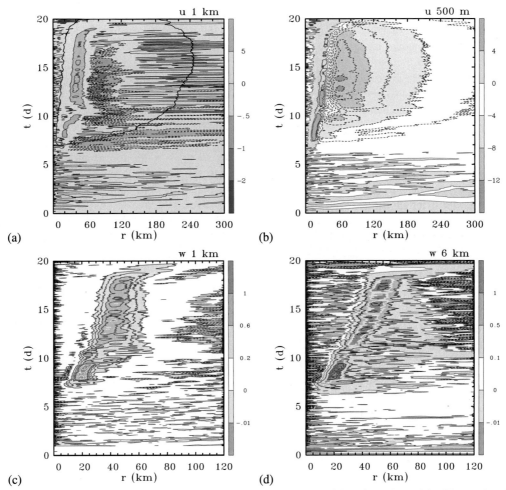

FIGURE 15.9 Hovmöller plots of the azimuthally-averaged radial velocity component at heights of (a) 1 km and (b) 500 m, and vertical velocity at heights of (c) 1 km and (d) 6 km in the life cycle simulation. Note the different contour values in the upper panels necessitated by the much stronger inflow at 500 m height and the different scales on the abscissa in the lower panels. Shading as indicated on the side bar. Units are m s^{-1}. The black contour in (a) shows the radius of gale-force winds, R_{gales}. The yellow contour in panels (c) and (d) highlights values of vertical velocity larger than 1 m s^{-1}. Adapted from Smith et al. (2021). Republished with permission of Wiley.

The degree to which the slab boundary layer model is able to replicate the radial distribution of vertical motion in the time-dependent numerical model is a measure of the degree to which boundary layer dynamics determine the low-level inflow. Kilroy et al.'s comparison of the two vertical velocity fields and their time evolution showed good qualitative agreement, except near and beneath the eyewall updraft, which led them to conclude that the boundary layer was the dominant mechanism in producing the inflow, except in the vicinity of the eyewall updraft, itself.

Even though the expansion of the vortex core in the present calculation is consistent with the ideas of boundary layer control discussed in Section 10.14, an explanation for the temporary decay and re-intensification of the vortex between $9\frac{1}{2}$ and 12 days requires a consideration of processes inside and near R_{vmax}. In this region, boundary layer dynamics may not be wholly applicable as noted earlier. We return to this issue in Section 15.5.4.

15.5.3 Boundary layer coupling

In Section 10.14, we proposed a way of thinking about the dynamical control that the boundary layer has on vortex evolution. At radii where there is inflow just above the boundary layer and where the radial advection of mean absolute vorticity dominates that of vertical advection, the tangential wind at that level will increase as absolute angular momentum surfaces are drawn inwards. If the flow at this level is approximately in gradient wind balance, the local radial pressure gradient will increase as well. Boundary layer theory tells us that this increased pressure gradient will be transmitted through the boundary layer (see Section 6.2.1), where it will locally accelerate the inflow. Thus, the boundary-layer convergence at inner radii will increase, as will the ascent at the top of the boundary layer.

Because the diabatic heating rate depends strongly on the vertical velocity of the ascending air (see Section 5.11) as well as on the thermodynamic properties of this air, and because the pattern of ascent is strongly tied to that at the top of the boundary layer, the boundary layer can be expected to exert a strong influence on the evolution of the vortex. This influence is embodied in the concept of boundary layer control, aspects of which are discussed in Section 10.14.

Although the boundary layer control mechanisms are underpinned by axisymmetric reasoning, they find broad support from the present calculations, despite the fact, that as shown in Fig. 15.4, the decay phase is accompanied by the formation of small-scale banded vorticity-convective structures. These structures are strongly asymmetric, but they are localized near the inner-edge of the eyewall updraft. The part played by these asymmetric structures on the overall decay of the vortex has yet to be determined, but their localized nature would suggest that they are not the major factor.

15.5.4 Ventilation of the boundary layer inflow

In Chapter 8, Section 8.4, we explored in a balance framework the interplay between the effects of deep convection and the frictional boundary layer in determining whether or not a vortex will intensify and in Section 9.9.1, we examined aspects of deep convection that are important to this interplay. Here we show how the ventilation ideas are useful for interpreting the vortex behavior in the life-cycle simulation.

To reiterate these ideas, if deep convection is strong enough to ventilate mass at a rate larger than the rate supplied by the boundary layer, it will draw air inwards above the boundary layer as well as the $\langle M \rangle$-surfaces, leading to a spin up of the tangential flow above the boundary layer. In turn, this spin up will boost the boundary-layer mass influx together with the rate that this mass is offered to the convection. On the assumption that the spatial gradients of heating rate do not change during this period of spin up, the foregoing processes will bring the rate at which boundary-layer mass is offered to the convection closer to the rate at which mass is being "sucked upwards" by the convection. The inflow above the boundary layer will then decrease, together with the rate at which the tangential wind at the top of the boundary layer increases. In reality, of course the spatial gradients of heating rate *will* change, partly because of possible changes to the radial structure of the vertical velocity at the top of the boundary layer and partly because of changes in the thermodynamic properties of the air that is ascending as a result, for example, of changes in the radial distribution of surface enthalpy fluxes.

If, for any reason, the sucking effect of the convection decreases in strength, i.e., the aggregate mass flux it can ventilate upwards decreases, or, if the boundary layer inflow increases because, for example, the radial pressure gradient at its top and within it increases, it is possible that the convection may no longer be able to ventilate mass at the rate it

is converging in the boundary layer. In this case, the residual mass will flow outwards just above the boundary layer.

When the vertical advection of $\langle M \rangle$ is small enough, the radial flow above the boundary layer will advect the $\langle M \rangle$-surfaces outwards and the tangential flow at the top of the boundary layer will spin down. On account of this spin down, the boundary layer inflow will weaken until, possibly, the convection is again able to suck mass upwards at a rate greater than the rate supplied by the boundary layer. Then inflow resumes at the top of the boundary layer leading again to spin up. The re-invigoration of convection would be facilitated by continuing surface enthalpy fluxes as the flow evolves.

When the vertical advection of $\langle M \rangle$ out of the boundary layer is not so small, especially at small radii where the tangential velocity in the boundary layer is larger than that above, $\langle M \rangle$-surfaces above the boundary layer can move inwards on account of the vertical advection of $\langle M \rangle$ from the boundary layer, including eddy advection. At such radii and altitudes, the vortex would spin up locally, even if the radial flow is outwards.

15.5.4.1 Quantifying ventilation

The foregoing ideas are applicable to interpreting aspects of the flow evolution illustrated in Figs. 15.2 and 15.8. The degree of ventilation by deep convection can be quantified by a *ventilation diagnostic*, ΔM_{flux}, defined as the difference in azimuthally-averaged radially-integrated mass flux between heights of 6 and 1 km over a radial distance R_{int}. Mathematically,

$$\Delta M_{flux} = 2\pi \int_0^{R_{int}} [\langle \rho w \rangle_{z=6km} - \langle \rho w \rangle_{z=1km}] r dr, \quad (15.1)$$

where $\langle \rho w \rangle$ is the azimuthally-averaged vertical mass flux with ρ the air density, w the vertical velocity and r the radius.

A Hovmöller diagram of ΔM_{flux} as a function of the radius of integration R_{int} and time t for the life-cycle simulation is shown in Fig. 15.10. Shown also are the 0.5 m s^{-1} contours of vertical velocity at heights of 1 km and 6 km. Positive values of ΔM_{flux} indicate times and radii of integration at which the mass flux at 6 km exceeds that emerging from the boundary layer at a height of 1 km within the radius R_{int}. These are locations in the diagram where deep convection is more than able to ventilate the mass converging in the boundary layer inside R_{int}. In contrast, negative values of ΔM_{flux} indicate times and radii of integration at which the convection is too weak to ventilate all the mass converging in the boundary layer inside R_{int}. Then, the fraction of mass not being ventilated flows outwards just above the inflow layer. As exemplified by Fig. 15.6d, this air may be ventilated by convection at larger radii. The latter may be

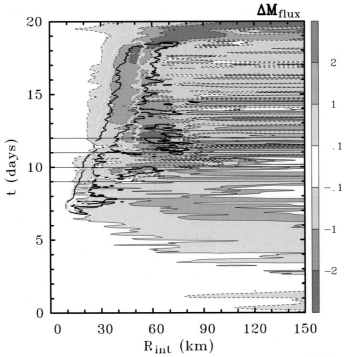

FIGURE 15.10 Hovmöller plots of the ventilation diagnostic, Eq. (15.1), i.e., the azimuthally-averaged radially integrated vertical mass flux difference between heights of 6 km and 1 km as a function of the integration radius, R_{int} in the life cycle simulation. Contour values of ΔM_{flux} should be multiplied by 10^9 kg m^{-1}. Shown also are contours of the 0.5 m s^{-1} azimuthally-averaged vertical velocity at a height of 1 km (yellow contour) and 6 km (black contour) to provide an indication of the updraft tilt below 6 km height. Adapted from Smith et al. (2021). Republished with permission of Wiley.

the case if the eyewall updraft has an appreciably outward tilt with height, which is why the two contours of vertical velocity are shown in the figure. The diagnostic should provide the most robust information on ventilation for values of R_{int} beyond the radius of the eyewall updraft at 6 km height. In the present case, the eyewall tilt *below this height* would not be judged appreciable, except perhaps for intervals of a few hours, until the decay phase, beyond about 16 days.

Important features to note in Fig. 15.10 are:

- Until almost 9 days, including the RI phase and early maturity, $\Delta M_{flux} > 0$ for all values of R_{int}. Up to this time, there is mostly inflow in the low to mid-troposphere as exemplified by Fig. 15.6b. However, as noted in Section 15.4.1, even at 9 days, a shallow tongue of radial outflow has formed just above the surface-based inflow layer. Animations of fields akin to those Fig. 15.6b, but without time averaging, show that the $\langle M \rangle$-surfaces continue to move slowly inwards in this tongue of outflow. It follows that there must be a vertical transfer of $\langle M \rangle$ from the inflow layer. Recall from Fig. 15.6 that the $\langle M \rangle$-surfaces have an inward-pointing nose within the inflow layer indicating that the maximum tangential wind speed resides there. This feature is reflected also in Fig. 15.6b.

- At 9 days, ΔM_{flux} has just become negative in the mean eyewall updraft and soon after 9 days, the radius R_{int} beyond which ΔM_{flux} becomes positive increases sharply.

Just before $9\frac{1}{2}$ days, the 0.5 m s^{-1} contour of vertical velocity extends rapidly outwards, a reflection of the formation of the secondary mean updraft feature beyond the inner eyewall updraft seen in Fig. 15.6c. As pointed out in Section 15.4, this updraft is related to the outer convective band seen in the horizontal cross section in Fig. 15.4c. From these figures, it is seen that the outer convective band is being fed by air that was not able to be ventilated by the eyewall convection, itself.

- As the vortex continues to decay between 10 and 11 days (Fig. 15.2a), the radius beyond which R_{int} first becomes positive extends beyond 110 km. During this time, the eyewall updraft at 1 km becomes quite narrow, and, as noted in Section 15.3.2, it becomes fragmented in the azimuthal direction.

- During the subsequent reintensification phase, the width of the mean eyewall updraft at a height of 1 km increases sharply at about $11\frac{1}{2}$ days. This increase is preceded a few hours earlier by a pulse of increased ascent at this level, between 40 and 50 km radius. Once more, there are frequent periods with closed 0.5 m s^{-1} contours of vertical velocity beyond the eyewall updraft that are most evident up to about 14 days. During these periods, $\Delta M_{flux} > 0$ at some radius inside 150 km. At other times, when the outer convection is weaker, ΔM_{flux} is never positive inside this radius.

- Beyond 11 days until well into the final decay phase, the inner radius of the 0.5 m s^{-1} contour of mean vertical velocity at 1 km moves progressively outwards as the inner eyewall expands, but the mean outer radius remains more or less constant. The width of the mean eyewall updraft at 6 km altitude expands and the mean radius moves outwards until shortly after 18 days, when deep convection ceases in an azimuthal mean sense and the mean eyewall updraft collapses (cf. Fig. 15.5d). From about 14 days onwards, deep convection becomes progressively unable to ventilate the mass ascending out of the boundary layer and there are many times when ΔM_{flux} never becomes positive inside 150 km. After the mean eyewall updraft collapses, ΔM_{flux} is strongly negative inside 100 km radius.

To summarize, the developing eyewall updraft is more than able to ventilate the mass of boundary-layer air being supplied to it during the first intensification phase. This ability ceases during the first mature phase, leading to low-level outflow above the boundary layer that initiates deep convection in a band beyond the mean eyewall updraft and the eyewall updraft weakens. As a result, the vortex decays for about a day before re-intensifying. The period of decay and re-intensification is characterized by a reorganization of deep convection in which the mean eyewall updraft reconsolidates during a 6-8 hour period at a somewhat larger radius than before.

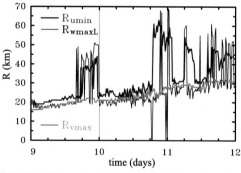

FIGURE 15.11 Time series of R_{vmax} (red (mid gray in print version) curve), R_{umin} (blue (dark gray in print version) curve) and R_{wmaxL} (black curve) during the period 9-12 days in the life cycle simulation, during which there is a major reorganization of deep convection. Adapted from Smith et al. (2021). Republished with permission of Wiley.

Aspects of the reorganization of deep convection are illustrated in Fig. 15.11, which shows time series of R_{vmax}, R_{umin}, and R_{wmaxL} during the period 9-12 days. Here, R_{umin} and R_{wmaxL} are the radii of maximum azimuthally-averaged radial inflow and vertical velocity in the lowest 3 km (hence the subscript L), respectively. Features to note are the sharp increase of R_{umin} at about 9 days, 16 hours and the large fluctuations in R_{wmaxL}. The latter indicate a substantial change in the character of the moist convection just before the rapid increase of R_{umin} a few hours later.

Both quantities return to near their former values at about 10 days, but begin to fluctuate again just after 10 days, 18 hours. At around this time, the mean eyewall updraft below 3 km moves outwards over a period of 6-8 hours from approximately 25 km to 55 km radius. Subsequently, the large fluctuations in R_{umin} and R_{wmaxL} become more persistent and sharp positive increases in these quantities are an indication that the roots of the outer band of deep convection are stronger collectively than those of the eyewall convection, itself. During the entire three day period, R_{vmax} increases comparatively steadily, the largest rate of increase being during the 6-8 hours that the mean eyewall updraft below 3 km moves outwards.

The reorganization of deep convection is a consequence of its progressive inability to ventilate the mass that is being ejected by the boundary layer as indicated by increasing radii with negative values of ΔM_{flux}. Presumably, this inability is partly a reflection of the decreasing convective instability evident in the weakening maximum local vertical velocity after 10 days, a feature seen in Fig. 15.2d. Following the reorganization of convection and the brief period of re-intensification between 11 and 12 days, the width of the mean eyewall updraft continues to expand and its mean radius increases. This updraft and the deep convection in spiral bands beyond it increasingly struggle to evacuate the mass being ejected by the boundary layer. Again, the decreasing convective instability suggested by the decline in w_{max} in Fig. 15.2d would explain this struggle. During this later period, first the rainbands decay and then the eyewall updraft, itself.

There are significant fluctuations of ΔM_{flux} on time scales of an hour or less. These fluctuations point to a general mismatch between the rate of boundary layer mass convergence and the rate at which this mass is being ventilated by deep convection. This mismatch serves to explain the rapid fluctuations in the evolution of V_{max} seen in Fig. 15.2a. It serves also to explain the pulses of positive radial flow near the top of the boundary layer seen in Fig. 15.9a. These pulses occur during periods of prolonged negative values of ΔM_{flux}.

It seems likely that the processes described above may operate also in realistic tropical cyclones and it is conceivable that they could be incorrectly interpreted as a result of ambient vertical wind shear, which is not present in the model discussed here.

15.5.5 Convection component of ventilation

The discussion of deep convection in tropical cyclones in Section 9.9 is pertinent to understanding why the deep convective mass flux might fluctuate over the lifetime of a tropical cyclone. As suggested in Subsection 9.9.1, one might guess that the ability of deep convection to ventilate the mass of air converging in the boundary layer depends in some way on the collective buoyancy of updrafts, which depends presumably on the degree of conditional instability, typically quantified by local values of *CAPE*. Here we explore this issues involved in a little detail.

Data from the research aircraft surveys of the convective environment of pre-genesis disturbances during the PREDICT field campaign in 2010 (Montgomery et al., 2012) show that there were large values of pseudo-adiabatically calculated *CAPE* available to support vigorous deep convective systems in these disturbances (Smith and Montgomery, 2012b) and therefore ample cloud buoyancy.

In contrast, Houze (2010) notes that, in an azimuthally-averaged view and to a first approximation, the eyewall convection in a mature tropical cyclone is close to a moist adiabat, in which case the eyewall would have only minimal buoyancy and hence minimal *CAPE*. This is because the computed buoyancy of the lifted air parcel would be relative to a reference density[6] that lies within the eyewall, itself. As is evident from the discussion in Section 9.2.4, in the center of such a moist environment, where the air is saturated and rising approximately along a moist adiabat, lifted air parcels would not be expected to have much local buoyancy. Even if the air ascending in the eyewall were warmer than that in its immediate environment and had more or less uniform properties, the relatively large width of the eyewall would imply that the effective buoyancy would be minimal, even not taking into account the rotational constraint (Section 5.12.1).

Significantly, in Section 7 of his review, Houze writes:

"While a circular outward-sloping eyewall cloud consisting of purely slantwise conditionally symmetric neutral motions (Section 6) is a useful idealization that explains a great deal about the typical structure of an eyewall cloud, the eyewall cloud does not exist in reality without containing superimposed cells of buoyant convective updrafts arising from the local release of conditional buoyant instability. The air masses in which tropical cyclones occur are not exactly slantwise neutral; they usually exhibit some degree of conditional instability."

The calculations in this chapter fully support this scenario, whereupon *CAPE* might provide a useful measure of local conditional instability within the eyewall region, itself.

Since the determination of *CAPE* requires a deep sounding through the troposphere, observational estimates of *CAPE* in tropical cyclones have been difficult to obtain. Only in the last few years have such deep dropsonde soundings become feasible from dedicated research aircraft flying in the upper troposphere or from unmanned aircraft flying in the lower stratosphere.

In an early observational study, Bogner et al. (2000) showed some evidence in hurricanes of a general increase in *CAPE* with radius from the center. Nevertheless, this inferred increase is based largely on data from Omega dropsonde soundings made at heights below the 500 mb level. These sounding data were extended to larger heights using composite reference soundings derived from radiosonde ascents from nearby land stations. Even so, Bogner et al.'s finding is supported by a more recent study by Molinari et al. (2012), based on data from dropsondes released from an altitude closer to the tropopause. Unfortunately, in neither studies were soundings obtained in the eyewall region of storms.

Molinari et al. maintain that *CAPE* values calculated with the normal assumptions,[7] even assuming reversible ascent with no fusion, are unrealistically large because the maximum vertical velocities inferred from these values are much too large. They proposed that entrainment[8] must be taken into account, stating that

"entraining CAPE was consistent with the observed radial distribution of convective intensity ... ".

They did acknowledge below their Eq. (1) the potential adverse effect of a negative perturbation pressure gradient force, but they did not calculate this force. Presumably, they did not think it would be a dominant effect. However, the reduction in *CAPE* values that take account of entrainment in the data shown in their figure 3 are comparable with the reduction in *CAPE* values shown in our Table 9.1. The latter arise solely because the decrease in effective buoyancy with the width of the buoyant column. This decrease reflects the effect of the adverse vertical perturbation pressure gradient force in regions of deep convection with widths on the order of 20 km. Such widths are comparable with the scale of convective rainbands observed in tropical cyclones (e.g., Houze, 2010; Bell et al., 2012b). This result calls into question the conclusion that accounting for entrainment alone accounts for "the observed radial distribution of convective intensity", at least without qualification. Molinari et al.'s conclusion might remain valid for the embedded convective structures in the eyewall region noted in the quote by Houze as these have a smaller horizontal scale than the eyewall region, itself, and therefore a larger effective buoyancy.

The tropical-cyclone life cycle simulation described in this chapter indicates that, as the vortex matures, the eyewall expands and its mean radius increases. Despite this increase in area, the eyewall becomes increasingly less capable of ventilating all the air that is being funneled in the frictional boundary layer. It is plausible that the reduction of effective

---

6. In the presence of a rapidly-rotating vortex one needs to decide whether or not to include or exclude the system buoyancy from the calculation, i.e., the part of the buoyancy field that would be in thermal wind balance with the vortex in the definition of a reference density (Section 5.12.1).

---

7. See Section 9.2.1.

8. An effect touched upon in Section 9.8.

buoyancy accompanying the expansion and areal growth of the eyewall would be an element of any explanation for the reduction of the areally-averaged mass flux carried by the eyewall.

To our knowledge, an exploration of this likelihood has not yet been carried out. This task would require, *inter alia*, an extension of the concept of buoyancy in a vortical flow discussed in Chapter 5 to one for effective buoyancy discussed in Chapter 9. It would require also an analysis of the dynamically-induced pressure gradient field where the boundary layer terminates, i.e., a solution of the $\nabla^2 p_d' = F_d$ part of Eq. (9.15) formulated in cylindrical coordinates with appropriate consideration of centrifugal forces.[9]

At the time of writing, the precise controls that determine the vertical mass flux carried by deep convection in a particular environment are not well understood. Such controls would call upon the theory of convective parameterization, which has a long history, but remains a formidable research challenge (Emanuel, 1994; Smith, 1997; Siebesma et al., 2020), especially in regions of strong rotation and related nonlinear boundary layer effects.

15.6 Life cycle summary

This chapter summarized the results of a recent analysis of a multi-week numerical simulation of a tropical cyclone in a quiescent environment on an f-plane using a relatively high-resolution, idealized, three-dimensional model. The aim of the simulation was to explore some fundamental dynamical and thermodynamical aspects of the cyclone's life cycle in the framework of the rotating-convection paradigm. Because the simulation has a homogeneous sea surface temperature and does not represent any ambient flow, the intensity and structure changes that occur represent internal modes of vortex variability that need be understood as a prerequisite to understanding observed or simulated tropical cyclone intensity change in more complex situations.

The simulation begins with a weak, axisymmetric, cloud-free vortex in thermal wind balance and lasts for 20 days. During this time, the vortex undergoes a life cycle that includes a gestation period leading to genesis, a rapid intensification phase, a first mature phase followed by a transient decay and re-intensification phase, a second mature phase and a final rapid decay phase. Throughout most of the life cycle, the flow evolution is generally asymmetric, but to a first approximation, much of the evolution can be interpreted in terms of an azimuthally-averaged framework that implicitly includes both mean and eddy processes.

During much of the life cycle, especially during the mature and decay phases, vortex evolution can be interpreted in terms of the boundary-layer control mechanism discussed in Section 10.14 and a ventilation diagnostic, discussed in

Section 15.5.4. The boundary-layer control mechanism provides an explanation for the gradual expansion of the inner-core of the vortex. The ventilation diagnostic characterizes the ability of deep convection within a given radius to evacuate the mass of air that is ascending out of the boundary layer within that radius and highlights the increasing difficulty of deep convection to ventilate the air at the rate that it exits the boundary layer within a particular radius, especially in the later stages of the life cycle.

The final rapid decay of the vortex is a consequence of the un-ventilated air flowing radially outwards in the lower troposphere. This outflow, in which mean absolute angular momentum is approximately materially conserved, leads to a rapid spin down and vortex collapse. As this decay process occurs in the absence of ambient vertical wind shear, it is possible that it operates at times in observed tropical cyclones. In such cases, the decay might be incorrectly attributed to the presence of ambient vertical wind shear.

The transient decay and re-intensification phase is not associated with an eyewall replacement cycle, but rather with a process in which the eyewall temporally fragments while a rain band complex forms beyond it. This process was shown to be an example of the interplay between the boundary layer inflow and ventilation. Like the rapid decay process described above, this process may operate in realistic tropical cyclones and, again, might be incorrectly attributed to the effects of ambient vertical wind shear.

The results of this study are in line with the discussion in Section 14.6, suggesting that even in a quiescent environment on an f-plane, isolated tropical cyclones are intrinsically transient systems and never reach a globally steady state in the presence of strong boundary layer control.

---

9. Coriolis forces should be weak in this inner region of the vortex.

Chapter 16

Applications of the rotating-convection paradigm

This chapter addresses several topics that build on and consolidate the material of previous chapters. In particular, it provides answers to some outstanding questions using concepts developed earlier, some under the umbrella of the rotating-convection paradigm.

The chapter begins with a review and extension of recent work by the authors (Montgomery and Smith, 2022) that explores the connection between the early axisymmetric theories of tropical cyclone intensification, based on the classical Eliassen balance theory and the air-sea interaction, or so called WISHE theories. One aim of this section is to interpret the purported WISHE feedback mechanism in terms of the classical balance theory and its extension to include boundary-layer friction and the unbalanced effects encapsulated in the boundary-layer spin up enhancement mechanism.

Subsequent sections address various attributes of vortex development in the tropical atmosphere. These attributes include a comparison of axisymmetric and three-dimensional intensification in an idealized configuration, the effects of latitude and sea surface temperature on tropical cyclone intensification, the effects of initial vortex size on genesis and intensification, the process of tropical cyclogenesis at and near the Equator, observational tests of the rotating convection model, and the formation of tropical lows, also called "landphoons", in monsoon environments as well as the formation of polar lows, sometimes called "arctic hurricanes" in polar environments. The penultimate section offers explanations of recent findings from other authors using the same framework. The chapter concludes with a sketch of some of the key elements required for a basic understanding of the interaction of a pre-cyclone disturbance or intensifying/mature cyclone with an environmental shear flow.

16.1 Minimal conceptual models for vortex intensification

Throughout the book we have progressively built up a framework for understanding tropical cyclone structure and evolution, ranging from the classical Eliassen balance theory in Chapters 5 and 8, the rotating-convection paradigm in Chapters 10 and 11 and the WISHE theories in Chapter 12.

As noted in Chapter 12, since the early 1990's, the WISHE feedback mechanism has been the widely accepted theory to explain tropical cyclone intensification (e.g., Zhang and Emanuel, 2016). Indeed, Ruppert et al. (2020) begin their study of cloud-infrared radiation feedback effects on tropical cyclone intensification by pointing out that:

"A long history of research indicates that TCs intensify through the WISHE feedback ..., whereby the rate of evaporation increases with surface wind speed."

However, nowhere in Ruppert et al. is the WISHE feedback process clearly articulated. Rather, WISHE appears to be largely associated with the formula for the evaporation of moisture at the air-sea interface, which is proportional to the near-surface wind speed. Similar ambiguities arise in studies of convective self-aggregation (e.g., Muller and Romps, 2018, p2), where the WISHE *mechanism* appears to be associated with surface flux feedbacks that

"connect enhanced surface winds to enhanced surface fluxes."

In their pioneering review of large-scale circulations in convective atmospheres, Emanuel et al. (1994) state that:

"Tropical cyclones are quasi-balanced, warm-core systems. They are premier examples of systems arising from WISHE."

They articulate vortex amplification by first noting that

"for amplification to occur, the troposphere must become nearly saturated on the mesoscale in the core. When this happens, the vertical entropy distribution no longer has a prominent minimum and there is no low entropy air for downdrafts to import to the sub-cloud layer. The effective stratification approaches zero in this case, and Ekman pumping becomes inefficient in cooling the core. ... Now the enhanced surface fluxes associated with strong surface winds near the core can actually increase the subcloud-layer entropy and thus the core temperature. The

Tropical Cyclones. https://doi.org/10.1016/B978-0-44-313449-4.00024-2

WISHE process results now in a positive feedback to the warm-core cyclone, and the system amplifies."

This argument is essentially an energetics argument wherein the increase in core temperature, in conjunction with the positive diabatic heating rate, is the source for available potential energy. Given that the rising motion in the core will be largely positively correlated with the temperature warming, it seems reasonable that the increase in available potential energy will simultaneously increase the kinetic energy of the vortex. The energetics concepts being employed here are presumably global energy quantities involving spatial integrals over the entire vortex. If this were not the case and these energy quantities were intended to represent local properties, then energy transport terms would need to be included to give a complete and consistent account of the system energetics (Chapter 14). Far less is known theoretically and/or observationally about the specifics of these transport terms during intensification. Therefore, the implied association of an increase in global kinetic energy with an increase of the maximum azimuthally-averaged tangential wind, V_{max}, is strictly an ad hoc assumption that will not apply in general. We hold that a physically consistent dynamical explanation for the *amplification of V_{max}* is required.

An explanation of the spin up of the maximum tangential wind speed using the WISHE feedback viewpoint is underpinned only by a rather complex model formulated in potential radius coordinates, a model that is based on some dubious assumptions (Section 12.2). In particular, spin up in the model depends crucially on the radial gradient of an *ad hoc* parameter, β, which is introduced to mimic the deleterious effects of convective downdrafts into the boundary layer yet is not derivable from the equations of motion. The parameter β seems to be a legacy from the invocation of convective downdrafts noted in the Emanuel et al. (1994) explanation above. Recognizing the *ad hoc* nature of this formulation, Emanuel (2012) reformulated the model to avoid the introduction of β, but this has not led to a revised theory for the WISHE feedback and the new model has its own issues (Section 12.3). An outstanding question remains as to whether a minimal dynamical model exists that can transparently explain the essential physics of the WISHE feedback mechanism?

A similar issue arises in Ruppert et al. (see their Fig. 2), where the "WISHE feedback effect" is quantified through a horizontal integral of a covariance between anomalies of vertically-averaged moist static energy and surface enthalpy flux, with anomalies defined as deviations relative to a large square domain. A positive co-variance integral is equated with a positive feedback effect and promotion of vortex development. However, these anomaly variables are not vortex specific and a positive integral and/or an increasing positive integral does not guarantee growth in V_{max}.

Besides becoming the widely accepted theory to explain the intensification of tropical cyclones, the WISHE feedback mechanism is extensively invoked to explain the intensification process of occasional tropical-cyclone-like low pressure systems over the Mediterranean, so-called "medicanes" (see Section 16.9).

A recent explanation by Miglietta and Rotunno (2019), which applies to tropical cyclones also, highlights the main issue. On page 1445 they write:

"All these categories of hybrid cyclones (medicanes: our insertion) share with tropical cyclones the mechanism of development in the "tropical-like" part of their lifetime, the so-called Wind Induced Surface Heat Exchange (WISHE: Emanuel, 1986; Rotunno and Emanuel, 1987); these storms are developed and maintained against dissipation entirely by self-induced sea-surface fluxes with virtually no contribution from pre-existing convective available potential energy (CAPE), so they result from an air-sea interaction instability."

Turning then to the summary of the air-sea interaction instability given on page 559 of Rotunno and Emanuel (1987), these authors state that:

"We have established, using a numerical model, that a hurricane-like vortex may grow as a result of a finite amplitude instability in an atmosphere which is neutrally stable to the model's moist convection. The mechanism (our emphasis), which is a form of air-sea interaction instability, operates in such a way that wind-induced latent heat fluxes from the ocean lead to locally enhanced values of θ_e in the boundary layer which, after being redistributed upward along angular momentum surfaces, lead to temperature perturbations aloft. These temperature perturbations enhance the storm's circulation, which further increases the wind-induced surface fluxes, and so on. The tropical cyclone will continue to intensify so long as boundary-layer processes permit steadily increasing values of θ_e near the core or until the boundary layer there becomes saturated."

Although the foregoing explanation may seem plausible on a first reading, after some reflection it raises a number of questions. First, how does a redistribution of locally enhanced values of θ_e in the boundary layer along (absolute) angular momentum surfaces lead to the inward movement of these surfaces above the boundary layer? According to the classical mechanism discussed in Chapter 8, this inward movement would be a necessary requirement for the tangential velocity component to increase. Second, how do the "temperature perturbations enhance the storm's circulation"?

Recall from Sections 12.5 and 13.7.2 that, in the later Emanuel intensification theories, the overturning circulation is assumed to be exactly moist neutral, i.e., ascending air parcels have effectively no local buoyancy. This assumption effectively prevents temperature perturbations from directly driving the overturning circulation via local buoyancy forces. As far as we are aware, none of the papers on medicanes invoking the term WISHE, including that of Emanuel, 2005a, have provided a satisfactory physical explanation of the WISHE feedback mechanism to increase V_{max}.

Zhang and Emanuel (2016) and Emanuel (2019) confound matters further by seemingly redefining the "WISHE feedback process" as simply the formula relating the increase of surface enthalpy flux to that of the surface wind speed and to the degree of thermodynamic disequilibrium near the surface, but they do not explain how the increased fluxes lead to an increase in V_{max} (see Fig. 12.4). While Zhang and Emanuel do present an example of a feedback process, it is unclear how this example relates to the purported WISHE process (see Section 12.2).

A first attempt to explain the WISHE feedback mechanism for the growth of V_{max} was presented in Section 12.2, but as noted there, this explanation has two significant weaknesses. Therefore, in the light of the foregoing review, we conclude that a clear articulation of the air-sea interaction feedback mechanism (WISHE) for the amplification of V_{max} is lacking. Without this articulation, it is unclear how other processes, such as cloud infrared radiation and small-scale turbulent mixing, interact with the WISHE mechanism.

As a way forward to resolving the issues concerning the WISHE feedback mechanism, it would appear fruitful to develop a framework to bridge the classical intensification theory of Chapter 8 to the Emanuel formulations in Chapter 12 and, in particular, to clarify the role of air-sea enthalpy fluxes in the amplification of V_{max}. Here, we review our efforts to do this by developing a small hierarchy of minimal conceptual models with increasing complexity, the objective being to describe the spin up of the system-scale tangential winds in a tropical cyclone (Montgomery and Smith, 2022). The present chapter extends the latter study to account for the slantwise ascent of rising air parcels in the eyewall region, which becomes more realistic as the tropical cyclone intensifies and matures.

The proposed intensification models described below are based on the axisymmetric Eliassen balance model discussed in Chapter 5, which assumes the flow to be in strict gradient wind and hydrostatic balance. This formulation is perhaps the simplest framework in which to represent the main elements of tropical cyclone intensification, at least in a first approximation of a slowly evolving vortex. Recall that the Emanuel models used to underpin the WISHE

theory adopt also an axisymmetric balance formulation. Of course, we have shown that the balance evolution framework does not capture the nonlinear spin up dynamics in the boundary layer for a realistic tropical cyclone (Section 6.6; Montgomery and Persing, 2020; Wang et al., 2020) and it has solvability issues as the vortex intensifies (see Sections 8.1.1.4 and 8.4.5).

16.1.1 A general prognostic balance model

The equations for a general prognostic balance model formulated in cylindrical r-z coordinates are reviewed at the beginning of Chapter 8, but are collected again here for the reader's convenience. The prognostic element is the tendency equation for the tangential wind component, v:

$$\frac{\partial v}{\partial t} = -u\frac{\partial v}{\partial r} - w\frac{\partial v}{\partial z} - \frac{uv}{r} - fu - \dot{V}, \qquad (16.1)$$

where u and w are the radial and vertical velocity components, t is the time, f is the Coriolis parameter (assumed constant), and $-\dot{V}$ is the azimuthal momentum sink per unit mass associated with the near-surface frictional stress divergence. The balanced density field is obtained from the thermal wind equation, which has the general form:

$$\frac{\partial \ln \chi}{\partial r} + \frac{C}{g}\frac{\partial \ln \chi}{\partial z} = -\frac{\xi}{g}\frac{\partial v}{\partial z}, \qquad (16.2)$$

where $\chi = 1/\theta$ is the inverse of potential temperature θ, $C = v^2/r + fv$ is the sum of centrifugal and Coriolis forces per unit mass, $\xi = f + 2v/r$ is the modified Coriolis parameter, i.e., twice the local absolute angular velocity, and g is the acceleration due to gravity. Eq. (16.2) is a first order partial differential equation for $\ln \chi$, which on an isobaric surface is equal to the logarithm of density ρ plus a constant, with characteristics $z_c(r)$ satisfying the ordinary differential equation $dz_c/dr = C/g$ (Section 5.2).

At the heart of the balance formulation is the Eliassen equation, which is a second-order partial differential equation for the streamfunction of the overturning circulation, ψ. As explained in Chapter 5, this equation determines the overturning circulation required to keep the primary (tangential) circulation in gradient wind balance and hydrostatic balance in the presence of diabatic heating and friction that would otherwise destroy such balances. The Eliassen equation (5.67) is:

$$\frac{\partial}{\partial r}\left[\gamma\left(N^2\frac{\partial \psi}{\partial r} - B\frac{\partial \psi}{\partial z}\right)\right] + \frac{\partial}{\partial z}\left[\gamma\left(I_g^2\frac{\partial \psi}{\partial z} - B\frac{\partial \psi}{\partial r}\right)\right] =$$
$$g\frac{\partial \dot{\Theta}}{\partial r} + \frac{\partial}{\partial z}\left(C\dot{\Theta}\right) + \frac{\partial}{\partial z}\left(\chi\xi\dot{V}\right), \qquad (16.3)$$

where $\gamma = \chi/(\rho r)$, $N^2 = -(g/\chi)\partial\chi/\partial z$ is the dry static stability, $I_g^2 = \xi\zeta_a + (C/\chi)\partial\chi/\partial r$ is the generalized centrifugal (inertial) stability, $B = (1/\chi)/\partial(\chi C)/\partial z$ is the baroclinicity, $\zeta_a = \zeta + f$ is the absolute vorticity, $\zeta = (1/r)\partial(rv)/\partial r$ is the vertical component of relative vorticity, $\dot{\Theta} = \chi^2\dot{\theta}$, $\dot{\theta} = D\theta/Dt$ is the diabatic heating rate, and $\dot{V}$ is defined above. The transverse velocity components u and w are given in terms of ψ by:

$$u = -\frac{1}{r\rho}\frac{\partial\psi}{\partial z}, \qquad w = \frac{1}{r\rho}\frac{\partial\psi}{\partial r}, \qquad (16.4)$$

which satisfy exactly the steady form of the continuity equation.

The discriminant of the Eliassen equation, Δ, is given by

$$\Delta = 4\gamma^2\left(I_g^2 N^2 - B^2\right), \qquad (16.5)$$

and the Eliassen equation is elliptic if $\Delta > 0$.

If the heating rate and frictional stress are given, and provided that the Eliassen equation is everywhere elliptic, the equation system (16.1)-(16.4) may be solved, in principle, as outlined at the beginning of Chapter 8. There, examples of such solutions are discussed, in which the heating rate and frictional stress are prescribed spatial functions, the location of the heating distribution being tied to a moving M-surface. Here, *we examine an even simpler situation in which the heating distribution is held fixed in time and space*, a situation that provides a basis for the zero-order model below. The motivation for examining this case is to pave the way for the first-order model discussed later, in which the heating distribution is determined as part of the solution.

While solutions for a prescribed heating rate are useful for developing basic understanding, a complete theory would require the heating rate to be linked intrinsically to the vortex dynamics and thermodynamics. One simple way to do this was outlined in Section 5.11. This method is generalized below to include slantwise ascent.

If an ascending air parcel becomes saturated, condensation of water vapor occurs at a rate proportional to the material rate-of-change of saturation mixing ratio, r_v^*. For a slowly evolving vortex, the local rate-of-change of r_v^* should be small compared with the advective rates of change. In this case, the rate of latent heat release is given by the approximate formula

$$\dot{Q} \approx -L_v\left(u\frac{\partial r_v^*}{\partial r} + w\frac{\partial r_v^*}{\partial z}\right), \qquad (16.6)$$

where L_v is the latent heat of condensation.[1]

For nearly upright ascent, the effect of $u\partial r_v^*/\partial r$ on $\dot{Q}$ is much smaller than that of $w\partial r_v^*/\partial z$, and since $\partial r_v^*/\partial z$ is related to θ_e^* of the ascending air, the prescription of $\dot{Q}$ is more or less equivalent to a prescription of w where the

air is ascending in a deep convective cloud. Typically, in such clouds, $\dot{Q}$ increases in the lower troposphere to reach a maximum somewhere in the mid-troposphere and decreases again in the upper troposphere. Thus, the prescription of such a vertical structure for $\dot{Q}$ implies a similar structure of w in the heated region, consistent with the fact that deep convective clouds are accompanied by mean horizontal convergence in the lower troposphere and mean horizontal divergence in the upper troposphere. It is for this reason that the use of a prescribed heating rate with this vertical structure to represent a region of deep convection produces results that have some degree of realism when used in a zero-order minimal model for tropical cyclone intensification (cf. Section 16.1.4.2, Case B for further discussion of this point in the context of the first order model).

16.1.2 Zero-order model

The prognostic models presented in Chapter 8 are arguably amongst the simplest for isolating the dynamical aspects of spin up in the foregoing balance framework. In these models, the amplitude and structure of the diabatic heating rate are held fixed in time, but the heating rate distribution is allowed to contract as the vortex contracts. Calculations were carried out there for the frictionless case with heating (Section 8.1.1) and for cases with a simple formulation of surface drag, with or without heating (Sections 8.1.3 and 8.1.2). For the zero-order model here, we consider first the case without friction in which the location of the heating distribution is held fixed in time. The spatial distribution of the heating is the same as that in Section 8.1.1.

The upper panels of Fig. 16.1 show radius-height cross sections of the diabatic heating rate and streamfunction of the overturning circulation at the initial time and after 24 hours for this calculation, while the lower panels show the radial and tangential velocity components at these times. In all figures, the M-surfaces are superimposed. Both times are before the regularization procedure is first required at about 30 hours.

As in the calculations in Section 8.1.1, where the heating region moves inwards with a particular M-surface, the flow outside the heating region is quasi-horizontal with inflow in the lower troposphere, below about 5 km, and outflow above this level. The M-surfaces are advected inwards in the lower troposphere, highlighting the classical spin-up mechanism for tropical cyclone intensification (Section 8.2), which, in an axisymmetric framework, attributes vortex spin up to the convectively-induced inward radial advection of absolute angular momentum M at levels where this quantity is materially conserved. Most of the inflow passes through the region of heating and at low to mid-levels there, the flow has a significant component across the M-surfaces in the direction towards decreasing M. In other words, the vertical advection of M becomes important in spinning up the flow there.

---

1. The diabatic heating rate, $\dot{\theta}$ and the latent heating rate $\dot{Q}$ are related by the formula $\dot{\theta} = \dot{Q}/(c_{pd}\pi)$, where π is the Exner function (see Chapter 5).

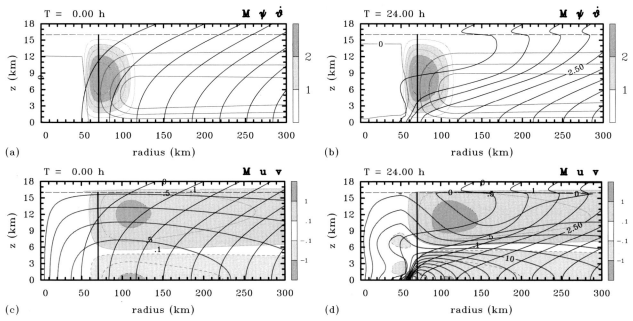

FIGURE 16.1 Radius-height cross sections of M-surfaces (thin black curves) superimposed on various quantities for the zero-order model without friction. These quantities include: (a, b) streamlines of the secondary circulation, ψ (blue (dark gray in print version) curves), and diabatic heating rate $\dot\theta$ (shaded); (c, d) contours of tangential velocity, v (blue (dark gray in print version) curves), and radial velocity u (shaded). The upper panels are at the initial time and the lower panels are at 24 hours. The dashed horizontal line at a height of 16 km indicates the model tropopause. The thick vertical line shows the axis of the maximum diabatic heating rate. Contour intervals are: for M, 5×10^5 m^2 s^{-1}; for ψ, 1×10^8 kg s^{-1}; for $\dot\theta$, 0.5 K h^{-1}; for u, 0.1 m s^{-1}, 0.5 m s^{-1}, and 1 m s^{-1} (negative values dashed); for v, 2 m s^{-1}. From Montgomery and Smith (2022). Republished with permission of Elsevier.

Again, in the case without friction, the maximum tangential wind speed occurs at or just above the surface where the inflow is a maximum. This is to be expected because V_{max} lies at the surface initially and the convectively-induced inflow remains a maximum close to the surface,[2] within 400 m of the surface at subsequent times. From Eq. (16.1), the nonzero vertical advection of M just above the surface accounts for the progressive, but small elevation of V_{max}. Thus, the largest inward advection of the M-surfaces occurs just above the surface. The asymmetry in the depths of the inflow and outflow layers is a consequence of mass continuity and the fact that density decreases approximately exponentially with height. The maximum outflow occurs at a height of about 12 km.

When the location of the distribution of diabatic heating rate is held fixed with time, spin up occurs at a slightly reduced rate compared with the case in Section 8.1.1 where this distribution moves inwards with a prescribed M-surface.

16.1.3 A minimal representation of friction

As explained in Section 8.4.4, the addition of friction in the balance model is somewhat pathological as friction is applied only in the tangential momentum equation. Moreover, while the application of a frictional force in this equation leads to radial inflow, the assumption of gradient wind balance in the friction layer requires frictional effects to be no more than a small perturbation to the balanced flow (Smith and Montgomery, 2008). In contrast, more realistic vortex boundary layers are intrinsically unbalanced (see Section 16.1.6 below). With the balance approximation, the inflow is forced by the need to maintain balance in the presence of a frictional torque that is trying to destroy balance. This means that the frictionally-driven inflow is present from the initial instant, whereas the boundary layer in the tangential direction is not and requires time to develop. For these reasons, one might describe the representation of friction in a balanced vortex as a *minimal representation of friction*.

16.1.3.1 Zero-order model with friction only

It is well known that the primary mechanism of spin down of realistic high-Reynolds number vortices is associated with the outward radial advection of absolute angular momentum above the surface-based friction layer. This advection is brought about by the secondary overturning circula-

---

2. In our prior publication of Smith et al. (2018b), the statement on p3174 (lc) that "U_{min} occurs at the surface at subsequent times" is not quite accurate. In fact, after several hours U_{min} occurs slightly above the surface and we retract that statement, even though the broad description of the results is still correct.

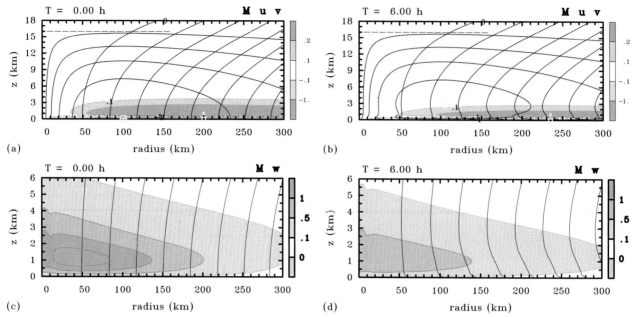

FIGURE 16.2 Radius-height cross sections of M-surfaces (thin black curves) superimposed on various quantities for the friction-only simulation. These quantities include: (a, b) contours of tangential velocity, v (blue (dark gray in print version) curves), and radial velocity u (shaded), and (c, d) contours of vertical velocity, w (shaded) up to an altitude of 6 km. The left panels are at the initial time and the right panels are at 6 hours. The dashed horizontal line at a height of 16 km indicates the model tropopause. Contour intervals are: for M, 5×10^5 m^2 s^{-1}; for v, 2 m s^{-1}; for w, 0.5 cm s^{-1} with the additional contour 0.1 cm s^{-1} shown also. Contours for u are 0.1 m s^{-1} and 0.2 m s^{-1} for positive values, 0.1 and 1 m s^{-1} for negative values. Positive contours are solid, negative contours dashed. The yellow (light gray in print version) symbols in (a) and (b) indicate the locations of maximum v, ($\oplus$), maximum u, ($\times$), and minimum u, ($\square$).

tion induced by friction and not by the direct diffusion of tangential momentum to the lower boundary (Section 6.1). This mechanism is an essential feature of the spin down of a balanced vortex in the absence of diabatic heating, solutions for which were shown in Section 8.1.2.

If there is no diabatic heating and if the atmosphere is stably stratified, the frictionally-driven inflow leads to a shallow layer of outflow just above the boundary layer (see Fig. 8.7). Here we use the same formulation of friction as in Section 8.1.2 and the simulation there provides the friction only simulation for the zero-order model. Fig. 16.2a is a repeat of Fig. 8.7a and Fig. 16.2b shows the same fields after 6 hours. As expected, the overturning circulation is weaker and shallower after 6 hours and the maximum tangential velocity has diminished in strength and has become elevated.

The lower panels of Fig. 16.2 show the corresponding vertical velocity at these two times in the lowest 6 km. Of interest is the fact that, in the domain shown, the vertical velocity is everywhere positive at both times with a broad maximum centered at an altitude of 700 m inside the radius of maximum tangential wind (initially 100 km). This outcome could be determined from the analysis in Section 6.9.2.

16.1.3.2 Zero-order model with heating and friction

Fig. 16.3 shows radius-height cross sections of the radial and tangential velocity components at the initial time and after 20 hours for the simulation with heating and friction. As in the case where the heating distribution moves with a prescribed M-surface (Section 8.1.3), the largest inflow in the presence of surface friction occurs within the frictional boundary layer, even in this case where gradient wind balance is assumed to hold approximately in the boundary layer as well. In fact, in this case, the maximum inflow remains at the surface.

In the presence of friction, the solution where the heating distribution is held fixed in time breaks down at 22 hours 45 min, about 3 hours earlier than in the case in Section 8.1.3 where the axis of maximum heating is allowed to move inwards. At the time step before breakdown, the maximum tangential wind speed has increased to 26.4 m s^{-1} from 10 m s^{-1} at the initial time and the maximum inflow has decreased to 1.3 m s^{-1} from 3.0 m s^{-1} at the initial time, presumably on account of the increase in inertial stability in a balance configuration. The Eliassen equation requires regularization after just under 8.5 hours.

What happens at breakdown in this case is that the shallow frictionally-induced return flow above the boundary layer, inside the heating region, is accentuated by the out-

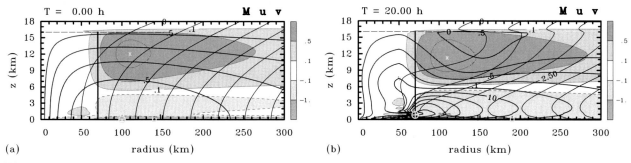

FIGURE 16.3 Radius-height cross sections of M-surfaces (thin black curves) superimposed on various quantities for the simulation with heating and friction. These quantities include: contours of tangential velocity, v, (blue (dark gray in print version) curves), and radial velocity u (shaded) at (a) the initial time, and (b) 20 hours. The dashed horizontal line at a height of 16 km indicates the model tropopause. The thick vertical line shows the location of the maximum diabatic heating rate, which is fixed in time. Contour intervals are: for M, 5×10^5 m^2 s^{-1}; for u, 0.5 m s^{-1}, 0.02 m s^{-1} (thin solid contours below 2.5 km height and inside 75 km radius), 1 m s^{-1} for negative values dashed contours; for v, 2 m s^{-1}. The yellow (light gray in print version) symbols indicate the locations of maximum v, ($\oplus$), maximum u, ($\times$), and minimum u, ($\square$).

flow there driven by the heating. As seen in Fig. 16.3b, this shallow outflow leads to a rapid increase in the radial gradient of M near the axis of heating and thereby a large radial gradient of tangential velocity and large vorticity as anticipated in Section 8.1.1.4. In turn, these large gradients appear to prevent the successive-over-relaxation method for solving the Eliassen equation from converging.

Despite the fact that the zero-order model does not introduce surface fluxes of moist enthalpy explicitly, the fact that fixing the diabatic heating rate in space and time still leads to spin up calls into question the need to invoke the WISHE feedback mechanism. In other words, if the thermodynamic processes that determine the diabatic heating rate were included in the formulation and if these were such as to produce a steady heating rate distribution, the vortex would still intensify. There would be no need to invoke a temporal increase in the amplitude of the surface enthalpy fluxes with near surface wind speed to explain intensification. All that is needed would be for these fluxes to be sufficiently large to maintain convective instability and thereby the diabatic heating rate distribution over a deep layer to maintain the lower tropospheric inflow.

16.1.4 First-order model

We consider now the possibility of relating the diabatic heating rate to the thermodynamics of the developing vortex, which would certainly enhance the realism of the zero-order models described above. Attempts to incorporate a parameterization for deep convection in an axisymmetric balance framework go back to Ooyama (1969), Sundqvist (1970a,b), and Emanuel (1989). However, the later models by Emanuel discussed in Section 12.1.2-12.1.3 adopt a different approach to representing the bulk effects of moist processes, one that is arguably more akin to the simple explicit approach adopted in the first-order model described below.

16.1.4.1 Mathematical formulation in cloudy regions

In cloudy regions $r_v = r_v^*(p, \chi)$, the substitution for $\dot{Q}$ using Eq. (16.6), with Eq. (16.4) used to relate u and w to ψ, $\dot{\Theta}$ is

$$\dot{\Theta} = -\frac{L_v \chi^2}{r \rho c_{pd} \pi} J(\psi, r_v^*) = -\gamma \frac{L_v}{c_{pd} T} J(\psi, r_v^*), \quad (16.7)$$

where the Jacobian operator is defined by

$$J(\psi, r_v^*) = \left(\frac{\partial \psi}{\partial r} \frac{\partial r_v^*}{\partial z} - \frac{\partial \psi}{\partial z} \frac{\partial r_v^*}{\partial r} \right). \quad (16.8)$$

When modified to incorporate the relationship (16.7) linking the heating rate and slantwise motion that supports the condensation process, the Eliassen equation takes form

$$\frac{\partial}{\partial r} \left[\gamma \left\{ N^2 \frac{\partial \psi}{\partial r} - B \frac{\partial \psi}{\partial z} + \frac{g L_v}{c_{pd} T} J(\psi, r_v^*) \right\} \right] + \frac{\partial}{\partial z} \left[\gamma \left\{ I_g^2 \frac{\partial \psi}{\partial z} - B \frac{\partial \psi}{\partial r} + \frac{C L_v}{c_{pd} T} J(\psi, r_v^*) \right\} \right] = \frac{\partial}{\partial z} (\chi \xi \dot{V}), \quad (16.9)$$

which, using Eq. (16.8), may be written

$$\frac{\partial}{\partial r} \left[\gamma \left\{ \left(N^2 + \frac{g L_v}{c_{pd} T} \frac{\partial r_v^*}{\partial z} \right) \frac{\partial \psi}{\partial r} - \left(B + \frac{g L_v}{c_{pd} T} \frac{\partial r_v^*}{\partial r} \right) \frac{\partial \psi}{\partial z} \right\} \right] + \frac{\partial}{\partial z} \left[\gamma \left\{ \left(I_g^2 - \frac{C L_v}{c_{pd} T} \frac{\partial r_v^*}{\partial r} \right) \frac{\partial \psi}{\partial z} - \left(B - \frac{C L_v}{c_{pd} T} \frac{\partial r_v^*}{\partial z} \right) \frac{\partial \psi}{\partial r} \right\} \right]$$
$$= \frac{\partial}{\partial z} (\chi \xi \dot{V}). \quad (16.10)$$

Within moist saturated ascent, the discriminant of this equation is

$$\Delta = 4\gamma^2 \left[\left(I_g^2 - \frac{C L_v}{c_{pd} T} \frac{\partial r_v^*}{\partial r} \right) \left(N^2 + \frac{g L_v}{c_{pd} T} \frac{\partial r_v^*}{\partial z} \right) - \right.$$

$$\left\{ B + \frac{L_v}{2c_{pd}T}\left(g\frac{\partial r_v^*}{\partial r} - C\frac{\partial r_v^*}{\partial z}\right)\right\}^2 \right].$$
$$(16.11)$$

In unsaturated regions, $r_v < r_v^*(p,\chi)$, the Eliassen equation in the form of Eq. (16.3) remains appropriate, since Eq. (16.10) reduces to Eq. (16.3) if the terms involving latent heat release, which are proportional to L_v, are set to zero. To determine whether or not the air is cloudy, we need to introduce a prediction equation for r_v in the first-order model. Further, the influence of surface moisture fluxes would enter through this equation.

Note that, if friction were not included as part of the first-order model (i.e., $\dot{V} = 0$), there would be no forcing term on the right-hand-side of Eq. (16.10) and the only solution would appear to be the trivial solution $\psi = 0$. This outcome is supported by physical considerations. Because $\dot{Q}$ is proportional to w, without frictional forcing, a solution starting with no heating initially would not be able to develop heating.

Consider first the case of cloudy regions where there is pseudo-adiabatic upright ascent. Then, $\partial r_v^*/\partial z < 0$ and the terms involving $\partial r_v^*/\partial r$ would be absent (Exercise 16.2). In these regions, the coefficient of $\partial^2\psi/\partial r^2$ in Eq. (16.10) may be written as

$$g\gamma\left(\frac{1}{\theta}\frac{\partial\theta}{\partial z} + \frac{L_v}{c_{pd}T}\frac{\partial r_v^*}{\partial z}\right) \approx \frac{\gamma g}{\theta_e^*}\frac{\partial\theta_e^*}{\partial z}, \qquad (16.12)$$

where θ_e^* is the saturation pseudo-equivalent potential temperature and the weak dependence of $1/T$ on z has been neglected. If air parcels were rising without mixing, θ_e^* would be materially conserved and in a steady flow with air parcels rising vertically, $\partial\theta_e^*/\partial z$ would be zero. In this case, the discriminant Δ would be zero or negative. Note that the quantity $(g/\theta_e^*)/(\partial\theta_e^*/\partial z)$ is the effective static stability in cloudy air.

In the more general case of slantwise ascent, the release of latent heat during radial outward motion has the effect of *increasing* the inertial stability as then $\partial r_v^*/\partial r < 0$, since the pressure increases with radius and, for a warm-cored vortex, the temperature decreases. The effect of latent heat release on I_g^2 is analogous to the effect on N^2, remembering that the sum of centrifugal and Coriolis forces is in the direction of radially outward motion whereas, in the case of ascent, the gravitational acceleration is in the opposite direction to ascent. Thus the generalized buoyancy force associated with an air parcel that is less dense than its environment is inwards and upwards (Section 5.3).

The term $B + L_v/(2c_{pd}T)\left(g\partial r_v^*/\partial r - C\partial r_v^*/\partial z\right)$ may be written as $B + L_v/(2c_{pd}T)\mathbf{h}\cdot\nabla r_v^*$, where $\mathbf{h}$ is the vector $(g, 0, -C)$, which is related to the generalized gravitational acceleration, $\mathbf{g}_* = (C, 0, -g)$ defined in Section 5.4: see Exercise 16.5. Thus, the effect of latent heat release on B

depends on the sign of B and on the angle between the vectors $\mathbf{h}$ and ∇r_v^* (see Exercise 16.4).

Typically, in the atmosphere, the angle between ∇r_v^* and $\mathbf{g}_*$ is less than 90 degrees and for a warm-cored vortex, $B < 0$. In this case, latent heat release leads to a reduction in the magnitude of the last squared term in Eq. (16.11). Therefore, the effect of latent heat release on the magnitude and sign of the discriminant Δ is not immediately obvious: one has to perform the calculation in a specific case.

16.1.4.2 Qualitative structure of the solutions

In view of the foregoing mathematical results, it is enlightening to examine the qualitative structure of solutions of Eq. (16.10) in two situations, depending on the character of the equation in cloudy air as determined by the sign of Δ.

Case A. Suppose that the effective static stability in saturated air becomes relatively small, but remains sufficiently positive to keep $\Delta > 0$ there, and that $\Delta > 0$ in unsaturated air also. In this case, the Eliassen equation (16.10) would be globally elliptic and the solutions thereto would be qualitatively similar to those in the case with no heating in the sense that there will be inflow in the layer with friction and outflow above (cf. Section 16.1.3.1). However, we would expect the reduced static stability in cloudy air to allow for a deeper layer of outflow than in the case with friction only (e.g., Montgomery et al., 2001). In support of this conjecture, we show in Fig. 16.4 a repeat of the calculation with no heating in Section 16.1.3.1, but with the coefficient of the second-order derivative $\partial^2\psi/\partial r^2$ multiplied by a factor 0.2 at each grid point in a region extending in radius from 20-90 km and from an altitude of 1 km to 14 km to crudely mimic the effect of reduced static stability on account of latent heat release (indicated by a rectangular box in the figure). The upper panels show similar fields as Fig. 16.2a, b at the initial time and at 6 hours.

As one might have anticipated by invoking the membrane analogy (see Sections 2.19 and 5.10.2), a significant reduction in the coefficient of $\partial^2\psi/\partial r^2$ in the Eliassen equation, i.e., a reduction in the effective static stability, would be expected to lead to an increase in the depth to which air would rise as it leaves the boundary layer and therefore a deepening of the layer of outflow above the boundary layer. The calculations shown in Fig. 16.4 support this supposition: compare for example the upper panels of Fig. 16.4 with those of Fig. 16.2. Comparing the lower panels of Fig. 16.4 with the upper panels shows that the overturning circulation has undergone significant decay over a 6 hour period.

Despite this decay, there is a slight, but temporary amplification of the maximum tangential wind speed (by less than 2 m s^{-1}) over this time and this maximun has moved a little inwards. It appears that the abrupt change in the coefficient of $\partial^2\psi/\partial r^2$ at a height of 1 km at inner radii leads to enhanced inflow near the top of the inflow layer to account

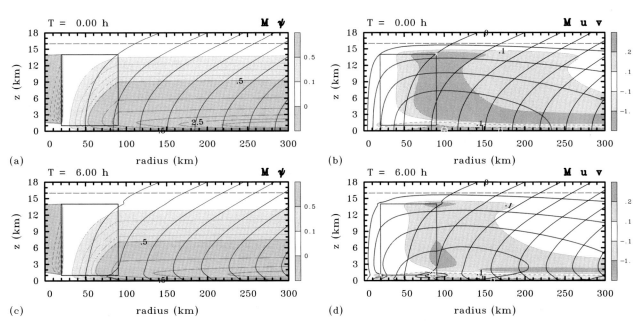

FIGURE 16.4 Radius-height cross sections of M-surfaces (thin black curves) superimposed on various quantities for the friction-only simulation, but with the coefficient of the term $\partial^2 \psi / \partial r^2$ in the Eliassen equation multiplied by a factor 0.2 in the rectangular region outlined in dark blue (dark gray in print version). These quantities include: (a, c) streamlines of the secondary circulation, ψ (red and blue (mid and dark gray in print version) contours and shading); (b, d) contours of tangential velocity, v (blue (dark gray in print version) curves), and radial velocity u (shaded). The upper panels are at the initial time and the lower panels are at 6 hours. The dashed horizontal line at a height of 16 km indicates the model tropopause. Contour intervals are: for M, $5 \times 10^5 \text{ m}^2 \text{ s}^{-1}$; for ψ, $5 \times 10^7 \text{ kg s}^{-1}$ thick positive contours, $1 \times 10^7 \text{ kg s}^{-1}$ thin positive contours. $2 \times 10^5 \text{ kg s}^{-1}$ negative contours; for v, 2 m s^{-1}. Contours for u are 0.1 m s^{-1} and 0.2 m s^{-1} for positive values, 0.1 and 1 m s^{-1} for negative values. Positive contours are solid, negative contours dashed. The yellow (light gray in print version) symbols in (b) and (d) indicate the locations of maximum v, ($\oplus$), maximum u, ($\times$), and minimum u, ($\square$).

for this amplification. Note that the inflow layer in the right panels of Fig. 16.4 is deeper than in the upper panels of Fig. 16.2 with the maximum depth inside the radius range of the reduced coefficient of $\partial^2 \psi / \partial r^2$.

Case B. In contrast to Case A, suppose the effective static stability becomes negative in cloudy air. In such a region, $\Delta < 0$ and Eq. (16.10) would be hyperbolic there. In these unstable regions, air parcels would be able to accelerate vertically under their local buoyancy so that the hydrostatic approximation and therefore thermal wind balance would break down. In this most likely scenario, the breakdown of the formulation as a globally elliptic problem is an indication that localized buoyant deep convection is seeking to develop. In particular, it suggests also that the simple parameterization encapsulated in the formula (16.7) in conjunction with the hydrostatic constraint is inadequate to represent such convection. With the hydrostatic approximation, and therefore within the balance framework, air parcels experience no net vertical force to accelerate them vertically and for this reason, the characteristic flow patterns of low-level convergence and upper-level divergence typical of deep convection are unable to develop.

The conclusion from the foregoing analysis is that the simple parameterization given by the formula (16.7), in conjunction with the imposition of hydrostatic and gradient balance, is unable to facilitate the low-level inflow

required to reverse the frictionally-induced outflow in the lower troposphere above the boundary layer. *Therefore, the parameterization is unable to provide the necessary convergence in this region for vortex spin up.*

16.1.5 Exercises

Exercise 16.1. Verify the derivation of Eqs. (16.10) and (16.11).

Exercise 16.2. Show that for saturated pseudo-adiabatic upright ascent, Eq. (16.11) reduces to

$$\Delta = 4\gamma^2 \left[I_g^2 \left(N^2 + \frac{gL_v}{c_{pd}T} \frac{\partial r_v^*}{\partial z} \right) - \left(B - \frac{CL_v}{2c_{pd}T} \frac{\partial r_v^*}{\partial z} \right)^2 \right]. \quad (16.13)$$

Exercise 16.3. Using vector notation, show that nonzero diabatic heating, $\dot{\theta}$, requires the transverse velocity, $\mathbf{u}$, to be in the opposite direction to ∇r_v^*.

Exercise 16.4. Show that in the vertical $r - z$ plane,

$$\nabla r_v^* = \frac{\partial r_v^*}{\partial p} \nabla p + \frac{\partial r_v^*}{\partial T} \nabla T, \quad (16.14)$$

where, here, $\nabla = \hat{\mathbf{i}}\frac{\partial}{\partial r} + \hat{\mathbf{k}}\frac{\partial}{\partial z}$ and $\hat{\mathbf{i}}$, $\hat{\mathbf{k}}$ are unit vectors in the radial and vertical directions, respectively.

Exercise 16.5. Show that the second term in the expression $B + \frac{1}{2}\frac{L_v}{c_{pd}T}\left(g\partial r_v^*/\partial r - C\partial r_v^*/\partial z\right)$ in Eq. (16.11) can be written in the vector form

$$\frac{1}{2}\frac{L_v}{c_{pd}T}\mathbf{h}\cdot\nabla r_v^*,$$

where $\mathbf{h}$ is the vector $(g, -C)$. Sketch the relationship between the generalized gravity, $\mathbf{g}^* = (C, 0, -g)$ and the vector $\mathbf{h} = (g, 0, -C)$ in relation to an isobaric surface.

Exercise 16.6. Using the approximate equation for saturation pseudo-equivalent potential temperature

$$\ln\theta_e^* = \ln\theta + \frac{L_v r_v^*}{c_{pd}T}, \qquad (16.15)$$

which treats L_v and $1/T$ as approximately constant (see Section 2.5.4), show that the axisymmetric moist saturation potential vorticity defined by

$$P_m^* = \frac{(\boldsymbol{\omega}+\mathbf{f})\cdot\nabla\theta_e^*}{\rho} \qquad (16.16)$$

is determined by the equation

$$\frac{g\rho\xi}{\theta_e^*}P_m^* = \left(I_g^2 - \frac{CL_v}{c_{pd}T}\frac{\partial r_v^*}{\partial r}\right)\left(N^2 + \frac{gL_v}{c_{pd}T}\frac{\partial r_v^*}{\partial z}\right) - \left(B - \frac{CL_v}{c_{pd}T}\frac{\partial r_v^*}{\partial z}\right)\left(B + \frac{gL_v}{c_{pd}T}\frac{\partial r_v^*}{\partial r}\right). \qquad (16.17)$$

Exercise 16.7. Using vector calculus, show that the saturation potential vorticity defined in Exercise 16.6 is equal to $\hat{\mathbf{j}}/(r\rho)\cdot\nabla\theta_e^*\wedge\nabla M$, where M is the absolute angular momentum per unit mass and θ_e^* are as defined also in Exercise 16.6 and $\hat{\mathbf{j}}$ is a unit vector in the azimuthal direction.

Exercise 16.8. Show that the discriminant of the Eliassen equation for cloudy air given by Eq. (16.11) is related to P_m^* by the equation

$$\Delta = 4g\gamma^2\left[\frac{\rho\xi P_m^*}{\theta_e^*} - \frac{g}{4}\left(\frac{L_v}{c_{pd}T}\right)^2\left(\frac{\partial r_v^*}{\partial T}\right)_p^2\left(\frac{\partial T}{\partial r}\right)_p^2\right]. \qquad (16.18)$$

Note that, if the M and θ_e^* surfaces are assumed to be congruent as in the later Emanuel intensification theories discussed in Chapter 12, $P_m^* = 0$ (see Exercise 16.7), then $\Delta < 0$, whereupon the Eliassen equation in a classical balance formulation would be hyperbolic.

16.1.6 Beyond the minimal representation of friction

Before examining the implications of the foregoing deductions from the first-order model, it is appropriate to reiterate the limitations of the representation of friction in a balance framework (Section 16.1.3). The scale analysis presented in Chapter 6 shows that the boundary layer is intrinsically unbalanced and the secondary circulation it induces is, in fact, a result of gradient-wind imbalance. For this reason, it might seem appropriate to apply the classical balance theory (Chapter 5) only to the free atmosphere above the boundary layer, treating the boundary layer as a separate component of the flow. In such an extended formulation, the strength and geometrical configuration of the secondary circulation above the boundary layer would be determined by the spatial distribution of diabatic heating rate throughout the troposphere, together with the radial distribution of vertical velocity at the top of the boundary layer, which would have to be supplied by a boundary layer model.

The above conceptualization would allow the vertical motion at the top of the boundary layer to be determined by a more realistic boundary layer model that incorporates unbalanced processes. By including a full radial momentum equation, important nonlinear effects arise in the ensuing dynamics of vortex spin up. As discussed in Section 6.6, these effects lead to the counter-intuitive result that the maximum storm-relative tangential wind in the azimuthally averaged flow occurs within, but near the top of the frictional layer, despite the tendency for surface stress at the air-sea interface to locally slow down the tangential winds underneath the vortex.

It is reasonable to expect that, for realistic values of the drag coefficient, the inclusion of a nonlinear boundary layer would increase the spin-up rate in comparison to the zero- and first-order models employing a minimal frictional boundary layer as described above.[3] In particular, the enhanced near-surface wind speeds will increase values of θ_e entering the central core convection, thereby enhancing the diabatic heating rate and its radial gradient.

There are two issues with the foregoing separation of a free atmosphere in which there is approximate thermal wind balance and a boundary layer that is not in balance. The first is that, in general, the tangential velocity expelled from the boundary layer at inner radii will not be in gradient balance as would be required as soon as it leaves the boundary layer. There would seem to be a need to incorporate an adjustment layer in this formulation.

The second issue is that the boundary layer model depends only on the tangential velocity profile at its top and it does not know about the suction effect of deep convection, which would be trying to draw air out of the boundary layer

---

3. See Montgomery and Smith (2022), Section 2.5, for a list of references in support of this statement.

into the convective region. We saw that, in the zero-order model without friction, the maximum inflow driven by the diabatic heating distribution occurs at, or very close to the surface. This is purely a manifestation of the suction effect of the region of heating. If friction is included in the zero-order model, the effect of suction is less transparent, but it is still there and, in this model, it adds linearly to the effects of friction.

In an extended formulation with a separate and unbalanced boundary layer representation, the boundary layer model would not represent the suction effect of the diabatic heating in the free atmosphere at all. However, since the net mass flux of the overturning circulation in the balance model is determined by the heating distribution, the suppressed ability of the heating to draw air inwards in the boundary layer has the unintended effect of drawing in the required air above the boundary layer, thereby leading to stronger spin up of winds above and at the top of the boundary layer than it would otherwise. Recent illustrations of this unintended effect are given in Montgomery and Persing (2020), their Fig. 5c, 5e, Appendix B and accompanying discussion.

16.1.7 Cumulus parameterization in minimal models

An important deduction from Case A in Section 16.1.4.2 is that, using the explicit representation of latent heat release encapsulated in Eq. (16.6) in conjunction with the balance vortex formulation, the flow above the boundary layer must be everywhere outwards if the Eliassen equation remains globally elliptic. In the alternative Case B, the Eliassen equation becomes hyperbolic in regions of conditional instability. These hyperbolic regions are where deep convective clouds would be expected to form. If strong enough, these clouds would have the ability to reverse the frictionally-induced outflow above the boundary layer, leading to inflow in the lower troposphere, which is necessary to converge absolute angular momentum to facilitate vortex spin up (Section 8.2).

The foregoing considerations explain why early attempts to simulate tropical cyclones using axisymmetric numerical models were frustrated by the emergence of deep convective clouds instead of a coherent tropical-cyclone-like vortex.[4] If these early studies had had the computational resources to simulate for a longer time period, they would have likely found a coherent vortex to develop (e.g., Rotunno and Emanuel, 1987), even though the deep convective rings that evolve in the early stages of development are not realistic and give erroneous characteristics and dependencies on the surface drag (see Section 16.2). As is well known to forecasters and highlighted in Chapter 10,

deep convection in tropical cyclones is mostly localized and highly asymmetric.

Because the early modeling studies lacked the computational ability to resolve deep convective clouds, parameterization schemes were developed to represent these clouds as sub-grid-scale phenomena, not only for idealized studies in the Tropics, but more generally for numerical weather prediction and climate models. In particular, many of the early minimal models for tropical cyclones incorporated a parameterization scheme for deep cumulus clouds: examples include Ooyama (1968), Ooyama (1969), Emanuel (1989), Zhu et al. (2001), Nguyen et al. (2002). An important feature of these schemes is that they facilitate a pattern of low-level convergence and upper-level divergence in regions of deep convection to enable the natural tendency for frictionally-driven outflow above the boundary layer to be reversed.

It is notable that the early Ooyama models listed above and that of Emanuel (1989) include a parameterization scheme for deep convection that would serve such a purpose, placing these models in the category of Case B in Section 16.1.4.2. In contrast, the later Emanuel models, i.e., those of Emanuel (1995), Emanuel (1997), Emanuel (2012), do not incorporate such parameterization schemes and these models would appear at first sight to fall into the category of Case A, since there is radial outflow present everywhere above the boundary layer (see e.g., Figs. 12.3, 12.5).[5] On the other hand, the result of Exercise 16.8 would appear to place them in Case B, since the assumption that the M and θ_e^* surfaces are congruent means that the Eliassen equation would be hyperbolic. At the time of writing, the consequences of these conflicting conclusions concerning Emanuel's later intensification theories remain to be understood, but we are led to tentatively conclude that they are incompatible with the classical balance theory of Eliassen.

The bottom line is that, *in order for a minimal balance model to represent the classical spin up mechanism, it would need to incorporate a parameterization scheme for deep cumulus convection to ensure that there is horizontal convergence in the lower troposphere and horizontal divergence in the upper troposphere.* This remark applies, in particular, to the first-order model outlined in Section 16.1.4.

---

4. A nice review of this early work is given by Emanuel (2018), page 15.6.

5. The analysis in Section 16.1.4.2 calls into question a statement made in our review of paradigms for tropical cyclone intensification (Montgomery and Smith, 2014) concerning the Emanuel 1997 (E97) theory. There, on page 48, we said: "The effects of latent heat release in clouds are implicit also in the E97 model, but the negative radial gradient of θ_e in the boundary layer is roughly equivalent to a negative radial gradient of diabatic heating in the interior, which, according to the balance concepts discussed earlier in "The overturning circulation" will lead to an overturning circulation with inflow in the lower troposphere". The more in-depth analysis of the Eliassen equation underpinning Case A above suggests that this is not the case. Thus, we are led to retract the statement that "the spin-up above the boundary layer (in E97) is entirely consistent with that in Ooyama's cooperative intensification theory".

However, there remain questions about the integrity of the later intensification models of Emanuel as true balance models in the sense of being able to invert an Eliassen equation for the streamfunction of the overturning circulation as an elliptic partial differential equation.

16.1.8 Role of the WISHE feedback?

An important feature of the Emanuel models for vortex intensification discussed in Chapter 12 is the recognition that deep convection outside the central cumulus zone and in rainbands in the outer region of the vortex will produce convective downdrafts that tend to lower θ_{eb}. Of course, the surface enthalpy fluxes act to increase θ_{eb} and to counter the dilution due to downdrafts. As the central cumulus zone is approached from the outside, the air throughout the troposphere will become progressively moistened by deep convection so that downdrafts will become weaker. At the same time, the near-surface winds will increase.

As long as the increase of sea surface enthalpy flux with wind speed is not outweighed by a decrease in the thermodynamic disequilibrium at the sea surface in the aerodynamic flux formula, the enthalpy fluxes will increase, leading to an increase in θ_{eb} with decreasing radius. In fact, at a given time, the decreasing surface pressure with decreasing radius will serve to increase the saturation mixing ratio, r_v^*, which would help to maintain the degree of thermodynamic disequilibrium, an effect argued to be important by Ooyama (1969).

The moist physics version of the prognostic balance model formulation in Section 16.1.1 was developed in an effort to interpret the purported WISHE feedback mechanism in terms of the classical balance theory discussed in Chapter 8. The essence of the WISHE theory would appear to be that an increase of V_{max} with time *is the result of* the increase in the surface enthalpy fluxes with time, which, in turn, increase with near-surface wind speed. However, as noted in Section 16.1, the mechanism by which the increase in the surface enthalpy fluxes lead to an increase of V_{max} remains to be articulated satisfactorily.

According to the first-order model *with some suitable parameterization of deep convection*, as long as the surface fluxes maintain a deep overturning circulation with inflow in the lower troposphere, and, as long as the air converging in the boundary layer can be ventilated by inner-core deep convection to the upper troposphere, the vortex will continue to intensify. *In this sense, there is no need for coupling between the increasing surface enthalpy fluxes and the increase in the tangential wind speed.*

Even though *an increase* in surface enthalpy fluxes with wind speed is not required for intensification, it is easy to see how a temporal increase of these fluxes might lead to an increase in the intensification rate and V_{max}. If the surface enthalpy fluxes increase with time, the increase of θ_{eb} will be augmented relative to a simulation in which the surface flux distribution is held constant-in-time. What would be required is for the increased θ_{eb} to increase the areally-integrated vertical mass flux of the convection in the developing eyewall updraft to strengthen the secondary circulation. This would lead to a strengthened inflow in the lower troposphere and an increase in the intensification rate as M-surfaces there are drawn inwards faster.

As a result, the increase of the surfaces fluxes with time will lead to a larger V_{max}. This process would be our interpretation of the WISHE feedback and is supported by the numerical modeling studies of Nguyen et al. (2008), Montgomery et al. (2009), Montgomery et al. (2015) and later experiments of Zhang and Emanuel (2016), all of whom showed that the intensification rate is most rapid and the maximum intensity is greatest when the wind-speed dependence of the surface fluxes is retained.

What is still missing in this chain of arguments is why the increase in θ_{eb} as a result of increased surface enthalpy fluxes would increase the areally-integrated vertical mass flux of the convection in the developing eyewall, an aspect that we return to in Item (7) of the next section.

16.1.9 Important caveats

We have used the zero-order and first-order conceptual models for tropical cyclone intensification in this section to outline the key elements contributing to the spin up of the maximum tangential winds. In doing so, we glossed over some important caveats relating to these models and we return to these now for completeness. One or two are consequences of the balance model formulation, but others are more subtle and require careful consideration in specific implementations of the first-order model.

(1) It was shown in Chapter 8 that a feature common to all explicit solutions of the zero-order model is that, within a relatively short space of time, on the order of half a day for the model parameters chosen, the M-surfaces fold over in the upper-troposphere and regions of inertial instability develop. In these regions, the Eliassen equation for the streamfunction of the overturning circulation becomes hyperbolic and strict balance solutions beyond this time are possible only if the coefficients of the equation are modified (or regularized) to remove the unstable region.

When regularization is needed, the balance solutions are only "weak solutions" and the details of the predicted flow, especially in the regularized regions are questionable. The regularization procedure is not only *ad hoc*, but is only a temporary fix. It does not eliminate instability completely because the prognostic equation for the tangential wind component still contains a term involving the negative radial gradient of M. In fact, after another half a day, the solutions break down completely. Nevertheless, the development of such unstable regions in the upper troposphere is a robust feature of three-dimensional model simulations as seen in Figs. 10.8c,d; Figs. 10.17d,f (see also Wang et al. (2020), Figs. 9c,d and Montgomery and Persing (2020), Fig. 4).

(2) The later Emanuel models conceived to underpin the WISHE feedback mechanism have the topology of the M-surfaces near the radius of maximum gradient wind hard-wired to ascend to the upper troposphere and flare out to large radius without bending downward or overturning. This assumption has the unintended effect of preventing the development of inertially unstable regions. Moreover, by definition, it precludes the possibility of layers of inflow developing above and below the upper tropospheric outflow layer, a feature commonly exhibited in many modeling studies (see Section 11.6.1). Even so, and perhaps more importantly, as noted in Section 16.1.7, the later axisymmetric balance models of Emanuel (1995), Emanuel (1997), and Emanuel (2012) appear at first sight to fall in the category of Case A in Section 16.1.4.2 as they do not include a parameterization of deep cumulus convection that would allow the classical spin up mechanism to be a feature of the dynamics. Nevertheless, the assumption in these models that the M and θ_e^* surfaces are congruent would imply a negative discriminant that would place them in case B. At this time of writing, we are unable to resolve this conundrum.

(3) The intensification process envisaged in Sections 16.1.2 and 16.1.4 is based on the idea that the diabatically-driven overturning circulation draws M-surfaces inwards at the radius R_{Vmax} of maximum gradient wind V_{max} in order to amplify the latter, even though, in general, R_{Vmax} will not remain fixed. This idea is consistent with the classical Shapiro and Willoughby (1982) explanation for spin up.

(4) In the first-order model, it is presumed that deep convection is strong enough to carry moist air into the upper troposphere. When surface friction is taken into account in the explanation, the deep overturning circulation produced by the diabatic heating would need to be strong enough to ventilate the moist air that is converging in the boundary layer. As discussed in Chapters 8 and 15, this ventilation cannot be taken for granted. In contrast, as noted above, in the E95, E97 and E12 models, all air converging in the boundary layer is constrained to ascend to the upper troposphere.

(5) To sustain a long-lived vortex, it is necessary that θ_{eb} be high enough to main deep convective instability in the inner-core region. Such an elevation of boundary layer θ_{eb} can come about only through moist enthalpy fluxes at the sea surface. These fluxes must outweigh the adverse effect of frictionally-induced, mesoscale and convective downdrafts that transport cool dry air into the boundary layer. The crucial role of the entropy fluxes is then to maintain deep convective instability in the face of the frictional subsidence and downdrafts. This role was importantly recognized by Emanuel (1986); Rotunno and Emanuel (1987); Emanuel (1989).

(6) In the first-order model, the congruence between θ_e^* and M surfaces has not been assumed as in the 1995, 1997 and 2012 Emanuel models discussed in Chapter 12. In this sense, the model is more general than the Emanuel models and, in principle, it would permit convective updrafts to significantly cross the M surfaces in the earlier stages of intensification.

(7) An important step in the arguments offered in the previous section for why an increase in surface enthalpy flux with time would lead to a strengthening of V_{max} was that the increased flux would be expected to strengthen the overturning circulation. The reasons for this expectation are subtle and call for an articulation of the key steps. Recall from Eq. (16.3) that the strength of the overturning circulation with the balance assumption is proportional to the magnitude of $|\partial\dot{\Theta}/\partial r|$, in the free troposphere, where $\dot{\Theta} = \chi^2\dot{\theta}$ and $\dot{\theta}$ is the diabatic heating rate. Thus, one needs to build a connection between the increase in θ_{eb} *with time* and the increase in $|\partial\dot{\Theta}/\partial r|$ *with time*, assuming that there is approximate thermal wind balance above the boundary layer.

First, one needs a relationship between $\partial\dot{\Theta}/\partial r$ and $\partial\dot{Q}/\partial r$ at a particular time. Referring to footnote 1 in this chapter, $\dot{\theta} = \dot{Q}/(c_{pd}\pi)$, where π is the Exner function so that, ignoring the weak pressure dependence and the weak radial variation of χ, $\partial\dot{\Theta}/\partial r$ is approximately proportional to $\partial\dot{Q}/\partial r$. In the lower half of the troposphere, where most of the condensation takes place, the ascent into the eyewall updraft is nearly upright, whereupon from Eq. (16.6), $\dot{Q} \approx -L_v w(\partial r_v^*/\partial z)$. As long as the boundary layer doesn't saturate, its moisture content, as characterized by r_v, will increase with decreasing radius. Then, where the air ascends out of the boundary layer and saturates, the release of latent heat, which is proportional to $-\partial r_v^*/\partial z$, will increase also with decreasing radius, the proviso being that the local w_{max} doesn't decrease with radius.[6]

Next, since θ_{eb} is a monotonic function of r_v, it follows that $\partial\theta_{eb}/\partial r$ will increase also with decreasing radius. Thus, combining the foregoing steps, $|\partial\dot{\Theta}/\partial r|$

---

6. The caveat regarding w_{max} seems hardly necessary since, according to the Eliassen balance theory as exemplified by the zero-order models, a negative radial gradient of $\dot{Q}$ (and corresponding negative radial gradient of $\dot{\Theta}$), would lead to a deep overturning circulation in which the vertical profile of w progressively amplifies with decreasing radius. Such an increase is supported by the explicit calculations shown in Section 8.1.1, Figs. 8.3e,f and Fig. 8.6c.

is a monotonically increasing function of $|\partial\theta_{eb}/\partial r|$. Finally, an increase of the surface enthalpy flux *with time* will lead to a similar increase in $|\partial\theta_{eb}/\partial r|$, an increase in $|\partial\tilde{\Theta}/\partial r|$ and therefore to an increase in the strength of the overturning circulation, characterized by the vertical mass flux carried by it.

(8) In the foregoing analysis it has been presumed that the bulk of the vortex dynamics and thermodynamics are governed to zero order by azimuthally-averaged flow quantities relative to a nominal, slowly-varying, center of circulation. However, we have seen in Chapter 11 that, in general, the eddy (non-axisymmetric) terms are not negligible. In specific circumstances, the eddy covariance terms may be comparable to, or even dominate the strictly mean terms in the inner-core region of the vortex and would need to be included explicitly in a more complete analysis. The rotating-convection paradigm is the proper framework in this circumstance.

16.1.10 Synthesis

In this section, we examined a hierarchy of minimal conceptual models for tropical cyclone intensification because popular explanations of intensification still appear to invoke the imprecisely articulated WISHE feedback process. The minimal models build on the prognostic balance theory of Chapter 8 with an extension to incorporate latent heat release explicitly and have added an explicit link to the moist enthalpy fluxes that are required to support the deep convection in the central region of the developing vortex. The minimal models serve to fill the gap between the popular explanations of the WISHE feedback mechanism noted in Section 16.1 and the classical theory of vortex intensification discussed in Section 8.2.

First, we noted that the popular explanations of the WISHE feedback mechanism are underpinned by the E95, E97, E12 intensification theories analyzed in Chapter 12. An important finding from the minimal model formulations is that these intensification theories are inconsistent with Eliassen balance dynamics, which includes the classical spin up mechanism as presented in Chapter 8.

Second, we demonstrated that *vortex intensification does not require the surface enthalpy fluxes to increase with time, but only that the surface enthalpy fluxes are sufficient to maintain convective instability and a deep region of latent heating with a negative radial gradient.* In this sense, a WISHE feedback mechanism is not necessary to explain vortex intensification.

Finally, we showed how *wind-speed dependent surface fluxes enhance the intensification rate of the vortex characterized by the increase in V_{max} in comparison to formulations that do not allow the fluxes to increase with time.* This enhancement of the intensification rate would be our interpretation of "the WISHE feedback".

16.2 Comparison between three-dimensional and axisymmetric tropical cyclone dynamics

The process of tropical cyclone genesis and intensification observed in reality and found in three-dimensional, cloud-representing, numerical model simulations is markedly asymmetric (Chapters 1, 10 and 11). Thus, the question naturally arises: how does the development process in a three-dimensional model compare with that in an otherwise identically configured axisymmetric model? This question is stimulated also by curiosity into the possible deficiency of axisymmetric convection in general, a likelihood suggested by a comparison of convection in slab-symmetric geometry with that in three-dimensions by Moeng et al. (2004). In particular, is an axisymmetric depiction of convection as concentric rings overly efficient in generating buoyancy fluxes compared to the 3D convection in isolated plumes? If true, such an outcome would have potentially significant ramifications for modeling tropical cyclones as it would implicate an intrinsic tendency towards excessive condensation heating and an overly rapid spin up in strictly axisymmetric formulations. We show below that these proffered implications are essentially correct.

The study by Persing et al. (2013) examined the dynamical differences between tropical cyclone evolution in three-dimensional (3D) and axially symmetric (AX) configurations for the prototype intensification problem defined in Chapter 10. The study identified a number of important differences between the two configurations, which are summarized here.

Many of the simulated differences are attributable to the dissimilarity of deep cumulus convection in the two model configurations. Specific distinguishing features are as follows:

1. There are important differences in convective organization. Recalling the vortex axisymmetrization process studied in Chapter 4, deep convection in the three-dimensional model tends to be sheared azimuthally by the differential angular rotation of the azimuthally-averaged system-scale circulation in the radial and vertical directions, quite unlike convective rings in the axisymmetric configuration.

2. Because convection is generally not organized into concentric rings during the spin up process, the azimuthally-averaged heating rate and its radial gradient is generally considerably less than that in the AX model (e.g., Fig. 16.5). For most of the time this lack of organization results in a slower spin up and leads ultimately to a weaker mature vortex.

As discussed in more detail by Persing et al., there is a short period of time when the rate of spin up in the 3D model exceeds the maximum spin up rate in the

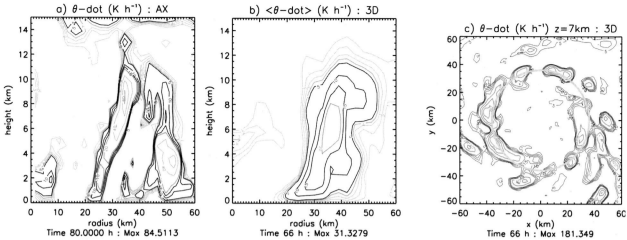

FIGURE 16.5 Comparing diabatic heating rates in identically configured AX and 3D simulations. Diagnosed latent heating rate during rapid intensification in the axisymmetric (AX) and three-dimensional (3D) simulations from Persing et al. (2013). (a) Radius-height contour plot of $\dot{\theta}$ from the AX simulation. (b) Radius-height contour plot of the azimuthally-averaged heating rate from the 3D simulation. In the panels, the bracket symbol $\langle ... \rangle$ denotes an azimuthal average on a constant height surface in the case of 3D. (c) Plan-view of $\dot{\theta}$ at 7 km height in the 3D simulation. Courtesy John Persing. Adapted from Persing et al., 2013. Republished with permission from Atmos. Chem. Phys./European Geosciences Union under the Creative Commons Attribution 4.0 License.

AX model. During this period the convection is locally more intense than that found in the AX model and the convection is organized in a ring-like structure resembling a developing eyewall. These regions of relatively strong updrafts have associated vertical eddy momentum fluxes that contribute significantly to the spin up of the azimuthally-averaged tangential velocity (Fig. 16.6), shown also in Sections 11.5.1 and 11.5.2, and to the contraction of the radius of maximum tangential velocity in the developing eyewall above the boundary layer.

The eddy processes provide a plausible explanation for the enhanced spin up rate in the 3D model, despite the relative weakness of the azimuthally-averaged heating rate in this model. Further, inspection of Fig. 16.6f confirms the interpretation offered in Section 11.5.3 that the vertical divergence of the vertical flux of eddy radial momentum, $- < u'w' >$ per unit mass, enhances the near-surface inflow just outside the developing eyewall updraft region. This eddy effect is the manifestation of the suction of low-level air by deep convection in the developing eyewall complex.

3. The findings of Persing et al. suggest that eddy processes associated with vortical plume structures in and around the developing eyewall region can assist the intensification process via vertical transport of eddy momentum and up-gradient momentum fluxes in the radial direction. This finding is in contrast to traditional ideas about the role of eddies and their treatment as agents of a downgradient diffusion process.

4. These vortical plumes contribute significantly also to the azimuthally-averaged heating rate as in Fig. 16.5c, and the corresponding overturning circulation in Fig. 16.6f.

5. The analysis of Persing et al. unveiled an important issue in the representation of subgrid scale parameterizations of eddy momentum fluxes in hurricane models. In particular, the structure of the resolved specific eddy momentum fluxes above the boundary layer differs significantly from that prescribed by the usual "downgradient" subgrid scale parameterizations in both the 3D and AX configurations (e.g., Fig. 16.6, panels a-c), with the exception perhaps of the resolved horizontal eddy momentum flux during the mature stages (not shown).

In addition to the foregoing differences, a notable difference in flow phenomenology that appears to be underappreciated is the fact that the flow in the AX configuration tends to be much noisier than in the more realistic 3D configuration. The larger flow variability in the AX configuration occurs because deep convection generates azimuthally-coherent, large-amplitude, inertia-gravity waves. Although deep convection in the 3D model generates inertia-gravity waves also, intense convection during intensification is typically confined to small ranges of azimuth and tends to be strained by the radial variation of the angular velocity of the system-scale vortex (as noted in item (1) above and discussed in Chapter 4). These two effects lead together to a reduced amplitude of variability in the azimuthally-averaged flow fields in the 3D simulations.

A particularly dramatic illustration of the differences between AX and 3D dynamics is highlighted by a series of calculations examining the sensitivity of vortex intensity to changes of the surface drag coefficient, C_D, for the same value of the surface enthalpy exchange coefficient, C_K. Fig. 16.7 shows time series of the maximum azimuthally-

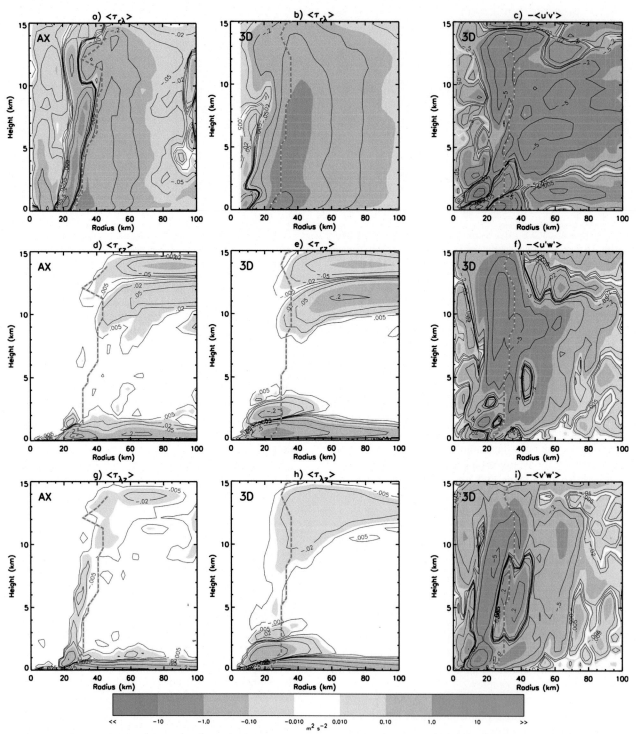

FIGURE 16.6 Comparing parameterized specific (subgrid scale) and resolved specific momentum fluxes (in m² s⁻²) in the AX and 3D simulations from a sample time interval of rapid intensification at comparable vortex intensities from the calculations presented in Persing et al. (2013). Each of the three rows show radius-height ($r-z$) contour plots of the time and azimuthally-averaged sub-grid scale stresses and resolved stresses from the companion AX simulation (77-83 h, left panel) and the 3D simulation (144-148 h, center and right panels). Top row shows time and azimuthally-averaged horizontal sub-grid scale stresses (left panel, AX; center panel, 3D) and resolved stresses (right panel, 3D). Middle and bottom rows show corresponding vertical subgrid scale stress plots during the same sub-interval of rapid intensification from the companion AX simulation (77-83 h, left); the 3D simulation (144-148 h, center); and the corresponding resolved eddy stresses in the 3D simulation (144-148 hours, right). The color bar shows that levels of red (blue) shading correspond to increasing magnitude of positive (negative) values in m² s⁻². Black contours have values ± {0.005, .02, .05, .2, .5, 2, 5, 20, 50} m² s⁻². Black railroad track curve in each panel depicts the radius of v_{max} up to a height of 15 km. The bracket symbol ⟨...⟩ denotes a temporal mean in AX and an azimuthal-temporal mean in 3D. Courtesy John Persing. [Comparison with the corresponding figure for the 3D results (their Fig. 15) suggests that Persing et al. incorrectly reported using data with 2-min spacing and rather used the hourly dataset for their plots. This has been confirmed by using their exact diagnostic code with the 2-min dataset.] Adapted from Persing et al., 2013. Republished with permission from Atmos. Chem. Phys./European Geosciences Union under the Creative Commons Attribution 4.0 License.

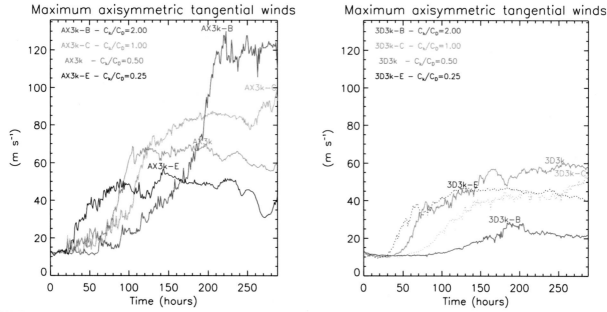

FIGURE 16.7 Sensitivity of vortex intensity in axisymmetric (AX, left) versus three-dimensional (3D, right) simulations to changes of surface drag coefficient, C_D, in the otherwise identical simulations conducted by Persing et al. (2013). Surface enthalpy exchange coefficient $C_K = 1.2 \times 10^{-3}$ is held constant for all these experiments. The "3k" nomenclature is shorthand for the 3 km horizontal (or radial) grid spacing employed in the vortex region. Courtesy John Persing. Adapted from Persing et al., 2013. Republished with permission from Atmos. Chem. Phys./European Geosciences Union under the Creative Commons Attribution 4.0 License.

averaged tangential velocity V_{max} for a suite of AX (left panel) and 3D (right panel) numerical experiments with progressively increasing values of surface drag coefficient as simulated over a 12 day period. Here, increasing surface drag is represented by a decreasing ratio C_K/C_D for fixed C_K. For the weakest drag configuration, the vortex struggles to intensify in 3D throughout the 12 day integration time. Fig. 16.8 shows an example of the highly asymmetric organization of convection at the time of rapid intensification in the 3D simulation with the lowest C_D (left panel) (i.e., $C_K/C_D = 2$, large) and the realistic C_D (right panel). The difference between the convection organization in these three-dimensional experiments is noteworthy and is due solely to the difference in C_D. Needless to say, as is evident from Fig. 16.7, the lowest drag solution, 3D3k-B, is considerably weaker than the corresponding AX solution after approximately 210 hours (8.75 days) (i.e., $V_{max} = 29$ m s^{-1} versus 120 m s^{-1}!).

The AX and 3D results summarized in Figs. 16.7 and 16.8 lend strong support to the idea that as the drag coefficient is increased from small to realistic values, surface friction significantly fosters convective organization in the 3D configuration. The importance of surface drag on the convective organization in the presence of sustained surface fluxes of moist enthalpy has been confirmed independently by other investigators (see Section 2.5 of Montgomery and Smith (2022), for a detailed list of references). This effect is

not significant in the AX configuration as the convection is already organized into concentric rings, by definition.

A useful insight from this facet of the Persing et al. study was the observation that there are competing effects in explaining the dependence of spin up on the surface drag and it is not possible to anticipate *a priori* which of the effects dominates the intensification process. An illustrative example of these competing effects was articulated by those authors (p12336):

"In terms of strictly axisymmetric boundary-layer dynamics, an increase of C_D in the radial momentum equation will lead to a larger inward agradient force, but also to a larger outward frictional drag on the inflow (cf. Chapter 6, our addition). Whether or not the inflow increases with increasing C_D depends on which of these effects dominates. If the boundary layer inflow increases, so will the inward advection of $\langle M \rangle$ surfaces. However, if the drag force increases, $\langle M \rangle$ will be lost at a greater rate because of the increased frictional torque."

The upshot is

"one has to do the calculation to determine which effect dominates in the radial and tangential momentum equations."

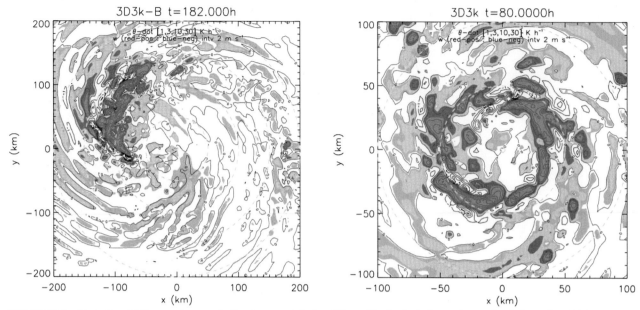

FIGURE 16.8 Illustration of 3D convective organization during period of rapid intensification in the lowest drag setting (left panel) at 182 hours versus the realistic drag setting (right panel) at 80 hours as found by Persing et al. (2013). Surface enthalpy exchange coefficient $C_K = 1.2 \times 10^{-3}$ is held constant for these experiments. Shown in left and right panel is a plan-view of vertical motion w at z = 5 km height with contour interval 2 m s^{-1} and diabatic heating rate $\dot{\theta}$ using shading for levels of 1, 3, 10, and 30 K h^{-1}. The dashed green (gray in print version) rings are constant radii spaced at 50 km from the storm center. The "3k" nomenclature is shorthand for the 3 km horizontal grid spacing employed in the vortex region. Courtesy John Persing. Adapted from Persing et al., 2013. Republished with permission from Atmos. Chem. Phys./European Geosciences Union under the Creative Commons Attribution 4.0 License.

A partial explanation of the foregoing calculations varying C_D is presented by Persing et al. and the reader is referred to their Section 6.7 for details.

16.2.1 Synthesis

The foregoing results illustrate intrinsic differences between 3D and strictly AX tropical cyclone dynamics. These differences highlight subtle, but important effects that surface drag has on the organization of convection and the concentration of ambient and local vertical vorticity. In particular, they constitute a warning against reliance on axisymmetric simulations to ascertain dependencies of tropical cyclone evolution on various parameters in operational forecast settings, or in research settings that may include variables comprising projected climate change scenarios, etc.

The findings concerning the role of eddies raise the practical and research question of whether a down-gradient turbulence closure for all predicted quantities is indeed appropriate in the rotating, convective turbulence region that pervades a rapidly intensifying tropical cyclone?

Overall, the results of this section appear sufficiently interesting to warrant further analysis using both higher spatial resolution and non-quiescent environments as a means to explore more completely the eddy momentum and eddy heat contributions to vortex spin up, as well as the influence of surface drag on convective organization in more realistic environments.

16.3 The effects of latitude on tropical cyclone intensification

Several previous studies have reported idealized numerical model simulations in the prototype problem for tropical cyclone intensification, discussed in Chapter 10, in which the effects of latitude on vortex evolution were examined (e.g., DeMaria and Pickle, 1988; Bister, 2001; Smith et al., 2011; Rappin et al., 2011; Li et al., 2012; Smith et al., 2015a). In every case, the model cyclones were found to intensify more rapidly and reach a higher intensity as the latitude in the model is reduced. This behavior is contrary to that found in the prognostic balance simulations described in Chapter 8 (see Section 8.3.3), but, as shown later, that is most likely to be a result of using the same prescribed heating distribution at different latitudes in these simulations.

The question then arises: is there observational evidence for an increased intensification rate and final intensity at lower latitudes? The answer is a qualified yes. A statistical analysis of Atlantic hurricanes by Kaplan and DeMaria (2003) indicates that a larger proportion of low-latitude storms undergo rapid intensification than storms at higher latitudes. Fig. 16.9 shows the distributions of the RI and

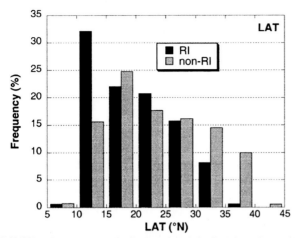

FIGURE 16.9 Frequency distribution of the latitudinal dependence of RI and non-RI samples. From Kaplan and DeMaria, 2003. Republished with permission of the American Meteorological Society.

non-RI samples as a function of latitude, taken from Fig. 5c of Kaplan and DeMaria. Quoting from these authors,

> *"Rapid intensification occurs most frequently from 10°N to 15°N, and the fraction of RI cases generally decreases with increasing latitude. In contrast, the non-RI cases occur most frequently from 15°N to 20°N and exhibit a much slower poleward decrease".*

The qualification is that, in a follow up study by Kaplan et al. (2010), latitude did not feature as an explicit predictor in their updated statistical prediction scheme. Nevertheless, as discussed by Kaplan and DeMaria (2003) and in a climatological study by Hendricks et al. (2010), there are several environmental properties affecting storm intensification. In particular, vertical shear and SST are additional effects that may be influencing the statistics. Since SST typically increases as the latitude decreases, the question arises as to whether this effect might overwhelm the latitudinal effects found in the numerical simulations, which normally assume a uniform SST. Clearly, *from a theoretical viewpoint, it is of interest to understand the effects of the individual processes separately.*

In this section, we review the findings of Smith et al. (2015a), which summarizes and appraises those of previous authors and offers an explanation for the latitude dependence of tropical cyclone evolution. Then, in the section that follows we examine the effects of SST at various latitudes. The explanation of these dependencies touches on practically all facets of the dynamics and thermodynamics of tropical cyclones, invoking the classical spin-up mechanism (Chapter 8, Section 8.2), the boundary layer spin up enhancement mechanism (Section 6.6) and the boundary-layer control mechanism (Section 10.14).

16.3.1 The Smith et al. (2015a) simulations

In brief, the numerical simulations of Smith et al. are three-dimensional and use the MM5 model referred to in Section 10.14. They are similar to those described in Nguyen et al. (2008) with a horizontal grid spacing of 5 km on an inner domain and 24 σ levels in the vertical, 7 of which are below 850 mb. Deep moist convection is resolved explicitly and represented by the warm-rain scheme discussed in Chapter 10. The boundary layer is represented by a parameterization scheme proposed by Blackadar.[7] The SST is a constant (27 °C). As the time period of the calculations is short (3 days), no representation of radiative cooling is implemented.

The initial vortex is axisymmetric with a maximum tangential wind speed of 15 m s^{-1} at the surface at a radius of 100 km. The tangential wind speed decreases sinusoidally with height, vanishing at the top model level. The temperature field is initialized to be in gradient wind balance with the wind field using the method of characteristics sketched in Section 5.2. The far-field temperature and humidity are based on Jordan's Caribbean sounding for the hurricane season (Jordan, 1958). The vertical structure of the initial velocity and temperature deviation fields at various latitudes are shown later in (Fig. 16.13).

16.3.2 Vortex evolution at different latitudes

Three simulations are performed on an f-plane centered at latitudes 10°N, 20°N, and 30°N. Figs. 16.10a-d show time series of various metrics of vortex evolution in these simulations, in the three simulations. These metrics are based on azimuthally-averaged quantities including the maximum tangential wind speed, V_{max}, and the radius at which it occurs, R_{Vmax}; the maximum radial inflow, U_{min}, and the maximum vertical velocity, w_{max}. The azimuthal average is calculated with respect to a vortex center defined as the location of the pressure minimum in a filtered surface pressure field as described at the start of Section 10.5. This center location is taken to be independent of height. Typically, at latitude 10°N, the maximum tangential wind speed occurs at a height of 750 m, while at latitude 30°N it is slightly lower, about 600 m.

As in the previous studies referred to above, the intensification rate increases with decreasing latitude so that, after a few days, the maximum intensity is achieved at the lowest latitude (Fig. 16.10a). However, this result is opposite to those in Sections 8.3.3 and 8.3.5 in which the maximum intensification rate increased slightly with latitude (see Figs. 8.14b and 8.16b), an indication that the convective forcing of the overturning circulation is an important feature of the latitudinal dependence.

---

7. The Blackadar scheme is one of several representations available in the MM5 model.

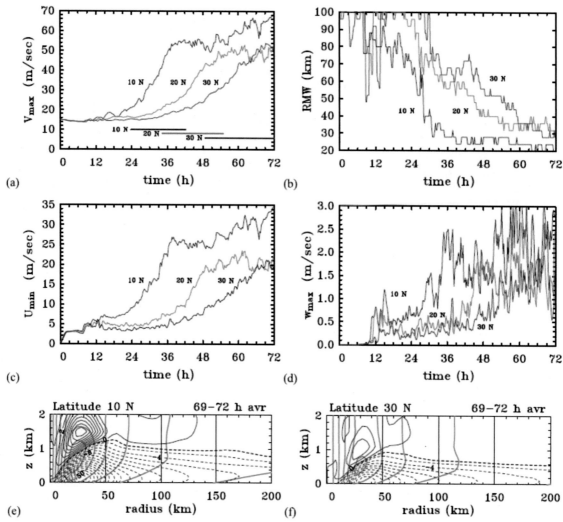

FIGURE 16.10 Time series of azimuthally-averaged quantities in calculations for the prototype intensification problem at latitudes 10°N, 20°N, and 30°N. (a) The maximum tangential wind speed, V_{max}; (b) the radius at which it occurs, R_{Vmax}; (c) the maximum radial inflow; and (d) the maximum vertical velocity. The horizontal lines in (a) show the period of rapid intensification (defined in the text) at each latitude and are referred to in Fig. 16.16. (e), (f) Vertical cross sections of the azimuthally averaged tangential (red (mid gray in print version) contours) and radial wind (blue (dark gray in print version) contours) in the lowest 2 km at 10°N and 30°N, respectively (contour intervals: 10 m s^{-1} for the tangential wind and 2 m s^{-1} for the radial wind). Dashed curves indicate negative values; the thick dashed blue (dark gray in print version) contour shows the -1 m s^{-1} contour of radial wind. From Smith et al., 2015a. Republished with permission of the American Meteorological Society.

In line with the idea that the closer air parcels can approach the axis, the faster they can spin (see e.g., Section 8.2), R_{Vmax} decreases also with decreasing latitude at any given time (Fig. 16.10b). The larger inward displacements of air parcels is consistent with the increase in the maximum radial wind speeds as the latitude decreases (Fig. 16.10c). As discussed in Section 6.6, the maximum tangential wind speeds occur near the top of the strong surface-based inflow layer, which broadly encompasses the boundary layer. Finally, the maximum vertical velocity is largest for the vortex at 10°N and this maximum decreases with increasing latitude (Fig. 16.10d). The maximum vertical velocity occurs typically at a height of between 10 and 14 km (not shown).

Figs. 16.10e,f show radius-height cross sections of the azimuthally-averaged radial and tangential wind components averaged during the period from 69 to 72 hours for the 10°N and 30°N calculations. Noteworthy features relevant to the present study are the deeper boundary layer (as characterized, for example, by the depth of appreciable inflow indicated by the -1 m s^{-1} radial velocity contour) at 10°N and the monotonic increase of boundary layer depth with decreasing radius to approximately the radius of maximum tangential wind speed, which, itself, occurs near the top of the layer of appreciable inflow. We show now that the stronger boundary layer inflow with decreasing latitude is consistent with that found in solutions of the simple slab boundary layer formulation presented in Chapter 6 and

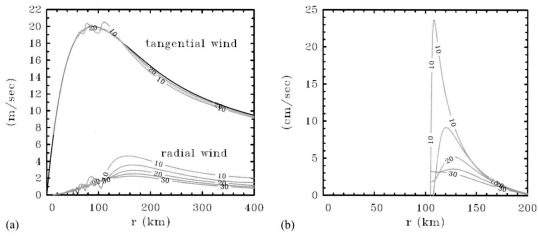

FIGURE 16.11 (a) Radial profiles of radial and tangential wind components, u_b and v_b, respectively, in the slab model boundary layer for different values of latitude for a fixed profile of gradient wind at the top of the boundary layer (the black line). For ease of plotting, the magnitude of the radial inflow has been plotted. (b) the corresponding profiles of vertical velocity at the top of the boundary layer (*note the difference in the radial extent of the abscissa in the two panels*). The calculations are based on the assumption of a fixed boundary layer depth of 1 km. From Smith et al., 2015a. Republished with permission of the American Meteorological Society.

also with the balance boundary layer calculations in Section 8.3.6.

16.3.3 Slab boundary layer solutions

As a further illustration of the dependence of frictionally-induced inflow on latitude, we present next a few solutions of the nonlinear slab boundary layer equations (6.46)-(6.48) for a range of latitudes between 10°N and 30°N and for a constant depth boundary layer. The radial profile[8] of gradient wind is given by the analytical formula:

$$v_g = v_1 s \exp(-\alpha_1 s) + v_2 s \exp(-\alpha_2 s), \qquad (16.19)$$

where $s = r/R_{Vmax}$, r is the radius, R_{Vmax} is the radius of maximum tangential wind speed, and $v_1, v_2, \alpha_1, \alpha_2$ are constants. The solutions for the boundary layer wind components u_b, v_b, and w_b are shown in Fig. 16.11 for a boundary layer with a depth of 1000 m and for a gradient wind profile given by Eq. (16.19). This profile has a maximum tangential wind speed of 20 m s^{-1} at a radius of 90 km, corresponding with a weak tropical storm that is marginally stronger than the initial vortex used for the numerical simulation and it has a slightly smaller value of R_{Vmax}. It is seen that at any given radius beyond about 130 km, the tangential wind falls increasingly below the gradient wind as the latitude decreases, whereas a little inside this radius, the tangential wind speed becomes supergradient ($v_b > v_g$). At these inner radii, the degree to which v_b exceeds v_g increases modestly with decreasing latitude.

At a fixed radius, the radial inflow increases also with decreasing latitude, the maximum inflow increasing from

barely 3 m s^{-1} at 30°N to more than 6 m s^{-1} at 10°N. Note also that the radius of maximum inflow decreases with decreasing latitude.

As pointed out in Section 6.5.7, when the tangential wind becomes supergradient in the boundary layer, all forces in the radial momentum equation there are directed radially outwards and the radial flow decelerates rapidly. The decrease in radial inflow acts as a brake to further the continued increase in v_b and leads also to a sharp increase in the vertical velocity at the top of the boundary layer (Fig. 16.11b). In the calculations shown, the radial inflow flow continues down to a radius of about 10 km and, inside R_{Vmax}, the tangential wind in the boundary layer oscillates about the gradient wind. We do not attribute much significance to these oscillations in reality and, in fact, Kepert (2012) showed that they were an artifact of certain approximations made in deriving the slab model. For this reason, one should not give much weight to the solutions at inner radii where the oscillations occur.

Note that, as the latitude decreases, the maximum vertical velocity increases sharply and its radial location moves inwards. As shown by Smith et al. (see their Fig. 3), this is not a feature of the corresponding linear slab boundary layer model, where the radial location of maximum vertical velocity moves outwards with decreasing latitude. The same is true also of the balance boundary layer formulation used in the prognostic balance model in Chapter 8 (not shown).

The features described above for the nonlinear slab boundary layer are notable ones of the numerical model simulations shown in Figs. 16.10e,f and, with certain caveats discussed in Chapter 6 (see Sections 6.7 and 6.7.2) are suggestive that the behavior of the model simulations can be attributed to the dynamics of the boundary layer.

8. This profile is labeled profile 4 detailed in Smith (2003) for which: $v_1 = 14.538$, $v_2 = -5.462$, $\alpha_1 = 1.263$, and $\alpha_2 = 0.3$ in Eq. (16.19).

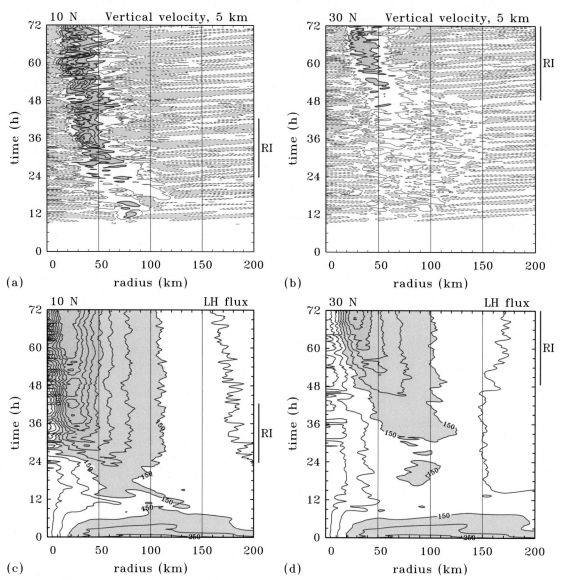

FIGURE 16.12 Time-radius plots of azimuthally-averaged vertical velocity at a height of 5 km and surface latent heat fluxes in the calculations in Section 16.3.1 at latitude 10°N (left column), and 30°N (right column). Contour intervals for vertical velocity: thick solid contours 0.5 m s^{-1}, values ≥ 0.5 m s^{-1} shaded (pink), thin solid contour 0.1 m s^{-1}, values ≤ -0.01 m s^{-1} shaded (light blue). Contour intervals for surface latent heat fluxes: 50 W m^{-2}, values ≥ 150 W m^{-2} shaded. Adapted from Smith et al., 2015a. Republished with permission of the American Meteorological Society.

Indeed, these features support the ideas of boundary layer control articulated in Section 10.14. There it is argued that the ability of the slab boundary layer model to produce a radial distribution of radial, tangential and vertical motion similar in behavior to those in the time-dependent numerical simulation provides a useful measure of the degree of boundary layer control on the dynamics of vortex evolution. The boundary layer exerts a degree of thermodynamic control on the vortex evolution as well, since this layer is the source of moist air needed to support prolonged deep convection. We examine this thermodynamic support in the next subsection.

16.3.4 Thermodynamic support for deep convection

We have seen in Chapter 15 that a crucial component of vortex spin up is the ability of deep convection to ventilate air at the rate that it is converging in the boundary layer. Therefore, as a first step, it is pertinent to examine the differences in the patterns of ascent in the calculations at 10°N and 30°N and the thermodynamical support for the convectively driven ascent in the form of the corresponding surface latent heat fluxes. To this end, we show in Fig. 16.12 Hovmöller diagrams of the azimuthally-averaged vertical velocity at an altitude of 5 km for the two calculations, a quantitative mea-

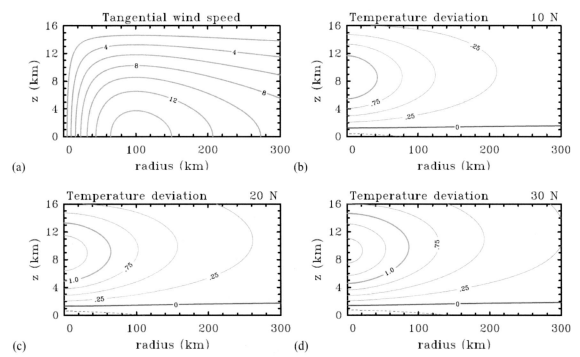

FIGURE 16.13 Radius-height cross sections of (a) the tangential wind speed in the idealized warm-cored vortex used as initial condition for the present simulations (contour interval 2 m s^{-1}), and corresponding balanced temperature perturbations at latitudes (b) 10°N, (c) 20°N, and (d) 30°N (contour interval: thick contours 1 °C, thin contours 0.25 °C). From Smith et al., 2015a. Republished with permission of the American Meteorological Society.

sure of deep convective vigor, and the azimuthally-averaged surface latent heat flux, an indication of support for the maintenance of the convection.

Consistent with the earlier intensification and stronger intensity, deep convection is seen to be more vigorous in the calculation at 10°N and it develops earlier than at 30°N. Moreover, consistent with the idea of boundary-layer control, this convection occurs at a smaller mean radius. The stronger near-surface winds in the 10°N calculation lead to larger latent heat fluxes at the ocean surface, providing support for maintaining the stronger deep convection. Smith et al. (2015a) showed that the larger latent heat fluxes do not lead to a larger near-surface moisture (see the middle panels of their Fig. 6). In this regard, it should be remembered that stronger convection will be accompanied by stronger downdrafts which will tend to decrease the boundary-layer moisture levels, especially during the early stages of vortex evolution when there is still some dry air aloft.

Attempts to relate the differences in convective strength in the two calculations throughout the entire 72 hour integrations, based on the azimuthally-averaged patterns of vertical velocity, to the patterns of $CAPE$ and CIN, based on azimuthally-averaged thermodynamic fields, were inconclusive. With hindsight, this is now not wholly surprising, bearing in mind the limitations of $CAPE$ in reference to effective buoyancy discussed in Chapter 9 and especially the fact that while the vertical velocity attained updrafts may

be related to the CAPE,[9] the vertical mass flux carried by the convection depends also on the area occupied by these updrafts. Even so, Smith et al. (2015a) pointed out that the upper-level temperature anomaly in thermal wind balance with a given tangential wind field is weaker and radially more confined at low latitudes (Fig. 16.13), an effect that should be an element in facilitating greater convective instability there.

16.3.5 Diabatically-forced overturning circulation

In the context of axisymmetric balance dynamics, it was shown in Chapter 5 that a negative radial gradient of diabatic heating rate, $\dot{\theta}$, associated with deep convection will produce inflow in the lower troposphere. This inflow is an essential feature of the classical spin-up mechanism (Section 8.2). Since the diabatic heating rate is proportional to the vertical velocity (Eq. (5.77)), the upper panels of Fig. 16.12 are suggestive that this heating rate is much larger in the simulation at 10°N compared with that at 30°N. This feature was verified by Smith et al. (2015a), who calculated the distribution of azimuthally-averaged diabatic heating

---

9. In particular, it was noted there that the reference density for a lifted air parcel is based on the (vertical) sounding through the air parcel. Moreover, because deep convection exhibits strong asymmetries during spin up as shown in earlier chapters, the azimuthally-averaged thermodynamic fields may be inappropriate for estimating effective buoyancy.

rate averaged over the period 14–16 hours in the calculations for these two latitudes. The maximum at 10°N was 26 K h^{-1}, compared with only 6 K h^{-1} at 30°N. The radial gradient of $\dot{Q}$ was found to be much larger at 10°N also. Smith et al. went on to solve the Eliassen equation (5.67) (without friction) for the streamfunction of the balanced overturning circulation based on these two heating rate distributions, confirming expectations that this circulation is much stronger in the 10°N calculation, as is the inflow in the lower troposphere. The inflow is determined from from Eq. (16.4). The maximum inflow at 10°N is 8.1 m s^{-1}, compared with only 2.5 m s^{-1} at 30°N.

It seems reasonable to presume that the large difference in diagnosed azimuthally-averaged vertical velocity at 5 km altitude during the foregoing time interval is associated, at least to a degree, with greater convective instability. Moreover, it seems likely that an element in this greater instability is the weaker upper-level temperature anomaly at lower latitudes in thermal wind balance with the tangential wind field. Other factors contributing the larger vertical velocity at 5 km would be the difference in the vertical velocity at the top of the boundary layer between the two calculations, which we have shown to be much larger at the lowest latitude, and the differing effects of rotational stiffness above the boundary layer, which at higher latitudes would suppress the inflow above the boundary layer to a larger degree and thereby, through continuity, inhibit the vertical velocity.

The above results show why the predictions of the three-dimensional simulations with explicitly-determined latent heat release differ from those of the prognostic balance simulations in Section 8.3.3. They show that the assumption of the same heating rate at different latitudes is unrealistic, even if the SST remains the same.

16.3.6 Quantifying the effects of rotational stiffness

Basic principles developed in Chapter 5 and explored further in Section 8.3 would suggest that the ability of deep convection to draw air inwards above the boundary layer would be constrained by the rotational stiffness of the vortex, as quantified by the inertial stability I_g^2. For a cyclonic vortex, this quantity increases with latitude through its dependence on the Coriolis parameter, f. At least for a fixed distribution of diabatic heating rate and for the same initial vortex, one would expect there to be a stronger inflow above the boundary layer at lower latitudes. This expectation is supported by the heating only simulations in Section 8.3.3, but, as seen in Fig. 8.13, the effect is not very large. In fact, if the vortex Rossby number is order 1 or larger, this effect would not be expected to be large enough to explain the significant difference in diabatic heating rates and associated radial gradients found above.

To isolate the effects of rotational stiffness, Smith et al. (2015a) calculated the balanced secondary circulation at 30°N, again solving the Eliassen equation (5.67) without friction, but with the diabatic heating rate at 10°N. As anticipated, there is only a small reduction in the maximum inflow velocity from 8.1 to 7.1 m s^{-1} (i.e., less than about 13%). The reduced inflow with increasing latitude will not necessarily lead to slower spin up, because the tendency of the tangential wind on account of radial influx of absolute vorticity is the product of the inflow velocity and the absolute vorticity, but the latter increases with increasing latitude. Basically, *one has to do the calculation to determine the net effect on the tangential wind tendency.*

Smith et al. (2015a) compared the isotachs of the tangential wind tendency at 10°N and 30°N for the two calculations with different heating rates and that for the calculation at 30°N with the same heating rate as at 10°N. In the first two calculations, with different heating rates and different latitudes, there is a much larger positive tendency at 10°N (5.6 m s^{-1} h^{-1} compared with 1.7 m s^{-1} h^{-1} at 30°N). However, for the same heating rate (i.e., at 10°N), the differences are small. In fact, the inviscid tendency at 30°N is actually slightly larger (6.2 m s^{-1} h^{-1}) than that at 10°N (5.6 m s^{-1} h^{-1}). Taken together, *these results demonstrate that the difference in intensification rates in the two simulations with different heating rates and latitudes cannot be attributed to the larger inertial stability at the higher latitude, leaving the difference in the heating rates and associated radial gradients as the reason. This conclusion underpins the limitations of assuming a fixed heating rate at different latitudes as in Section 8.3.3.*

16.3.7 Flow asymmetries

At this stage, the explanations of the higher rate of intensification at low latitudes have been based on axisymmetric concepts. However, as shown in Chapters 10, 11, 14, and 15, the flow evolution during the intensification phase is distinctly nonaxisymmetric, with rotating convective structures and their progressive aggregation being a dominant feature. As an illustration, Fig. 16.14 shows vertical velocity and relative vorticity at 5-km height for the 10°N experiment at 24 and 36 hours. At 24 hours, the vortex is just beginning to intensify (Fig. 16.14a), and the vertical velocity field consists of a few isolated convective towers, with nearby patches of mainly cyclonic relative vorticity. At 36 hours (Fig. 16.14b), which is beyond the middle of the rapid intensification period, the vorticity and vertical velocity have consolidated to some degree, but still exhibit significant asymmetry. The situation at 30°N is qualitatively similar (not shown), but in that case the intensification process proceeds more slowly. As a result, the vorticity structures consolidate some 18-20 hours later compared to the 10°N simulation.

During the intensification phase, the azimuthally-averaged vertical velocity, vertical vorticity, and diabatic heating

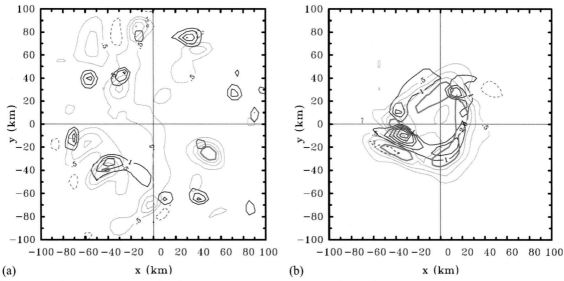

FIGURE 16.14 Horizontal cross sections of vertical velocity (black contours) and vertical relative vorticity (positive values are red (mid gray in print version); negative values are blue (dark gray in print version)) at a height of 5 km in the 10°N calculation at (a) 24 and (b) 36 hours. Contour interval for vertical velocity (black): 1 m s^{-1}; solid contours are positive and dashed contours are negative. Contour intervals for vertical vorticity: thin contours 5×10^{-4} s^{-1} from 5×10^{-4} to 1.5×10^{-3} s^{-1} and thick contours 2×10^{-3} s^{-1}. Solid (red (mid gray in print version)) contours are positive, and dashed (blue (dark gray in print version)) contours are negative. From Smith et al., 2015a. Republished with permission of the American Meteorological Society.

rate fields are dominated by local features, or "eddies". As shown in Chapter 11, eddy processes are a necessary part of a complete dynamical explanation of the intensification process, but the results of that chapter should be applicable qualitatively to the present simulations (see also Section 16.2).

16.3.8 Summary of latitudinal dependence

The analyses presented above offer a plausible explanation for the findings of numerical model simulations of tropical cyclones that the lower the latitude, the more rapid is the intensification rate and the mature intensity is larger. Comparison between the steady, slab boundary layer model and the time-dependent numerical solutions suggest that the nonlinear frictional boundary layer is a key controlling element in the latitudinal dependence of the intensification rate in the numerical simulations. In essence, boundary layer dynamics lead to stronger low-level convergence and stronger and radially more confined ascent as the latitude is decreased. At the radii where air is lofted out of the boundary layer, the convective inhibition of the initial sounding is reduced so that these radii become favorable ones for the initiation of deep convection. The stronger the ascent, the sooner deep convection will be initiated. The earlier onset of deep convection and its occurrence in a smaller band of radii are prominent features of the solution for 10°N.

In an azimuthally-averaged view of the problem, the ensuing deep convection will drive an overturning circulation with inflow in the lower troposphere. Above the boundary layer, the inflow will converge absolute angular momen-

tum leading to vortex spin up. This is the classical spin up mechanism at work as discussed in Section 8.2. The earlier development of the overturning circulation at 10°N leads to earlier intensification of the vortex, to stronger near surface winds and stronger surface moisture fluxes, which are required to maintain instability in the inner convective region.

The stronger the overturning circulation, the larger will be the diabatic heating rate and its radial gradient in the ascending branch. The more radially confined the region of heating will be an additional factor contributing to a larger radial gradient of the heating rate, assuming of course that there is adequate thermodynamic support for the heating in the form of evaporation from the ocean. The stronger overturning circulation in the simulation and its more radially-confined nature at 10°N is a prominent feature of the solutions.

The spin up of the tangential winds above the boundary layer via the classical mechanism is proportional to the strength of the low-level inflow and the radial gradient of absolute angular momentum. The inflow is larger as the latitude is decreased and the radial gradient of absolute angular momentum increases with latitude. The calculations indicate that the low level inflow is the dominant effect in the tangential wind tendency equation in producing stronger intensification on the system scale. In other words, *the difference between the intensification rates in the 10°N and 30°N simulations cannot be attributed to the larger inertial stability at the higher latitude: it is the difference in diabatic heating rates and their associated negative radial gradients that dominates in producing stronger inflow at low latitudes.*

The maximum tangential wind speed in the 10°N and 30°N simulations occurs in the surface-based inflow layer, an indication that the boundary-layer spin up enhancement mechanism is operative in both cases.

The foregoing summary reaffirms the statement made earlier that the explanation for the dependence of tropical-cyclone spin up with latitude touches on essentially all facets of the dynamics and thermodynamics of tropical cyclones. These facets comprise the classical spin-up mechanism, the boundary-layer spin up enhancement mechanism and the boundary-layer control mechanism.

16.4 The effects of sea surface temperature on intensification

As discussed in Section 16.3, observational data relating to the effects of latitude on tropical cyclone intensification are likely to be contaminated by other effects such as SST and vertical wind shear. In this section, we review the findings of a study by Črnivec et al. (2015), which sought to extend the findings discussed in the last section to isolate and understand the dependence of the tropical cyclone intensification rate on the SST at different latitudes.

The Črnivec et al. calculations focused on the same prototype problem described above, but used the CM1 numerical model as in the majority of calculations in Chapters 10, 11, 14, and 15. Again the warm-rain scheme was used to represent deep convection explicitly, but with a horizontal grid spacing of 3 km instead of 5 km used by Smith et al. (2015a). Other model details are to be found in Section 2 of Črnivec et al.

Fig. 16.15 compares time series of maximum azimuthally-averaged[10] tangential wind speed, V_{max} and intensification rate, IR, in a series of calculations for one of three different latitudes (10°N, 20°N, 30°N), combined with one of three different SSTs (26 °C, 28 °C, 30 °C). The intensification rate at time t is defined as the 24 hour change in V_{max}, i.e., $IR = V_{max}(t + 24h) - V_{max}(t)$. The black solid horizontal line in each right panel indicates the threshold value of 15 m s^{-1} day^{-1}. Values of IR exceeding this threshold are used to characterize RI defined in Footnote 2 of Chapter 10.

Figs. 16.15a and b show the time series of V_{max} and IR in the calculations with the lowest SST (26 °C). It is seen that the intensification is more rapid when the latitude decreases. At 72 hours, V_{max} is approximately 37 m s^{-1} for the vortex at 10°N, whereas at 20°N and 30°N it is much less, with values of $\sim$ 27 and $\sim$ 20 m s^{-1}, respectively. Only the vortex at 10°N has an intensification rate exceeding the criterion for RI. This RI period begins at about 28 hours and lasts approximately 20 hours. The intensification rates

at 20°N and 30°N reach maxima of 11 and 5 m s^{-1} day^{-1}, respectively.

Figs. 16.15c-f show the corresponding time series for SSTs of 28 °C and 30 °C. At these SSTs, all vortices undergo a period of RI and, as for the lowest SST, IR decreases with increasing latitude. Moreover, the time at which RI begins occurs later. At either SST, the vortices at 28 °C and 30 °C reach approximately the same intensity after 72 hours and this intensity increases with SST (about 50 m s^{-1} at 28 °C and 70 m s^{-1} at 30 °C). These results confirm and extend those discussed in Section 16.3, where an explanation for the latitudinal dependence of vortex evolution was provided. In particular, the results show that the differences in the evolution of V_{max} at different latitudes for a given SST become less as the SST is increased.

Next, we examine the dependence of vortex intensification on SST at a fixed latitude. Although the information on this dependence is contained in Fig. 16.15, it is insightful to reorder the curves to highlight the dependence on SST, as in Fig. 16.16. In contrast to the dependence on latitude at a fixed SST, it is seen that there is a strong SST dependence of intensification rate and the time at which RI commences at all latitudes. As might be expected, the intensification is more rapid as the SST is increased (see Figs. 16.16a, c, and e). Moreover, the onset of RI occurs earlier (when it occurs at all) and the duration of RI is longer as the SST is increased. For this reason, the final intensity after 72 hours is greater for larger SSTs, although the difference is less at 10°N than at higher latitudes (possibly because the vortices at 10°N have achieved a greater degree of maturity). In the next section, we explore the reasons for the foregoing behavior.

16.4.1 Interpretation of the SST dependence

In the previous section, we quantified the dependence of IR on SST at different latitudes compared with the dependence of IR on latitude at a fixed SST. We showed that IR is substantially larger, and that intensification begins sooner, for larger values of SST. Here we examine why there is this strong dependence of intensification rate on SST.

16.4.1.1 Near-surface moisture and θ_e

In seeking an explanation, a starting point might be to examine the differences in the azimuthally-averaged surface latent heat fluxes as a function of radius and time. This is because the saturation mixing ratio increases exponentially with temperature, implying a potentially strong dependence of surface evaporation on the SST and possibly a greater disequilibrium in moisture at the sea surface. Note that here we are not invoking the WISHE feedback mechanism discussed in Section 12.2 and the modification discussed in Section 16.1. Even so, an elevation of the latent heat flux does not automatically lead to an increase in low-level

---

10. The center location for calculating the azimuthal average is as defined in Section 16.3.2.

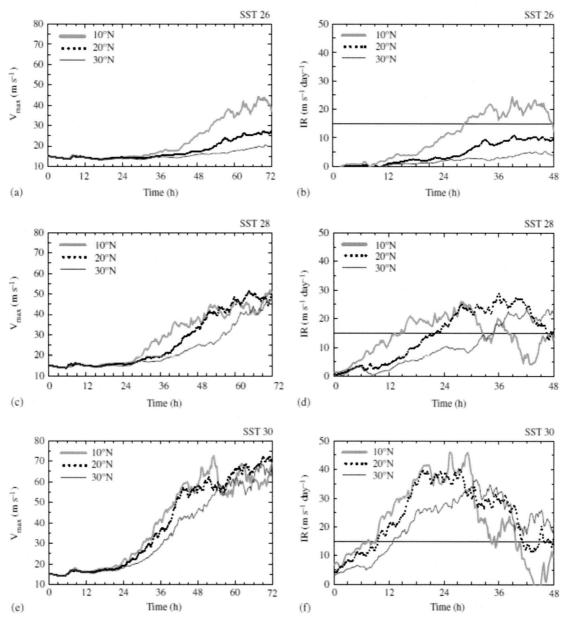

FIGURE 16.15 Time series of (a, c, e) maximum azimuthally-averaged tangential wind speed and (b, d, f) intensification rate in the nine CM1 calculations at three different latitudes (10°N - thick solid red (mid gray in print version) curve, 20°N - dotted black curve, 30°N - thin solid blue (dark gray in print version) curve), combined with three different SSTs: (a, b) 26 °C, (c, d) 28 °C, and (e, f) 30 °C. Note the difference in time period on the abscissa between the left and right panels. From Črnivec et al., 2015. Republished with permission of Wiley.

moisture and thereby an increase in low-level θ_e, because this elevation may be mitigated by the effects of frictionally induced subsidence of dry air and precipitation cooled downdrafts into the boundary layer (see e.g., Section 9.3.4). For this reason, it might be better to examine azimuthally-averaged distribution of low-level θ_e, itself.

As it turns out, at a particular latitude, the amount of moisture is still appreciably larger in the calculations with higher SSTs on account of the larger latent heat flux (not shown). These higher values of low-level moisture are reflected in larger values of θ_e, as indicated in Fig. 16.17.

This figure compares Hovmöller diagrams of azimuthally-averaged θ_e at a height of 1.1 km for all nine experiments. The most striking feature of this comparison is that inner-core values of θ_e at this level increase with increasing SST, but another feature is that, at a fixed SST, the radial extent of larger values of θ_e increases with latitude.

16.4.1.2 Diabatic heating rate and secondary circulation

It is shown in Section 5.11, that the value of r_v^* and hence θ_e as air exits the boundary layer is one factor influencing the

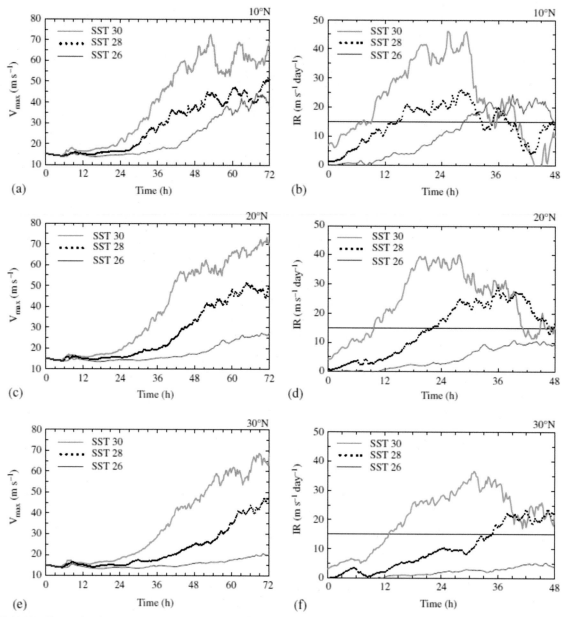

FIGURE 16.16 Time series of (a, c, e) maximum azimuthally-averaged tangential wind speed and (b, d, f) intensification rate in the nine CM1 calculations with three different SSTs (26 °C - thin solid blue (dark gray in print version) curve, 28 °C - dotted black curve, 30 °C - thick solid red (mid gray in print version) curve) combined with three different latitudes: (a, b) 10°N, (c, d) 20°N, and (e, f) 30°N. Note the difference in time period on the abscissa between the left and right panels. From Črnivec et al., 2015. Republished with permission of Wiley.

material diabatic heating rate of ascending air parcels, the other being the radial distribution of vertical velocity. A detailed radius-time distribution of the azimuthally-averaged diabatic heating rate, $\langle \dot{\theta} \rangle$, in the middle troposphere (at a height of 6 km) in the three simulations at latitude 20°N is shown in Črnivec et al. (2015) and this figure indicates a strong dependence of $\langle \dot{\theta} \rangle$ on SST. For the purposes of this subsection it suffices to show that radius-height distribution of the diabatic heating rate around the time of maximum IR for three different SSTs.

Črnivec et al. (2015) show that the magnitude of $\langle \dot{\theta} \rangle$ and its radial gradient, $\partial \langle \dot{\theta} \rangle / \partial r$, are largest when the SST is 30 °C and weaken considerably as the SST is decreased. In accordance with the classical axisymmetric balance theory for intensification above the boundary layer (Section 8.2), a stronger radial gradient of $\langle \dot{\theta} \rangle$ (a measure of the convective forcing in the balance theory for vortex intensification) leads to a stronger secondary circulation. In particular, a negative radial gradient of the diabatic heating rate leads to radial inflow in the lower troposphere at radii outside

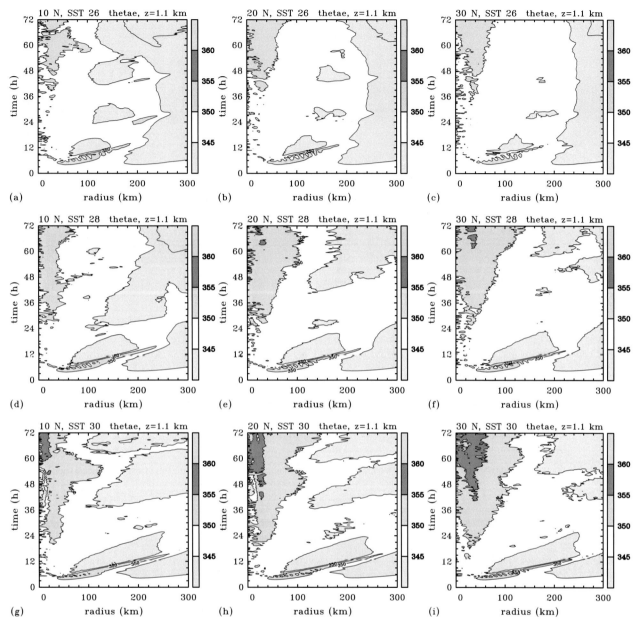

FIGURE 16.17 Radius-time plots of azimuthally-averaged equivalent potential temperature, $\langle \theta_e \rangle$, at a height of 1.1 km in the nine CM1 calculations at (a, d, g) 10°N, (b, e, h), 20°N and (c, f, i) 30°N for SSTs of (a–c) 26 °C, (d–f) 28 °C, and (g–i) 30 °C. The contour interval is 5 K. Shading is as indicated in the color bar. Adapted from Črnivec et al., 2015. Republished with permission of Wiley.

the axis of maximum heating rate (Section 5.10.5). This inflow draws absolute angular momentum surfaces inwards above the boundary layer and, because $\langle M \rangle$ is approximately materially conserved above the boundary layer, the inflow leads to a spin-up of tangential wind speed there.

To quantify the foregoing effect, we diagnose the spatial distribution of the time average of $\langle \dot{\theta} \rangle$ during a 3 hour time interval surrounding the maximum value of IR for each of the three experiments (here the maximum IR is judged from a centered difference from $t - 12$ hours to $t + 12$ hours). The radius-height distributions of this average are shown in

Figs. 16.18a-c for the three calculations at 20°N with different SSTs. Again it is clear that there is a large difference in the mean heating rates between the calculations. The maximum values of $\langle \dot{\theta} \rangle$ reached in the calculations with SSTs of 26 °C, 28 °C, and 30 °C are 11 K h^{-1}, 17 K h^{-1}, and 22 K h^{-1}, respectively. The radial gradient of diabatic heating rate is considerably larger and its maximum is located closer to the storm axis in the calculation with the highest SST (30 °C), compared with other two calculations. Moreover, the vertical extent of high $\langle \dot{\theta} \rangle$ values (e.g., those exceed-

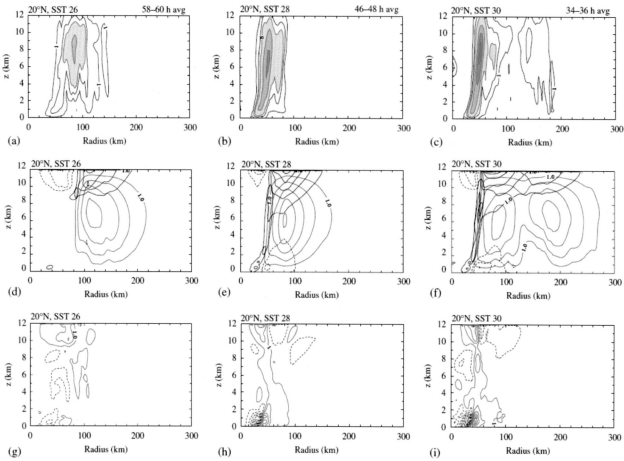

FIGURE 16.18 Radius-height plots of the time-averaged diabatic heating rate, based on azimuthally-averaged fields from the CM1 calculations at 20°N for SST values of (a) 26 °C, (b) 28 °C, and (c) 30 °C. The contour interval for blue (dark gray in print version) contours is 2 K h^{-1} starting with the contour 1 K h^{-1}. The contour interval for red (mid gray in print version) contours is 5 K h^{-1}, values between 5 and 10 K h^{-1} are shaded light pink (light mid gray in print version), values between 10 and 15 K h^{-1} are shaded light red (mid gray in print version) and values greater than 15 K h^{-1} are shaded darker red (dark mid gray in print version). Panels (d), (e), and (f) show the corresponding streamfunction of the balanced secondary circulation obtained by solving the Eliassen equation (5.67) with these heating rates as forcing terms (red (mid gray in print version) contours, contour interval 1 × 10^{-8} kg s^{-1}). These panels show also the contours of radial wind speed (contour interval 1 m s^{-1}, positive values are solid blue (dark gray in print version), negative values are dashed blue (dark gray in print version)) and vertical velocity (black contours, contour interval 0.5 m s^{-1}). Panels (g), (h), and (i) show the corresponding tendencies of the balanced tangential wind (contour interval 2 m s^{-1} h^{-1}, positive values are solid red (mid gray in print version), negative values are dashed blue (dark gray in print version)). Adapted from Črnivec et al., 2015. Republished with permission of Wiley.

ing 10 K h^{-1}) increases as the SST increases, leading to a deeper secondary circulation.

Figs. 16.18d,e,f show isotachs of radial and vertical velocity in the balanced secondary circulation obtained by solving the frictionless form of the Eliassen equation (5.67) for the corresponding vortex and with the corresponding heating rate distribution from panels (a), (b), and (c) as forcing. For simplicity, we neglect both frictional forcing and the 'eddy terms' that would be present in this case. The streamfunction contours of the overturning circulation are shown in this figure also. Confirming expectations, the balanced secondary circulation is strongest in the calculation with the highest SST (30 °C) and weakest in the calculation with the SST of 26 °C. In particular, the strongest vertical motion amounts to 1.1 m s^{-1} at 30 °C, while it is only

0.5 m s^{-1} at 26 °C. Similarly, the maximum low-level inflow velocity is 3.0 m s^{-1} for an SST of 30 °C, compared with 0.99 m s^{-1} in the calculation with the lowest SST (26 °C).

Figs. 16.18g,h,i show isotachs of the tangential wind tendency in the three calculations. As might be expected, the largest positive low-level (below 2 km) tendency of the tangential wind is found in the calculation with the SST of 30 °C and amounts to 13.2 m s^{-1} h^{-1}. For an SST of 28 °C, it is approximately 9.0 m s^{-1} h^{-1}, while at 26 °C it is only 1.6 m s^{-1} h^{-1}. Collectively, the panels in Fig. 16.18 affirm the applicability of the classical spin-up mechanism for broadly interpreting the dependence of the spin up on the SST.

16.4.1.3 Dependence of ambient thermodynamic profile on SST

As pointed out by Črnivec et al. (2015), there is a potential issue in carrying out experiments at a fixed latitude when varying the SST, but keeping the ambient sounding the same. In reality, the sounding itself would be expected to have some dependence on the SST. This issue is investigated at length in Section 5.1 of Črnivec et al. and shown to be a small effect for the short 3-day model integration time which captures the entire spin up of each vortex: see especially Fig. 9b therein.

16.4.2 Summary of SST effects

We have examined the effects of changing the SST on the intensification rate of tropical cyclones at different latitudes. The study complements and extends the results of Section 16.3 that investigates only the latitudinal dependence. For a given SST, the results confirm the findings of Section 16.3 that intensification begins earlier and the intensification rate increases as the latitude is decreased. However, they show important quantitative differences in the latitudinal dependence of intensification rate as the SST is changed. The dependence on latitude is largest when the SST is marginal for tropical cyclone development (26 °C) and becomes smaller as the SST is increased. For example, at 20°N or 30°N latitude, the intensification rate is more strongly dependent on the SST than at 10°N. Furthermore, at a fixed latitude, vortex intensification begins earlier and is more rapid as the SST is increased. An explanation for this behavior is offered in terms of the classical axisymmetric mechanism for tropical cyclone spin-up (Section 8.2).

In brief, an increase in the SST is accompanied by a significant increase in the surface water vapor fluxes and corresponding increases in the low-level moisture and equivalent potential temperature. Where boundary-layer air is lofted into the vortex above, these increases lead to a larger negative radial gradient of diabatic heating in the low to middle troposphere and thereby to a stronger diabatically-forced overturning circulation. The stronger inflow draws absolute angular momentum surfaces inwards at a higher rate, leading to faster spin-up.

The foregoing explanations are based on an azimuthally averaged perspective, but the flow exhibits significant departures from axial symmetry, especially during the rapid intensification phase. The non-axisymmetric features, which are a prominent feature of real storms also, are discussed in Chapter 11 as well as in Sections 16.2 and 16.3.7.

16.5 The effects of initial vortex size on genesis and intensification

Observations show that tropical cyclones have a wide range of sizes, as measured, for example, by the radius of near-surface gale force winds or the radius of the outermost closed isobar at the surface (Merrill, 1984; Weatherford and Gray, 1988a; Liu and Chan, 1999; Kimball and Mulekar, 2004; Rudeva and Gulev, 2007; Yuan et al., 2007; Chavas and Emanuel, 2010; Lu et al., 2011; Chan and Chan, 2012). At one end of the size spectrum are the so-called "midget storms", in which even the extent of gale-force winds is no more than 100 km from the storm center (an example is Tropical Cyclone Tracy that devastated the Australian city of Darwin on Christmas Day in 1974, in which the radius of gales was a mere 50 km). At the other end of the spectrum are storms such as Typhoon Tip (1979), which had gales as far as 1100 km from the center (Fig. 16.19). Observations have shown also that there is little correlation between the storm size and its intensity as measured by the maximum near-surface wind speed or the minimum surface pressure (Merrill, 1984; Weatherford and Gray, 1988a; Chavas and Emanuel, 2010).

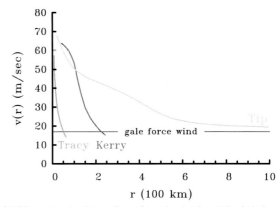

FIGURE 16.19 Radial profiles of low-level azimuthal winds in Australian region Tropical Cyclones Tracy (1974) and Kerry (1979), and Supertyphoon Tip (1979). Adapted from Holland and Merrill, 1984. Republished with permission of Wiley.

The extent of gale force winds is an important component of a tropical cyclone forecast as it provides an indication of the area over which damaging winds may be expected. Typically, the radius of gales will vary in different sectors of the storms and forecasts try to convey such structural information.

There have been numerous theoretical attempts to account for the size differences between storms, including the size changes that may occur during the lifetime of an individual storm (Rotunno and Emanuel, 1987; DeMaria and Pickle, 1988; Xu and Wang, 2010; Smith et al., 2011; Chan and Chan, 2014; Frisius, 2015; Kilroy et al., 2016a; Tsuji et al., 2016). Some of these studies have pointed to a monotonic relationship between the final size of storms and the size of the initial disturbance from which they form. Most of the explanations proffered for size change focus on angular momentum budgets and do not quantify the collective effects of deep convection in drawing these angular momen-

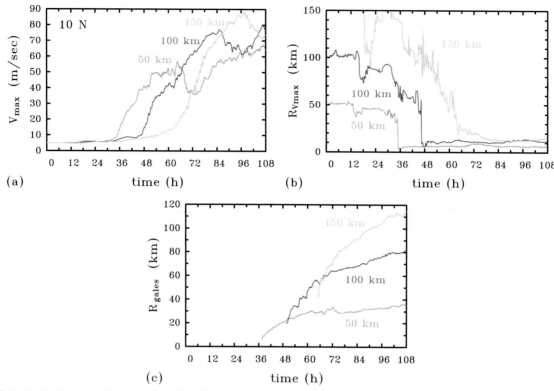

FIGURE 16.20 Evolution metrics in the three simulations E1, E2, and E3 defined in the text. Time series of (a) maximum azimuthally averaged tangential wind speed (V_{max}), (b) radius R_{Vmax} at which V_{max} occurs, and (c) radius at which gale force winds (R_{gales}) occur. Here, R_{gales} is calculated at a height of 1 km and corresponds to the outer radius at which the tangential wind speed is 17 m s^{-1}. Adapted from Kilroy and Smith, 2017. Republished with permission of Wiley.

tum surfaces inwards. Kilroy and Smith (2017) suggested that:

"robust explanations for controls on size require consideration of the controls on deep convection that drive the inflow to produce size changes. They require also a quantification of the convection in terms of, say, the diabatic heating rate and its spatial gradients and/or the convective mass flux."

They noted also that, important controls on deep convection are rooted in the dynamics and thermodynamics of the boundary layer.

The most basic processes controlling vortex size may be understood in the context of an axisymmetric vortex as discussed in Section 8.3. In the present section we review briefly a study by Kilroy and Smith (2017), which extended previous numerical modeling studies (e.g., Chan and Chan, 2014) investigating the dependence of storm size on that of the initial disturbance to include the genesis process. The study sought also to provide an explanation for storm size evolution in the context of the rotating-convection paradigm.

16.5.1 Numerical experiments on vortex size

Kilroy and Smith (2017) repeated the warm-rain experiment of Chapter 10 with the initial radius of maximum tangential wind speed moved to 50 km or 150 km, retaining the 500 m horizontal grid spacing. The calculations were carried out on an f-plane with the Coriolis parameter $f = 2.53 \times 10^{-5}$ s^{-1}, corresponding to a latitude of 10°N. It was shown that vortex evolution in these new experiments is similar to that in the warm-rain experiment, with a progressive organization of convectively-induced, cyclonic relative vorticity into a monopole structure. The results are summarized in Fig. 16.20, which compares the evolution of V_{max}, the radius at which this occurs, R_{Vmax}, and the radius of gale force winds, R_{gales}, for all three experiments designated E1 (initial $R_{Vmax} = 50$ km), E2 (initial $R_{Vmax} = 100$ km), and E3 (initial $R_{Vmax} = 150$ km). As the initial vortex size is increased, the organization of relative vorticity occurs later. In addition, the sizes of the vorticity monopole, the sizes of the inner and outer core tangential wind circulations and the lifetime intensity of the vortex all become larger. Some of these features are illustrated in Figs. 16.21 and 16.22.

Fig. 16.21 shows horizontal cross sections of relative vertical vorticity, horizontal velocity vectors and the 1 m s^{-1} contour of vertical velocity at altitudes of 2 and 6 km

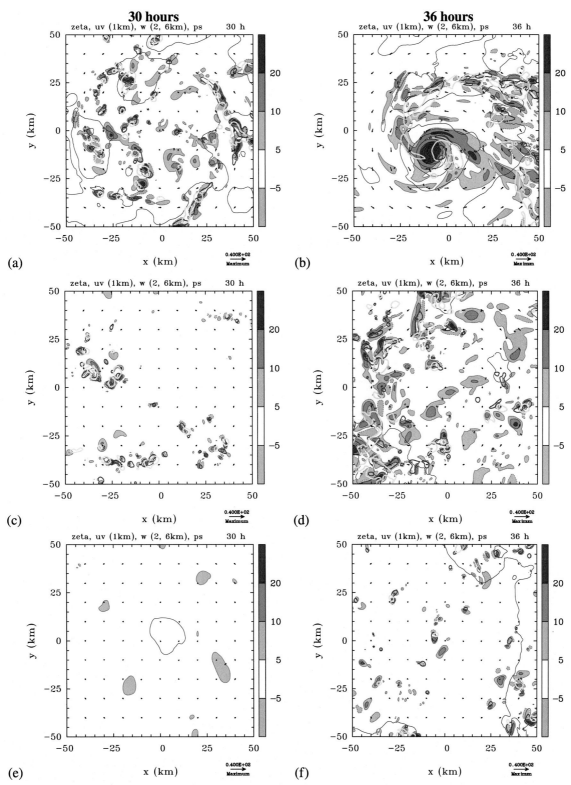

FIGURE 16.21 Horizontal cross sections of relative vertical vorticity, ζ, and wind vectors, **u** at various times for experiments E1 (top row), E2 (middle row), and E3 (bottom row) at 30 hours and 36 hours. Shown also are contours of vertical velocity, w, at heights of 2 km (blue) and 6 km (yellow), and black contours of surface pressure, contoured every 0.5 mb. Values for the shading of vertical vorticity are given in the color bar, multiplied by 10^{-4} s^{-1}. Vertical velocity contour is 1 m s^{-1} for both heights. The wind vectors are in relation to the maximum reference vector (15 m s^{-1}) at the bottom right, while on the bottom left the maximum total wind speed in the domain plotted is given in m s^{-1}. From Kilroy and Smith (2017) (the colors of the vertical velocity contours in this paper were labeled incorrectly, being the reverse of those shown here). Republished with permission of Wiley.

48 hours   **72 hours**

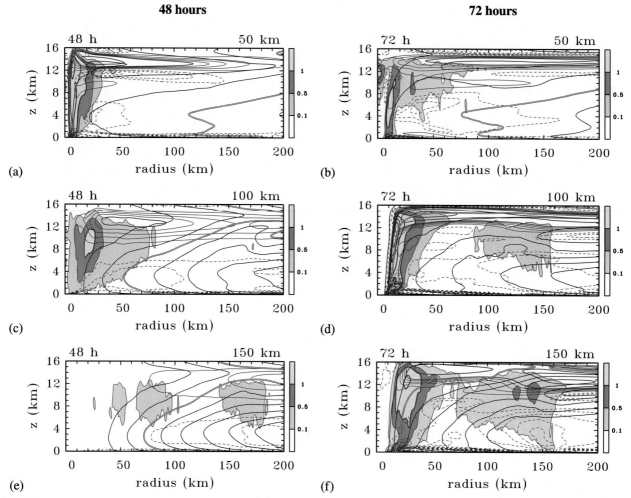

FIGURE 16.22 Vertical cross-sections of the azimuthally averaged, 3 h time-averaged radial (contoured) and vertical (shaded and contoured) velocity components centered at (a, c, e) 48 hours and (b, d, f) 72 hours. Panels (a, b) initial $R_{Vmax} = 50$ m s^{-1}, (c,d) initial $R_{Vmax} = 100$ m s^{-1} and (e, f) initial $R_{Vmax} = 150$ m s^{-1}. Contour intervals are as follows. Radial velocity: blue contours - solid thick, interval 5 m s^{-1}, thin, interval 1 m s^{-1} to 4 m s^{-1}, negative thick, interval 5 m s^{-1}, thin, interval 0.5 m s^{-1} from -0.5 m s^{-1} to -4 m s^{-1}. Vertical velocity: red contours - solid thick, interval 1 m s^{-1}, thin 0.1 and 0.5 m s^{-1}, negative thin dashed contour -0.2 m s^{-1}. Shading as indicated on the side bar. Superimposed on each panel are contours of absolute angular momentum: black contours every 2×10^5 m^2 s^{-1}, with the 2×10^5 m^2 s^{-1} contour highlighted in thick yellow and black. Zero contours of all quantities are not plotted. From Kilroy and Smith, 2017. Republished with permission of Wiley.

at early times for the three simulations at 30 hours and 36 hours. Similar cross sections at additional times are shown in Kilroy and Smith. At 30 hours, deep convection is already present in the inner core region in E1 (panel (a)) and there are many patches of enhanced cyclonic vertical vorticity within 50 km of the circulation center. As discussed in Section 10.3.2, the enhanced vorticity is generated by convective clouds that stretch and tilt the background vorticity. Tilting of horizontal vorticity generates vorticity dipoles which are evident, for example near $(-30, 27.5$ km), (40, 30 km), and $(-5, -35$ km). The vertical vorticity produced near the circulation center is mostly cyclonic on account of the lack of vertical wind shear (and, thus, a lack of horizontal vorticity) (Kilroy and Smith, 2016).

At the same time in E2, there is much less deep convection (panel (c)) and what there is tends to be further from the vortex center. As a result, there are fewer patches of enhanced cyclonic vorticity present. In E3 there is no active convection at 30 hours, although there are some remnants of enhanced cyclonic vertical vorticity originating from prior deep convection (panel 3(e)). As shown by Montgomery et al. (2006b) and Wissmeier and Smith (2011), the vorticity generated by deep convection generally outlives the convection, itself.

The vortex in E1 begins to spin up around 33 hours (Fig. 16.20a) and by 36 hours a clear monopole of strong cyclonic vertical vorticity (values exceeding 2×10^{-3} s^{-1}) has emerged near the circulation center (Fig. 16.21b). There is some deep convection near the vorticity monopole at this time and there are two spiral bands of deep convection emanating from the eastern and western sides of the vortex core. The western band contains patches of both cyclonic and anticyclonic vertical vorticity. At the same time in E2 there has been a noticeable increase in the number of deep convective cells present and there are now some patches of cyclonic vorticity located close to the circulation center (Fig. 16.21d). In E3 at 36 hours, there are a few sparsely separated deep convective cells and a few patches of convectively-induced cyclonic vorticity exceeding 1×10^{-3} s^{-1} in magnitude (Fig. 16.21f).

The subsequent evolution in each experiment is similar to that described in Section 10.3.2, but, as seen in Fig. 16.20a, the larger the initial vortex, the more prolonged is the gestation period before RI commences.

The differences in vortex structure as time proceeds are highlighted by vertical cross sections of the azimuthally-averaged, 3 hour time-averaged, radial velocity $\langle u \rangle$, vertical velocity $\langle w \rangle$, and absolute angular momentum $\langle M \rangle$ at 48 and 72 hours in the three experiments (Fig. 16.22). The time averaging is centered on the time indicated and is based on data output every 15 minutes. Again, similar cross sections are shown in Kilroy and Smith (2017) for earlier times.

At 48 hours, the vortex in E1 has many features of a mature tropical cyclone, including an outflow region located just above the strong near-surface inflow layer. There is still weak inflow in the low to mid-troposphere and, as expected, the $\langle M \rangle$ surfaces show a prominent inward-pointing nose in this region. The vertical velocity field is compact with significant ($\langle w \rangle > 0.5$ m s^{-1}) mean ascent confined inside a radius of about 30 km.

At this time in E2, the near surface inflow is again a prominent feature and the overturning circulation shows inflow in the lower troposphere and outflow above. Even at this early time, there is a layer of marked inflow above and below the upper tropospheric outflow layer. Further, the vertical velocity field is broader than in E1 with significant mean ascent extending now to 60 km radius. In contrast, in E3, the radial and vertical velocity fields are still relatively weak and the most prominent features of the overturning circulation lie beyond a radius of 80 km. In both E2 and E3, the $\langle M \rangle$ surfaces have developed a nose-like feature in the lower troposphere as in E1, and, of course, the radial gradient of $\langle M \rangle$ is an increasing function of the initial vortex size.

By 72 hours, there is a mature cyclone in all three experiments. In E1, the radial inflow in much of the lower troposphere has strengthened, but the updraft has contracted somewhat and weakened as reflected also in a weakening of the upper-level outflow and the inflow layer below the outflow layer. In E2, the updraft has contracted, but strengthened considerably as has the overturning circulation. In E3, the strongest inflow now occurs near the surface at a radius of about 20 km, although there is still relatively strong

(> 2.5 m s^{-1}) mid-level inflow centered at a radius of 150 km and a height of about 3 km, stronger indeed than in E2 at that radius. This inflow contributes to spinning up the broader outer circulation. In contrast, in E1 at this time, the radial inflow is relatively weak at all heights beyond a radius of 150 km.

Notably, the vertical velocity is stronger and covers a broader range of radii as the initial vortex size is increased and the inflow in the low to mid-troposphere beyond a radius of 100 km is stronger also. The stronger inflow in combination with the larger radial gradient of $\langle M \rangle$ leads to a faster spin up of the tangential winds in the outer part of the vortex.

The foregoing behavior is consistent with the findings of Fudeyasu and Wang (2011) and is as one would expect. When deep convection is located at large radii, it will induce inflow on its outer flank leading to an inward displacement of the $\langle M \rangle$ surfaces and thereby a spin up of the tangential winds above the frictional boundary layer, where $\langle M \rangle$ is approximately materially conserved. If there is little or no outer convection, the inflow at large radii will be weaker and the spin up there will be correspondingly weaker.

Kilroy and Smith offered an explanation for the differences in behavior described above in terms of a boundary-layer control of the overturning circulation discussed in Section 10.14. In essence, a simple slab boundary-layer model like that described in Section 6.5 shows that, the smaller the initial vortex, the stronger is the ascent at the top of the boundary layer and this ascent is located closer to the circulation center. The results of the slab boundary-layer calculations are supported by a comparison of time-radius diagrams of vertical velocity near the top of the boundary layer from the three experiments (see Kilroy and Smith, 2017, Fig. 7). Even though the ascent out of the boundary layer is relatively weak in all cases before deep convection begins (on the order of a few mm s^{-1}), it is sufficient to provide a location with the lowest CIN, where deep convection can focus and amplify the vertical vorticity locally. The stronger the ascent, the more rapid is the removal of the CIN and the sooner is the onset and subsequent organization of deep convection. The earlier the onset of convection explains also the earlier the onset of rapid intensification.

When the initial vortex is broader, so is the radial extent of deep convection and the outer convection leads to stronger inflow at large radii. The stronger inflow, in combination with the larger absolute vorticity there, leads to a more rapid spin-up of the tangential winds in the outer part of the vortex, but not necessarily in the inner-core region where V_{max} is located. *This latter fact would explain the observation in Fig. 1.15 that there is not a strong relationship between intensity and size.*

The increase in lifetime intensity of the vortex with increasing initial vortex size may be explained, in part, by the more extensive reservoir of absolute vorticity for the inner-core convection to concentrate. However, there are many factors at play affecting V_{max}, including those that affect the strength of the inner-core convection. These factors are explored in some detail by Smith et al. (2011) and Kilroy and Smith (2017).

16.5.2 Synthesis

The foregoing results serve to underpin a frequently cited observation that, when environmental conditions are favorable, small storms tend to spin up much earlier than larger ones. Importantly, the timing of RI is sensitive to the initial vortex size. Moreover, while observations do not show a clear relationship between intensity and size, the idealized experiments do suggest that tropical cyclones which develop within a large initial circulation have the capacity to become stronger than those originating from small initial disturbances. These results may be of interest to forecasters, although we are conscious of the fact that the idealized experiments neglect the effects of vertical wind shear, which may limit their immediate applicability to real-world situations that forecasters have to contend with.

16.6 Tropical cyclogenesis at and near the Equator

It was thought for a long time that tropical cyclones do not form within about 5 deg. of the Equator on the grounds that the Earth's background rotation is insufficient at such low latitudes to support vortex spin up. This thought was proved to be incorrect by the occurrence of Typhoon Vamei, which was first classified as a tropical depression at about 1.5°N latitude on 26 Dec. 2001 over the South China Sea and made landfall one day later along the southeastern coastline of the Malaysian Peninsula. However, the question remains, from where do such storms acquire their rotation?

The underlying dynamics of the formation of tropical cyclones within a few degrees latitude of the Equator was investigated by Steenkamp et al. (2019) using European Centre for Medium Range Weather Forecasts (ECMWF) analyses of some prominent cyclogenesis events, including Typhoon Vamei. The possibility of formation at the Equator was demonstrated also using idealized model simulations, similar to the warm-rain experiment in Chapter 10, but with the Coriolis parameter set to zero.

In the real events investigated by Steenkamp et al., vortex formation occurred within a broad-scale counterclockwise flow that encompasses a region of predominantly positive absolute vertical vorticity extending typically more than 5 deg. south of the Equator. Patches of enhanced vertical vorticity form within this region as a result of vorticity stretching by deep convection. These vorticity patches are organized by the convection, the collective effects of which are to produce an overturning circulation that fluxes vortic-

ity at low levels towards some center within the convective region. The process of spin up is much as that described for the warm-rain experiments north of the Equator in Chapter 10.

As an illustration, Fig. 16.23 shows the evolution of Typhoon Vamei from a vorticity perspective, exemplified by selected fields from the ECMWF analyses. These fields include the absolute vorticity distribution, the geopotential height field and the horizontal wind vectors at 850 mb. Within 5 degrees of the Equator, values of f are relatively small, less than 1.3×10^{-5} s^{-1}, so that the absolute vorticity, ζ_a, is dominated by relative vorticity, ζ. In some localized regions, ζ can exceed 5×10^{-4} s^{-1} in magnitude, long before any named tropical depression has formed (not shown). Of course, as the latitude increases, the contribution of f to ζ_a becomes progressively more important.

An increase in the strength of the northeasterly winds from 00 UTC 23 Dec. 2001 is accompanied by the development of a low-level cyclonic (counter-clockwise) circulation northwest of Borneo (not shown). This cyclonic circulation is indicative of the development of the so-called "Borneo vortex" and is seen at 18 UTC 24 Dec. in Fig. 16.23a at around 108°E, 3°N. The low develops just north of the Equator in an extensive region of positive absolute vorticity that extends well into the Southern Hemisphere at this time.

A prominent feature of the absolute vorticity in Fig. 16.23a is a comma shaped region of enhanced positive ζ_a ($> 1 \times 10^{-4}$ s^{-1}), starting near the Equator and curving round to the northeast. This region is one also of low geopotential height. At the Equator, itself, positive values of ζ_a extend from approximately 104°E to 110°E and the winds are northerly west of about 108°E and southerly to the east of this longitude, at least as far as 113°E. Clearly, there is a horizontal flux of positive absolute vorticity into the Southern Hemisphere between 104°E and 108°E and back from the Southern Hemisphere into the Northern Hemisphere between 108°E and 110°E. At longitudes west of 104°E, there is a flux of negative absolute vorticity from the Northern Hemisphere to the Southern Hemisphere and a similar return flux east of about 110°E.

During the 24 hour period to 18 UTC 25 Dec. (Fig. 16.23b) the comma shaped region of positive ζ_a has consolidated into a monopole of enhanced vorticity centered at about 108°E, 2°N with a comma-shaped tail and the geopotential heights have fallen in the region of the monopole. Positive absolute vorticity continues to extend several degrees into the Southern Hemisphere.

By 12 UTC 26 Dec. (Fig. 16.23c), the vorticity has continued to consolidate into a monopole. The geopotential field has now a sharp minimum within, but slightly to the southwest of the monopole center. The winds have increased in strength and the counter-clockwise circulation has become focused around the vorticity monopole, extending well into the Southern Hemisphere. However, the exten-

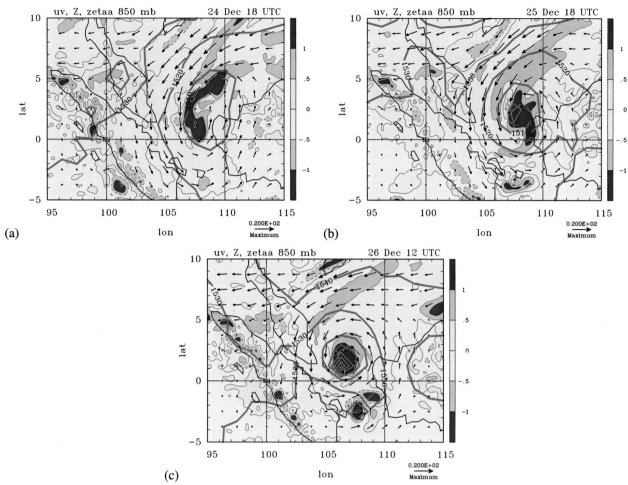

FIGURE 16.23 Horizontal cross sections of wind vectors and geopotential height at 850 mb together with 850 mb absolute vorticity, ζ_a during the genesis and evolution of Typhoon Vamei (2001) in the ECMWF analyses at the times shown. Scale for wind vectors in m s^{-1} at bottom right. Geopotential height in geopotential meters (gpm) (solid green contours, interval 10 gpm). (The geopotential of an air parcel at height h is defined as $\Phi = \int_0^h g(z, \lambda) dz$, where $g(z, \lambda)$ is the acceleration due to gravity, z is the geometric height above mean sea level and λ is the latitude. The geopotential height is defined as Φ/g^*, where $g^* = 9.80665$ m s^{-2} is the standard gravity at mean sea level. In the present context, it is sufficient to think of geopotential height and geometric height as synonymous.) The red shading indicates counterclockwise vorticity (e.g., cyclonic in the Northern Hemisphere, anticyclonic in the Southern Hemisphere) and blue shading clockwise vorticity (shading values multiplied by 1×10^{-4} s^{-1}) From Steenkamp et al. (2019). Republished with permission of Wiley.

sive region of positive absolute vorticity that was south of the Equator has now retreated to the Northern Hemisphere.

In this case as well as the others investigated by Steenkamp et al. there was an apparently sufficient environment of low-level counterclockwise vorticity to enable a vortex to be amplified and there was sustained "deep convection" within this environment to help concentrate the vorticity. This convection is parameterized in the ECMWF modeling system and its collective effects may be judged by regions of enhanced ascent at the 500 mb level, about 5.5 km high. It may have been fortuitous that all four vortices studied were counterclockwise since there seems no reason why vortex growth in a region spanning the Equator of clockwise absolute vorticity of Southern Hemisphere origin could not occur on occasion.

16.6.1 An idealized numerical study

Kilroy et al. (2020) examined further the idealized model simulations referred to above that are similar to the warmrain experiment in Chapter 10, but with the Coriolis parameter set to zero. Setting the Coriolis parameter to zero will answer the curiosity-driven question whether the formation process described in the prior section can still occur without the direct influence of the Earth's rotation.[11] Three simulations were carried out in which the maximum tangential wind speed (5 m s^{-1}) is specified at an initial radius of 50, 100, or 150 km, essentially repeating the calculations of Section 16.5 with $f = 0$. Fig. 16.24 summarizes the metrics

---

11. At least one other study has suggested an affirmative answer to this question (Montgomery et al. (2006b), their Expt. E1 and accompanying summary on page 365).

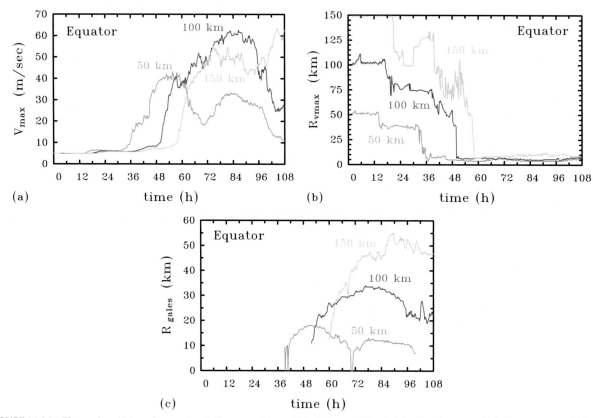

FIGURE 16.24 Time series of (a) maximum azimuthally averaged tangential wind speed (V_{max}), (b) radius R_{vmax} at which V_{max} occurs, and (c) radius at which gale force winds (R_{gales}) occur in the three idealized numerical simulations at the Equator with initial vortices having maximum tangential wind radii of 50 km (red (mid gray in print version) curves), 100 km (blue (dark gray in print version) curves), and 150 km (aqua curves). Here, R_{gales} is calculated at a height of 1 km and corresponds to the outer radius of 17 m s^{-1} tangential wind outside at that level. These should be compared with the corresponding panels of Fig. 16.20, which are for analogous simulations at 10°N. From Kilroy et al., 2020. Republished with permission of Wiley.

of vortex evolution used for these simulations and compares them with those for the simulations at 10°N shown in Section 16.5.

After a period of gestation lasting between 30 hours and 60 hours, all the vortices intensify rapidly, the evolution being similar to that for vortices away from the Equator. In particular, the larger the initial vortex size, the longer the gestation period, the larger the maximum intensity attained, and the longer the vortex life time. Beyond a few days, however, the vortices decay as the cyclonic vorticity source provided by the initial vortex is depleted and negative vorticity surrounding the vortex core is drawn inwards by the convectively driven overturning circulation.

The vortex evolution during the mature and decay phases differs from that in simulations at 10°N, where inertially-unstable regions are much more limited in area. In these negative vorticity regions, the flow is inertially/centrifugally unstable (Chapter 5, Sections 5.7.1, 5.7.3). Vortex decay in the $f = 0$ simulations appears to be related intimately to the development of inertial instability, which is accompanied by an outward-propagating band of deep con-

vection. The degree to which this band of deep convection is realistic is unknown.

16.6.2 Synthesis

The foregoing findings support the hypothesis that tropical cyclogenesis on the Equator, where $f = 0$, is possible, in principle, provided there is a sufficient reservoir of relative vorticity to concentrate. Vortex development continues for some time until the vorticity reservoir is exhausted and a decay phase ensues, dominated by centrifugal instability.

16.7 Observational tests of the rotating-convection paradigm

Field experiments have offered the opportunity to test various aspects of the rotating-convection paradigm. In particular, experimental data collected using airborne Doppler radar, dropsondes and aircraft flight level data provide the ability to interrogate both the classical and boundary-layer spin up enhancement mechanisms that comprise essential elements of the rotating-convection paradigm summarized in Section 11.7. The 2008 field experiment, TCS08, and the 2010 experiments GRIP and IFEX discussed in Section 1.5.3 collected a wealth of such data. Here, we highlight a few of the prominent tests of the rotating-convection paradigm that emerged from these experiments in regards the dynamics and thermodynamics of the tropical cyclone spin up process.

The first observational support of the boundary-layer spin up enhancement mechanism for an intensifying tropical cyclone was presented by Sanger et al. (2014), who examined *inter alia* the azimuthally averaged boundary-layer structure during the intensification of Typhoon Jangmi, which was observed as part of the TCS08 experiment. That study documented also rotating convective structures that comprised a sample of the latent heat forcing. The aggregate effect of this forcing is responsible, in part, for the overturning circulation required to sustain the spin up process as envisaged in the classical spin up mechanism. Details of both aspects are presented in Sanger et al. and for brevity we summarize here only some of the key boundary layer findings.

Fig. 16.25 shows radial distributions of the aircraft dropsonde data at various levels in the inner-core boundary layer in a cylindrical coordinate system that moves with the center of the storm. The left column depicts the observed pressure data at the indicated levels alongside simple polynomial fits to these data during the tropical storm, typhoon and supertyphoon stages of Jangmi. Since the pressure field in an intense vortex is close to axisymmetric, the observed pressure data in and around the high wind region may be assumed to have an approximate axisymmetric distribution. The right column depicts the observed, storm-relative, tangential velocity (v) alongside the gradient wind velocity (v_g) deduced from solving the quadratic gradient wind equation, given the inferred pressure gradient obtained from the polynomial fit shown in the left column and the observed air density. The second curve in panels (b) and (f) represent a second estimate of the derived gradient wind using the extrapolated surface pressure from the high density aircraft flight level data (HDOB) available on the Air Force C-130 Hurricane Hunter Aircraft (see Sanger et al. for further details).

The findings demonstrate the presence of significant supergradient winds *within the boundary layer and near and within the radius of maximum tangential wind* in all three stages of the storm evolution. Referring to Fig. 16.25, the authors concluded that (p26):

> *"The largest supergradient wind speeds occurred near and just inside the radius of maximum tangential wind speed during the supertyphoon stage of Jangmi, though the highest supergradient winds relative to the local gradient wind occurred in the tropical storm phase".*

The study corroborates the predictions of the rotating-convection paradigm that unbalanced boundary layer dynamics in the inner-core region are an important component in determining the maximum axisymmetric radial (not shown here) and tangential flow at all times during the evolution of the storm.

An even more detailed data set for testing the rotating-convection paradigm was obtained in Hurricane Earl (2010) during four days of intensive measurements based on airborne Doppler radar and dropsonde data released from the upper and lower troposphere during the collaborative GRIP-IFEX experiments. A few of the key findings from Montgomery et al. (2014) pertaining to Earl's intensification and evolution are summarized here.

One of the unique findings from the study was the first observational confirmation of the progressive inward movement of absolute angular momentum surfaces during Earl's intensification (see their Fig. 5). This result, while to be expected based on the physical arguments presented in this book, is significant because it demonstrates unequivocally the operation of the classical spin up mechanism articulated in Chapter 8.

An analysis similar to that of Sanger et al. was carried out for the azimuthally averaged boundary-layer structure during the intensification, secondary eyewall formation and re-intensification stages of Earl (see their Fig. 11). The findings corroborate the finding of Sanger et al. that

> *"significant supergradient winds are observed within the boundary layer and near and within the radius of maximum tangential wind during storm spin up and mature evolution".*

In particular,

> *"during spin up and maturity, the maximum tangential winds occur persistently within the layer of strong boundary layer inflow (< 1 km depth)".*

The quantitative analyses showed that

> *"the average maximum tangential winds beneath the eyewall exceed the gradient wind by between 20 and 60%, with the largest excess occurring during the re-intensification period following the eyewall replacement".*

An analysis of the possible departures from gradient wind balance at 2 km altitude indicated also that

> *"the gradient balance approximation in the low-level vortex interior above the strong inflow layer may not be as accurate as has been widely held in the inner-core region of a tropical cyclone during its intensification"*

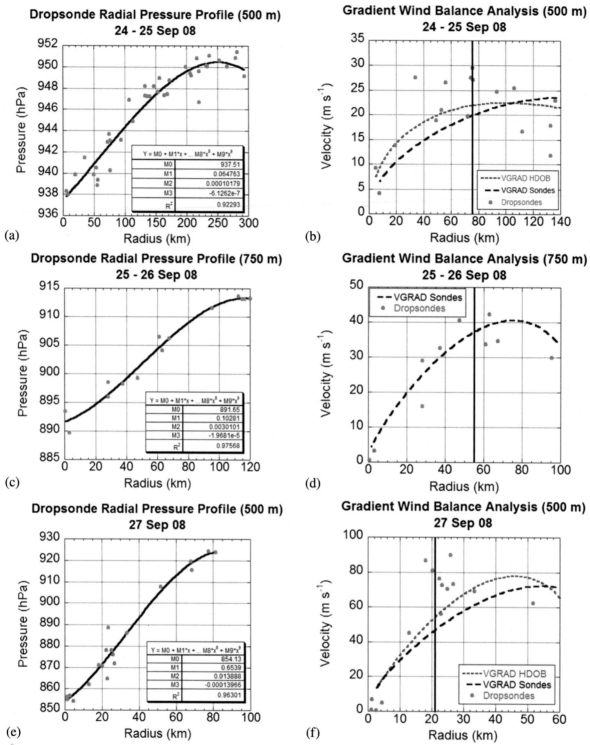

FIGURE 16.25 Observed pressure (mb) from aircraft dropsondes (solid red (mid gray in print version) circles) and third-degree polynomial fit (solid blue (dark gray in print version) line) at (a) 500 m for Tropical Storm Jangmi, (c) 750 m for Typhoon Jangmi, and (e) 500 m for Supertyphoon Jangmi. Inset table shows curve fit coefficients and r^2 values. Tangential wind (ms^{-1}) at (b) 500 m for Tropical Storm Jangmi, (d) 750 m for Typhoon Jangmi, and (f) 500 m for Supertyphoon Jangmi from aircraft dropsondes (solid red (mid gray in print version) circles) and gradient wind from aircraft High Density Observation (HDOB) data (short-dashed green (gray in print version) curve) and dropsonde data (long-dashed blue (dark gray in print version) line). The solid black vertical lines in (b), (d), (f) denote the mean radius of maximum tangential wind speed. From Sanger et al. (2014). Republished with permission of the American Meteorological Society.

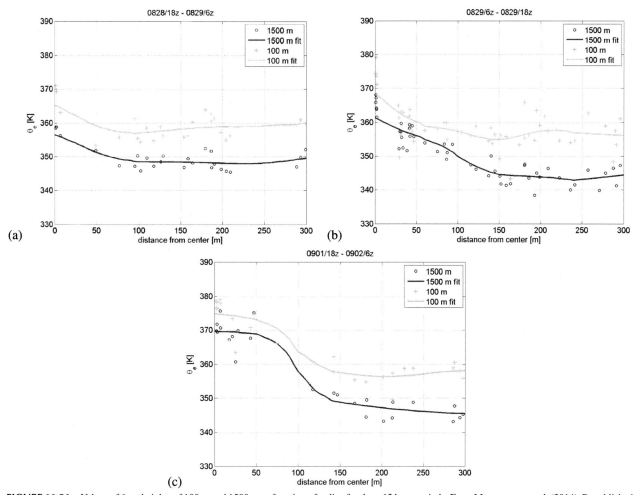

FIGURE 16.26 Values of θ_e at heights of 100 m and 1500 m as function of radius for three 12 hour periods. From Montgomery et al. (2014). Republished with permission of Wiley.

(cf. Section 11.5). Finally, the dropsonde composites show that

"the maximum radial inflow is very close to the sea surface, which is consistent with fluid dynamical consideration for a rapidly rotating vortex adjacent to a frictional boundary layer"

as discussed in Chapter 6.

We conclude this section by summarizing a particularly interesting and surprising (at least to us) thermodynamic finding from the Earl study that would appear to have deep implications on the dynamics of tropical cyclone intensification in the real world. Fig. 16.26 shows values of θ_e at heights of 100 m and 1500 m as a function of radius for three 12 hour periods during hurricane Earl's spin up and mature evolution. We see that the axisymmetric θ_e structure at 1.5 km altitude is consistently less than the corresponding near-surface value at all radii, even where the air is ascending into the eyewall. Quantitatively speaking, in the innermost 150

km, the maximum difference is approximately 10 K, while the minimum is about 5 K.

These observations appear to be the first to suggest that the air ascending into the eyewall has significantly lower values of θ_e than those near the surface. This discovery is significant because it is not consistent with the axisymmetric eruption of the boundary layer into the eyewall unless there are non-conservative (eddy) processes acting to modify the entropy of ascending air. It was pointed out to us by D. Raymond (personal communication) that

"the expected high value of θ_e at 1.5 km may be concentrated in isolated updrafts that were missed by the dropsondes, whereas these θ_e values are more spread out at the surface."

As noted by Montgomery et al.:

"If the pilots were deviating around high-reflectivity areas as they penetrated the eyewall, this would almost certainly be the case. These latter consider-

ations implicate an important role of localized up-drafts and associated eddy processes in the eyewall region during the intensification of a tropical cyclone".

The foregoing findings independently corroborate the importance of understanding the role of the vortical convection and eddy dynamics during tropical cyclone spin up and evolution as summarized in Section 16.2.

The fact that the azimuthally averaged boundary-layer θ_e is not well-mixed for several hundred kilometers from the center of the vortex is contrary to the well-mixed assumption for θ_e invoked in axisymmetric theoretical formulations of the hurricane boundary layer presented in Chapters 12 and 13. The Montgomery et al. study offered an interpretation of these findings based on an analogy with the "shear sheltering" concept that has been proffered to explain "anti-mixing" in strongly sheared boundary-layer flows and also, in part, eddy transport barriers. Those authors hypothesized that the strong vertical shear of the tangential and radial winds in the vortex boundary layer plays an important role in limiting vertical mixing of θ_e in the boundary layer across the broad scale of the hurricane vortex. The ramifications of these thermodynamic findings and future tests of the "shear-sheltering" concept would seem to be fruitful avenues of future research at both basic and practical levels.

16.8 Tropical lows over land

A prominent feature of the Australian "wet season" (November-April) is a trough of low pressure that lies south of the Equator. At the beginning of the wet season, this monsoon trough lies to the north of the Australian continent, but later it moves southwards, sometimes lying over the continent. It is common for tropical lows to develop within the trough, sometimes two or more at the same time at spatial intervals along the trough. *We use the term "tropical low" to describe a non-frontal, warm-core vortex with cyclonic flow at the surface and with embedded moderate or deep moist convection to distinguish it from a dry heat low that is a common feature of the flow over the continent.* Lows over land are sometimes referred to locally as monsoon lows or monsoon depressions, or even "landphoons".

Under suitable conditions,[12] when the trough lies over ocean, the lows may act as seeds for tropical cyclone development, but when the trough lies close to the coast or over the continent, lows may still intensify into significant vortical rain producing systems, even while moving inland and polewards. If the inland lows move over the sea, they often develop rapidly into tropical cyclones (e.g., Tory et al., 2007; Smith et al., 2015b). Other inland lows originate

from landfalling tropical cyclones, which normally weaken after landfall as the supply of latent heat from the ocean surface is cut off and the surface friction increases. Whatever their origin, tropical lows over land are a major source of monsoonal rainfall and frequently lead to major flooding. An early landmark study of tropical cyclogenesis during the Australian monsoon is that of McBride and Keenan (1982).

Even when the lows intensify over land, they do so where there is relatively sparse conventional data coverage and, until the turn of the 21st century, little was known about their mechanism of formation or their structure. Two early case-studies of lows that initially formed over land are those of Foster and Lyons (1984) and Davidson and Holland (1987). There have been a few more studies more recently, including cases of storms re-intensifying over land after initially weakening (Emanuel et al., 2008; Arndt et al., 2009; Brennan et al., 2009; Evans et al., 2011) and storms that formed over land (Smith et al., 2015b; Kilroy et al., 2017b, 2016b; Tang et al., 2016; Zhu and Smith, 2020).

Inland lows pose significant challenges to forecasters, who need to anticipate whether or not an incipient low will develop and how rapidly it might develop. While answering this question would inevitably require the use of guidance from numerical forecast systems, an improved understanding of the basic mechanisms would be surely useful for forecasters. *An important question is to what extent the formation, structure and intensification of these lows over land are similar to those of tropical cyclones over sea?*

Emanuel et al. (2008) hypothesized that inland intensification of lows over the northern part of the Australian continent is made possible by the special type of sandy soils that are found there. These soils are argued to become hot prior to rain and are such that their thermal conductivity is greatly enhanced after being moistened by the first rains of an approaching storm so that they are able to supply sufficient surface enthalpy fluxes to support storm intensification. The ideas were demonstrated in simulations using a simple, axisymmetric, coupled soil-atmosphere model applied to tropical cyclone Abigail (2001) which made landfall from the Gulf of Carpentaria and appeared to undergo two cycles of re-intensification as it progressed over land. However, the question remains as to whether this special soil type is crucial to the intensification of real storms.

In a related study, Evans et al. (2011) carried out numerical simulations of Northern Hemisphere tropical cyclone Erin (2007), which made landfall from the Gulf of Mexico and re-intensified over the central United States. Their analysis showed, inter alia, that the along-track rainfall feedback mechanism proposed by Emanuel et al. was

"of minimal importance to the evolution of the vortex".

Instead, they found that

---

12. SSTs in the monsoon region to the north of Australia can be especially favorable for cyclogenesis, exceeding 30 deg C, but vertical wind shear associated with the monsoon may be detrimental.

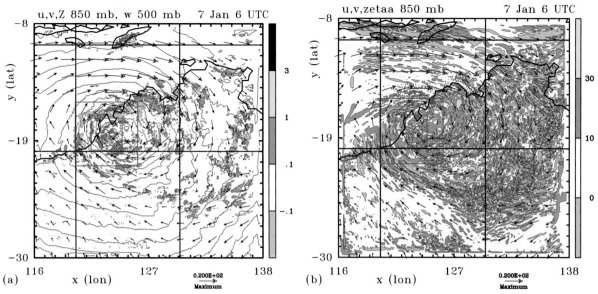

FIGURE 16.27   (a) Vertical velocity fields at 500 mb (color shaded) with wind fields and geopotential fields at 850 mb superimposed and (b) minus the absolute vorticity (color shaded) and wind fields at 850 mb from the 4 km grid UM forecast at 0600 UTC 7 Jan. Contour intervals: for vertical velocity as indicated on the color bar in m s^{-1}; for absolute vorticity as indicated on the color bar $\times 10^{-5}$ s^{-1}, for geopotential height, Z: 10 gm. Adapted from Zhu and Smith (2020). Republished with permission of Wiley.

"the final intensity of the simulated (and presumably observed) vortex appears to be closely linked to the maintenance of boundary-layer moisture over pre-existing near-climatological soil moisture content along the track of the vortex and well above climatological soil moisture content".

They argued that

"variations in soil moisture content result in impacts upon the boundary-layer thermodynamic environment via boundary-layer mixing. Greater soil moisture content results in weaker mixing, a shallower boundary layer, and greater moisture and instability. Differences in the intensity of convection that develops and its accompanying latent heat release aloft result in greater warm core development and surface vortex intensification within the simulations featuring greater soil moisture content".

More recently, a series of case-studies examined the formation and evolution of inland lows seen in ECMWF analyses (Smith et al., 2015b; Kilroy et al., 2016b, 2017b). The upshot of these studies suggested that the lows are warm-cored, convectively driven systems, similar in structure to tropical cyclones. The fact that inland intensification occurs in the ECMWF analysis and forecast system and in the MM5 model used by Tang et al. (2016), neither of which represent the special soil types hypothesized to be necessary by Emanuel et al., supports the finding of Evans et al.

that the along-track rainfall feedback mechanism proposed by Emanuel et al. is not of major importance in real storms.

16.8.1   A tropical low case study

One drawback of the ECMWF analyses for investigating the detailed evolution of lows is the relatively coarse temporal resolution as the analyses are available only every six hours. Another limitation is the fact that the ECMWF forecasts that contribute most to the analyses in data-sparse regions use a parameterization of deep convection. For this reason, Zhu and Smith (2020) carried out a simulation of an intensifying low that moved inland over Western Australia using a convection-permitting version of the United Kingdom Unified Model (UM) with a 4 km horizontal grid spacing, aspects of which were compared with the corresponding ECMWF analyses. Here we summarize briefly the main findings from that study. The reader is referred to the published paper for details of the simulation and a more complete description of the low and its evolution. The aim here is show a few basic features that highlight the relevance of the rotating-convection paradigm to understanding such events.

Three figures set the scene. Fig. 16.27 shows the vertical velocity fields at 500 mb and absolute vorticity fields at 850 mb at 0600 UTC 7 Jan. 2015 with the geopotential isopleths and wind vectors at 850 mb superimposed in each panel. Fig. 16.28 shows the track of the low beginning one day earlier and ending at 0000 UTC 11 Jan. and Fig. 16.21 compares the intensity as characterized by the minimum geopotential at 850 mb, Z_{min}, between the UM forecast, the ECMWF analyses and a UM forecast in which the surface latent heat flux is set to zero over land. The latter forecast will be touched on later.

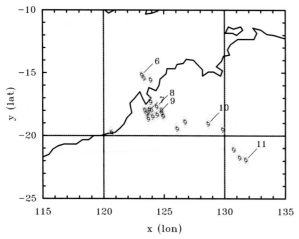

FIGURE 16.28 Locations of minimum surface pressure every six hours denoted by a cyclone symbol in the simulation starting at 00 UTC 6 Jan. The locations at 00 UTC are indicated by the day of the month. From Zhu and Smith (2020). Republished with permission of Wiley.

The fields in Fig. 16.27 are reminiscent in character to those in the idealized simulations of Chapters 10 (Figs. 10.4, 10.12, 10.15) and 15 (Fig. 15.3). In essence, deep convective updrafts are localized in scale and rather randomly distributed and the absolute vorticity field has much fine-scale structure reflecting the nature of its local amplification by convection. The close coupling between the vorticity and convection is underscored by the near overlap between the enhanced absolute vorticity and the vertical motion fields. These fine-scale structures are much less apparent in the wind and geopotential fields.

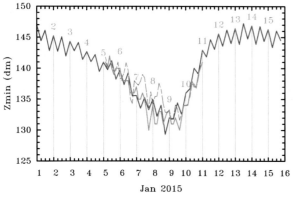

FIGURE 16.29 Comparison of intensities as characterized by the minimum geopotential at 850 mb, Z_{min}, between the ECMWF analyses (blue (dark gray in print version)), the UM forecast with 4 km grid spacing (red (mid gray in print version)), and a UM forecast in which the surface latent heat flux is set to zero over land (yellow/green (light gray/gray in print version) dashed). From Zhu and Smith (2020). Republished with permission of Wiley.

As seen in Figs. 16.28 and 16.29, the low continued to deepen as it moved inland, reaching its maximum intensity around 9 Jan., with excellent agreement between Z_{min} in the UM simulation and the ECMWF analyses.

16.8.1.1 Dynamical aspects of spin up

As we have seen in Chapters 8 and 10, the key element of tropical cyclone spin up is the concentration of absolute vorticity by the convectively-induced low-level convergence above the frictional boundary layer, the classical mechanism discussed in Section 8.2. This concentration of absolute vorticity leads to an increase in the low-level circulation of the vortex. To highlight this process in the present case study, we show in Fig. 16.30 the evolution of two bulk properties of the low characterizing the strength and location of the convection and the changes in circulation. Specifically, the figure shows time-height cross-sections of system-averaged vertical mass flux within latitude-longitude columns 3° latitude × 3° longitude and 6° latitude × 6° longitude in cross-section, centered on the location of the minimum wind speed at 850 mb as well as the corresponding cross-sections of normalized circulation (circulation divided by the length of the circuit) around these columns. The latter calculation quantifies the average tangential wind around the closed circuit.

In the smaller area column (Fig. 16.30a), there are regular "bursts" of deep upward mass flux until about 0600 UTC on 9 Jan., a time at which the storm was near its maximum intensity (Fig. 16.29). These bursts have a tendency to occur in the afternoon or evening (0300 UTC-1200 UTC) as would be expected over land. At most later times, the bursts do not extend through a deep layer.

In the larger area column (Fig. 16.30b), the strongest bursts occur in the 12 hour period starting 1800 UTC 7 Jan. and mass flux values tend to be larger than in the smaller area column, at least until about 1200 UTC 8 Jan., indicating that much of the deep convection in the low lies outside the inner 3° latitude × 3° longitude area.

Figs. 16.30c and 16.30d show the evolution of the normalized circulation around the two columns with time and height. The normalized circulation around the innermost column shows a sharp increase through a deep layer starting a little before 18 UTC 7 Jan., but the next prominent and sustained increase occurs about 18 hours later, one that is confined mostly to heights below the 700 mb level. In other words, the low is a "bottom up" development as opposed to "top down", a finding similar to that in the cases described by (Kilroy et al., 2016b, Fig. 9f) and (Kilroy et al., 2017b, Figs. 5f, 10f, and 14f). Even so, the enhanced cyclonic circulation extends through most of the troposphere during the intensification and mature stages.

The normalized circulation around the 6° × 6° column shows a more prolonged enhancement, especially at altitudes below the 600 mb isobar, with a significant increase

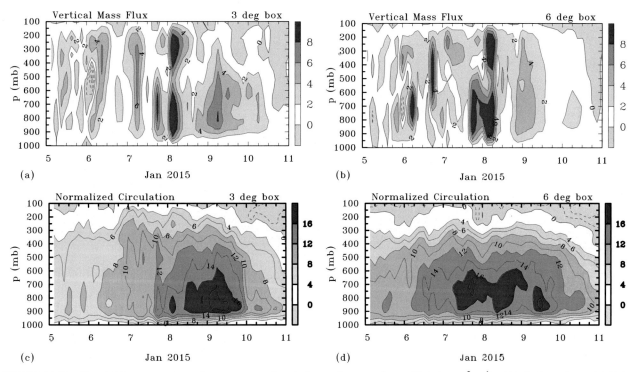

FIGURE 16.30 Time-height cross-sections of system-averaged vertical mass flux per unit area (Unit: kg m^{-2} s^{-1}) within lat-lon columns (a) 3° lat × 3° lon, and (b) 6° lat × 6° lon, centered on the location of the minimum wind speed at 850 mb (Contour interval 2 units as indicated on the color bar). Panels (c) and (d) show the corresponding time series of the magnitude of system-averaged normalized circulation (m s^{-1}) around these columns. Contour interval 2 units as indicated on the color bar. From Zhu and Smith (2020). Republished with permission of Wiley.

beginning about 0900 UTC 7 Jan., about a day earlier than in the 3° × 3° column. The cyclonic circulation around the 6° × 6° column does not extend quite so high as in the 3° × 3° column. This feature would be expected and is consistent with the idea that as air parcels move outwards at upper levels, the Coriolis force they experience leads to an anticyclonic circulation beyond a certain radius.

Further insight into the role of convectively-induced low-level convergence in leading to an increase in the low-level vortex circulation is provided by the Hovmöller (radius-time) plots of the azimuthally-averaged radial and tangential velocity components relative to the moving vortex center at 850 mb shown in Fig. 16.31.

In general, tangential winds exceeding 20 m s^{-1} do not occur until 3 UTC 8 Jan., about 9 hours after the minimum geopotential (cf. Fig. 16.29). These values are located beyond a radius of 150 km and persist until late on 9 Jan. The maximum tangential velocity (23.6 m s^{-1}) occurs at 15 UTC 9 Jan. at a radius of 177 km. Thus, unlike a typical tropical cyclone in the region, *the low did not have a concentrated core of high rotational winds* within, say, 100 km of the center. There were significant fluctuations in the radius, R_{Vmax}, of maximum tangential wind speed, V_{max}, as illustrated by the yellow (light gray in print version) line in panel (a). In broad terms, there is a broad correspondence between the "bursts" of inflow in panel (b) and increases

in tangential wind speed in panel (a), bearing in mind that an increase in tangential wind speed is an integrated effect of the inflow over time (and thus such tangential wind maxima follow radial inflow maxima chronologically). This correspondence is a reflection of the classical intensification mechanism. In turn, the bursts of inflow are a consequence of the system-scale pulsing of deep convection, which, over land, has a marked diurnal component.

Zhu and Smith (2020) showed a more detailed azimuthally-averaged analysis of fields to cement the interpretation that the spin up of the low is a consequence of the classical spin up mechanism. This analysis included Hovmöller diagrams of the advective and non-advective vorticity fluxes (cf. Section 11.1) to the vorticity tendency at 850 mb and their contribution to the total vorticity tendency.

16.8.1.2 Thermodynamical aspects of spin up

Through much of its intensification stage, the low center was over land, albeit not too far inland, no more than about 300 km at its most intense stage. However, it was embedded in a large-scale monsoonal circulation with abundant moisture supply to the north and west from the warm ocean, where sea surface temperatures were in excess of 29 °C.

The question that immediately arises is whether the ocean moisture source was sufficient to support the intensification process over land, or whether soils, moistened by

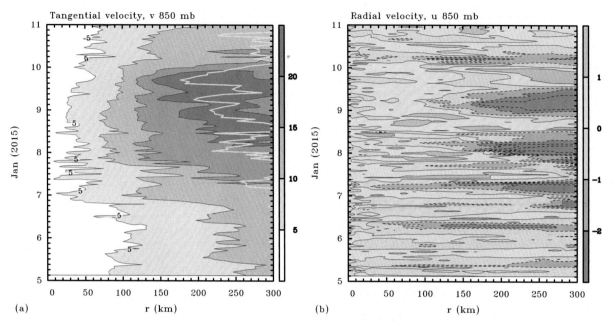

FIGURE 16.31 Hovmöller radius-time cross-sections of the azimuthally-averaged (a) tangential, and (b) radial velocity components at 850 mb from the UM forecasts. Contour intervals as indicated on the color bar in m s^{-1}. From Zhu and Smith (2020). Republished with permission of Wiley.

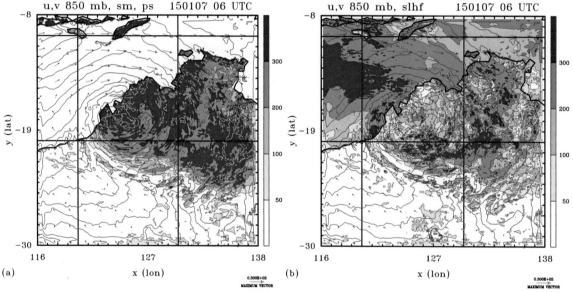

FIGURE 16.32 Distribution of (a) soil moisture (shaded) and (b) surface latent flux (shaded) at 0600 UTC on 7 Jan. while the low was still strengthening. Both panels show also the surface pressure and surface wind vectors. Unit for soil moisture: kg m^{-3} as indicated on the color bar. Contour interval for surface pressure 1 mb, wind vector scale in m s^{-1} indicated by the arrow below the figure. Unit for heat flux: W m^{-2} as indicated on the color bar. From Zhu and Smith (2020). Republished with permission of Wiley.

previous monsoonal rainfall events, were an important factor also? To set the scene for answering this question, we show in Fig. 16.32a the distribution of soil moisture at 0600 UTC 7 Jan. At this time, the soil within 1-2° latitude of the circulation center was relatively moist with large patches having values exceeding 300 kg m^{-3}. These moist soils extend over much of the region shown, to the east of the low and to about 21°S, while soils beyond a few degrees to the

south and south west of the circulation center are comparatively dry. The region of moist soils had extended a little southwards since the start of the UM simulation and, as would be expected, the moisture has increased a little in the vicinity of the low. On the other hand, the soils to the south and southwest have become drier.

To examine the potential influence of soil moisture on the intensification of the low, Fig. 16.32b shows the spatial distribution of surface latent flux at the same time, when the low in the model was still strengthening. It turns out that the latent heat fluxes have a strong diurnal variation, being a maximum during the daytime (e.g., at 0600 UTC, 2 pm local time in Western Australia) and falling to zero over night (e.g., at 1800 UTC, 2 am local time). Typical values are in the range 100-200 W m^{-2}, comparable with values found by Kilroy et al. (2016b), Fig. 10, and Kilroy et al. (2017b), p774. Typical values over the ocean during the monsoon are somewhat higher (from 200 W m^{-2} to more than 400 W m^{-2} in the case examined by Kilroy et al. (2016b), their Fig. 10).

To examine further the impact of soil moisture on the intensification of the low over land, Zhu and Smith (2020) carried out an additional simulation in which the surface latent heat flux is set to zero over land. The intensification curve for this simulation is included in Fig. 16.29. Not surprisingly, the intensity is lower in this simulation than in the control simulation, but only slightly and intensification still occurs. *Thus, there is sufficient moisture available from the sea in the suppressed moisture flux simulation to sustain deep convection and thereby the convectively-driven overturning circulation.* However, as shown by Zhu and Smith (see their Fig. 16b), the center of the suppressed moisture flux simulation remains a little closer to the coast until after 10 Jan. and at early times the center is frequently over the sea.

It should be noted also that the foregoing behavior is different from that postulated by Emanuel et al. (2008) and that found by Evans et al. (2011), presumably due to the proximity of the relatively warm sea in our case.

16.8.2   Synthesis

The dynamics of low intensification in the UM simulation is shown to be consistent with that in the classical paradigm for tropical cyclone intensification outlined in Chapter 8, supporting the interpretations of earlier studies that were based on coarser resolution ECMWF analyses. In particular, the UM simulation indicates that high soil moisture or special soil types are not essential to vortex intensification over land, at least in a monsoonal environment, which can provide adequate moisture to sustain an overturning circulation driven by deep convection.

16.9   Polar lows, medicanes and tropical cyclones

Other types of lows have been suggested to have some characteristics of tropical cyclones. These include certain low pressure systems that develop over the Mediterranean Sea, sometimes referred to as "medicanes", and some that develop over the polar oceans, so-called "polar lows". Both types are characterized by deep convection near their center

and both types sometimes develop cloud-free eye-like features near their center. Up-to-date reviews of medicanes are given by Michaelides et al. (2018) and Pytharoulis (2018), while polar lows are the subject of a popular scientific article by Businger (1991) and a book by Rasmussen and Turner (2003). Medicanes are referred to in Section 16.1.

Medicanes have a variety of forms, but strong smallscale vortices may form near the center of an existing low following an outbreak of deep convection near the center. The same kind of development can occur in the case of polar lows with deep convection being triggered by moist instability arising from the destabilization of an extremely cold air mass as it flows over a much warmer sea. In both cases, airsea interaction involving enhanced surface enthalpy fluxes has been proposed as a major feature of these types of lows (Emanuel, 2005a; Emanuel and Rotunno, 1989). An alternative explanation of polar low formation that is not restricted to axisymmetric dynamics was proposed by Montgomery and Farrell (1992). Their work hypothesized and demonstrated plausibly that polar low formation could be captured by a minimal moist Eady model in which the surface enthalpy fluxes act only to support deep moist convection in a baroclinic environment that is otherwise nearly neutral along absolute momentum surfaces.

In a recent paper, Kilroy et al. (2022) carried out a case study of a medicane that occurred in the eastern Mediterranean in December 2020 and showed that many aspects of this case could be interpreted in terms of the rotatingconvection paradigm. It is possible that this paradigm might be applicable also for interpreting the formation of at least some polar lows.

16.10   The rotating-convection paradigm in the research of others

A stringent test of a new theory is whether it can explain observed or simulated features not previously understood or predict and verify altogether new properties not previously anticipated. In this section, we draw attention to an independent study that validates many key features of the rotating-convection paradigm and go on to demonstrate the versatility of this paradigm by offering explanations or constructive suggestions of some interesting findings documented recently by other authors, but not yet explained under one overarching framework.

16.10.1   An idealized numerical study

Many aspects of the rotating-convection paradigm have been independently confirmed in a comprehensive numerical study by Wang et al. (2019). In brief, in their simulation of the genesis and intensification of a model tropical cyclone, they find that the spin up process above the boundary is largely captured by axisymmetric balance dynamics, at

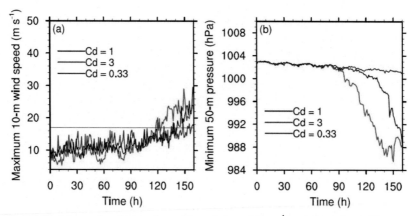

FIGURE 16.33 Numerically simulated time series of (a) maximum 10-m wind speed (m s^{-1}) and (b) minimum 50-m pressure (mb) from simulations with different drag coefficients ($C_D = 1.0 \times 10^{-3}$, 3.0×10^{-3}, 0.33×10^{-3}) and a fixed surface entropy exchange coefficient ($C_K = 1.2 \times 10^{-3}$). Adapted from Wang et al. (2019). Republished with permission of the American Meteorological Society.

least beyond the deep convective region. This result affirms the importance of the classical mechanism (Section 8.2) in increasing the tangential wind speed above the boundary layer outside of the eyewall region. In regard to the role of the boundary layer they write:

"The results suggest that even though friction in the boundary layer reduces the near-surface wind speed everywhere, it induces an imbalance of forces in the radial direction when there is a pressure gradient imposed through the depth of the boundary layer. The resultant frictional convergence associated with the pressure gradient promotes the process of vortex spinup in TC genesis from the top of the boundary layer to the near surface. The spinup role of the frictional boundary layer is more important than its direct reduction of wind speed in TC formation."

This deduction supports the subtle, but nonetheless important role of surface friction as discussed in Section 16.2. Their Fig. 17 (shown here as Fig. 16.33) affirms that frictional drag importantly supports the organization of moist convection in the central region of the incipient circulation and increased drag enhances the spin up rate of the emergent vortex (as suggested long ago by Ooyama (1968) and more recently by Schecter (2011) and Kilroy et al. (2017a)). Reinforcing this point, Wang et al. write in their conclusion:

"... our results, consistent with Kilroy et al. (2017a), imply that convection is organized by the boundary layer to some degree and that important influences of friction can occur in the earliest stages of development."

These findings add support to the position advocated in this book that boundary layer dynamics play a hitherto underappreciated role, even during the early phase of cyclogen-

esis and during rapid intensification. In particular, *the findings strengthen the view that further understanding of the transition of a weak disorganized system to a rapidly intensifying cyclone may hinge to an important degree on the boundary layer.*

16.10.2 Formation of a thermodynamic shield in a Category 5 hurricane, but not in a Category 3 hurricane

In an interesting paper, Wadler et al. (2021a) examined the thermodynamic effect of downdrafts on the boundary layer and nearby updrafts in idealized simulations of Category-3 and Category-5 tropical cyclones. These simulations were designated Ideal3 and Ideal5, respectively. The authors showed that in the stronger storm, Ideal5,

"downdrafts underneath the eyewall pose no negative thermodynamic influence because of eye-eyewall mixing below 2-km altitude. Additionally, a layer of higher θ_e between 1 and 2 km altitude associated with low-level outflow that extends 40 km outward from the eyewall region creates a "thermodynamic shield" that prevents negative effects from downdrafts".

In the weaker storm, Ideal3, no such outflow occurred. The impact of downdrafts beyond the main eyewall in the two simulations were compared in a cartoon schematic shown in Fig. 16.34 (their Fig. 17, reproduced here for convenience).

In the case of Ideal5, a lower-tropospheric radial outflow jet was found to advect enhanced θ_e air outwards from the interior eyewall/eye region. Presumably, θ_e is the pseudo-equivalent potential temperature. The radial outflow jet of enhanced θ_e air was argued to act as a thermodynamic shield that modifies downdraft air descending into the boundary layer. As a downdraft attempts to pass through the shield,

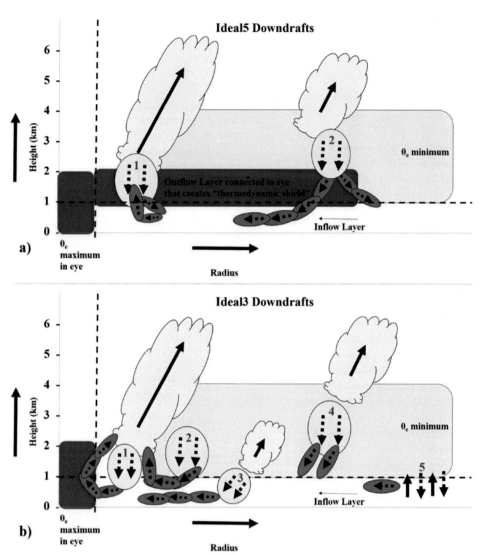

FIGURE 16.34 Radius-height cross-section schematic of the thermodynamic impact that downdrafts have in (a) Ideal5 and (b) Ideal3. In both panels updrafts are indicated by clouds, downdrafts are indicated by dashed down arrows in a light blue circle, and parcel trajectory locations are indicated by dashed arrows in dark blue circles. The downdrafts are numbered for reference in the text. The mid-tropospheric minimum in θ_e is indicated in light blue and the high θ_e in the eye and low-level outflow in (a) is indicated in red. From Wadler et al. (2021a). Republished with permission of the American Meteorological Society.

downdraft air would be warmed and moistened through mixing with the air in the shield. The net effect of the shield is to elevate the downdraft θ_e before it is drawn inwards by the frictional inflow. In contrast, for the case of Ideal3, there is no appreciable persistent outflow jet to shield the vortex from downdraft influences.

In the conclusions, Wadler et al. wrote:

"The presence of a high-θ_e air above the inflow layer in Ideal5, which is also discussed in previous observational studies (e.g., Barnes, 2008; Wadler et al., 2021b), highlights the importance of storm structure in determining the thermodynamic effect of downdrafts. However, it remains unknown the exact mech-

anisms which lead to the formation of the high-θ_e above the boundary layer in the TC and why this layer formed in Ideal5, but not in Ideal3. This will be a topic of future work."

Here we follow our comment (Smith and Montgomery, 2022a) and offer an explanation for this feature, which we believe to be a common occurrence as a storm matures and decays, especially in more intense storms with a broad tangential wind circulation. The underlying reasons are rooted in the ventilation concept first introduced in Chapter 8 and discussed in greater detail in Chapter 15. In essence, the evolution of the inner-core region of a tropical cyclone at a particular stage of its life cycle depends broadly on the rate at which moist air is funneled by the boundary-layer inflow towards the inner region, where the deepest convection prevails, and the rate at which this mass can be carried to the upper troposphere by the aggregate effects of the convection. Smith and Montgomery (2022a) write

"Typically, in the early stages of tropical cyclone formation and intensification, the boundary layer inflow is relatively weak and deep convection is more than capable of removing mass at the rate at which it is funnelled inwards. However, as the storm intensifies and the tangential wind field expands, the rate at which air is funnelled inwards increases, while the progressive warming of the upper troposphere tends to reduce the degree of convective instability and thereby the ability of inner-core deep convection to ventilate the mass converging in the boundary layer. This reduced convective instability may be accompanied by the reduction of effective buoyancy in the eyewall as the eyewall broadens in size (Smith and Montgomery, 2022b). As soon as the boundary layer inflow begins to dominate, the residual mass that cannot be ventilated by deep convection flows outwards in a shallow layer just above the boundary layer. Typically, the tangential velocity component of air ascending out of the boundary layer in the inner-core region is supergradient (Smith and Vogl, 2008; Smith et al., 2008, 2009) and has a natural tendency to flow outwards. Unless this air reaches a level of free convection, it remains stably stratified which accounts for the outflow occurring in a shallow layer."

An important finding of Smith and Montgomery (2022a), p2011, was that

"as a tropical cyclone matures, the low-level radial outflow becomes more and more prevalent and leads ultimately to the decay of the vortex, even in a quiescent environment. At radii where the radial outflow is sufficiently large so that the radial advection of absolute angular momentum, M, exceeds the vertical advection of M out of the boundary layer, the tangential flow at the top of the boundary layer will spin down.."

The foregoing evolution is illustrated, for example, in Figs. 15.6 and 15.7.

Smith and Montgomery (2022a) used the boundary layer control and ventilation concepts, which are important elements of the rotating-convection model, to explain the difference between the two simulations Ideal3 and Ideal5. They argued that

"the difference is presumably because, at the time of analysis, Ideal3 is not yet in a state where the mass influx in the boundary layer exceeds the rate at which this flux can be ventilated by deep convection."

They suggested then a way to validate this conjecture by using the ventilation diagnostic introduced in Chapter 15: see Eq. (15.1) and Fig. 15.10.

Although balance dynamics was not invoked by Wadler et al. to explain the origin of the thermodynamic shield, it is worth noting here that axisymmetric balance model solutions of tropical cyclone evolution, like those presented in Chapter 8, can mimic the same effect, even though the winds in the boundary layer do not become supergradient in such a model by definition. This correspondence was demonstrated in Section 8.4 in a series of simulations with a prognostic balance model in which the diabatic heating rate was varied in strength in relation to the frictional forcing. As already noted in Chapter 8, the unbalanced forces in the radial and vertical directions do not exist in such a balance model.

16.10.3 Invocation of WISHE-like positive feedback mechanism to explain the rapid intensification of Hurricane Michael (2018)

A study by Zhu et al. (2021) found that intensification and track forecasts of recent North Atlantic Hurricanes are significantly improved when the Hurricane Analysis and Forecast System (HAFS) is improved to better estimate the static stability in clouds associated with the intense turbulent mixing in the eyewall and rainband regions. Specifically, the study found that:

"... sub-grid scale (SGS) turbulent transport above the PBL (Planetary Boundary Layer, our insertion) in the eyewall plays a pivotal role in initiating a positive feedback among the eyewall convection, mean secondary overturning circulation, vortex acceleration via the inward transport of absolute vorticity, surface evaporation, and radial convergence of moisture in the PBL."

The finding that an improved SGS scheme above the PBL helps support the vigor of eyewall convection is not implausible and may highlight a useful way forward to being able to better predict rapid intensification events in real-world cases. If this is indeed the case, then it would seem important to understand the underlying reason for the improved predictions. The authors attribute this improved performance (p17) to the "kicking off" of a "WISHE-like positive feedback mechanism" in the updated modeling system. However, it is unclear precisely what the authors mean by the "WISHE-like positive feedback mechanism" and it seems mysterious, at least to us, why the WISHE-like mechanism kicks in only when the representation of the SGS processes influencing the vertical stability in clouds has been updated. Following closely the findings of Sections 16.1.8 and 16.1.9, especially Caveat [(7)], we offer some suggestions on the foregoing issues.

The study by Zhu et al. (2021) invokes this "kicking off" mechanism to explain the differences between two highlighted simulations of Hurricane Michael (2018), one with an improved representation of sub-grid-scale turbulence.

On p17 they show that a key difference in the simulation using the improved scheme is the larger influx of cyclonic absolute vorticity in the planetary boundary layer. They argue that the large positive tangential wind tendency implied by this influx of cyclonic vorticity and

> *"the resultant increase of tangential wind enhance surface evaporation and radial convergence of moisture in the PBL ..., which further fosters stronger eyewall convection evidenced from the increase of vertical velocity ... . The enhanced eyewall convection in turn causes the further increase of radial inflow ... ."*

In contrast, in the old HAFS model, the positive wind tendency associated with the influx of vorticity is found to remain relatively weak and insufficient to

> *"generate the needed acceleration of tangential wind to kick off the WISHE-like feedback mechanism underlying the RI (rapid intensification, our insertion) of Michael".*

In the new HAFS model, the presumed chain of effects listed in the foregoing description is thought to be an affirmation of the initiation of the purported WISHE mechanism. However, in view of the results presented in Sections 16.1.8 and 16.1.9, especially Caveat [(7)], the conflation of this chain of effects with a WISHE feedback seems problematic.

The question remains whether a similar result would emerge had the surface heat fluxes been frozen in time during the RI period. Certainly, an increase of tangential wind will lead through boundary layer dynamics to an increase in the boundary layer inflow and to an increase in the radial convergence of moisture even if the fluxes are frozen in time. Another point to consider is the proposed link between the increased moisture convergence and the stronger eyewall convection as evidenced by the increase of azimuthally-averaged vertical velocity at the 900 mb level. However, an increased vertical velocity at this level does not guarantee an increase in the mean convective mass flux in the middle troposphere. It could happen that the increased

boundary layer inflow could not be all ventilated by the developing eyewall convection.

Recalling the results of the life cycle simulation of Chapter 15, the ventilation of the boundary layer inflow is easily accomplished during the early phase of intensification, but in the later phases, this is not the case. In real-world cases, it should not be taken for granted that all the mass being funneled to the base of the eyewall by the boundary layer will be ventilated by the convection. In fact, moisture convergence *per se* is the wrong quantity to examine, since, as discussed in Section 12.2, the equivalent potential temperature of the boundary layer θ_{eb} is arguably the more relevant quantity. Hence the presumed linkage between moisture convergence and convective strength seems tenuous.

All else being equal, based on the ideas developed in this book, one might plausibly associate an increase in θ_{eb} in time with an increase in local buoyancy in the developing eyewall and thus an increase in the local mean vertical velocity. However, the *effective buoyancy* and therefore the vertical acceleration it produces depends *inter alia* on the horizontal scale of the updraft (Section 9.6). Even if in Eq. (16.7) the vertical gradient of the saturation mixing ratio were unchanged in time, an increase in the mean vertical velocity would imply an increase in the mean heating rate $\dot{Q}$ and its associated radial gradient. According to the balance model of Section 16.1.1, an increased magnitude of $\partial \dot{Q}/\partial r$ would imply a strengthening of the overturning circulation and hence an increased spin up tendency of mean tangential velocity via the increased mean influx of cyclonic absolute vorticity.

In a related issue, the authors suggest on p17 (also p21) that

> *"regardless of the strength of individual convective elements the azimuthal-mean eyewall convection must exceed a critical level so that the induced mean secondary overturning circulation can generate sufficiently large inward transport of absolute vorticity needed for vortex intensification."*

We are perplexed about the meaning of this suggestion because as long as there is mean inflow above the boundary layer, the vortex will intensify. Why is a critical level of radial influx of cyclonic vorticity necessary to initiate intensification?

16.10.4 Synthesis

We have reviewed recent independent research that corroborates some of the key ideas developed in this book. In particular, this research points to the important role of the boundary layer even in the early stages of cyclone development.

16.11 Vertical shear regimes

The additional complexities associated with the influence of horizontal and vertical shear deformation on a candidate tropical storm or tropical disturbance within an embedded wave-pouch structure should be amenable to analysis with many of tools developed herein. A first step in this direction was taken by Smith et al. (2017), who sought to interpret the dynamics of intensification in an HWRF simulation of Hurricane Earl (2010). Nevertheless, there are subtleties associated with the asymmetric dynamics of the emergent vortex that require further analysis outside the scope of this textbook. Related to this problem is the possible intrusion of relatively dry air ("anti-fuel") into the pouch region of a pre-cyclone disturbance or a well defined tropical cyclone, which requires a careful analysis of the sources and sinks of θ_e in and around the pouch and incipient vortex (e.g., Riemer and Montgomery, 2011; Nolan and McGauley, 2012; Freismuth et al., 2016; Rutherford et al., 2018; Wadler et al., 2021a; Smith and Montgomery, 2022a).

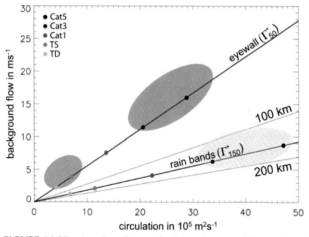

FIGURE 16.35 A regime diagram for the environmental interaction of tropical cyclones (TCs) in vertical wind shear. The sloping lines denote the closest radius (in km) of the dividing streamline for a combination of relative circulation (Γ, abscissa) and storm-relative flow (U, ordinate). For eyewall interaction, the TC categories are defined by the relative circulation for a circular disk of 50 km radius, Γ_{50}; for rain band interaction, a circular disk of radius 150 km is used, Γ_{150} (see Section 3.2.1 of Riemer and Montgomery for details). The TC categories are highlighted by the shaded dots. The red (mid gray in print version) shading indicates the regime of eyewall interaction for weak and mature TCs (as defined in Table 1 of Riemer and Montgomery), respectively. Yellow (light gray in print version) shading indicates the regime of rain band interaction for mature TCs. The eyewall radius (50 km), 100 km radius, the rain band radius (150 km), and 200 km radius are depicted (from top to bottom). From Fig. 11 of Riemer and Montgomery (2011)). Republished with permission from Atmos. Chem. Phys./European Geosciences Union under the Creative Commons Attribution 4.0 License.

To be specific, but without going into too many details, the "regime diagram" of Fig. 16.35 helps illustrate some of the key issues that arise in the vortex-environment interaction problem associated with an impinging vertical shear

flow. The figure summarizes a truncated view of the problem using just two regimes of environmental interaction:

- a direct interaction of environmental air with the eyewall; and
- an interaction with rain bands.

Here we summarize a zero-order estimate of the potential environmental interaction based on the closest approach of the dividing streamline between environmental air and vortex air. For this purpose, we represent a tropical cyclone as a non-divergent point vortex with a 1/radius decay of tangential velocity (like that employed in the toy, wave-vortex, model of Chapter 4, at least outside $r = a$). In this model it is assumed that for a vertically sheared tropical cyclone, the environmental air is located outside of the dividing streamline.[13] The storm-relative flow for the point vortex immersed in a uniform westerly environmental flow U is depicted in Fig. 16.36.

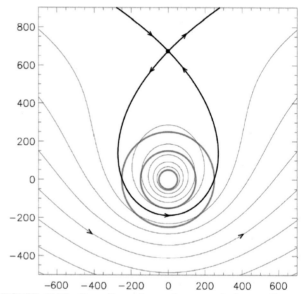

FIGURE 16.36 Streamlines of a stationary, cyclonic, non-divergent point vortex with a circulation of 33.7×10^5 m^{2}s^{-1} (Category 3, see Table 1 of Riemer and Montgomery, 2011) in 5 ms^{-1} westerly background flow. The strength of the vortex and the storm-relative environmental flow are chosen to be similar to the idealized, three-dimensional, numerical experiment studied in Riemer and Montgomery (2011) at 2 km height (cf. their Fig. 3). The streamlines passing through the stagnation point (black dot), the flow's manifolds, are highlighted. The horizontal scale is in km. The gray concentric circles denote radii of 50 km (nominal eyewall radius), 150 km (nominal rainband radius) and 200 km, respectively. Adapted from Fig. 7 of Riemer and Montgomery (2011). Republished with permission from Atmos. Chem. Phys./European Geosciences Union under the Creative Commons Attribution 4.0 License.

If the closest approach of the dividing streamline to the center is within the rain band (150 km) or eyewall (50 km)

---

13. This assumption may be relaxed by including the effect of weak convergence above the frictional inflow layer. The results of the improved model broadly support the zero-order model presented here. See Riemer and Montgomery, Section 4.2 and supporting subsections for details.

radius, then pronounced interaction with rain bands and eyewall convection, respectively, would be expected. The closest approach of the dividing streamline is based on the solution of the transcendental equation (16) of Riemer and Montgomery:

$$\alpha + \ln\alpha = -1 \qquad (16.20)$$

where $\alpha = R_{nd}U/\Gamma$, with R_{nd} denoting the closest approach of the dividing streamline south of the vortex center in Fig. 16.36, Γ denoting the relative circulation of the vortex[14] normalized by 2π and U denoting the storm-relative environmental flow, averaged, say, between the top of the frictional inflow layer and 5 km height. The solution to (16.20) is found numerically: $\alpha \approx 0.278$. Hence we obtain

$$R_{nd} = \alpha\frac{\Gamma}{U}. \qquad (16.21)$$

This simple model and its corresponding regime diagram serve several purposes. First, it provides theoretical support for the hypothesis advanced by Riemer et al. (2010) that the "character" of shear interaction scales with the ratio of tropical cyclone intensity and vertical shear magnitude. Here, the storm "intensity" is measured by the relative circulation Γ and the shear magnitude is represented by the storm-relative environmental flow, U. Second, we see from Eq. (16.21) that the storm-relative flow that is necessary to move the dividing streamline to a specific radius increases linearly with Γ. Third, the regime diagram (Fig. 16.35) suggests that direct eyewall interaction seems likely for the tropical depression (TD) and tropical storm (TS) vortices for storm-relative environmental flow starting at approximately 3 m s^{-1}, while environmental flow well above 10 m s^{-1} is necessary to move the dividing streamline to a radius of 50 km for the stronger Category 3 (Cat3) and Category 5 (Cat5) vortices.

If one makes the reasonable assumption that the low-level, storm-relative flow is approximately half of the wind difference between 850 mb and 200 mb, i.e., the deep-layer vertical "shear", it is concluded that direct interaction of environmental air with the eyewall of vortices of tropical storm intensity and below is likely in weak to moderate deep-layer "shear" of 6 m s^{-1} and above. On the other hand, for mature tropical cyclones of Category 3 or stronger, a direct interaction of environmental air with the eyewall is unlikely even in strong deep-layer "shear" of 20 m s^{-1}.

As noted by Riemer and Montgomery, the foregoing model used Γ_{50} to derive the above results. Due to the slower radial decay of the tangential winds in real storms (e.g., Mallen et al., 2005), these estimates represent a lower bound of the requisite storm-relative flow. Using larger circulation values to represent the respective cyclone category,

e.g., Γ_{150}, the differences between the weak and the strong cyclone categories become more pronounced. For mature storms (Category 3 and above), moderate to strong deep-layer shear is likely to promote pronounced interaction of environmental air with rain bands. As intimated in Chapters 9, 10, and 13, such interaction is expected to produce persistent, vortex-scale downdrafts leading to a considerable weakening of the storm as demonstrated in Fig. 5 of Riemer and Montgomery and Riemer et al. (2010).

The foregoing model led Riemer and Montgomery to draw what appears to be an important insight for forecasters and for future tropical cyclone research in regards to the interaction of tropical cyclones with their real environment:

"it is not only the dynamic resiliency of a TC that increases with intensity (Jones, 1995; Reasor et al., 2004). The ability of a TC to isolate itself from adverse thermodynamic interaction with environmental air increases with intensity also. A similar point can be made for vortex size. A broad TC with a relatively large radius of gale force winds has evidently a larger circulation than a TC with the same maximum intensity (as defined by V_{max}, addition ours) but a smaller radius of gale force winds. Thus, it can be expected that the ability of a TC to thermodynamically isolate itself from the environment increases with TC size. Very distinct scenarios of possible environmental interaction in vertical shear are indicated for values of U and Γ that represent realistic TC conditions in this simplified framework."

Recalling the material presented in Section 16.10.2 regarding the formation of a thermodynamic shield in a Category 5, but not a Category 3 cyclone, in conjunction with the dynamical interpretation provided therein, points to an important new resilience element that needs to be understood as part of the vertical shear interaction problem.

16.11.1 Synthesis

We have provided here a brief overview of what we think are some of the key elements required for a basic understanding of the interaction of a pre-cyclone disturbance or intensifying/mature cyclone with an environmental shear flow. There are several aspects to be explored beyond the cited references, but the analytical model summarized, its accompanying regime diagram, together with an improved understanding of the dynamics of the boundary layer, convective ventilation and related formation of a downdraft shield for intense cyclones seems adequate to begin a systematic analysis in a realistic three-dimensional configuration.

---

14. $\Gamma = \oint_C \mathbf{u} \cdot \mathbf{dl}/2\pi$ where C denotes a closed horizontal contour surrounding the vortex at a suitably large radius.

Chapter 17

Epilogue

In the Preface we intimated the need for a modern textbook on tropical cyclones that develops the foundational fluid dynamics and thermodynamics necessary to understand a range of basic processes involved in their observed behavior. We mentioned also that a particular scientific aim was to provide an understanding of how tropical cyclones form from weak tropical disturbances in a moist, non-saturated environment and how they subsequently intensify, mature and decay. In this textbook, we have presented what we think to be a robust set of physical ideas and tools that provide an elementary understanding of these aspects. The tools developed are rooted in the three-dimensional dynamics of the moist tropical atmosphere and relate to an emerging tropical depression within a favorable environment for sustained development.

While the axisymmetric dynamics of a tropical cyclone provides a useful first step for understanding particular problems beyond the genesis phase, the genesis problem, itself, is far from axisymmetric and even the azimuthally-averaged dynamics fall well short of providing a complete description of the phenomenon. That being said, there is much to be learned from the behavior of axisymmetric vortices and several early chapters have focused on this simplification in order to establish a basic foundation for understanding vortex behavior. Nevertheless, because coherent wave and vortex structures are recognized to be key ingredients of geophysical turbulence, including the tropical cyclogenesis problem, later chapters have adopted a three-dimensional (non-axisymmetric) approach that incorporates the fluid dynamics of some of the most pertinent coherent structures in the formation, intensification, mature evolution and decay of a tropical cyclone.

The material presented in this textbook, along with its supporting exercises, provides plausible scientific answers to the specific questions raised in the Preface. This material constitutes what we think is a physically sound foundation for continued work on these questions.

17.1 Examples of recent events

The case of the formation and intensification of Hurricanes Fiona and Ian in the Caribbean sea during the 2022 Atlantic hurricane season serve to highlight some applications of the rotating-convection paradigm in conjunction with the marsupial paradigm.

17.1.1 Formation and intensification of Hurricane Fiona (2022)

Fig. 17.1 shows the analyzed wave-pouch structure for the tropical disturbance at 00 UTC 14 September, which became Tropical Storm Fiona approximately 15 hours after this analysis time. The data are derived from the US GFS global forecast model output[1] and have been processed following the Lagrangian methodology that proved highly successful during the PREDICT experiment as summarized in Chapter 1. The analysis methodology for the wave-pouch structure has been updated to a fully Lagrangian analysis system following Rutherford et al. (2018) and has become automated also. The left panel depicts the non-dimensional Lagrangian vorticity field, as defined by Rutherford et al. (2018), at the 850 mb level, while the right panel depicts the horizontal wind streamlines and relative vertical vorticity in a frame of reference moving with the wave-pouch structure at the 925 mb level before it was a named system by NHC forecasters. Both diagnostic analyses depict a quasi-closed circulation of approximately several hundred kilometers in horizontal scale around the so-called "sweet spot" of the pre-Fiona disturbance (as defined in Chapter 1, by the intersection of wave trough and critical line in the right panel) at these levels. These analyses support the new necessary condition for storm formation discussed in Chapter 1 of a sub-synoptic-scale pouch region with an approximately closed cyclonic circulation.

As discussed in Chapter 1, such a pouch provides a favorable environment for subsequent convective organization, provided the thermodynamical conditions remain favorable to support deep convection and a pathway of vortical-convective organization as illustrated in Chapter 10. Around the time of the images in Fig. 17.1, the vertical wind shear is relatively modest also (not shown) and, given the circumstances, the disturbance appears poised to develop.

In reality, the pouch flow did subsequently support continuing deep convection near the sweet spot and a tropical storm was declared 15 hours after these analyses. Analyses like these are routinely available online,[2] making further in-depth analysis of the vortical-convective organization that

---

1. https://www.ncei.noaa.gov/products/weather-climate-models/global-forecast.
2. The url is https://marsupialpouches.nps.edu/global.php?m=gfs&p=OWBtree700&dtg=2022111800&fh=loop&r=atl&.

Tropical Cyclones. https://doi.org/10.1016/B978-0-44-313449-4.00025-4

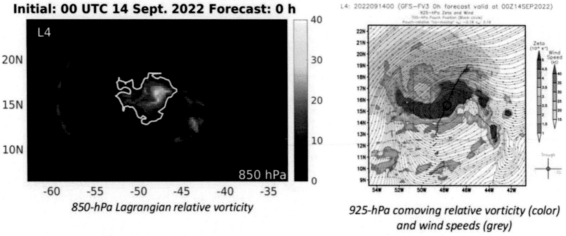

First indication of Fiona by the GFS-based automated pouch tracker is a complete 120-hour forecast initiated at 00 UTC 14 September 2022.

850-hPa Lagrangian relative vorticity

925-hPa comoving relative vorticity (color) and wind speeds (grey)

FIGURE 17.1 Analyzed wave-pouch structure for the tropical disturbance at 00 UTC 14 September, which became Tropical Storm Fiona approximately 15 hours after this analysis time. The data come from the U.S. GFS global forecast model output and have been processed following the Lagrangian methodology as summarized in Chapter 1. The analysis includes both a semi-Lagrangian analysis (right panel) and a fully Lagrangian analysis (left panel, following Rutherford et al., 2018). The left panel depicts the nondimensional Lagrangian vorticity field at the 850 hPa level, while the right panel depicts the horizontal wind streamlines and relative vertical vorticity in a frame of reference moving with the wave-pouch structure (before it was a named system by NHC forecasters) at the 925 hPa level. Courtesy Mark Boothe.

much easier. We think such analyses should prove useful to forecasters on the bench and offer compelling evidence to reverse the entrenched reluctance of analyzing the wave-pouch kinematics and vortical-convective organization in a Lagrangian framework as advocated and demonstrated by Dunkerton et al. (2009), Davis and Ahijevych (2012), and Asaadi et al. (2017).

Fig. 17.2 shows an example of early-cycle intensity guidance for the emergent tropical storm Fiona at 00 UTC 18 September. The time series displays intensity forecasts from some of the main weather centers and forecast models around the world. Superimposed on the intensity forecast guidance plot is the NOAA-NHC public advisory forecast at the time of the guidance (red (mid gray in print version) curve, labeled as "OFCI"), and the actual storm intensity ("X" symbols) as adjudicated by NHC forecasters based on their analysis of hurricane reconnaissance aircraft data and satellite data. Between 12 UTC 19 September and 00 UTC 20 September, the storm intensity increased by 20 knots in 12 hours (from 75 to 95 knots). This intensification rate exceeds the official forecaster definition of "rapid intensification" (30 knots in 24 hours).

From the assorted forecasts shown, we see clearly the challenge of forecasting storm intensity. The forecasters try to consult all available satellite, reconnaissance and numerical model data to assess the degree of convective organization and the likelihood for further strengthening. In this case, at 72 hours, the discrepancy between the 18 September forecast and the analyzed observations is two full storm

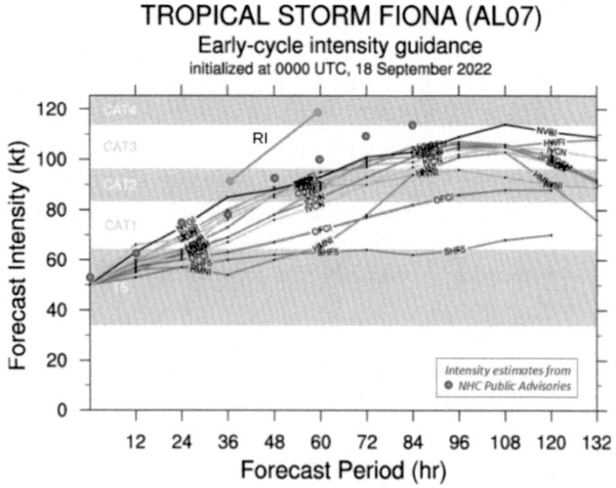

FIGURE 17.2 Early-cycle intensity guidance for tropical storm Fiona on 18 September 00 UTC as presented by NCAR-RAL's Tropical Cyclone Guidance Project (J. Vigh, https://hurricanes.ral.ucar.edu/realtime/plots/northatlantic/2022/al072022//intensity_early/aal07_2022091800_intensity_early.png). The time series displays intensity forecasts from some of the main weather centers and forecast models. The model abbreviations are defined at: https://hurricanes.ral.ucar.edu/guide/intensity/index.php. Superposed on the intensity forecast guidance plot is NOAA-NHC's official forecast at the time of this guidance (red (mid gray in print version) curve, labeled as "OFCI"), and the actual storm intensity ("X" symbols) as adjudicated by NHC forecasters based on their analysis of hurricane reconnaissance aircraft data and satellite data. The sloping orange (gray in print version) line marked RI has the gradient 30 knots/per day, which is the forecasters' criterion for rapid intensification. (These actual intensities were obtained from the NOAA-NHC public advisories: https://www.nhc.noaa.gov/archive/2022/FIONA.shtml?)

categories. This is not to be critical of forecasters on the bench, who do not have the benefit of hindsight.

To highlight the multi-scale nature of the intensity problem, Fig. 17.3 shows satellite images of Fiona's convective organization late on 19 September before sunset (visible) and during nighttime, approximately 7 hours later (infrared). For the times shown, the eye region (the dark region near the center of circulation) has contracted to a near "pinhole" configuration, which is barely resolved in the second satellite image (Fig. 17.3b).

(a)

(b)

FIGURE 17.3 NOAA GOES16-East GEOCOLOR composite satellite images of Hurricane Fiona (2022) on 19 and 20 September. Top panel: visible satellite image at 22:00 UTC 19 September. Bottom panel: Infrared satellite image at 04:50 UTC 20 September. Courtesy of NOAA/NESDIS/STAR. Real-time GEOCOLOR composite satellite imagery available at: https://cdn.star.nesdis.noaa.gov/GOES16/ABI/ SECTOR/taw/GEOCOLOR/GOES16-TAW-GEOCOLOR-900x540.gif. Archived satellite imagery available at: https://inventory.ssec.wisc.edu/ inventory/?date=2011/10/18&time=&satellite=GOES-16.

As discussed in Chapters 6, 8, 10, and 15, the radial contraction of the vortex is the result of the aggregate effect of the deep convection that draws in absolute vorticity above the boundary layer. This same mechanism accounts

for the increase in vortex size as explained in the next subsection. Both classical and boundary layer spin up enhancement mechanisms are at work to yield an eye less than 10 nautical miles in radius, like the pinhole indicated in the second satellite image. Forecast models employing a horizontal grid spacing of 2 or 3 km will be challenged to capture accurately this nonlinear structure in the inner core. Furthermore, forecasters typically do not have access to this type of information in real time and, moreover, are handicapped by the limited effective resolution of the current forecast models for this purpose. The foregoing discussion serves to illustrate that we still have a notable journey ahead before we can claim success on the intensity forecast problem.

17.1.2   Increasing size of Hurricane Fiona

The evolution of storm size is another important facet of the forecasting problem. Fig. 17.4 shows the evolution of the intensity (V_{max}) and the radius of gale force winds (17 m s^{-1}, 34 knots) taken from the best track data that were available at the time of writing. It is seen that during the hurricane phase, the radius of gales increases with time at a progressively increasing rate and there is little obvious correlation between the rate of size increase and the rate of intensity increase. Notably, the size increases markedly during the extra-tropical transition phase as the intensity decreases rapidly. This behavior exemplifies that described in Section 1.2.6.

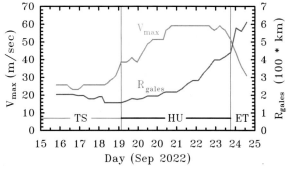

FIGURE 17.4 Intensity V_{max} and mean radius of gale force winds R_{gales} for Hurricane Fiona (2022) for the period 15-25 September, based on best track data available at the time of writing. The thin vertical lines delineate the period of hurricane intensity (HU), marked by the thicker horizontal line. The tropical storm (TS) and extra-tropical (ET) periods are indicated also. R_{gales} is an arithmetic average of the values given in four sectors of the storm in the best track data. Before 00 UTC 18 September, the value is the arithmetic average of three values as there was no radius of gales given in the southwest quadrants.

In Chapters 8, 11, and 15, a framework was articulated in which the general tendency of storm size to increase may be understood. As long as the convectively-induced overturning circulation is maintained, the low-tropospheric inflow will continue to flux cyclonic absolute vorticity, both mean and eddy components, inwards, thereby increasing the circulation at a given radius and the mean tangential wind at that radius. The spin up of the tangential wind field includes the inward flux of cyclonic eddy vorticity by the axisymmetrization of vorticity anomalies in the storm environment as discussed in Section 4.5.2.

17.1.3 Formation and intensification of Hurricane Ian (2022)

As in the case of Fiona, the formation of Ian was found to be associated with a quasi-closed pouch circulation in the middle to lower troposphere more than a day before the storm was declared a tropical depression by hurricane forecasters.

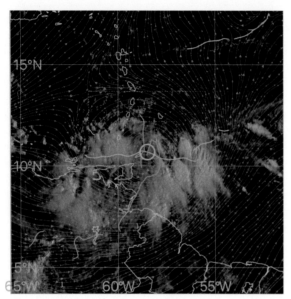

FIGURE 17.5 NOAA GOES16 visible satellite image using Channel 2 at 12 UTC, 21 September 2022.

Fig. 17.5 shows visible satellite imagery of the tropical convection overlaid with the analyzed co-moving streamline flow at the 850 mb level at 12 UTC, 21 September. In this analysis, the translation velocity (approximately 6 m s^{-1} almost due west) of the pre-Ian disturbance as calculated by hurricane forecasters is used to calculate the co-moving circulation. The figure depicts a cluster of thunderstorms, which are located within and around the pouch-like circulation.

For reasons explained in Sections 1.5.3-1.5.6, the intersection of Ian's pre-cursor wave trough line (gray curve) and critical line (purple curve) indicates the preferred location for convective organization, tropical cyclogenesis and intensification. This is, indeed, the case at subsequent times (not shown). To highlight the advantage of this co-moving framework and the related facets that come with the marsupial paradigm, Fig. 17.6 shows the wind barbs and streamline flow at 850 mb in the co-moving and earth-relative reference frames.

In the earth-relative framework, the flow is closed with a center of circulation over land near 61°W and 8°N. This center is more than 3 degrees southwest of the center over the ocean in the co-moving reference frame, where Ian eventually becomes a tropical depression almost two days later. On the basis of this analysis, we would suggest that it would be misleading to focus on the earth-relative center for predicting the preferred zone for convective organization.

Given the presence of deep convective activity in and around the pouch and the warm ocean water underneath the pouch at this time in the Caribbean region, for reasons discussed in Sections 1.5.5, 8.2, 10.10, and 11.7, we would anticipate continued convective activity in the pouch region and a progressive amplification of the swirling wind component in association with the aggregate effect of the cumulus convection that draws cyclonic absolute vorticity inwards in the lower troposphere above the boundary layer.

17.1.4 Increasing size of Hurricane Ian

Fig. 17.7 shows the evolution of the intensity (V_{max}) and the mean radius of gale force winds (17 m s^{-1}, 34 knots) for Hurricane Ian, analogous to Fig. 17.4 for Hurricane Fiona. In contrast to Fiona, which remained over the ocean, Ian crossed the Florida peninsula from the Gulf of Mexico and made a further landfall over the coast of South Carolina after a brief period of reintensification over the Atlantic Ocean. Despite these differences, the radius of gales increases progressively with time as in Fiona. During the period before the first landfall at about 20:30 UTC 28 September, the growth in size accompanied the storm's progressive intensification, but shortly after landfall the mean radius of gales expanded dramatically at a time when the intensity was falling rapidly. Again, based on the idea that the radial extent of the convectively-generated overturning circulation is related to that of the convection itself, it is likely that this size increase is associated with an increase in the areal extent of deep convection after landfall as a result of the classical intensification mechanism (Section 8.2).

While Wikipedia currently attributes the spin down after landfall to an increase of vertical wind shear,[3] a plausible alternative scenario is that the increase in the low-level frictionally-driven inflow could no longer be ventilated locally by deep convection, despite the increase in the areal extent of convection. In turn, the inadequate ventilation would lead to low-level outflow above the boundary layer inflow as, for example in the life-cycle simulation in Chapter 15. This enhanced outflow would lead, in turn, to a decrease in the tangential wind speed above the boundary

---

3. Specifically, Wikipedia notes "Continual land interaction resulted in the frictional displacement of the system, and that coupled with high vertical wind shear caused Ian to quickly degrade to a tropical storm by 09:00 UTC as it moved north-northeast off of the eastern Florida coastline."

IAN 2022-09-22-12Z GFS Analysis

850 mb wind vectors

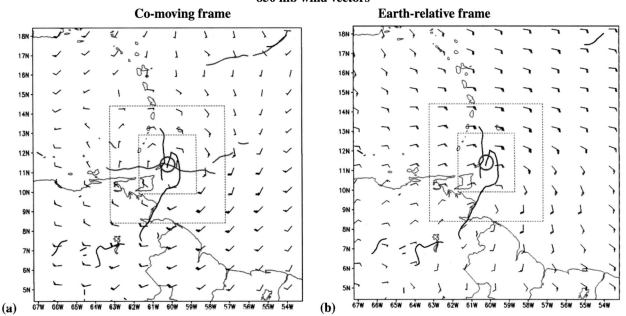

850 mb streamlines, wind speed and relative vorticity

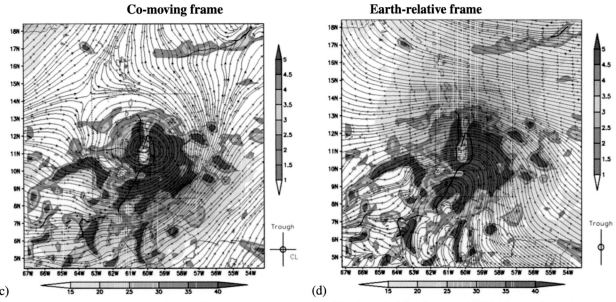

FIGURE 17.6 As in Fig. 17.4: Analyzed pre-Ian flow structure at the 850 mb level for the tropical disturbance at 12 UTC, 21 September, which became Tropical Depression Ian approximately 42 hours after this analysis time. The top row shows the wind barbs, the trough axis and critical line in the co-moving (left) and earth-relative (right) frames. The bottom row shows the streamline flow, isotachs (gray shading) and relative vertical vorticity (color) in co-moving frame of reference (left) and earth-relative frame of reference (right). Wind barbs in top row follow meteorological convention, with single barb 10 knots and flag 50 knots. Relative vertical vorticity in bottom row depicted using color shading (and color bar) in units of 1×10^{-5} s^{-1}. Isotachs in gray shading (and bar) in units of knots (e.g., lightest shading equals range 15-20 knots). In all panels, 3 by 3 degree and 6°by 6°boxes are centered at intersection (open circle) of wave trough (solid black curve) and critical line (purple curve). Boxed areas are used to quantity average flow properties on proto-vortex and pouch-scale. Courtesy Mark Boothe.

layer. These features merit further study with the rotating-convection paradigm offering a suitable framework for such an investigation.

17.2 Applications and future directions

This book has sought to articulate the building blocks and a framework required for understanding tropical cyclone

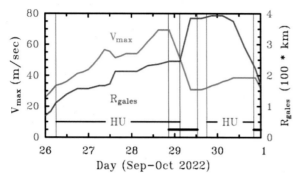

FIGURE 17.7 Intensity, V_{max} and mean radius of gale force winds, R_{gales}, for Hurricane Ian (2022) for the period 26 September to 1 October, based on best track data available at the time of writing. The thin vertical lines delineate the periods of hurricane intensity (shown by the thicker horizontal line marked HU) and the periods that the storm was over Florida and later over South Carolina (marked by the thick horizontal lines below this line). R_{gales} is an arithmetic average of the values given in northeast, southeast, southwest, and northwest quadrants of the storm in the best track data. Before 06 UTC 26 September and after 18 UTC 30 September, the value is the arithmetic average of three values as there was no radius of gales given in the southwest quadrants.

behavior. This framework, which continues to undergo refinement is referred to as the rotating-convection paradigm. The framework has been used to provide interpretations of many observed features of tropical cyclone behavior, noting that the tight nonlinear coupling between deep convection and the frictional boundary layer often makes it difficult to offer cause and effect statements. In these circumstances, all one can do is to articulate the elements of the coupling. Beyond that, one has to do the calculation.

As noted in Section 16.10, a stringent test of any theory is whether it can explain experimental features not previously understood, or predict and verify new properties not yet conceived in prior experiments. Based on the evidence presented herein, we suggest that the rotating convection paradigm passes this test in relation to tropical cyclones.

It is hard to exaggerate the importance of having a conceptual model for tropical cyclone behavior for interpreting observations and numerical simulations. It is our experience that many studies involving numerical model simulations continue to be carried out "in the dark" without any conceptual framework for interpretation and it is worth reiterating the quote of Davis and Emanuel in the Preface that:

> *"A proper integration of the equations is not synonymous with a conceptual grasp of the phenomenon being predicted".*

There is clearly limited value in carrying out numerical simulations in which the model is run as a "black box" and the results simply reported with little attempt at interpretation. But interpretation requires a conceptual framework from the outset. The same remark applies to the interpretation of observations, although the observational data sets are arguably never as complete as those from numerical models. It seems hardly necessary to stress the importance to forecasters of having a firm conceptual understanding of how storms form, intensify, mature and decay, even though quantitative calculations using complex numerical prediction models are ultimately necessary to apply this understanding.

Some highlights of the material presented here that may be relevant to forecasters are:

- In the rotating-convection paradigm, there is no distinction between the dynamical mechanisms of tropical cyclogenesis and tropical cyclone intensification and, furthermore, there is no distinction between the processes of intensification and rapid intensification of storms in a protected pouch region.

- For a traveling precursor wave disturbance, a Lagrangian framework that moves with the disturbance is necessary to deduce the true flow boundaries in and around the low pressure region of the wave. These flow boundaries help protect a disturbance, including the subsequent tropical cyclone if the genesis process is complete, from dry air intrusions that are forced by the relative flow associated with an impinging vertical shear. As long as these flow boundaries are maintained and the thermodynamic conditions at the surface remain favorable, a vortical convective organization process will likely ensue. Whether or not there is a closed surface isobar or westerly surface winds equatorwards of the wave is immaterial.

- The incorporation of the boundary layer spin up enhancement mechanism in the rotating-convection paradigm has led to improved estimates of turbulent diffusion coefficients in the boundary layer parameters, which have been shown to improve the performance of the NOAA forecasting model with respect to intensification.

- The rotating-convection paradigm has drawn attention to the need to determine the validity of using a down-gradient turbulence closure for all predicted quantities in the rotating, convective turbulence region of the inner core that pervades a rapidly intensifying tropical cyclone.

Some readers may be disappointed at the lack of a thorough treatment given here to the role of vertical wind shear. To do this topic justice would have required a significant expansion of the text and a considerable amount of more time to complete the book. As discussed in the Preface, it is our intention to write some form of sequel to this book that focuses on the effects of vertical wind shear and other more advanced topics, including the hypothesized effects of global warming on tropical cyclone behavior.

References

Aberson, S., Montgomery, M.T., Bell, M., Black, M., 2006. Hurricane Isabel (2003): new insights into the physics of intense storms. Part II. Extreme localized wind. Bull. Am. Meteorol. Soc. 87, 1349–1354.

Anderson, J.D., 2005. Ludwig Prandtl's boundary layer. Phys. Today 58, 3042–3060.

Andreas, E.L., Mahrt, L., Vickers, D., 2012. A new drag relation for aerodynamically rough flow over the ocean. J. Atmos. Sci. 69, 2520–2537.

Anthes, R.A., 1974. The dynamics and energetics of mature tropical cyclones. Rev. Geophys. Space Phys. 12, 495–522.

Anthes, R.A., 1982. Tropical Cyclones: Their Evolution, Structure and Effects. Meteor. Monogr., vol. 41. American Meteorological Society, Boston, Mass., USA. 208 pp.

Aref, H., 1983. Integrable chaotic and turbulent vortex motion in two-dimensional flows. Annu. Rev. Fluid Mech. 15, 345–389.

Arndt, D.S., Basara, J.B., McPherson, R.A., Illston, B.G., McManus, G.D., Demkos, D.B., 2009. Observations of the reintensification of tropical storm Erin (2007). Bull. Am. Meteorol. Soc. 99, 1079–1093.

Asaadi, A., Brunet, G., Yau, M.K., 2016a. On the dynamics of the formation of the Kelvin cat's-eye in tropical cyclogenesis. Part I: climatological investigation. J. Atmos. Sci. 73, 2317–2338.

Asaadi, A., Brunet, G., Yau, M.K., 2016b. On the dynamics of the formation of the Kelvin cat's-eye in tropical cyclogenesis. Part II: numerical simulation. J. Atmos. Sci. 73, 2339–2359.

Asaadi, A., Brunet, G., Yau, M.K., 2017. The importance of critical layer in differentiating developing from nondeveloping easterly waves. J. Atmos. Sci. 74, 409–417.

Baines, P.G., 1995. Topographic Effects in Stratified Flow. Cambridge University Press, 40 West 20th Street, New York, NY, 10011-4211, USA. 482 pp.

Barnes, G.M., 2008. Atypical thermodynamic profiles in hurricanes. Mon. Weather Rev. 136, 631–643.

Barnes, G.M., Zipser, E.J., Jorgensen, D., Marks Jr., F., 1983. Mesoscale and convective structure of a hurricane rainband. J. Atmos. Sci. 40, 2125–2137.

Bell, M.M., Montgomery, M.T., 2008. Observed structure, evolution, and potential intensity of Category 5 Hurricane Isabel (2003) from 12 to 14 September. Mon. Weather Rev. 136, 2023–2046.

Bell, M.M., Montgomery, M.T., 2019. Mesoscale processes during the genesis of Hurricane Karl (2010). J. Atmos. Sci. 76, 2235–2255.

Bell, M.M., Montgomery, M.T., Emanuel, K.A., 2012a. Air-sea enthalpy and momentum exchange at major hurricane wind speeds observed during CBLAST. J. Atmos. Sci. 69, 3197–3222.

Bell, M.M., Montgomery, M.T., Lee, W.C., 2012b. An axisymmetric view of concentric eyewall evolution in Hurricane Rita (2005). J. Atmos. Sci. 69, 2414–2432.

Bister, M., 2001. Effect of peripheral convection on tropical cyclone formation. J. Atmos. Sci. 58, 3463–3476.

Bister, M., Emanuel, K.A., 1997. The genesis of Hurricane Guillermo: TEXMEX analyses and a modeling study. Mon. Weather Rev. 125, 2662–2682.

Bister, M., Emanuel, K.A., 1998. Dissipative heating and hurricane intensity. Meteorol. Atmos. Phys. 50, 233–240.

Bister, M., Renno, N., Paulius, O., Emanuel, K.A., 2011. Comment on Makarieva et al. "A critique of some modern applications of the Carnot heat engine concept: the dissipative heat engine cannot exist". Proc. R. Soc. A 467, 1–6. https://doi.org/10.1098/rspa.2010.0087.

Black, M.L., Burpee, R.W., Marks, F.D., 1996. Vertical motion characteristics of tropical cyclones observed with airborne Doppler radial velocities. J. Atmos. Sci. 53, 1887–1909.

Black, P.G., D'Asoro, E.A., Drennan, W.M., French, J.R., Niller, P.P., Sanford, T.B., Terril, E.J., Walsh, E.J., Zhang, J.A., 2007. Air-sea exchange in hurricanes. Synthesis of observations from the coupled boundary layer air-sea transfer experiment. Bull. Am. Meteorol. Soc. 88, 357–374.

Bogner, P.B., Barnes, G.M., Franklin, J.L., 2000. Conditional instability and shear for six hurricanes over the Atlantic Ocean. Weather Forecast. 15, 192–207.

Bohren, C.F., Albrecht, B.R., 1998. Atmospheric Thermodynamics. Oxford University Press, New York, USA. 402 pp.

Bolton, D., 1980. The computation of equivalent potential temperature. Mon. Weather Rev. 108, 1046–1053.

Boyce, W.E., DiPrima, R.C., 1986. Elementary Differential Equations and Boundary Value Problems, fourth edition. John Wiley and Sons, New York, USA. 654 pp.

Bracken, E.W., Bosart, L.F., 2000. The role of synoptic-scale flow during tropical cyclogenesis over the North Atlantic Ocean. Mon. Weather Rev. 128, 353–376.

Braun, S.A., 2002. A cloud resolving simulation of Hurricane Bob (1991): storm structure and eyewall buoyancy. Mon. Weather Rev. 130, 1573–1592.

Braun, S.A., 2006. High-resolution simulation of Hurricane Bonnie (1998): Part II: water budget. J. Atmos. Sci. 63, 43–64.

Braun, S.A., Kakar, R., Zipser, E., Heymsfield, G., Albers, C., Brown, S., Durden, S.L., Guimond, S., Halverson, J., Heymsfield, A., Ismail, S., Lambrigtsen, B., Miller, T., Tanelli, S., Thomas, J., Zawislak, J., 2013. NASA'S genesis and rapid intensification processes (GRIP) field experiment. Bull. Am. Meteorol. Soc. 94, 345–363.

Braun, S.A., Montgomery, M.T., Mallen, K.J., Reasor, P.D., 2010. Simulation and interpretation of the genesis of Tropical Storm Gert (2005) as part of the NASA Tropical Cloud Systems and Processes Experiment. J. Atmos. Sci. 67, 999–1025.

Brennan, M.J., Knabb, R.D., Mainelle, M., Kimberlain, T.B., 2009. Atlantic hurricane season of 2007. Mon. Weather Rev. 137, 4061–4088.

Bretherton, C.S., Smolarkewicz, P.K., 1989. Gravity waves, compensating subsidence and detrainment around cumulus clouds. J. Atmos. Sci. 46, 740–759.

Brunet, G., Montgomery, M.T., 2002. Vortex Rossby waves on smooth circular vortices. Part I: theory. Dyn. Atmos. Ocean. 35, 153–177.

Bryan, G.H., 2013. Notes and correspondence comments on "sensitivity of tropical-cyclone models to the surface drag coefficient". Q. J. R. Meteorol. Soc. 139, 1957–1960.

Bryan, G.H., Fritsch, J.M., 2002. A benchmark simulation for moist non-hydrostatic numerical model. Mon. Weather Rev. 130, 2917–2928.

Bryan, G.H., Rotunno, R., 2009a. Evaluation of an analytical model for the maximum intensity of tropical cyclones. J. Atmos. Sci. 66, 3042–3060.

Bryan, G.H., Rotunno, R., 2009b. The influence of near-surface, high entropy air in hurricane eyes on maximum hurricane intensity. J. Atmos. Sci. 66, 148–158.

Bryan, G.H., Rotunno, R., 2009c. The maximum intensity of tropical cyclones in axisymmetric numerical model simulations. Mon. Weather Rev. 137, 1770–1789.

Bryan, G.H., Rotunno, R., Chen, Y., 2009. The effects of turbulence on hurricane intensity. In: 29th Conference on Hurricanes and Tropical Meteorology, vol. 66, pp. 3042–3060.

Bureau of Meteorology, A., 1977. Report on cyclone Tracy, December, 1974. Australian Govt. Publ. Service, P.O. Box 84, Canberra, A.C.T. 2600. 82 pp.

Businger, S., 1991. Arctic hurricanes. Am. Sci. 79, 18–33.

Caillol, P., 2017. A singular vortex Rossby wave packet within a rapidly rotating vortex. Phys. Fluids 29, 046601-1–046601-30.

Caillol, P., 2019. A singular vorticity wave packet within a rapidly rotating vortex: spiralling versus oscillating motions. J. Fluid Mech. 873, 688–741.

Callaghan, J., Smith, R.K., 1998. The relationship between maximum surface wind speeds and central pressure in tropical cyclones. Aust. Meteorol. Mag. 47, 191–202.

Camargo, S.J., Tippett, M.K., Sobel, A.H., Vecchi, G.A., Zhao, M., 2014. Testing the performance of tropical cyclone genesis indices in future climates using the HIRAM model. J. Climate 27, 9171–9196.

Carnevale, G.F., McWilliams, J.C., Pomeau, Y., Weiss, J.B., Young, W., 1991. Evolution of vortex statistics in two-dimensional turbulence. Phys. Rev. Lett. 66, 2735–2738.

Carr, L.E., Elsberry, R.L., 1990. Observational evidence for predictions of tropical cyclone propagation relative to environmental steering. J. Atmos. Sci. 47, 542–546.

Carr, L.E., Williams, R.T., 1989. Barotropic vortex stability to perturbations from axisymmetry. J. Atmos. Sci. 46, 3177–3191.

Carrió, D.S., Homar, V., Jansa, A., Romero, R., Picornell, M.A., 2017. Tropicalization process of the 7 November 2014 Mediterranean cyclone: numerical sensitivity study. Atmos. Res. 197, 300–312.

Chan, J.C.L., Williams, R.T., 1987. Analytical and numerical studies of the beta effect in tropical cyclone motion. Part I: zero mean flow. J. Atmos. Sci. 44, 1257–1265.

Chan, K.T.F., Chan, J.C.L., 2012. Size and strength of tropical cyclones as inferred from QuikSCAT data. Q. J. R. Meteorol. Soc. 140, 811–824.

Chan, K.T.F., Chan, J.C.L., 2014. Impacts of vortex intensity and outer winds on tropical cyclone size. Q. J. R. Meteorol. Soc. 141, 525–537.

Charney, J.G., Eliassen, A., 1964. On the growth of the hurricane depression. J. Atmos. Sci. 21, 68–75.

Chavas, D.R., Emanuel, K., 2010. A QuikSCAT climatology of tropical cyclone size. Geophys. Res. Lett. 37, L18816.

Chen, S.S., Frank, W.M., 2001. A numerical study of the genesis of extratropical convective mesovortices. Part I: evolution and dynamics. J. Atmos. Sci. 50, 2401–2426.

Chen, Y., Brunet, G., Yau, M.K., 2003. Spiral bands in a simulated hurricane. Part II: wave activity diagnosis. J. Atmos. Sci. 60, 1239–1256.

Chorin, A., 1994. Vorticity and Turbulence. Springer-Verlag Publishing Company, Inc., 175 Fifth Avenue, New York, N.Y. 10010. 174 pp.

Chorin, A., Marsden, J.E., 2000. A Mathematical Introduction to Fluid Mechanics, fourth printing. Springer-Verlag Publishing Company, Inc., 175 Fifth Avenue, New York, N.Y. 10010. 181 pp.

Cione, J.J., Black, P.G., Houston, S.H., 2000. Surface observations in the hurricane environment. Mon. Weather Rev. 128, 1550–1561.

Cione, J.J., Bryan, G.H., Dobosy, R., Zhang, J.A., de Boer, G., Aksoy, A., Wadler, J.B., Kalina, E., Dahl, B.A., Ryan, K., Neuhaus, J., Dumas, E., Marks, F.D., Farber, A.M., Hock, T., Chen, X., 2020. Eye of the storm: observing hurricanes with a small unmanned aircraft system. Bull. Am. Meteorol. Soc. 101, E186–E205.

Coles, D., 1985. The uses of coherent structure (Dryden lecture). In: 23rd Aerospace Sciences Meeting. 14-17 January 1985. Reno, NV, USA. Published Online: 17 Aug 2012.

Cram, T.A., Persing, J., Montgomery, M.T., Braun, S.A., 2007. A Lagrangian trajectory view on transport and mixing processes between the eye, eyewall, and environment using a high resolution simulation of Hurricane Bonnie (1998). J. Atmos. Sci. 64, 1835–1856.

Črnivec, N., Smith, R.K., Kilroy, G., 2015. Dependence of tropical cyclone intensification rate on sea surface temperature. Q. J. R. Meteorol. Soc. 140, 1618–1627.

Curcic, M., Haus, B.K., 2020. Revised estimates of ocean surface drag in strong winds. Geophys. Res. Lett. 47, 1–8.

D'Asaro, E., Sanford, T.B., Niller, P., Terrill, E., 2007. Cold wake of Hurricane Frances. Geophys. Res. Lett. 34, l15609. https://doi.org/10.1029/2007GL030160.

Davidson, N.E., Holland, G.H., 1987. A diagnostic analysis of two intense monsoon depressions over Australia. Mon. Weather Rev. 115, 380–392.

Davies-Jones, R., 2003. An expression for effective buoyancy in surroundings with horizontal density gradients. J. Atmos. Sci. 60, 2922–2925.

Davis, C.A., Ahijevych, D.A., 2012. Mesoscale structural evolution of three tropical weather systems observed during predict. J. Atmos. Sci. 69, 1284–1305.

Davis, C.A., Emanuel, K.A., 1991. Potential vorticity diagnostics of cyclogenesis. Mon. Weather Rev. 119, 1929–1953.

DeMaria, M., Kaplan, J., 1994. Sea surface temperature and the maximum intensity of Atlantic tropical cyclones. J. Climate 7, 1324–1334.

DeMaria, M., Pickle, J.D., 1988. A simplified system of equations for simulation of tropical cyclones. J. Atmos. Sci. 45, 1542–1554.

Donelan, M.A., Haus, B.K., Reul, N., Plant, W.J., Stiassnie, N., Graber, H.C., Brown, O.B., Saltzman, E.S., 2004. On the limiting aerodynamic roughness of the ocean in very strong winds. Geophys. Res. Lett. 31, L18306.

Dunion, J.P., 2011. Rewriting the climatology of the tropical North Atlantic and Caribbean Sea atmosphere. J. Climate 24, 893–908.

Dunkerton, T.J., Montgomery, M.T., Wang, Z., 2009. Tropical cyclogenesis in a tropical wave critical layer: easterly waves. Atmos. Chem. Phys. 9, 5587–5646.

Dunnavan, G.M., Diercks, J.W., 1980. An analysis of super typhoon Tip (October 1979). Mon. Weather Rev. 108, 1915–1923.

Edson, J.B., Jampana, V., Weller, R.A., Bigorre, S.P., Plueddemann, A.J., Fairall, C.W., Miller, S.D., Mahrt, L., Vickers, D., Hersbach, H., 2013. On the exchange of momentum over the open ocean. J. Phys. Oceanogr. 43, 1589–1610.

Eliassen, A., 1951. Slow thermally or frictionally controlled meridional circulation in a circular vortex. Astrophys. Nor. 5, 19–60.

Elsberry, R.L., Harr, P.A., 2008. Tropical Cyclone Structure (TCS08) field experiment: science basis, observational platforms, and strategy. Asia-Pac. J. Atmos. Sci. 44, 1–23.

Emanuel, K., 2005a. Genesis and maintenance of "Mediterranean hurricanes". Adv. Geosci. 2, 217–220.

Emanuel, K., Callaghan, J., Otto, P., 2008. A hypothesis for the redevelopment of warm-core cyclones over northern Australia. Mon. Weather Rev. 136, 3863–3872.

Emanuel, K.A., 1986. An air-sea interaction theory for tropical cyclones. Part I: steady state maintenance. J. Atmos. Sci. 43, 585–604.

Emanuel, K.A., 1988. The maximum intensity of hurricanes. J. Atmos. Sci. 45, 1143–1155.

Emanuel, K.A., 1989. The finite amplitude nature of tropical cyclogenesis. J. Atmos. Sci. 46, 3431–3456.

Emanuel, K.A., 1994. Atmospheric Convection. Oxford University Press. 580 pp.

Emanuel, K.A., 1995. Sensitivity of tropical cyclone to surface exchange coefficients and a revised steady-state model incorporating eye dynamics. J. Atmos. Sci. 52, 3969–3976.

Emanuel, K.A., 1997. Some aspects of hurricane inner-core dynamics and energetics. J. Atmos. Sci. 54, 1014–1026.

Emanuel, K.A., 1999. Thermodynamic control of hurricane intensity. Nature 401, 665–669.

Emanuel, K.A., 2003. Tropical cyclones. Annu. Rev. Earth Planet. Sci. 31, 75–104.

Emanuel, K.A., 2004. Tropical cyclone energetics and structure. In: Fedorovich, E., Rotunno, R., Stevens, B. (Eds.), Atmospheric Turbulence and Mesoscale Meteorology: Scientific Research Inspired by Doug Lilly. Cambridge University Press, pp. 165–192.

Emanuel, K.A., 2005b. Divine Wind: The History and Science of Hurricanes. Oxford University Press, New York. 296 pp.

Emanuel, K.A., 2012. Self-stratification of tropical cyclone outflow. Part II: implications for storm intensification. J. Atmos. Sci. 69, 988–996.

Emanuel, K.A., 2017. Will global warming make hurricane forecasting more difficult? Bull. Am. Meteorol. Soc. 98, 495–501.

Emanuel, K.A., 2018. 100 years of progress in tropical cyclone research. Meteorol. Monogr. 59, 15.1–15.68.

Emanuel, K.A., 2019. Inferences from simple models of slow, convectively coupled processes. J. Atmos. Sci. 76, 195–208.

Emanuel, K.A., Neelin, J.D., Bretherton, C.S., 1994. On large-scale circulations of convecting atmospheres. Q. J. R. Meteorol. Soc. 120, 1111–1143.

Emanuel, K.A., Nolan, D.S., 2004. Tropical cyclone activity and global climate. In: Preprints, 26th Conf. on Hurricanes and Tropical Meteorology. Miami, FL. Amer. Meteor. Soc., pp. 240–241.

Emanuel, K.A., Rotunno, R., 1989. Polar lows as Arctic hurricanes. Tellus, Ser. A Dyn. Meteorol. Oceanogr. 41, 1–17.

Emanuel, K.A., Rotunno, R., 2011. Self-stratification of tropical cyclone outflow. Part I: implications for storm structure. J. Atmos. Sci. 68, 2236–2249.

Enagonio, J., Montgomery, M.T., 2001. Tropical cyclogenesis via convectively forced vortex Rossby waves in a shallow water primitive equation model. J. Atmos. Sci. 58, 685–705.

Evans, C., Schumacher, R.S., Galarneau, T.J., 2011. Sensitivity in the overland reintensification of tropical cyclone Erin (2007) to near-surface soil moisture characteristics. Mon. Weather Rev. 139, 3848–3870.

Fang, J., Zhang, F., 2011. Evolution of multiscale vortices in the development of Hurricane Dolly (2008). J. Atmos. Sci. 68, 103–122.

Fiorino, M., Elsberry, R.L., 1987. Some aspects of vortex structure related to tropical cyclone motion. J. Atmos. Sci. 46, 975–990.

Foster, I.J., Lyons, T.J., 1984. Tropical cyclogenesis: a comparative study of two depressions in the northwest of Australia. Q. J. R. Meteorol. Soc. 110, 105–119.

Frank, N.L., 1970. Atlantic tropical systems of 1969. Mon. Weather Rev. 98, 307–314.

Frank, W.M., 1987. Tropical cyclone formation. In: Elsberry, R.L. (Ed.), A Global View of Tropical Cyclones. Office of Naval Research, pp. 53–90.

Franklin, J.L., 1990. Dropwindsonde observations of the environmental flow of Hurricane Josephine (1984): relationships to vortex motion. Mon. Weather Rev. 118, 2732–2744.

Franklin, J.L., Black, M.L., Valde, K., 2003. GPS dropwindsonde wind profiles in hurricanes and their operational implications. Weather Forecast. 18, 32–44.

Franklin, J.L., Feuer, S.E., Kaplan, J., Aberson, S.D., 1996. Tropical cyclone motion and surrounding flow relationships: searching for the beta gyres in Omega dropwindsonde datasets. Mon. Weather Rev. 124, 64–84.

Franklin, J.L., Lord, S.J., Feuer, S.E., Marks Jr., F.D., 1993. The kinematic structure of Hurricane Gloria (1985) determined from nested analyses of dropwindsonde and Doppler radar data. Mon. Weather Rev. 121, 2433–2451.

Freismuth, T.M., Rutherford, B., Boothe, M.A., Montgomery, M.T., 2016. Why did the storm ex-Gaston (2010) fail to redevelop during the PREDICT experiment? Atmos. Chem. Phys. 16, 8511–8519.

French, J.R., Drennan, W.M., Zhang, J.A., Black, P.G., 2007. Turbulent fluxes in the hurricane boundary layer. Part I: momentum flux. J. Atmos. Sci. 64, 1089–1102.

Frisch, U., 1995. Turbulence: The Legacy of A.N. Kolmogorov. Cambridge University Press, Press Syndicate of the University of Cambridge, The Pitt Building, Trumpington Street, Cambridge, CB2 1RP. 296 pp.

Frisius, T., 2015. What controls the size of a tropical cyclone? Investigations with an axisymmetric model. Q. J. R. Meteorol. Soc. 141, 2457–2470.

Fudeyasu, H., Wang, Y., 2011. Balanced contribution to the intensification of a tropical cyclone simulated in TCM4: outer-core spinup process. J. Atmos. Sci. 68, 430–449.

Garner, S., 2015. The relationship between hurricane potential intensity and CAPE. J. Atmos. Sci. 72, 141–163.

Gent, P.R., McWilliams, J.C., 1986. The instability of barotropic circular vortices. Geophys. Astrophys. Fluid Dyn. 35, 209–233.

Gill, A.E., 1982. Atmosphere - Ocean Dynamics, fourth edition. Academic Press, New York. 662 pp.

Glatz, A., Smith, R.K., 1996. Vorticity asymmetries in hurricane Josephine (1984). Q. J. R. Meteorol. Soc. 122, 391–413.

Goody, R., 1995. Principles of Atmospheric Physics and Chemistry. Oxford University Press, New York, Oxford. 324 pp.

Gray, W.M., 1968. A global view of the origin of tropical disturbances and storms. Mon. Weather Rev. 96, 669–700.

Gray, W.M., 1975. Tropical Cyclone Genesis. Bluebook, vol. 234. Colorado State University, Department of Atmospheric Science, pp. 1–121.

Gray, W.M., 1998. The formation of tropical cyclones. Meteorol. Atmos. Phys. 67, 37–69.

Greenspan, H.P., Howard, L.N., 1963. On a time-dependent motion of a rotating fluid. J. Fluid Mech. 17, 385–404.

Guimond, S.R., Heymsfield, G.M., Turk, F.J., 2010. Multiscale observations of hurricane *Dennis* (2005): the effects of hot towers on rapid intensification. J. Atmos. Sci. 67, 633–654.

Haller, G., Beron-Vera, F., 2013. Coherent Lagrangian vortices: the black holes of turbulence. J. Fluid Mech. 731, 1–10.

Haus, B.K., Jeong, D., Donelan, M.A., Zhang, J.A., Savelyev, I., 2010. Relative rates of sea-air heat transfer and frictional drag in very high winds. Geophys. Res. Lett. 37, l07802. https://doi.org/10.1029/2009GL042206.

Hausman, S.A., Ooyama, K.V., Schubert, W.H., 2006. Potential vorticity structure of simulated hurricanes. J. Atmos. Sci. 63, 87–108.

Hawkins, H.F., Imbembo, S.M., 1976. The structure of a small, intense hurricane Inez 1966. Mon. Weather Rev. 104, 418–442.

Haynes, P., McIntyre, M.E., 1987. On the evolution of vorticity and potential vorticity in the presence of diabatic heating and frictional or other forces. J. Atmos. Sci. 44, 828–841.

Haynes, P., McIntyre, M.E., 1990. On the conservation and impermeability theorems for potential vorticity. J. Atmos. Sci. 47, 2021–2031.

Helmholtz, H.V., 1958. Über Integrale der hydrodynamischen Gleichungen, welche den Wirbelbewegungen entsprechen. Walter de Gruyter GmbH and Co. KG, Berlin, pp. 25–55.

Hendricks, E.A., Peng, M.S., Fu, B., Li, T., 2010. Quantifying environmental control on tropical cyclone intensity change. Mon. Weather Rev. 138, 3243–3271.

Hendricks, E.A., Schubert, W.H., 2010. Adiabatic rearrangement of hollow PV towers. J. Adv. Model. Earth Syst. 2. 19 pp.

Hendricks, E.A., Schubert, W.H., Taft, R.K., 2009. Life cycles of hurricane-like vorticity rings. J. Atmos. Sci. 66, 705–722.

Hirji, Z., Aldhous, P., 2020. The Busiest Hurricane Season on Record is Officially over. BuzzFeed News.

Hock, T.F., Franklin, J.L., 1999. The NCAR GPS dropwindsonde. Bull. Am. Meteorol. Soc. 80, 407–420.

Hogan, R., Bozzo, A., 2018. Nonlinear response of a tropical cyclone vortex to prescribed eyewall heating with and without surface friction in tcm4: implications for tropical cyclone intensification. J. Adv. Model. Earth Syst. 10, 1990–2008.

Holland, G.H., Merrill, R.T., 1984. On the dynamics of tropical cyclone structural changes. Q. J. R. Meteorol. Soc. 110, 723–745.

Holthuijsen, L.H., Powell, M.D., Pietrzak, J.D., 2012. Wind and waves in extreme hurricanes. J. Geophys. Res. 117, c09003. https://doi.org/10.1029/2012JC007983.

Holton, J.R., 2004. An Introduction to Dynamic Meteorology, 4th edition. Academic Press, London. 535 pp.

Holton, J.R., Hakim, G., 2013. An Introduction to Dynamic Meteorology, 5th edition. Elsevier, London. 532 pp.

Hoskins, B.J., Bretherton, F.P., 1972. Atmospheric frontogenesis models: mathematical formulation and solution. J. Atmos. Sci. 29, 11–37.

Hoskins, B.J., McIntyre, M.E., Robertson, A.W., 1985. On the use and significance of isentropic potential vorticity maps. Q. J. R. Meteorol. Soc. 111, 877–946.

Houze, R.A., 1993. Clouds Dynamics. Academic Press, San Diego. 570 pp.

Houze, R.A., 2010. Clouds in tropical cyclones. Mon. Weather Rev. 138, 293–344.

Houze, R.A., 2014. Clouds Dynamics, 2nd edition. Academic Press, London. 496 pp.

Houze, R.A., Chen, S.S., Smull, B.F., Lee, W.-C., Bell, M.M., 2007. Hurricane intensity and eyewall replacement. Science 315, 1235–1239.

Houze, R.A., Lee, W.C., Bell, M.M., 2009. Convective contribution to the genesis of Hurricane Ophelia (2005). Mon. Weather Rev. 137, 2778–2800.

Jakob, C., Singh, M.S., Jungandreas, L., 2019. Radiative convective equilibrium and organised convection - an observational perspective. J. Geophys. Res. 124, 5418–5430.

James, I.N., 1994. Introduction to Circulation Atmospheres. Cambridge Atmospheric and Space Science Series. Cambridge University Press. 422 pp.

James, R.P., Markowski, P.M., 2010. A numerical investigation of the effects of dry air aloft on deep convection. Mon. Weather Rev. 138, 140–161.

Jeong, D., Haus, B.K., Donelan, M.A., 2012. Enthalpy transfer across the air–water interface in high winds including spray. J. Atmos. Sci. 69, 2733–2748.

Johnson, R.H., Rickenbach, T.M., Rutledge, S.A., Ciesielski, P.E., Schubert, W.H., 1999. Trimodal characteristics of tropical convection. J. Climate 12, 2397–2418.

Jordan, C.L., 1958. Mean soundings for the West Indies area. J. Meteorol. 15, 91–97.

Jorgensen, D.P., Zipser, E.J., LeMone, M.A., 1985. Vertical motions in intense hurricanes. J. Atmos. Sci. 42, 839–856.

Judt, F., 2018. Insights into atmospheric predictability through global convection permitting model simulations. J. Atmos. Sci. 75, 1477–1497.

Kaplan, J., DeMaria, M., 2003. Large-scale characteristics of rapidly intensifying tropical cyclones in the North Atlantic basin. Weather Forecast. 18, 1093–1108.

Kaplan, J., DeMaria, M., Knaff, J.A., 2010. A revised tropical cyclone rapid intensification index for the Atlantic and Eastern North Pacific basins. Weather Forecast. 25, 220–241.

Kármán, T. von, 1956. Some aspects of the turbulence problem. Collect. Works III, 120–155.

Kasahara, A., Platzmann, G.W., 1963. Interaction of a hurricane with the steering flow and its effect upon the hurricane trajectory. Tellus 15, 321–335.

Kelvin, Lord, 1880. On the vibrations of a columnar vortex. Philos. Mag. 10, 155–168.

Kepert, J.D., 2001. The dynamics of boundary layer jets within the tropical cyclone core. Part I: linear theory. J. Atmos. Sci. 58, 2469–2484.

Kepert, J.D., 2006a. Observed boundary-layer wind structure and balance in the hurricane core. Part I. Hurricane Georges. J. Atmos. Sci. 63, 2169–2193.

Kepert, J.D., 2006b. Observed boundary-layer wind structure and balance in the hurricane core. Part II. Hurricane Mitch. J. Atmos. Sci. 63, 2194–2211.

Kepert, J.D., 2010a. Comparing slab and multi-level models of the tropical cyclone boundary layer. Part I: comparing the simulations. Q. J. R. Meteorol. Soc. 136, 1700–1711.

Kepert, J.D., 2010b. Slab- and height-resolving models of the tropical cyclone boundary layer. Part II: why the simulations differ. Q. J. R. Meteorol. Soc. 136, 1700–1711.

Kepert, J.D., 2012. Choosing a boundary-layer parameterisation for tropical cyclone modelling. Mon. Weather Rev. 140, 1427–1445.

Kepert, J.D., 2013. How does the boundary layer contribute to eyewall replacement cycles in axisymmetric tropical cyclones? J. Atmos. Sci. 70, 2808–2830.

Kepert, J.D., Wang, Y., 2001. The dynamics of boundary layer jets within the tropical cyclone core. Part II: nonlinear enhancement. J. Atmos. Sci. 58, 2485–2501.

Khain, A., Sutyrin, G., 1983a. Part I: Tropical Cyclones, Their Structure, Climatology, Energetics, Numerical Modeling. Hydro-Meteo-Press, Leningrad, USSR. 147 pp.

Khain, A., Sutyrin, G., 1983b. Part II: Interaction of Tropical Cyclones with the Ocean. Observations and Modeling Studies. Hydro-Meteo-Press, Leningrad, USSR. 124 pp.

Kieu, C., 2015. Revisiting dissipative heating in tropical cyclone maximum potential intensity. Q. J. R. Meteorol. Soc. 139, 1255–1269.

Kilroy, G., Montgomery, M.T., Smith, R.K., 2017a. The role of boundary-layer friction on tropical cyclogenesis and subsequent intensification. Q. J. R. Meteorol. Soc. 143, 2524–2536.

Kilroy, G., Smith, R.K., 2013. A numerical study of rotating convection during tropical cyclogenesis. Q. J. R. Meteorol. Soc. 139, 1255–1269.

Kilroy, G., Smith, R.K., 2015. Tropical-cyclone convection: the effects of a vortex boundary layer wind profile on deep convection. Q. J. R. Meteorol. Soc. 141, 714–726.

Kilroy, G., Smith, R.K., 2016. A numerical study of deep convection in tropical cyclones. Q. J. R. Meteorol. Soc. 142, 3138–3151.

Kilroy, G., Smith, R.K., 2017. The effects of initial vortex size on tropical cyclogenesis and intensification. Q. J. R. Meteorol. Soc. 143, 2832–2845.

Kilroy, G., Smith, R.K., Montgomery, M.T., 2016a. Why do model tropical cyclones grow progressively in size and decay in intensity after reaching maturity? J. Atmos. Sci. 73, 487–503.

Kilroy, G., Smith, R.K., Montgomery, M.T., 2017b. Tropical low formation and intensification over land as seen in ECMWF analyses. Q. J. R. Meteorol. Soc. 143, 772–784.

Kilroy, G., Smith, R.K., Montgomery, M.T., 2017c. A unified view of tropical cyclogenesis and intensification. Q. J. R. Meteorol. Soc. 143, 450–462.

Kilroy, G., Smith, R.K., Montgomery, M.T., 2018. The role of heating and cooling associated with ice processes on tropical cyclogenesis and intensification. Q. J. R. Meteorol. Soc. 144, 99–114.

Kilroy, G., Smith, R.K., Montgomery, M.T., 2020. An idealized numerical study of tropical cyclogenesis and evolution at the equator. Q. J. R. Meteorol. Soc. 146, 685–699.

Kilroy, G., Smith, R.K., Montgomery, M.T., Lynch, B., Earl-Spurr, C., 2016b. A case study of a monsoon low that intensified over land as seen in the ECMWF analyses. Q. J. R. Meteorol. Soc. 142, 2244–2255.

Kilroy, G., Zhu, H., Chang, M., Smith, R.K., 2022. Application of the rotating-convection paradigm for tropical cyclones to interpreting medicanes: an example. Trop. Cycl. Res. Rev. 11, 131–145.

Kim, S.-H., Kwon, H.J., Elsberry, R.L., 2009. Beta gyres in global analysis fields. Adv. Atmos. Sci. 26, 984–994.

Kimball, S.K., Mulekar, M.S., 2004. A 15-year climatology of North Atlantic tropical cyclones. Part I: size parameters. J. Climate 17, 3555–3575.

Kleinschmidt, E., 1951. Gundlagen einer Theorie des tropischen Zyklonen. Arch. Meteorol. Geophys. Bioklimatol. 4, 53–72.

Knutson, T.R., Chung, M.V., Vecchi, G., Sun, J., Hsieh, T.-L., Smith, A.J.P., 2021. Climate change is probably increasing the intensity of tropical cyclones. In: Le Quéré, P.L.C., Forster (Eds.), Critical Issues in Climate Change Science. Science Brief, pp. 1–8. https://zenodo.org/record/4570334#.Yk1Zv4VBxdg.

Kossin, J.P., 2015. Hurricane wind-pressure relationship and eyewall replacement cycles. Weather Forecast. 30, 177–181.

Kraus, A.B., Smith, R.K., Ulrich, W., 1995. The barotropic dynamics of tropical cyclone motion in a large-scale deformation field. Contrib. Atmos. Phys. 68, 249–261. https://www.meteo.physik.uni-muenchen.de/%7Eroger/Publications/CAP_1995_Krauss_etal.pdf.

Kundu, P.K., Cohen, I.M., 2010. Fluid Mechanics, fourth edition. Elsevier, Amsterdam. 872 pp.

Ladyzhenskaya, O., 1969. The Mathematical Theory of Viscous Incompressible Flow. English Translation. Gordon and Breach, New York, New York. 224 pp.

Lajoie, F., Walsh, K., 2008. A technique to determine the radius of maximum wind of a tropical cyclone. Weather Forecast. 23, 1007–1015.

Landau, L.D., Lifshitz, E.M., 1966. Fluid Mechanics, third revised English edition. Pergamon Press. 536 pp.

Lander, M.A., 1994. Description of a monsoon gyre and its effects on the tropical cyclones in the western North Pacific during August 1991. Weather Forecast. 9, 640–654.

Large, W.G., Pond, S., 1981. Open ocean momentum flux measurements in moderate to strong winds. J. Phys. Oceanogr. 11, 324–336.

Leipper, D.F., Volgenau, D., 1972. Hurricane heat potential of the Gulf of Mexico. J. Phys. Oceanogr. 2, 218–224.

Lemone, M.A., Zipser, E.J., 1980. Cumulonimbus vertical velocity events in GATE. Part I: diameter, intensity and mass flux. J. Atmos. Sci. 37, 2444–2457.

Li, T., Ge, X., Peng, M., Wang, W., 2012. Dependence of tropical cyclone intensification on the Coriolis parameter. Trop. Cycl. Res. Rev. 1, 242–253.

Lindzen, R.S., 1990. Dynamics in Atmospheric Physics. Cambridge University Press, 40 West 20th Street, New York, NY, 10011-4211, USA. 310 pp.

Lindzen, R.S., Christy, J.R., 2020. The Global Mean Temperature Anomaly Record: How It Works, and Why It Is Misleading. Co2 Coalition, 1621 North Kent Street, Suite 603 Arlington, Virginia 22209, pp. 1–14.

Liu, K.S., Chan, J.C.L., 1999. Size of tropical cyclones as inferred from ERS-1 and ERS-2 data. Mon. Weather Rev. 127, 2992–3001.

Long, R.R., 1953. Some aspects of the flow of stratified fluids. I. A theoretical investigation. Tellus 5, 42–58.

Lorenz, E.N., 1963. Deterministic nonperiodic flow. J. Atmos. Sci. 20, 130–141.

Lorenz, E.N., 1969. The predictability of a flow which possesses many scales of motion. Tellus 21, 289–307.

Lu, X., Yu, H., Lei, X., 2011. Statistics for size and radial wind profile of tropical cyclones in the western North Pacific. Acta Meteorol. Sin. 25, 104–112.

Lucas, C., Zipser, E.J., LeMone, M.A., 1994. Vertical velocity in oceanic convection off tropical Australia. J. Atmos. Sci. 51, 3183–3193.

Lussier, L.L., Rutherford, B., Montgomery, M.T., Boothe, M.A., Dunkerton, T.J., 2015. Examining the roles of the easterly wave critical layer and vorticity accretion during the tropical cyclogenesis of Hurricane Sandy. Mon. Weather Rev. 143, 1703–1722.

Majda, A.J., Bertozzi, A.L., 2002. Vorticity and Incompressible Flow. Cambridge University Press, The Pitt Building, Trumpington Street, Cambridge CB2 IRP. 545 pp.

Makarieva, A.M., Gorshkov, G., Li, B.L., Nobre, A.D., 2011. A critique of some modern applications of the Carnot heat engine concept: the dissipative heat engine cannot exist. Proc. R. Soc. A 466, 1893–1902.

Makarieva, A.M., Gorshkov, V.G., Nefiodov, A.V., Chikunov, A.V., Sheil, D., Nobre, A.D., Nobre, P., Li, B.L., 2019. Hurricane's maximum potential intensity and surface heat fluxes. ArXiv, arXiv:1810.12451v2.

Makarieva, A.M., Nefiodov, A.V., Sheil, D., Nobre, A.D., Chikunov, A.V., Plunien, G., Li, B.L., 2020. Comments on "An evaluation of hurricane superintensity in axisymmetric numerical models" by Raphaël Rousseau-Rizzi and Kerry Emanuel. ArXiv, arXiv:2005.11522 [physics.ao-ph].

Malkus, J.S., Riehl, H., 1960. On the dynamics and energy transformations in steady-state hurricanes. Tellus 12, 1–20.

Mallen, K.J., Montgomery, M.T., Wang, B., 2005. Reexamining the near-core radial structure of the tropical cyclone primary circulation: implications for vortex resiliency. J. Atmos. Sci. 62, 408–425.

Mapes, B.E., 1993. Gregarious tropical convection. J. Atmos. Sci. 50, 2026–2037.

Mapes, B.E., Houze, R.A., 1995. Diabatic divergence profiles in western Pacific mesoscale convective systems. J. Atmos. Sci. 52, 1807–1828.

Mapes, B.E., Zuidema, P., 1996. Radiative-dynamical consequences of dry tongues in the tropical troposphere. J. Atmos. Sci. 53, 620–638.

Marks, F.D., Black, P.G., Montgomery, M.T., Burpee, R.W., 2008. Structure of the eye and eyewall of Hurricane Hugo (1989). Mon. Weather Rev. 136, 1237–1259.

McBride, J.L., 1981. Observational analysis of tropical cyclone formation. Part I. Basic definition of data sets. J. Atmos. Sci. 38, 1117–1131.

McBride, J.L., 1995. Tropical cyclone formation. In: Elsberry, R.L. (Ed.), A Global View of Tropical Cyclones. IWTC. World Meteorological Organization, Geneva, pp. 63–105. Chapter 3.

McBride, J.L., Keenan, T.D., 1982. Climatology of tropical cyclone genesis in the Australian region. J. Climate 2, 13–33.

McIntyre, M.E., 2015. The atmospheric wave-turbulence jigsaw. In: Diamond, P.H., Garbet, X., Ghendrih, P., Sarazin, Y. (Eds.), Rotation and Momentum Transport in Magnetized Plasmas. World Scientific Publishing Co. Pte. Ltd., Singapore, pp. 1–43.

McWilliams, J.C., 1984. The emergence of isolated coherent vortices in turbulent flow. J. Fluid Mech. 146, 21–43.

McWilliams, J.C., 2011. Fundamentals of Geophysical Fluid Dynamics. Cambridge University Press, Cambridge. ISBN 978-1107404083. 89 pp.

McWilliams, J.C., 2019. A perspective on the legacy of Edward Lorenz. Earth Space Sci. 6, 336–350.

McWilliams, J.C., Graves, L.P., Montgomery, M.T., 2003. A formal theory for vortex Rossby waves and vortex evolution. Geophys. Astrophys. Fluid Dyn. 97, 275–309.

Menalou, K.D., Schecter, D.A., Yau, M.K., 2016. On the relative contribution of inertia–gravity wave radiation to asymmetric instabilities in tropical cyclone–like vortices. J. Atmos. Sci. 73, 3345–3370.

Merrill, R.T., 1984. A comparison of large and small tropical cyclones. Mon. Weather Rev. 112, 1408–1418.

Merrill, R.T., 1988. Environmental influences on hurricane intensification. J. Atmos. Sci. 45, 1678–1687.

Michaelides, S., Karacostas, T., Sánchez, J.L., Retalis, A., Pytharoulis, I., Homar, V., Romero, R., Zanis, P., Giannakopoulos, C., Bühl, J., Ansmann, A., Merino, A., Melcón, P., Lagouvardos, K., Kotroni, V., Bruggeman, A., López-Moreno, J.I., Berthet, C., Katragkou, E., Tymvios, F., Hadjimitsis, D.G., Mamouri, R.-E., Nisantzi, A., 2018. Reviews and perspectives of high impact atmospheric processes in the Mediterranean. Atmos. Res. 208, 4–44.

Miglietta, M.M., Carnevale, D., Levizzani, V., Rotunno, R., 2020. Role of moist and dry air advection in the development of Mediterranean tropical-like cyclones (medicanes). Q. J. R. Meteorol. Soc. 146, 876–899.

Miglietta, M.M., Rotunno, R., 2019. Development mechanisms for Mediterranean tropical-like cyclones (medicanes). Q. J. R. Meteorol. Soc. 145, 1444–1460.

Möller, J.D., Shapiro, L.J., 2002. Balanced contributions to the intensification of hurricane Opal as diagnosed from a GFDL model forecast. Mon. Weather Rev. 130, 1866–1881.

Moeng, C.H., McWilliams, J.C., Rotunno, R., Sullivan, P.P., Weil, J., 2004. Investigating 2D modeling of atmospheric convection in the PBL. J. Atmos. Sci. 61, 889–903.

Moffatt, H.K., 2011. A brief introduction to vorticity dynamics and turbulence. In: Moffatt, H.K., Shuckburgh, E. (Eds.), Environmental Hazards-The Fluid Dynamics and Geophysics of Extreme Events, Lecture Notes Series. Institute for Mathematical Sciences, National University of Singapore, World Scientific Publishing Co. Pte. Ltd., pp. 1–27.

Molinari, J., Romps, D.M., Vollaro, D., Nguyen, L., 2012. CAPE in tropical cyclones. J. Atmos. Sci. 69, 2452–2462.

Montgomery, M.T., Bell, M.M., Aberson, S.D., Black, M.L., 2006a. Hurricane Isabel (2003): new insights into the physics of intense storms. Part I: mean vortex structure and maximum intensity estimates. Bull. Am. Meteorol. Soc. 87, 1335–1348.

Montgomery, M.T., Brunet, G., 2002. Vortex Rossby waves on smooth circular vortices. Part II: idealized numerical experiments for tropical cyclone and polar vortex interiors. Dyn. Atmos. Ocean. 35, 179–204.

Montgomery, M.T., Davis, C., Dunkerton, T., Wang, Z., Velden, C., Torn, R., Majumdar, S., Zhang, F., Smith, R.K., Bosart, L., Bell, M.M., Haase, J.S., Heymsfield, A., Jensen, J., Campos, T., Boothe, M.A., 2012. The pre-depression investigation of cloud systems in the tropics (PREDICT) experiment: scientific basis, new analysis tools, and some first results. Bull. Am. Meteorol. Soc. 93, 153–172.

Montgomery, M.T., Farrell, B.F., 1992. Polar low dynamics. J. Atmos. Sci. 49, 2484–2505.

Montgomery, M.T., Kallenbach, R.J., 1997. A theory for vortex Rossby waves and its application to spiral bands and intensity changes in hurricanes. Q. J. R. Meteorol. Soc. 123, 435–465.

Montgomery, M.T., Kilroy, G., Smith, R.K., 2020. Contribution of mean and eddy momentum processes to tropical cyclone intensification. Q. J. R. Meteorol. Soc. 146, 3101–3117.

Montgomery, M.T., Lussier III, L.L., Moore, R.W., Wang, Z., 2010a. The genesis of Typhoon Nuri as observed during the Tropical Cyclone Structure 2008 (TCS-08) field experiment. Part 1: The role of the easterly wave critical layer. Atmos. Chem. Phys. 10, 9879–9900.

Montgomery, M.T., Nguyen, S.V., Smith, R.K., Persing, J., 2009. Do tropical cyclones intensify by WISHE? Q. J. R. Meteorol. Soc. 135, 1697–1714.

Montgomery, M.T., Nichols, M.E., Cram, T.A., Saunders, A.B., 2006b. A vortical hot tower route to tropical cyclogenesis. J. Atmos. Sci. 63, 355–386.

Montgomery, M.T., Persing, J., 2020. Does balance dynamics well capture the secondary circulation and spin-up of a simulated tropical cyclone? J. Atmos. Sci. 77, 75–95.

Montgomery, M.T., Persing, J., Smith, R.K., 2015. Putting to rest WISHE-ful misconceptions. J. Adv. Model. Earth Syst. 7, 92–109.

Montgomery, M.T., Persing, J., Smith, R.K., 2019. On the hypothesized outflow control of tropical cyclone intensification. Q. J. R. Meteorol. Soc. 145, 1846–1863.

Montgomery, M.T., Shapiro, L.J., 1995. Generalized Charney-Stern and Fjørtoft theorems for rapidly rotating vortices. J. Atmos. Sci. 52, 1829–1833.

Montgomery, M.T., Smith, R.K., 2011. Tropical cyclone formation: theory and idealized modelling. In: Proceedings of Seventh WMO International Workshop on Tropical Cyclones (IWTC-VII). La RéUnion, Nov. 2010. (WWRP 2011-1). World Meteorological Organization, Geneva, Switzerland, p. 23.

Montgomery, M.T., Smith, R.K., 2012. The genesis of Typhoon Nuri as observed during the Tropical Cyclone Structure 2008 (TCS08) field experiment - Part 2: observations of the convective environment. Atmos. Chem. Phys. 12, 4001–4009.

Montgomery, M.T., Smith, R.K., 2014. Paradigms for tropical cyclone intensification. Aust. Meteorol. Ocean. Soc. J. 64, 37–66.

Montgomery, M.T., Smith, R.K., 2017a. On the applicability of linear, axisymmetric dynamics in intensifying and mature tropical cyclones. Fluids 2, 1–15.

Montgomery, M.T., Smith, R.K., 2017b. Recent developments in the fluid dynamics of tropical cyclones. Annu. Rev. Fluid Mech. 49, 541–574.

Montgomery, M.T., Smith, R.K., 2018. Comments on: "Revisiting the balanced and unbalanced aspects of tropical cyclone intensification", by J. Heng, Y. Wang and W. Zhou. J. Atmos. Sci. 75, 2491–2496.

Montgomery, M.T., Smith, R.K., 2019. Towards understanding the dynamics of spinup in Emanuel's tropical cyclone model. J. Atmos. Sci. 76, 3089–3093.

Montgomery, M.T., Smith, R.K., 2020. Comments on: "An evaluation of hurricane superintensity in axisymmetric numerical models" by Raphaël Rousseau-Rizzi and Kerry Emanuel. J. Atmos. Sci. 77, 1877–1892.

Montgomery, M.T., Smith, R.K., 2022. Minimal conceptual models for tropical cyclone intensification. Trop. Cycl. Res. Rev. 11, 61–75.

Montgomery, M.T., Snell, H.D., Yang, Z., 2001. Axisymmetric spindown dynamics of hurricane-like vortices. J. Atmos. Sci. 58, 421–435.

Montgomery, M.T., Vladimirov, V.A., Denissenko, P.V., 2002. An experimental study on hurricane mesovortices. J. Fluid Mech. 471, 1–32.

Montgomery, M.T., Wang, Z., Dunkerton, T.J., 2010b. Coarse, intermediate and high resolution numerical simulations of the transition of a tropical wave critical layer to a tropical storm. Atmos. Chem. Phys. 10, 10803–10827.

Montgomery, M.T., Zhang, J.A., Smith, R.K., 2014. An analysis of the observed low-level structure of rapidly intensifying and mature Hurricane Earl (2010). Q. J. R. Meteorol. Soc. 140, 2132–2146.

Moon, Y., Nolan, D.S., 2015. Spiral rainbands in a numerical simulation of Hurricane Bill (2009). Part II: propagation of inner rainbands. J. Atmos. Sci. 72, 191–215.

Morrison, H., Curry, J.A., Khvorostyanov, V.I., 2005. A new double-moment microphysics parameterization for application in cloud and climate models. Part I: description. J. Atmos. Sci. 62, 1665–1677.

Morton, B.R., 1966. Geophysical vortices. In: Küchemann (Ed.), Prog. Aeronaut. Sci. 7, 145–194.

Muller, C.J., Romps, D.M., 2018. Acceleration of tropical cyclogenesis by self-aggregation feedbacks. Proc. Natl. Acad. Sci. 115, 2930–2935.

Naylor, J., Schecter, D.A., 2014. Evaluation of the impact of moist convection on the development of asymmetric inner core instabilities in simulated tropical cyclones. J. Adv. Model. Earth Syst. 6, 1027–1048.

Nguyen, C.M., Smith, R.K., Zhu, H., Ulrich, W., 2002. A minimal axisymmetric hurricane model. Q. J. R. Meteorol. Soc. 128, 2641–2661.

Nguyen, V.S., Smith, R.K., Montgomery, M.T., 2008. Tropical-cyclone intensification and predictability in three dimensions. Q. J. R. Meteorol. Soc. 134, 563–582.

Nicholls, M., Montgomery, M.T., 2013. An examination of two pathways to tropical cyclogenesis occurring in idealized simulations with a cloud resolving numerical model. Atmos. Chem. Phys. 13, 5999–6022.

Nikitina, L.V., Campbell, L.J., 2015a. Dynamics of vortex Rossby waves in tropical cyclones, Part 1: linear time-dependent evolution on an f-plane. Stud. Appl. Math. 135, 377–421.

Nikitina, L.V., Campbell, L.J., 2015b. Dynamics of vortex Rossby waves in tropical cyclones, Part 2: nonlinear time-dependent asymptotic analysis on a β-plane. Stud. Appl. Math. 135, 422–446.

Nolan, D.S., 2007. What is the trigger for tropical cyclogenesis? Aust. Meteorol. Mag. 56, 241–266.

Nolan, D.S., Farrell, B.F., 1999. The intensification of two-dimensional swirling flows by stochastic asymmetric forcing. J. Atmos. Sci. 56, 3937–3962.

Nolan, D.S., McGauley, M.G., 2012. Tropical cyclogenesis in wind shear: climatological relationships and physical processes. In: Oouchi, K., Fudeyasu, H. (Eds.), Cyclones: Formation, Triggers, and Control. Nova Science Publishers, Happauge, New York. 270 pp.

Nolan, D.S., Montgomery, M.T., 2000. The algebraic growth of wavenumber one disturbances in hurricane-like vortices. J. Atmos. Sci. 57, 3514–3538.

Nolan, D.S., Montgomery, M.T., 2002. Three-dimensional, nonhydrostatic perturbations to balanced, hurricane-like vortices. Part I: linearized formulation, stability, and evolution. J. Atmos. Sci. 59, 2989–3020.

Nolan, D.S., Montgomery, M.T., Grasso, L., 2001. The wavenumber-one instability and trochoidal motion of hurricane-like vortices. J. Atmos. Sci. 58, 3243–3270.

Ooyama, K.V., 1968. Numerical simulation of tropical cyclones with an axi-symmetric model. New York University Technical Report. pp. 1–8.

Ooyama, K.V., 1969. Numerical simulation of the life cycle of tropical cyclones. J. Atmos. Sci. 26, 3–40.

Ooyama, K.V., 1982. Conceptual evolution of the theory and modeling of the tropical cyclone. J. Meteorol. Soc. Jpn. 60, 369–380.

Ooyama, K.V., 1987. Numerical experiments of steady and transient jets with a simple model of the hurricane outflow layer. In: Extended Abstracts, 17th Conf on Hurricanes and Tropical Meteorology. Miami, Fl. American Meteorological Society, Boston, pp. 318–320.

Palmén, E., 1948. On the formation and structure of tropical hurricanes. Geophysica 3, 26–38.

Palmén, E., 1956. Formation and development of tropical cyclones. In: Proc., Trop. Cyclone Symp. Brisbane. Australian Bureau of Meteorology, Melbourne, Australia 3001, pp. 213–231.

Palmén, E., Newton, C.W., 1969. Atmospheric Circulation Systems: Their Structure and Physical Interpretation. Academic Press, New York. 602 pp.

Pauluis, O., Zhang, F., 2017. Reconstruction of thermodynamic cycles in a high-resolution simulation of a hurricane. J. Atmos. Sci. 74, 3367–3381.

Parks, J.P., Montgomery, M.T., 2001. Hurricane maximum intensity: past and present. Mon. Weather Rev. 129, 1704–1717.

Pedlosky, J., 1987. Geophysical Fluid Dynamics. Springer-Verlag, New York. 710 pp.

Peng, K., Rotunno, R., Bryan, G.H., 2018. Evaluation of a time-dependent model for the intensification of tropical cyclones. J. Atmos. Sci. 75, 2125–2138.

Peng, K., Rotunno, R., Bryan, G.H., Fang, J., 2019. Evolution of an axisymmetric tropical cyclone before reaching slantwise moist neutrality. J. Atmos. Sci. 76, 1865–1884.

Persing, J., Montgomery, M.T., 2003. Hurricane superintensity. J. Atmos. Sci. 60, 2349–2371.

Persing, J., Montgomery, M.T., McWilliams, J., Smith, R.K., 2013. Asymmetric and axisymmetric dynamics of tropical cyclones. Atmos. Chem. Phys. 13, 12299–12341.

Persing, J., Montgomery, M.T., Smith, R.K., McWilliams, J., 2019. Quasi-steady hurricanes revisited. Trop. Cycl. Res. Rev. 8, 1–17.

Peters, J.M., 2016. The impact of effective buoyancy and dynamic pressure forcing on vertical velocities within two-dimensional updrafts. J. Atmos. Sci. 73, 4531–4551.

Powell, M.D., 1990. Boundary layer structure and dynamics in outer hurricane rainbands. Part I: mesoscale rainfall and kinematic structure. Mon. Weather Rev. 118, 891–917.

Powell, M.D., Vickery, P.J., Reinhold, T.A., 2003. Reduced drag coefficient for high wind speeds in tropical cyclones. Nature 422, 279–283.

Press, W.H., Flannery, B.P., Teukolsky, S.A., Vetterling, W.T., 2007. Numerical Recipes: The Art of Scientific Computing, 3rd ed. Cambridge University Press, Cambridge, UK. 1234 pp.

Pytharoulis, L., 2018. The hurricane-like Mediterranean cyclone of January 1995. Atmos. Res. 208, 261–279.

Rappaport, E.N., Franklin, J.L., Avila, L.A., Baig, S.R., Beven II, J.L., Blake, E.S., Burr, C.A., Jiing, J.-G., Juckins, C.A., Knabb, R.D., Landsea, C.W., Mainelli, M., Mayfield, M., McAdie, C.J., Pasch, R.J., Sisko, C., Stewart, S.R., Tribble, A.N., 2009. Advances and challenges at the National Hurricane Center. Weather Forecast. 24, 395–419.

Rappin, E.D., Morgan, M.C., Tripoli, G.J., 2011. The impact of outflow environment on tropical cyclone intensification and structure. J. Atmos. Sci. 68, 177–194.

Rasmussen, E.A., Turner, J., 2003. Polar Lows: Mesoscale Weather Systems in the Polar Regions. Cambridge University Press, Cambridge, England. 624 pp.

Rayleigh, Lord, 1880. On the stability, or instability, of certain fluid motions. Proc. Lond. Math. Soc. 11, 57–70.

Rayleigh, Lord, 1916. On the dynamics of revolving fluids. Proc. R. Soc. A 93, 148–154.

Raymond, D.J., 1995. Regulation of moist convection over the West Pacific warm pool. J. Atmos. Sci. 52, 3945–3959.

Raymond, D.J., Carillo, C.L., 2011. The vorticity budget of developing Typhoon Nuri (2008). Atmos. Chem. Phys. 11, 147–163.

Raymond, D.J., Gjorgjievska, S., Sessions, S.L., Fuchs, Z., 2014. Tropical cyclogenesis and mid-level vorticity. Aust. Meteorol. Ocean. Soc. J. 64, 11–25.

Reasor, P.D., Montgomery, M.T., 2001. Three-dimensional alignment and corotation of weak, TC-like vortices via linear vortex Rossby waves. J. Atmos. Sci. 58, 2306–2330.

Reasor, P.D., Montgomery, M.T., 2015. Evaluation of a heuristic model for tropical cyclone resilience. J. Atmos. Sci. 72, 1765–1782.

Reasor, P.D., Montgomery, M.T., Bosart, L.F., 2005. Mesoscale observations of the genesis of Hurricane Dolly (1996). J. Atmos. Sci. 62, 3151–3171.

Reasor, P.D., Montgomery, M.T., Marks, F.D., Gamache, J.F., 2000. Low-wavenumber structure and evolution of the hurricane inner core observed by airborne dual-Doppler radar. Mon. Weather Rev. 128, 1653–1680.

Reed, R.J., Norquist, D.C., Recker, E.E., 1977. The structure and properties of African wave disturbances as observed during Phase III of GATE. Mon. Weather Rev. 105, 317–333.

Reeder, M.J., Smith, R.K., Lord, S.J., 1991. The detection of large scale asymmetries in the tropical cyclone environment. Mon. Weather Rev. 1199, 848–854.

Ren, S., 1999. Further results on the stability of rapidly rotating vortices in the asymmetric balance formulation. J. Atmos. Sci. 56, 475–482.

Renno, N.O., Williams, E.R., 1995. Quasi-Lagrangian measurements in convective boundary layer plumes and their implications for the calculation of CAPE. Mon. Weather Rev. 123, 2733–2742.

Richter, D.H., Stern, D.P., 2010. Evidence of spray-mediated air-sea enthalpy flux within tropical cyclones. Geophys. Res. Lett. 41, 2997–3003.

Richtmeyer, R.D., Morton, K., 1967. Difference Methods for Initial Value Problems. Interscience Publishers, New York, USA. 405 pp.

Riehl, H., 1963. Some relations between wind and thermal structure of steady state hurricanes. J. Atmos. Sci. 20, 276–287.

Riehl, H., 1967. Tropical Meteorology. McGraw-Hill, New York, USA. 1234 pp.

Riehl, H., Malkus, J.S., 1961. Some aspects of Hurricane Daisy, 1958. Tellus 13, 181–213.

Riemer, M., Montgomery, M.T., 2011. Simple kinematic models for the environmental interaction of tropical cyclones in vertical wind shear. Atmos. Chem. Phys. 11, 9395–9414.

Riemer, M., Montgomery, M.T., Nicholls, M.E., 2010. A new paradigm for intensity modification of tropical cyclones: thermodynamic impact of vertical wind shear on the inflow layer. Atmos. Chem. Phys. 10, 3163–3188.

Riemer, M., Montgomery, M.T., Nicholls, M.E., 2013. Further examination of the thermodynamic modification of the inflow layer of tropical cyclones by vertical wind shear. Atmos. Chem. Phys. 13, 327–346.

Ritchie, E.A., Holland, G.J., 1997. Scale interactions during the formation of Typhoon Irving. Mon. Weather Rev. 125, 1377–1396.

Rogers, R.F., Aberson, S., Bell, M.M., Cecil, D.J., Doyle, J.D., Kimberlain, T.B., Morgerman, J., Shay, L.K., Velden, C., 2017. Rewriting the tropical record books: the extraordinary intensification of Hurricane Patricia. Bull. Am. Meteorol. Soc. 98, 2091–2112.

Rogers, R.F., Aberson, S., Black, M., Black, P., Cione, J., Dodge, P., Dunion, J., Gamache, J., Kaplan, J., Powell, M., Shay, N., Surgi, N., Uhlhorn, E., 2006. The Intensity Forecasting Experiment: a NOAA multiyear field program for improving tropical cyclone intensity forecasts. Bull. Am. Meteorol. Soc. 87, 1523–1538.

Rogers, R.F., Fritsch, J.M., 2001. Surface cyclogenesis from convectively driven amplification of midlevel mesoscale convective vortices. Mon. Weather Rev. 129, 605–637.

Romps, D.M., Kuang, Z., 2011. A transilient matrix for moist convection. J. Atmos. Sci. 68, 2009–2025.

Rotunno, R., 2022. Supergradient winds in simulated tropical cyclones. J. Atmos. Sci. 79, 1234.

Rotunno, R., Bryan, G.H., 2012. Effects of parameterized diffusion on simulated hurricanes. J. Atmos. Sci. 69, 2284–2299.

Rotunno, R., Emanuel, K.A., 1987. An air-sea interaction theory for tropical cyclones. Part II: evolutionary study using a nonhydrostatic axisymmetric numerical model. J. Atmos. Sci. 44, 542–561.

Rozoff, C.M., Schubert, W.H., McNoldy, B.D., 2006. Rapid filamentation zones in intense tropical cyclones. J. Atmos. Sci. 63, 325–340.

Rudeva, I., Gulev, S.K., 2007. Climatology of cyclone size characteristics and their changes during the cyclone life cycle. Mon. Weather Rev. 135, 2568–2587.

Ruppert, J.H., Wing, A.A., Tang, X., Duran, E.L., 2020. The critical role of cloud–infrared radiation feedback in tropical cyclone development. Proc. Natl. Acad. Sci. 45, 27884–27892.

Rutherford, B., Boothe, M.A., Dunkerton, T.J., Montgomery, M.T., 2018. Dynamical properties of developing tropical cyclones using Lagrangian flow topology. Q. J. R. Meteorol. Soc. 144, 218–230.

Rutherford, B., Dunkerton, T.J., Montgomery, M.T., 2015. Lagrangian vortices in developing tropical cyclones. Q. J. R. Meteorol. Soc. 141, 3344–3354.

Sampson, C.R., Fukada, E.M., Knaff, J.A., Strahl, B.R., Brennon, M.J., Marchok, T., 2017. Tropical cyclone gale wind radii estimates for the Western North Pacific. Weather Forecast. 32, 1029–1040.

Sanger, N.T., Montgomery, M.T., Smith, R.K., Bell, M.M., 2014. An observational study of tropical-cyclone spin-up in supertyphoon Jangmi from 24 to 27 September. Mon. Weather Rev. 142, 3–28.

Schecter, D.A., 2011. Evaluation of a reduced model for investigating hurricane formation from turbulence. Q. J. R. Meteorol. Soc. 137, 155–178.

Schecter, D.A., Dubin, D.H.E., 1999. Vortex motion driven by a background vorticity gradient. Phys. Rev. Lett. 83, 2191–2194.

Schecter, D.A., Montgomery, M.T., 2006. Conditions that inhibit the spontaneous radiation of spiral inertia-gravity waves from an intense mesoscale cyclone. J. Atmos. Sci. 63, 435–456.

Schecter, D.A., Montgomery, M.T., 2007. Waves in a cloudy vortex. J. Atmos. Sci. 64, 314–337.

Schlichting, H., 1968. Boundary Layer Theory, seventh edition. McGraw-Hill, New York. 817 pp.

Schmidt, C., Smith, R.K., 2016. Tropical cyclone evolution in a minimal axisymmetric model revisited. Q. J. R. Meteorol. Soc. 142, 1505–1516.

Schubert, W.H., Hausman, S., Garcia, M., Ooyama, K.V., Kuo, H.-C., 2001. Potential vorticity in a moist atmosphere. J. Atmos. Sci. 58, 3148–3157.

Schubert, W.H., Montgomery, M.T., Taft, R.K., Guinn, T.A., Fulton, S.R., Kossin, J.P., Edwards, J.P., 1999. Polygonal eyewalls, asymmetric eye contraction, and potential vorticity mixing in hurricanes. J. Atmos. Sci. 56, 1197–1923.

Shapiro, L.J., Montgomery, M.T., 1993. A three-dimensional balance theory for rapidly-rotating vortices. J. Atmos. Sci. 50, 3322–3335.

Shapiro, L.J., Ooyama, K.V., 1990. Barotropic vortex evolution on a beta plane. J. Atmos. Sci. 47, 170–187.

Shapiro, L.J., Willoughby, H., 1982. The response of balanced hurricanes to local sources of heat and momentum. J. Atmos. Sci. 39, 378–394.

Shin, S., Smith, R.K., 2008. Tropical-cyclone intensification and predictability in a minimal three dimensional model. Q. J. R. Meteorol. Soc. 134, 1661–1671.

Siebesma, A.P., Bony, S., Jakob, C., Stevens, B. (Eds.), 2020. Clouds and Climate: Climate Science's Greatest Challenge. Cambridge University Press, Cambridge, UK, p. 409.

Sinclair, P.C., 1969. General characteristics of dust devils. J. Appl. Meteorol. 8, 32–45.

Slocum, C.J., Williams, G.J., Taft, R.K., Schubert, W.H., 2014. Tropical cyclone boundary layer shocks. arXiv 1234, 1–19.

Smith, G.B., Montgomery, M.T., 1995. Vortex axisymmetrization: dependence on azimuthal wave-number or asymmetric radial structure changes. Q. J. R. Meteorol. Soc. 121, 1615–1650.

Smith, R.A., Rosenbluth, M.N., 1990. Algebraic instability of hollow electron columns and cylindrical vortices. Phys. Rev. Lett. 64, 649–652.

Smith, R.K., 1991. An analytic theory of tropical cyclone motion in a barotropic shear flow. Q. J. R. Meteorol. Soc. 117, 685–714.

Smith, R.K. (Ed.), 1997. The Physics and Parameterization of Moist Atmospheric Convection. NATO ASI Series. Kluwer Academic Publishers, Dordrecht. 498 pp.

Smith, R.K., 2003. A simple model of the hurricane boundary layer revisited. Q. J. R. Meteorol. Soc. 129, 1007–1027.

Smith, R.K., Kilroy, G., Montgomery, M.T., 2015a. Why do model tropical cyclones intensify more rapidly at low latitudes? J. Atmos. Sci. 72, 1783–1804.

Smith, R.K., Kilroy, G., Montgomery, M.T., 2021. Tropical cyclone life cycle in a three-dimensional numerical simulation. Q. J. R. Meteorol. Soc. 147, 3373–3393.

Smith, R.K., Leslie, L.M., 1976. Thermally driven vortices: a numerical study with application to dust devils. Q. J. R. Meteorol. Soc. 102, 791–804.

Smith, R.K., Leslie, L.M., 1978. Tornadogenesis. Q. J. R. Meteorol. Soc. 104, 189–198.

Smith, R.K., Montgomery, M.T., 2008. Balanced depth-averaged boundary layers used in hurricane models. Q. J. R. Meteorol. Soc. 134, 1385–1395.

Smith, R.K., Montgomery, M.T., 2012a. How important is the isothermal expansion effect to elevating equivalent potential temperature in the hurricane inner-core? Q. J. R. Meteorol. Soc. 138, 75–84.

Smith, R.K., Montgomery, M.T., 2012b. Observations of the convective environment in developing and non-developing tropical disturbances. Q. J. R. Meteorol. Soc. 138, 1721–1739.

Smith, R.K., Montgomery, M.T., 2014. On the existence of the logarithmic surface layer in hurricanes. Q. J. R. Meteorol. Soc. 140, 72–81.

Smith, R.K., Montgomery, M.T., 2015. Towards clarity on tropical cyclone intensification. J. Atmos. Sci. 72, 3020–3031.

Smith, R.K., Montgomery, M.T., 2016. Understanding hurricanes. Weather 71, 219–223.

Smith, R.K., Montgomery, M.T., 2020. The generalized Ekman model for the tropical cyclone boundary layer revisited: the myth of inertial stability as a restoring force. Q. J. R. Meteorol. Soc. 146, 3435–3449.

Smith, R.K., Montgomery, M.T., 2021a. Correction to: the generalized Ekman model for the tropical cyclone boundary layer revisited. Q. J. R. Meteorol. Soc. 147, 2082–2085.

Smith, R.K., Montgomery, M.T., 2021b. The generalized Ekman model for the tropical cyclone boundary layer revisited: addendum. Q. J. R. Meteorol. Soc. 147, 1471–1476.

Smith, R.K., Montgomery, M.T., 2022a. Comments on "Thermodynamic characteristics of downdrafts in tropical cyclones as seen in idealized simulations of different intensities". J. Atmos. Sci. 79, 2011–2012.

Smith, R.K., Montgomery, M.T., 2022b. Effective buoyancy and CAPE: some implications for tropical cyclones. Q. J. R. Meteorol. Soc. 148, 1–14.

Smith, R.K., Montgomery, M.T., Braun, S.A., 2018a. Azimuthally-averaged structure of Hurricane Edouard (2014) just after peak intensity. Q. J. R. Meteorol. Soc. 145, 211–216.

Smith, R.K., Montgomery, M.T., Bui, H., 2018b. Axisymmetric balance dynamics of tropical cyclone intensification and its breakdown revisited. J. Atmos. Sci. 75, 3169–3189.

Smith, R.K., Montgomery, M.T., Kilroy, G., 2018c. The generation of kinetic energy in tropical cyclones revisited. Q. J. R. Meteorol. Soc. 144, 2481–2490.

Smith, R.K., Montgomery, M.T., Kilroy, G., Tang, S., Müller, S., 2015b. Tropical low formation during the Australian monsoon: the events of January 2013. Aust. Meteorol. Ocean. Soc. J. 65, 318–341.

Smith, R.K., Montgomery, M.T., Nguyen, S.V., 2009. Tropical cyclone spin up revisited. Q. J. R. Meteorol. Soc. 135, 1321–1335.

Smith, R.K., Montgomery, M.T., Persing, J., 2014. On steady-state tropical cyclones. Q. J. R. Meteorol. Soc. 140, 2638–2649.

Smith, R.K., Montgomery, M.T., Vogl, S., 2008. A critique of Emanuel's hurricane model and potential intensity theory. Q. J. R. Meteorol. Soc. 134, 551–561.

Smith, R.K., Montgomery, M.T., Zhu, H., 2005. Buoyancy in tropical cyclones and other rapidly rotating vortices. Dyn. Atmos. Ocean. 40, 189–208.

Smith, R.K., Schmidt, C.W., Montgomery, M.T., 2011. Dynamical constraints on the intensity and size of tropical cyclones. Q. J. R. Meteorol. Soc. 137, 1841–1855.

Smith, R.K., Ulrich, W., 1990. An analytical theory of tropical cyclone motion using a barotropic model. J. Atmos. Sci. 47, 1973–1986.

Smith, R.K., Ulrich, W., 1993. Vortex motion in relation to the absolute vorticity gradient of the vortex environment. Q. J. R. Meteorol. Soc. 119, 207–215.

Smith, R.K., Ulrich, W., Dietachmayer, G., 1990. A numerical study of tropical cyclone motion using a barotropic model. Part I. The role of vortex asymmetries. Q. J. R. Meteorol. Soc. 116, 337–362.

Smith, R.K., Vogl, S., 2008. A simple model of the hurricane boundary layer revisited. Q. J. R. Meteorol. Soc. 134, 337–351.

Smith, R.K., Wang, S., 2018. Axisymmetric balance dynamics of tropical cyclone intensification: diabatic heating versus surface friction. Q. J. R. Meteorol. Soc. 144, 2350–2357.

Smith, R.K., Weber, H.C., 1993. An extended analytic theory of tropical cyclone motion. Q. J. R. Meteorol. Soc. 119, 1149–1166.

Smith, R.K., Zhang, J.A., Montgomery, M.T., 2017. The dynamics of intensification in an HWRF simulation of Hurricane Earl (2010). Q. J. R. Meteorol. Soc. 143, 293–308.

Sneddon, I.N., 1957. Elements of Partial Differential Equations. McGraw Hill, New York. 327 pp.

Steenkamp, S.C., Kilroy, G., Smith, R.K., 2019. Tropical cyclogenesis at and near the equator. Q. J. R. Meteorol. Soc. 145, 1846–1864.

Stern, D.P., Bryan, G.H., Aberson, S., 2016. Extreme low-level updrafts and wind speeds measured by dropsondes in tropical cyclones. Mon. Weather Rev. 144, 2177–2204.

Stern, D.P., Kepert, J.D., Bryan, G., Doyle, J.D., 2020. Understanding atypical mid-level wind speed maxima in hurricane eyewalls. J. Atmos. Sci. 77, 1531–1557.

Stern, D.P., Vigh, J.L., Nolan, D.S., Zhang, F., 2015. Revisiting the relationship between eyewall contraction and intensification. J. Atmos. Sci. 72, 1283–1306.

Stevens, B., 2005. Atmospheric moist convection. Annu. Rev. Earth Planet. Sci. 33, 1283–1306.

Sullivan, P.P., McWilliams, J.C., Patton, E.G., 2014. Large-eddy simulation of marine atmospheric boundary layers above a spectrum of moving waves. J. Atmos. Sci. 71, 4001–4027.

Sundqvist, H., 1970a. Numerical simulation of the development of tropical cyclones with a ten-level model. Part I. Tellus 4, 359–390.

Sundqvist, H., 1970b. Numerical simulation of the development of tropical cyclones with a ten-level model. Part II. Tellus 5, 505–510.

Sutyrin, G.G., 1989. Azimuthal waves and symmetrization of an intense vortex. Sov. Phys. Dokl. 34, 104–106.

Sutyrin, G.G., 2019. On vortex intensification due to stretching out of weak satellites. Phys. Fluids 31, 075103505.

Sutyrin, G.G., Radko, T., 2019. On the peripheral intensification of two-dimensional vortices in smaller-scale randomly forcing flow. Phys. Fluids 31, 101701.

Tang, B., Emanuel, K.A., 2010. Midlevel ventilation's constraint on tropical cyclone intensity. J. Atmos. Sci. 67, 1817–1830.

Tang, S., Smith, R.K., Montgomery, M.T., Gua, M., 2016. Numerical study of the spin-up of a tropical low over land during the Australian monsoon. Q. J. R. Meteorol. Soc. 142, 2021–2032.

Terwey, W., Montgomery, M.T., 2003. Vortex Waves and Evolution in Sharp Vorticity Gradient Vortices. Bluebook, vol. 734. Colorado State University, Dept. of Atmospheric Science, pp. 1–97.

Thompson, P.D., 1957. Uncertainty of initial state as a factor in the predictability of large scale atmospheric flow patterns. Tellus 9, 275–295.

Tory, K.J., Davidson, N.E., Montgomery, M.T., 2007. Prediction and diagnosis of tropical cyclone formation in an NWP system. Part III: diagnosis of developing and nondeveloping storms. J. Atmos. Sci. 64, 3195–3213.

Tory, K.J., Dare, R., Davidson, N.E., McBride, J.L., Chand, S.S., 2013. The importance of low-deformation vorticity in tropical cyclone formation. Atmos. Chem. Phys. 13, 2115–2132.

Tsuji, H., Itoh, H., Nakajima, K., 2016. Mechanism governing the size change of tropical cyclone-like vortices. J. Meteorol. Soc. Jpn. 94, 219–236.

Turner, J.S., Lilly, D.K., 1963. The carbonated-water tornado vortex. Q. J. R. Meteorol. Soc. 20, 468–471.

Uhlhorn, E.W., Black, P.G., 2003. Verification of remotely sensed sea surface winds in hurricanes. J. Atmos. Ocean. Technol. 20, 99–116.

Ulrich, W., Smith, R.K., 1991. A numerical study of tropical cyclone motion using a barotropic model. II: motion in spatially-varying large-scale flows. Q. J. R. Meteorol. Soc. 117, 107–124.

Vickery, P.J., Wadhera, D., Powell, M.D., Chen, Y., 2009. A hurricane boundary layer and wind field model for use in engineering applications. J. Appl. Meteorol. Climatol. 48, 381–405.

Vigh, J.L., Schubert, W.H., 2009. Rapid development of the tropical cyclone warm core. J. Atmos. Sci. 66, 3335–3350.

Wadler, J.B., Nolan, D.S., Zhang, J.A., Shay, L.K., 2021a. Thermodynamic characteristics of downdrafts in tropical cyclones as seen in idealized simulations of different intensities. J. Atmos. Sci. 79, 3503–3524.

Wadler, J.B., Zhang, J.A., Rogers, R.F., Jaimes, B., Shay, L.K., 2021b. The rapid intensification of Hurricane Michael (2018): storm structure and the relationship to environmental and air-sea interactions. Mon. Weather Rev. 149, 245–267.

Wang, B., Murakami, H., 2020. Dynamic genesis potential index for diagnosing present-day and future global tropical cyclone genesis. Environ. Res. Lett. 15, 1–10.

Wang, S., Smith, R.K., 2019. Consequences of regularizing the Sawyer-Eliassen equation in balance models for tropical cyclone behaviour. Q. J. R. Meteorol. Soc. 145, 3766–3779.

Wang, S., Smith, R.K., Montgomery, M.T., 2020. Upper tropospheric inflow layers in tropical cyclones. Q. J. R. Meteorol. Soc. 146, 3466–3487.

Wang, Y., 2002. An explicit simulation of tropical cyclones with a triply nested movable mesh primitive equation model: TCM3. Part II: model refinements and sensitivity to cloud microphysics parameterization. Mon. Weather Rev. 130, 3022–3036.

Wang, Y., Davis, C.A., Huang, Y., 2019. Dynamics of lower-tropospheric vorticity in idealized simulations of tropical cyclone formation. J. Atmos. Sci. 76, 707–727.

Wang, Z., 2012. Thermodynamic aspects of tropical cyclone formation. J. Atmos. Sci. 69, 2433–2451.

Wang, Z., Montgomery, M.T., Dunkerton, T.J., 2010. Genesis of pre-hurricane Felix (2007). Part I: the role of the easterly wave critical layer. J. Atmos. Sci. 67, 1711–1729.

Wang, Z., Montgomery, M.T., Fritz, C., 2012. A first look at the structure of the wave pouch during the 2009 PREDICT-GRIP dry runs over the Atlantic. Mon. Weather Rev. 140, 1144–1163.

Weatherford, C.L., Gray, W.M., 1988a. Typhoon structure as revealed by aircraft reconnaissance. Part I: data analysis and climatology. Mon. Weather Rev. 116, 1032–1043.

Weatherford, C.L., Gray, W.M., 1988b. Typhoon structure as revealed by aircraft reconnaissance. Part II: structural variability. Mon. Weather Rev. 116, 1044–1056.

Weber, H.C., Smith, R.K., 1995. Data sparsity and the tropical cyclone analysis and prediction problem: some simulation experiments with a barotropic model. Q. J. R. Meteorol. Soc. 121, 631–654.

Wei, D., Blyth, A.M., Raymond, D.J., 1998. Buoyancy of convective clouds in TOGA COARE. J. Atmos. Sci. 55, 3381–3391.

Weiss, J., 1991. The dynamics of enstrophy transfer in two-dimensional hydrodynamics. Physica D 48, 273–294.

Williams, E.R., Renno, N.O., 1993. An analysis of the conditional instability of the tropical atmosphere. Mon. Weather Rev. 121, 21–36.

Willoughby, H.E., 1988. The dynamics of the tropical cyclone core. Aust. Meteor. Mag. 36, 183–191.

Willoughby, H.E., 1990. Gradient balance in tropical cyclones. J. Atmos. Sci. 47, 465–489.

Willoughby, H.E., 1995. Mature structure and motion. In: Elsberry, R.L. (Ed.), A Global View of Tropical Cyclones. World Meteorological Organization, Geneva, pp. 21–62. Chapter 2.

Willoughby, H.E., 1998. Tropical cyclone eye thermodynamics. Mon. Weather Rev. 126, 1653–1680.

Willoughby, H.E., Clos, J.A., Shoreibah, M.G., 1982. Concentric eye walls, secondary wind maxima, and the evolution of the hurricane vortex. J. Atmos. Sci. 39, 395–411.

Wirth, V., Dunkerton, T.J., 2006. A unified perspective on the dynamics of hurricanes and monsoons. J. Atmos. Sci. 63, 2529–2547.

Wissmeier, U., Smith, R.K., 2011. Tropical-cyclone convection: the effects of ambient vertical vorticity. Q. J. R. Meteorol. Soc. 137, 845–857.

WMO, 1993. Global Guide to Tropical Cyclone Forecasting. WMO/TD 560, Report No. TCP-31. World Meteorological Organization, Geneva, Switzerland.

Xu, J., Wang, Y., 2010. Sensitivity of tropical cyclone inner-core size and intensity to the radial distribution of surface entropy flux. J. Atmos. Sci. 67, 1831–1852.

Yano, J.-I., Emanuel, K.A., 1991. An improved model of the equatorial troposphere and its coupling with the stratosphere. J. Atmos. Sci. 48, 377–389.

Yang, B., Wang, Y., Wang, B., 2007. The effect of internally generated inner-core asymmetrics on tropical cyclone potential intensity. J. Atmos. Sci. 64, 1165–1188.

Yuan, J.D., Wang, D., Wan, Q., Liu, C., 2007. A 28-year climatological analysis of size parameters for northwestern Pacific tropical cyclones. Adv. Atmos. Sci. 24, 24–34.

Zagrodnik, J.P., Jiang, H., 2014. Rainfall, convection, and latent heating distributions in rapidly intensifying tropical cyclones. J. Atmos. Sci. 71, 2789–2809.

Zhang, D.-L., Liu, Y., Yau, M.K., 2002. A multiscale numerical study of Hurricane Andrew (1992). Part V: inner-core thermodynamics. Mon. Weather Rev. 130, 2745–2763.

Zhang, F., Emanuel, K.A., 2016. On the role of surface fluxes and WISHE in tropical cyclone intensification. J. Atmos. Sci. 73, 2011–2019.

Zhang, F., Sun, Y.Q., Magnusson, L., Buizza, R., Lin, S., Chen, J., Emanuel, K., 2019. What is the predictability limit of midlatitude weather? J. Atmos. Sci. 76, 1077–1091.

Zhang, J.A., 2010. Estimation of dissipative heating using low-level in situ aircraft observations in the hurricane boundary layer. J. Atmos. Sci. 67, 1853–1862.

Zhang, J.A., Drennan, W.M., 2012. An observational study of vertical eddy diffusivity in the hurricane boundary layer. J. Atmos. Sci. 69, 3223–3236.

Zhang, J.A., Marks, F.D., Montgomery, M.T., Lorsolo, S., 2011a. An estimation of turbulent characteristics in the low-level region of intense Hurricanes Allen (1980) and Hugo (1989). Mon. Weather Rev. 139, 1447–1462.

Zhang, J.A., Montgomery, M.T., 2012. Observational estimates of the horizontal eddy diffusivity and mixing length in the low-level region of intense hurricanes. J. Atmos. Sci. 69, 1306–1316.

Zhang, J.A., Rogers, R.F., Nolan, D.S., Marks, F.D., 2011b. On the characteristic height scales of the hurricane boundary layer. Mon. Weather Rev. 139, 2523–2535.

Zhang, J.A., Nolan, D.S., Rogers, R.F., Tallapragada, V., 2015. Evaluating the impact of improvements in the boundary layer parameterization on hurricane intensity and structure forecasts in HWRF. Mon. Wea. Rev. 143, 3136–3155.

Zhang, J.A., Rogers, R.F., Reasor, P.D., Gamache, J., 2023. The mean kinematic structure of the tropical cyclone boundary layer and its relationship to intensity change. Mon. Wea. Rev. 151, 63–84.

Zhu, H., Smith, R.K., 2020. A case-study of a tropical low over northern Australia. Q. J. R. Meteorol. Soc. 146, 1702–1718.

Zhu, H., Smith, R.K., Ulrich, W., 2001. A minimal three-dimensional tropical cyclone model. J. Atmos. Sci. 58, 1924–1944.

Zhu, P., Hazelton, A., Zhang, Z., Marks, F.D., Tallapragada, V., 2021. The role of eyewall turbulent transport in the pathway to intensification of tropical cyclones. J. Geophys. Res., Atmos. 126, e2021JD034983. https://doi.org/10.1029/2021JD034983.

Zipser, E.J., 1969. The role of organized unsaturated convective downdrafts in the structure and rapid decay of an equatorial disturbance. J. Appl. Meteorol. 8, 799–814.

Zipser, E.J., 1977. Mesoscale and convective-scale downdrafts as distinct components of squall-line structure. Mon. Weather Rev. 105, 1568–1588.

Zipser, E.J., 2003. Some views on "hot towers" after 50 years of tropical field programs and two years of TRMM data. In: Tao, W.-K., Adler, R. (Eds.), Cloud Systems, Hurricanes, and the Tropical Rainfall Measuring Mission (TRMM) a Tribute to Dr. Joanne Simpson. American Meteorological Society, pp. 49–58.

Zipser, E.J., Lemone, M.A., 1980. Cumulonimbus vertical velocity events in GATE. Part II: synthesis and model core structure. J. Atmos. Sci. 37, 2458–2469.

Index

Printed in the United States
by Baker & Taylor Publisher Services